MANUAL OF METEOROLOGY

MANUAL OF METEOROLOGY

VOLUME II

COMPARATIVE METEOROLOGY

BY

SIR NAPIER SHAW, LL.D., Sc.D., F.R.S.

Late Professor of Meteorology in the Imperial College of Science and Technology and Reader in Meteorology in the University of London; Honorary Fellow of Emmanuel College, Cambridge; sometime Director of the Meteorological Office, London, and President of the International Meteorological Committee

WITH THE ASSISTANCE OF

ELAINE AUSTIN, M.A.

of the Meteorological Office
formerly of Newnham College, Cambridge

SECOND EDITION

CAMBRIDGE
AT THE UNIVERSITY PRESS
MCMXXXVI

TO

THE DIRECTOR OF THE METEOROLOGICAL OFFICE

Sir George Simpson, K.C.B., F.R.S.

this volume is gratefully inscribed

CAMBRIDGE
UNIVERSITY PRESS

University Printing House, Cambridge CB2 8BS, United Kingdom

Cambridge University Press is part of the University of Cambridge.

It furthers the University's mission by disseminating knowledge in the pursuit of education, learning and research at the highest international levels of excellence.

www.cambridge.org
Information on this title: www.cambridge.org/9781107475472

© Cambridge University Press 1936

This publication is in copyright. Subject to statutory exception and to the provisions of relevant collective licensing agreements, no reproduction of any part may take place without the written permission of Cambridge University Press.

First published 1936
First paperback edition 2014

A catalogue record for this publication is available from the British Library

ISBN 978-1-107-47547-2 Paperback

Cambridge University Press has no responsibility for the persistence or accuracy of URLs for external or third-party internet websites referred to in this publication, and does not guarantee that any content on such websites is, or will remain, accurate or appropriate.

PREFACE TO THE FIRST EDITION

FROM the study of Meteorology in History as set out in the introductory volume of this Manual the conclusion was arrived at that the primary need of the science was a sufficient knowledge of the facts about the atmosphere in its length and breadth and thickness to furnish a satisfactory representation of the general circulation and its changes. The present volume is intended to provide the reader with the means of making himself acquainted with the nature and extent of the material which is available for satisfying that need. In the first place the normal general circulation and its seasonal changes are represented by maps, tables and diagrams, with references to the original sources for further details. Next the transitory changes are dealt with, first by considering the results of the purely arithmetical methods of chronology, periodicity and correlation, and secondly by the results of the graphic methods of the weather-map in combination with the autographic records of the meteorological observatory. Incidentally we should like to call attention to remarkable evidence of the effectiveness of the maps in the case of pressure, notwithstanding their small size. It will be found on p. 213.

The tables and diagrams which are added to supplement the information contained in the maps are only samples of many possible compilations. Their number and extent are necessarily limited by the considerations which control the size of the volume. In so far as those tables and diagrams are successful they will recall the suggestion of the preface to the first volume, of an encyclopaedia or dictionary which would embody information of that character for all parts of the world. If each year a part were brought up to date the whole would constitute a permanent work of reference to which a student could turn for information now accessible only to the few privileged persons who live within easy access of a meteorological library. The advantage of some provision of that kind will be apparent to the reader who notices the discontinuous and unceremonious introduction of certain information in the form of supplementary tables. At the last moment they seemed indispensable as additions to material which had already been included in the text, and room had to be found for them.

The maps of normal distribution, which form the original foundation of the whole work, are reproduced from blocks prepared by the Cambridge Press under the supervision of the late Mr J. B. Peace. The original drawings were made for me in the Meteorological Office during the years 1919 and 1920 by the late Mr Charles Harding and Dr C. E. P. Brooks.

The incorporation of these charts into the body of what is intended to serve as a text-book would not have been possible without the initiative enterprise of Mr Peace and the skilful management of the material by his successor, Mr Lewis, and the staff of the University Press.

For further help of various kinds in relation to the illustrations I make the following grateful acknowledgment:

To His Majesty's Stationery Office for permission to reproduce Figs. 4 (inset), 5, 6 Ice in the N. Atlantic and in the Southern Ocean, 55 Curves of normal conditions in the upper air of England, 156, 161–2 Wind-roses of the equatorial belt, 181 Diurnal variation of wind at St Helena, 184 One hundred years' rainfall over London, 198 Waterspouts (Capt. J. Allan Mordue), 205 Isanakatabars (Solar Physics Observatory), 207 and 213 Meteorograms of British observatories, 208 Meteorogram for July 27, 1900, and 209 Embroidery of the barogram August 14–15, 1914. Fig. 216 Early life-history of a secondary, and the entablatures of pp. 244 and 246 are redrawn from the *Life History of Surface Air-Currents* and Fig. 217 from *The Weather Map*.

To the Lords Commissioners of the Admiralty and the Astronomer Royal for the charts of magnetic elements Figs. 9 and 10.

To the Meteorological Office for the original copy for Fig. 3 Influence of orographic features on weather, 12 and 13 Thunderstorms, 15 Weekly weather, 182 Wind-velocity at the Eiffel Tower and the Bureau Central, 185–8 Quarterly weather, 217 Weather in relation to the centres of cyclones, also blocks for Fig. i Temperature scales, and Fig. 60 Glass models of temperature July 27 and 29, 1908.

To Professor H. H. Turner and Mr J. J. Shaw for permission to reproduce Fig. 7, Earthquakes.

To Dr S. Fujiwhara for details of the distribution of Japanese volcanoes incorporated in Fig. 8.

To Dr W. J. S. Lockyer for the original unpublished charts from which Figs. 204 and 206 have been prepared.

To the Royal Meteorological Society for Figs. 61, 62, 210, 211, 212 and 214.

To the Scottish Geographical Society for the block of Fig. 163.

To the Royal Society for permission to reproduce Colonel Gold's article on "Forecasting," from the Catalogue of the Society's exhibit in the British Empire Exhibition 1924, and to Colonel Gold for his concurrence in that permission. Also, with the permission of the Society, we have made use of the diagram of Electro-magnetic-waves in *Phases of Modern Science*.

To the Royal Astronomical Society and to Mr J. H. Reynolds for photographs of Jupiter, pp. 169–171.

To Dr J. Bjerknes for the stereo-photograph of his model of the circulation of a cyclone (Fig. 219).

Other borrowings are acknowledged in the text with equal gratitude; and foremost among my obligations I must regard that to the authors of *Les Bases de la Météorologie dynamique* from which I have ventured to take several illustrations.

The maps of the distribution of pressure over the Northern and Southern Hemispheres for February 14, 1923 (Figs. 189, 190), were prepared for the International Commission for the Exploration of the Upper Air. In a different form they have appeared in a folio of charts of the distribution of pressure over the globe on the international days of the year 1923 for the illustration of a specimen volume of the results of the observations of the upper air of that year. Thirty-six days are represented in the volume which was presented to a meeting of the Commission held at Leipzig at the end of August 1927 with the title *Comptes rendus des jours internationaux*, 1923. The maps are believed to be the first examples of synchronous charts of pressure for the whole globe.

I am again indebted to Captain D. Brunt and Commander L. G. Garbett, R.N., for the reading of the proof-sheets and for valuable suggestions of various kinds.

Finally I am indebted to the Meteorological Office for Miss Elaine Austin's continued assistance.

Prefixed to the text of this volume is a collection of definitions of quantities and ideas which are often employed or referred to in meteorological literature and may be wanted in the reading of this book.

The volume concludes as the previous volume did with a summary chapter on the atmosphere, but in a form which may be unfamiliar to the reader. Certain assumptions, principles and conclusions are set out in very definite wording, and in the scholastic manner, so that they may challenge assent or dissent. The chapter forms, in fact, the brass plate, bell-pull and knocker of the house which is to be represented by the remaining volumes.

NAPIER SHAW

November 4, 1927

PREFACE TO THE SECOND EDITION

In the eight years that have elapsed since the completion of the original volume many notable additions have been made to the information available for a meteorological gazetteer of the world. On the table adjoining that on which I am writing is a collection of data for the upper air which covers 196 pages for 1923, 185 for 1924, 1058 for 1925, 1228 for 1926, 1674 for 1927 with a volume of 140 pages for one country for 1932.

A complete revision of the presentation of meteorological data from the point of view of 1935 instead of that of 1927 is not possible. The effort has been limited to correcting errors and omissions, substituting new figures in place of Nos. 43–4, 57–9, 72, 184–8, 192, and, at the end of the volume, a chapter of notes in place of the "New Views about Cyclones and Anticyclones" and the forty-five articles of the syllogistic summary of conditions of the middle atmosphere.

For additional information as to matters to be added in the supplement we are indebted to many colleagues, to Col. Gold, Dr Whipple, Mr R. Corless, Prof. Brunt, Capt. Garbett, Dr Brooks and Miss L. D. Sawyer, to Dr C. W. B. Normand and his colleagues of the Indian Meteorological Department, to Mr D. C. Archibald of the Canadian Service and to Prof. F. Eredia of the Italian Aeronautical Department.

And to the Director of the Meteorological Office for Miss Austin's assistance in the revision of the volume.

NAPIER SHAW

March 4, 1936

TABLE OF CONTENTS

VOLUME II. COMPARATIVE METEOROLOGY

LIST OF ILLUSTRATIONS

Physics *Meteorology*

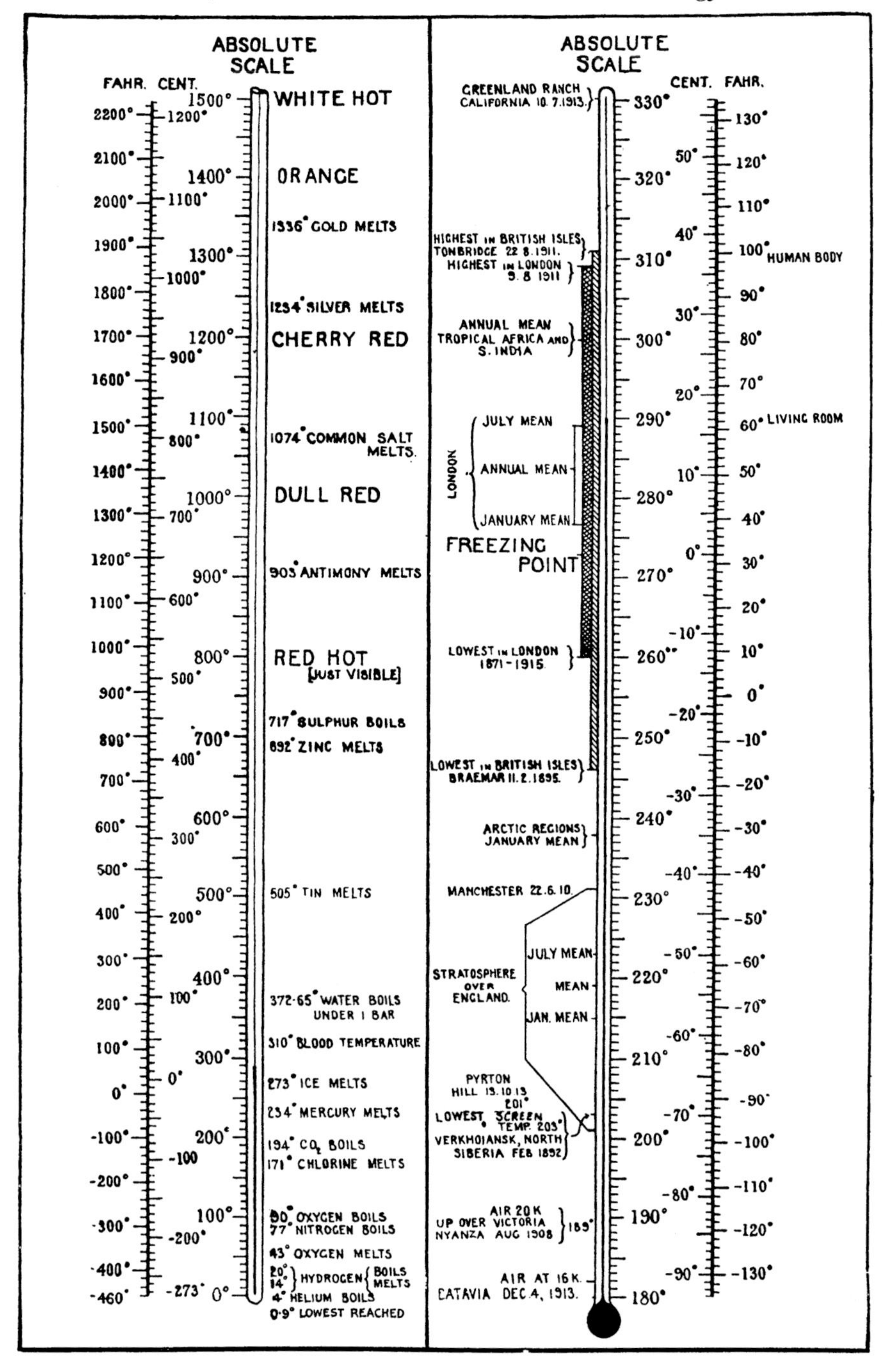

Fig. i. Comparison of the absolute scale of temperature with the Fahrenheit and Centigrade scales. Extreme and mean air-temperatures, and a selection of physical thermometrical constants. From *The Weather of the British Coasts*, M.O. 230, London, 1918. The "lowest reached" at Leiden is now ·08 (July 1933).

DEFINITIONS AND EXPLANATIONS OF CERTAIN TECHNICAL TERMS

Absolute: a word that is frequently used with respect to temperature to indicate the graduation of a scale of temperature which is independent of the nature of the "thermometric substance." Thermometers based upon the expansion of solids or of liquids or gases in glass enclosures are not expected to agree exactly all over their scales although they may be constructed to agree at any two fixed points. An absolute scale can be made out in various ways as, for example, by the saturation pressure of vapour of a liquid of known composition; but a better known way of arriving at an absolute temperature is an indirect one based upon the thermodynamic principles of the relation between heat and work. A gas-thermometer of constant volume gives a very close approximation to the absolute scale; and a thermometer with a perfect gas as thermometric substance, agreeing with the absolute scale at two points, would agree throughout the range. Hence a hydrogen-gas-thermometer is the medium by which a practical realisation of the absolute scale is arrived at. If the freezing-point of water be taken as 273 on the absolute scale, and the boiling-point of water under standard pressure as 373, the position of the absolute zero is found to be 273·1 below the freezing-point of water.

At the absolute zero the hydrogen of a gas-thermometer would have no pressure, water or any other liquid would have no vapour-pressure, and the limit of conversion of heat into work would have been reached. It has therefore been regarded hitherto as unattainable; and from the point of view of measurements is theoretical, though temperature has been carried down to ·08 absolute, 273·02t below the freezing-point.

The readings of a well-made mercury thermometer, such as those used at meteorological stations, differ little throughout the range of meteorological measurements from the absolute scale of temperature counting 273 as the freezing-point of water and 373 as the boiling-point. Consequently it is customary to convert the meteorological centigrade measurements of temperature into "absolute" temperature by adding 273. That does not give, strictly speaking, the absolute temperature, though it is near enough for meteorological purposes. In this work we avoid the inaccuracy by calling the centigrade reading increased by 273, the tercentesimal temperature, with the symbol tt; but we use the resulting figure in formulae based upon absolute temperature, generally without correction, because the difference is too small to be of importance.

Adiabatic: a compound Greek word meaning "no road" or "impassable." It is applied to a gas or liquid which being enclosed in a cylinder or other environment (q.v.) is subject to changes of pressure and consequent changes of temperature when the environment is such that no heat can pass either way across the boundary between the substance operated upon and its surroundings.

In a laboratory under ordinary experimental conditions adiabatic changes cannot be examined because there the thermal isolation of the substance operated on is far from complete. Heat flows more or less freely between the substance and its environment to make up for any loss or gain of temperature due to rarefaction or compression; and the conditions approximate more nearly to those known as isothermal (i.e. of the same temperature) than to the adiabatic (i.e. no heat transference). In order to approximate to adiabatic conditions the operations must be within the limits of the protected enclosure of a vacuum flask, or the changes must be very rapid and the observations correspondingly quick.

But in the free air when a very large volume of air is undergoing change of pressure the exchange of heat across the boundary is small and has no immediate effect upon any part of the mass except that quite close to the boundary. The only way in which the internal mass of free air can lose heat or gain it is by radiation from or to those of its constituents which are capable of emitting or receiving radiation: this effect is so small that it is customary to regard even slow changes of pressure in the free atmosphere as operating under adiabatic conditions.

The word "adiabatic" is also used as an abbreviation for adiabatic curve which shows the relation between the pressure and temperature of a mass of air when the adiabatic condition is rigorously maintained. The curves are derived by computation from other known properties of air, not (up to the present) from direct observation. Two sets of adiabatics have been computed, namely dry adiabatics, applicable to those changes which do not cause the air operated upon to pass below its point of saturation, or dew-point, and saturation adiabatics that are appropriate to air which is always saturated, any changes of pressure being associated with evaporation (if the air is compressed) or condensation (if the air is expanded) so that the condition of saturation is always maintained.

Such curves are often called **isentropic** because what is called the "entropy" of the substance cannot change when the conditions are adiabatic. Isentropic is contrasted with isothermal as representing essentially different conditions connected with the transformation of heat into mechanical energy. If a gas is working against its environment under isothermal conditions the energy which is transformed is derived from, or carried to, the environment; but if the conditions are isentropic the energy transformed is derived entirely from the heat-store of the working substance itself, at the sacrifice of its temperature. If atmospheric air were homogeneous the word "isentropic" would be as applicable as adiabatic; but, in consequence of the peculiar properties of water-vapour and the latent heat which it contains, saturated air can make use of the latent heat of water-vapour to carry out its work instead of drawing entirely upon its own supply; it saves thereby some of the loss of temperature which would be inevitable if the air were not saturated. Adiabatic, as expressing the condition with reference to environment, is in consequence a more useful term than isentropic; the latter refers to the substance under operation, which can change its composition during the operation.

Aerology: a word that has been introduced to denote the modern study of the atmosphere, which includes the upper air as well as the more conventional studies which have been connoted by the older word "meteorology." Sometimes it is used as limiting the study to the upper air. In like manner **Aerography** is a modern word for the record of the structure of the atmosphere and its changes.

Altimeter: an aneroid barometer which is graduated to read "heights" instead of barometric pressure in millibars or inches or millimetres of mercury. The word height in this case is necessarily used in a conventional sense because the instrument is actuated by pressure alone (with possibly some influence of temperature and mechanical lag) and the height is not determinable by observing change of pressure alone, but requires also a knowledge of the density of the air between the ground-level and point of observation.

There is a good deal of laxity about the use of the word height, of the same kind as that of the aeronauts who graduate a pressure-instrument to read what they call height. For example, V. Bjerknes and others would express the height of a point in the atmosphere by the geopotential at the point, calling the quantity expressed the dynamic height. We reproduce an extract from the *Avant Propos* of the *Comptes rendus des jours internationaux*, 1923.

"The relation of the geopotential Γ at any position to the geometric height of that position h and the gravitational acceleration g is $\Gamma = \int g dh$. The value is governed accordingly by the local value of gravity depending on the attraction of gravitation and the rotation of the earth; but not to any appreciable extent upon the condition of the atmosphere at the time of observation. We will refer to this measure for the time being as the geodynamic height. The dimension of its measure in the absolute system of units is l^2/t^2 and the special unit employed by Bjerknes is represented by 10 m²/sec².

"There are, however, other ideas in meteorological practice which may be associated with the word *height* and which are useful in their several ways, although they depend upon the state of the atmosphere. They refer to the pressure, specific volume of air, and entropy of unit mass of air (potential temperature or megatemperature[1]) respectively. The association of these ideas with height is based upon the fact that the atmosphere cannot be in equilibrium unless the rate of change of pressure with height is negative, the rate of change of specific volume with height is positive and the rate of change of entropy or potential temperature or megatemperature is positive.

"We may consider these several ideas of height in turn.

1. *Surfaces of equal pressure, isobaric levels expressed by the equation* $p_0 - p = \int g\rho dh$.

It is in this manner, by universal practice, that altimeters are graduated for determining the height attained by aircraft. What is actually indicated on the instrument is pressure but it is read as height. The height expressed in this manner is also 'dynamic' and might by analogy be called the *barodynamic height*; its dimensions are mass divided by length and the square of time and its unit on the C.G.S. system is gramme/(centimetre second²).

2. *Surfaces of equal specific volume: the isosteric surfaces of Bjerknes (or of equal density: isopycnic surfaces).* The expression of height by means of this measurement might be called the *elastic height*, the dimension of its measure on the absolute system is length³/mass and the unit in the C.G.S. system is cm³/gramme. These surfaces form the correlative of the isobaric surfaces in the dynamic representation of the atmosphere.

3. *Surfaces of equal entropy or potential temperature.* These surfaces are determined by the temperature of the air and its pressure (as the surfaces of equal specific volume are) but the measure is arrived at by expressing the temperature which would be attained if the pressure were increased from its local value to a standard pressure. As a standard pressure the estimated mean pressure at sea-level, 760 mm of standard mercury, has often been used and is sometimes so defined. Our instructions in the règlement of the Commission are to use a pressure of 1000 millibars as the standard and we therefore need a special name for the potential temperature with that pressure as standard. We have adopted 'megatemperature' as a suitable name because the measurement is given as temperature with a megadyne per square centimetre as standard pressure; the name will, therefore, automatically remind the user of the standard employed. The 'height' connoted in this manner may be called the 'thermodynamic height'; it is related to the other quantities by the equation

$$T = \int_p^{p_0} \partial t/\partial p\ (\phi \text{ const.})\, dp = \int_p^{p_0} \frac{\gamma - 1}{\gamma} \frac{t}{p}\, dp.$$

We use this conception of height in the diagrams which are referred to as tephigrams. In the form of entropy (which is proportional to the logarithm of potential temperature) the dimensions of thermodynamic height are (mass × velocity²)/temperature, and its unit on the C.G.S. system is gramme × (cm/sec)² per degree of absolute or tercentesimal scale.

"The idea of height is implicit in the values of p and t. Thus, while the use of the word height without qualification must always be understood to mean the geometrical or geographic height, we could, without much straining of language, speak of the 'geodynamic height,' the 'barodynamic height,' the 'elastic height' or the 'thermodynamic height' of a particular portion of the atmosphere if we wished to lay stress upon the general idea of height which is inherent in the several measures employed."

Amplitude: the extent of the excursion (on either side) of the value of a quantity which is subject to periodic change. (See **Phase**.)

[1] We have learned to prefer megadyne temperature.

Anabatic (ἀνά up, βαίνω I go). A word used to describe the character of a wind which is found flowing up a hill-side in consequence of the warming of the slope by the sun's rays or otherwise: more generally the travel of air upward through its environment. It need not be regarded as related to the distribution of pressure.

"Atmospheric." See **Stray.**

Atmospheric shells. A word is required to denote a layer of atmosphere between two surfaces having some defined characteristic on the analogy of a number of other words, to indicate different and more or less independent layers of the earth's structure: the solid earth is called the *lithosphere*, the water lying upon it the *hydrosphere*, the whole body of air lying upon the two the *atmosphere*, though by derivation the word refers to water-vapour rather than to air. The atmosphere is divided into the *stratosphere* in which convection is not possible and the *troposphere* which is permeated by convection when circumstances are favourable thereto. When circumstances are not favourable and we have a layer of atmosphere which has stability and consequent identity we may call the bounding surfaces thermodynamic surfaces, and the shell a thermodynamic shell; on the other hand, the shell between two surfaces of equal geopotential may be called a geodynamic shell. The *tropopause* is the name given to the boundary between the stratosphere and troposphere; it is not a thermodynamic surface nor a geodynamic surface. A thermodynamic shell which cuts across the tropopause is much thinner in the portion within the stratosphere than in the portion within the troposphere.

Autoconvection: the readjustment of equilibrium in a column of fluid when the density of a lower layer is less than that above it in consequence of the warming of the base or of differences of dynamical cooling (W. J. Humphreys). See **Stability.**

Azimuth (Bearing): a convenient but rather uncouth Arabic word used to indicate the horizontal angular deviation of a more or less distant object from the true North and South line. Bearing is a nautical term which means the same thing, when the bearing is described as "true." In practice the bearing is often given as referred to magnetic North and South. Moreover, bearing may be given in compass points, WSW, N, E, or whatever the point of the compass may be; but azimuth is almost invariably given in "degrees from true North," going round "clockwise." An azimuth of 221°, for example, is 41° past the true South line.

Boiling: a familiar process with water and some other liquids when heated. As temperature rises, by the heating of the vessel which contains the liquid, bubbles of vapour are formed, first at the hottest part of the container. They break away and float upwards to the surface carrying away part of the liquid as vapour. So far as meteorology is concerned the word is sometimes used metaphorically to describe the commotion of the atmosphere which is apparent when celestial objects are seen through a telescope; more frequently perhaps in relation to "boiling-point," by which we are reminded that the temperature at which any liquid boils depends on the pressure to which the surface is subjected by the atmosphere above it. So water boils at a lower temperature on a mountain than in the valley beneath because the pressure is less.

C.G.S.: the initials of Centimetre, Gramme, Second, the fundamental units selected, half a century ago, for the systematic expression of all physical quantities connected with electricity and magnetism, and consequently also all dynamical quantities such as velocity, acceleration, force, pressure, work, power.

The systematic nature of the practice is expressed by calling the series of units based upon the fundamentals centimetre, gramme, second, the C.G.S. *system*.

All the quantities which are used for meteorological measurements are included in the scheme of the C.G.S. system except temperature, and even that comes in when regarded from the thermodynamical point of view; and it would appear therefore that, in so far as the science is to be regarded as systematic, meteorology should come into line with other systematic sciences and join in a common and convenient "system of units."

It is hardly conceivable, but nevertheless true, that large numbers of people, including even scientific people of high distinction, are disposed to regard the use or otherwise of a system of units as a matter to be disposed of by personal habit and convenience, and in some quarters the idea of having to learn what is meant by a centimetre, a gramme and a second, is so appalling as to obliterate all other considerations, even the interests of future generations.

Still it seems probable that the millibar as a systematic unit for pressure will hold its own, and possibly the kilowatt per square dekametre as the unit of strength of sunshine; others may come in time.

Circulation: the displacement of air along lines forming closed curves.

When a solid body forces its way through a fluid medium such as air it has first to push the air in front of it out of the way, and the air at the back will travel after the departing mass to fill the space which would otherwise be left void. We may thus picture to ourselves a series of curves indicating the direction in which the air is moving, away from the body in front, towards it at the back and intermediately leading from the front round to the back.

When the body has a solid boundary and moves rapidly through the fluid the circulation which is set up is confined to the immediate neighbourhood of the moving body. The circulation produced by the passage of a motor-car is a familiar example. The study of such circulations is a part of the important question of the resistance which a fluid offers to a solid travelling through it.

When the moving body is part of the air itself the circulation may be very complicated and may require a very wide range for its completion. A horizontal circulation may extend all round the earth and a vertical circulation may be localised in its ascending part and distributed over a very large area in the descending part that is required to complete it, but it must always be completed and all motion of air through the air must be treated as circulation, not as the motion of a body projected through empty space.

Convection: a Latin word which means simply "conveying" or "carrying": applied to the redistribution of heat in a gas or liquid, it implies that the different parts of the fluid, when readjusting their relative positions, carry their own heat with them.

Thermal convection is the redistribution of heat by the redistribution of mass consequent upon difference of temperature of adjacent masses. In the atmosphere the physical process of convection is complicated by the inevitable fact that the temperature of air varies automatically with its pressure. The law of thermal convection as stated in *Principia Atmospherica* runs as follows: "In the atmosphere convection is the descent of colder air in contiguity with air relatively warmer." The statement is generally appropriate because the motive power is derived from the greater density of the colder air, and the most typical example of convection is the flow of air down the slope of a hill where there is spontaneous cooling by radiation. But by convection meteorologists often have in mind the opposite aspect of the process, viz. the ascent of the warmer air pushed upward, it is true, by the descent of the colder but appearing to rise by its own levity instead of by the gravity of its environment. The ascent of warm air by thermal convection in this way is complicated not only by the automatic response of temperature to change of pressure, which takes place in dry air too, but also by the secondary effects due to the condensation of water-vapour when the pressure is reduced sufficiently.

Ordinarily the relative displacement of warm air in juxtaposition with cold air or *vice versa* is very limited in consequence of the automatic changes of temperature with pressure; but if the rising air becomes saturated the condensation of water-vapour may enable it to penetrate upward beyond the limit for dry air, and in the case of air going downward the journey can be prolonged if heat can be taken from the descending air by the cooled surface. In either of the cases we have examples of **penetrative convection** because a limited mass of air penetrates successive layers above or beneath, as distinguished from the ascent by mere accumulation of warm air in juxtaposition with colder air when the convection is said to be cumulative.

Convective equilibrium: the automatic variation of temperature of air with change of pressure introduces peculiar complications into the question of the equilibrium of successive layers of the atmosphere. A layer of air which is of uniform temperature throughout its depth is not like a layer of water in similar condition. The water is in "convective equilibrium," that is to say any portion may be exchanged for any other equal portion without any further adjustment. But in the layer of air a portion from the bottom transferred to the top would be too cold for its environment and the equal portion from the top brought down in exchange would be too warm. If, however, the successive layers are properly adjusted in temperature, being colder as one ascends by about 1t for 100 metres of height, a portion from any one layer can be exchanged for an equal portion of any other layer (provided the air is dry) without any adjustment. No change of temperature would be observable if the air were mixed up mechanically. The air is then said to be in convective equilibrium or in a "labile" state, provided the air remains unsaturated. Convective equilibrium has also a meaning when applied to saturated air, but it is a much more complicated one.

Corona: the name given to a coloured ring round the sun or moon, or indeed round any bright light, caused by the *diffraction* of the light by particles forming a cloud, if they are sufficiently regular in size. The best example of corona is that formed artificially by interposing in front of a bright light a plate of glass which has been dusted over with lycopodium grains. Each point of light gives its own corona, so that to get a good effect the source should be concentrated as nearly as possible in one point.

Débâcle: the name given to the resumption of summer conditions in rivers which are frozen in winter: a very important seasonal feature in continental countries. The ice gives way irregularly over a long stretch so that the flow of the river becomes blocked; in time the blocks give way suddenly and may produce floods and other disasters in the lower levels.

Diffraction: the secondary effect of large or small obstacles upon a beam of light. When a point of light throws a shadow of the straight edge of an obstacle the margin of the shadow will show colours. If a greasy finger be drawn across a plate of glass and the plate held up to the light there is a brilliant display of colours, like mother of pearl; iridescent clouds and coronas are phenomena of like nature. They are explained by supposing that the light spreads out from a point as a "wave front," and if the uniformity of the front is broken there is a formation of new wave fronts from the edge. A cloud of small particles may remake the front altogether.

Eddy: a word of unknown history. "The water that by some interruption in its course runs contrary to the direction of the tide or current (Adm. Smyth); a circular motion in water, a small whirlpool"—according to the *New English Dictionary*.

Eddies are formed in water whenever the water flows rapidly past an obstacle. Numbers of them can be seen as little whirling dimples or depressions on the surface close to the side of a ship which is moving through the water. In the atmosphere similar eddies on a larger scale are shown by the little whirls of dust and leaves sometimes formed at street corners and other places which present suitable obstacles. The peculiarity of these wind-eddies is that they seem to last for a little while with an independent existence of their own. They sometimes attain considerable dimensions and, in fact, they seemed to pass by insensible degrees from the corner-eddy to the whirlwind, the dust-storm, the waterspout, the tornado, the hurricane, and finally the cyclonic depression. It is not easy to draw the line and say where the mechanical effect of an obstacle has been lost, and the creation of a set of parallel circular isobars has begun, but it serves no useful purpose to class as identical phenomena the street-corner-eddy, twenty feet high and six feet wide, and the cyclonic depression, a thousand miles across and three or four miles high.

The special characteristic of every eddy is that it must have an axis to which the circular motion can be referred. The axis need not be straight nor need it be fixed in shape or position. The best example of an eddy is the vortex-ring or smoke-ring, which can be produced by suddenly projecting a puff of air, laden with smoke to make the motion visible, through a circular opening. In that case the axis of the eddy is ring-shaped; the circular motion is through the ring in the direction in which it is travelling and back again round the outside. The ring-eddy is very durable, but the condition of its durability is that the axis should form a ring. If the continuity of the ring is broken by some obstacle the eddy rapidly disappears in irregular motion.

It is on that account that the eddy-motion of the atmosphere is so difficult to deal with. When air flows past an obstacle a succession of incomplete eddies is periodically formed, detached, disintegrated and reformed. There is a pulsating formation of ill-defined eddies. The same kind of thing must occur when the wind blows on the face of a cliff, forming a cliff-eddy with an axis, roughly speaking, along the line of the cliff and the circular motion in a vertical plane.

Whenever wind passes over the ground, even smooth ground, the air near the ground is full of partially formed, rapidly disintegrating eddies, and the motion is known as turbulent, to distinguish it from what is known as stream-line motion, in which there is no circular motion. The existence of these eddies is doubtless shown on an anemogram as gusts, but the axes of the eddies are so irregular that they have hitherto evaded classification. Irregular eddy-motion is of great importance in meteorology, because it represents the process by which the slow mixing of layers of air takes place, an essential feature of the production of thick layers of fog. Moreover, all movements due to convection must give rise to current and return current which at least simulates eddy-motion. (*Meteorological Glossary*, M.O. 225 ii, Fourth Issue, 1918, p. 90.)

Eddy-viscosity: every liquid or gas has a certain viscosity, or redistribution of momentum with dissipation of energy, in consequence of the relative motion of the fluid on the two sides of a surface of separation within it. If the surface of separation be horizontal, one effect of viscosity is that there is an exchange of molecules across the surface, even if the exchange requires molecules of a heavier fluid to ascend and those of a lighter fluid to descend. If the discontinuity of the motion is sufficiently marked to cause eddies, as it always is between a solid surface and the natural wind, there is an exchange of mass between the surface-layers and the layers above, which follows a law similar to that of molecular viscosity but is five hundred thousand times more effective in causing a redistribution of the mass. The process as thus enhanced is described as eddy-viscosity. The numerical magnitude is dependent upon many conditions such as vertical distribution of temperature, the nature of the surface, whether water or land, and so on.

Entropy: a term introduced by R. Clausius to be used with temperature to identify the thermal condition of a substance with regard to a transformation of its heat into some other form of energy. It involves one of the most difficult conceptions in the theory of heat, about which some confusion has arisen.

The transformation of heat into other forms of energy, in other words, the use of heat to do work, is necessarily connected with the expansion of the working substance under its own pressure, as in the cylinder of a gas engine, and the condition of a given quantity of the substance at any stage of its operations is completely specified by its volume and its pressure. Generally speaking (for example, in the atmosphere) changes of volume and pressure go on simultaneously, but for simplifying ideas and leading on to calculation it is useful to suppose the stages to be kept separate, so that when the substance is expanding the pressure is maintained constant by supplying, in fact, the necessary quantity of heat to keep it so; and, on the other hand, when the pressure is being varied the volume is kept constant; this again by the addition or subtraction of a suitable quantity of heat. While the change of pressure is in progress, and generally, also, while the change of volume is going on, the temperature is changing, and heat is passing into or out of the substance. The question arises whether the condition of the substance cannot be specified by the amount of heat that it has in store and the temperature that has been acquired, just as completely as by the pressure and volume.

To realise that idea it is necessary to regard the processes of supplying or removing heat and changing the temperature as separate and independent, and it is this step that makes the conception useful and at the same time difficult.

For we are accustomed to associate the warming of a substance, i.e. the raising of its temperature, with supplying it with heat. If we wish to warm anything we put it near a fire and let it get warmer by taking in heat, but in thermodynamics we separate the change of temperature from the supply of heat altogether by supposing the substance to be "working." Thus, when heat is supplied the temperature must not rise; the substance must do a suitable amount of work instead; and if heat is to be removed the temperature must be kept up by working upon the substance. The temperature can thus be kept constant while heat is supplied or removed. And, on the other hand, if the temperature is to be changed it must be changed dynamically, not thermally; that is to say, by work done or received, not by heat communicated or removed.

So we get two aspects of the process of the transformation of heat into another form of energy by working; first, alterations of pressure and volume, each independently, the adjustments being made by adding or removing heat as may be required, and secondly, alterations of heat and temperature independently, the adjustments being made by work done or received. Both represent the process of using heat to perform mechanical work or *vice versa*.

In the mechanical aspect of the process, when we are considering an alteration of volume at constant pressure, $p\,(v-v_0)$ is the work done; and in the thermal aspect of the process $H-H_0$ is the amount of heat disposed of. There is equality between the two.

But if we consider more closely what happens in this case we shall see that quantities of heat ought also to be regarded as a product, so that $H-H_0$ should be expressed as $T\,(\phi-\phi_0)$, where T is the absolute temperature and ϕ the entropy.

The reason for this will be clear if we consider what happens if a substance works under adiabatic conditions, as we may suppose an isolated mass of air to do if it rises automatically in the atmosphere into regions of lower pressure, or conversely if it sinks. In that case it neither loses nor gains any heat by simple transference across its boundary; but as it is working it is drawing upon its store of heat, and its temperature falls. If the process is arrested at any stage, part of the store of heat will have been lost through working, so in spite of the adiabatic isolation part of the heat has gone all the same. From the general thermodynamic properties of all substances, it is shown that it is not H, the store of heat, that remains the same in adiabatic changes, but H/T, the ratio of the store of heat to the temperature at which it entered. We call this ratio the ***entropy***, and an adiabatic line which conditions thermal isolation and therefore equality of entropy is called an isentropic. If a new quantity of heat h is added at a temperature T the entropy is increased by h/T. If it is taken away again at a lower temperature T' the entropy is reduced by h/T'.

In the technical language of thermodynamics the mechanical work for an elementary cycle of changes is $\delta p\,.\,\delta v$ and the element of heat, $\delta T\,.\,\delta\phi$. The conversion of heat into some other form of energy by working is expressed by the equation

$$\delta T\,.\,\delta\phi=\delta p\,.\,\delta v$$

when heat is measured in dynamical units.

It is useful in meteorology to consider these aspects of the science of heat although they may seem to be far away from ordinary experience because, in certain respects, the problem of dynamical meteorology seems to be more closely associated with these strange ideas than those which we regard as common. For example, it may seem natural to suppose that if we could succeed in completely churning the atmosphere up to, say, 10 kilometres (6 miles) we should have got it uniform in temperature or isothermal throughout. That seems reasonable, because if we want to get a bath of liquid uniform in temperature throughout we stir it up; but it is not true. In the case of the atmosphere there is the difference in pressure to deal with, and, in consequence of that, complete mixing up would result, not in equality, but in a difference of temperature of about 100t between top and bottom, supposing the whole atmosphere dry. The resulting state would not, in fact, be isothermal; the temperature at any point would depend upon its level and there would be a temperature difference of 1t for every hundred metres. But it would be perfectly isentropic. The entropy would be the same everywhere throughout the whole mass. And its state would be very peculiar, for if one increased the entropy of any part of it by warming it slightly the warmed portion would go right to the top of the isentropic mass. It would find itself a little warmer, and therefore a little lighter specifically than its environment, all the way up. In this respect we may contrast the properties of an isentropic and an isothermal atmosphere. In an isentropic atmosphere each unit mass has the same entropy at all levels, but the temperatures are lower in the upper levels. In an isothermal atmosphere the temperature is the same at all levels, but the entropy is greater at the higher levels.

An isothermal atmosphere represents great stability as regards vertical movements, any portion which is carried upward mechanically becomes colder than its surroundings and must sink again to its own place; but an isentropic atmosphere is in the curious state of neutral equilibrium which is called "labile." So long as it is not warmed or cooled it is immaterial to a particular specimen where it finds itself, but if it is warmed, ever so little, it must go to the top, or cooled, ever so little, to the bottom.

In the actual atmosphere above the level of ten kilometres (more or less) the state is not far from isothermal, below, in consequence of convection, it tends towards the isentropic state, but stops short of reaching it by a variable amount in different levels. The condition is completely defined at any level by the statement of its entropy and its temperature, together with its composition which depends on the amount of water-vapour contained in it.

Speaking in general terms the entropy increases, but only slightly, as we go upward from the surface through the troposphere until the stratosphere is reached, and from the boundary upwards the entropy increases rapidly.

If the atmosphere were free from the complications arising from the condensation of water-vapour the definition of the state of a sample of air at any time by its temperature and entropy would be comparatively simple. High entropy and high level go together; stability depends upon the air with the largest stock of entropy having found its level. In so far as the atmosphere approaches the isentropic state, results due to convection may be expected, but in so far as it approaches the isothermal state, and stability supervenes, convection becomes unlikely. (From the *Meteorological Glossary*, M.O. publication, no. 225 ii, H.M. Stationery Office.)

Environment: the material immediately surrounding a mass which is moving or undergoing other physical changes. When the behaviour and motion of a limited portion of air are under consideration the surrounding air forms its environment. There is always action and reaction between the limited mass and its environment which will affect its behaviour and its motion. The behaviour of the mass will therefore be controlled partly by its own condition and changes, and partly by the nature and condition of its environment.

Equiangular spiral: called also **Logarithmic spiral**: a curve described by a point which moves so that every step along its path is at a fixed angle to the straight line drawn from its position to a fixed point which is called the pole of the spiral. The equation of the spiral is $r=ae^{k\theta}$. A circle is a limiting case, being an equiangular spiral of ninety degrees. If the angle between the direction of motion and the radius passing outward is greater than ninety degrees, the point moves along the spiral towards the pole; if, on the other hand, the angle is less than ninety degrees the point goes outward with the radius and takes a widening sweep.

LENGTHS OF ELECTRO-MAGNETIC VIBRATIONS, VISIBLE AND INVISIBLE,

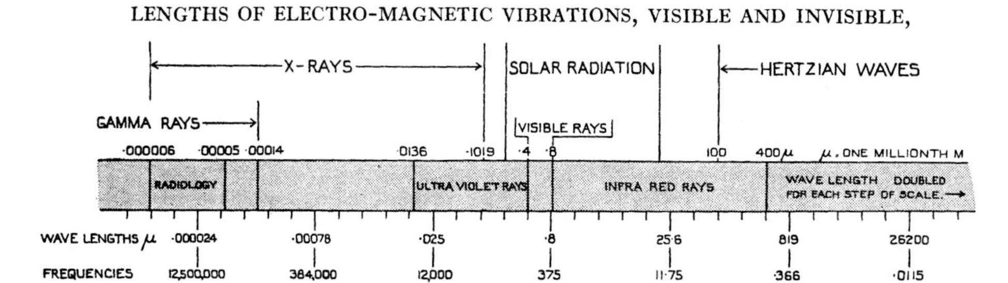

Fig. ii. Exhibiting the relation in respect of wave-length and of frequency of gamma rays, X-rays, ultra-violet rays, visible rays, infra-red rays, Hertzian waves (including

Frequency is employed in meteorology as a means of dealing statistically with non-instrumental observations as of winds from the several directions, gales, snow-storms, rain-days, fogs, etc. It is also extended to instrumental observations as a device for representing climate, for example, the frequency of occurrence of temperatures or relative humidities within specified limits. In this way curves of frequency can be constructed which are one of the implements of the statistical treatment of observations; the greatest ordinate of such a curve, representing the most frequent occurrence within the selected limits, identifies what is called the *mode* for the variation of that particular element.

Frequency is also used to mean the number of complete oscillations in a second or other interval, as the correlative of period, with reference to oscillations or vibrations such as those of sound, light or the electrical vibrations of the ether. The period is the time of one complete oscillation; the wave-length the space travelled by the waves in one period. When the frequency is doubled and the period and consequently the wave-length halved the new vibration is said to be the octave of the first. The formula connecting wave-length λ, period of oscillation τ and velocity of transmission v is $v\tau=\lambda$. The methods of generation of the respective groups of waves and the means by which they are detected are enumerated in the diagram in *Phases of Modern Science* prepared by a Committee of the Royal Society for the British Empire Exhibition, 1925. Taking all kinds of vibration of the ether into account they range over sixty octaves. One of them covers the range of vibration of visible light; four cover the infra-red of solar radiation and one the ultra-violet. The rest are produced by electrical radia-

* * *

Friction is a difficult word to deal with from the point of view of meteorology. The idea is derived from the frictional force between two solid surfaces which are made to slide one over the other—a very complicated question in itself. It is extended to the frictional forces between air and the sea or land, and thence to the effect of obstacles on the surface of either, which is not entirely frictional, and also to the interaction of two masses of air moving relatively the one to the other. In the last case molecular diffusion between the two masses comes in under the name of molecular viscosity with accompanying eddy-motion, and that again is complicated by other physical differences such as those of temperature and humidity as between a cold surface-layer and warm upper layer or *vice versa*, which may alter the extent of the interchange so much that the general character is changed. Friction in the atmosphere is indeed a special subject of undoubted importance but of great complexity.

Gale: a surface-wind exceeding force 7 on the Beaufort scale. When wind-velocities were measured exclusively by the Robinson anemometer, which smooths out the gustiness, the lower limit for a gale for the British coasts was set at 39 miles per hour, about 17·5 m/s; now that wind is recorded by anemometers which show the rapid fluctuations in the velocity of the wind, the word can only be used with much

WHICH CAN TRAVEL 30,000 MILLION CM PER SEC.—AND THEIR FREQUENCIES

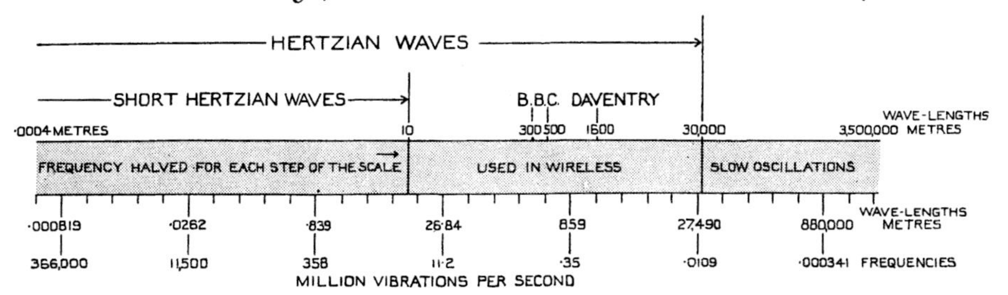

wireless waves) and "slow" oscillations for wave-lengths exceeding 30,000 metres—less than 10 kilocycles, 10,000 oscillations per second.

tion of various kinds. The arrangement of the whole series is set out in fig. ii, which is adapted from the diagram of wave-lengths exhibited at the British Empire Exhibition in 1925. They show a range from ·000006μ (millionths of a metre) on the left-hand side, to 3,500,000 m on the right.

The band representing wave-lengths in the figure extends across the double page. A scale is shown at the bottom and has to be read in a peculiar manner. Each division of it represents the range of an "octave," by which is meant that the wave-length of the vibration at any division-line is double that of the vibration at the next line on the left, and one-half of the wave-length at the line next on the right. In that way the whole range of vibrations from 100 million billions per second on the extreme left to 85 per second on the extreme right can be brought within the limits of the double page.

The frequencies which are of importance in meteorology are primarily the solar radiation which has been recognised as including not only the visible rays but also the shorter waves known as ultra-violet and the longer waves of infra-red. The range in million million vibrations per second is from 1500 to 16. The infra-red rays of terrestrial radiation range from 375 to about 1 million million. There are besides the ionising waves of short wave-length which are related to aurora, and the Hertzian waves of wireless telegraphy which are often manifest in lightning flashes.

In recent practice of radio-telegraphy a complete vibration has become a "cycle" and the frequency is indicated by the number of "kilocycles" instead of by the wave-length in metres.

* * *

less confidence. On June 1, 1908, a solitary gust or squall which indicated a transient velocity of 27 m/s at Kew Observatory brought down part of the avenue of chestnuts in Bushey Park; it is excusable to call the catastrophe the result of a gale of wind; but on the basis of hourly velocity there was no gale.

Geopotential: the potential energy per unit mass at a point above the earth's surface consequent upon the separation of the mass from the earth under the influence of gravity and the rotation of the earth. It is expressed algebraically as $\int_0^H g dh$, where g represents the acceleration of gravity and is consequently variable with height and geographical position. The dimensions of geopotential on the C.G.S. system are cm^2/sec sec, a convenient practical unit is a square dekametre per second per second (10 m^2)/sec sec.

Surfaces of equal geopotential are "level surfaces" and are horizontal in the technical sense. The height of a point can therefore be expressed with advantage for scientific purposes in terms of the geopotential of the "level surface" which passes through the point.

The shapes of level surfaces of a homogeneous fluid like water or air can be calculated from the law of gravitation and the angular velocity of the earth's rotation.

The result is a surface of revolution round the polar axis with a polar diameter smaller than the equatorial diameter by about one-third of one per cent.

The "level surface" for the sea, or "sea-level," is the level to which all heights above or below are referred. The shape of the fundamental level surface is called the "geoid" (*see* C. F. Marvin, Washington, *Monthly Weather Review*, October 1920).

A table of the equivalents of geopotential in terms of geometric height is given on p. 260.

Geostrophic: a compound Greek word coined to carry a reference to the earth's spin; thus the geostrophic wind for a certain pressure-gradient is the wind as adjusted in direction and velocity to balance and therefore maintain the existing distribution of pressure by the aid of the earth's rotation in the absence of all external forces except gravity. The direction of the geostrophic wind is along the isobar and its velocity G is equal to $bb/(2\omega\rho \sin\phi)$, where bb is the gradient of pressure, ω the angular rotation of the earth, ρ the density of the air and ϕ the latitude of the place to which the gradient refers. A uniform system of units must be employed to give a proper numerical result. In latitude 50° the velocity in metres per second is $7{\cdot}45 \times$ pressure difference in millibars per 100 km.

For certain reasons it is supposed that the geostrophic wind is a very good representation of the actual wind when the isobars do not diverge appreciably from great circles on the earth's surface, and when the point of observation of wind and corresponding gradient of pressure is so high above the surface that the direct effect of the friction at the surface is inappreciable. The effect of friction, whether due to the earth or to air, is always to reduce the velocity and divert the wind from the line of the isobar towards the lower pressure.

Gradient (of pressure, temperature, etc.): the change in the pressure, temperature, etc. corresponding with a unit step along a horizontal surface. The unit step on the C.G.S. system is a centimetre and it is desirable to use that step in any formula that includes a considerable number of variables, but for some practical purposes in meteorology a unit step of 100 kilometres (10^7 cm) is more convenient.

Gradient wind: the wind tangential to the isobaric line and computed from the gradient bb by the formula $bb = 2\omega V\rho \sin\phi \pm \frac{V^2\rho}{E} \cot r$, where V is the velocity in C.G.S. units, ω the angular rotation of the earth, ρ the density of the air, ϕ the latitude, r the angular radius of curvature of the path of the air, and E the radius of the earth. The formula may be written $bb = bb\frac{V}{G} \pm \rho V^2 \cot r/E$, where G represents the geostrophic wind. The upper sign is to be used when the circulation is cyclonic and the lower sign when the circulation is anticyclonic.

Homogeneous as applied to a material or substance is generally understood to mean that every finite portion of the substance, however small, into which the substance may be supposed to be divided is composed of the same constituents in the same proportions by weight, thus every separate chemical substance is homogeneous. But in its particular application in the phrase "homogeneous atmosphere" we have to understand an atmosphere consisting of vertical columns of limited height each of which has the same density throughout its height as the air at its base and produces the same pressure at its base as the column of the actual atmosphere.

Horizontal: We have already explained in chapter XI of vol. I that the word must be understood as meaning perpendicular to the force of gravity, whence it follows that the surface of still water is horizontal. In the upper air the word must have the same meaning, a horizontal surface or level surface is everywhere perpendicular to the force of gravity. Successive horizontal or level surfaces are consequently not equidistant the one from the other over their whole extent, they are closer together at the poles than at the equator.

Hyperbar: a term proposed by A. McAdie to indicate a region of high pressure. Thus what are sometimes called the anticyclones of the tropical region of the oceans would be the oceanic hyperbars. As the correlative of these, the areas of low pressure, McAdie has suggested the word "infrabar," but the guardians of literary propriety are critical of a word made up of a Greek root with a Latin prefix. Such words are apt to be classed as hybrid.

Inversion: as applied to temperature, the name given to the phenomenon of a rise in place of the usual lapse of temperature with height in the atmosphere. The change of temperature with height used to be called the "gradient of temperature," and inversion was an abbreviation of inversion of temperature-gradient. But the use of the word gradient (adapted from "barometric gradient") to denote change in the vertical is liable to cause confusion because barometric gradient originally referred to the change experienced in a horizontal step; and the corresponding expression for temperature, though less often required than for pressure, is still necessary; "lapse-rate" is now used in English in place of vertical temperature-gradient and an inversion means inverted lapse-rate. Counterlapse would be a good English word for it.

Great attention is paid in meteorological literature to "inversions" as being separated by the isothermal condition from the states of positive temperature-gradient or lapse-rate. It is difficult to justify the picking out of the isothermal condition as a critical condition or dividing line. The air is fundamentally stable when the lapse-rate of saturated air passing through it, say, ·4t per 100 metres, is greater than that of the existing environment; from that critical condition the stability increases continuously without any limit as the lapse-rate of the environment diminishes, the isothermal condition marks only a stage in a continuous process.

Ion: see p. 38 and vol. I, p. 323.

Isobar: or isobaric line, a line drawn on a map through all the points at which the pressure when "reduced to sea-level," or to any given level, is the same. An isobaric surface is a surface drawn through space connecting the points at which the pressure is the same. An isobaric surface will cut the "sea-level" or any other level surface in an isobar.

Isothermal line: a line drawn on a map through the points at which the temperature is the same. For maps of large areas, with great differences of level, the temperatures are "reduced to sea-level," but for smaller areas and for daily synoptic charts they are left unreduced. In like manner an isothermal surface is a surface drawn in space connecting points which have the same temperature.

Katabatic, *κατά* downward, *βαίνειν* go. A name chosen to describe the wind which is formed by the downward flow of air on a slope that is cooled by radiation. It is frequent in the form of valley-winds or fjord-winds which may at times reach alarming velocities. The importance of having a separate word arises from the fact that these radiation winds occur especially in calm weather, and are the expression of the simple physical process of the downward convection of cooled air. They are quite inexorable. They are not related to the rotation of the earth as geostrophic winds are, nor to centrifugal force as cyclostrophic winds are, and consequently they defy the laws of winds that apply in the free atmosphere. They are controlled by radiation and orographic features alone. They belong especially to surfaces which are exposed to radiation and are not warmed by the sun, and consequently they are specially noticeable at night in places where there are day and night, and in the winter night of the polar regions.

For these winds, slopes are necessary and mountain-slopes are very effective. High land in the polar regions and high mountains at night and in winter, and clear sky for radiation, may be regarded as their necessary and sufficient causes.

On July 20, 1926, the author watched from the sea the transformation of a vigorous katabatic wind down the slope behind Funchal, Madeira, into an irregular anabatic wind between the hours of 6 a.m. (before the sun had risen above the hill) and 9 a.m.

when the slope had been solarised for about two hours. The anabatic wind of 9 a.m. which was maintained as a sea-breeze throughout the day, joined the permanent north-easter above the ridge of the island and was carried away to the South.

Kilobar: a name which it is proposed to substitute for the word "millibar" to express a pressure of 1000 dynes per square centimetre, on the ground that the word "bar" had been used in Physical Chemistry to denote the C.G.S. unit of pressure (1 dyne per square centimetre) before it was employed in meteorology to express a million dynes per square centimetre.

Kilograd: a scale of temperature proposed by A. McAdie for international use and employed by him in *Aerography*. On the kilograd scale the zero is that of the absolute thermodynamic scale and the freezing-point of water under standard conditions is marked 1000; the length of a degree of the scale on a mercury-thermometer is approximately ·273 of a degree centigrade. Temperatures below the freezing-point are expressed by numbers less than 1000. The lowest temperature reached by Kamerlingh Onnes would be about 3 kilograd.

Lapse-rate: applied to the temperature of the air to indicate the rate at which temperature "lapses" with increase of vertical height. The expression "vertical temperature-gradient," often abbreviated to "temperature-gradient," has been employed to define the rate of change of temperature with height, but confusion arises if the same word "gradient" is used for the changes along a horizontal surface, and also for changes in the vertical line. Both are frequently required with temperature, and whereas pressure-gradient had already acquired a conventional meaning, temperature-gradient was beginning to mean something of a different character altogether.

In this and other publications we endeavour to restrict the use of the word "gradient" to changes along a *horizontal* surface and to find some other word for changes along a vertical line. L. F. Richardson has the word "upgrade" for this purpose.

Megadyne temperature: potential temperature (q.v.) referred to a standard pressure of 1000 mb.

Nucleus: a name given to the particle or molecular aggregate upon which drops of water can form when air is cooled sufficiently below the dew-point. The precise nature of the nuclei in different circumstances is still undetermined. It is understood from the work of John Aitken, C. T. R. Wilson and others that without a nucleus condensation cannot occur, and with only "ions" for nuclei the reduction of temperature below the dew-point must be sufficient to represent at least "fourfold saturation." It has been supposed that the dust-particles visible as motes in a sunbeam and impaled upon a glass-slide by the Owens dust-counter were nuclei for condensation, but that is now doubtful. The tendency is to seek for hygroscopic nuclei or molecular aggregates too small to form particles which can be seen in a microscope.

Phase: the position of a planet in its orbit: from the Greek word φάσις. The use is familiar in connexion with the phases of the moon. But as is common in such circumstances the meaning has become specialised. New moon, first quarter, full moon and last quarter, had better be called the principal phases, because the intermediate phases have their place in the sequence of the moon's revolution in its orbit.

From the motion of the moon and planets the idea is transferred to any periodic motion, and especially to periodic motion resolved into harmonic components as explained in vol. I, p. 273. As in that chapter we take a component of the periodic motion as represented by $x = A\cos\left(\frac{2\pi nt}{\tau} - \alpha\right)$, x is the displacement, A the amplitude or one-half the range of the oscillation, t the time at which the displacement is x, τ the period of a complete oscillation, n an integer giving the number of the component, and the angle $\left(\frac{2\pi nt}{\tau} - \alpha\right)$ is the phase by which the position in the oscillation is determined: α is the angle $\left(\text{at a time } t \text{ from the datum point, equal to } \frac{\alpha\tau}{2\pi n}\right)$ which corresponds with the time of maximum of the quantity, and is called the phase of the maximum, because when $t = \frac{\alpha\tau}{2\pi n}$ the phase angle of the oscillation is zero, the cosine is unity, and the displacement is A.

These rather complicated uses of the word phase, which are understood by practised workers in harmonic analysis, are incidental to the fact that in the study of the analysis of a complex curve the datum point for the time-measurement has to be arbitrarily chosen. On that account the angle α has to be introduced. An artifice is adopted to get rid of it in practice.

The most useful mental picture of harmonic oscillation to have in mind is the one which Maxwell uses in *Matter and Motion*. A body P is revolving in a circular orbit $PA'OA$, and we watch the motion of the point M at the foot of the perpendicular from P on the diameter AA'. The point M describes harmonic motion. Starting from the extreme point A its distance CM varies according to the cosine law, and reaches zero when P is over the centre C, then it passes on to A' and returns thence through C to the starting point A. The phase is indicated by the position of P in its orbit, and therefore by the angle AP if the time-datum is when the body is at A. But if the time-datum is taken arbitrarily when the body is at some point O, then OP is the phase and OA is the phase of maximum.

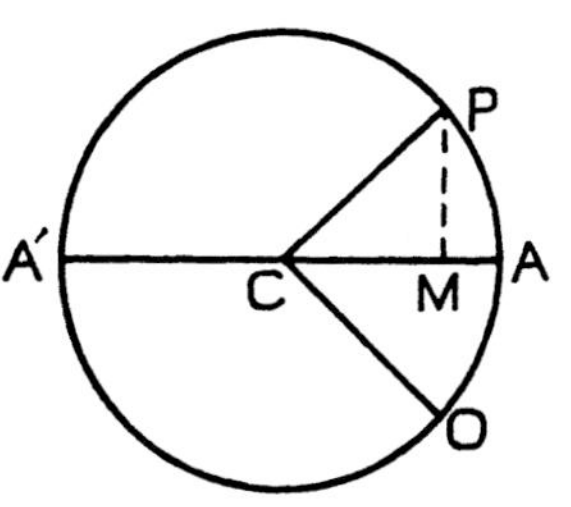

Phenology. "The study of the times of recurring natural phenomena especially in relation to climatic conditions," *New English Dictionary*. The word "phenological" was used by C. Fritsch in *Jahrb. d. k. k. Central-Anstalt für Meteorologie*, 1853, Vienna, 1858. Instructions for the Observation of Phenological Phenomena were published by the Council of the Meteorological Society in 1875. The observations included the dates of the first flowering of certain plants, the arrival and departure of migrant birds and so on. The study is included in the "crop-weather scheme" of the Ministry of Agriculture and Fisheries and is in process of organisation upon an international basis with Nijhoff's *Acta Phaenologica* as its official journal.

A notable difference of practice between phenological observations and observations of weather is that phenologists make use of the days of observation of certain specified phenomena by which the progress of the seasons is indicated, whereas meteorologists generally express the same ideas by values of the meteorological elements for months or in some cases for weeks. In order that the facts may be effectively collated it seems necessary to agree upon a time-unit to which both sets of phenomena shall be referred. The month covers too long an interval for observations of plants and animals and it is probable that in course of time the week will come to be regarded as the most effective unit for both aspects of the cycle of the seasons.

Potential temperature of the air at any point of the atmosphere (where the pressure is below a fixed standard) is the temperature which the air at the point would assume if its pressure were increased to the standard pressure without any communication of heat or evaporation of water. A convenient standard pressure for the purpose is 1000 mb. The potential temperature t_0 at the standard pressure p_0 is calculated from the actual temperature t and the actual pressure p by the formula:

$$\log t_0 = \log t + \cdot 286\,(\log p_0 - \log p).$$

The numerical constant expresses the ratio $(\gamma - 1)/\gamma$, where γ is the ratio of the specific heat at constant pressure to the specific heat at constant volume (1·40). The ratio $(\gamma - 1)/\gamma$ may also be expressed as the gas-constant divided by the dynamical equivalent of heat and by the specific heat at constant pressure.

Pressure: the distributed force exerted by the atmosphere. It is expressed numerically in millibars. A pressure of a thousand millibars is approximately equivalent to 14½ lb. per square inch, but the actual figure of the equivalent depends upon latitude; it is also the pressure of 750·076 mm or 29·5306 in. of mercury at the freezing-point of water in latitude 45°. A millibar is approximately the pressure of 1 cm. of water.

Radiation is the name used for the transmission of energy from a body through space or through a transparent medium by which the radiating body or radiator is surrounded. According to the undulatory theory of radiation the energy which is stored in the radiation takes the form of wave-motion in the space traversed and appears again in the form of heat in the body by which the radiation is ultimately absorbed. The waves travel in straight lines, or the wave-surfaces are spherical, so long as the medium traversed is homogeneous and uniform in structure; but the wave-surface is deformed, and the rays consequently bent, by any change in the nature of the material traversed, or any defect in its uniformity. The natural procedure in respect of radiation is very complicated. There is first the radiation received from the sun, which depends upon the intensity of the original emission from the sun (the solar constant), the nature of the atmosphere traversed in relation to the various wave-lengths, the nature of the receiving surface and finally the angle at which the rays impinge. Then there is the radiation from the earth outward which depends upon the nature of the radiating surface and on its temperature and "aspect." A perfectly black body at temperature t loses heat at the rate of σt^4 from each square centimetre of its surface, where σ is Stefan's constant.

The waves set up by the radiation from the surface penetrate the atmosphere if the sky is clear; but if there are opaque clouds the loss of heat by radiation is diminished by the radiation which comes from the clouds, and even if there are no clouds the water-vapour and carbonic acid gas of the atmosphere and its casual impurities, consisting of solid or liquid particles, absorb some of the radiation on its passage and, on the other hand, radiate some to earth.

The difference in the behaviour of the atmosphere in respect of radiation of different wave-lengths is clearly shown by the difference in the character of the photographs taken with screens that limit the rays to different parts of the spectrum. A photograph of a landscape with infra-red light is notably more distinct than one taken with ordinary light, and still more so than one taken with ultra-violet. It would appear that, to an eye which was sensitive only to violet light, mist or fog would be much more frequent and conspicuous than to an eye which was sensitive only to the extreme red waves.

The radiation of heat, which alone is considered here, is only one of many phenomena of similar nature which are differentiated by various properties associated with differences of wave-length. There is thus a whole series of radiations—from the long waves of wireless telegraphy, the Hertzian waves, through the dark infra-red rays to light rays (from red to violet), and the ultra-violet rays, and finally through shorter X-rays to gamma rays, which are the shortest rays known to science. See fig. ii.

A new departure in science has arisen from the suggestion of Planck that radiation is emitted in indivisible units or quanta.

Refraction: the bending of a beam of light as it passes through transparent media of different density or of different composition. In meteorology refraction comes in to explain how the sun may be visible when the disk itself is actually below the horizon, and consequently the length of the day at the equinox is increased by 7 minutes in latitude 50° and by 5 minutes at the equator. In that case the bending is due to the increased density of the air as the rays come very obliquely towards the Eastern or Western horizon.

The other notable effect of refraction is the formation of haloes, rings round the sun and moon which mark the maximum concentration of light that passes through a cloud of ice-crystals. The formation of the halo-ring is most easily suggested experimentally by having a glass prism or crystal attached to a string and spinning it (by twisting the string) in the path of a lantern-beam; of the light that gets through the prism, none can ever get nearer to the centre of the beam on the screen than the angle of minimum deviation, and that angle is very sharply defined in the experiment.

Réseau: the French word for a net, applied also to what is called in English a network; in that way it is commonly used as a convenient term for the group of stations which form a meteorological system. The term has come into general use in recent years for a network of stations extending over the whole globe under the title "réseau mondial."

Resilience: "Springiness." The distribution of forces in a continuous material which is called into play when the material is deformed, and which causes the material to return to its original shape if the deforming forces are gradually removed, or which sets up oscillations about the original configuration if the disturbing forces suddenly cease to act.

Saturation: a word applied to solutions of salts in water on the one hand, and to the limit of suspension of water-vapour in the air on the other. By analogy it can be applied in many other connexions. We have used it in vol. I with reference to the limit of population which can be supported by a given area under specified economic conditions. The really effective implication in all these cases is that, after saturation has been reached, any further change in the direction of still more restricted conditions, as by cooling a saturated solution or saturated air, or some adverse change of climate in the economic parallel, means that some part of the material has to be eliminated by crystallisation or condensation or annihilation.

Stable—Stability: words derived from the Latin *stare* (to stand) and meaning, capable of standing, capacity for standing, without auxiliary support. In technical language an arrangement of mass can stand when the forces acting upon it (gravity and the supporting resistance of the ground, for example) are in equilibrium. But for an ordinary structure, exposed to casual disturbing forces, to stand, it must not only have the forces which are acting upon it in equilibrium but the structure itself must be so supported that it will return to its position of equilibrium after it is displaced by any slight disturbing force: then the equilibrium is said to be stable and the structure is in stable equilibrium, or briefly, stable. Otherwise, though it is capable of standing in one particular unique position, any slight displacement dislocates it beyond recovery: it is in unstable equilibrium, or briefly, unstable. Thus a lead pencil can stand without other support on its flat end in spite of slight disturbance due to ordinary shaking, the equilibrium is stable; but there is no practical standing on its sharpened end without support. Though there is force enough to balance the weight the equilibrium is unstable and there is always disturbance enough to upset it. Technically it can stand, practically it falls over on account of the want of stability.

Thus stability and instability ought strictly speaking to be associated with the results which follow small displacements. In meteorology it is a column of air which we think about as being supported, and such a column can stand without auxiliary support, other than that of its environment, so long as there is a continuous diminution of its density with height. The density depends partly on pressure, which always diminishes with height, and partly on temperature, which may diminish or increase with height. If the temperature, by loss of heat through radiation or otherwise, becomes so low that in spite of the lower pressure the air above is denser than that beneath, equilibrium is not possible in still air and must readjust itself by the descent of the cold air.

W. J. Humphreys (*Physics of the Air*, p. 102) has calculated that in a column of air heated at the base the density of any layer will not be less than that of the layer immediately above it so long as the lapse-rate of temperature does not exceed 1t for a step of R/g or 29·27 metres. That rate is more than three and a half times the lapse-rate of an atmosphere in convective equilibrium. The ratio is $\gamma/(\gamma-1)$. "In order that the lower layer shall be distinctly lighter than the upper the temperature decrease with increase of altitude must be four or five times the adiabatic rate."

Thereby a distinction is drawn between the limit of equilibrium of a column of air heated at the base and the upward convection of air slightly warmer than its environment at the same level. It is the peculiar condition of the heated column that is held to account for certain types of mirage. Though there is equilibrium it is essentially unstable. The column is sistible (see below) but not stable.

It is customary to describe a condition of that kind in the atmosphere as instability, though really the forces are not in equilibrium. There is, however, no name for the condition of the atmosphere when it is not in static equilibrium. In a footnote on p. 235 of vol. I, we have suggested the word "sistible," derived from *sistere*, as a word that might be useful for the purpose. Some word is wanted because the atmosphere, in conditions when static equilibrium is not possible, might still be stable through the influence of its motion.

Stratification: as applied to the atmosphere is the separation of the atmosphere into more or less nearly horizontal layers which mark separate steps in the change of some particular element with height. Evidence of stratification in the atmosphere, apart from the continuous variation of pressure, is shown by layers of cloud as well as by notable changes in curves representing the variation with height of water-vapour, temperature or wind-velocity.

Stratosphere: the name given to the layer of the atmosphere beyond the limit of thermal convection. See **Atmospheric shells**.

Stray. "A natural electromagnetic wave in the ether. The term is used in reference to the effects of such waves in producing erratic signals in radio-telegraphic receivers. Strays are known collectively as **static**," C. F. Talman, *Meteorology*, p. 382. In England these disturbances in radio-telegraphic transmission are most frequently called **atmospherics**. Many are attributed to lightning flashes, sometimes hundreds of miles away, and some have been found associated with rainfall irrespective of the observation of lightning. A good deal of attention is now being paid to the nature and occurrence of "atmospherics" or to "static," which promises to become a new meteorological element of importance.

Stream-function: a quantity which is related to velocity in the same way that pressure is related to pressure-gradient; thus pressure-gradient is inversely proportional to the shortest distance between consecutive isobars, and velocity is inversely proportional to the shortest distance between consecutive lines of stream-function; but the direction of the velocity is at right angles, across the lines of shortest distance, whereas the direction of the pressure-gradient is along the line by which the differences are taken.

Tephigram: a name given to the graph of a sounding of the air by balloon, kite or aeroplane upon a temperature-entropy diagram. If abscissae drawn from right to left on a linear scale represent temperature, and ordinates drawn upwards represent megadyne temperature (potential temperature referred to 1000 mb as standard pressure) on a logarithmic scale, an area on the diagram will represent energy or work done.

On such a diagram adiabatic lines for dry air are horizontal lines, and isothermal lines are vertical. Adiabatic lines for saturated air which can be set out by means of the equation that represents their relation to pressure and temperature are curved lines.

The area of the diagram between the graph of the sounding and the line of zero temperature represents the amount of energy which would be necessary to account for the increase in the entropy of unit mass of air passing along the graph; and the area between the graph and the adiabatic for saturated air drawn through the the starting-point represents the "superfluous" energy of unit mass of saturated air at the starting-point.

The superfluous energy is available for the production of dynamical or electrical effects.

Tercentesimal: a name chosen to denote the scale of temperature which is measured in centigrade degrees counting from a zero 273 degrees below the freezing-point of water instead of from the freezing-point itself. The numbers for temperatures in the tercentesimal scale are very nearly identical with those in the so-called absolute or thermodynamic scale, but the technical definition of the absolute scale is different and hence a different name is required.

Term-hours: "Prescribed hours for taking meteorological observations," C. F. Talman, *Meteorology*, p. 383. By national or international agreement the fixed hours for observations for telegraphic reports in Europe are 01h, 07h, 13h, 19h G.M.T. (18h is actually used) and in North America 08h and 20h 75th Meridian time.

Self-recording instruments are set and climatological observations are arranged generally according to local mean time, and the term-hours may be different in different countries; in Britain 09h and 21h are usual, with 15h (3 p.m.) for an intermediate reading. On the Continent 08h, 14h and 20h; and 07h, 13h and 21h are sometimes used. It is not always easy to decide whether the observations in a meteorological table refer to local time or to standard time.

Thermodynamics: the science which treats of the forces called into play by heat or cold; or heat regarded as a source of motive power. The main object of the science is to explore and formulate the conditions under which heat is spent in the production of mechanical work, or is produced as the equivalent of work. The evidence of the working power of heat may be either kinetic energy of motion, as in a train, a ship, or an aeroplane, or natural forces overcome, such as gravity on a railway incline or in an aeroplane, or friction on a railway or in an aeroplane. We have to put these two together because, in all ordinary cases, when the limiting speed is reached the power of the heat expended is all used up in overcoming the friction and maintaining the motion without making any addition to it, and in fact the result of the whole process is a flow of heat by a somewhat complicated route from the region of combustion to the region of friction.

The atmosphere itself is the seat of an extremely complicated transformation of heat into work and *vice versa*, with the sun supplying heat by radiation and the earth losing it by a similar process at a different temperature. So there is an interesting analogy between the thermodynamic process of the atmosphere and those of a steam-vessel at full speed. During the flow of heat, and in consequence thereof, all sorts of displays of energy occur, and it is the business of the thermodynamics of the atmosphere to describe them and explain them on the same thermodynamical principles as those which are operative in the case of a steam-boat or a locomotive.

Tropopause: the layer of the atmosphere which marks the outer limit of the troposphere and the lower limit of the stratosphere. Subject to reservations, the tropopause may be regarded as a surface; but the transition is not always so abrupt as to produce real discontinuity, and it is therefore convenient to use the word tropopause to connote the phenomena of the region of transition from the troposphere to the stratosphere. The phenomena may include a sudden transition to nearly isothermal conditions, a counterlapse leading to isothermal conditions, or a gradual transition from a lapse-rate which is near the adiabatic to a condition approximately isothermal.

Troposphere: the name given to the lower layer of the atmosphere which is characterised by the effects of thermal convection. See **Atmospheric shells**.

Turbulence: the general name for the state of any portion of the atmosphere which is disturbed by eddies. It is not necessary for turbulence that the eddies should be complete vortices or even nearly so.

The motion of air, or of any other fluid, in which eddies are not being formed is called "stream-line" motion, as distinguished from the turbulent motion with which eddies are associated. When a current of air is passing over a lower layer, slow motion is represented as stream-line motion; but if the velocity of the passing current is increased beyond a certain limit, eddies are produced and the motion of the current becomes turbulent. The effect is dependent upon the difference of density of the current and the air over which it moves.

Vector: "a quantity having direction as well as magnitude denoted by a line drawn from its original to its final position" (*New English Dictionary*). A Latin word, used in geometry, meaning a *carrier*. The **radius vector** of a curve is the line, or radius, that carries or reaches from a chosen centre to a point on the curve. Vectors are required to describe displacement, velocity, angular velocity, acceleration, angular acceleration, force, but not to describe energy, mass or pressure, which are distinguished as scalar quantities.

Vortex: a Latin word meaning an eddy of water, wind or flame, a whirlpool, whirlwind, in modern scientific use a rapid movement of particles of matter round an axis, a whirl of atoms, fluid or vapour. A violent eddy or whirl of air, a whirlwind or cyclone, or the central portion of this (*New English Dictionary*).

A vortex, eddy or whirl is a stream which either returns into itself or moves in a spiral course towards or from an axis, in the latter case two or more successive turns of the same vortex may touch each other laterally without the intervention of any solid partition (W. J. M. Rankine, *Applied Mechanics*, p. 412).

Rankine (p. 574 *et seqq.*) refers to the following different kinds of vortex: (i) **Free circular vortex**, a system of circular currents in which the velocity at any point is inversely proportional to the distance of the point from the axis. "If owing to a slow radial movement particles should find their way from one circular current to another they would assume freely the velocities proper to the several currents entered by them outside the action of any force but weight or fluid-pressure." "A free vortex is a condition towards which every vortex not acted upon by external forces tends because of the tendency of intermixture of particles of adjoining circular currents." (ii) A **free spiral vortex**, a free circular vortex combined with a radiating current to or from the axis of rotation; so the actual motion of each particle is along an equiangular spiral. (iii) A **forced vortex** is one in which the velocity of revolution of the particles follows any law different from that of the free vortex: the kind of forced vortex which is most useful to consider is one in which the particles revolve with equal angular velocities of revolution as if they belonged to a rotating solid body. The energy of any particle is half actual and half potential. (iv) A **combined vortex** consists of a free vortex outside a given cylindrical surface, and a forced vortex within. Conditions must be so adjusted that the velocities of rotation agree at the circle of maximum velocity which is common to the two circulations.

Bjerknes gives a number of vortices all called simple and with a general formula for velocity proportional to the nth power of the distance from the axis and all lying between $n = -1$ which corresponds with Rankine's free vortex and $n = +1$ the special example of his forced vortex. Bjerknes also cites a number of combined vortices the profile of which can be expressed by single equations.

It is to be noted that in all these descriptions the motion described is in a plane section. Any diagram representing one or other of the vortices, for example the diagram of lines of variation of pressure with distance from the axis, is very suggestive of depth as well as breadth, but depth is not really represented and a vortex as here considered is two-dimensional.

To obtain an idea of reality with finite thickness we may suppose the moving fluid to be confined between two parallel plane surfaces, above or below which other and slightly different vortices may be circulating: if the axes are more or less continuous the final boundaries of a persistent atmospheric vortex must be the ground on the one hand and some substitute for a corresponding rigid surface overhead. The resilience of the atmosphere is a sufficient substitute.

It would make for simplicity if we could regard all the two-dimensional vortices into which an actual vortical system in the atmosphere may be supposed divided by horizontal planes, as being all similar with identical laws of velocity and continuous axes, but it is to be feared that the natural phenomena do not permit of simplification in that way.

***vr* Vortex, *v/r* Vortex**: the system of motion represented by Rankine's free vortex, where the velocity is inversely proportional to the distance from the axis, we call a

vr vortex, and that of the special type of forced vortex in which rotation takes place as a solid we call a v/r vortex.

It is further to be noted that in a v/r vortex the vorticity, defined mathematically as $\left(\frac{\partial v_y}{\partial x} - \frac{\partial v_x}{\partial y}\right)$, is everywhere constant and equal to twice the angular velocity of rotation, but in a simple vr vortex it is everywhere zero. Bjerknes[1] points out that the system has zero vorticity except at the axis where it is infinite. The motion expressed as a vr vortex relative to the earth's surface is not in reality a simple vortex free from vorticity, because the motion with the earth round the earth's axis provides the necessary vorticity.

With reference to the dynamical theory, attention ought to be drawn to Rankine's statement that a free vortex, or, as we should call it, a vr vortex, is the condition towards which every vortex tends in consequence of the viscosity of the fluid. The points to notice are (1) that an atmospheric vortex can generally be regarded as a vr vortex with a v/r core, (2) that a vr vortex cannot persist but must always be in process of degrading in consequence of the relative motion of adjacent layers, whereas, on the other hand, (3) a v/r vortex is free from any intrinsic elements of degradation and may persist when the vr portion of the system has become negligible.

Waves: any regular periodic oscillations, the most noticeable case being that of waves on the sea. The three magnitudes that should be known about a wave are the amplitude, the wave-length and the period. The amplitude is half the distance between the extremes of the oscillations, in a sea-wave it is half the vertical distance between the trough and crest, the wave-length is the distance between two successive crests, and the period is the time-interval between two crests passing the same point. In meteorological matters the wave is generally an oscillation with regard to time, like the seasonal variation of temperature, and in such cases the wave-length and the period become identical. The motion of the material in which the wave is travelling may be either in the direction of the travel, as in sound waves, or transverse to it, as in light waves; a particle taking part in the transmission of the oscillation may describe an orbit, circular or elliptical, round its undisturbed position as in the case of a water-wave.

If a quantity varies so as to form a regular series of waves it is usual to express it by a simple mathematical formula of the form $y = a \sin(nt + a)$ or $a \cos(nt + \alpha)$. The method of expressing periodic oscillations by one or more terms of either form is known as "putting into a sine curve," "into a Fourier series," or as "harmonic analysis."

Any periodic oscillations either of the air, water, temperature, or any other variable, recurring more or less regularly, may be referred to as waves. During the passage of sound-waves the pressure of the air at any point alternately rises above and falls below its mean value at the time. A pure note is the result of waves of this sort that are all similar, that is to say, that have the same amplitude and wave-length. The amplitude is defined in this simple case as the extent of the variation from the mean, while the wave-length is the distance between successive maximum values. The period is defined as the time taken for the pressure to pass through the whole cycle of its variations and return to its initial value. Another good example of wave-form is provided by the variations in the temperature of the air experienced in these latitudes on passing from winter to summer. This is not a simple wave-form because of the irregular fluctuations of temperature from day to day, and the amplitude of the annual wave cannot be determined until these have been smoothed out by a mathematical process. Fourier has shown that any irregular wave of this sort is equivalent to the sum of a number of regular waves of the same and shorter wave-length. In America "heat waves" and "cold waves" are spoken of. These are spells of hot and cold weather without any definite duration, and do not recur regularly.

[1] V. Bjerknes, 'On the Dynamics of the Circular Vortex,' *Geofysiske Publikationer*, vol. II, no. 4, Kristiania, 1921, p. 35.

LEST WE FORGET

1 therm = 100,000 British Thermal Units, equivalent to 100,000 lb. of water raised 1° F., or $2{\cdot}52 \times 10^7$ gramme calories, or $1{\cdot}053 \times 10^{15}$ ergs.

1 foot-ton ($g = 981$) = $3{\cdot}097 \times 10^7$ g.cm = $3{\cdot}0380 \times 10^{10}$ ergs.

1 foot-pound ($g = 981$) = $1{\cdot}3562 \times 10^7$ ergs.

1 joule = 10^7 ergs.

1 kilowatt-hour (1 Board of Trade Unit of electricity) = $3{\cdot}6 \times 10^{13}$ ergs.

1 gramme calorie = 4·18 joules.

Capacity for heat of 1 gramme of air at constant pressure 1·010 joules = $1{\cdot}010 \times 10^7$ ergs.

1 watt = 10^7 ergs per second.

1 horse-power = 746 watts = $\frac{3}{4}$ kilowatt (nearly).

1 kilowatt = 10^{10} ergs per second = $1\frac{1}{3}$ horse-power.

1 kilowatt per square dekametre = $1{\cdot}43 \times 10^{-2}$ g.cal per cm^2 per min.

1 flash of lightning is of the order of 10^{10} joules (C. T. R. Wilson).

1 hour of strong summer sunshine in England is about 100 kilowatt-hours per square dekametre ($3{\cdot}6 \times 10^8$ joules).

1 cm of water in evaporating uses $2{\cdot}5 \times 10^{16}$ ergs per square dekametre.

"Surplus" energy of 1 ton of air saturated before a summer thunderstorm may be of the order of 2 kilowatt-hours.

1 cubic dekametre of air at standard pressure and temperature weighs $1\frac{1}{4}$ tons.

The kinetic energy of a ton of:

a light air (force 0–3)	$< 3 \times 10^{11}$ ergs,
a breeze (force 4–5)	3×10^{11} to 10^{12} ergs,
a strong wind (force 6–7)	10^{12} to 3×10^{12} ergs,
a gale (force 8–9)	3×10^{12} to 6×10^{12} ergs,
a storm (force 10–11)	6×10^{12} to $1{\cdot}2 \times 10^{13}$ ergs,
a hurricane (force 12)	$1{\cdot}2 \times 10^{13}$ to [10^{14}] ergs.

The potential energy of 1 ton of rain or hail at the level of 1000 geodynamic metres is 10^{14} ergs;

1 ton of rain is 1 centimetre over 1 square dekametre.

1 pound weight ($g = 981$) = $4{\cdot}45 \times 10^5$ dynes.

1 pound = 453·59243 g. 1 kg = 2·2046223 lb.

g at the equator 978·03 cm/sec²; at the poles 983·21 cm/sec²; at latitude 45° 980·617 cm/sec² or 32·172 ft/sec².

Energy of geostrophic motion of a layer of air 100m thick, contained within a square lying between consecutive 2mb isobars = $5 \times 10^9 \operatorname{cosec}^2 \phi / \rho\omega^2$, or 10^{21} ergs (28 million kilowatt-hours) in lat. 60°; $1{\cdot}5 \times 10^{21}$ ergs in lat. 45°; 3×10^{21} ergs in lat. 30°.

⁂ A list of conversion tables is given in the Index.

NOMENCLATURE AND UNITS OF MEASUREMENT FOR METEOROLOGICAL QUANTITIES

The preceding list of definitions and explanations of technical terms invites a note on the importance of appropriate nomenclature for the success of a science like that of meteorology which must make its appeal for co-operation to the whole world, and must therefore use technical terms which are definite in meaning and commend themselves for their appropriateness to the purpose for which they are employed. It is not the only science which depends on words. Medicine is largely dependent upon words of Greek origin. Botanists and zoologists find the solution of their difficulty in the adoption of Latin names for the genera and species of plants or animals. In meteorology also the practice has been, generally, to use words of Latin or Greek origin as "technical" terms, in some cases with marked success as with Luke Howard's names for the forms of clouds and Teisserenc de Bort's contribution of "troposphere" and "stratosphere" for distinguishing the region of convection from the upper layer beyond the range of convection. The word "tropopause," suggested to the writer in 1918 by E. L. Hawke as a name for the region where ordinary convection ceases, has also found general acceptance. It has been found useful by Professor Pettersson for the corresponding transition between upper and lower layers in the hydrosphere.

But the nomenclature in general use is not always appropriate. Who, without due warning, would suppose that when we speak of a depression we recognise that it is the isobaric surface that is depressed with *relaxation of surface-pressure* not *depression* as a consequence. In the list of definitions objection is taken to the use of the word gradient for the rate of change of temperature in the vertical, because it is required for various rates of change in the horizontal; lapse-rate, suggested as an alternative, is already in common use. Objection is also taken to inversion as adequate for "reversal of temperature-gradient in the vertical", and since this volume was published the word **counterlapse**, which is as good Latin as inversion, has been proposed for the duty but not generally accepted.

Potential temperature is another common expression which needs some consideration. Initially the potential temperature of a sample of air means the temperature which would be attained by the sample if its pressure were increased to some adopted standard without loss or gain of heat. The usual standard is the pressure of 760mm of mercury at the freezing-point of water in lat. 45°, not exactly a happy specification; but the adoption of the millibar in place of the millimetre for the measurement of pressure leads naturally enough to the use of 1000mb as the standard pressure for potential temperature and something more explicit than the general term potential temperature is required. On p. 117 we have suggested the name **megatemperature** but it has not been appreciated, perhaps **megadyne temperature** would be better, of which megatemperature may be regarded as an abbreviation. Or, since the pressure-unit is the millibar, the potential temperature at the **bar** standard might be the **bar-temperature.** For English-speaking people there is the difficulty that the word bar has such a great variety of meanings; but it is not insurmountable.

A term **pseudo-potential-temperature** has been employed to indicate what is in fact the *limiting value of the entropy* of a mass of air which contains water-vapour when the pressure is reduced adiabatically to an extreme low limit. The figure expressing the limit of entropy is the logarithm of the bar-temperature. It is difficult to see that there is anything pseudo about it. The condition reached is quite real.

Other adjectives are used to indicate something different from what is ordinarily understood by *temperature*. V. Bjerknes uses **virtual temperature** as a correction to the gas equation for moisture, defining it as "the temperature which dry air ought to have in order to get the same specific volume as the assumed moist air of temperature *tt*." In England **equivalent temperature** is used for a scale to indicate the degree of human comfort which the air of a room affords, while on the Continent equivalent temperature is in use in order to bring under one number the expression of the whole amount of thermal energy carried by unit mass of air as estimated by the difference of its temperature from the freezing-point of water and augmented by the latent heat of the water-vapour which it contains. The whole is expressed in terms of the range of air-temperature that might correspond therewith. And further we find the expression **equivalent potential temperature** used in like manner to convey the idea of the total thermal energy of temperature and water-vapour which would be available in a specimen of air if it were brought down to sea-level from its present elevation.

From the consideration of quantities of this kind conclusions have been drawn by W. Schmidt, M. Robitzsch and others. They may serve to establish the identity of special air-masses in the atmosphere, and Schmidt has drawn the somewhat surprising conclusion that, contrary to the prevalent idea that the heated surface is the locus of supply of the sun's heat to the atmosphere, the real **Austausch** is the other way (vol. IV, p. 69)—the surface derives heat by convection from the upper air. But the question is to some extent dependent upon whether we have in view the comparative quiet of the daily life of the atmosphere expressed by mean values or the occasional occurrence of strong convection in heavy rain or thunderstorm. There is a certain amount of rashness in assuming that the mean result of the individual causes is the actual result of the mean cause.

Whether it will ever be possible to adopt a word to indicate the equilibrium of balanced forces in the atmosphere without borrowing the dynamical terms stable and stability, which were introduced in the science of dynamics in order to distinguish between different "kinds of equilibrium," stable and unstable, we cannot say; but that desirable end has certainly not yet been achieved. A boy on a moving bicycle, regarded as subject to balancing forces, is in equilibrium, but it would be wrong to call him stable.

Universal units.

When we consider the units in which the measures of the meteorological elements are expressed we must allow that progress is distinctly slow towards that uniformity which the world at large can hardly fail to regard as a desirable end and only waits for some wave of psychological appreciation. The slowness is indeed not surprising when one appreciates the meaning of the association of meteorological work with political and administrative agencies. When, in discharge of his official duty of time-keeper for the nation, with the object of securing the accurate determination of longitude at sea, the British Astronomer Royal advises the adoption of a 24-hour clock, the British Government disregards the advice because there is no apparent demand for it on the part of the general public! So it is not surprising that when the International Meteorological Organisation prescribes for the results of the soundings of the upper air the expression of height in geodynamic metres, pressure in millibars, temperature in "absolute (273° + C)" and humidity in percentage of saturation, the contributor of a report on the state of our knowledge of the upper air in an encyclopaedic work on meteorology should explain that he has disregarded the prescription because other parts of the book express their results in the more familiar metres, millimetres of mercury and centigrade degrees with the discrimination of + or −. And, for reasons which are ultimately identical, the British Meteorological Office reports observations of the upper air to the Air Ministry expressed in feet, millibars and Fahrenheit, and the British Navy gives wind-velocity in knots while its allied services use miles per hour or kilometres per hour. A public that understands what an astronomer *really* means by 12.29 p.m. as the exact time of a disaster may well be proud.

Humidity.

There is a specious uniformity about the expression of humidity as a percentage; but the information is really uninforming until we know of what? And it is hardly reasonable to deny that the proper expression for humidity is a "proper fraction" of which the numerator is the actual pressure or vapour-density and the denominator the greatest possible at the temperature of observation. Here is a table of the state of the air at South Farnborough expressed in that way.

Humidity at South Farnborough recorded in the 28th week of 1934, *July* 9–15.

(Vapour-pressure and saturation-pressure to the nearest millibar)

July	9	10	11	12	13	14	15	Aggregate
7h	$\frac{15}{18}$	$\frac{14}{18}$	$\frac{15}{20}$	$\frac{17}{18}$	$\frac{19}{20}$	$\frac{14}{18}$	$\frac{13}{17}$	$\frac{107}{129}$
13h	$\frac{14}{39}$	$\frac{10}{39}$	$\frac{14}{40}$	$\frac{16}{19}$	$\frac{17}{20}$	$\frac{12}{26}$	$\frac{13}{29}$	$\frac{96}{212}$
16h	—	$\frac{6}{37}$	—	—	—	—	—	$\frac{6}{37}$
18h	$\frac{8}{31}$	$\frac{8}{33}$	$\frac{12}{34}$	$\frac{17}{20}$	$\frac{19}{22}$	$\frac{11}{24}$	$\frac{13}{23}$	$\frac{88}{187}$

For percentage humidity these fractions are reduced to the common denominator 100.

There is of course a difficulty about taking the mean of a number of proper fractions; but that only raises the question whether the mean of the relative humidity for the day, or the year, has anything more than an arithmetical meaning. To substitute for the figures in the table the mean value of the quotients of the several terms would be nothing less than heedless massacre of the innocent facts upon which the future must depend for progress in the science.

The numerical aggregates given in the table show the totals of the seven numerators and denominators of the fractions in the separate rows, and some hint of the natural sequence of events is given by comparing the totals for the several hours of observation. The surface-air, apparently, loses 20 per cent. of its water in passing from 7h to 18h. The mean for the 24 hours would be a parable without a meaning.

Something more nearly intelligible is expressed if we consider the **drying power of the air**, that is the difference from saturation at the time of observation or the difference between the denominator and the numerator in each fraction. In round numbers they run as follows:

Drying power of air at South Farnborough, July 1934.

July	9	10	11	12	13	14	15
				millibars			
7h	3	5	5	1	1	5	4
13h	25	29	26	3	3	14	16
16h	—	32	—	—	—	—	—
18h	23	25	22	3	3	13	11

The differences are responsible for the evaporation of water into the atmosphere and it can hardly be doubted that the best way of summarising the state of the air as regards humidity within any period, a day, a week, a month or a year is by recording the evaporation of water within the period under specified conditions of exposure, leaving to the observations of humidity the enjoyment of their separate utility. Evaporation can obviously be aggregated whereas the relative humidity can not nor indeed the drying power.

The transformations of energy in meteorological processes.

It is perhaps in the expression of the different forms of energy that we find the greatest need for some universal understanding. Whatever modern mathematical analysis may suggest to the contrary we cannot disregard the conclusion of the first half of the nineteenth century about the conservation of energy. Provided that we include heat and electricity among the possible forms, energy in any form is convertible but indestructible by the processes which we observe. And in the phenomena of weather we can find, and must find, only the transformations, never the annihilation of the energy that comes within our powers of observation.

Meteorology is in fact the study of the dynamical and physical processes of weather and must naturally bring its system of measurement into relation with that which the physicists have adopted for the physical quantities which we employ. The international system for electrical and magnetic measurements is based upon the metric system with its conventions about decimal multiples and fractions, deka, hecto, kilo mega: deci, centi, milli, micro. Hence we have the C.G.S. system, based upon the centimetre, gramme and second as fundamental units, which includes force and energy as primary considerations, with the dyne and the erg for units. It is desirable that meteorology should use the system where possible.

The main source of supply of energy which we can recognise in meteorological processes is the radiation which reaches us from the sun, one of the forms of expression of the mysterious law that every body radiates its energy to every other body within sight, according to its temperature. And, with that, we have to associate the other mysterious influence of every body upon every other body in the universe, whether within sight or not, namely the influence of gravitation. These two mysterious influences acting together upon the solid earth and the atmosphere, with its load of that versatile substance water, exhibit our weather as the effect of their co-operation in the transformation of the energy which they have in charge, and heat is one of the forms of energy employed in the process.

In common life we have gone so far as to treat energy as a commercial product, and sell it in therms, 100,000 British Thermal Units (each $1{\cdot}053 \times 10^{10}$ ergs) in the form of combustible coal gas, or as electrical energy in Board of Trade Units (kilowatt-hours, $3{\cdot}6 \times 10^{13}$ ergs). And the diversity of practice is regarded as excused because thermal energy is only convertible into work or motion under thermodynamical conditions, whereas electrical energy can be converted with only resistance-losses. And yet the physicist would suggest to us that heat itself is a form or "mode" of motion, perhaps of the ultimate particles of the body which we call warm.

From the meteorologist's point of view it would seem to be important to lay stress on the ultimate identity of energy rather than its superficial diversity. It would be a boon for the science if we might have ergs with some power of ten in which to express it while we are thinking about the details of its transformations.

Temperature in meteorological practice.

The one feature of common practice which any recording angel (or meteorological observer) must regard with derision is that which uses a positive quantity for one part of a continuous scale of magnitude and a negative quantity for the other part of the same scale. We get the result that a physical quantity is increasing when its numerical expression is decreasing, and particularly for that element which is observed by everybody who contributes to our stock of information about weather. Even the most rudimentary observer has to become aware, when he estimates a fraction of a degree on his scale, Fahrenheit or centigrade, that he must count downwards when he looks at one part of his instrument and upwards when he looks at the other part. It is true that one of the points of transition, the freezing-point of water, is a very important point, but the recognition of that fact does not remove the difficulty. It is true, too, that the same convention prevails with angles, as in latitude and longitude, or with heights referred to sea-level; but then we give a separate name to the readings of the scale above and below the point of transition—for latitude we have North and South, for longitude East and West, the one side of the vertical scale is height and the other depth; but we have no such convention about the parts of the scale above and below the freezing-point of water, except that newspapers sometimes express low temperatures as degrees of frost or as degrees below zero without any indication of which. The distinction is by the symbols $+$ and $-$; and $-$ is so much in demand by the typist and the printer for other purposes that it is already overworked.

Various plans have been adopted to avoid a change of sign at the freezing-point. Fahrenheit's suggestion was to call the freezing-point of water 32°, with the zero point at what was thought to be the lowest observable temperature. A temperature of 272·92° centigrade below the freezing-point has now been attained and we may even await with some interest an experiment which will indicate a temperature below what is now accepted as the absolute zero, in accordance with nature's incurable habit of laughing at the locksmiths of natural philosophy. An extension of Fahrenheit's principle is advisable. That prescribed by the International Organisation is to graduate the instrument uniformly in centigrade degrees and mark the freezing-point 273°.

That is the plan which has been adopted in this work; with the suggestion already referred to that the scale so graduated should be called **tercentesimal** because the graduation 300 is not far from the extreme, and the symbol for the unit thus employed should be tt. The prescription of the International Organisation calls the scale absolute, but the comparison of the graduation of any thermometer with the real absolute scale is an elaborate process which meteorologists would not often undertake. The practice of naming the scale absolute can only be accepted *cum grano salis* which a large number of meteorological observers would fail to understand. The prescription gets over the difficulty by putting (273° + C) in a bracket after the word absolute.

The use of the degree sign, *the symbol of angular measurement*, to indicate steps in a scale is not really appropriate for temperature. The magnitude of an angle indicates an angular interval from a zero line which is necessarily arbitrary. So with temperature, so long as we understand that the scale of degrees represents intervals from an arbitrary zero, the practice has some justification; but modern meteorology cannot regard the magnitude assigned to temperature simply as recording the position of the mercury on an arbitrary scale. We must regard temperature as controlling many of the physical processes upon which the science of meteorology depends. Temperature controls the saturation pressure of the vapour of water, the amount of radiation from every body in the universe and so on. Hence for us temperature takes on the rôle of a physical quantity closely concerned with the transformation of energy and should certainly be measured from a zero at which evaporation and radiation are zero, that is the absolute zero.

In the interest of the rising generation of meteorologists it would perhaps be best to make a present of the word temperature to those who cannot dissociate it from its relation to the freezing-point of water and to use another name for the index of thermal energy. Thermancy has been suggested in vol. III. Meanwhile we may use the tercentesimal scale of temperature with a *quantity symbol* instead of a *degree sign* for its expression. The quantity symbol which we have adopted is tt, with the understanding that the symbol t may be used for the number of units of quantity corresponding with a *range* of temperature.

Temperature or thermancy as a measure of energy.

The graduation of our thermometers in practice marks equal increments in the volume of the mercury in its containing tube and the physicists have taught us that the same graduations mark equal increments in the volume of a mass of dry air at constant pressure, and indeed of moist air too, and that the changes correspond also with equal steps of temperature on the absolute scale. The actual volume of the mercury in a glass vessel is not of direct importance in meteorological work but the volume of a gramme of dry air is an important meteorological quantity especially in respect of evaluation of the thermal energy; and here we quote from p. 223 of vol. III with the correction of a casual error.

> The algebraical representation of the energy of a gramme of air in C.G.S. units is $3pv/2$ which is proportional to *tt*. At 278·2 tt the arithmetical value is approximately $1{\cdot}2 \times 10^9$ ergs, at 273tt $1{\cdot}17 \times 10^9$ ergs and at 300tt $1{\cdot}29 \times 10^9$ ergs.... We ought to find another name for that aspect of temperature which is of vital importance for the science and which indicates the molecular energy of a gas, "how fast the atoms are moving." Perhaps the development of the subject might proceed more smoothly if we regarded "tt" or absolute temperature as indicating the **thermancy** of the gas instead of its temperature.

Will the time ever arrive when the meteorologists of the world agree to adopt an instrument which marks the graduation of temperature by the corresponding volume of unit mass of dry air at the pressure of 1000mb, when the freezing-point would read 784cm^3 and 300tt 861cm^3, or by the corresponding equivalents in the energy of a gramme of dry air, $1{\cdot}174 \times 10^9$ ergs at the freezing-point and $1{\cdot}29 \times 10^9$ ergs at 300? On such a scale of thermancy the graduation marks would be: 1000 megalergs at 232·3tt, 1100 at 255·5tt, 1200 at 278·7tt and 1300 at 302·0.

The question is really worth consideration.

Symbols for units.

One small point may be noticed with reference to the printing of the expressions of meteorological quantity. The unit is not always written in full, as ergs and dynes often are but is expressed by a symbol. It is a printer's rule to regard an abbreviation as a separate word and follow it with a full stop of the same type as that which marks our decimal point; but that is not necessary when the unit is indicated by a symbol. It is better to print the symbol close to the number as, for example, 50g, and leave the figures and the unit-symbol to be a single word. That practice has been followed in this book, and is perhaps successful, except in the case of in as a symbol for inch. The symbols are printed in roman type, leaving italics for the variables of equations. If the plan were generally adopted we might ask some good friend with physical influence to suggest for us symbols for the dyne and the erg.

Units of time for meteorological purposes.

The natural time-intervals to which meteorological observations must be related are the mean solar day based on the earth's rotation round its axis in relation to the sun, and subdivided into twenty-four hours, fourteen hundred and forty minutes, or 86,400 seconds, and the year, the period of revolution of the earth round the sun which, strictly speaking, covers 365·2422 mean solar days. In practice we make use of a nominal year of 365 days with an additional day for leap year every fourth year, century years generally excepted, in order to keep an exact number of days.

Some time-unit intermediate between the day and the year is necessary. Meteorological practice has adopted for that purpose the so-called months of our Calendar as established by Julius Caesar, revised by his nephew Augustus and finally revised by Pope Gregory XIII in 1582. There are many objections to the calendar months as time-units, they are not of equal length and for many purposes they are too long for the registration of the sequence of weather. Also in the lapse of time they have become very uncomfortably related to the solstices and equinoxes which are the really cardinal points in the kalendar. Some slight discrepancy within the limits of a day in the position of these events is unavoidable on account of the extra quarter of a day in the mean solar year which is corrected by leap year; but, quite inexcusably, the festivals selected for the celebration of the solstices and equinoxes, Christmas Day, Lady Day, St John's Day (our Midsummer Day) and Michaelmas Day which are our quarter-days, have been allowed to wander from their proper places in the kalendar: 22nd or 23rd December to 25th, 20th or 21st March to 25th, 21st or 22nd June to 24th, and 23rd or 24th September to 29th.

There is another objection to the months as time-units. Their name implies that they are intended to be related to the changes in the moon, but as the lunar period is 29½ days there is room in the year for 12⅓ months, and the reduction to twelve, the adjustment of months to the signs of the Zodiac, has been the source of much trouble in the arrangement of our Calendar.

Some ancient kalendars were real lunar kalendars; the Hebrew and the Muslim kalendars are so still. It is easy to understand that when the only practical illumination of the night-hours was moonlight it was necessary to have regard to the moon's position in arranging a kalendar. We must not therefore be surprised at the attention paid to moonlight up to the time of the invention of gas and other substitutes for sunlight. In our civil Calendar the Zodiac, the twelve successive constellations through which the sun appears to pass, has won and we are left with mere names for months.

But for meteorological purposes it is sunlight, directly related to the solstices and equinoxes, that we have to keep in mind rather than moonlight, and the question which we have now to consider is whether we can find a group of days to serve as a time-unit intermediate between the day and the year that can be related directly to the duration of sunlight.

The week of seven days is already in use for many social and economic purposes and is not unsuitable in length for meteorological records. All that remains for us to consider is whether it can be utilised as the basis of a meteorological kalendar.

There would be 52 weeks in the year with one day over in each year, two days in leap year. One advantage of the 52, not possessed by groups of five days or six days, which are sometimes used as units, is that it is divisible by four and the year can be built up of four quarters, of thirteen weeks each, centred approximately at the solstices and equinoxes. The extra days, after the completion of 52 weeks, would have to be separately allowed for in the meteorological accounts, we might regard them as audit days.

The seventh week of each quarter would contain a solstice or an equinox and each quarter can be arranged to consist of three chapters, one of five weeks centred at the solstice or equinox and one of four weeks on either side of the central chapter of the quarter.

To bring this arrangement into relation with the existing Calendar, and for other reasons too, it would be convenient to begin the year which has been called the Grower's Year with 6 November. That gives a quarter extending to 4 February and centred at the winter solstice; eight of the weeks would be in the old year and five in the new year beginning with 1 January. So, if we count by the total number of weeks elapsed, transition from the grower's year to the civil year, a difference of eight weeks, would be an easy matter.

The week has been used for meteorological purposes, notably in the *Weekly Weather Report* of the Meteorological Office, but the rule adopted for that publication was that a week must begin with Sunday, so that the first day of a new year varied from 29 December to 4 January and in some cases there were 53 weeks in the year; consequently it was never possible to compare the week of one year with those of other years having the same duration of daylight. If the duration of daylight is our guiding principle we must allow the weeks to be in strict relation to solstices and equinoxes and waive the claim for the week to begin with Sunday. That is not inappropriate for meteorological observations because the weather pays no attention to the names of the days of the week.

The reckoning of meteorological time by counting weeks and days from a fixed zero is illustrated in a paper with the title of "The Natural History of Weather" read before the Royal Meteorological Society. A kalendar setting out the arrangement may be called a "daylight kalendar." The relation of the grower's year and the daylight kalendar to the civil year as at present arranged is set out for Greenwich in the following table.

Counting by days is sometimes desirable; and, with weeks beginning at a fixed date, the transition is easy, as the table shows.

THE DAYLIGHT KALENDAR

Short-days quarter				Lengthening-days quarter				Long-days quarter				Shortening-days quarter			
The weeks begin:			Daylight hrs.	The weeks begin:			Daylight hrs.	The weeks begin:			Daylight hrs.	The weeks begin:			Daylight hrs.
	No. 1	Nov. 6	64¾		No. 14	Feb. 5	66¾		No. 27	May 7	107¼		No. 40	Aug. 6	105
Ch.	2	13	62¼	Ch.	15	12	69¾	Ch.	28	14	109¾	Ch.	41	13	102¼
I	3	20	59¾	IV	16	19	73	VII	29	21	112	X	42	20	99¼
	4	27	57¾		17	26	76¼		30	28	113¾		43	27	96¼
	5	Dec. 4	56¼		18	Mar. 5	79¼		31	June 4	115		44	Sept. 3	93
	6	11	55¼		19	12	82½		32	11	116		45	10	90
II	7	18	*55	V	20	19	†85¾	VIII	33	18	*116½	XI	46	17	†86¾
	8	25	55		21	26	89		34	25	116¼		47	24	83½
	9	Jan. 1	55¾		22	April 2	92		35	July 2	115¼		48	Oct. 1	80¼
	10	Jan. 8	57¼		23	April 9	95¼		36	July 9	114		49	Oct. 8	77¼
III	11	15	59¼	VI	24	16	98½	IX	37	16	112½		50	15	74
	12	22	61½		25	23	101½		38	23	110¼	XII	51	22	71
	13	29	64		26	30	104¼		39	30	107¾		52	29	68
													Appendix	Nov. 5	9½

* Solstice. † Equinox.

The daylight hours are for London's weeks.

For the application of meteorology to agriculture, daylight is obviously a leading consideration, and in many other aspects of the influence of weather daylight is an essential factor. Consequently the daylight kalendar for an agricultural station is of some importance; but the length of daylight depends on the latitude and can be expressed by an astronomical formula. So our "frontispiece" for this volume of meteorological facts is a diagram (fig. iii) which exhibits the duration of daylight at every latitude throughout the year, and on it we have marked the limits of the twelve chapters of the grower's year as well as the months of the Calendar as at present arranged.

DURATION OF DAYLIGHT IN RELATION TO LATITUDE

Fig. iii. Lines of equal numbers of hours of daylight in different latitudes during the several quarters of the grower's year (beginning with 6 November), and in the months of the civil year.

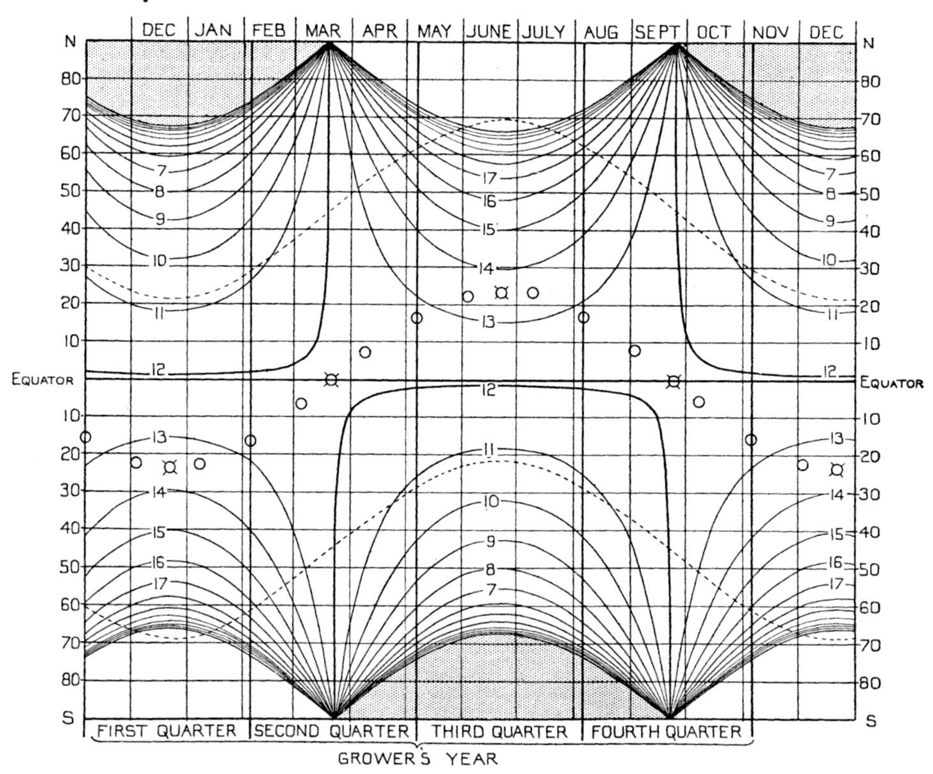

Horizontal lines are lines of latitude in steps of ten degrees North and South of the equator.

Thin vertical lines separate the months of the civil year; the thicker ones mark the quarters of the grower's year: 6 Nov. to 4 Feb., 5 Feb. to 6 May, 7 May to 5 Aug., 6 Aug. to 4 Nov. with 5 Nov. as audit day.

To find the duration of daylight at any time of the year—look out the line of latitude of the station, e.g. Bergen 60° N and read along the line, just over $8\frac{1}{4}$ hr./day on 6 Nov., $6\frac{1}{2}$ hr. on 1 Dec., 6 hr. on 1 Jan., $7\frac{3}{4}$ hr. on 1 Feb., $10\frac{1}{4}$ hr. on 1 Mar., 13 hr. on 1 April, $15\frac{3}{4}$ hr. on 1 May, 18 hr. on 1 June, $18\frac{3}{4}$ hr. on 1 July, 17 hr. on 1 Aug., 14 hr. on 1 Sept., $11\frac{1}{2}$ hr. on 1 Oct., and $8\frac{3}{4}$ hr. on 1 Nov. At the commencements of the four quarters we find 6 Nov. $8\frac{1}{4}$ hr., 5 Feb. $8\frac{1}{4}$ hr. 7 May $16\frac{1}{4}$ hr., 6 Aug. $16\frac{1}{2}$ hr. [5 Nov. $8\frac{1}{4}$ hr.]

The row of circles crossing the equator shows the latitude at which the sun is vertical at noon, and parallel curves to the North and South mark the latitudes at which the altitude of the sun at noon is 45°. Then for a latitude n degrees *North* of the 45° curve in the Northern Hemisphere the altitude of the sun at noon will be n degrees less than 45°. For the Southern Hemisphere a corresponding curve is shown; and so the altitude of the sun at noon for any part of the earth's surface at any time of the year is determinable.

CHAPTER I

THE INFLUENCE OF SUN AND SPACE

SOME PRELIMINARY FIGURES

Conditions of balance between solar and terrestrial radiation for a horizontal black surface and a perfectly transparent atmosphere.

Temperature (a) of a black horizontal surface ...	200	210	220	230	240	250	260	270	280	290	300
Sun's altitude for balance	3½°	4¼°	5°	6°	7¼°	8½°	10°	11¾°	13¾°	15¾°	18°
Temperature (a) of a black horizontal surface ...	310	320	330	340	350	360	370	380	390	400	402
Sun's altitude for balance	20¾°	23¾°	27°	31°	35¼°	40°	46°	53¼°	62¼°	79¼°	90°

Temperature (a) = tercentesimal temperature (tt) + ·10 ± ·05 (*Dict. App. Phys.* vol. 1).

Mean solar constant 135 kilowatts per square dekametre, 1·93 gramme calories per square centimetre per minute.

Stefan's constant of radiation from unit area of a black body $5{\cdot}72 \times 10^{-9}$ kilowatts per square dekametre per degree of absolute temperature, or 82×10^{-12} gramme calories per square centimetre per minute.

Constant of gravitation $6{\cdot}6576 \times 10^{-8}$ $cm^3/(g.\ sec^2)$.

Mean distance of earth from sun 149,500,000 kilometres.

Minimum distance of earth from sun 146,700,000 kilometres.

Maximum distance of earth from sun 152,100,000 kilometres.

In the fifteen chapters of our historical introduction we have sketched the evolution of modern methods of obtaining current information about the condition of the atmosphere and the facilities for dealing with the information thus obtained. In the present volume we offer for the reader's consideration a representation of the structure of the atmosphere, and some indication of the general circulation which is based upon observations collected in the manner described.

The representation cannot pretend to take account of all the observations which are, in one way or other, pertinent to the subject under consideration. The author, when he was director of the Meteorological Office in London, endeavoured to bring together in compact form the information about the weather of the British Isles which was collected in the ordinary course of duty and found himself responsible for about 2000 maps and as many pages of tables, expressing the data for one year. This was exclusive of the work on the meteorology of the sea for which, strangely enough, the publication of current data remains without adequate organisation on an international basis. The number of "significant figures" on a page, many of which are themselves summaries, may run to 5000. A year's output on this scale is quite beyond the capacity of any human being to keep in mind. In ten years a corresponding output for 50 countries of similar meteorological importance would provide 1,000,000 pages and would occupy some 100 metres' run of a substantial book-shelf. In the common jargon of the geophysicist the number of data to be dealt with is of the order 10^{10} to 10^{12}. The publication of such a vast number of facts for the several countries can only be justified on the understanding that the data are available for a large variety of economic and scientific purposes apart from the special study of the atmosphere; the general solution of the atmospheric problem can only be approached by adopting some systematic plan of dealing with the vast accumulation of data.

By general consent the first step is to deal with "normals" for selected periods based upon the co-ordination of data extending over a long series of years. In accordance with international agreement, confirming a practice already established, the year and, wisely or unwisely, the calendar months are the selected periods.

Our first endeavour is therefore to represent the "normal" structure of the atmosphere for the month or the year and the normal circulation which corresponds therewith.

The atmosphere has thickness as well as length and breadth. The length and breadth at the surface constitute the base of the structure; and, for these, vast collections of data are available, though there are still many regions for which no adequate monthly data exist. For the thickness comparatively few data are available. We must therefore deal with the thickness in a much more sketchy manner than the base.

By the presentation month by month of the base of the normal structure and circulation, seasonal changes are disclosed which are the main features of climate all over the globe; and one of the primary problems of meteorology is to study these changes and if possible to ascertain their causes. On that account we have judged it necessary to represent the conditions month by month. In meteorological text-books, for purposes of illustration, it is often deemed sufficient to present the extremes of seasonal conditions as represented by the months of January and July. But for the purposes of study the transition months are indispensable because it is precisely the course and causes of transition that we seek.

When in this way the normal structure and the normal circulation have been briefly indicated, we shall endeavour to represent the data upon which we must rely for insight into the nature of those local deviations of the normal circulation which constitute the sequence of weather.

SOLAR RADIATION AND SUNSPOTS

While we set out here as clearly as we can the general features of the problem of the normal circulation of the atmosphere and its local variations, we must reserve for a subsequent volume the consideration of the progress that has been achieved in tracing the physical and dynamical relationships between the several features, and the contributions made thereby towards the explanation of the sequence of weather as the natural effects of ascertained physical causes.

The fundamental causes have to be sought in solar and terrestrial radiation. That aspect of meteorology we must consider in due course; but the details of the physical processes by which, for example, radiation is related quantitatively to temperature or its possible alternative vapour-pressure are still in the stage of development that belongs rather to the meteorological laboratory than to the normal observatory, and we must accordingly postpone the detailed consideration of the subject until we come to deal with our knowledge of the physical processes which are operative in the atmosphere.

Yet even here we think it desirable to enable our readers to keep in mind some of the information about the sun and its radiation, and about radiation from the earth into free space, the use of which may be called for at any time and does not require any expert knowledge of the details of the physical processes involved. The information includes, first, the distribution of solar energy over the different regions of the globe as computed by A. Angot[1] assuming a value for the "constant" of solar radiation, that is the amount of energy which would reach unit-area of receiving surface at right angles to the sun's rays at the outer limit of the atmosphere; we shall take a square dekametre as the unit area and in a table on pp. 4, 5 express the energy of radiation received by it in the ordinary unit of power, the kilowatt.

SOLAR ENERGY, SUNSHINE AND SOLAR HEAT

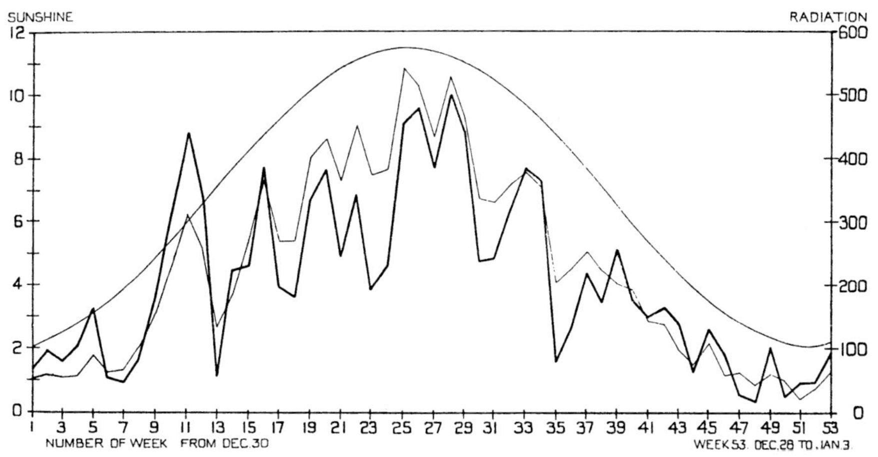

Week 1. Dec. 30 to Jan. 5. Week 2. Jan. 6 to Jan. 12, etc.

Fig. 1. Thick line —— mean daily duration of sunshine in hours per day. Thin line —— mean daily total of radiation received from sun and sky in kilowatt-hours per square dekametre of horizontal surface at Rothamsted week by week in 1924.

The smooth curve represents, on the same scale as that of radiation received, 50 per cent. of the solar radiation per square dekametre of horizontal surface.

The numbers for the smooth curve of solar energy in fig. 1 are the halves of those given in the column for 50° in the table of p. 4.

By way of comparison we have included in the same figure the amount of energy received by a Callendar recorder at Rothamsted in 1924 as daily averages for successive weeks; daily duration of sunshine for corresponding weeks is also shown. The latitude of Rothamsted is 51° 48′ N.

[1] *Ann. bur. cent. météor.*, Paris, 1883, part 1 (1885), pp. B. 121–169.

Energy, in kilowatt-hours, which would be received, if there were no atmosphere, upon a hundred square metres of horizontal surface by direct radiation from the sun with a solar constant of 135 kilowatts per square dekametre.

Totals for the middle day of successive weeks of the year.
Multiply by $3{\cdot}6 \times 10^{13}$ to express the result in ergs per sq. dekametre.
Multiply by $3{\cdot}6$ to express the result in joules per sq. centimetre.

NORTHERN HEMISPHERE

		Date	90°	80°	70°	60°	50°	40°	30°	20°	10°	0°
	22° 45′ S	Jan. 4	—	—	—	68	217	385	554	714	859	987
	21° 50′	,, 11	—	—	—	80	234	401	568	728	869	990
	20° 35′	,, 18	—	—	—	99	255	424	590	745	882	998
	19° 01′	,, 25	—	—	4	124	285	452	616	767	898	1006
	17° 10′	Feb. 1	—	—	19	155	320	487	647	791	915	1015
	15° 4′	,, 8	—	—	45	193	360	525	680	818	934	1023
	12° 45′	,, 15	—	—	78	238	406	568	717	846	952	1030
	10° 16′	,, 22	—	—	123	289	456	614	755	875	971	1037
	7° 40′	Mar. 1	—	20	174	344	509	660	794	903	987	1041
SPRING	4° 58′	,, 8	—	63	234	404	564	709	833	932	1002	1042
	2° 14′ S	,, 15	—	123	298	467	621	757	871	957	1014	1041
	† 0° 32′ N	,, 22	30	196	370	532	679	805	906	980	1025	1038
	3° 17′	,, 29	186	281	444	599	736	852	941	1002	1033	1033
	5° 59′	Apr. 5	338	377	524	667	792	896	973	1021	1038	1025
	8° 36′	,, 12	482	482	603	733	846	938	1002	1037	1041	1014
	11° 5′	,, 19	617	608	683	798	898	976	1027	1050	1041	1002
	13° 26′	,, 26	743	732	763	860	946	1013	1050	1061	1040	990
	15° 36′	May 3	857	844	841	918	991	1045	1072	1069	1038	977
	17° 33′	,, 10	957	944	915	971	1031	1072	1088	1076	1034	964
	19° 16′	,, 17	1045	1029	984	1018	1065	1096	1102	1080	1029	952
	20° 43′	,, 24	1116	1100	1050	1058	1095	1115	1112	1083	1025	940
	21° 53′	,, 31	1175	1157	1103	1092	1118	1131	1122	1084	1021	930
SUMMER	22° 44′	June 7	1215	1196	1142	1115	1134	1142	1127	1085	1017	923
	23° 15′	,, 14	1239	1220	1165	1130	1143	1148	1130	1085	1014	918
	*23° 27′	,, 21	1249	1229	1173	1135	1148	1150	1130	1085	1013	915
	23° 18′	,, 28	1239	1222	1165	1130	1143	1148	1129	1084	1013	917
	22° 49′	July 5	1215	1197	1142	1115	1133	1139	1125	1083	1014	919
	22° 01′	,, 12	1176	1157	1104	1091	1115	1127	1116	1080	1015	925
	20° 54′	,, 19	1119	1102	1052	1057	1091	1111	1107	1076	1018	933
	19° 30′	,, 26	1049	1033	987	1017	1062	1091	1095	1072	1021	942
	17° 50′	Aug. 2	964	949	918	971	1027	1067	1081	1067	1025	953
	15° 56′	,, 9	865	852	845	918	988	1040	1064	1060	1027	965
	13° 50′	,, 16	756	744	770	861	945	1007	1044	1050	1029	977
	11° 32′	,, 23	633	624	691	801	898	972	1021	1040	1029	988
	9° 6′	,, 30	504	500	612	737	846	934	995	1027	1029	999
AUTUMN	6° 33′	Sept. 6	365	393	533	672	794	894	967	1011	1026	1008
	3° 54′	,, 13	217	297	456	606	738	851	936	994	1021	1017
	† 1° 12′ N	,, 20	68	212	382	541	683	805	903	972	1013	1023
	1° 32′ S	,, 27	—	138	312	477	628	759	868	950	1004	1027
	4° 15′	Oct. 4	—	77	246	414	571	713	832	926	992	1029
	6° 56′	,, 11	—	30	186	355	517	666	794	899	979	1027
	9° 32′	,, 18	—	1	135	300	466	620	757	872	963	1025
	12° 1′	,, 25	—	—	90	250	416	575	720	845	946	1021
	14° 21′	Nov. 1	—	—	54	205	370	533	684	818	930	1015
	16° 30′	,, 8	—	—	27	166	329	494	651	792	913	1008
	18° 26′	,, 15	—	—	8	132	293	460	621	768	896	1000
	20° 05′	,, 22	—	—	—	107	263	431	594	747	882	994
	21° 27′	,, 29	—	—	—	85	239	406	572	729	868	987
WINTER	22° 28′	Dec. 6	—	—	—	72	221	387	556	716	859	981
	23° 8′	,, 13	—	—	—	62	209	377	545	706	853	979
	*23° 26′	,, 20	—	—	—	58	204	371	540	702	851	977
	23° 21′	,, 27	—	—	—	59	207	373	543	705	852	979
	23° 7′	,, 31	—	—	—	62	211	378	547	709	855	981

* Solstice. † Equinox.

The table shows the year divided into fifty-two weeks with one day, namely, December 31, over. The weeks are grouped in fours and fives. The groups of five introduce the seasons of the Farmers' year, spring, summer, autumn, winter, and are so chosen that their middle weeks contain the solstices and the equinoxes. The first days of the seasons would be March 5, June 4, September 3, December 3 respectively. Each group of five with a group of four before and after forms a quarter of the "May-year" which is arranged in accordance with the sun's declination, the dates of commencement being February 5, May 7, August 6 and

Energy, in kilowatt-hours, which would be received, if there were no atmosphere, upon a hundred square metres of horizontal surface by direct radiation from the sun with a solar constant of 135 kilowatts per square dekametre.

Totals for the middle day of successive weeks of the year.

Multiply by ·86 to express the result in gramme calories per square centimetre.

SOUTHERN HEMISPHERE

0°	10°	20°	30°	40°	50°	60°	70°	80°	90°	Week of May-year		Orbit factor
987	1084	1157	1202	1218	1210	1189	1218	1277	1296	I	9	·9832
990	1085	1153	1192	1202	1188	1160	1170	1227	1246		10	·9834
998	1087	1148	1179	1180	1157	1118	1106	1158	1177		11	·9839
1006	1087	1138	1161	1153	1119	1067	1027	1072	1089		12	·9846
1015	1085	1127	1139	1121	1075	1008	945	969	984		13	·9855
1023	1084	1114	1114	1081	1023	944	857	852	865	II	1	·9866
1030	1079	1098	1083	1040	967	872	767	721	733		2	·9879
1037	1072	1077	1050	992	907	799	676	581	590		3	·9895
1041	1064	1054	1014	944	846	725	587	454	440		4	·9911
1042	1052	1029	975	892	783	651	501	344	285		5	·9929
1041	1037	1000	934	840	720	576	419	248	128		6	·9948
1038	1019	971	891	786	656	506	340	166	—		7	·9968†
1033	1000	938	848	732	594	437	270	97	—		8	·9988
1025	979	906	805	679	535	374	207	45	—		9	1·0008
1014	957	872	761	628	478	316	151	7	—		10	1·0028
1002	934	838	720	581	425	263	104	—	—		11	1·0048
990	911	806	679	535	378	216	65	—	—		12	1·0067
977	888	776	643	494	335	176	36	—	—		13	1·0084
964	868	748	609	456	297	142	15	—	—	III	1	1·0101
952	848	722	579	424	265	113	3	—	—		2	1·0116
940	832	701	555	398	239	92	—	—	—		3	1·0130
930	817	683	535	377	219	74	—	—	—		4	1·0141
923	806	670	520	360	204	63	—	—	—		5	1·0151
918	799	662	510	351	194	57	—	—	—		6	1·0158
915	797	657	506	347	192	54	—	—	—		7	1·0164*
917	797	660	509	350	193	56	—	—	—		8	1·0167
919	803	667	516	358	201	62	—	—	—		9	1·0167
925	811	678	529	373	215	73	—	—	—		10	1·0166
933	824	694	548	392	234	88	—	—	—		11	1·0162
942	838	714	571	417	259	109	1	—	—		12	1·0157
953	857	737	599	447	289	135	12	—	—		13	1·0149
965	876	764	630	482	325	167	31	—	—	IV	1	1·0137
977	898	792	666	521	366	207	59	—	—		2	1·0125
988	919	824	703	564	412	251	95	—	—		3	1·0110
999	941	855	744	612	462	301	139	4	—		4	1·0094
1008	963	887	784	660	516	356	192	35	—		5	1·0077
1017	983	918	826	710	572	417	251	82	—		6	1·0059
1023	1002	949	869	761	632	482	319	146	—		7	1·0040†
1027	1019	980	911	814	694	551	393	223	86		8	1·0020
1029	1034	1008	952	867	756	622	473	313	240		9	1·0000
1027	1046	1034	991	918	818	695	556	416	393		10	·9980
1025	1057	1058	1027	967	879	768	643	535	541		11	·9960
1021	1065	1079	1061	1014	938	841	732	672	683		12	·9941
1015	1072	1098	1092	1058	995	911	821	803	815		13	·9922
1008	1076	1114	1121	1098	1048	979	907	923	938	I	1	·9905
1000	1077	1126	1145	1133	1096	1040	992	1031	1048		2	·9889
994	1080	1137	1165	1164	1137	1095	1072	1123	1141		3	·9874
987	1080	1145	1181	1189	1172	1141	1145	1200	1219		4	·9861
981	1080	1152	1193	1208	1199	1176	1199	1257	1276		5	·9851
979	1081	1156	1203	1220	1216	1200	1235	1295	1314		6	·9843
977	1081	1158	1207	1227	1224	1211	1251	1311	1331		7	·9837*
979	1083	1158	1207	1227	1223	1210	1247	1308	1328		8	·9833
981	1083	1158	1204	1223	1218	1202	1237	1296	1316		x	·9832

* Solstice. † Equinox.

November 5, these comply with the specification given in chapter III of vol. I. In the table the groups of four and five weeks are shown by spaces between the lines. Large spaces separate the quarters of the May-year, the smaller spaces the seasons, or quarters of the Farmers' year. The quarters of the present kalendar year are somewhat deranged, each one of them being in its turn a week late, that is to say, a quarter begins with Jan. 8, April 9, July 9, October 8. There is no great disadvantage about this so far as statistical meteorology is concerned; indeed it leads us to a kalendar adjusted to the duration of daylight.

Secondly, we give a table of the accepted mean values of the solar "constant" during the period 1912–24 as determined by Dr C. G. Abbot and his colleagues of the Smithsonian Institution of Washington.

Mean values of the "solar constant" 1912–20[1]
in kilowatts per square dekametre.

Mount Wilson		kw/(10 m)²	Hump Mountain		kw/(10 m)²
1912	May to September	135·6	1917	June to December	133·7
1913	July to November	132·7	1918	January to March	134·3
1914	June to October	136·4			
1915	June to October	136·0			
1916	June to October	135·6			
1917	July to October	136·5		**Calama**	
1918	June to October	135·7	1918	July to December	135·7
1919	June to September	135·9	1919	January to December	135·7
1920	July to September	134·6	1920	January to July	136·0

For the period 1918–24 we give the following provisional values of the solar constant for each month at the three stations Calama, Harqua Hala and Montesuma[2]. A definitive table for Montesuma (1921–30) is given in chap. x.

Year	kw/(10 m)²	Ja	F	Mr	Ap	My	J	Jy	Au	Se	Oc	No	De
		Thousandths of gramme calories per square centimetre per minute											
1918	135·4	Calama					1900 +	21	54	44	39	41	62
1919	135·8	43	49	41	53	40	55	54	53	39	53	53	50
1920	136·0	64	56	45	52	53	39	45	—	—	—	—	—
Means	1900 + 48	54	53	43	53	47	47	40	54	42	46	47	56 g. cal./1000
	130 + 5·8	6·2	6·1	5·4	6·1	5·7	5·7	5·2	6·2	5·4	5·6	5·7	6·3 kw/(10 m)²
1920	135·8	Harqua Hala								1900 +	43	52	48
1921	135·7	64	49	44	48	54	35	39	37	43	44	58	48
1922	134·1	41	47	30	24	28	20	12	20	05	19	15	30
1923	134·0	26	17	18	17	23	18	—	—	25	33	26	25
1924	134·0	24	18	13	12	20	16	23	26	23	34	34	—
Means	1900 + 30	39	33	26	25	31	22	25	28	24	35	37	38 g. cal./1000
	130 + 4·5	5·1	4·7	4·2	4·2	4·6	4·0	4·2	4·4	4·1	4·9	5·0	5·1 kw/(10 m)²
1920	135·6	Montesuma						1900 +	30	47	44	48	57
1921	135·7	55	56	49	44	43	39	47	35	53	46	50	52
1922	134·2	47	42	37	30	24	13	11	18	22	26	28	14
1923	134·0	30	12	12	12	16	18	26	31	34	30	31	23
1924	134·1	31	22	19	17	22	29	22	18	20	29	30	—
Means	1900 + 31	41	33	29	26	26	25	27	26	35	35	37	37 g. cal./1000
	130 + 4·6	5·3	4·7	4·5	4·2	4·2	4·2	4·3	4·2	4·9	4·9	5·0	5·0 kw/(10 m)²

Thirdly, since there is at least some reason to consider that the intensity of solar radiation which reaches the earth is dependent upon the activity of the sun's surface as indicated by the spots to be noticed upon it, we give a table of mean annual frequency of sunspots during the past 185 years, estimated according to a regular plan devised at Zürich by Professor Wolf and now continued regularly by Professor Wolfer.

The mean period of frequency of spots is 11·1 years and anything with a period approximating to 11 years or a multiple or sub-multiple thereof, may suggest a connexion with sunspots. The most recent and most effective relation that has come to the knowledge of the Meteorological Office is the direct relation between the sunspot-number and the variation of level of the water in Lake Victoria at Port Florence. The correlation in this case is + ·8.

[1] *Annals of the Astrophysical Observatory of the Smithsonian Institution*, vol. IV, p. 193. Washington, 1922. The values at Hump Mountain are of inferior weight.

[2] *Smithsonian Misc. Coll.*, vol. LXXVII, No. 3. Washington, 1925.

Table of sunspot-numbers[1], 1750–1934.

Years	0	1	2	3	4	5	6	7	8	9
	n	n	n	n	n	n	n	n	n	n
175–	83	48	48	31	12	10	10	32	48	54
176–	63	86	61	45	36	21	11	38	70	106
177–	101	82	66	35	31	7	20	92	154	126
178–	85	68	38	23	10	24	83	132	131	118
179–	90	67	60	47	41	21	16	6	4	7
180–	14	34	45	43	48	42	28	10	8	2
181–	0	1	5	12	14	35	46	41	30	24
182–	16	7	4	2	8	17	39	50	62	67
183–	71	48	28	8	13	57	122	138	103	86
184–	63	37	24	11	15	40	62	98	124	96
185–	66	65	54	39	21	7	4	23	55	94
186–	96	77	59	44	47	30	16	7	37	74
187–	139	111	102	66	45	17	11	12	3	6
188–	32	54	60	64	64	52	25	13	7	6
189–	7	36	73	85	78	64	42	26	27	12
190–	10	3	5	24	42	64	54	62	49	44
191–	19	6	4	1	10	47	57	104	81	64
192–	38	26	14	6	17	44	64	69	78	65
193–	36	21	11	6	9	—	—	—	—	—

TERRESTRIAL RADIATION

Finally, we quote in advance, as "Stefan's law," the formula for the heat emitted by radiation from a square centimetre of surface at temperature t in a perfectly transparent medium as σT^4, where T is the absolute temperature for which the tercentesimal temperature *tt* can be substituted with sufficient accuracy for meteorological work. The symbol σ represents a "constant" independent of the temperature but dependent upon the nature of the surface. It is called Stefan's constant[2]. "According to Professor Millikan, who has recently reviewed the literature of the subject, the most probable value of σ is $5{\cdot}72 \times 10^{-12}$" watts per square centimetre.

Radiation from a square dekametre of black earth at various temperatures, computed according to the formula $\Sigma = \sigma t^4$.

$\sigma = 5{\cdot}72 \times 10^{-9}$ kw per square dekametre $= 82 \times 10^{-12}$ g. cal. per square cm per minute

tt	0	1	2	3	4	5	6	7	8	9
					kilowatts per square dekametre					
26–	26·1	26·5	27·0	27·4	27·8	28·2	28·6	29·1	29·5	30·0
27–	30·4	30·9	31·3	31·8	32·2	32·7	33·2	33·7	34·2	34·7
28–	35·2	35·7	36·2	36·7	37·2	37·7	38·3	38·8	39·4	39·9
29–	40·5	41·0	41·6	42·2	42·7	43·3	43·9	44·5	45·1	45·7
30–	46·3	47·0	47·6	48·2	48·8	49·5	50·2	50·8	51·5	52·1

It is almost needless here to say that no material medium is perfectly transparent, the atmosphere certainly not, and from that fact arises the necessity of considering in some detail the subject of radiation before we can bring it into the quantitative explanation of meteorological phenomena.

[1] *Meteorological Glossary*, p. 245, H.M.S.O. 1916. The table has been extended by the addition of data for the years 1915–34 published in the *Astronomische Mitteilungen*, Zürich.

[2] *Meteorological Glossary*, s.v. Radiation. A more recent value (National Research Council Washington) is 5·709 for absolute temperature, equivalent to 5·717 for tercentesimal.

THE RELATION OF SOLAR AND TERRESTRIAL RADIATION

In the heading to this chapter we have given a short table in order to indicate a relation between the intensities of solar and terrestrial radiation. The figures are really hypothetical because they assume that the atmosphere which intervenes between the sun and earth and between the earth and free space is perfectly transparent both to the solar and terrestrial radiation. With that gratuitous assumption the table indicates the altitude at which the sun would have to be in order to compensate for the loss of heat by radiation from a horizontal surface of "black" earth. The final figure of the table indicates that a black surface at the temperature of about 402 tt would be in balance with a vertical sun.

We are bold enough to introduce this table with its assumption of a transparent atmosphere for the simple reason that it is precisely the part which the atmosphere plays in affecting the incoming and outgoing radiation which is a matter of chief interest for the science of meteorology.

Actual observations of the conditions of balance between incoming radiation and outgoing radiation must be included in the facts upon which any effective physical theory of atmospheric changes is to be built. In considering this subject, on account of the lack of perfect transparency of the atmosphere, we have to deal with radiation not only from the sun and the solid earth or the sea but also from the atmosphere itself which for this purpose is called "the sky." Not much progress has yet been made in the effective study of the relation of terrestrial radiation to the radiation from sun and sky. The subject is however coming gradually into our knowledge through the efforts of Anders Ångström, C. Dorno and W. H. Dines. During the years 1921 to 1927 the last-mentioned compared the loss of energy from a grass meadow with that received from the whole sky near the time of sunset. We quote a summary of some figures given by the author in a kalendar for 1925 based on the figures published in the *Meteorological Magazine*. They are arranged according to the quarters of the May-year (see pp. 4, 5).

Summary of inward long-wave radiation from the sky, and outward radiation from a grass field on cloudless days near sunset at Benson in 1924.

		On cloudless days near sunset. Kilowatts per square dekametre		Total daily income from sun and sky. Rothamsted
		Gain from sky	Loss from field	kw-hr/(10 m)^2
First quarter	May 6 to Aug. 5	30	37	423
Second quarter	Aug. 6 to Nov. 4	28	35	210
Third quarter	Nov. 5 to Dec. 31 Jan. 1 to Feb. 4	24	31	57
Fourth quarter	Feb. 5 to May 5	25	33	212

It appears that whatever may be the time of the year, with clear sky about sunset, the earth is losing heat at the rate of 7 kilowatts per square dekametre.

Further details of the physical aspects of the behaviour of radiation are given in chapters IV and V of volume III.

CHAPTER II

LAND, WATER AND ICE. OROGRAPHIC FEATURES AND OTHER GEOPHYSICAL AGENCIES

Continental masses and unknown seas.

Asia	44,680,000 square kilometres	Europe ...	9,710,000 square kilometres
Africa	29,840,000 ,,	Australia ...	7,633,000 ,,
North America ...	20,018,000 ,,	[Greenland]	2,142,000 ,,
Central and South America ...	18,461,000 ,,	Antarctic (summer)	14,170,000 ,,
		[,, (winter) about	45,000,000] ,,

Unknown Sea: Arctic 3,445,000 square kilometres, Antarctic 2,200,000 square kilometres

after W. S. Bruce, *The Scottish Geographical Magazine*, July 1906.

Area of land-surface 145,000,000, water-surface 367,000,000 square kilometres
Equatorial diameter ... 12,755 kilometres. Polar diameter 12,712 kilometres
Volume of the earth ... 1,082,000,000,000 cubic kilometres
Mass of the earth ... $5{\cdot}98 \times 10^{21}$ metric tons
Mass of the atmosphere $5{\cdot}34 \times 10^{21}$ grammes
Acceleration of polar gravity 983·21, equatorial 978·03, lat. 45° 980·62 cm/sec^2

At this stage of our inquiry we have no wish to enter into details, our aim is to give a general idea of the structure and circulation, as accurate as circumstances permit, but always capable of improvement in detail. We conceive that, of the several schools of meteorology, each will have its own set of basic maps and diagrams corresponding with those in this volume but on a scale which is quite beyond the capacity of its modest page. The basic maps can be improved as detailed maps of the several countries and the several oceans are developed. They can be used as a kind of note-book into which new information can be incorporated from time to time by superposition.

Furthermore, for adequate conception of the actual behaviour of the atmosphere, the idea of circulation is fundamental, and circulation is related either to the earth's axis, which is permanent, and in that sense meteorologically normal, or to some local axis which is meteorologically speaking transitory. Every portion of an isobaric line on the earth's surface has what mathematicians call a centre of curvature, and air which moves along it has an "instantaneous axis of rotation." We have therefore chosen for the ground-plan of the great majority of our maps hemispheres in pairs, Northern and Southern. A simple geometrical representation of a hemisphere upon a plane surface by projection is an insoluble problem and the scheme which we have used cannot be said technically to be a projection, or even a map; it is a diagram or working picture, in which meridians are radii drawn through the poles, and lines of latitude are concentric circles at distances proportional to the co-latitude, that is, the angular distance from the pole. There is accordingly much exaggeration of the distance between the consecutive meridians at the equator compared with the polar regions, represented by a ratio of $\pi : 2$. Wind-directions, as represented on these charts, are not "true" in the sense that applies to winds on Mercator's projection, that is to say, an East wind is not always represented by a line drawn from right to left nor a West wind by a line from left to right; indeed the line for each wind has every direction in turn according to the position, on the map, of the point to which it refers; but, taken with due regard to the meridians which all pass through the pole of the chart, a true idea of convergence or divergence of direction is obtained which is more effective for many purposes than the idea that the directions of winds of the same name are everywhere parallel.

An obvious disadvantage of our charts is that the plan of computing the direction and velocity of the geostrophic wind from the separation of consecutive isobars with the aid of a common geostrophic scale is not applicable on account of the difference of scale along parallels from that along meridians. In the first edition of this volume we indicated a method of meeting that difficulty by means of scales which give separately the components of velocity along parallels of latitude and along meridians respectively.

But our maps are unsuitable for the study of the circulation in the equatorial regions, partly on account of the distortion and partly because of the separation at the equator. For subjects which are related especially to these regions we have accordingly used a special map drawn on Mercator's projection and extending from 30° N to 30° S of the equator. The division is appropriate for the region so specified because, as we shall see, the general idea of the normal atmospheric motion in the upper air near the equator is a circulation from East to West, whereas for middle latitudes the circulation is from West to East.

OROGRAPHIC FEATURES

We shall begin our representation of the atmospheric structure and circulation with maps of the main orographic features of the two hemispheres (figs. 2*a* and 2*b*) which are arranged to show: (1) the coast-lines together with the summer and winter boundaries of sea-ice in the two hemispheres, (2) the contour of 200 metres, and (3) the contour of 2000 metres. The contour of 200 metres is chosen because the orographic features below that level offer comparatively little obstruction to atmospheric currents and the contour may thus be regarded as a sort of secondary coast-line in considering such questions as the travel of cyclonic depressions, whatever may prove to be the final result of the analysis of the phenomena which are connoted by that term. The reader may be interested to notice that in this sense there is an atmospheric coast-line for North Western Europe which runs from the Pyrenees to the Ural Mountains and which circumscribes the great plain of Northern Europe enclosing the Baltic Sea and leaving the Scandinavian Peninsula as a huge island.

Even that line is broken through by a gap in Western Russia and by a large area between the Northern coast of Russia and the basins of the Caspian Sea and the Sea of Aral. There is also a large area under 200 metres in Siberia East of the Ural Mountains. Other notable areas under that level are marked by the valleys of the Nile, the Niger, the Indus, the Ganges, the Chinese rivers, the Mississippi and Hudson Bay in the Northern Hemisphere, by the

FIGURE 2*a*

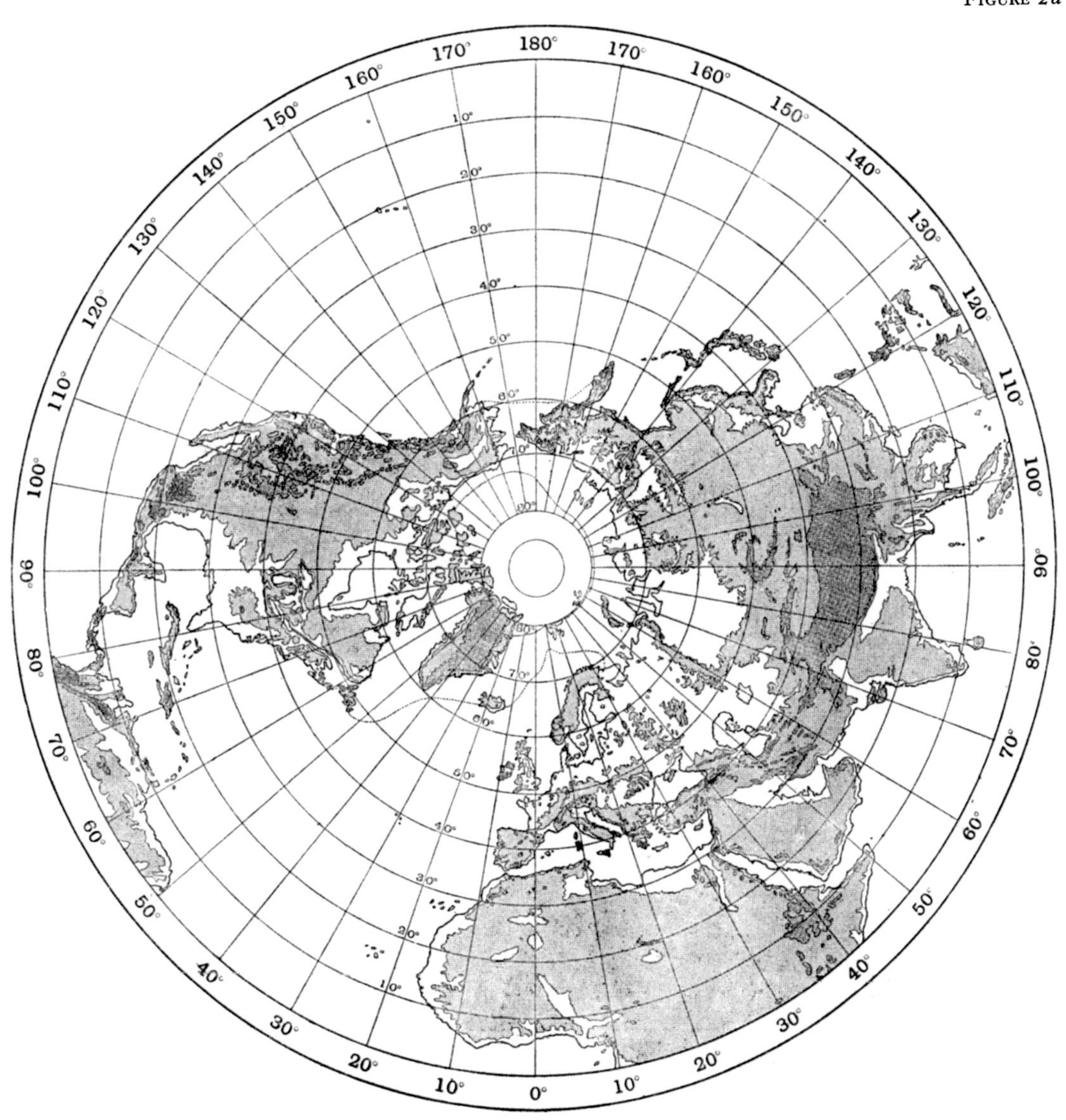

Snow-line is used to indicate roughly the level at which snow may be found throughout the year. The approximate height in different parts of the world is as follows:

S. Alaska	1500 metres	Latitude 78° N	Sea-level
Cascade Mts.	2500–3500 ,,	S. Greenland	600 metres
Glacier Park, Montana	2750 ,,	N. Scandinavia	900 ,,
Yellowstone Park	3200 ,,	S. Norway	1500 ,,
Colorado	3800 ,,	Alps	2600–2750 ,,
Mexico	4500 ,,	Pyrenees	2000 ,,
Bolivian Andes	4500–5500 ,,	Himalayas on the Northern side about 5500 metres but lower on the Southern side	
Sierra Nevada	3350–4000 ,,		
S. Chile	600 ,,	Tropics	4500–5800 metres

The summer and winter lines of sea-ice are dotted on the orographic maps.

[The seasonal distribution of snow at lower levels is given in chap. x, note 2.]

FIGURE 2*b* SOUTHERN HEMISPHERE

(see chap. x, note 3)

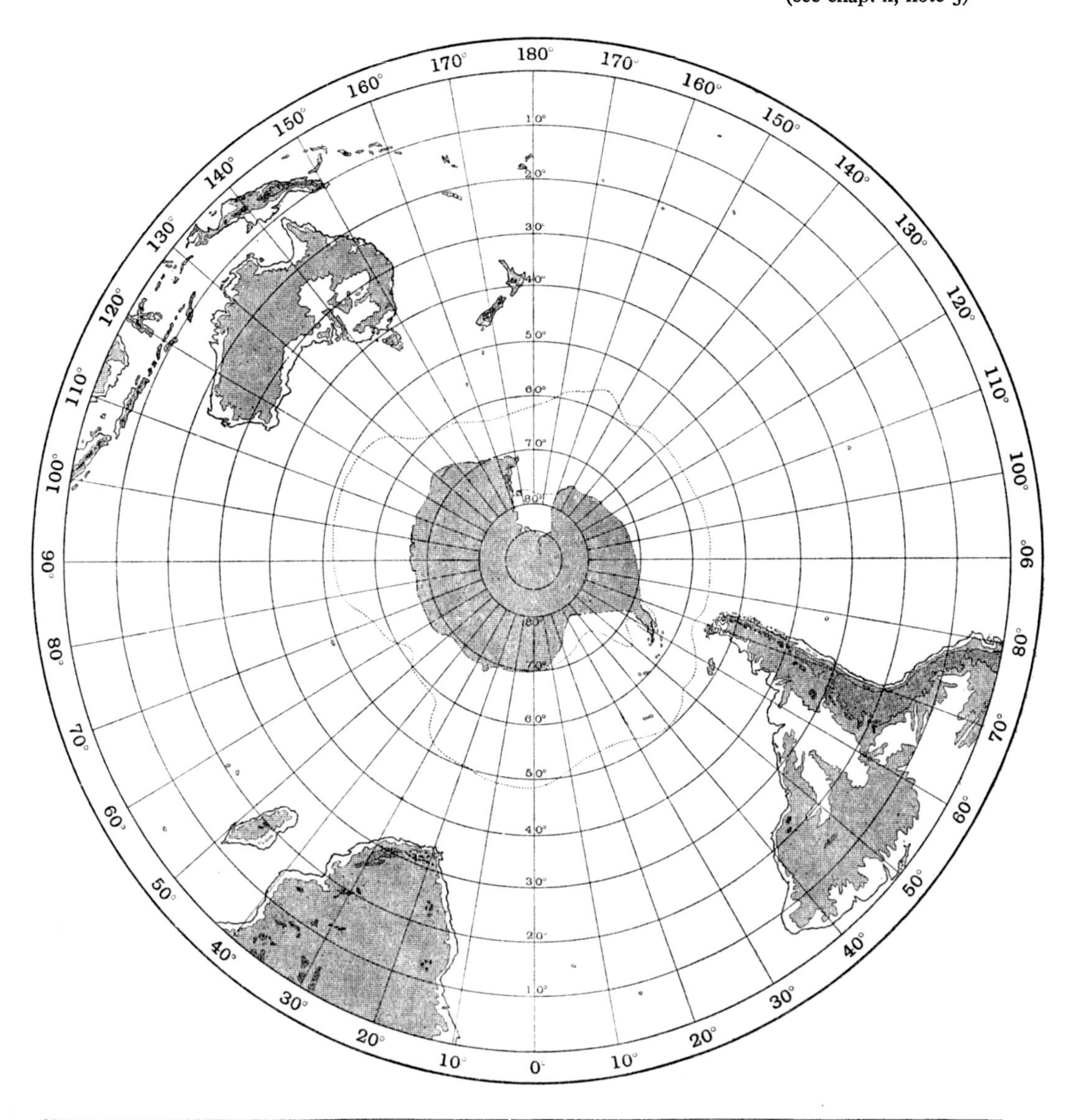

The highest mountains

	Metres		Metres		Metres
Everest, Himalaya	8840	Chimborazo, Andes	6248	Mt St Elias, Rockies	5494
Godwin-Austen (K. 2),		McKinley, Alaska	6187	Charles Louis, New Guinea	5486
Himalaya	8610	Llullaillaco, Andes	6170	Popocatepetl, Mexico	5420
Kanchanganga 1, Himalaya	8579	Cotopaxi, Andes	5978	Sangay, Ecuador	5323
Kanchanganga 2, Himalaya	8474	Mt Logan, Rockies	5955	Kenya, Kenya	5194
Makalu, Himalaya	8470	Kilima-Njaro, Tanganyika	5889	Ararat, Armenia	5157
Aconcagua, Andes	6970	Demavend, Caucasus	5669	Koshtan Tau, Caucasus	5143
Illampu (Sorata), Andes	6550	Elbruz, Caucasus	5630	Ruwenzori, Uganda	5121
Sahama, Andes	6416	Tolima, Cordilleras	5584	Kazbek, Caucasus	5013
Illimani, Andes	6410	Citlaltepetl, Mexico	5582	Mt Blanc, Alps	4810

Samen mountains, Abyssinia, 4600 m., snow-line about 4000 m.

The values are taken from Bartholomew's *Handy Reference Atlas* supplemented by the *Encyclopaedia Britannica*. The heights are apparently subject to considerable uncertainty.

Amazon and the La Plata in South America. These regions are the flood-areas of the earth. We must not however omit to notice that there are considerable areas below 200 metres in the interior of Australia which are not subject to floods but, on the contrary, are dry.

For heights below the level of 200 metres the reduction of pressure to sea-level may be trusted to a degree of accuracy which is not strictly applicable to higher levels, and consequently the computation of the gradient-wind for sea-level has a reality of meaning which does not hold for mountainous districts. Local variations of contour may affect local rainfall but the details of the surface wind are as a rule beyond calculation.

The contour of 2000 metres is chosen because the level is not very different from that of the snow-line for Europe, and the regions above that level are suggestive of permanent snow-fields which must exercise a great influence upon the atmospheric conditions.

The most notable area included within this contour comprises the large mountain-masses of Central Asia which protrude with some interruptions through the Caucasus to the Alps and Pyrenees. The curious knot of Abyssinia with isolated points in equatorial Africa is an important meteorological feature, as are also the irregular massif of the Rocky Mountains, Mexico and Central America and the more continuous massif of the Andes. All these form serious obstacles to the circulation of the atmosphere, easily recognised in the distribution of rainfall but not less important in the study of winds. In subsequent maps, as in those for the more restricted region of the Mediterranean in chap. II of vol. I, the area enclosed by the contour is more boldly marked in black, though on a smaller scale; and in like manner the contours of land at 4000 metres and at 6000 metres are shown in the maps which illustrate chap. VI. A table of estimates of the height of the snow-line in different parts of the world is given as an entablature of the map of fig. 2*a*, and a corresponding table of the greatest heights attained in the various areas of elevation occupies the same position with regard to fig. 2*b*.

These two entablatures are the first of a large number, some consisting of diagrams and others of numerical tables. By these the information contained in the maps is supplemented in order to embody information about the general circulation which is available in meteorological libraries. It should be noted that the choice of tables and diagrams is a very wide one and only few can find places in this book. We have selected those which may be regarded as typical either from the meteorological or geographical point of view with the hope that students of meteorology may be encouraged to amplify the collection for themselves by noting any pertinent examples that they may find in the course of their reading.

We leave the maps which are here reproduced and the notes set out upon them to tell the story of their own contribution to the structure of the atmosphere in their own way without much verbal description. Any features which are thought to be of importance in discovering the relations between the structure and the circulation will be referred to when occasion arises.

METEOROLOGICAL SECTIONS ROUND THE GLOBE IN DIFFERENT LATITUDES

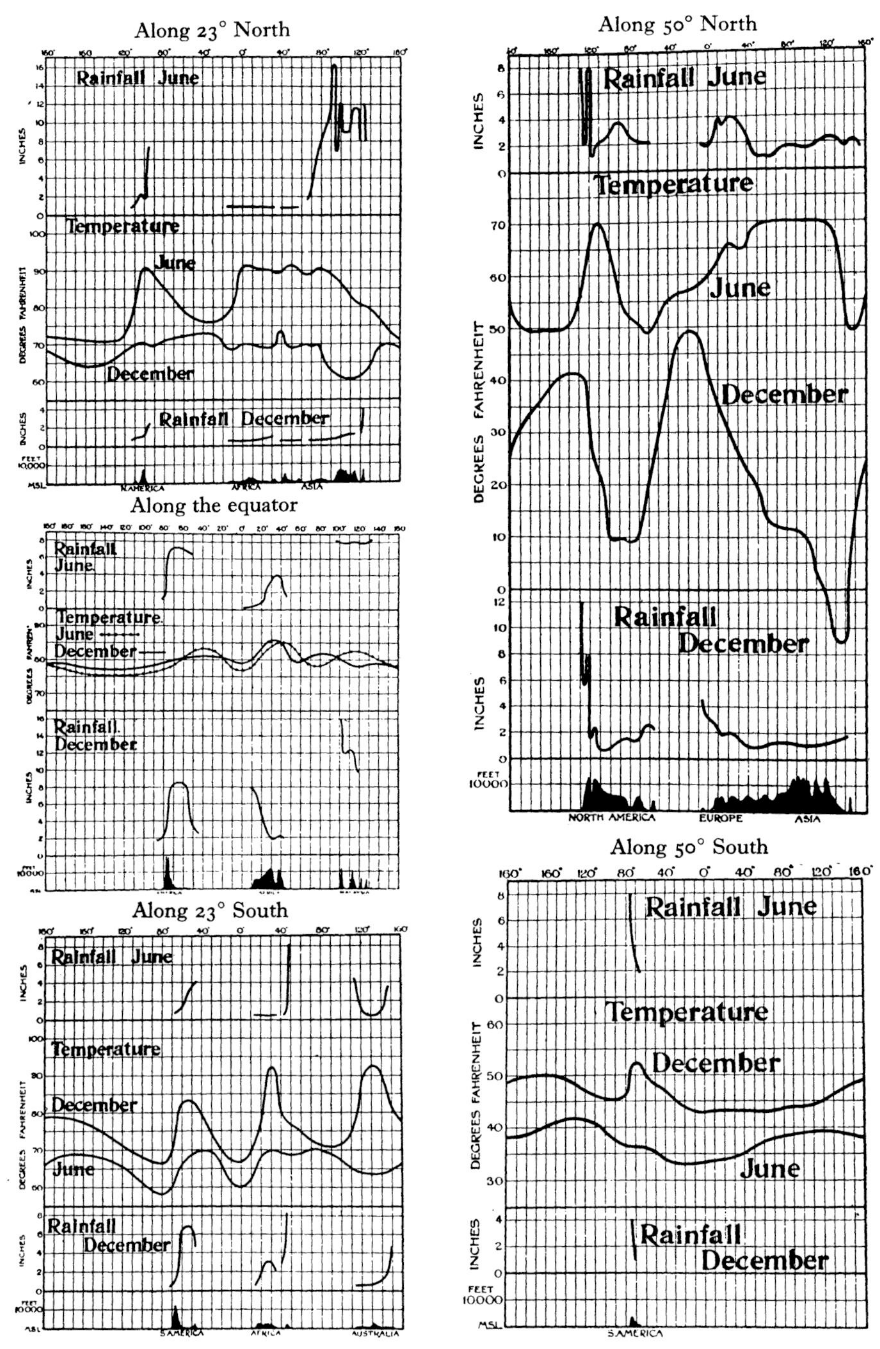

Fig. 3. The normal influence of orographic features on weather and climate.

The figures at the head of the sections indicate the longitude from 160° E Eastward through 180° to 160° E.

The scales at the side show temperature in degrees F and rainfall in inches.

The influence of orographic features upon weather.

The extent to which the orographic features of the globe control the weather is effectively illustrated by some "meteorological sections" round the globe which were prepared for the opening of the new building of the Meteorological Office in London in December 1910, and are still exhibited there. The sections represented are for the equator, for 23° North and South latitude, and for 50° North and South. They are reproduced in fig. 3.

At the bottom of each diagram is a little block section representing the orographic features which are intersected. The particular countries can be identified by the scale of longitude which is common to all the sections. In the middle portion of each diagram are graphs of the variation of mean temperature with longitude, for June and December, which show enormous differences over the continental areas compared with the oceanic portions of the sections, except the section round the equator which shows little variation but yet some interesting details. The continents are thus picked out in quite a remarkable manner.

Rainfall is in every case very notably affected by the orographic features and in a way that will be better understood from the diagrams than from many words of description. The most noteworthy effect is perhaps that of the double range of mountains, the Coast Range and the Rocky Mountains, in the section round 50° North latitude, but that is only one of many striking influences.

One of the chief lessons to be drawn from these diagrams is the lamentable want of information over the sea. Gaps in the run of the graphs are inevitable so long as rainfall is not adequately measured at sea, and the reader cannot help wondering how the broken ends of the curves should be joined and what the real distribution of rainfall over the sea can be.

SEA-ICE AND ICEBERGS (see also chap. x, note 4)

The influence of ice upon climate and weather is of profound importance to the atmospheric circulation whether it is represented by snow-fields in the higher levels of the land-areas or by the freezing of the surface of the sea. So long as the surface remains liquid the temperature of the air in immediate contact with it cannot pass much below 273tt, the freezing-point of ordinary fresh water. Sea-water of normal composition freezes at about 271tt, but water in the immediate neighbourhood of ice-fields is often much diluted by surface-water derived from the melting of ice or the flow of rivers, and freezing takes place before the freezing-point of normal sea-water is reached[1]. In the course of each cycle of the seasons the area of frozen surface extends outwards from either pole and recedes again. We have shown the approximate limits

[1] The relation between the salinity of water, its maximum density and its freezing-point is given in the following table derived from Schokalsky, *Oceanography*, Petrograd, 1917.

Salinity %	0	·5	1·0	1·5	2·0	2·4695	2·5	3·0	3·5	4·0
Temperature max. density	276·98	275·9	274·9	273·8	272·7	271·668	271·6	270·5	269·5	268·5
Temperature freezing-pt.	273·0	272·7	272·5	272·2	271·9	271·668	271·65	271·4	271·1	270·8

of the winter and summer ice-fields on the orographic maps (figs. 2*a* and *b*). The range in the Southern Hemisphere amounts in area to about one-eighth of the area of the hemisphere and if the range for the Northern Hemisphere, which is more difficult to estimate, be taken at the same figure, we may express the effect of the seasonal variation as the transition of one-eighth of the earth's surface from water to ice and back again from ice to water in the course of the year.

ICE IN THE NORTH POLAR REGIONS

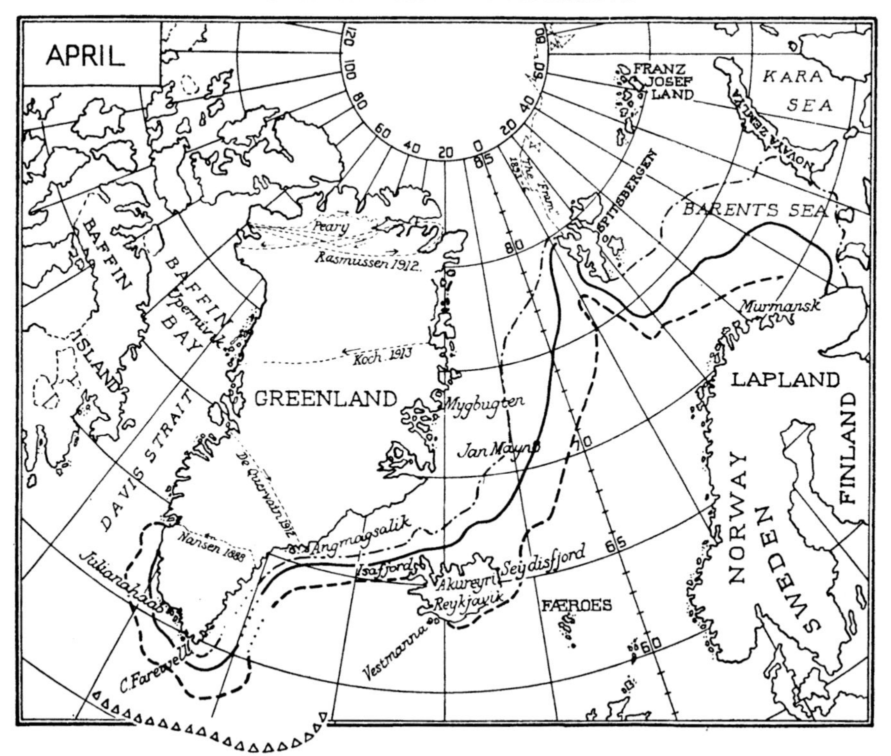

Fig. 4. Average limit of field-ice (full line) and extreme limits (broken lines) in the month of April. The extreme limit of icebergs is shown by a chain of triangles.
From *Isforholdene i de Arktiske Have*, Det Danske Met. Inst., Kjøbenhavn, 1917.

We ought therefore to give some expression of the sequence of changes in the course of the seasons. In the Northern Hemisphere the state of the ice in polar regions has been the subject of special study by the Danish Meteorological Office, and from their publications we have taken the material for the representation of the fluctuation of the boundary of the field-ice in the North Atlantic in three maps for the months of April, June and August respectively (fig. 4). Insets in the latter show the limits of field-ice and the limits of icebergs in the North Western region. A more recent chart of ice in this region is given in *The Marine Observer* for April 1935. The distinction between icebergs and field-ice is no longer drawn.

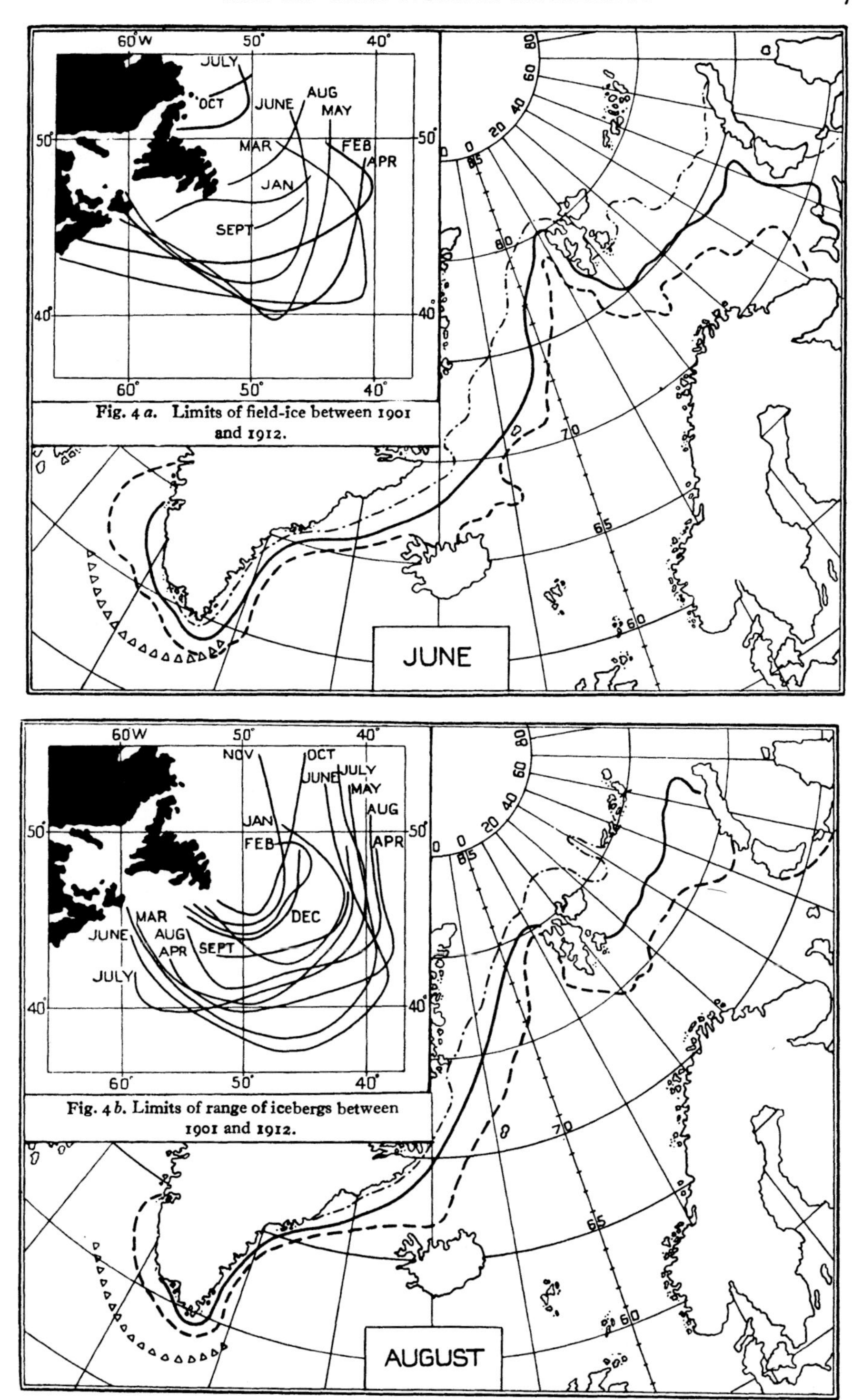

Fig. 4. Ice-limits in June and August to compare with those for April.

To the information which is thus represented we require to add that collected about the icebergs which are detached from the coastal ice-fields of the Northern and Southern Hemispheres and are carried by the oceanic currents to the warmer waters of the globe.

In the Northern Hemisphere the supply of ice in this special manner comes mainly from the ice-fields of Greenland. The icebergs off the Eastern shore are carried Southward along the Greenland coast and rounding Cape Farewell pass up Northward along the Western coast, whereas those off the North Western coast of Greenland are carried Southward along the Western shores of Davis Strait and Baffin Bay and thence by the Labrador current to the Strait of Belle Isle, Newfoundland and the Great Bank. There they become merged in the warmer water of the Gulf Stream. They are a cause of so much anxiety in respect of the navigation of those waters that since 1914, following the loss of ss. *Titanic* in 1912, they have become the subject of an international patrol which is carried out for the maritime nations of the world by the Government of the United States.

ICEBERGS IN THE ATLANTIC

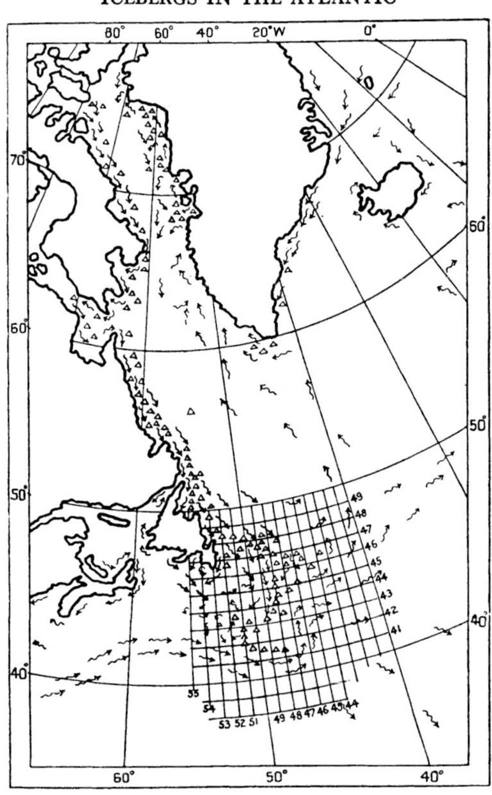

Fig. 5. Icebergs and the currents in which they drift. 'The International Ice Patrol,' *Met. Mag.* (M.O. 278), Nov. 1925.

From a recent account of the proceedings of this patrol we have transcribed a map (fig. 5) of the course of the icebergs which find their way to the Great Bank. It expresses with clearness the genesis of these dangers to navigation in one of the most interesting regions of the globe for the student of meteorology. There or thereabout equatorial and polar air and water are in a perpetual condition of action and reaction. Information about the region is at present insufficient; if it could be extended to give a clear account of the action it would add materially to our comprehension of the circulation.

The Pacific Ocean is much less subject to the menace of icebergs than its companion area in the Northern Hemisphere, the Atlantic. We have little information on this subject; but it is evident that the access to the Pacific from the polar regions through the Behring Strait is very much more restricted than that which is afforded to the Atlantic by Baffin Bay and Davis Strait, and moreover the general drift from the Behring Strait is towards the Arctic Ocean rather than towards the Pacific.

For the Southern Hemisphere we reproduce, from the *Seaman's Handbook*, a map showing the limits of the field of icebergs round the Antarctic ice-fields according to information collected since 1772, and in addition the localities of

ICE IN THE SOUTHERN OCEAN

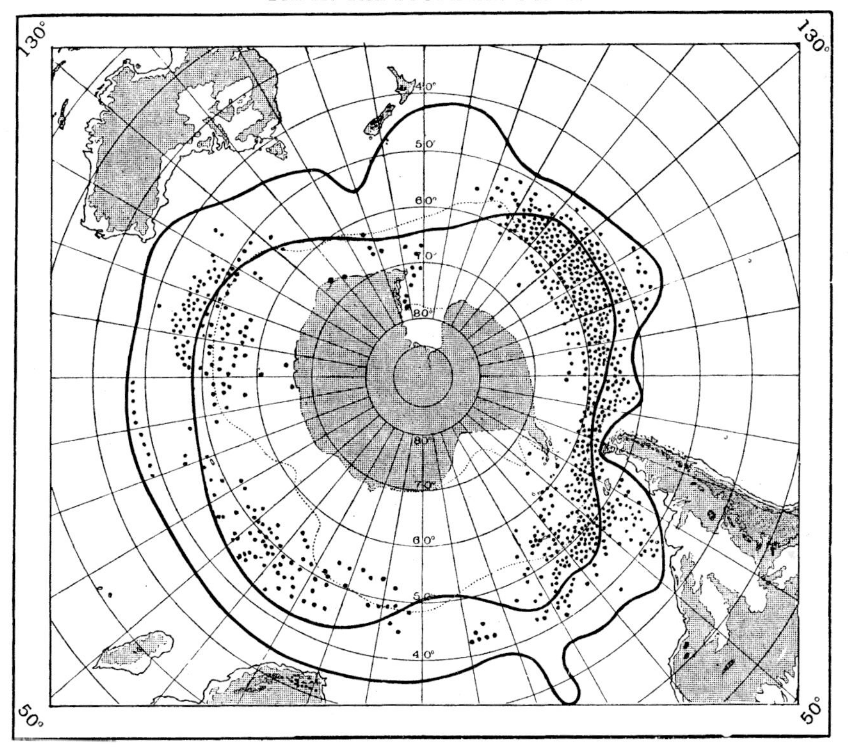

Fig. 6. Lines marking the range of the Northern limit of ice throughout the year in the South polar regions since 1772.

The dots indicate the positions of icebergs reported during the years 1902–16.

icebergs which have been notified to the Meteorological Office in London during the years 1902–16. The distribution has some remarkable features, notably the comparative scarcity of icebergs in the range of longitude between 140° E and 170° W overlapping the Victoria quadrant on the one side and the Ross quadrant on the other.

In the original diagram separate symbols are used for the icebergs observed in the several months and lines are drawn to indicate the relative positions of the ice-pack for each month. Our illustration is upon too small a scale for these details; reference may be made to the *Seaman's Handbook* (M.O. 215). We call attention however to the limits of ice throughout the year as represented by the bold lines of fig. 6. It will be noticed that the extreme Northern limit of ice since 1772 for many meridians is far beyond the range of icebergs noted in the years 1902 to 1916. The winter-limit of field-ice is indicated by a faint dotted line hardly visible in the crowds of dots.

OCEAN-CURRENTS

It will be evident from the charts which disclose the presence of icebergs in the different regions of the globe that a description of the main ocean-currents ought not to be omitted from the geographical information to be placed at the disposal of students of the circulation of the atmosphere. We have referred already to the Labrador Current and the Gulf Stream, or its extension past the Great Bank on to the coasts of the British Isles and Norway with a branch to the Bay of Biscay and North West Africa, and also to the drift of icebergs down the East coast, and up the West coast of Greenland, which derives from the Arctic regions North of Spitsbergen and Franz Josef Land. Other noteworthy currents are the Kurosiwo Drift of the North Pacific that corresponds with the extension of the Gulf Stream across the North Atlantic, the Equatorial Currents North and South which pass, with some interruptions, from East to West and form indeed the Southern and Northern margins respectively of well-marked circulations round the anti-cyclonic areas of the tropical oceans. The Eastern members of these circulations are the well-known cold currents on the West coasts of California, South America, South Africa and Australia, and their Western branches are the warm currents of the Gulf Stream, the Kurosiwo Drift past Japan, the Brazilian Current, the Mozambique Current between Madagascar and South Africa, and the East Australian Current.

Further South we have the continuous current of the Southern Ocean, the so-called West Wind Drift of the "roaring forties" extending as far as a trough of low pressure about latitude 60° S.

The waters of the ocean are subject to physical and dynamical laws which are similar in many ways to those which govern the atmospheric circulation itself, only the coast-line occupies a more rigorous position in relation to water than it does in relation to air. Upward and downward convection play an important rôle. The whole system of ocean-currents is accordingly so complicated that something more than a page is necessary to give any adequate account of it. Instead of attempting that we will ask the reader to infer from what we have said about the notable currents that the analogy between the circulation of surface-winds and that of surface-currents is sufficiently close for him to regard the surface-circulation of winds, which we shall represent in a subsequent chapter (figs. 157–160), as a general indication of the flow of ocean-currents. This he can confirm by noting in figs. 47–54 the deflections of the isotherms of the sea-surface caused by the flow of water. The ideas can be corrected if necessary by reference to the detailed maps of the ocean-currents, which the Hydrographic Office provides for all oceans, based upon the observations co-ordinated by the Meteorological Office. A number of summaries appear as charts in the *Marine Observer*.

OTHER GEOPHYSICAL AGENCIES

Among the geophysical phenomena with which the meteorological features of the general circulation of the atmosphere may have to be brought into association we may mention, firstly, the subsidences or elevations of the earth's crust above or below sea-level by earthquakes which are the result of the seismic activity of the earth's interior; secondly, the activity of volcanoes which may fill the whole atmosphere with very fine dust; thirdly, the hitherto unexplained observations of terrestrial magnetism, which may be attributed partly to the solid earth and partly to the atmosphere, and with which, by general consent, are now associated the observations of aurora polaris in either hemisphere; and fourthly, the observations of the state of the atmosphere in respect of atmospheric electricity whether in ordinary quiet electrical conditions, or during the impressive manifestations of thunder-storms, or the less impressive but not less interesting displays of St Elmo's fire.

Earthquakes and volcanic eruptions.

In order to give the reader some idea of the distribution of earthquakes, we are fortunate in being able to make use of a map (fig. 7) prepared by H. H. Turner and J. J. Shaw for the British Empire Exhibition of 1924–5.

We are coerced into the consideration of volcanoes as a notable terrestrial influence upon weather by the voluminous reports of the dust-clouds and sunset-glows which followed the immense eruption of Krakatoa in 1883, and those of the diminished intensity of European sunshine in 1903 and 1912 which was attributed to the violent eruptions of Mont Pelée and Mount Katmai, and the stress laid upon similar events by W. J. Humphreys, who has been led to regard the dust in the atmosphere as a meteorological agency of great importance. We have accordingly put together maps of the distribution of active and recent volcanoes (fig. 8) in order to indicate the positions of the most notable centres of eruption, and we have also given a list of important volcanic eruptions recorded since A.D. 1500[1].

Important eruptions between 1500 *and* 1800.

1500	Java	1660	Katla	1712	Miyakeyama	1766	Mayon
1536	Etna	1660	Teon	1717	Kirishimayama	1766	Hekla
1550	Tala	1660	Omate	1717	Fuego	1768	Cotopaxi
1586	Keloet	1660	Pichincha	1721	Kötlugia	1772	Papandajan
1593	Ringgit	1673	Gamma Kunorra	1725–29	Leirhnukur	1779	Sakurashima
1597	Hekla	1680	Celebes	1727	Oeraefajökull	1783	Skaptar Jokull, Laki
1598	Oeraefajökull	1680	Krakatoa	1730–36	Lanzarote		
1598	Grimsvötn	1693	Hekla	1730	Raoen	1783	Asamayama
1614	Kl. Sunda Is.	1693	Seroea	1742–44	Cotopaxi	1783	Eldeyar
1631	Vesuvius	1694	Celebes	1749	Taal	1785	Vesuvius
1636	Hekla	1694–6	Gunong Api	1752	Kl. Sunda Is.	1786	Paulow
1638	Raoen	1694	Amboyna	1754	Taal	1786	Amukhta
1640	Komagatake	1700	Sawaii	1755	Kötlugia	1793	Tuxtla
1641	Awoe	1707	Fuji	1755	Katla	1794	Vesuvius
1641	G. Adiksa	1707	Vesuvius	1759	Jorullo	1795	Pogrumnoj
1646	Makjan	1707	Santorin	1760	Makjan	1796	Bogosloff
1650	Santorin					1799?	Fuego

[1] Karl Sapper, 'Beiträge zur Geographie der tätigen Vulkane,' *Zeit. für Vulkanologie*, Band III, 1916–17, Berlin, p. 148. W. J. Humphreys, *Physics of the Air*, Philadelphia, 1920.

OROGRAPHIC FEATURES AND GEOPHYSICAL AGENCIES

A new map of earthquake distribution by N. H. Heck based on instrumental records from 1899–1933 is reproduced in the *Geographical Review*, 1935, p. 125.

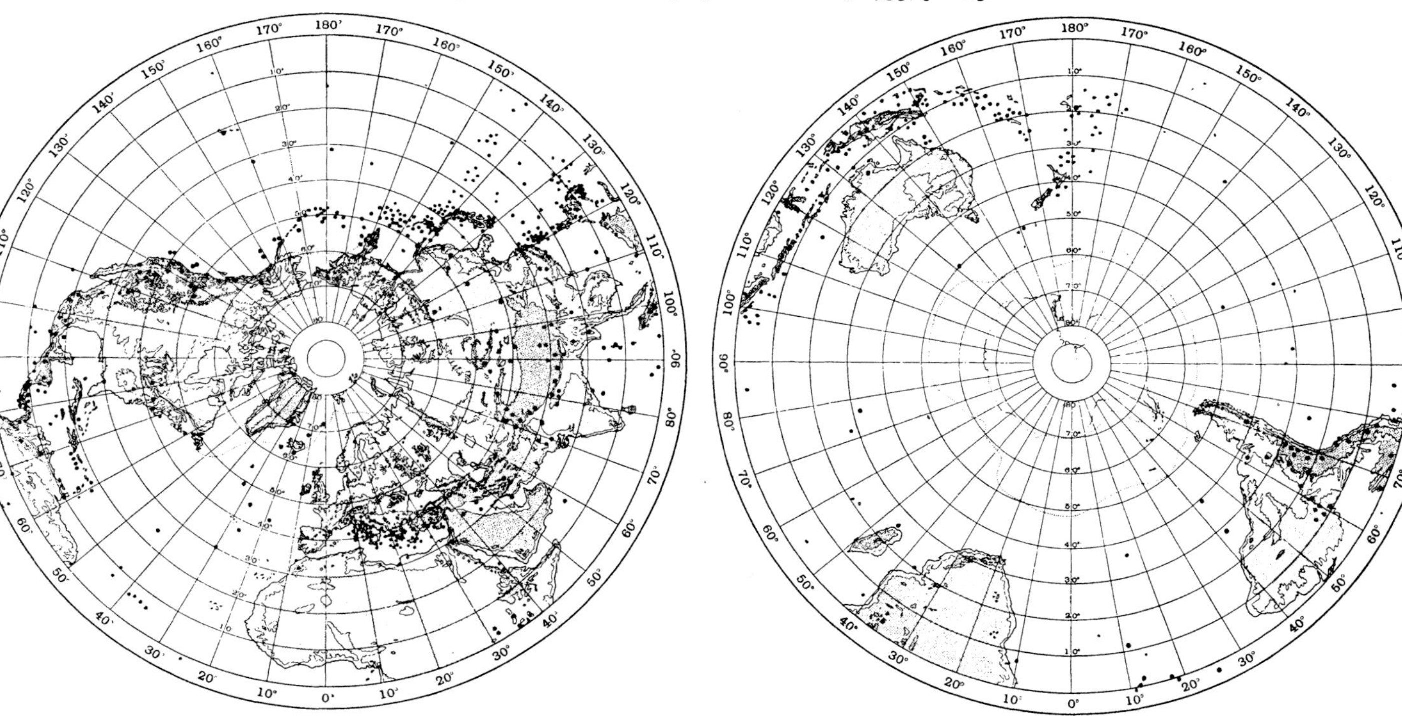

Fig. 7. The localities of earthquakes recorded in the years 1913 to 1919 as determined at Shide, Isle of Wight, or Oxford. (H. H. Turner and J. J. Shaw.) British Empire Exhibition, 1924–5.

It is pointed out that the earthquake centres lie upon two or three well-defined lines on the earth's surface.

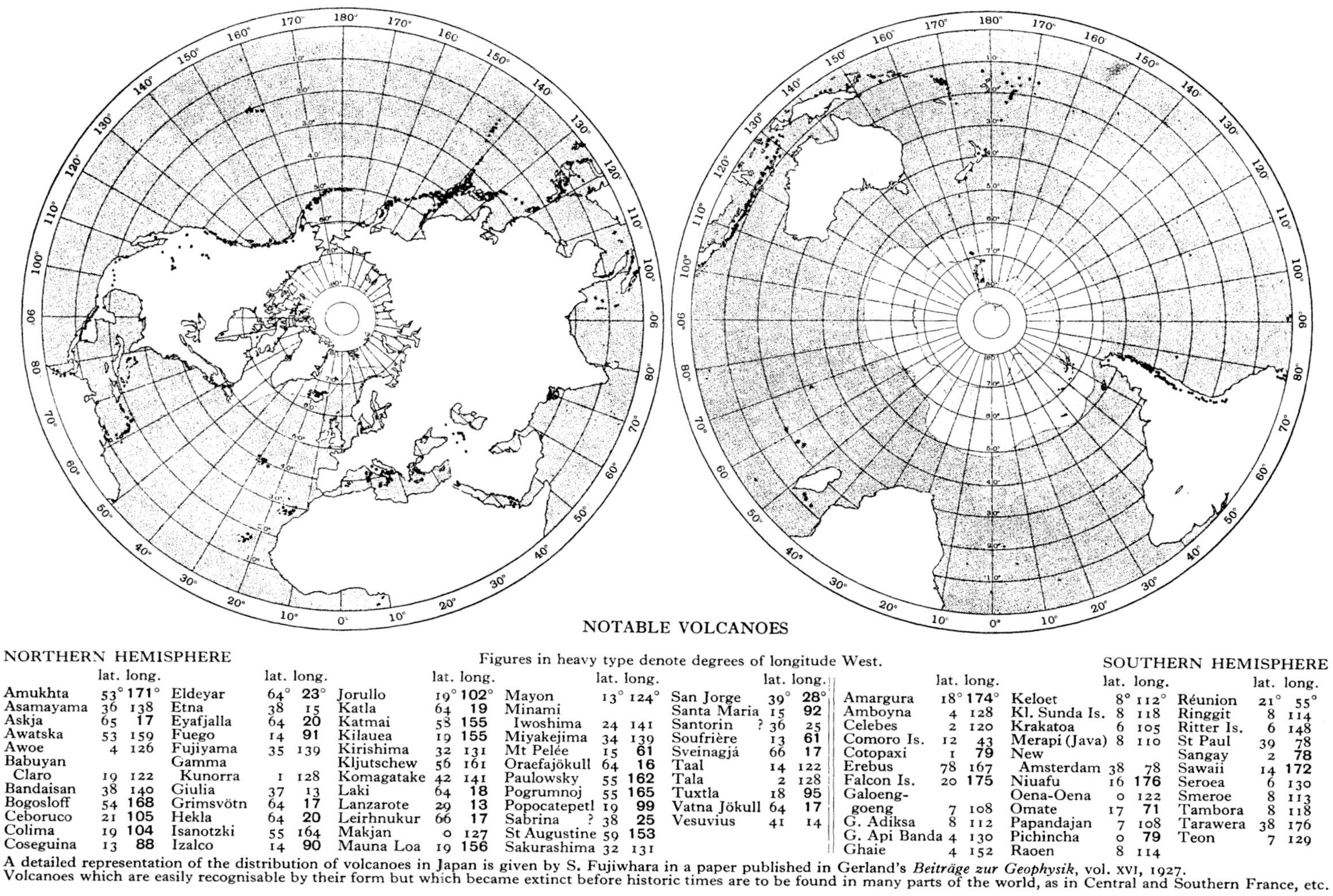

NOTABLE VOLCANOES

Figures in heavy type denote degrees of longitude West.

NORTHERN HEMISPHERE

	lat.	long.
Amukhta	53°	**171°**
Asamayama	36	138
Askja	65	**17**
Awatska	53	159
Awoe	4	126
Babuyan Claro	19	122
Bandaisan	38	140
Bogosloff	54	**168**
Ceboruco	21	**105**
Colima	19	**104**
Coseguina	13	**88**
Eldeyar	64°	**23°**
Etna	38	15
Eyafjalla	64	**20**
Fuego	14	**91**
Fujiyama	35	139
Gamma Kunorra	1	128
Giulia	37	13
Grimsvötn	64	**17**
Hekla	64	**20**
Isanotzki	55	164
Izalco	14	**90**
Jorullo	19°	**102°**
Katla	64	**19**
Katmai	58	**155**
Kilauea	19	**155**
Kirishima	32	131
Kljutschew	56	161
Komagatake	42	141
Laki	64	**18**
Lanzarote	29	**13**
Leirhnukur	66	**17**
Makjan	0	127
Mauna Loa	19	**156**
Mayon	13°	124°
Minami Iwoshima	24	141
Miyakejima	34	139
Mt Pelée	15	**61**
Oraefajökull	64	**16**
Paulowsky	55	**162**
Pogrumnoj	55	**165**
Popocatepetl	19	**99**
Sabrina ?	38	**25**
St Augustine	59	**153**
Sakurashima	32	131
San Jorge	39°	**28°**
Santa Maria	15	**92**
Santorin ?	36	25
Soufrière	13	**61**
Sveinagjá	66	**17**
Taal	14	122
Tala	2	128
Tuxtla	18	**95**
Vatna Jökull	64	**17**
Vesuvius	41	14

SOUTHERN HEMISPHERE

	lat.	long.
Amargura	18°	**174°**
Amboyna	4	128
Celebes	2	120
Comoro Is.	12	43
Cotopaxi	1	**79**
Erebus	78	167
Falcon Is.	20	**175**
Galoeng-goeng	7	108
G. Adiksa	8	112
G. Api Banda	4	130
Ghaie	4	152
Keloet	8°	112°
Kl. Sunda Is.	8	118
Krakatoa	6	105
Merapi (Java)	8	110
New Amsterdam	38	78
Niuafu	16	**176**
Oena-Oena	0	122
Omate	17	**71**
Papandajan	7	108
Pichincha	0	**79**
Raoen	8	114
Réunion	21°	55°
Ringgit	8	114
Ritter Is.	6	148
St Paul	39	78
Sangay	2	**78**
Sawaii	14	**172**
Seroea	6	130
Smeroe	8	113
Tambora	8	118
Tarawera	38	176
Teon	7	129

A detailed representation of the distribution of volcanoes in Japan is given by S. Fujiwhara in a paper published in Gerland's *Beiträge zur Geophysik*, vol. XVI, 1927.
Volcanoes which are easily recognisable by their form but which became extinct before historic times are to be found in many parts of the world, as in Central and Southern France, etc.

Fig. 8. Red dots show the positions of active volcanoes.

MAGNETIC LINES FOR THE GLOBE

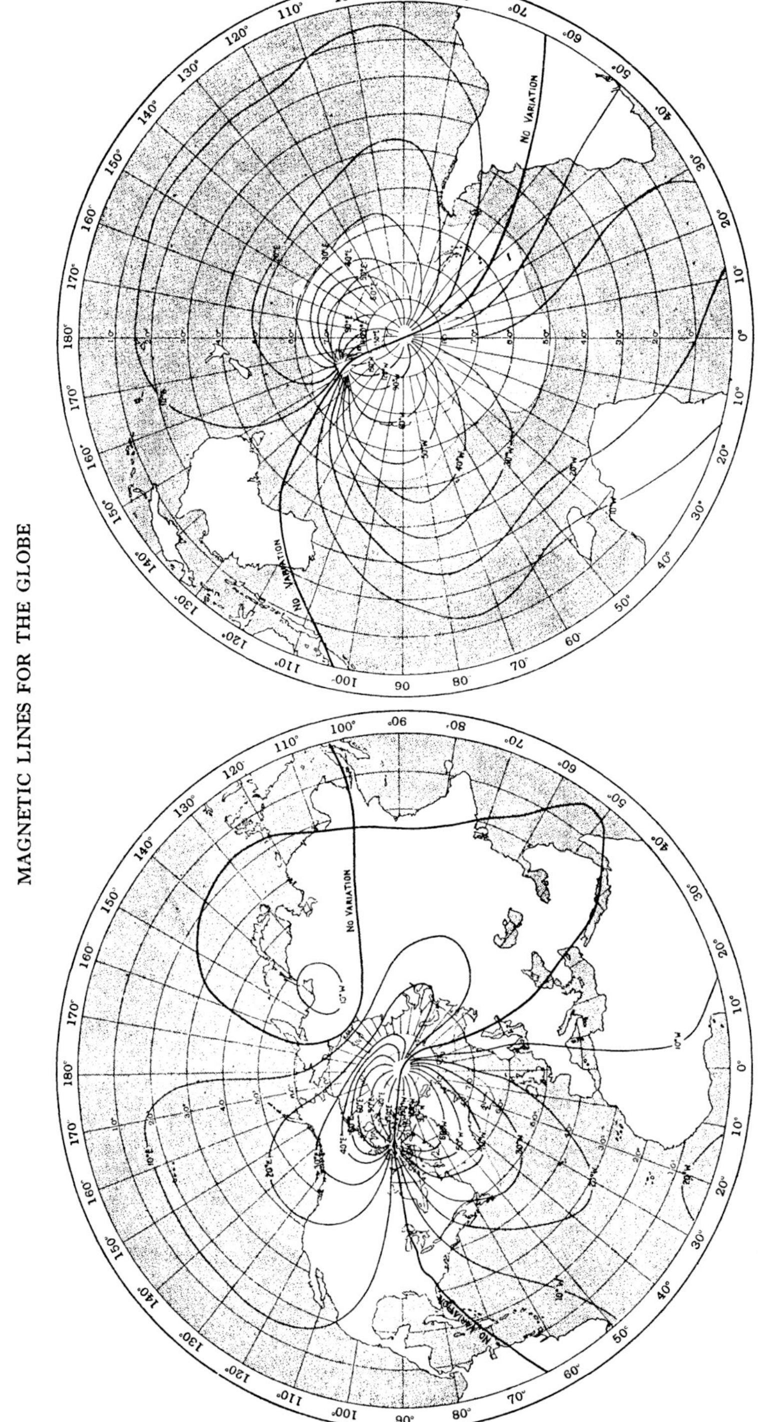

Fig. 9. Charts of lines of equal magnetic variation or declination for the epoch 1922 based upon Admiralty Chart No. 2598, compiled at the Royal Observatory, Greenwich. The variation of the compass in degrees East or West of true North is marked upon the several lines. It is subject to secular change which is expressed by a movement of the North end of the needle amounting to 15 minutes per annum towards West over the South Indian Ocean and 9 minutes per annum towards East over the British Isles.

Volcanic eruptions since A.D. 1800 (see also chap. X, note 5).

1808?	S. Jorge	1831	Pichincha	1872	Merapi (Java)	1892	Awoe
1809?	Etna	1831	Giulia	1875	Vatna Jökull	1898	Oena-Oena
1811	Sabrina	1835	Coseguina	1875	Askja	1899	Mauna Loa
1812	Soufrière, St Vincent	1837	Awatska	1875	Sveinagjá	1902	St Vincent
1812	Awoe	1840	Kilauea	1877	Cotopaxi	1902	Mt Pelée
1814	Mayon	1845–46	Hekla	1878	Ghaie	1902	Santa Maria
1815	Tambora	1846–47	Amargura	1881–2	Mauna Loa	1903	Colima
1817	Raoen	1852?	Etna	1883	St Augustine	1904	Minami Iwoshima
1821	Eyafjalla	1852	Mauna Loa	1883	Bogosloff	1905–6	Sawaii
1822	Galoenggoeng	1855	Mauna Loa	1883	Krakatoa	1906	Vesuvius
1822	Vesuvius	1855–56	Cotopaxi	1885	Falcon Is.	1906–7	Bogosloff
1825–31	Isanotzki	1856	Awoe	1886	Tarawera	1907	Mauna Loa
1826	Keloet	1859	Mauna Loa	1886	Niuafu	1911	Taal
1829	Kljutschew	1861	Makjan	1888	Bandaisan	1912	Katmai
1831	Graham's Is.	1868	Mauna Loa	1888	Ritter Is.	1913	Colima
1831	Babuyan Claro	1870	Ceboruco	1890	Bogosloff	1914	Sakurashima
		1872	Vesuvius	1891–99	Vesuvius	1914	Minami Iwoshima

Terrestrial magnetism and aurora.

The importance of the magnetic compass in navigation has secured close attention to the phenomena of terrestrial magnetism for the past three centuries and more. The lines of equal magnetic declination of the magnetic needle from the geographical meridian, called officially the variation of the compass, and the intensity of the earth's magnetic field are the subject of official inquiry on the part of the Admiralty, and are also subjects of research by the few organisations for the study of the physics of the globe which have an establishment independent of the requirements of Government services. The most notable of these, so far as terrestrial magnetism is concerned, is the Department of Terrestrial Magnetism and Atmospheric Electricity of the Carnegie Institution of Washington under the direction of Dr L. A. Bauer.

We give transformations of the charts of variation (fig. 9) and of horizontal force (fig. 10), compiled by the Royal Observatory, Greenwich, for the epoch 1922. Thereon are clearly indicated the positions of the magnetic poles.

The duality of the pole as indicated by the double convergence of lines of equal variation, a striking feature of the map, arises from the method of presentation of the facts. All directions co-exist at the geographical pole. Hence while the direction of the magnetic needle near the pole remains the same the direction of geographical or true North, and consequently the variation of the compass, would change through the whole range of 360° along a small circle surrounding the pole. It follows that all lines of variation must be represented there. All lines of magnetic variation must also converge to the magnetic pole. The duality has no magnetic significance.

The magnetic poles, in the North of Canada and the Antarctic respectively, are the points of convergence of the lines of horizontal magnetic force, they would be reached by following continuously the direction, North or South, of the compass-needle.

Aurora, now attributed to the bombardment of the earth by particles emitted from the sun, is much more definitely concerned with the atmosphere than the conventional elements of terrestrial magnetism. It is true that the brilliant

MAGNETIC LINES FOR THE GLOBE

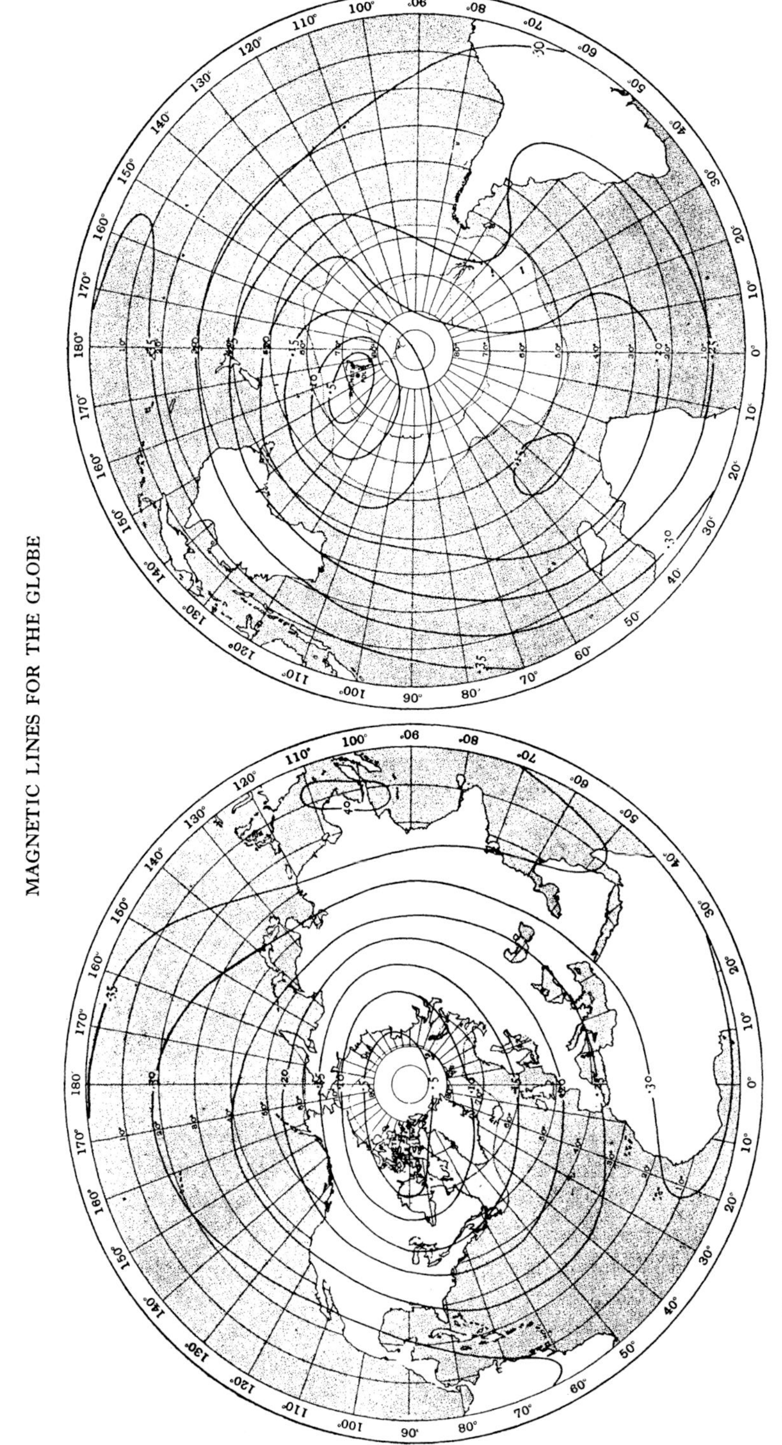

Fig. 10. Charts of lines of equal horizontal magnetic force for the epoch 1922 based upon Admiralty Chart No. 3603, compiled at the Royal Observatory, Greenwich. The intensity of the magnetic force expressed in C.G.S. units is marked on the several lines. It is subject to secular change. The figure for the innermost line should be ·05, not ·5.

lower parts of the auroral displays belong to a region of about 100 kilometres above the earth's surface but, on the other hand, a "green line" which is attributed to auroral light is almost as regular a characteristic of the atmosphere at night as the blue is of the daylight sky. Displays of aurora in the more popular sense, seen without the aid of a spectroscope or other physical apparatus, are more frequently visible the higher the latitude up to about 60° or 70°. Lines of frequency of observation have been prepared by H. Fritz which show a series of nearly circular and nearly concentric ovals, the common centre being on the extreme North West of Greenland about latitude 80°, and consequently some degrees North of the magnetic pole of the Northern Hemisphere. We give a reproduction of this map (fig. 11) on the lines of which the number of days of observation of aurora within a year are entered, and side by side with it we give a table of observations of aurora in the Southern Hemisphere by the various expeditions that visited the Antarctic about 1900[1]. The polar year 1932–3 should provide many additional observations.

Aurora Borealis.

Map of the distribution of frequency of observation.

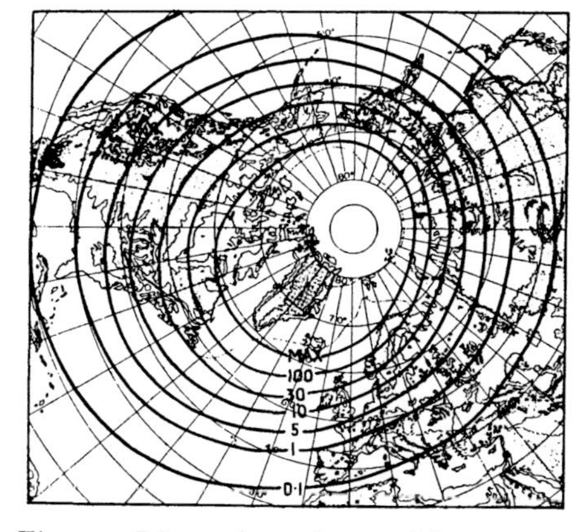

Fig. 11. Lines of equal annual frequency of observation of aurora (Isochasms) in the Northern Hemisphere. From curves by H. Fritz, *Das Polarlicht*, Leipzig, 1881. Data in *M. W. Rev.*, 1928, p. 401.

Aurora Australis.

Number of aurora observed in the several Antarctic expeditions, 1898–1903.

	Gauss 66° S 90° E 1902–3	C. Adare 71° S 170° E 1899	*Belgica* 70° S 87° W* 1898	Ross I. 78° S 167° E 1902	Ross I. 78° S 167° E 1903
	n	n	n	n	n
Feb.	4	—	—	—	—
Mar.	10	4	12†	0	2
Apr.	17	6	12	10	18
May	10	12	6	8	14
June	19	12	7	11	18
July	12	20	12	10	22
Aug.	10	16	7	9	14
Sept.	14	5	5‡	4	2
Oct.	12	1	—	—	—
Year	108	[76]	[61]	[52]	[90]

* Varying between 69° 51′ and 71° 36′ S and 82° 35′ and 92° 21′ W.
† 20 days of observation.
‡ 10 days of observation.

Later Expeditions.

1911, May 13 to July 11. Cape Evans 78° S. 36 per cent. of possible occasions.

Cape Adare 71° S. 64 per cent. of possible occasions.

1912–13. Cape Denison (Adelie Land) 67° S. 52 per cent. of possible occasions.

The South Magnetic Pole is located at 72½° S, 155° E (Shackleton).

Authorities:

Deutsche Südpolar Expedition, 1901–3. Band III. Meteorologie, I, I. Meteorologische Ergebnisse der Winterstation des "Gauss," 1902–1903. Von W. Meinardus. S. 162.

Magnetic and meteorological observations made by the *Southern Cross* Antarctic Expedition 1898–1900 under the direction of C. E. Borchgrevink. London, Royal Society, 1902.

Expédition Antarctique Belge. Résultats du Voyage du S. Y. *Belgica* en 1897–1898–1899. Météorologie, aurores australes. Par H. Arctowski. Anvers, 1901.

National Antarctic Expedition, 1901–1904. Physical Observations, p. 126. London, Royal Society, 1908.

[1] For a list of observations of aurora in the Southern Hemisphere from 1640–1895, see 'Das Südlicht' by Dr W. Boller, *Beiträge zur Geophysik*, Band III, Leipzig, 1898, pp. 56 and 550.

For the later expeditions C. S. Wright[1] gives the following in his account of the "Observations on the Aurora."

At Cape Evans,

Strictly comparable numbers are available for the period May 13th to July 31st [1911], inclusive, counting only observations at exact hours between 4 p.m. and 8 a.m. For this period, there were 656 occasions on which conditions for observation were favourable, while aurorae were recorded on only 236 occasions, or 36 per cent. of the whole.

At Cape Adare,

It is, in fact, estimated that aurorae were seen on 64 per cent. of the possible occasions, i.e. on about two in three observations when the meteorological conditions were favourable, corresponding with one in three (about) in the case of the Cape Evans observations.

Sir Douglas Mawson[2] gives the following table of analysis of the records of Cape Denison (Adelie Land, lat. 67°); the figures refer to the percentage of hours in which some form of auroral manifestation appeared, meteorological conditions permitting observation:

	March	April	May	June	July	Aug.	Sept.	Oct.	Mean
1912	54	54	36	54	63	54	46	54	51·9
1913	61	35	57	50	66	59	64	24	52
1912–13	57·5	44·5	46·5	52	64·5	56·5	55	39	52 (approx.)

This means that on 52 per cent. of all hours of moderate twilight and darkness, and including all moonlight hours, when the sky was clear for observation, auroral lights would be seen at least some time within the hour.

Atmospheric electricity (see also chap. x, note 6).

Observations of potential gradient obtained from self-recording electrometers at fixed observatories with radioactive collectors or water-droppers are the fundamental observations of atmospheric electricity. They have been supplemented by observations of ionisation and conductivity of the air and also, with the aid of balloons, by observations of the potential in the atmosphere at higher levels. These observations have found little application in the study of the general circulation of the atmosphere, and their consideration belongs more directly to the physics of the atmosphere. But the exceptional displays of atmospheric electricity which belong to thunder-storms are closely related to the general features of the circulation, being significant of atmospheric convection upon a vast scale.

Notes of the occurrence of thunder and lightning are indeed part of the routine of all meteorological stations and of observing ships at sea; consequently a large amount of information as to the frequency of these phenomena

[1] *British* (*Terra Nova*) *Antarctic Expedition*, 1910–1913, "Observations on the Aurora," by C. S. Wright, London, 1921.

[2] *Australasian Antarctic Expedition*, 1911–14, Scientific Reports, Series B, vol. II, Terrestrial Magnetism and Related Observations, Part I, Records of the Aurora Polaris, by Douglas Mawson, Sydney, 1925.

exists in ordinary climatic or statistical summaries and is available in any good meteorological library.

Accordingly, we indicate the distribution of thunder by the aid of maps compiled by C. E. P. Brooks for Geophysical Memoir, No. 24. The maps may seem to be more naturally associated with cloud, rainfall, pressure and winds and to be out of place here; but audible thunder, the phenomenon referred to in the maps, is in fact a geophysical phenomenon, the only one so far as we know of which the evidence is natural sound. The laws of its association with the recognised evidences of local atmospheric disturbance are not entirely self-evident nor even fully understood. In view of the recent work upon the disturbances of wireless transmission known as "atmospherics," or in France as "parasites," it seems likely that the localisation of thunder-storms, or other sources of disturbance, will be associated with the more definitely physical side of meteorological work, we are therefore including the maps of frequency of thunder with the orographical and general physical features.

In the entablatures of the two maps, on small scale, we are setting a collection of diagrams which were prepared in the Meteorological Office in 1919 for the *Advisory Committee for Aeronautics*. They represent, so far as the great majority are concerned, the seasonal distribution of the days of thunder at selected stations in the two hemispheres. There are however two diagrams of the diurnal distribution of the observations of thunder at Edinburgh and at Batavia respectively. Others are given in chap. x, note 6.

We extract from the memoir referred to the following particulars. The information is derived from the records taken over a number of years at 3265 stations of which 2680 are in Europe, together with a large number of observations on board ship. Assuming that a station will record thunder if a thunder-storm occurs within a surrounding area of 500 square kilometres there will be on the average 16 thunder-storms a year within each of such areas into which the earth's surface can be divided. That makes a total of 16,000,000 thunder-storms a year, or 44,000 a day. If an hour is taken as the average duration of a thunder-storm, we may infer that over the earth's surface 1800 thunder-storms are in progress at any moment throughout the year.

The following table taken from the Report of H.M.S. *Challenger* (p. 31) shows the distribution through the hours of the day of the number of occurrences of thunder-storms (thunder with lightning) over the open sea, and of lightning alone:

Hour	0–2	2–4	4–6	6–8	8–10	10–12	12–14	14–16	16–18	18–20	20–22	22–24
Thunder-storms	4	**7**	5	3	1	0	0	2	0	0	1	3
Lightning only	42	36	11	0	0	0	1	2	7	25	**46**	39

We gather therefrom that thunder-storms at sea are nocturnal phenomena, whereas on land they belong chiefly to the later afternoon following the greatest warmth of the day.

FIGURE 12

ANNUAL FREQUENCY OF DAYS OF THUNDER

Authority: Geophysical Memoir, No. 24.

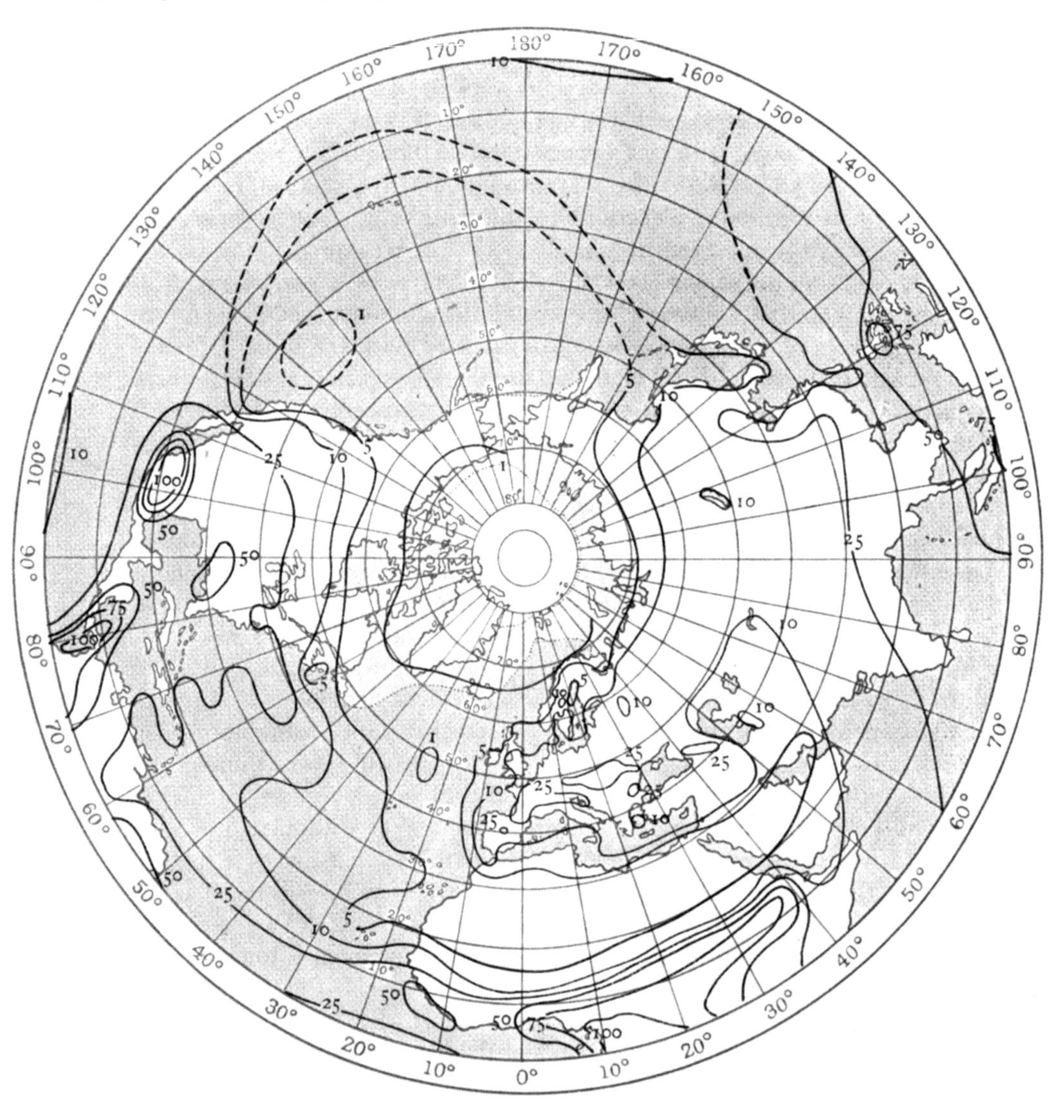

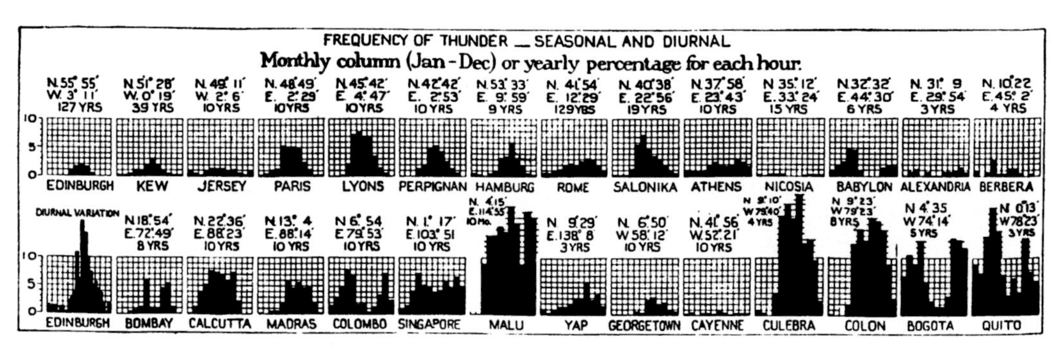

The scales at the side show the normal number of days of thunder indicated by the monthly columns.

ANNUAL FREQUENCY OF DAYS OF THUNDER

FIGURE 13

Authority: Geophysical Memoir, No. 24.

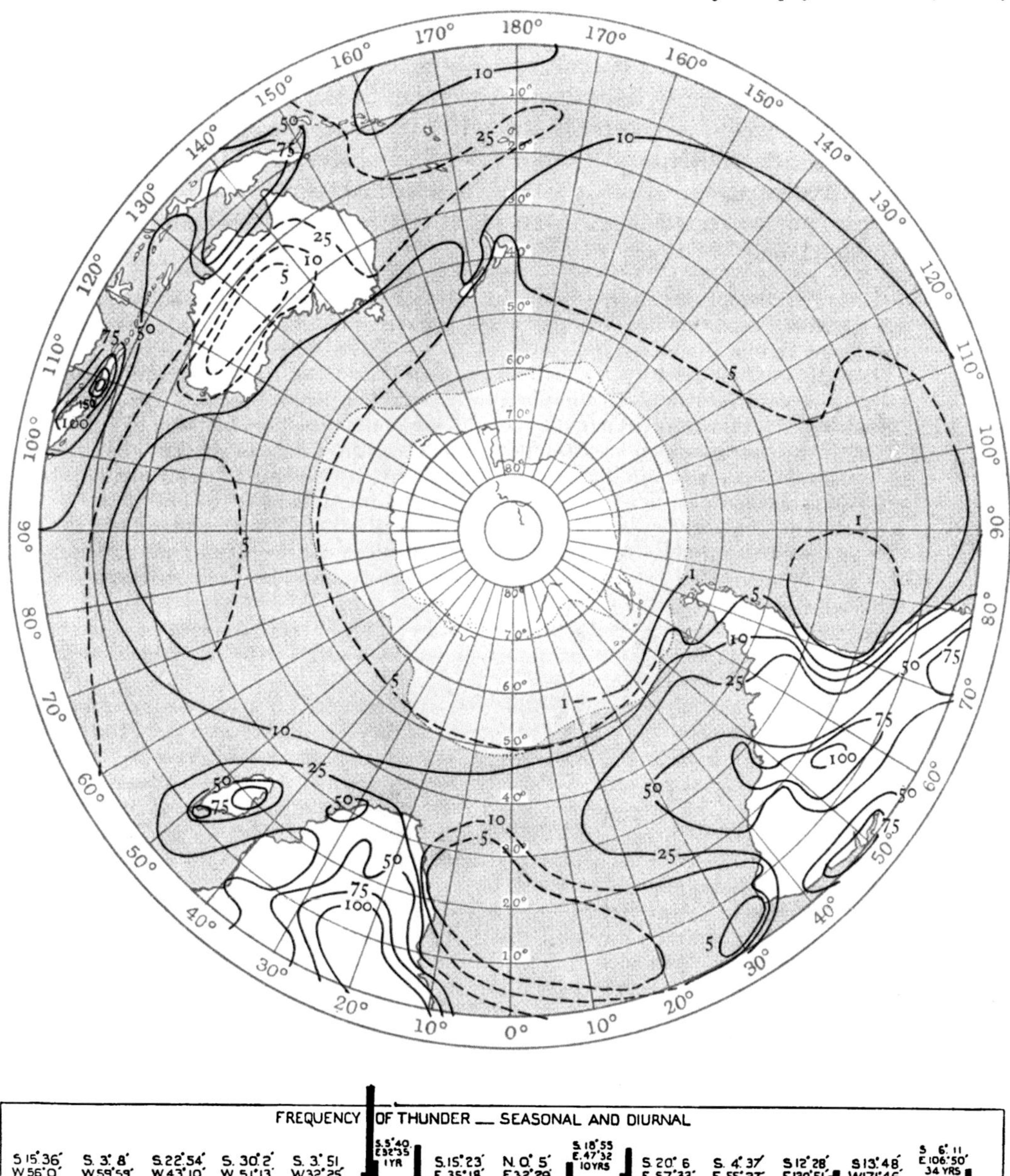

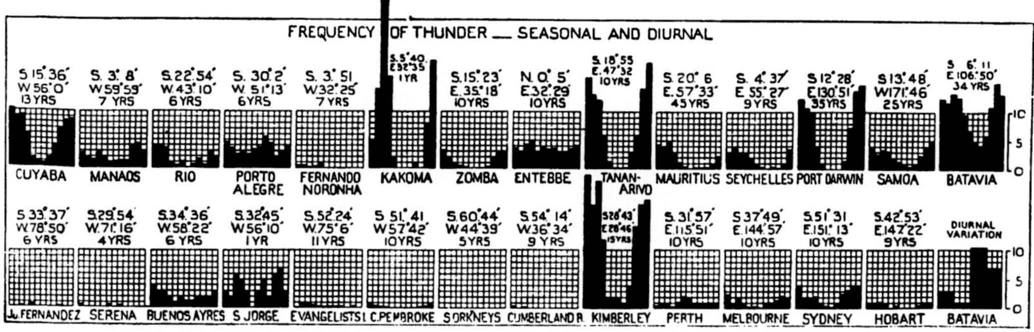

For diurnal variation the scale gives each hour's percentage of the total number of days of thunder.

St Elmo's fire.

The more normal phenomena of atmospheric electricity are not observed at a sufficient number of stations to afford material for the general survey of this volume, but we append a note concerning St Elmo's fire, corposants or *corpo santo*, which is sometimes observed as a meteorological phenomenon. It may be remembered that Dampier noted an observation of St Elmo's fire as antecedent to the typhoon which he describes. Our note is translated from a volume edited by E. Mathias, Director of the Observatory of the Puy-de-Dôme[1].

The difference of the electric potential between the surface of the earth and the atmosphere increases rapidly with the height, even in times of calm weather and when there is no sign of thunder-storm.

This difference of potential, which is a cause of loss of electricity, increases on the mountains particularly at their summits, and especially if they are pointed. Moreover, it is also seen on the top of the pyramids in Egypt and on the high plateaux of Persia. There is then such a difference of potential between the surface of the earth and the layers of air in contact with it that the loss of electricity becomes a tremendous and noisy phenomenon instead of proceeding slowly and invisibly as in the low plains.

During the day the discharge is expressed by an electric wind which crackles; the hair of travellers and the coats of their mounts stand on end when they are brushed. At night luminous aureoles stand out round the head and form at the finger-tips when the hand is raised above the head.

In times of thunder-storm the phenomena are exaggerated; luminous pencils are formed at the points of lightning-conductors, on the weather-cocks of houses and on the yards and tops of masts of ships.

The phenomenon has been recognised from long antiquity. It inspired superstitious terror in soldiers when, like Caesar's legionaries, they saw flames dart from the tips of their pikes and from their swords, which jumped from one point to another. Their leaders forced themselves on the contrary to draw from it a favourable augury for their enterprises. The ancients recognised these prodigies under the name of Castor and Pollux; in modern times the sailors call them St Nicholas' fire or St Elmo's fire; when they appear in the course of a tempest they are seen with joy because they announce its approaching end.

We will close this short exposition by referring to the phenomena that are observed on the two mountain meteorological observatories in France.

St Elmo's fires are extremely frequent at the Pic du Midi on lightning-conductors; it is sometimes enough, as I have been able to verify myself, merely to raise the hand in the air to see luminous sheaves at the finger-tips, accompanied by a characteristic whistling. (A. Angot.)

On June 25, 1888, the Prince of Wales (later King Edward) accompanied by his secretary, was able to observe the phenomenon of St Elmo's fire on the gallery of the platform of the tower at the summit of the Puy-de-Dôme. When he took off his hat his hair stood up on end on his head. Lifting his arms above his head a discharge of electricity took place from the finger-tips. When the Prince of Wales pointed his walking-stick to the sky the crackling manifested itself at the extremity of his stick. (Ch. Plumandon.)

In the only three cases known in which electrical manifestations were produced near the ground during aurora borealis, St Elmo's fire was visible. (A. Angot.)

[1] *Traité d'électricité atmosphérique et tellurique*, publié sous la direction de E. Mathias, Comité Français de Géodésie et de Géophysique, Paris, 1924, p. 293.

Dampier drew a distinction between the fixity of a corposant on the mast-head, and the washing about of the same phenomenon on deck. It seems possible that the luminosity of the latter may have been due to phosphorescence, which in tropical seas can be very impressive.

The observations of St Elmo's fire on Ben Nevis were summarised and discussed by A. Buchan in one of the volumes on the meteorology of that observatory. From the report[1] we make the following extract:

> During the five winters from 1883 to 1888, fifteen cases of St Elmo's fire were recorded. These all occurred during the night-time, and during the winter months from September to February. It is difficult, if not impossible, to observe this meteor in ordinary daylight, or in strong twilight, and this consideration perhaps accounts for the absence of recorded cases during day, and during the summer months. On one occasion it was *heard* in the day-time, being identified by the peculiar and unmistakable hissing sound which accompanies it.
>
> These fifteen cases have been discussed by Mr Rankin[2] in connection with the observations of pressure, temperature and rainfall for 30 hours before and 24 hours after the time of occurrence of St Elmo's fire. It is shown that the weather which precedes, accompanies and follows it has very definite characteristics, not only on Ben Nevis, but also over the West of Europe generally. Indeed, so well-marked is the weather and so notorious for its stormy character, that the observers regard it as a distinct type of weather, and call it by the name of St Elmo's weather....
>
> Thunder and lightning occurred at lower levels over the British Islands on several of the nights that St Elmo's fire appeared on Ben Nevis. Only on one occasion, on 4th Feb. 1887, was thunder and lightning observed on Ben Nevis about the times of occurrence of St Elmo's fire. On that occasion the thunder-storm was two hours subsequent to St Elmo's fire.

These remarks are quoted because the process of electrical discharge which forces itself upon the attention in its more "tremendous" manifestations must in fact be going on continuously, though with less intensity, as silent electrical discharge from all the projecting points of grass, shrubs, trees and buildings and mountain tops. It must be one of the most ubiquitous and persistent of all atmospheric phenomena. What its relation to weather may be has not yet been ascertained; but it is not safe on that account to assume that the transformation of energy which it represents is of no meteorological importance.

The normal potential gradient.

The phenomena of the silent discharge of electricity become exaggerated into St Elmo's fire when the atmospheric potential gradient at the surface is increased by the passage of an electrified cloud overhead, or by the development of a strong local field through the agency of blown snow or some other cause of electrical disturbance. We ought therefore to give some indication of the strength of the normal field. The ordinary potential gradient, as indicated by the records at permanent or temporary observatories, is of the order of 100 volts per metre "in the open." It is subject to very considerable

[1] 'The Meteorology of Ben Nevis,' *Trans. Roy. Soc. Edin.*, vol. XXXIV, Edinburgh, 1890, p. xliv.

[2] *Jour. Scot. Met. Soc.*, 3rd series, vol. VIII, 1889, p. 191.

variations both diurnal and seasonal, the normal range of variation may be nearly 100 per cent. of the normal value.

The typical diurnal curve is one which has a minimum in the early morning and a maximum in the late afternoon, the maximum being about twice the minimum.

At a number of the stations of observation, especially in summer, a semi-diurnal variation is indicated with minima in the early morning and early afternoon, and maxima about 8 o'clock in the morning and evening. Such semi-diurnal variations are conspicuous in the summer at Upsala, Paris and Tortosa, and in both summer and winter at Kew, Tokio, Davos, Kremsmünster and Munich.

According to observations at Kew Observatory there is strong correlation between the variation of potential gradient and that of the number of solid particles of the atmosphere which are measured as atmospheric pollution. The question of the origin of the variation of field which is thus raised appears again with regard to the potential gradient at Simla, which has so marked a minimum in the afternoon in June that the potential gradient changes sign and becomes negative. It is suggested that this may also be connected with the dust in the atmosphere.

According to observations made on the surveying ship *Carnegie* of the Magnetic Institute of Washington, the potential gradient at sea also has a marked diurnal variation with a single minimum in the early "morning" and maximum in the "evening" of Greenwich time.

It must be understood that the regular fluctuations which are described here are related to the quiet days of atmospheric electricity when there are no violent disturbances. The days of thunder and possibly of other occasional disturbing causes are marked by violent and rapid fluctuations from positive to negative gradients and *vice versa* of which the ordinary recording instrument does not give sufficient detail to enable the student to follow the course of the variations. That part of the subject is however of great interest in relation to the atmospherics of wireless telegraphy and will doubtless in course of time be more fully elucidated from that point of view.

The recognition of electrical wireless radiation began with a spark for the transmitter and a spark-gap for the receiver. In these days electro-magnetic arrangements, with no more exciting cause than the human voice, can set up radiation which will reproduce the voice in the listener's receiver, and lightning insists upon its own voice being appreciated by the listener. Fifty years ago we were curious to notice that an approaching thunderstorm might be heralded by the occasional tinkle of the telephone bell, but then there was the whole network of telephone wires to work upon; and now the flash, perhaps hundreds of miles away, without any network announces its achievement by the sharp cracks called strays, static or atmospherics. The "make" and "break" of the switch of an electric lamp can claim similar recognition and the electric lift of a neighbour's house may monopolise the conversation. There is still much to learn about electricity and its discharge; but it is not for this chapter. Whether all flashes produce atmospherics and all atmospherics are produced by lightning was the subject of a discussion[1] before the British Association in Bristol in 1930.

[1] See vol. I, p. 247; also R. A. Watson Watt, *Jour. Inst. Elec. Eng.*, vol. LXIV, 1926, pp. 596–610; and *Q. J. Roy. Meteor. Soc.*, vol. LVII, 1931, pp. 221–37.

CHAPTER III

THE COMPOSITION OF THE ATMOSPHERE

Pressure in millibars at high levels, in an atmosphere of uniform composition and isothermal above 12 km (192 mb at 12 km):

Height,	50	100	200	250	500	750 km
			Millibars			
tt = 180	$1{\cdot}41 \times 10^{-1}$	$1{\cdot}07 \times 10^{-5}$	$6{\cdot}12 \times 10^{-14}$	$4{\cdot}60 \times 10^{-18}$	$1{\cdot}13 \times 10^{-38}$	$2{\cdot}78 \times 10^{-59}$
tt = 200	$2{\cdot}91 \times 10^{-1}$	$5{\cdot}68 \times 10^{-5}$	$2{\cdot}16 \times 10^{-12}$	$4{\cdot}22 \times 10^{-16}$	$1{\cdot}19 \times 10^{-34}$	$3{\cdot}37 \times 10^{-53}$
tt = 220	$5{\cdot}25 \times 10^{-1}$	$2{\cdot}23 \times 10^{-4}$	$4{\cdot}01 \times 10^{-11}$	$1{\cdot}70 \times 10^{-14}$	$2{\cdot}33 \times 10^{-31}$	$3{\cdot}20 \times 10^{-48}$

Height of the Heaviside-Kennelly layer 50 km to 80 km; wireless waves of length 400 m are "reflected or refracted at night-time by a region about 80 km in height." (see chap. X, note 7.)

Layer of most frequent appearance of meteors 110 km; level of most frequent disappearance 80 km with a secondary maximum at 45 km. Lindemann and Dobson conclude that the temperature remains at 220tt up to 50 or 60 km and rises to about 300tt at about 140 km.

Lower limit of the height of aurora 80 km to 150 km. Wave-length of the auroral green line $\cdot 5577\mu = 5577$ Å.

Average height of luminous night clouds 82 km (Jesse). Last twilight arc not less than 700 km (Wegener).

"It is probable that the ozone would be formed chiefly at a height of about 50 km" (see chap. X, note 7).

BEFORE proceeding to represent the distribution of the ordinary meteorological elements which specify the condition of the atmosphere, we must pause to consider its composition.

The intrinsic constituents of the atmosphere are its permanent gases, nitrogen, oxygen, ozone, argon, carbon dioxide, with the rarer gases, neon, krypton, helium and, irregularly, hydrogen.

The proportions of these constituents are so nearly fixed that we are able to present the gaseous composition of the atmosphere throughout the region of the troposphere, and the stratosphere so far as it has been investigated experimentally, in the form of a table. Water-vapour has been included.

Some particulars of the gases of the atmosphere[1].

Gas	Specific gravity		Density at 1000 mb 290tt	Gas-constant R $p = R.t.\Delta$		Boiling-point (1000 mb)	Proportional composition in the troposphere	
	O, 16	Dry air, 1	Δ	R	γ		By volume or pressure	By weight
			g/m³	C.G.S. units		tt	%	%
Dry air	14·48	1·00	1201	$2{\cdot}870 \times 10^6$	1·4	—	100	100
Water-vapour	9·01	·6221	—	$4{\cdot}61 \times 10^6$	1·3	372·6	0 to 4	0 to 2·5
Nitrogen	14·01	·967	1162	$2{\cdot}966 \times 10^6$	1·4	78	78·03	75·48
Oxygen	16·00	1·105	1327	$2{\cdot}597 \times 10^6$	1·4	90	20·99	23·18
Argon	19·94	1·377	1653	$2{\cdot}083 \times 10^6$	5/3	87	·94	1·29
Carbon dioxide	22·15	1·529	1836	$1{\cdot}878 \times 10^6$	1·3	193	·03	·045
Hydrogen	1·008	·0696	83·6	$41{\cdot}22 \times 10^6$	1·4	20	·01?	·0007?
Neon	10·1	·697	837	$4{\cdot}12 \times 10^6$	5/3	34	·0012	·0008
Helium	1·99	·137	165	$20{\cdot}80 \times 10^6$	5/3	6	·0004	3×10^{-5}
Krypton	41·5	2·87	3440	$1{\cdot}00 \times 10^6$	5/3	121	$\frac{1}{2} \times 10^{-5}$	$1{\cdot}5 \times 10^{-5}$
Xenon	65	4·5	5400	$\cdot 63 \times 10^6$	5/3	164	$\frac{1}{2} \times 10^{-6}$	2×10^{-6}
Geocoronium	0·2?	·01?	17?	200×10^6?	?	?	?	?

The ratio of the capacities for heat is denoted by γ. The capacities at constant pressure and at constant volume are $c_p = R\gamma/(\gamma - 1)$ and $c_v = R/(\gamma - 1)$ respectively.

From the consideration of the conditions of equilibrium of gases of different densities and Dalton's law of the mutual independence of the composition of a gaseous mixture in respect of pressure, several physicists have calculated the variation in the composition of the atmosphere that must presumably be discovered whenever experimental investigations can be extended to a sufficient height. The basis of the calculation is set out in a paper by S. Chapman and

[1] *Computer's Handbook*, M.O. publication, No. 223, Section I, London, 1916.

E. A. Milne[1]. Similar formulae have been used by previous investigators; the results differ according to the composition which the investigators have assumed to represent the atmosphere at the surface. We reproduce from *The Air and its Ways* three diagrams representing the composition of the air by volume, at different heights up to 140 kilometres and 240 kilometres respectively, which illustrate very forcibly the important differences which may be displayed in the final result in consequence of very small differences in the basic assumptions. The first diagram (fig. 14*a*, Humphreys), which ranges up to 140 kilometres, shows the gradual change in proportion of nitrogen and oxygen, and an upper atmosphere above 100 kilometres consisting of hydrogen with a slight but diminishing admixture of helium. This is based on the assumption that hydrogen is a definite constituent of the lower atmosphere with a "normal" value for its proportion to the other gases.

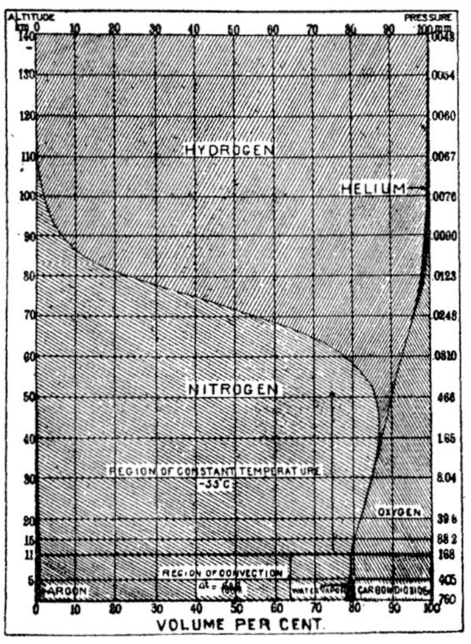

Fig. 14*a*. Percentage composition of air by volume (Humphreys).

The second diagram (fig. 14*b*, Chapman and Milne), on the other hand, shows in like manner a gradual change in the proportion of nitrogen to oxygen and an upper atmosphere, above 150 kilometres, consisting almost entirely of helium, with no hydrogen. This distribution is based on the assumption that the free hydrogen measured in the lower layers of the atmosphere is occasional and subject to combination with oxygen. It is therefore not a "normal" constituent and consequently cannot be used as a base upon which to build a hydrogen atmosphere.

Meanwhile Dr A. Wegener has suggested the existence of a new gas, still rarer even than hydrogen, which he associates with a green line in the spectrum of the solar corona and therefore calls "geocoronium." In his diagram (fig. 14*c*) he divides the atmosphere above 200 kilometres about equally between geocoronium and hydrogen with a relative increase of the former in still higher levels.

It is however supposed that gravity is not strong enough to keep an atmosphere so light as geocoronium or even as hydrogen attached to the earth[2], so perhaps we may regard Chapman and Milne's diagram as giving the best guide to the chemical composition of the atmosphere above the level of direct exploration. [Further information is given in chap. X.]

[1] 'The composition, ionisation and viscosity of the atmosphere at great heights,' *Q.J. Roy. Meteor. Soc.*, vol. XLVI, 1920, p. 357.

[2] Johnstone Stoney, 'On the escape of gases from planetary atmospheres according to the kinetic theory,' *Astrophys. J.*, 1900, pp. 251–8, 357–72, and subsequent papers in *Nature*, vol. LXI, 1899–1900, p. 515; LXII, 1900, pp. 78–9; *Proc. Roy. Soc.*, LXVII, 1900, pp. 286–91. Jeans, *Dynamical Theory of Gases*, 2nd edition, Cambridge, 1916, p. 361.

Some light is thrown upon this interesting but rather remote meteorological question by the results of inquiries into the aurora polaris. The lower limit of the aurora has been set out by Vegard and Krogness[1] at from 80 kilometres to 150 kilometres with maxima at 100 kilometres and 106 kilometres, and its spectrum is characterised by a very definite green line. Controversy has been very active about the origin of this green line. Vegard ascertained that a line very close to the auroral line was produced by the fluorescence under bombardment of solid nitrogen, but the wave-length seems now to have been identified by J. C. McLennan[2] as obtainable from a mixture of helium with oxygen; and in so far the conclusion is confirmatory of Chapman and Milne's diagram. Further researches by the same author indicate a closer agreement with the green line in a spectrum of a mixture of nitrogen and oxygen; and that conclusion, combined with the observation that the same green line is shown by the auroral light as high as 750 kilometres, suggests that the normal mechanical mixture of gases extends far beyond the limits set out in fig. 14.

PERCENTAGE COMPOSITION OF AIR BY VOLUME (see chap. x, note 8)

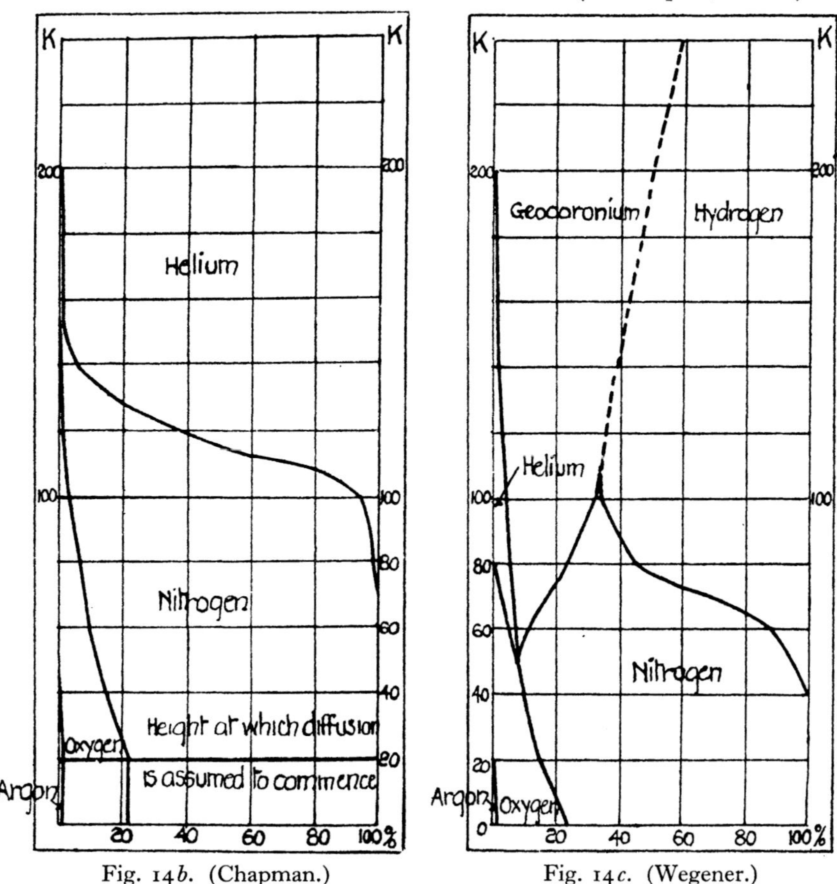

Fig. 14*b*. (Chapman.) Fig. 14*c*. (Wegener.)

[1] 'The Position in Space of the Aurora Polaris,' *Geofysiske Publikationer*, vol. 1, No. 1, Kristiania, 1920, p. 101.

[2] *Proc. Roy. Soc.* A, vol. CVIII, 1925, p. 501.

Apart from that which is provided by the study of the aurora, information as to the condition of the atmosphere, beyond the reach of the sounding balloons described in chapter XI of volume I, is obtained by observations of meteors and their trails, by the influence of the atmosphere upon the solar radiation as examined in a spectroscope, or by the behaviour of electric radiation in relation to the atmosphere. From the first, interesting conclusions have been drawn by Lindemann and Dobson[1] as to density and temperature to which we shall refer later, but nothing definite has been elicited in that way about the chemical composition. On the other hand, the spectroscope applied to the ultra-violet portion of the spectrum is affording a considerable amount of information about the presence of ozone, which is produced from atmospheric oxygen by the action of ultra-violet light. In recent years much attention has been paid to the detection of ozone and its distribution in height and latitude (see note 7). After Schönbein had discovered ozone and had ascertained that it would reduce iodine from potassium iodide it was supposed that the blue coloration of starch papers soaked in potassium iodide when exposed to the atmosphere was due to free ozone. The papers, which are still known as ozone papers, were frequently employed at meteorological stations to measure the amount of ozone. The action is however now attributed to hydrogen peroxide or to traces of the oxides of nitrogen[2].

The action of ultra-violet light in producing ozone must be associated with the phenomena of the aurora in the general effect of the solar radiation upon the constituents of the atmosphere. Included therein is the Heaviside layer which is of primary importance in radiotelegraphy. It is a conducting layer, and the electrical conductivity in gases is attributed to "ions," the atomic constituents of the molecules of chemical compounds which are supposed to be "dissociated" from their chemical combination by some electrical or physical process, in the present case by the impact of solar radiation of one kind or other. This is not the place for particulars of the action so far as they are known, but the subject must be mentioned here because the effects observed are held to imply a change in the composition of the atmosphere.

WATER IN THE ATMOSPHERE

The vapour of water has been included among the gaseous constituents of the atmosphere, but as a numerical value the average water-content of the atmosphere has very little effective meaning. Water in the atmosphere is constantly changing its form from the gaseous state in which it is known as water-vapour to the liquid form as water-drops with a great range of size, or to the solid form of ice-crystals or snow. As may be inferred from the pictures in chapter XI of vol. I, both the crystals and the drops may form visible clouds. Rainfall and snowfall are the common consequences of the condensation. Hence the composition of the atmosphere as regards water is subject to

[1] 'A Theory of Meteors, and the Density and Temperature of the Outer Atmosphere to which it leads,' *Proc. Roy. Soc.* A, vol. CII, 1922, p. 411.

[2] 'The Occurrence of Ozone in the Upper Atmosphere,' is discussed by J. N. Pring, *Proc. Roy. Soc.* A, vol. XC, 1914, p. 204.

incessant variation both in time and in space. We have already treated of water-vapour in the atmosphere and the method of measuring it. Limiting ourselves to water-vapour, or water in the gaseous form, we can give some idea of the condition at the surface by maps representing the normal distribution of water-vapour as derived from observations of humidity; and here we need only remind the reader that the amount of water-vapour which can be carried by a kilogramme of air is limited by the temperature of the air. The amount only exceeds 5 per cent. of the weight of air in the warmest climates and only then when they are at the same time very moist[1].

Water-drops and snow-crystals are carried by the atmosphere over great distances and for long periods by the eddy-motion which is always present to some extent in the air even when the general motion is very slow. The earlier natural philosophers exercised their ingenuity in finding an explanation of the floating of clouds. But as a matter of fact clouds do not float any more than the muddy constituents of turbid water can be said to float. It is the motion combined with the viscosity, or resistance to relative motion, of the air that prevents the particles "settling" as they would do if left undisturbed. If a fog, which consists of liquid drops, is watched, the motion of the particles can be seen as an indescribable turmoil. From that point of view the atmosphere is hardly ever at rest. When it is at rest, as in a closed globe in a laboratory, the particles settle in accordance with a certain law which will be referred to later.

SOLID PARTICLES

The same process of turbulent motion enables the atmosphere to carry solid particles as well as liquid-drops. The particles which are thus conveyed are of very varying character and size. From the point of view of the general circulation of the atmosphere the finer dust of volcanic eruptions is the most important kind of solid particle. It has been charged with weakening sunlight over the whole globe or a great portion of it, as in the eruption of Krakatoa near Java in 1883, in the eruption of La Soufrière and Mont Pelée in the West Indies in 1902, and in the eruption of Katmai in Alaska in 1912. We have already referred to Humphreys' treatise from which we have borrowed fig. 112 of volume I; a considerable part of the treatise is devoted to the influence of volcanic dust upon climate and weather and it appears that Humphreys would even refer the recurring ice-ages to that cause. (See vol. III, p. 126.)

Such fine dust as is here referred to may be regarded as the cause of the crimson colour of sunset which becomes very conspicuous when the atmosphere is charged with more than the wonted quantity of volcanic dust.

Coarser dust, which is raised in enormous quantities by the wind-storms of the tropical deserts, can also travel with the atmosphere for thousands of miles and is recognised in Europe in the production of "red rain." A notable example is described as occurring on February 21 and 22, 1903[2], in which

[1] See p. 131. [2] *Q.J. Roy. Meteor. Soc.*, vol. XXX, 1904, p. 57.

the source of the dust was tracked by reference to the meteorological conditions and located in North West Africa. The path travelled carried the sand round the Spanish peninsula to North Western Europe, round a high-pressure centre over Southern Europe. The amount of dust which fell in England on that occasion was computed to be about 10,000,000 tons.

Vast quantities of dust are conveyed from the deserts of Western Africa by the harmattan wind, which is thus described in the *Africa Pilot*:

> Off the West coast of Africa, between Cape Verde and Cape Lopez, a very dry Easterly wind, known as the harmattan, sometimes blows in December, January and February; it occasionally lasts five or six days, and has been known to continue as long as a fortnight, blowing with moderate force. It is always accompanied by a thick haze, which extends twelve or fifteen miles from the shore. From Sierra Leone Northward its direction is from ESE; on the Gold Coast NE; and at Cape Lopez NNE.
>
> At Fernando Po, 3° 50′ N to 1° 30′ S and 9° E to 5° 30′ E, the weather is often so thick and hazy that the land cannot be seen, and as the Easterly current is strong, vessels might run past the island.

The dust of the harmattan is relied upon to explain thick haze accompanied by a deposit of dust which is sometimes experienced far away on the Western side of the equatorial Atlantic.

The subject of dust-storms is treated in chap. x, note 9.

Smoke.

To the solid impurities carried into the atmosphere from volcanoes and deserts must be added that which comes as smoke from industrial, domestic and forest fires. The amount thus contributed may be unimportant in respect of the atmosphere as a whole, but it is in fact sufficient to affect the climate of the great industrial cities of Britain through the disturbance of the normal composition of the atmosphere by the addition of carbon, tar and sulphur compounds, and by the cutting off of a considerable fraction of sunlight or daylight from the areas in which the great cities are located.

The importance of these effects has led to the devising of means for the investigation of the amount and nature of the solid particles in the atmosphere, which has already added considerably to our knowledge of the subject under the guidance of the Advisory Committee for Atmospheric Pollution. The results of the investigation are given in twenty published reports and they are summarised in a recent work on *The Smoke Problem of Great Cities*[1]. One of the primary conclusions is that the acid products of combustion, as indicated by the amount of sulphuric acid which comes down with the rain, are spread much more widely and evenly than the solid particles. Of these, the larger generally fall in the near neighbourhood of the source. Occasionally, in exceptional atmospheric conditions appropriate for thunderstorms, the large aggregate soot-particles are carried bodily to great distances and fall as black rain. Within a mile or two of the source the smoke-particles are nearly uniform and about ·8 micron in diameter. It is doubtless mainly particles of this kind and of average size which form the dust-horizons in this country as represented

[1] Constable and Co., Ltd., 1925.

in fig. 24 of volume 1, but in other countries, such as India, the dust-horizon marks the boundary of the fine dust of blown sand.

In the investigation of the dust of the atmosphere of any particular locality all solid particles must be included whether they come from volcanoes, sandy deserts, coke ovens, factories or domestic fires. They may however be differentiated in the course of the investigation. The process consists in impaling the solid particles of a measured volume upon a glass plate and counting them with the aid of a microscope. In view of the general meteorological interest of the inquiry the instrument has been used in many countries and the following results have been obtained. *Smoke Problem*, p. 196.

Comparison of dust-counts in different countries.

Locality	Duration of observations	Smallest count (per cubic centimetre)	Largest count (per cubic centimetre)
Athens	Jan. to April, 1924	40. March 7	4300. January 2. Count only approximate
London	December, 1924	300. Very clear day	60,000. Fog-day
Melbourne	April, 1923 to August, 1924	23. March 6. Fine-weather type, clear atmosphere; frequent thunderstorms during previous day	2012. April 5, 1924. During April the atmosphere was unusually still, and consequently smoky
Stocksund (small villa town 5 km North of Stockholm)	1923	40 to 120. October 16. Few and uniform particles	480 to 800. July 5. Visibility 2 to 4 km
Warsaw	May to September, 1924	19. May 23, 7 h. Wind SE 5	854. July 8. Wind SW 1
Washington	1923 and 1924	59. September 21, 1923. Wind SW 6; raining	1785. November 12, 1924. Wind S. Visibility 3 miles

Wind-borne dust is not entirely confined to the lower atmosphere. A notable case has been disclosed by observations in aeroplanes and balloons in the United States. On April 6, 1923, the dust-content in the neighbourhood of Washington was observed to be at its maximum between 2000 and 3000 metres and the wind at the same level, as disclosed by a pilot-balloon, was moderately strong from North West and changing just above to a considerably stronger wind from the West. In this instance the lower layers were comparatively free from dust, and the larger load of dust in the higher levels was thought to be carried from the hot arid regions far to the West of the point of observation.

The normal amount of solid matter in the atmosphere of London varies between 0·1 milligramme and 1 milligramme per cubic metre. The latter figure would give 150,000 tons in a kilometre of height over the 150,000 square kilometres which represent the area of England. It is therefore very much less than the dust which came down as red rain in February, 1903.

Hygroscopic nuclei.

The examination of particles of solid or liquid matter in the atmosphere is of some importance to meteorology on account of the conditions which govern the formation of water-drops and consequently of clouds in the free atmosphere.

Many years ago John Aitken called attention to the fact that drops could not form, even in supersaturated air, in the absence of nuclei upon which molecules of water could collect to form a drop. The fundamental basis of his experiments was that air could be completely freed from nuclei for condensation under ordinary conditions of adiabatic expansion by passing it through a tube packed with cotton-wool. Subsequently C. T. R. Wilson pushed the inquiry further and ascertained the degree of supersaturation to which air must be reduced, if it is free from ordinary nuclei, in order to make use of negative ions or positive ions as nuclei for the purpose of condensation; but as the degree of supersaturation even for negative ions is "fourfold saturation" it is far beyond anything that is probable in the atmosphere, which is generally supplied with an ample number of suitable nuclei. Here therefore we need only consider the ordinary nuclei and leave out of account the possible condensation upon positive or negative ions.

Aitken provided an instrument which could be used to count them by the simple process of rarefaction of the air in a closed space after adding filtered air to ensure a suitable dilution. In that way the enclosed mass of air is cooled below its dew-point and drops are formed, they settle on a reticle and can be counted with the aid of a magnifying glass.

Aitken called the instrument a dust-counter and the impression became general that it was the solid particles in the atmosphere, ordinary dust, visible as motes in a sunbeam, that formed the nuclei for condensation. Recent investigations have, however, shown it to be doubtful whether the solid particles of dust have anything to do with ordinary condensation. The numbers obtained in an Aitken dust-counter are apparently quite unrelated to those obtained in an Owens' dust-counter, and also unrelated to the obvious dustiness of an atmosphere artifically loaded with visible dust.

We have therefore to find room in our specification of the composition of the atmosphere for another class of substance in addition to permanent gases, water-vapour, ions, and solid or liquid particles of dust or smoke. These we call hygroscopic nuclei, partly because they are in many cases formed of substances like common salt or other chlorides, or sulphuric acid, which are known to take up water from air which would not show "saturation" over a level surface of water, and partly because it is upon them that deposits of water-molecules are found to occur in preference to the sides of the vessel or a flat water-surface. These are the particles which can be counted in an Aitken dust-counter; they form the nuclei of water-drops in the atmosphere investigated by Köhler and noted in vol. III. They seem to be different from the particles of dust and smoke. An Atlantic fog consists of drops of condensed water-vapour, but shows no solid particles in the field of an Owens' dust-counter. The distinction may be found in the size of the hygroscopic particles when dry. The solid particles which count in the Owens' dust-counter are those which when dry can be seen with an object-glass of one-twelfth-inch focus with oil immersion; but with that high magnification there are still some particles (less therefore than ·2 micron) which are not clearly visible because they are below the limit of the resolving power of the microscope. Consequently we must imagine aggregations of molecules of hygroscopic salts too small to be focussed in a microscope and yet capable of forming visible aggregates of water when the humidity of the air which contains them is sufficiently high. Sulphur compounds and the chlorides of sodium and magnesium are specially suitable.

It is possible that we may recognise the difference of these solid particles and liquid aggregates in the colours of the sky. The deep blue of the Mediterranean sky is attributed to the "scattering" of the light of the shorter wave-lengths of the solar spectrum by the gaseous molecules of air, but many miles of thickness of solarised air are required in order to realise the blueness, and the transmitted light is not noticeably tinged with red. Smoke of the very finest solid particles, graded by long travel, may give the beautiful whitey-blue of peat-smoke seen against a dark background, and such particles tinge the sun's rays with a correlated red or crimson colour. The grosser coal-smoke in the neighbourhood of its source is too coarse to leave the red comparatively unscattered; it therefore looks white or grey.

The hygroscopic particles smaller than the finer smoke-particles are within the limits of size of the wave-lengths of the solar spectrum and must produce some differentiation between blue and red; but they easily load themselves with water and become large enough to scatter red light as well as the smaller wave-lengths, and they give us a white nebula when illuminated by the sun's rays (vol. I, fig. 93). They may therefore be regarded as responsible for the whiter skies of Northern latitudes.

J. S. Owens has ascertained that particles of salt collected in samples of haze at sea, or near thereto, are so hygroscopic that they become liquid-drops in an atmosphere with about 75 per cent. humidity. They may accordingly form the basis of fogs of considerable density in circumstances which are not otherwise suggestive of fog. In relation to these phenomena we may make the following quotations:

Walfisch Bay to Orange River. A thick haze generally hangs over the whole of this coast during the early part of the day, particularly at the distance of from four to six miles off-shore; but it generally clears off about 3 or 4 p.m. The breakers on the beach are frequently seen under the haze, while the land is barely discernible.

There is no rain during the Southerly winds, but generally very heavy dews at night, with occasionally a very dense fog, with large drops of dew-like heavy rain. A thick fog-bank on the Western horizon, with a well-defined line between it and the sky, is a sure indication of a strong Southerly gale. (*Africa Pilot*, Part II, Sixth edition, 1910, p. 327.)

Caribbean Sea. A peculiar phenomenon which has spread over the whole Caribbean from Barbados to St Kitts and extending South almost to Demerara has been the prevalence of a low-hanging mist which has shut off the horizon. For some time the idea prevailed that it was dust due to volcanic eruption, but this was removed by the reports from vessels arriving, and from advices from the neighbouring islands.

Captains trading in these waters for years state that they have never before experienced such continued low visibility at this time of the year. No scientific explanation of the phenomenon has yet been offered. (*Barbados Advocate*, May 23, 1922.)

These thick hazes and the curious sea-fogs of summer, which occasionally spread from sea to shore on British coasts and suddenly change the whole aspect of a summer's day, may thus be due to the formation of a cloud of hygroscopic nuclei by the spray of breakers on the shore.

The question of the supersaturation of air in clouds is considered in chap. X, note 10.

Clouds in the stratosphere.

Among the phenomena which are dependent upon some form of solid or liquid particles in the atmosphere are the luminous clouds which are occasionally seen far above the polar horizon of temperate latitudes during the summer period, from the middle of May to the middle of August in the Northern hemisphere, and from November to February in the South. "The characteristic feature of luminous clouds appears to be, as indicated by O. Jesse, their unchangeable altitude over the earth's surface; on the average 82 km[1]." It has been suggested by A. Wegener that they are formed by the condensation of water, but the general view is that they consist of dust-particles, and as they cannot always be attributed to volcanic action they are looked upon as evidence of the passage of the earth through a comet's tail.

Further investigation of clouds above the tropopause has been pursued but attention is usually concentrated on the explanation of the physical appearances whereas this volume limits itself to the appearances themselves.

Two types of such clouds have been observed. The higher noctilucent clouds have already been referred to. They are described as cirrus in form with a definite wave-structure and of shining blue white silvery colour. The lower type, known as Perlmutter or iridescent clouds, show brilliant prismatic colours and occur at heights between 20 and 30 kilometres. They differ from noctilucent clouds by appearing almost exclusively in winter.

The investigation of noctilucent clouds goes back to the observations of Jesse in 1885–91, that of Perlmutter clouds to Mohn in 1893. Information of recent observations will be found in the following references:

Noctilucent clouds. C. Störmer, 'Height and velocity of luminous night-clouds observed in Norway, 1932,' University Observatory, Oslo, Publication No. 6, 1933.

E. H. Vestine, 'Noctilucent clouds,' *Journ. R. Ast. Soc. Canada,* vol. XXVIII, 1934, p. 249. (The paper gives a summary of 290 observations between 1885 and 1933.)

Iridescent clouds. C. Störmer, 'Remarkable clouds at high altitudes,' *Nature,* vol. CXXIII, 1929, p. 260; 'Mother-of-pearl clouds over Scandinavia in January and February, 1932,' *Q. J. Roy. Met. Soc.,* vol. LVIII, 1932, pp. 307–9; *Geofys. Publ.,* vol. IX, No. 4, Oslo, 1932; 'Nuages dans la stratosphère,' *C. R. Acad. Sci.,* vol. CXCVI, 1933, p. 1824; 'Höhenmessungen von Stratosphärenwolken,' *Beitr. Phys. fr. Atmosph.,* Band XXI, 1933.

[1] V. Malzev, *Nature,* vol. CXVIII, 1926, p. 14.

CHAPTER IV

COMPARATIVE METEOROLOGY: TEMPERATURE

Some limiting values	Maximum		Minimum	
Mean temperature for a year	Massawa	303tt	Antarctic Ice-barrier	247tt
			Greenland (A. Wegener)	241tt
Mean temperature for a month	Death Valley	312tt	Verkhoïansk	222tt
Mean seasonal variation	Verkhoïansk	66t	Marshall Islands (Pacific)	0·4t
Mean daily range of temperature	Calama	23t	Indian Ocean and Polar region	0·5t
Absolute extremes	Death Valley	329tt	Verkhoïansk	205tt

['Grenzwerte der Klimaelemente auf der Erde,' by G. Hellmann, *Sitz. Preuss. Akad. Wiss.*, 1925, pp. 200–215]

Temperature over the Poles.

North Pole. "The minimum temperature recorded [by Commander Peary on April 6–7, 1909] during the thirty hours was −33° [237] and the maximum −12°?F [249]." (*Nature*, vol. LXXXI, 1909, p. 339.)

South Polar Plateau [ca 3000 m].

Month	Observer	Mean observed temperature	Maximum observed temperature	Minimum observed temperature
December	Amundsen	250·4	258·3	244·5
January	Scott	244·8	253·4	238·7

(G. C. Simpson, *British Antarctic Expedition* 1910–1913, vol. I, p. 41.)

Mount Everest. Camp III [ca 6400 m]. Lowest air-temperature observed by the expedition 254tt, 10 May, 1924.

HAVING now before us the main features of the base upon which the atmosphere lies, the earth's surface, land and sea, and a sufficient sketch of the composition of the superincumbent atmosphere in which the meteorological changes are produced, we may call attention to the normal values of the meteorological elements by which the working conditions of the atmosphere are defined.

When the science of meteorology attains its maturity the fortunate author will begin his survey of the meteorological elements with solar and terrestrial radiation, the proximate causes of the whole sequence of weather. We cannot do that. We have not yet attained sufficient knowledge of the quantitative relations of solar and terrestrial radiation to atmospheric changes; we must first describe the meteorological condition of the atmosphere in the form which the customary meteorological instruments make possible. The consideration of the influence of solar and terrestrial radiation beyond that which has been given already in chapter I must be regarded as a subject rather of inquiry than of knowledge and treated accordingly in volume III.

We begin our presentation of the meteorological conditions at the base of the atmospheric structure with the distribution of temperature.

The first section of this part of the subject is a set of monthly and annual maps of normal temperature at sea-level. These are supplemented by maps of the average daily range of temperature throughout the year and the range of normal temperature within the cycle of the seasons in different parts of the world.

With the notes which are inserted in the vacant spaces under the maps these will be a sufficient indication of the average conditions at the surface. The material available for the representation of the conditions of the upper air is too scanty for us to treat the upper levels in the same manner as sea-level.

We have chosen normal mean values as the basis of our representation. A more adequate representation of temperature would be afforded if the maps showed separate lines for normal maximum temperature, the temperature of about two hours after noon, when presumably the loss of heat by radiation to the environment from the surface-air and ground begins to exceed the amount that comes by radiation from that environment aided by the sun, and the normal minimum temperature just before sunrise when the radiation from the sun begins to affect the temperature of the surface-layer. If such a mode of representation were practicable it would bring out clearly an important fact about temperature to which attention is not sufficiently directed; that is the difference in the diurnal range of temperature over the sea and over the land. There is in fact hardly any diurnal range over the open sea. Observations of the *Challenger* expedition allowed a range of about 1 t (see p. 84); and, in accepting that, we must remember that every casual circumstance tends to exaggerate the apparent difference. As mounted in the ordinary way, a thermometer on board ship can hardly be screened from the effect of solar radiation by day, or of terrestrial radiation at night, and it would be a very slight exaggeration to say that there is no diurnal variation at all over the sea. That is the conclusion one arrives at naturally after an examination of the few thermographic records that have come to us from the sea. It is curious to note how easily one realises in those circumstances that, over the sea, pressure is the element which maintains a diurnal variation, while temperature takes up that rôle as soon as land is reached or the ship comes to rest. The question wants examining with the closely allied question of the range of temperature of the sea-water which should show some diurnal variation in calm weather but very little when there is a moderate wind (chap. X, note 11).

We should have been glad to show the isothermal lines as running in single line over the sea indicating both normal maximum and normal minimum (because they are practically coincident) and bifurcating upon reaching the coast-line, the northern branch marking out the places which had the normal temperature of the sea-air for maximum and the southern branch indicating in like manner the line of equal minimum temperatures. In that way the difference between temperatures of day and night would be very neatly shown by the bifurcation at or near the coast-line and subsequent separation until they came together again on the next coast-line. The thermal conditions proved however to be too complicated for this plan and we have had to content ourselves with maps of mean daily temperature over land and sea with an additional pair for the seasonal range and a collection of examples of the diurnal range of temperature in the twelve months of the year at selected stations in different parts of the world expressed in the form of step-diagrams.

It may be noted that we have been careful to have the lines of mean temperature over the sea drawn quite independently of those over the land without indicating any continuity between them. Continuity of course there must be; but whether the joining lines should run actually along the coast-line, within it or outside it, the available observations do not show.

For drawing the lines over the land the mean values of temperature have been "reduced to sea-level" according to rules of which particulars are given in the International Tables[1]. Geographers sometimes demur to this procedure because, as they rightly contend, the part of a line which crosses a mountainous region has no geographical application. Those who look for a mean temperature at a high-level station in accordance with the isotherms shown on the map will be disappointed. But it is obvious that a map of a whole hemisphere, or even of a whole country, is not a practical method of co-ordinating the details of the facts about temperature in mountainous regions; an impracticably large scale would be necessary. In fact the information which can be expressed by values reduced to sea-level or even to means is a very rudimentary step in the approach to a real comprehension of the behaviour of the atmosphere.

SEASONAL VARIATION

The representation of air temperature by maps of the normal monthly distribution is supplemented by tables which will give the reader some idea of the range of the information covered by the data obtained at meteorological stations in different parts of the world, from which the normal course of the seasons with regard to temperature can be made out. That does not however exhaust the representation of the data. We can go into further detail if necessary by taking the normal *daily* temperatures, maximum, minimum or mean, in a manner precisely similar to that in which the monthly normals have been obtained. Plotted as a graph we can get a curve of daily temperature showing the normal variation throughout the year. Graphs of that kind are available for many stations: those for the four chief observatories of the Meteorological Office, Aberdeen, Cahirciveen (Valencia), Falmouth and Richmond (Kew) for the years 1871–1900 are reproduced in a volume of *Temperature-tables for the British Islands* (M.O. 154). Richmond shows notable oscillations of mean temperature during May and also during December. The former is suggestive of the common incidence of low temperatures over Europe associated with the ice-saints Mamertius, Pancras and Gervais, May 11–13. It has been supposed that these notable lapses of temperature with their corresponding recoveries are not merely accidents of weather but are real features of the normal seasonal development of the general circulation[2]. Some years ago the idea of the normality of these occurrences was controverted because in the long sequence of the records of temperature at Greenwich Observatory they failed to preserve their distinctness. If however the general circulation should be habitually unpunctual with its sequence the impression of the sequence may be lost by being distributed over too long an interval.

[1] *International Meteorological Tables*, chapter III, section II, Paris, Gauthier Villars, 1890.

[2] A. Buchan, *Handy Book of Meteorology*. William Blackwood and Sons, London, 2nd edition, 1868.

THE NORMAL VICISSITUDES OF WEATHER WEEK BY WEEK

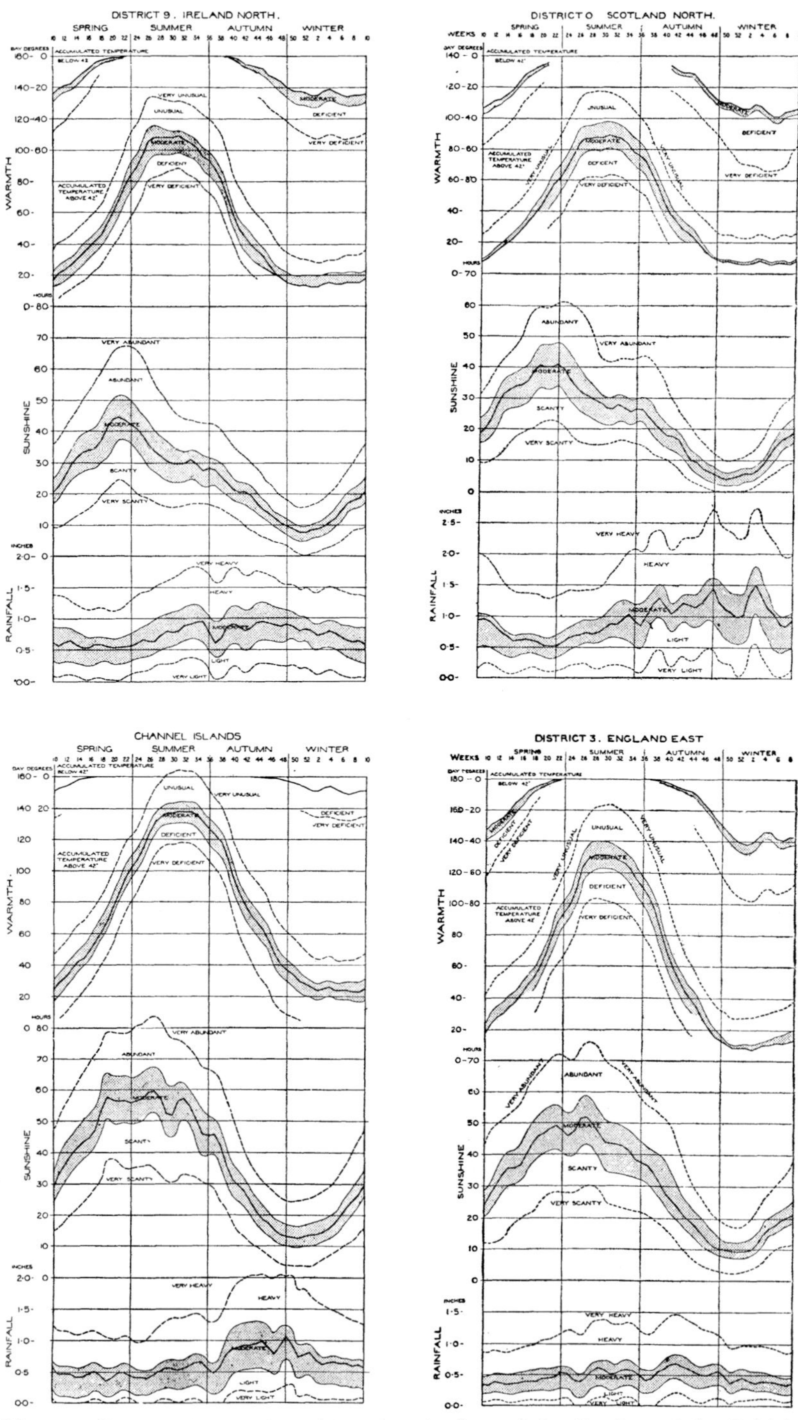

Fig. 15. Curves of normal weekly values in four of the districts of the British Isles, of accumulated temperature in day-degrees, above 42° F and below 42° F, the weekly duration of sunshine in hours and the rainfall in inches; with the limits of variations which have a frequency of one in three and one in twelve respectively.

Seasonal variation in weekly values.

Intermediate between the month and the day is the week which was chosen for a Report of the Meteorological Office specially designed for agricultural and hygienic application. A novel feature of this weekly series of data as compared with monthly climatological tables is that the values are taken out for districts, twelve in all for the British Isles and Channel Islands, and not only for individual stations. This valuable feature would have been more directly effective if the districts had agreed with those chosen by the establishments which gather agricultural, vital or commercial statistics.

Another special feature is that warmth and cold are not dealt with by the simple process of noting temperatures but, at the suggestion of leading authorities in agricultural science, by "accumulated temperature" above or below 42° F expressed in "day degrees." Rules were drawn up for making the computation for the weeks when the range of temperature passed across the limit.

An incidental disadvantage of the weekly system as arranged in the weekly report is that so far as concerns the relation of the events to the sun's declination, which is the fundamental climatic factor, the commencement of the first week ranges from 29 December to 4 January, and its last day from 4 January to 10 January, with corresponding ranges for all the other weeks. Hence the period represented by the means has one day of "umbra" with 6 days of "penumbra" on either side, and draws its values from 13 days of the cycle of the seasons instead of 7 days.

Nevertheless the Weekly Report of the Meteorological Office affords the best material known to the writer for the study of the features of climate in relation to its effect upon the sequence of events in agriculture, or the facts concerning health and disease, which are set out week by week in the tables of the Registrars General of the three kingdoms. The weekly records can easily be summarised to preserve in very short form for the community in general an effective memory of past weather. The purpose is achieved by classifying the records as described in the following extract:

WEEKLY WEATHER REPORT. Various changes were made in this publication at the commencement of the year 1907. The most important among these is the inclusion on the front page of the *Report* of a table in which the week's warmth, rainfall and sunshine, for each district are characterised by a selection of adjectives. The terms used are, for warmth: "unusual," "moderate," "deficient"; for rainfall: "heavy," "moderate," "light"; for sunshine: "abundant," "moderate," "scanty." For fixing the limits within which each of these adjectives should apply, the statistics of weekly values, for districts, of each of the five elements, accumulated temperature, above and below 42° F (278·6 tt), rainfall and sunshine, which have been published in the weekly reports for the years 1881–1905, have been examined from the point of view of their frequency distribution. The limits have been so selected, that of the total number of weekly values for each element included in the period mentioned above, one-third are characterised as moderate, one-third fall on the side of excess, and one-third on the side of defect. A further subdivision is effected by prefixing the adverb "very" to the description in the case of one-twelfth of the values reckoned from either extreme. In the *Report* the definitions of the terms are expressed as probabilities as follows:

Warmth. The week's warmth is called "*unusual*" if it is so much above the average for the time of year that, in the long run, it is likely to occur, for that week, only once in three years, and it is marked "*very unusual*" if it is likely to occur, for that week, only once in twelve years; similarly it is called "*deficient*" if it is so much below the average for the time of the year that it is only likely to occur, for that week, once in three years, and "*very deficient*" if it is likely to occur, for that week, only once in twelve years. Otherwise it is called "*moderate*."...

[In like manner the characteristics "very heavy," "heavy," "moderate," "light" and "very light," are applied to rainfall; and "very abundant," "abundant," "moderate," "scanty" and "very scanty," to the measures of sunshine. When the week has been without rainfall the word "nought" has been inserted in the column for the district, and similarly when the week has been without sunshine.]

The first step in the process of fixing the limits consisted in the determination of the average value for each week for each element, and for each district. The average values, found by taking the mean in the usual manner, were then smoothed by Bloxam's formula $(A+2B+C)/4$, and the results adopted as "average for the time of year." Subsequently the frequency distributions of the divergences from these smoothed averages were determined for each element. From these, working diagrams were prepared for each district.

[Four are reproduced in fig. 15.] The thickened line in the centre of the shaded belt in each section of the diagram shows the "average for the time of year." The shaded belts themselves show the regions of "moderate" value, and they are so drawn that one-third of the total number of values in the 25-year period falls within them, and that one-third falls above and one-third below them. The dotted lines are the limits for differentiating the values to which the adverb "very" is to be prefixed. They are so drawn that one-twelfth of the total number of values lies above the upper and one-twelfth below the lower line.

In the case of the quantity warmth, during the warm season of the year the classification is based entirely on the consideration of accumulated temperature above 42° F; but in the colder months the weeks of unusual warmth only are determined from this quantity, and those of deficient warmth are classified from the amount of accumulated temperature below 42°.

(*Report of the Meteorological Committee to H.M. Treasury*, 1907, p. 31.)

The descriptive adjectives for warmth: very unusual, unusual, moderate, deficient and very deficient, allow themselves to be abbreviated to the initial letters U, u, m, d, D; in like manner for rainfall we may use the initial letters H, h, m, l, L; and for sunshine A, a, m, s, S, with O for no rainfall or no sunshine if required. These abbreviations enable us to present the climatic results of a whole year in a statement of weekly values which requires only one line of print for each of the elements, temperature, rainfall, sunshine, and to these can be added, by anyone who is interested in that aspect of climate, weekly representation of the winds characterised as "polar" or "equatorial" with the adverbial supplement if desired, making use for that purpose of a table of the frequency and strength of the winds from the cardinal points which is based upon the records of anemometers.

We shall illustrate the facilities of this mode of procedure by a table in chapter VII, p. 304, which gives on one page the results for temperature, rainfall and sunshine for each week of twenty years.

We invite the reader to infer that for the purposes of a weekly report it is well to commence the statistical year on 8 January, to group the weeks in fours and fives, which by analogy may be called months, so arranged that the four groups of five weeks have the solstices and equinoxes in their middle weeks. The grouping makes it easy to display the quarters of the May-year and the seasons of the Farmers' year, autumn, winter, spring and summer, as shown in the distribution of radiation, pp. 4 and 5.

TEMPERATURE OF THE SEA

Following the monthly maps of mean temperature, mean daily range, and mean seasonal range, we have placed four pairs of maps of mean temperature of the sea-surface in the four months February, May, August and November. These maps are transformed from the maps of temperature of sea-surface in the *Barometer Manual for the Use of Seamen*, which were themselves compiled from the maps of the separate oceans in previous publications of the Meteorological Office. Probably these maps could now be improved[1].

In the meantime we may note the result of some observations of temperature and salinity of the sea, described by W. H. Harvey[2], which were carried out at a station 20 miles South West of Plymouth where the depth is 70 metres.

From the changes of temperature from month to month is derived the net daily loss or gain of heat of a column of 0·1 square cm cross-section.... It is concluded that the changes in temperature of the sea were controlled to a marked extent by evaporation. A very interesting observation was that, in the absence of windy weather and consequent mixing by waves, the upper layers may be heated by solar radiation in early May, giving a shallow warm layer separated from the cold water below by a sharp surface of discontinuity. Several days of rough sea are necessary to disturb materially this distribution of temperature. It is also pointed out that in fine clear weather with only light winds the upper inch or two of water become very hot. The normal method of sampling sea-water in a bucket represents the surface 6 inches, more or less, so that the sample is considerably cooler than the actual surface temperature of the sea.

In this connexion it is of interest to refer to the results of Knudsen's experiments on the intensity of light at various depths in the ocean[3].

Knudsen's results show that with increasing depth in the ocean the intensity of illumination decreases comparatively rapidly at the ends of the visible spectrum (i.e. in the red and violet) and that the rate of decrease is least in the green portion of the spectrum. The number of observations discussed in the publication is too small to provide conclusive data, but the following table based on them is instructive:

Wave length in microns, μ	·65	·60	·55	·50	·45	·40
Depth in metres at which light is reduced to one-thousandth of its intensity at the surface	18	25	38	37	23	14
Fraction to which the intensity is reduced in passing through a layer of 10 metres	$\frac{1}{45}$	$\frac{1}{16}$	$\frac{1}{7}$	$\frac{1}{7}$	$\frac{1}{20}$	$\frac{1}{148}$

[1] Helland Hansen and Nansen, 'Temperatur-schwankungen des Nordatlantischen Ozeans und in der Atmosphäre. Einleitende Studien über die Ursachen der klimatologischen Schwankungen,' Kristiania *Videnskaps. Skr.*, I, 1916, No. 9; tr. in *Smithsonian Misc. Coll.*, vol. LXX, No. 4, Washington, 1920. Monthly maps of mean sea-surface temperatures of the North Atlantic, compiled from all available sources during the period 1855 to 1917, are published in *The Marine Observer*, London, H.M.S.O., 1926. (Additional references in chap. x.)

[2] *Journal of the Marine Biological Association*, 1925, noted in *Nature*, vol. CXV, p. 778.

[3] *Conseil Permanent International pour l'Exploration de la Mer. Publications de Circonstance*, No. 76. The quotation is from *The Meteorological Magazine*, vol. LVIII, 1923, p. 75.

TEMPERATURE OF THE SEA AND AIR

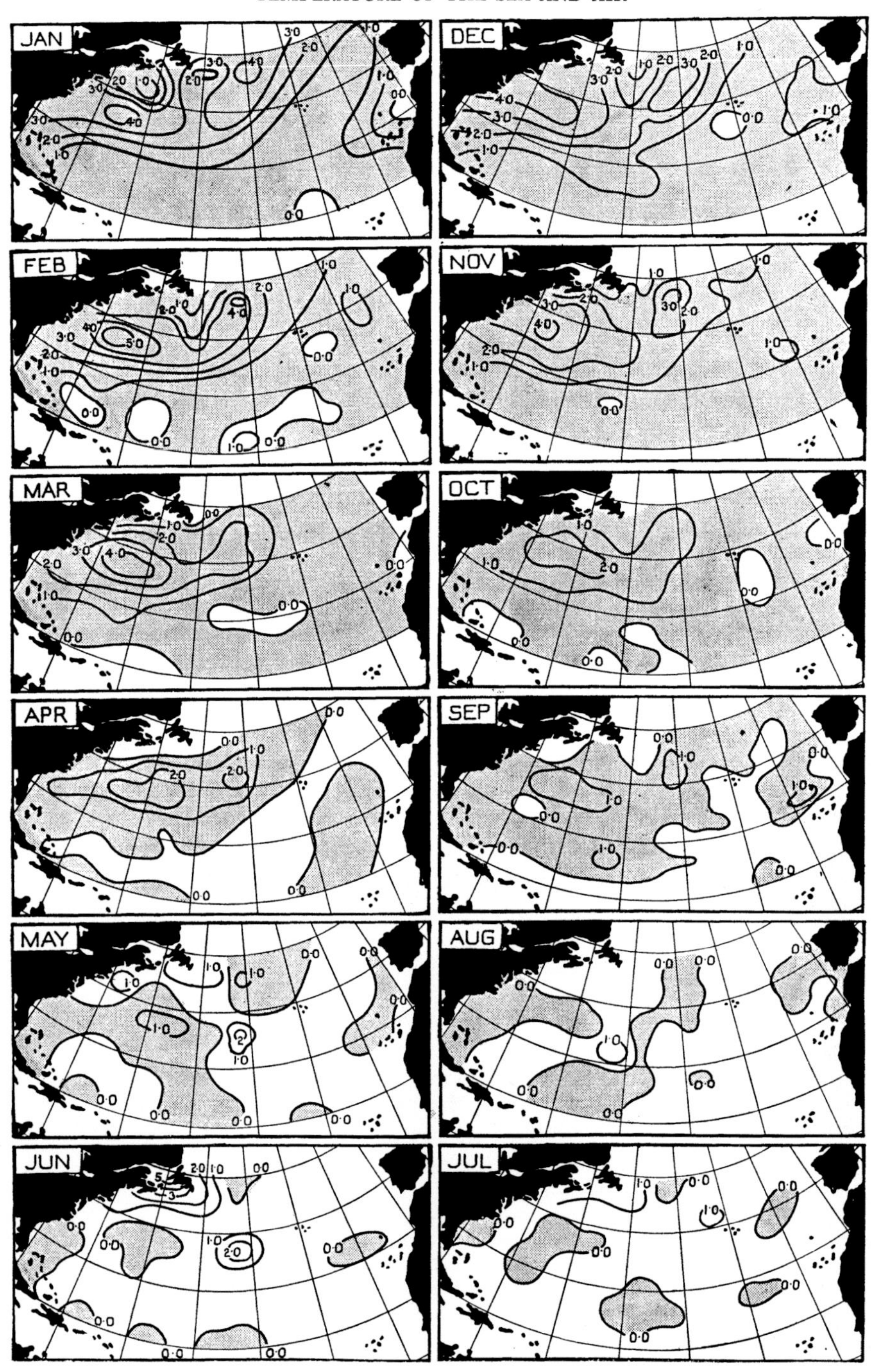

Fig. 16. Seasonal variation of the difference of temperature of the sea and air over the N. Atlantic 20°–50° N in tercentesimal degrees. Areas over which the temperature of the sea is greater than that of the air are stippled.

Difference of temperatures of the sea and air. North Atlantic Ocean.

From the nature of the case, the difference of mean temperature between sea and air represents one of the most important of the physical agencies which affect the atmosphere and must give rise to vigorous convective action if the sea is warmer than the air; or, on the other hand, the sea colder than the air is recognised as a contributory cause of fog. We call attention to the seasonal variation of such conditions for a few "squares" in the illustrations appended to the maps of sea-temperature (pp. 88–95) to which we shall refer, subsequently, when we consider the question of centres of action in chap. VII. The normal differences of temperature of sea and air, month by month, are represented in fig. 16. They are eloquent as evidence in relation to this important subject. February is the most exciting month.

Difference of temperature of sea and air. Mediterranean Sea.

Figures which indicate excess of sea-temperature over air-temperature are underlined.

		Jan.	Feb.	Mar.	Apr.	May	June	July	Aug.	Sept.	Oct.	Nov.	Dec.
West of Sardinia		° C	° C	° C	° C	° C	° C	° C	° C	° C	° C	° C	° C
36–38 N	5–0 W	0·6	0·1	0·5	0·7	0·3	1·0	1·4	1·0	0·2	0·8	0·3	0·2
40–42	0–5 E	1·9	1·0	0·5	0·2	1·3	0·5	0·2	0·1	0·2	0·9	1·5	1·6
38–40	0–5	1·3	0·1	0·4	0·0	0·8	0·5	0·2	0·1	0·4	1·1	0·6	1·0
36–38	0–5	1·3	0·3	0·2	0·5	0·4	0·5	1·4	0·9	0·4	0·3	0·4	1·3
42–44	5–10	2·6	1·7	0·7	0·2	0·3	0·4	0·1	0·1	0·2	1·1	1·5	2·4
36–40	5–10	1·7	0·9	0·3	0·5	0·2	0·6	1·1	0·4	0·2	0·5	0·9	2·0
Tyrrhenian Sea													
42–44 N	10–15 E	2·3	1·1	0·3	0·4	0·5	0·7	0·0	0·0	0·2	0·6	1·6	2·5
40–42	10–15	2·3	1·2	0·5	0·5	0·6	0·4	0·0	0·3	0·2	0·6	1·5	2·0
38–40	10–15	2·0	0·6	0·3	0·6	0·2	1·1	0·2	0·5	0·1	1·3	2·1	2·1
38–40	15–Coast	2·1	1·0	0·2	0·2	0·9	1·0	0·9	0·3	0·1	0·0	1·0	1·4
Ionian Sea													
36–38 N	10–15 E	1·1	0·5	0·5	0·3	0·2	0·1	1·1	0·2	0·4	0·0	0·9	1·6
36–38	15–20	1·5	0·8	0·0	0·1	0·9	0·4	0·8	0·1	0·3	0·1	1·0	1·1
34–36	15–20	1·2	0·5	0·7	0·1	0·4	0·0	1·3	0·4	0·0	0·3	1·0	1·2
34–38	20–25	1·2	0·5	0·0	0·3	0·6	0·5	0·9	0·5	0·2	0·0	1·1	1·7
Egyptian Waters													
32–34 N	20–25 E	0·7	0·9	0·4	0·5	0·3	0·6	1·0	0·6	0·5	0·0	1·0	1·8
32–34	25–30	0·9	0·7	0·1	0·3	0·8	0·1	0·7	0·4	0·2	0·4	1·2	1·7
32–34	30–32	1·0	0·6	0·2	0·2	0·2	0·1	0·1	0·1	0·4	1·0	1·2	1·8

For the Mediterranean Sea we have extracted the figures for the preceding table of the difference of temperature between air and sea, from *Annalen der Hydrographie und Maritimen Meteorologie*, 1905.

EARTH-TEMPERATURES

The temperature of the earth's surface is as important in the thermodynamics of meteorology as that of the sea-surface and it is much more irregular and more fluctuating. Its effect is easily recognised in the variations of temperature in the thermometer-screen under standard conditions such as are represented in the sections round the earth in different latitudes in fig. 3. But that information is not by any means sufficient for physical and dynamical purposes; it is perhaps more obviously insufficient for agricultural and biological purposes. It is not so much the mean values of the temperatures at the surface that are effective as the variations of temperature during the day; we ought to get behind the isogeotherms which correspond with the isotherms for the air. From observations in Japan[1] at different depths down to 7 metres it appears that the mean temperature for the year is practically the same at all levels and the same is perhaps generally true of other stations; but there is notable difference in the mean temperature of the day at different levels and in the range of temperature within the day and the year.

A few examples will illustrate the problems which present themselves.

J. H. Field, Director of the Meteorological Service of India, writes:

> The lapse rate in Agra is often extreme near the ground even on a winter mid-day...while the corresponding night-inversion of great stability would be impossible, I suppose, in Britain. The ground is known to reach a temperature of 156° F [342 tt] and may even go higher still[2]. In England you do not get a soil-surface of 156° F nor lapses averaging 17° C through the year and reaching 25° C [45° F] for the difference between the ground and a height of 4 feet on hot occasions.

In a paper on the biological aspects P. A. Buxton[3] remarks that as far North as Palestine the sand-surface temperature commonly reaches 60° C, 333 tt, at mid-day; and J. G. Sinclair[4] gives 71·5° C, 344·5 tt as the temperature for a depth of 4 mm at 1 p.m. on 21 June, 1915, at Tucson, Arizona. These are very high values of surface-temperature but still higher might be obtained with special arrangements of the receiving surface. Solar radiation can be used for cooking if suitably concentrated. On the other hand travellers across deserts often refer to frost at night even in summer months, so that the range of temperature between day and night for the surface-soil may be of the order of 70 t. In the former of the two references attention is called to the biological importance of the large range of temperature, which is common to all regions of clear sky, in the consequent change of humidity in the surface-air. The dried herbage which drifts about the surface takes from the atmosphere a considerable fraction of its own weight of water when the humidity is above 80 per cent., and furnishes a possible supply of water for the insects which are to be found in those dry regions (see chap. X, note 12).

For the lower limit of temperature at the surface in Greenland we note some observations of J. P. Koch and A. de Quervain quoted by W. H. Hobbs[5] which refer to temperatures at different depths of snow. They show about 247 tt at a depth of half a metre diminishing to 240 tt at about 4 metres, and increasing from that level downwards to more than 242 tt at 7 metres. The same kind of change would be shown in the earth during summer, in consequence of seasonal difference of mean temperature at the surface by which

[1] *The Bulletin of the Central Meteorological Observatory of Japan*, vol. I, No. 1, 1904.

[2] *Indian Met. Mem.*, XXIV, part 3, p. 40. Observations by Foureau showing the sand of the Sahara hotter than the shade temperature by as much as 29 t (52° F) are quoted by H. Wiszwianski in a paper on 'Die Faktoren der Wüstenbildung,' Geogr. Inst., Berlin, 1906. Surface temperatures of different kinds of ground are discussed by N. K. Johnson, *Q.J. Roy. Met. Soc.*, vol. LIII, 1927.

[3] *Proc. Roy. Soc.* B, vol. XCVI, 1924, p. 128.

[4] *Monthly Weather Review*, Washington, vol. L, 1922, p. 142.

[5] *The Glacial Anticyclones*, New York, 1926, p. 79.

a wave of variation is transmitted downwards. It is especially notable when there is great seasonal difference of temperature at the surface as in Northern Siberia, where the ground is permanently frozen not far below the surface.

In the temperate regions the seasonal range of temperature diminishes with increasing depth and the Japanese observations, to which we have already referred, indicate a change from a range of 22 t at the surface to less than ·5 t at a depth of 7 metres.

In view of the importance of earth-temperature for agriculture and public health it is customary to take readings at climatological stations of thermometers at various depths, of which that of 4 feet has been regarded as of importance in consequence of a relation to certain diseases which was brought out by Sir A. Mitchell many years ago. We give the values for a number of British stations as recorded in the *Weekly Weather Report*, 1907–20.

(The routine for British agricultural stations is now for 4 in, 8 in, 2 ft.)

Table of temperatures at a depth of 4 feet.
Maximum, minimum and range of weekly averages.

Station	Max.	Week	Min.	Week	Range	Station	Max.	Week	Min.	Week	Range
Scotland W.	tt	no.	tt	no.	t	Scotland E.	tt	no.	tt	no.	t
Rothesay	286·5	33	278·3	7	8·2	Aberdeen	285·5	33	277·0	7	8·5
Dumfries	287·8	33–4	277·5	5–8	10·3	Crathes	286·3	33	276·1	4–6	10·2
England NW and Wales.						England NE.					
Aspatria	286·2	33–4	278·4	8	7·8	York	287·3	34	277·9	7–8	9·4
Southport	289·3	33	277·3	7	12·0	Lincoln	287·8	33	277·7	7, 9	10·1
Manchester	288·7	34	278·4	7–8	10·3	England E.					
Cardiff	288·3	34	278·9	7–8	9·4	Yarmouth	288·9	33–4	278·8	6–7	10·1
Ireland.						Cambridge	288·7	33–4	278·5	6–7	10·2
Markree	287·2	34	279·3	7–8	7·9	England SE.					
Armagh	287·4	33	278·9	7	8·5	Wisley	288·6	33	278·6	6–7	10·0
Dublin	287·7	33–4	279·4	6–7	8·3	Richmond	288·0	33–4	278·9	6–7	9·1
Midland Counties.											
Birmingham	285·4	34	279·2	11–12	6·2	Harrogate	286·8	33	277·2	7–8	9·6
Nottingham	288·3	33	276·9	6–7	11·4	Buxton	286·3	34	277·4	9	8·9
Channel Is.											
Guernsey	289·9	34	280·4	7–8	9·5						

There are some remarkable points about the table which ask for further investigation. The method of mounting the thermometer is not yet beyond question. The only considerable table which we have available for comparison is a special one for a number of forest-stations in the United States[1] the specification of which is too elaborate for quotation here.

We give particulars for a few isolated stations in various parts of the world.

Earth-temperatures at depths of about 4 feet (1·219 m).

Station and depth	Maximum tt	Maximum Month	Minimum tt	Minimum Month	Range t	Station and depth	Maximum tt	Maximum Month	Minimum tt	Minimum Month	Range t
Bremen: 1·0 m	290·4	Aug.	276·1	Feb.	14·3	New Year Is.: 1·0 m	280·9	Feb.	275·7	Aug.	5·2
2·0 m	289·6	Aug.	277·5	Mar.	12·1	2·0 m	279·4	Apr.	277·3	Oct.	2·1
Dresden: 1·0 m	289·1	Aug.	275·6	Feb.	13·5	Kimberley: 4 ft	298·4	Feb.	287·6	July	10·8
1·5 m	288·3	Aug.	276·8	Feb.	11·5	Entebbe: 4 ft	297·6	Feb.	296·8	Aug.	0·8
Tiflis: 0·84 m	299·2	Aug.	278·5	Feb.	20·7	Java: 1·1 m	302·8	Oct.	301·8	Feb.	1·0
1·65 m	295·2	Aug.	281·3	Feb.	13·9	Cordoba: 1·2 m	294·0	Feb.	286·7	Aug.	7·3
Tokyo: 1·2 m	295·5	Sept.	281·8	Feb.	13·7	Helwan: 1·15 m	301·9	Aug. Sept.	291·1	Feb.	10·8
Allahabad: 3 ft	306·2	June	293·8	Jan.	12·4						

[1] *Forest types in the Central Rocky Mountains as affected by climate and soil*, by C. G. Bates. U.S. Department of Agriculture, Department Bulletin 1233, Washington, 1924.

There are also examples of continuous records of underground temperatures at fixed depths; the series best known to us is that of the Radcliffe Observatory, Oxford, under the direction of Dr A. A. Rambaut, for which the records were made by Callendar's electrical thermometers. The observations have an intrinsic value in relation to the physics of the earth but are not immediately available for direct meteorological application because meteorology is essentially a comparative science. An isolated example has to wait until other examples of the same or similar kind are available for comparison.

Such records form an ideal subject for the application of Fourier analysis either from the point of view of the transmission downwards of a periodic wave of heat or for the expression of the changes at the several levels in terms of harmonic series. Tokyo gives us a useful specimen of the latter. Using the form

$$y = A_0 + A_1 \cos\left(\frac{2\pi t}{\tau} - \alpha_1\right) + A_2 \cos\left(\frac{4\pi t}{\tau} - \alpha_2\right) + \text{etc.}$$

we have the following table for the seasonal variation.

Harmonic analysis of five-day means of earth-temperatures at Tokyo.

Depth	A_0	A_1	α_1	A_2	α_2	A_3	α_3	A_4	α_4
m	tt	tt	°	tt	°	tt	°	tt	°
0	288·6	12·9	205	1·2	152	0·6	314	0·5	159
0·3	288·7	10·7	214	0·9	166	0·5	317	0·3	161
0·6	288·7	9·1	225	0·7	170	0·4	21	0·2	179
1·2	289·0	6·7	244	0·4	188	0·2	57	0·1	179
3·0	288·8	2·5	307	0·1	276	0·1	159	< ·05	47
5·0	288·5	0·6	31	< ·05	308	< ·05	346	< ·05	30
7·0	288·4	0·2	109	< ·05	249	< ·05	104	< ·05	27

The origin of time precedes 1 January by $\frac{\tau}{2 \times 73}$ or 2·5° or $2\frac{1}{2}$ days.

A scheme of representation of the seasonal variation of temperature in the water of the Elbe at Hamburg, for which the annual range is greatest, as compared with that of the air and the earth at different depths down to 5 metres, was worked out by W. van Bebber. The diagram is reproduced in Willis Moore's *Descriptive Meteorology*.

Perhaps the fullest representation of the exploration of the distribution of temperature in the interesting case of an area covered by snow in the winter is that described by J. Keränen, 'Ueber die Temperatur des Bodens und der Schneedecke in Sodankyla (Finland) nach Beobachtungen mit Thermoelementen[1].' The paper includes charts of "geoisotherms" to a depth of 160 cm, and isotherms for a snow layer up to 90 cm in the winter 1915–16, and many other details which cannot be dealt with in the space available.

[1] *Annals of the Finnish Academy*, Ser. A, vol. XIII, No. 7, 1920.

THE ENTABLATURES

The maps of normal temperature of the air and sea are contained on 38 pages, but each map occupies little more than two-thirds; there is in consequence at the bottom of each of the 38 pages a space which is available for displaying some of the multitudinous facts about temperature—diurnal and seasonal range for example. We have accordingly arranged in the form of "panels" or "entablatures," to occupy the spaces below the maps, some specimens of the kind of information which can be found in various publications, mostly official. For the temperature maps these entablatures are as follows:

Figs. 17–18. Conversion tables from temperature on the Fahrenheit scale to the tercentesimal scale, in order that the reader may make the conversion whenever he so desires.

Figs. 19–20. Diurnal and seasonal variation in the Arctic regions.

Figs. 21–22. Diurnal and seasonal variation at Aberdeen.

Figs. 23–24. Diurnal and seasonal variation at Calcutta.

Figs. 25–26. Diurnal and seasonal variation at Batavia.

Figs. 27–28. Diurnal and seasonal variation at Cape of Good Hope.

Fig. 29. Diurnal and seasonal variation in the Antarctic.

Fig. 31. Maximum temperature in each month of the years 1910–18 recorded in the *Réseau Mondial* at stations in the Northern Hemisphere.

Fig. 32. Minimum temperature in each month of the years 1910–18 recorded in the *Réseau Mondial* at stations in the Southern Hemisphere.

Figs. 33–40. Comparison of variation at high and low level stations.

Figs. 33–34. Fort William and Ben Nevis.

Figs. 35–36. Paris (Parc St Maur) and the Eiffel Tower.

Figs. 37–38. Clermont Ferrand and Puy de Dôme.

Figs. 39–40. Lahore and Leh.

Fig. 41. Minimum temperature in each month of the years 1910–18 recorded in the *Réseau Mondial* at stations in the Northern Hemisphere.

Fig. 42. Maximum temperature in each month of the years 1910–18 recorded in the *Réseau Mondial* at stations in the Southern Hemisphere.

Figs. 43–44. Isopleth diagrams of diurnal and seasonal variation of temperature at Barnaoul, Agra, St Helena and in the Antarctic; with a table for the diurnal variation of temperature over the sea and a conversion table from range in degrees Fahrenheit to range on the tercentesimal scale.

Figs. 45–46. The temperature of the lower air of the Arctic and Antarctic.

Figs. 47–54. Seasonal variation of temperature of the sea and air in the Atlantic and Indian Oceans.

The authorities for these tables are enumerated on pp. 126, 127.

[A note on details of the temperature-charts for Greenland is given in chap. X, note 13, and material for a possible revision of the whole series of charts is indicated in the bibliography of note 24.]

FIGURE 17 MEAN TEMPERATURE OF THE AIR AT SEA-LEVEL

Fahrenheit scale.

Authority: see p. 124

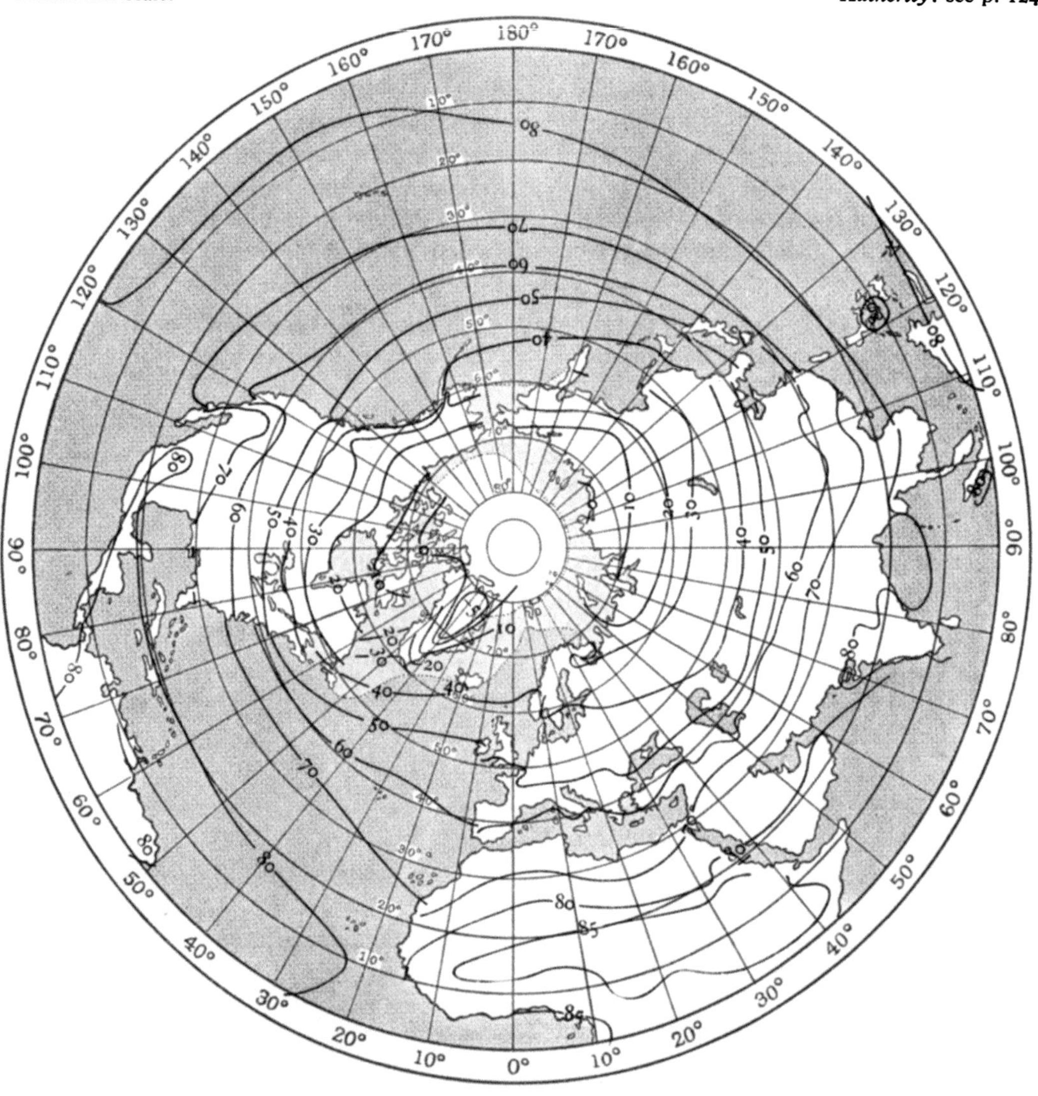

Table for Conversion from the Fahrenheit scale, °F to tt, the equivalent tercentesimal temperature

°F	0 tt	1 tt	2 tt	3 tt	4 tt	5 tt	6 tt	7 tt	8 tt	9 tt
−60	221·9	221·3	220·8	220·2	219·7	219·1	218·6	218·0	217·4	216·9
−50	227·4	226·9	226·3	225·8	225·2	224·7	224·1	223·6	223·0	222·4
−40	233·0	232·4	231·9	231·3	230·8	230·2	229·7	229·1	228·6	228·0
−30	238·6	238·0	237·4	236·9	236·3	235·8	235·2	234·7	234·1	233·6
−20	244·1	243·6	243·0	242·4	241·9	241·3	240·8	240·2	239·7	239·1
−10	249·7	249·1	248·6	248·0	247·4	246·9	246·3	245·8	245·2	244·7
−0	255·2	254·7	254·1	253·6	253·0	252·4	251·9	251·3	250·8	250·2
+0	255·2	255·8	256·3	256·9	257·4	258·0	258·6	259·1	259·7	260·2
10	260·8	261·3	261·9	262·4	263·0	263·6	264·1	264·7	265·2	265·8
20	266·3	266·9	267·4	268·0	268·6	269·1	269·7	270·2	270·8	271·3
30	271·9	272·4	273·0	273·6	274·1	274·7	275·2	275·8	276·3	276·9

MEAN TEMPERATURE OF THE AIR AT SEA-LEVEL

FIGURE 18

Authority: see p. 124.

Fahrenheit scale.

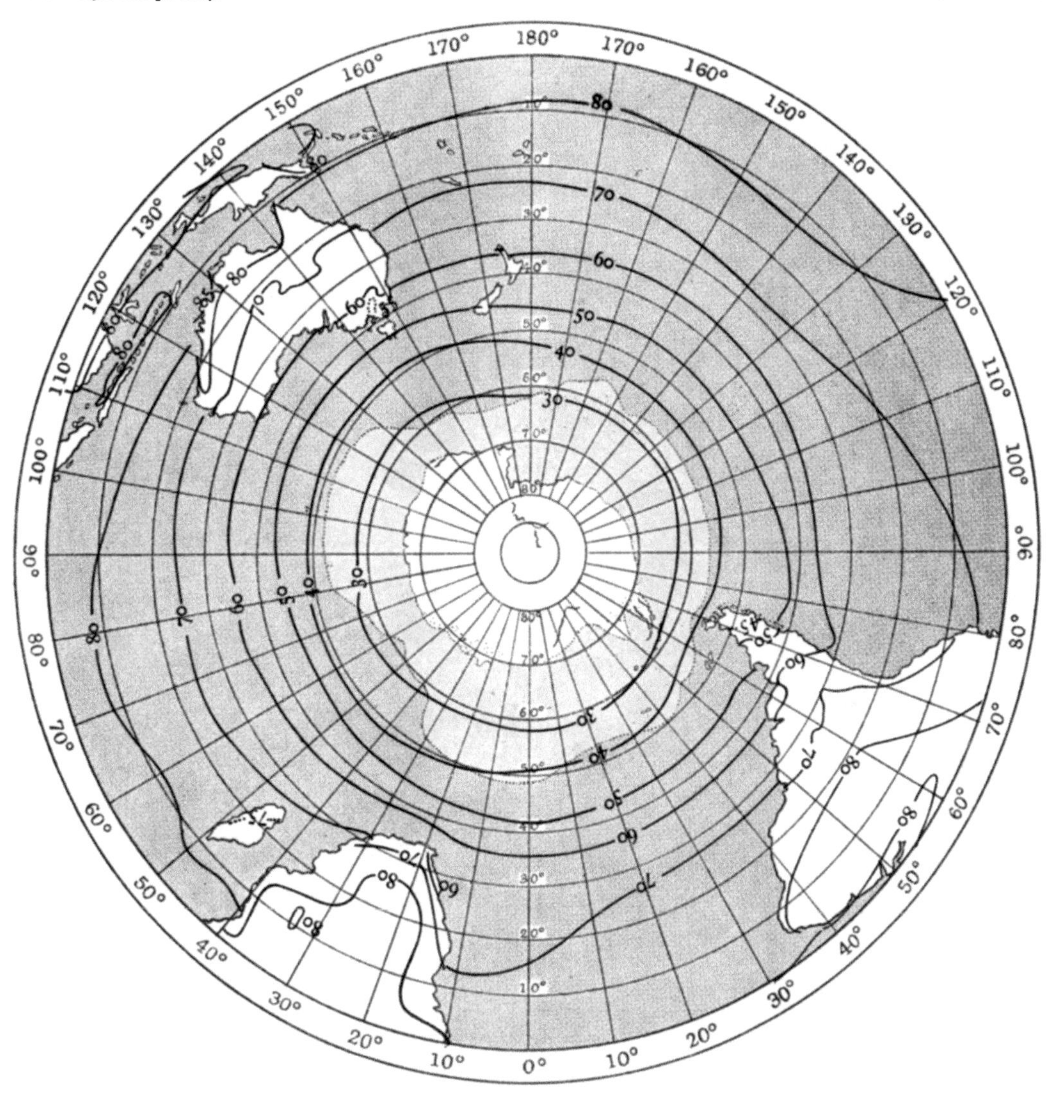

°F	0	1	2	3	4	5	6	7	8	9
	tt	tt	tt	tt	tt	tt	tt	tt	tt	tt
40	277·4	278·0	278·6	279·1	279·7	280·2	280·8	281·3	281·9	282·4
50	283·0	283·6	284·1	284·7	285·2	285·8	286·3	286·9	287·4	288·0
60	288·6	289·1	289·7	290·2	290·8	291·3	291·9	292·4	293·0	293·6
70	294·1	294·7	295·2	295·8	296·3	296·9	297·4	298·0	298·6	299·1
80	299·7	300·2	300·8	301·3	301·9	302·4	303·0	303·6	304·1	304·7
90	305·2	305·8	306·3	306·9	307·4	308·0	308·6	309·1	309·7	310·2
100	310·8	311·3	311·9	312·4	313·0	313·6	314·1	314·7	315·2	315·8
110	316·3	316·9	317·4	318·0	318·6	319·1	319·7	320·2	320·8	321·3
120	321·9	322·4	323·0	323·6	324·1	324·7	325·2	325·8	326·3	326·9
130	327·4	328·0	328·6	329·1	329·7	330·2	330·8	331·3	331·9	332·4
140	333·0	333·6	334·1	334·7	335·2	335·8	336·3	336·9	337·4	338·0

Table for Conversion from the Fahrenheit scale, °F to tt, the equivalent tercentesimal temperature

FIGURE 19

Fahrenheit scale.

MEAN TEMPERATURE OF THE AIR AT SEA-LEVEL

Authority: see p. 124.

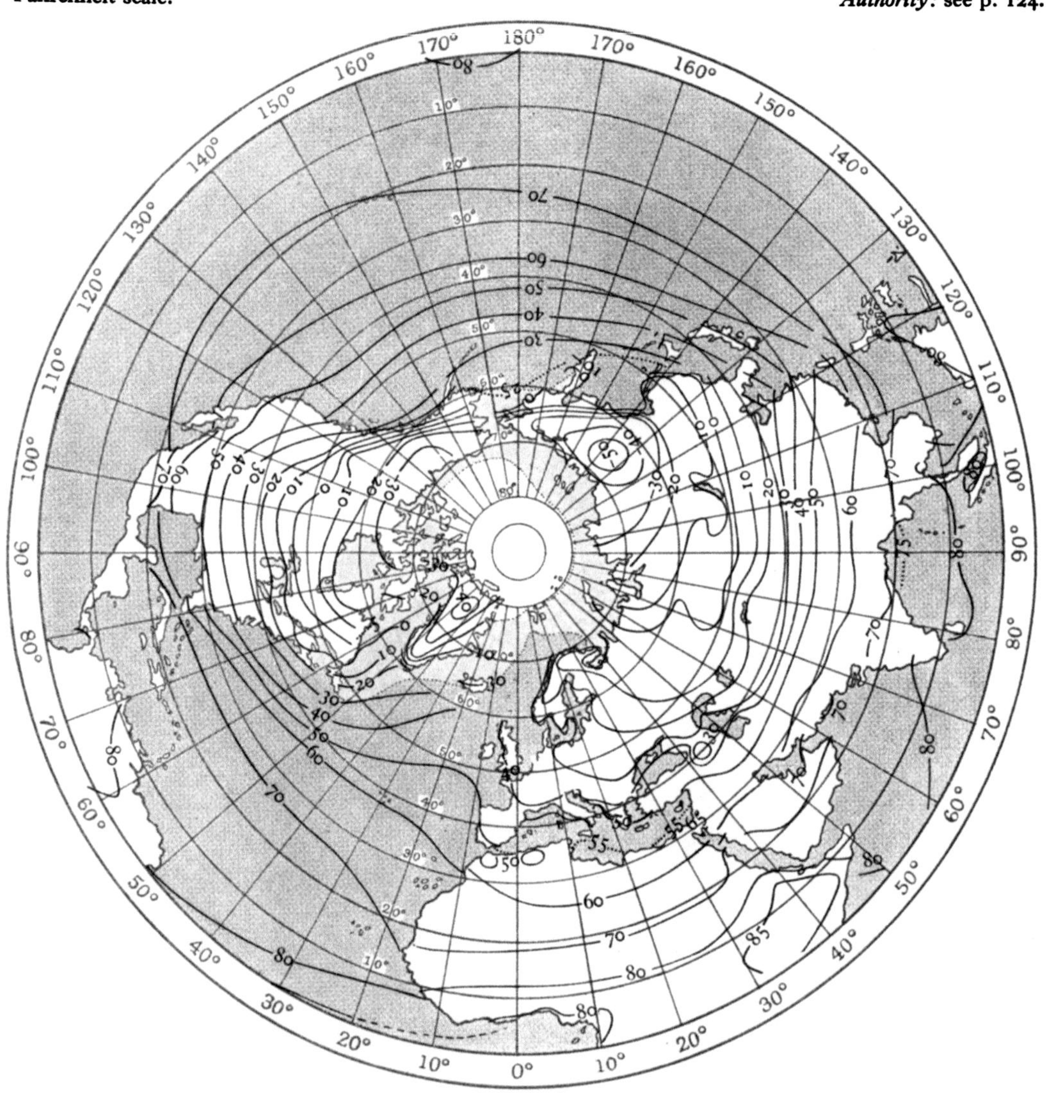

Diurnal variation of temperature (tercentesimal scale) in the Arctic. 1894–96.

Corrected for seasonal variation. The values underlined are negative, they indicate departures from the mean value downward.

Month	Mean	Mdt	1	2	3	4	5	6	7	8	9	10	11
Jan.	237·41	·14	·20	·25	·30	·32	·34	·34	·29	·21	·08	·01	·04
Feb.	237·17	·01	·00	·03	·05	·06	·09	·14	·16	·12	·05	·01	·04
Mar.	242·67	·41	·40	·36	·31	·26	·22	·20	·11	·02	·10	·14	·24
Apr.	250·22	1·30	1·52	1·63	1·62	1·48	1·24	·95	·57	·13	·30	·71	1·06
May	261·98	·62	·76	·84	·85	·80	·70	·58	·39	·13	·11	·29	·43
June	271·17	·55	·62	·68	·69	·62	·52	·40	·24	·06	·11	·26	·40
July	273·05	·34	·34	·34	·34	·30	·22	·13	·02	·08	·17	·22	·27
Aug.	271·24	·53	·51	·46	·42	·38	·28	·15	·01	·13	·19	·26	·34
Sept.	264·01	·33	·39	·42	·34	·20	·11	·05	·03	·02	·11	·15	·20
Oct.	251·24	·20	·25	·25	·14	·00	·13	·23	·29	·25	·15	·09	·12
Nov.	244·28	·28	·24	·19	·20	·25	·27	·26	·21	·12	·02	·07	·13
Dec.	240·77	·16	·21	·20	·17	·12	·05	·03	·12	·18	·19	·17	·13

MEAN TEMPERATURE OF THE AIR AT SEA-LEVEL

FIGURE 20

Authority: see p. 124.

Fahrenheit scale.

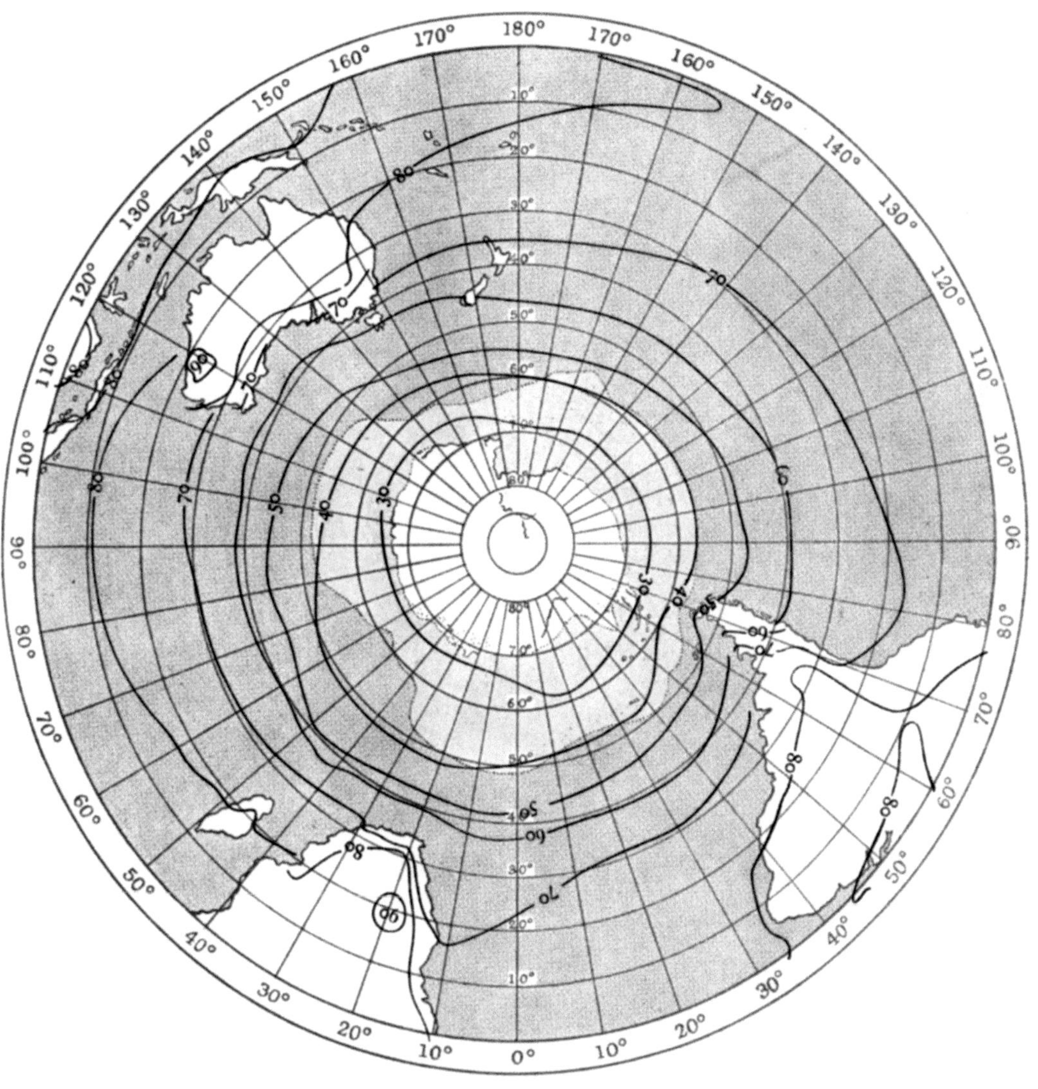

Diurnal variation of temperature (tercentesimal scale) in the Arctic. 1894–96.

Local mean time. The values underlined are negative, they indicate departures from the mean value downward.

12	13	14	15	16	17	18	19	20	21	22	23	Range	Month
·07	·16	·29	·40	·39	·33	·28	·24	·19	·12	·05	·04	·74	Jan.
·08	·08	·07	·03	·00	·00	·00	·08	·06	·07	·05	·01	·24	Feb.
·37	·45	·48	·48	·41	·30	·19	·06	·07	·20	·32	·39	·89	Mar.
1·31	1·47	1·55	1·52	1·43	1·27	·99	·60	·17	·23	·60	·98	3·29	Apr.
·57	·71	·82	·84	·80	·71	·55	·33	·08	·10	·29	·46	1·69	May
·52	·60	·63	·62	·57	·51	·39	·22	·06	·08	·25	·43	1·32	June
·33	·35	·35	·34	·32	·25	·14	·04	·07	·18	·26	·32	·69	July
·43	·50	·51	·50	·48	·35	·17	·02	·17	·28	·37	·47	1·07	Aug.
·29	·35	·37	·34	·29	·24	·15	·05	·03	·03	·06	·21	·78	Sept.
·13	·02	·10	·15	·10	·02	·04	·03	·00	·06	·10	·14	·55	Oct.
·19	·30	·40	·42	·35	·24	·14	·06	·01	·06	·16	·25	·71	Nov.
·05	·02	·04	·03	·02	·01	·05	·07	·07	·08	·03	·09	·40	Dec.

FIGURE 21

Fahrenheit scale.

MEAN TEMPERATURE OF THE AIR AT SEA-LEVEL

Authority: see p. 124.

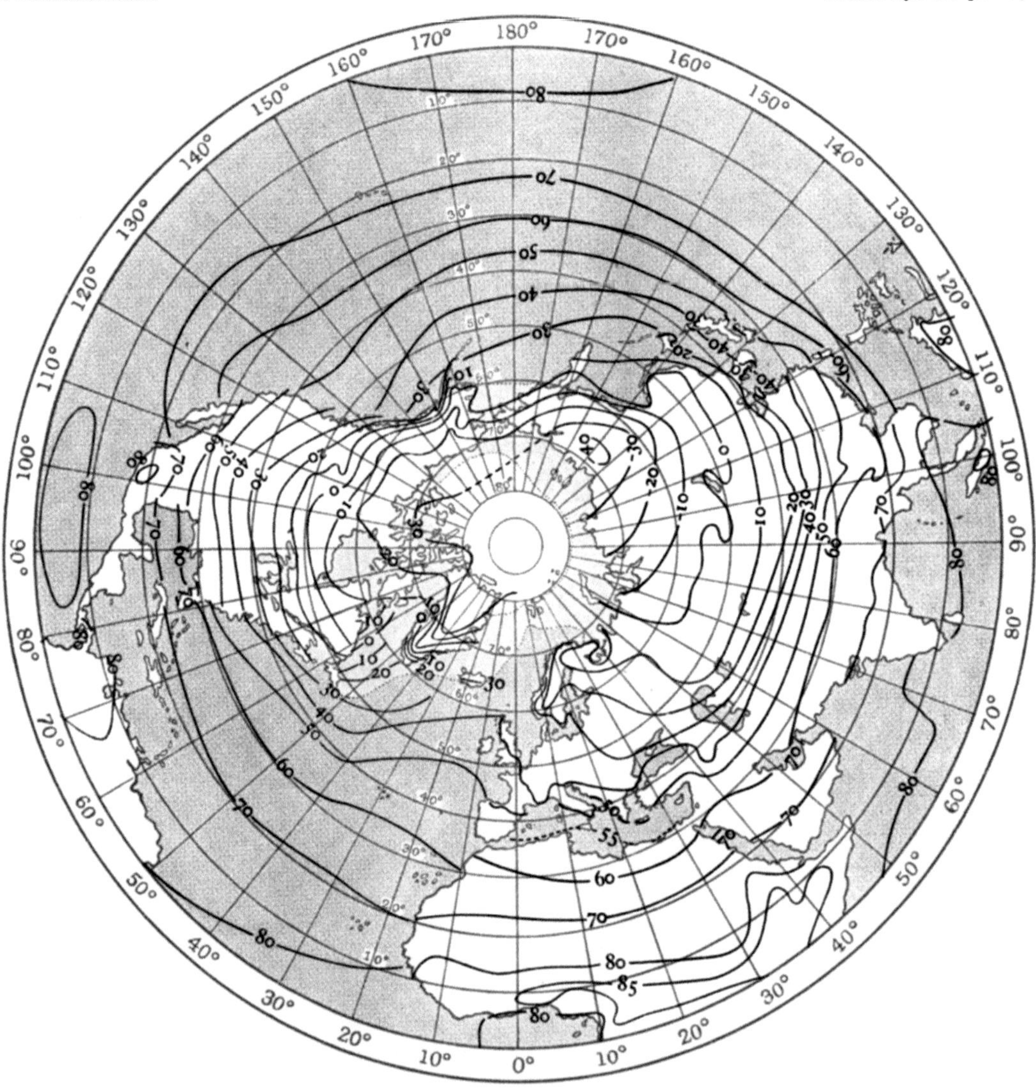

Diurnal variation of temperature (tercentesimal scale) at Aberdeen, 57° 10′ N, 2° 6′ W, 26·5 m. 1871–1915.

The values underlined are negative. They indicate departures from the mean value downward.

Month	Mean	G.m.t. 0	1	2	3	4	5	6	7	8	9	10	11
Jan.	276·46	·30	·33	·39	·41	·48	·48	·52	·51	·50	·36	·14	·32
Feb.	276·65	·52	·59	·67	·74	·83	·86	·89	·89	·85	·52	·03	·59
Mar.	277·34	·89	1·00	1·11	1·18	1·30	1·37	1·42	1·28	·86	·10	·54	1·14
Apr.	279·16	1·25	1·43	1·61	1·73	1·86	1·92	1·71	·94	·23	·56	1·07	1·51
May	281·55	1·53	1·71	1·93	2·11	2·26	1·90	1·19	·31	·20	·73	1·09	1·46
June	284·53	1·61	1·89	2·14	2·30	2·34	1·76	·91	·10	·35	·81	1·15	1·49
July	286·37	1·49	1·71	1·92	2·12	2·25	1·85	1·19	·34	·21	·73	1·09	1·46
Aug.	286·17	1·32	1·53	1·75	1·93	2·10	2·11	1·65	·76	·03	·69	1·11	1·55
Sept.	284·38	1·17	1·36	1·57	1·69	1·83	1·93	1·94	1·35	·51	·48	1·13	1·67
Oct.	281·39	·74	·86	·98	1·06	1·14	1·19	1·25	1·21	·85	·12	·60	1·21
Nov.	278·70	·41	·48	·53	·59	·65	·67	·70	·65	·59	·29	·13	·65
Dec.	276·75	·23	·28	·30	·33	·37	·35	·38	·37	·39	·29	·07	·30

MEAN TEMPERATURE OF THE AIR AT SEA-LEVEL

FIGURE 22

Authority: see p. 124.

Fahrenheit scale.

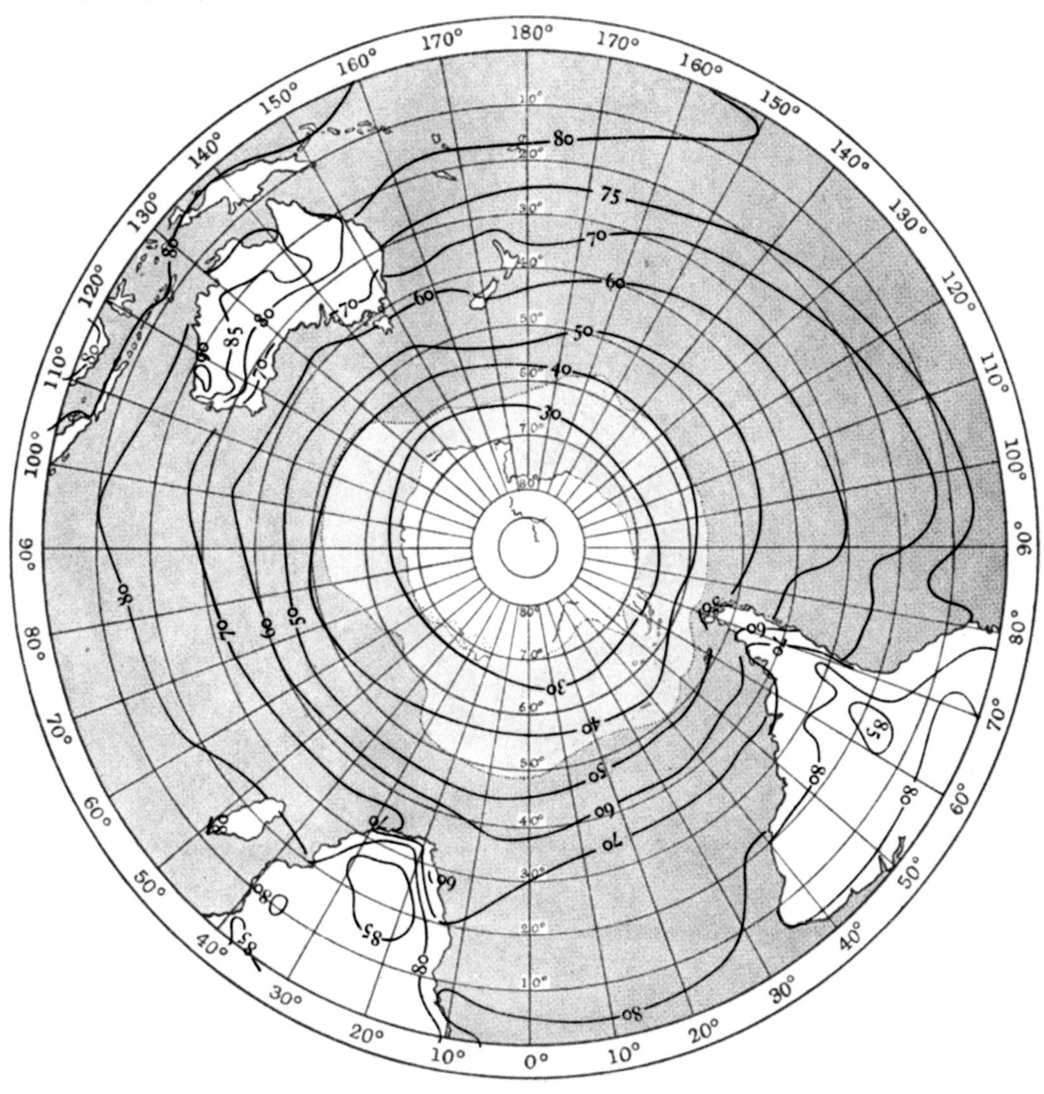

Diurnal variation of temperature (tercentesimal scale) at Aberdeen, 57° 10′ N, 2° 6′ W, 26·5 m. 1871–1915.

The values underlined are negative. They indicate departures from the mean value downward.

12	13	14	15	16	17	18	19	20	21	22	23	Mdt	Range t
·65	·92	·97	·87	·59	·34	·16	·08	·06	·12	·19	·23	·30	1·49
1·04	1·41	1·51	1·45	1·17	·74	·37	·12	·08	·22	·36	·44	·52	2·40
1·52	1·75	1·79	1·77	1·56	1·22	·67	·19	·12	·32	·54	·69	·85	3·21
1·75	1·93	1·91	1·86	1·59	1·33	·95	·42	·10	·39	·70	·97	1·19	3·85
1·67	1·84	1·80	1·76	1·54	1·40	1·04	·64	·04	·43	·82	1·17	1·43	4·10
1·60	1·79	1·74	1·69	1·50	1·43	1·08	·70	·17	·42	·83	1·18	1·51	4·13
1·62	1·81	1·79	1·77	1·53	1·42	1·07	·69	·11	·43	·87	1·22	1·48	4·06
1·79	2·00	1·99	1·94	1·71	1·44	1·05	·56	·08	·51	·88	1·11	1·36	4·11
1·93	2·15	2·16	2·05	1·77	1·41	·82	·27	·20	·49	·76	·98	1·20	4·10
1·59	1·83	1·86	1·72	1·33	·83	·36	·00	·24	·40	·59	·70	·85	3·11
1·05	1·27	1·27	1·08	·68	·34	·14	·01	·12	·18	·29	·38	·50	1·97
·57	·79	·77	·59	·35	·21	·09	·03	·06	·07	·14	·17	·25	1·18

FIGURE 23

Fahrenheit scale.

MEAN TEMPERATURE OF THE AIR AT SEA-LEVEL

Authority: see p. 124.

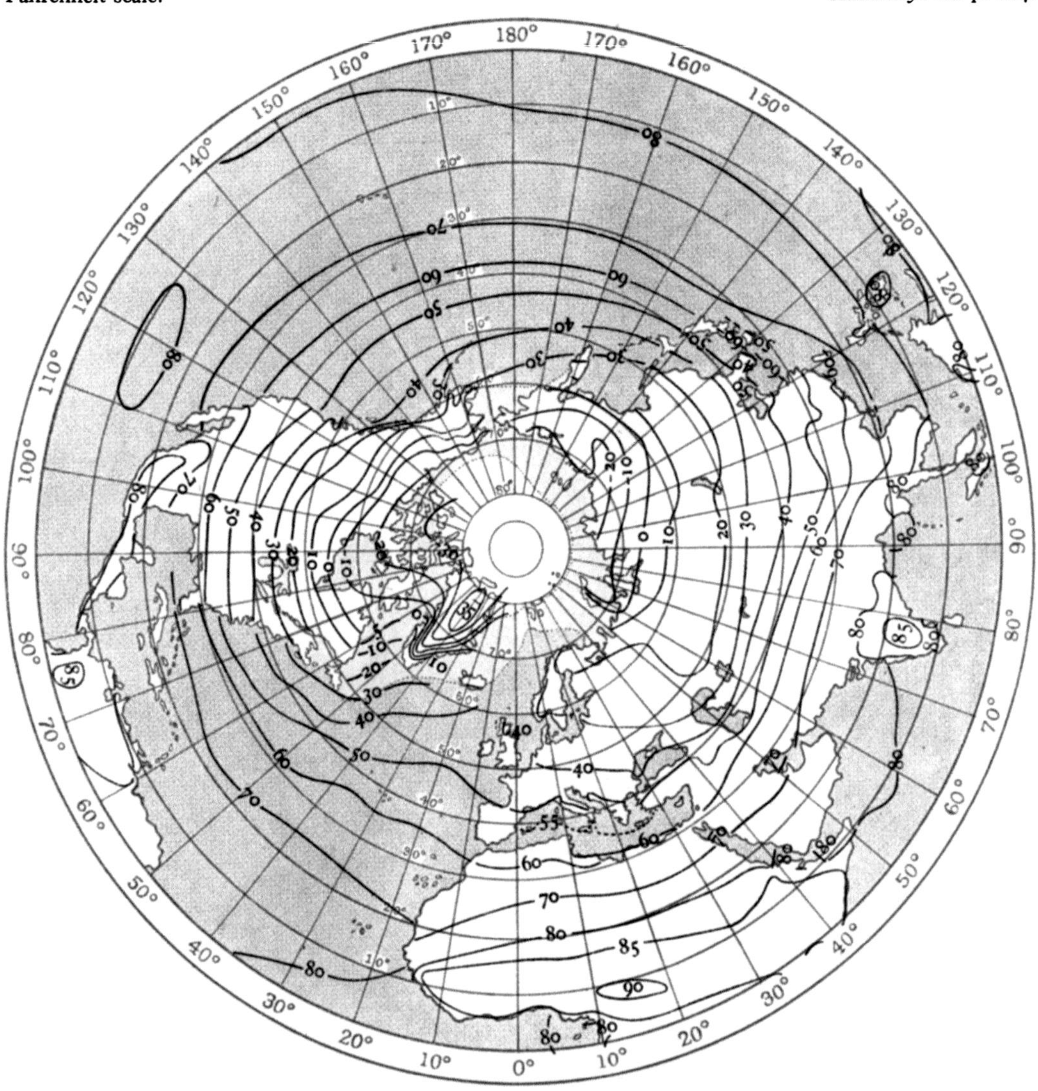

Diurnal variation of temperature (tercentesimal scale) at Calcutta (Alipore), 22° 32′ N, 88° 20′ E, 6·5 m. 1881–93.

Local mean time. The values underlined are negative, they indicate departures from the mean value downward.

Month	Mean	0	1	2	3	4	5	6	7	8	9	10	11
Jan.	291·4	2·4	2·8	3·0	3·3	3·5	3·7	3·9	4·0	2·9	·5	1·2	2·9
Feb.	293·6	2·8	3·1	3·3	3·4	3·6	3·8	4·0	4·1	2·7	·4	1·2	2·7
Mar.	299·0	2·5	2·8	3·1	3·3	3·6	3·7	3·9	3·5	1·9	·2	1·2	2·6
Apr.	302·5	2·6	2·8	3·1	3·3	3·4	3·7	3·7	2·7	1·2	·4	1·6	2·9
May	302·6	2·3	2·4	2·6	2·7	2·8	2·9	2·8	1·6	·3	·9	1·9	2·8
June	302·3	1·4	1·7	1·7	1·8	1·9	2·0	1·9	1·0	·1	·8	1·4	2·1
July	301·3	1·0	1·1	1·2	1·3	1·4	1·4	1·4	·8	·2	·6	1·0	1·5
Aug.	301·0	·9	1·0	1·1	1·2	1·3	1·3	1·4	·8	·2	·6	1·1	1·4
Sept.	301·0	1·0	1·1	1·2	1·3	1·4	1·4	1·5	·8	·1	·7	1·1	1·6
Oct.	299·6	1·4	1·6	1·7	1·9	2·0	2·1	2·1	1·4	·3	·8	1·5	2·1
Nov.	295·2	1·9	2·1	2·3	2·5	2·7	2·9	3·0	2·7	1·2	·4	1·7	2·8
Dec.	291·2	2·3	2·5	2·7	2·9	3·0	3·3	3·5	3·5	2·2	·2	1·4	2·8

MEAN TEMPERATURE OF THE AIR AT SEA-LEVEL

FIGURE 24

Authority: see p. 124.

Fahrenheit scale.

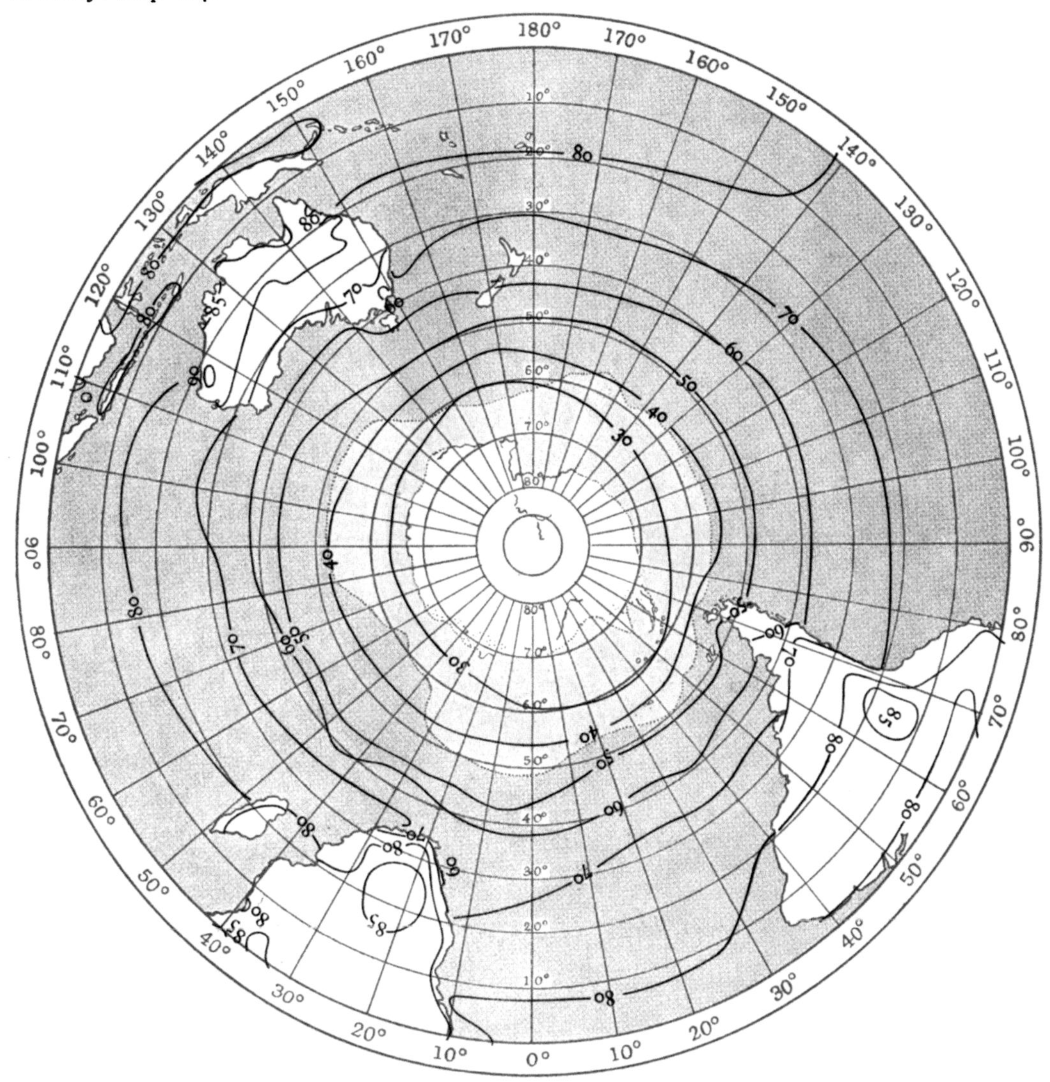

Diurnal variation of temperature (tercentesimal scale) at Calcutta (Alipore), 22° 32′ N, 88° 20′ E, 6·5 m. 1881–93.

The values underlined are negative, they indicate departures from the mean value downward.

12	13	14	15	16	17	18	19	20	21	22	23	Range	Month
3·8	4·7	5·2	5·4	5·3	4·5	2·2	·5	·4	1·2	1·6	2·2	9·4	Jan.
3·6	4·5	5·0	5·3	5·3	4·7	2·8	·8	·2	·9	1·4	1·9	9·4	Feb.
3·5	4·3	4·8	5·2	5·1	4·1	2·4	·6	·3	1·0	1·4	1·9	9·1	Mar.
3·8	4·5	4·9	5·0	4·6	3·3	1·8	·2	·8	1·5	1·9	2·2	8·7	Apr.
3·3	3·7	3·9	3·8	3·2	2·3	1·2	·1	·8	1·4	1·8	2·0	6·8	May
2·4	2·6	2·5	2·3	1·9	1·4	·7	·1	·6	·9	1·2	1·4	4·6	June
1·6	1·8	1·7	1·6	1·3	1·0	·6	·0	·3	·5	·7	·9	3·2	July
1·7	1·7	1·7	1·5	1·2	·8	·3	·1	·4	·5	·7	·8	3·1	Aug.
1·7	1·8	1·8	1·6	1·4	·9	·3	·2	·5	·7	·8	·9	3·3	Sept.
2·4	2·7	2·7	2·7	2·4	1·7	·4	·3	·7	·9	1·2	1·4	4·8	Oct.
3·3	3·8	4·0	4·1	3·8	2·4	·9	·1	·7	1·2	1·6	1·9	7·1	Nov.
3·7	4·3	4·7	4·9	4·7	3·6	1·4	·1	·7	1·3	1·7	2·2	8·4	Dec.

FIGURE 25

Fahrenheit scale.

MEAN TEMPERATURE OF THE AIR AT SEA-LEVEL

Authority: see p. 124.

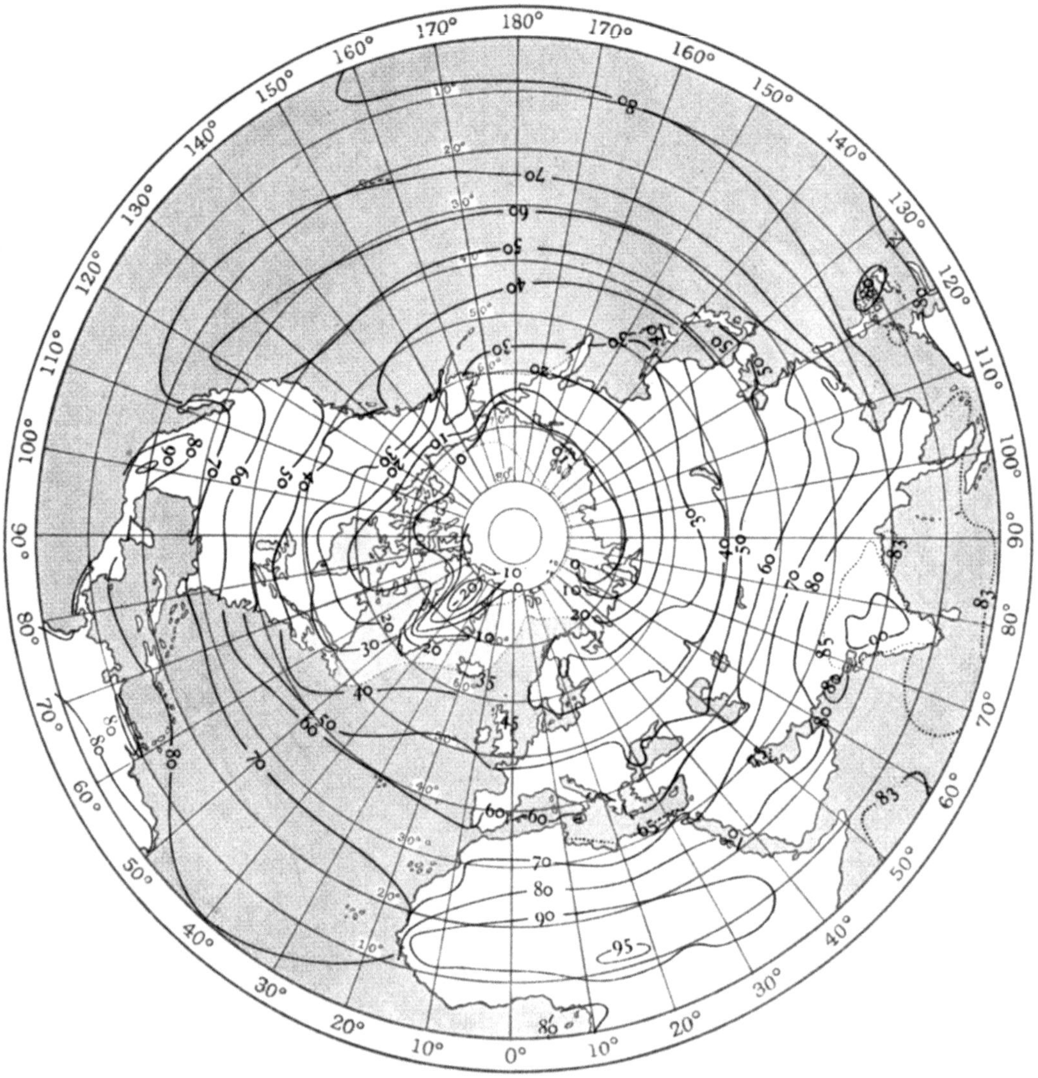

Diurnal variation of temperature (tercentesimal scale) at Batavia, 6° 11′ S, 106° 50′ E, 8 m. 1866–1915.

Local mean time. The values underlined are negative, they indicate departures from the mean value downward.

Month	Mean	0	1	2	3	4	5	6	7	8	9	10	11
Jan.	298·55	1·20	1·41	1·60	1·78	1·92	2·06	2·16	1·86	·96	·14	1·08	1·77
Feb.	298·53	1·17	1·39	1·56	1·73	1·88	2·00	2·11	1·89	1·02	·06	1·00	1·73
Mar.	298·96	1·36	1·59	1·79	1·97	2·14	2·30	2·45	2·22	1·14	·26	1·38	2·22
Apr.	299·41	1·57	1·80	2·00	2·19	2·37	2·54	2·69	2·39	1·18	·35	1·58	2·47
May	299·51	1·62	1·86	2·09	2·30	2·50	2·70	2·88	2·61	1·44	·14	1·45	2·41
June	299·14	1·60	1·84	2·08	2·31	2·51	2·71	2·90	2·73	1·64	·05	1·28	2·31
July	298·88	1·61	1·91	2·19	2·45	2·69	2·92	3·14	3·03	1·89	·17	1·29	2·41
Aug.	299·11	1·76	2·12	2·42	2·69	2·95	3·20	3·43	3·22	1·86	·04	1·59	2·79
Sept.	299·44	1·85	2·20	2·49	2·77	3·01	3·24	3·43	2·95	1·34	·53	2·01	3·11
Oct.	299·55	1·96	2·24	2·49	2·71	2·91	3·11	3·23	2·45	·76	·97	2·35	3·29
Nov.	299·25	1·78	1·98	2·18	2·36	2·52	2·68	2·75	1·97	·49	1·01	2·20	2·99
Dec.	298·79	1·44	1·66	1·85	2·03	2·18	2·33	2·41	1·85	·72	·54	1·56	2·37

MEAN TEMPERATURE OF THE AIR AT SEA-LEVEL

FIGURE 26

Authority: see p. 124.

Fahrenheit scale.

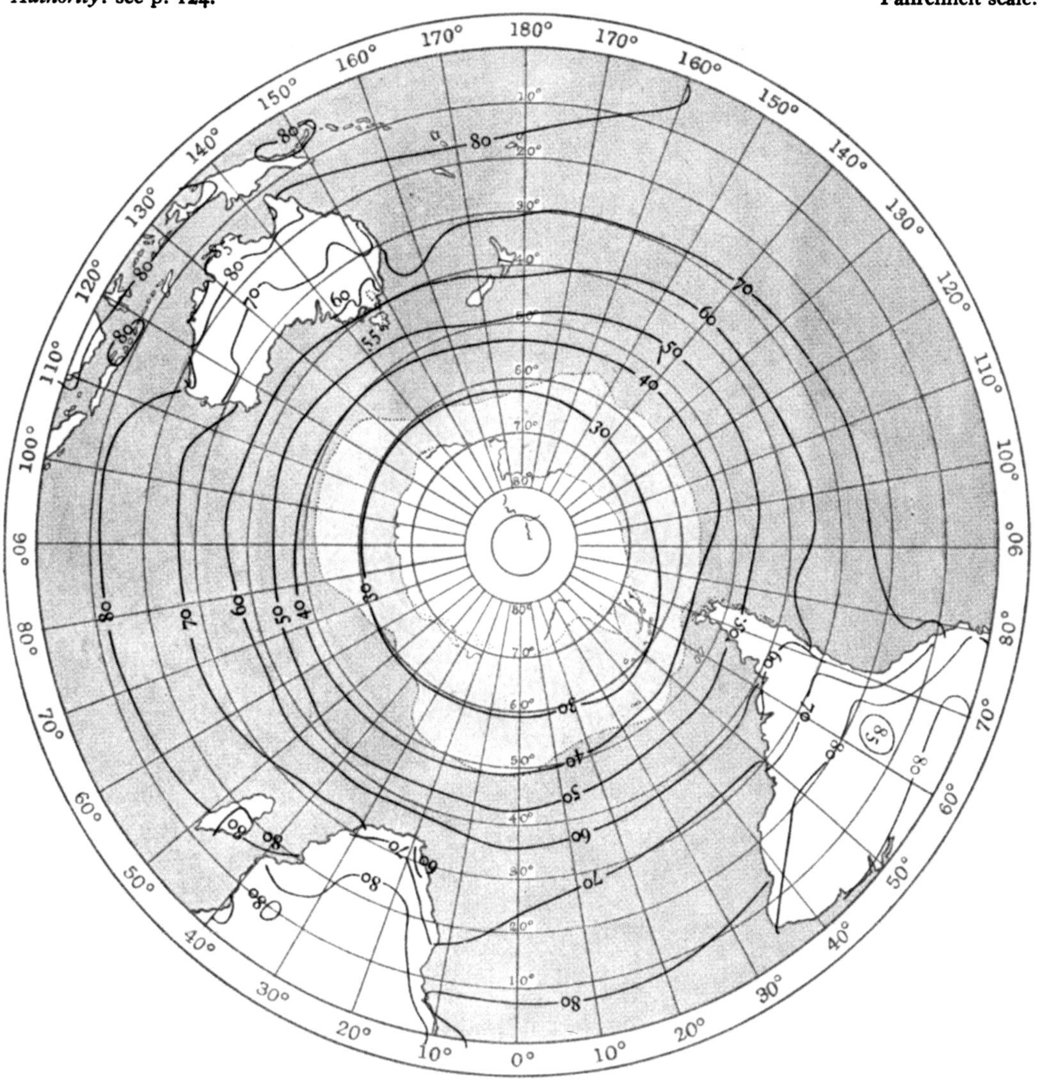

Diurnal variation of temperature (tercentesimal scale) at Batavia, 6° 11′ S, 106° 50′ E, 8 m. 1866–1915.

The values underlined are negative, they indicate departures from the mean value downward.

12	13	14	15	16	17	18	19	20	21	22	23	Range	Month
2·27	2·47	2·49	2·32	1·98	1·52	·93	·31	·12	·45	·73	·98	4·65	Jan.
2·22	2·48	2·50	2·32	2·01	1·54	·93	·30	·13	·43	·71	·95	4·61	Feb.
2·72	2·93	2·88	2·65	2·17	1·60	·91	·23	·24	·59	·89	1·14	5·38	Mar.
3·03	3·22	3·16	2·92	2·42	1·77	·95	·21	·31	·72	1·05	1·33	5·91	Apr.
3·05	3·36	3·36	3·12	2·67	2·04	1·16	·40	·18	·65	1·03	1·34	6·24	May
3·03	3·40	3·45	3·25	2·79	2·15	1·29	·49	·13	·61	1·00	1·32	6·35	June
3·20	3·59	3·61	3·40	2·95	2·33	1·46	·63	·02	·52	·94	1·30	6·75	July
3·55	3·83	3·78	3·51	2·99	2·32	1·47	·69	·00	·55	1·00	1·40	7·26	Aug.
3·70	3·78	3·65	3·30	2·74	2·04	1·24	·51	·12	·66	1·10	1·49	7·21	Sept.
3·64	3·63	3·46	3·09	2·55	1·84	1·01	·22	·41	·92	1·32	1·65	6·87	Oct.
3·32	3·36	3·20	2·83	2·24	1·51	·66	·07	·60	·99	1·29	1·55	6·11	Nov.
2·76	2·88	2·79	2·46	1·96	1·40	·74	·12	·31	·66	·95	1·21	5·29	Dec.

FIGURE 27

Fahrenheit scale.

MEAN TEMPERATURE OF THE AIR AT SEA-LEVEL

Authority: see p. 124.

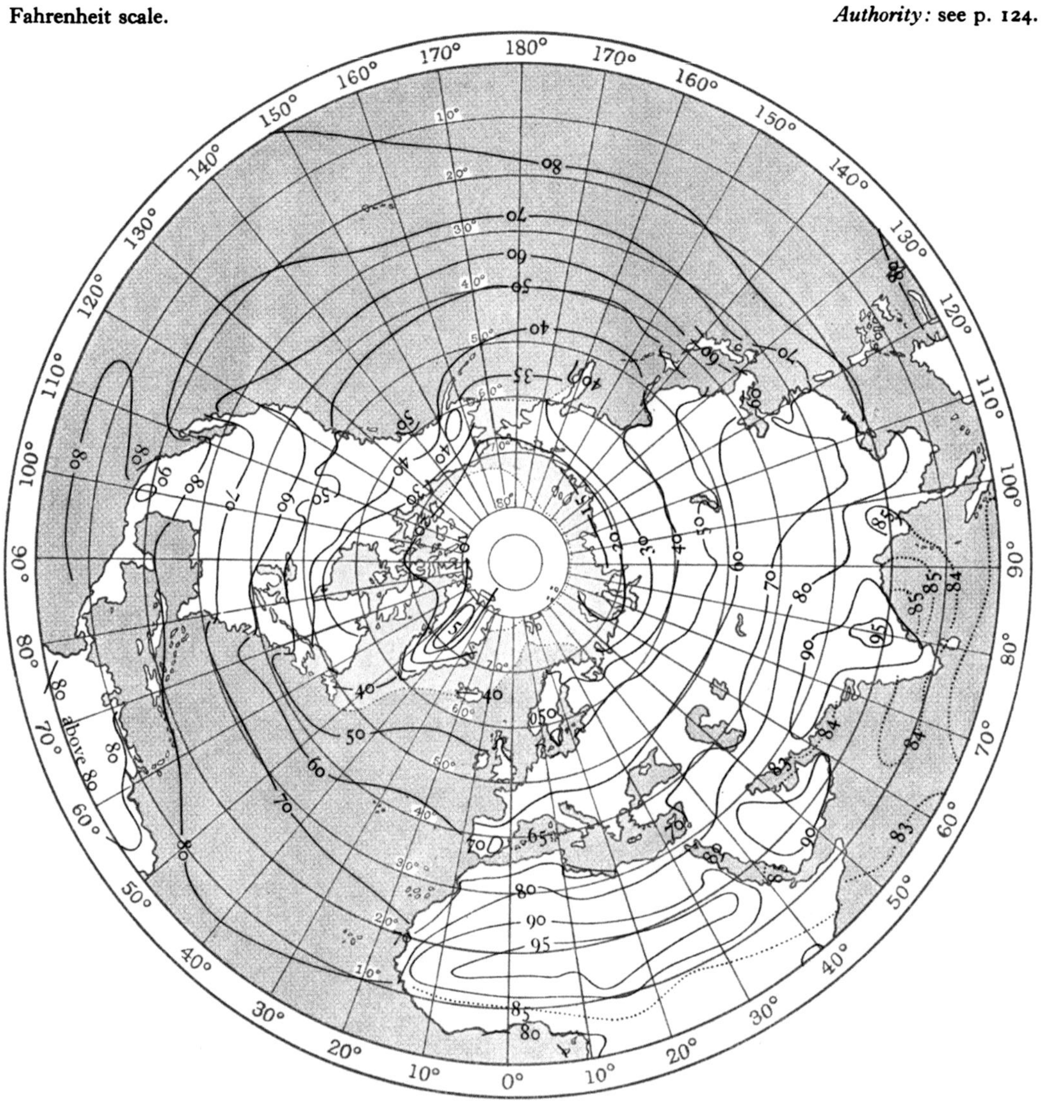

Diurnal variation of temperature (tercentesimal scale) at the Royal Observatory, Cape of Good Hope, 33° 56′ S, 18° 29′ E. 1841–6.

Local mean time. The values underlined are negative, they indicate departures from the mean value downward.

Month	Mean 1841–52	0	1	2	3	4	5	6	7	8	9	10	11
Jan.	292·83	2·01	2·18	2·32	2·44	2·57	2·61	2·06	·84	·16	1·00	1·77	2·49
Feb.	293·19	1·75	1·94	2·11	2·28	2·44	2·59	2·44	1·45	·48	·43	1·43	2·39
Mar.	291·71	1·75	1·96	2·14	2·32	2·48	2·60	2·66	2·11	·96	·32	1·51	2·48
Apr.	289·76	1·59	1·78	1·96	2·11	2·26	2·40	2·49	2·28	1·17	·12	1·33	2·22
May	287·01	1·24	1·32	1·40	1·51	1·63	1·77	1·91	1·98	1·29	·17	·94	1·70
June	285·42	1·09	1·22	1·34	1·47	1·59	1·74	1·84	1·91	1·61	·54	·66	1·49
July	285·26	1·19	1·38	1·54	1·68	1·78	1·92	2·07	2·12	1·62	·44	·83	1·73
Aug.	285·60	1·07	1·20	1·36	1·51	1·64	1·80	1·89	1·77	1·13	·06	·91	1·67
Sept.	286·53	1·30	1·46	1·65	1·85	2·01	2·13	2·16	1·77	·48	·52	1·38	2·06
Oct.	288·52	1·80	1·93	2·07	2·19	2·34	2·44	2·17	1·18	·04	1·01	1·87	2·55
Nov.	289·96	1·93	2·13	2·34	2·54	2·72	2·74	1·89	·58	·39	1·16	1·87	2·50
Dec.	291·63	2·20	2·41	2·60	2·78	2·96	2·96	2·06	·32	·57	1·30	2·00	2·56

MEAN TEMPERATURE OF THE AIR AT SEA-LEVEL

FIGURE 28

Authority: see p. 124.

Fahrenheit scale.

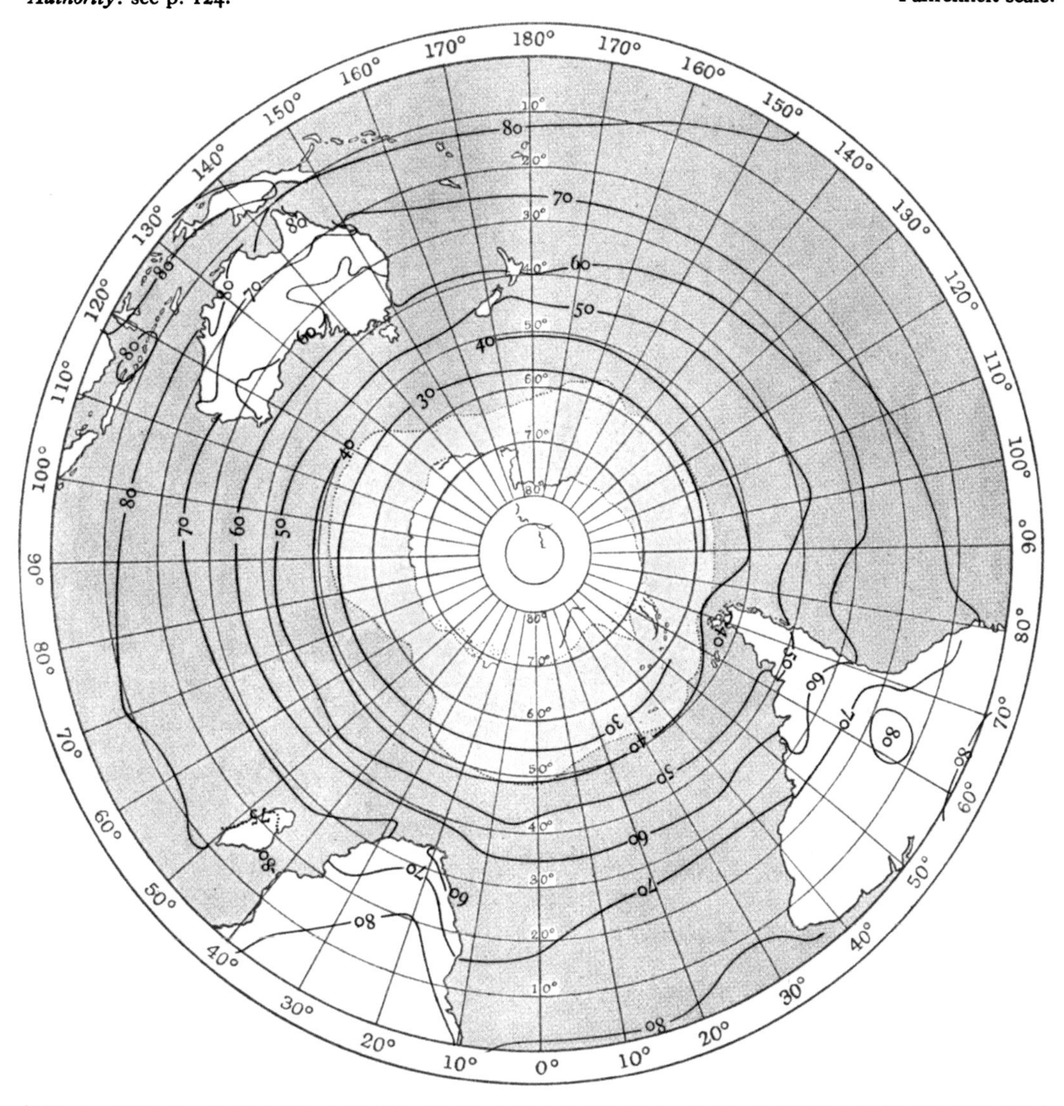

Diurnal variation of temperature (tercentesimal scale) at the Royal Observatory, Cape of Good Hope, 33° 56′ S, 18° 29′ E. 1841–6.

The values underlined are negative, they indicate departures from the mean value downward.

12	13	14	15	16	17	18	19	20	21	22	23	Range	Month
2·96	**3·16**	3·14	2·90	2·52	1·93	1·00	·39	·98	1·29	1·56	1·80	5·77	Jan.
3·16	**3·45**	3·36	3·00	2·49	1·78	·84	·20	·81	1·09	1·29	1·53	6·04	Feb.
3·09	3·51	**3·69**	3·41	2·84	1·93	·63	·13	·58	·94	1·24	1·51	6·35	Mar.
2·89	3·37	**3·57**	3·33	2·72	1·67	·47	·09	·43	·73	1·04	1·33	6·06	Apr.
2·34	2·74	**2·79**	2·58	2·06	1·31	·49	·06	·30	·61	·89	1·11	4·77	May
2·16	2·57	**2·64**	2·49	2·09	1·23	·59	·23	·07	·35	·62	·87	4·55	June
2·40	2·79	**2·93**	2·85	2·43	1·39	·64	·18	·16	·46	·73	·98	5·05	July
2·17	2·46	**2·54**	2·37	1·98	1·13	·42	·02	·31	·54	·73	·90	4·43	Aug.
2·52	2·76	**2·79**	2·57	2·06	1·13	·36	·15	·48	·73	·95	1·13	4·95	Sept.
2·93	3·13	**3·16**	2·90	2·33	1·47	·30	·49	·88	1·20	1·45	1·67	5·60	Oct.
2·93	**3·10**	3·04	2·74	2·33	1·78	·77	·39	·91	1·23	1·50	1·73	5·84	Nov.
2·97	**3·22**	**3·22**	2·94	2·59	2·09	1·09	·39	·99	1·35	1·67	1·94	6·18	Dec.

FIGURE 29

MEAN TEMPERATURE OF THE AIR AT SEA-LEVEL

Fahrenheit scale.

Authority: see p. 124.

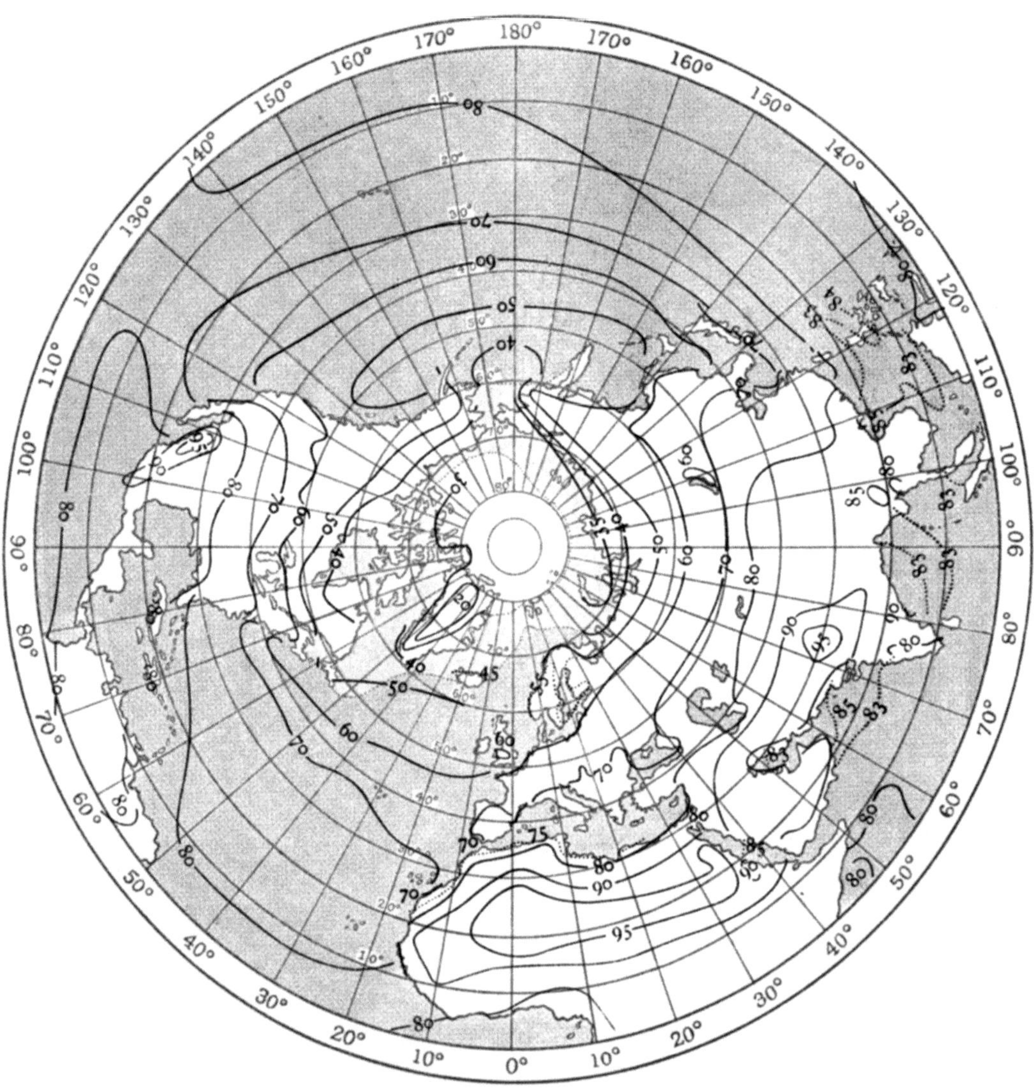

Diurnal variation of temperature (tercentesimal scale) at Hut Point and Cape Evans, 77½° S, 167° E. 1902–4, 1911–12. *Local time.*

Corrected for non-periodic change. The values underlined are negative, they indicate departures from the mean value downward.

Month	Mean	0	2	4	6	8	10	12	14	16	18	20	22	Range
Jan.	268·1	·96	1·37	1·57	·74	·27	·32	·70	·99	1·21	·99	·50	·16	2·78
Feb.	263·3	·49	·93	·86	·51	·10	·19	·53	·67	·71	·49	·35	·06	1·72
Mar.	257·6	·07	·18	·44	·42	·12	·27	·36	·34	·20	·02	·08	·02	·83
Apr.	250·6	·12	·03	·29	·14	·03	·05	·19	·03	·06	·18	·17	·04	·50
May	248·6	·07	·33	·22	·28	·16	·14	·36	·39	·38	·07	·19	·24	·72
June	247·9	·26	·41	·13	·04	·08	·04	·13	·01	·11	·34	·17	·06	·72
July	247·4	·14	·02	·44	·02	·04	·06	·17	·08	·02	·23	·08	·04	·67
Aug.	247·3	·18	·04	·33	·29	·09	·17	·39	·22	·22	·07	·01	·08	·72
Sept.	246·6	·24	·10	·11	·39	·49	·04	·24	·48	·42	·27	·13	·08	1·00
Oct.	251·8	·64	·46	·66	·78	·32	·63	·74	·83	·75	·29	·03	·47	1·61
Nov.	262·5	·99	·70	·89	·68	·19	·38	·77	1·18	·92	·60	·08	·56	2·17
Dec.	268·4	·35	·87	·72	·54	·40	·26	·53	·84	·82	·61	·21	·35	1·78

MEAN TEMPERATURE OF THE AIR AT SEA-LEVEL

FIGURE 30

Authority: see p. 124.

Fahrenheit scale.

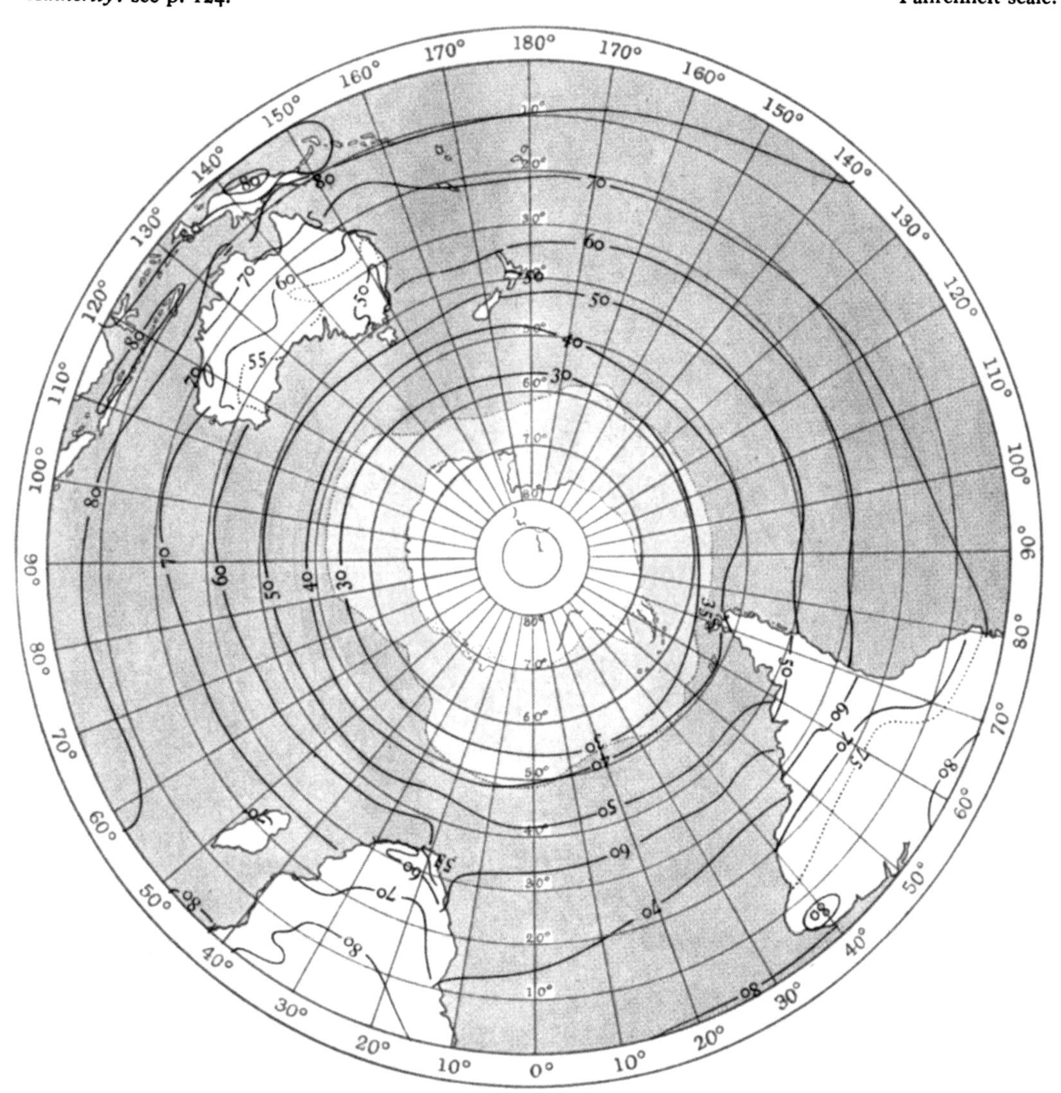

Mean daily temperature.

The mean temperature of a day is most correctly represented by the mean ordinate of the trace of a thermograph for that day. In practice the mean of twenty-four equally spaced hourly readings is treated as a correct representation, and there are many variants. The mean of observations at 9 h and 21 h, or of other term-hours, is also employed; but the commonest approximation is one-half of the sum of the maximum and minimum within the twenty-four hours, or briefly the mean of the max. and min. Correction-factors can be computed to represent the differences of these various methods of forming means. See vol. I, p. 194.

Diurnal variation and negative quantities.

In meteorological tables negative quantities should, if possible, be avoided because they are not only inconvenient and expensive to print but intractable and even dangerous in computation, and of no special arithmetical benefit even in the search for harmonic coefficients. Unfortunately it is customary to represent diurnal variations as positive or negative differences from the mean values for the months. We have quoted a number of tables of diurnal and seasonal variation arranged in that manner, asking the reader to accept a substitute for the negative sign. Other examples we have arranged in the form of the differences of the several hours from the hour of *lowest mean value* in each month. In those tables all the values are positive and the largest in each line is the mean daily range for the month.

FIGURE 31

Fahrenheit scale.

MEAN TEMPERATURE OF THE AIR AT SEA-LEVEL

Authority: see p. 124.

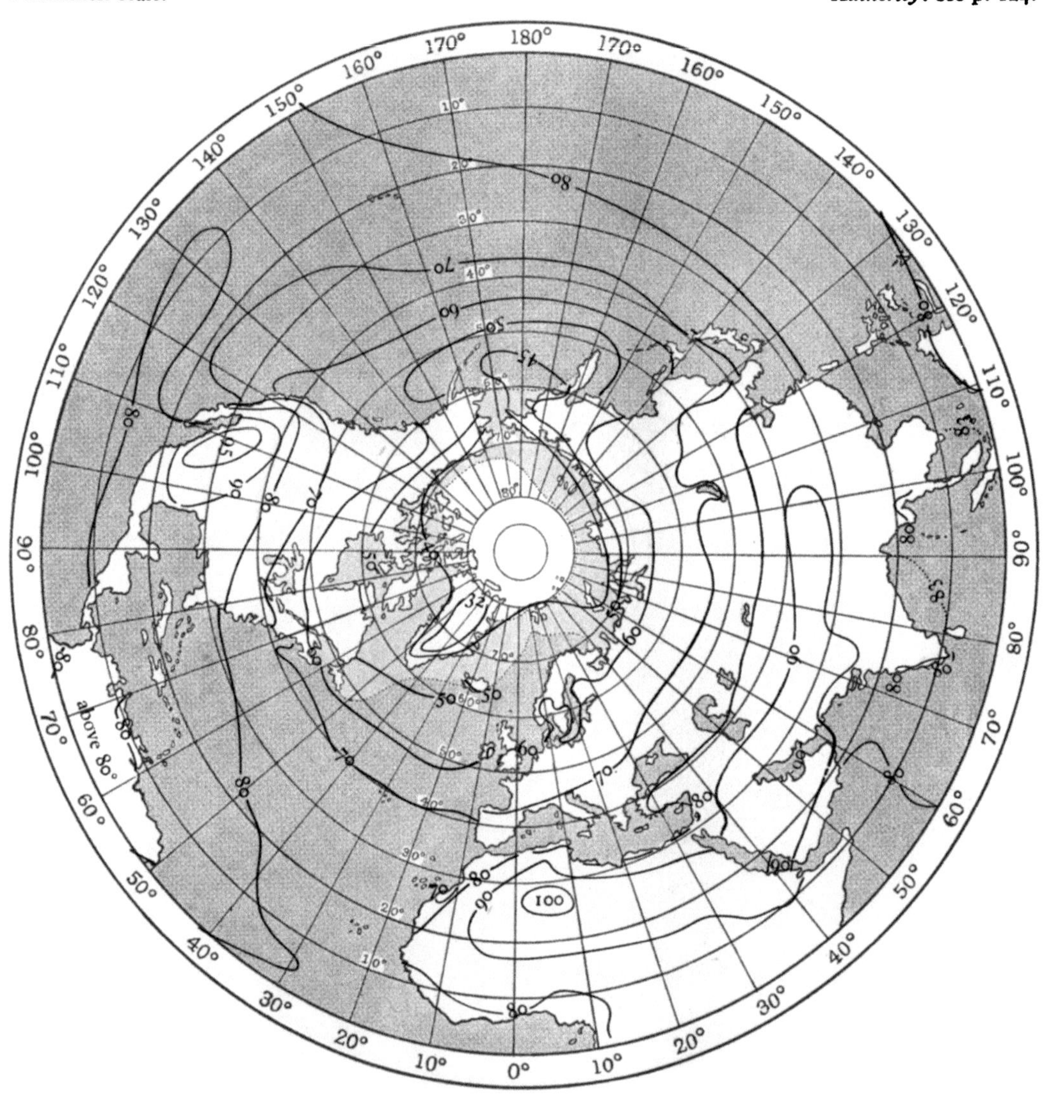

	Jan.	Feb.	Mar.	Apr.	May	June	July	Aug.	Sept.	Oct.	Nov.	Dec.	Station of maximum
1910	312·2	313·6	318·6	319·1	320·4	321·4	324·6	323·5	320·2	318·0	315·0	313·0	Insalah
1911	312·1	314·0	316·9	319·0	319·9	320·5	323·0	321·0	318·5	317·0	313·0	311·3	Insalah
1912	313·5	314·1	319·0	319·1	321·0	321·6	323·9	322·4	319·0	317·0	315·0	311·0	Insalah
1913	314·1	316·3	319·0	321·3	322·0	322·2	323·0	319·6	319·7	317·5	313·0	312·4	Insalah
1914	314·0	315·8	318·0	319·1	327·6	325·3	329·3	323·3	322·2	315·0	314·7	313·0	Insalah
1915	314·7	314·7	318·0	319·9	323·6	324·5	325·1	324·4	322·0	315·4	313·6	311·9	Jacobabad
1916	314·0	314·1	318·6	318·0	322·2	322·6	324·0	323·0	320·1	314·2	312·4	313·6	Timimoun
1917	313·0	314·7	318·6	319·1	318·8	323·4	323·4	320·9	320·0	314·0	314·1	311·9	See note
1918	312·4	315·2	318·6	319·1	323·7	323·0	324·2	323·0	319·3	315·5	?314·0	314·0	Insalah
Station of maximum	Maiduguri	McCarthy Island	Timbouctoo	Timbouctoo	Insalah	Insalah	Insalah	Timimoun	Insalah	Timbouctoo	Dakhla Oasis, Timbouctoo	Mongalla	Insalah

Highest temperature recorded in the "Réseau Mondial" at stations in the Northern Hemisphere. 1910–18.

The highest temperature for 1917, 323·4 tt, is shared by Baghdad and Jacobabad.

MEAN TEMPERATURE OF THE AIR AT SEA-LEVEL

FIGURE 32

Authority: see p. 124.

Fahrenheit scale.

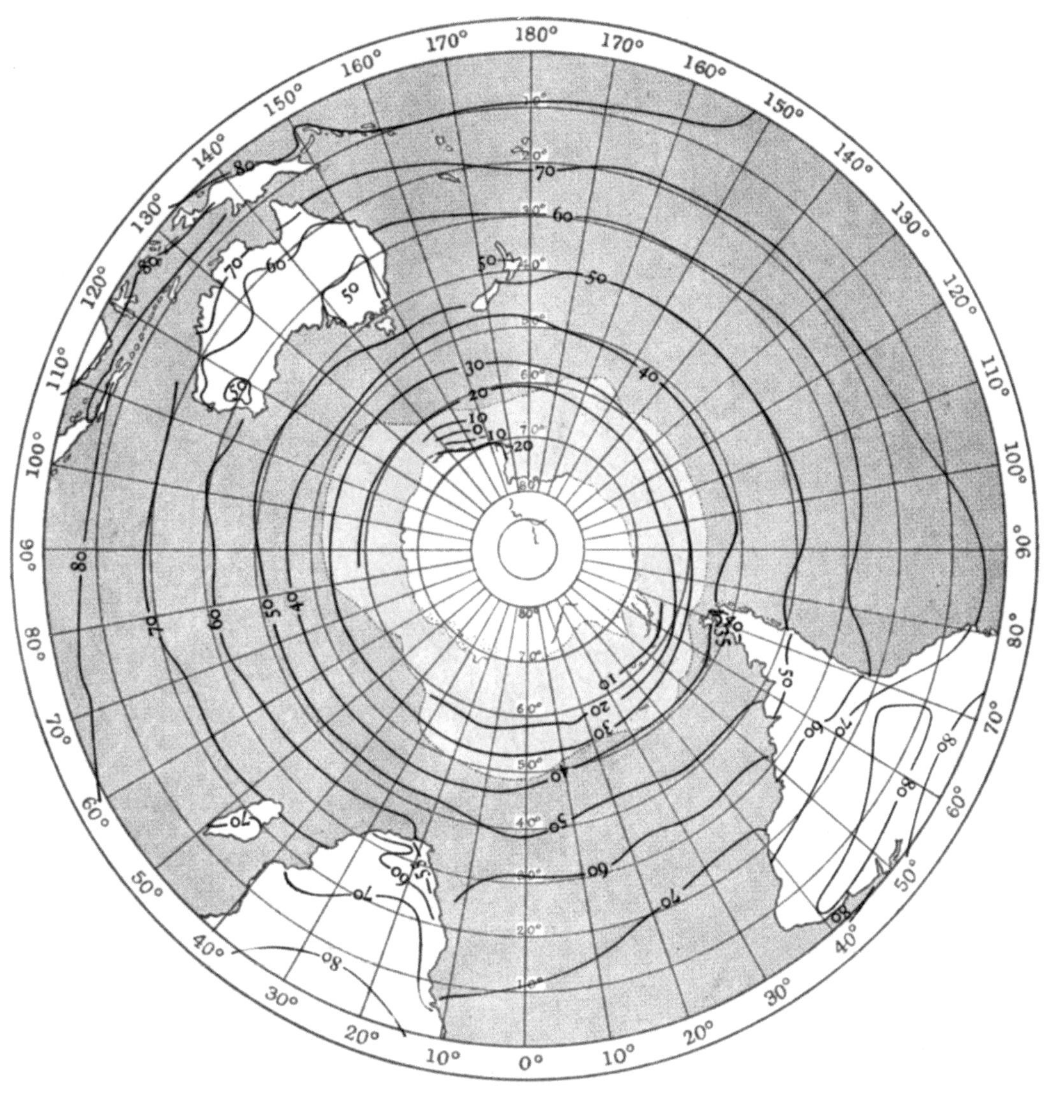

Lowest temperatures recorded in the "Réseau Mondial" at stations in the Southern Hemisphere. 1910–18.

	Jan.	Feb.	Mar.	Apr.	May	June	July	Aug.	Sept.	Oct.	Nov.	Dec.	Station of minimum
1910	269·9	269·8	269·0	260·5	261·7	257·8	244·4	241·3	251·2	254·1	265·8	266·6	S. Orkneys
1911	257·9	251·6	250·7	225·0	222·4	214·8	219·0	214·5	220·0	232·8	245·0	256·2	Framheim
1912	255·8	253·1	247·1	240·2	237·2	234·4	235·9	238·6	239·9	244·2	251·6	263·0	C. Evans
1913	267·4	268·5	265·0	258·8	253·9	237·0	241·5	243·0	251·1	252·9	256·6	269·2	S. Orkneys
1914	267·5	267·9	265·7	265·7	253·5	246·0	240·4	248·0	251·0	254·3	259·5	267·5	S. Orkneys
1915	260·8	246·3	243·6	240·8	241·6	238·6	236·3	238·6	238·0	241·9	?254·1	?259·7	"Endurance"
1916	269·6	265·5	262·0	251·9	245·9	246·8	241·7	249·0	246·4	249·5	265·5	268·7	S. Orkneys
1917	269·5	269·5	266·8	263·5	259·0	247·2	250·5	245·1	240·4	262·3	268·3	269·5	S. Orkneys
1918	270·4	268·0	269·3	260·5	251·3	243·5	246·2	239·8	242·5	267·0	266·7	268·2	S. Orkneys
Station of minimum	Framheim	"Endurance"	"Endurance"	Framheim	Framheim	Framheim	Framheim	Framheim	Framheim	Framheim	Framheim	Framheim	Framheim

AUGUST

FIGURE 33

MEAN TEMPERATURE OF THE AIR AT SEA-LEVEL

Fahrenheit scale.

Authority: see p. 124.

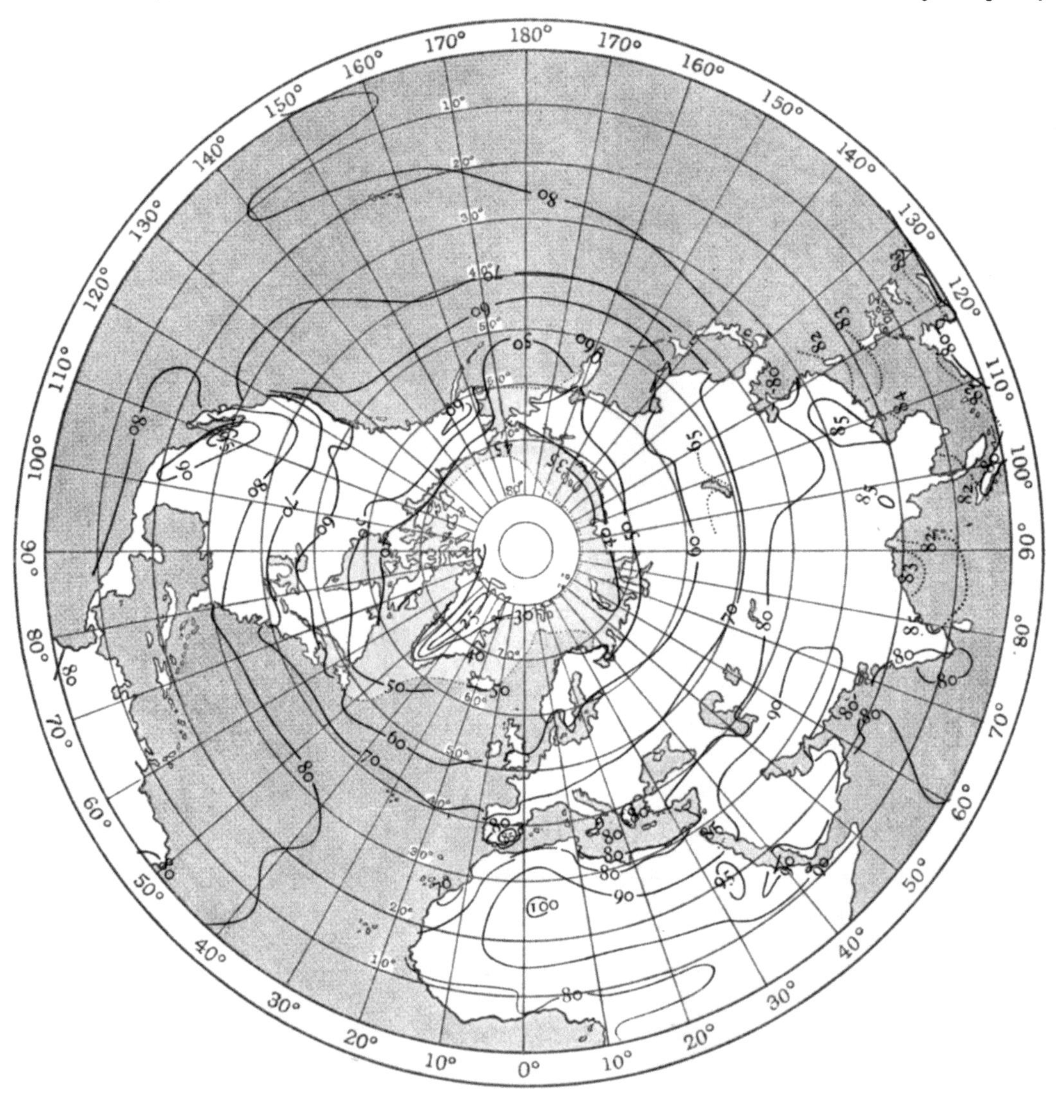

Diurnal variation of temperature (tercentesimal scale) at Fort William and Ben Nevis. 1891–1903.

Fort William, 56° 48′ N, 5° 5′ W. Corrected for seasonal change. The figures give the values *in excess of the minimum* for the respective months. Those underlined are below the mean value for the month.

Month	Min.	Mdt	2	4	6	8	10	12	14	16	18	20	22	Mean
Jan.	276·38	·23	·13	·10	·04	·00	·14	·68	·98	·67	·38	·30	·28	·35
Feb.	275·92	·41	·26	·13	·04	·00	·53	1·54	2·24	2·24	1·37	·90	·58	·87
Mar.	276·14	·76	·46	·21	·00	·32	1·60	2·75	3·37	3·43	2·57	1·65	1·13	1·54
Apr.	277·83	·95	·46	·12	·03	1·38	3·06	4·26	4·90	5·06	4·33	2·67	1·68	2·43
May	279·83	1·12	·44	·00	·43	2·17	3·84	5·02	5·79	5·88	5·26	3·49	2·05	2·98
June	282·98	1·12	·41	·00	·77	2·36	3·84	4·91	5·65	5·93	5·47	3·73	2·11	3·05
July	284·58	·82	·33	·00	·44	1·71	3·03	3·91	4·60	4·74	4·26	2·78	1·55	2·37
Aug.	284·49	·79	·41	·09	·08	1·33	2·74	3·63	4·18	4·38	3·72	2·18	1·35	2·09
Sept.	282·97	·79	·41	·11	·00	·79	2·23	3·33	3·93	4·07	3·00	1·81	1·23	1·82
Oct.	279·91	·52	·31	·11	·00	·16	1·23	2·52	3·03	2·79	1·68	1·12	·81	1·21
Nov.	279·13	·21	·06	·03	·01	·00	·41	1·17	1·53	1·19	·75	·48	·33	·53
Dec.	277·19	·19	·10	·03	·00	·09	·23	·67	·89	·66	·41	·30	·25	·33

MEAN TEMPERATURE OF THE AIR AT SEA-LEVEL

FIGURE 34

Authority: see p. 124.

Fahrenheit scale.

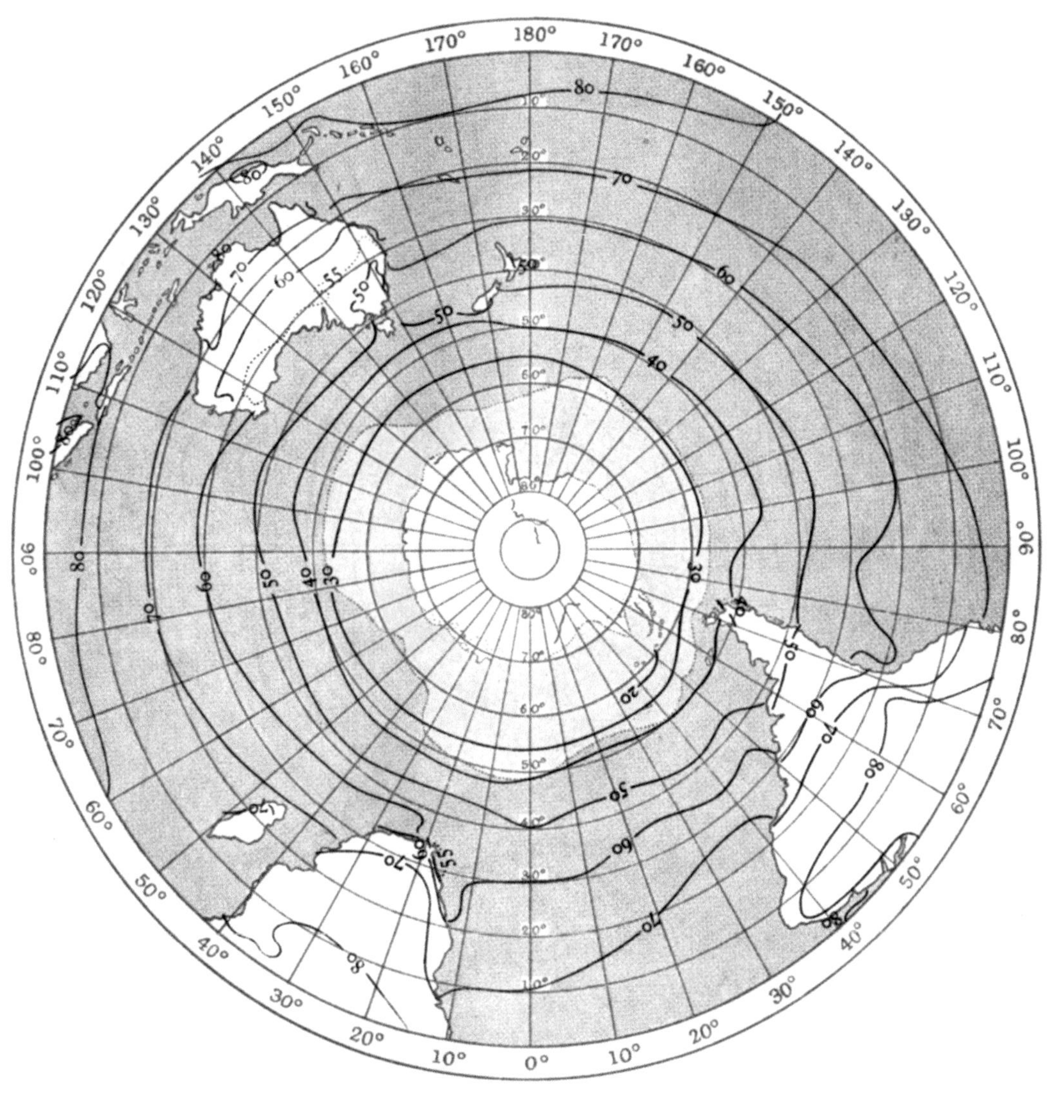

Diurnal variation of temperature (tercentesimal scale) at Fort William and Ben Nevis. 1891–1903.

Ben Nevis, 56° 48′ N, 5° W, 1343 m. Greenwich mean time. The figures give the values *in excess of the minimum* for the respective months. Those underlined are below the mean value for the month.

Month	Min.	Mdt	2	4	6	8	10	12	14	16	18	20	22	Mean
Jan.	268·07	·08	·09	·09	·03	·02	·16	·36	·36	·15	·06	·08	·07	·13
Feb.	268·23	·28	·17	·13	·07	·03	·43	·82	·94	·77	·36	·36	·27	·38
Mar.	268·11	·34	·20	·11	·00	·31	·86	1·31	1·43	1·34	·74	·56	·47	·63
Apr.	270·15	·26	·15	·05	·09	·51	1·19	1·62	1·83	1·68	1·14	·62	·42	·79
May	272·57	·41	·20	·00	·23	·83	1·52	2·14	2·39	2·25	1·70	·99	·65	1·12
June	276·27	·45	·19	·00	·27	·82	1·45	2·09	2·52	2·45	1·98	1·33	·77	1·19
July	277·31	·47	·23	·00	·14	·57	1·22	1·83	2·23	2·20	1·77	1·19	·73	1·06
Aug.	277·01	·39	·23	·06	·03	·37	·99	1·58	1·91	1·81	1·42	·92	·63	·87
Sept.	275·72	·29	·15	·02	·01	·30	·84	1·27	1·49	1·34	·85	·59	·40	·63
Oct.	272·23	·18	·19	·11	·03	·12	·58	·93	1·02	·76	·48	·36	·31	·42
Nov.	271·44	·04	·03	·11	·05	·04	·27	·46	·47	·21	·09	·08	·06	·16
Dec.	269·37	·06	·07	·04	·00	·00	·13	·36	·28	·18	·16	·14	·11	·13

FIGURE 35

MEAN TEMPERATURE OF THE AIR AT SEA-LEVEL

Fahrenheit scale.

Authority: see p. 124.

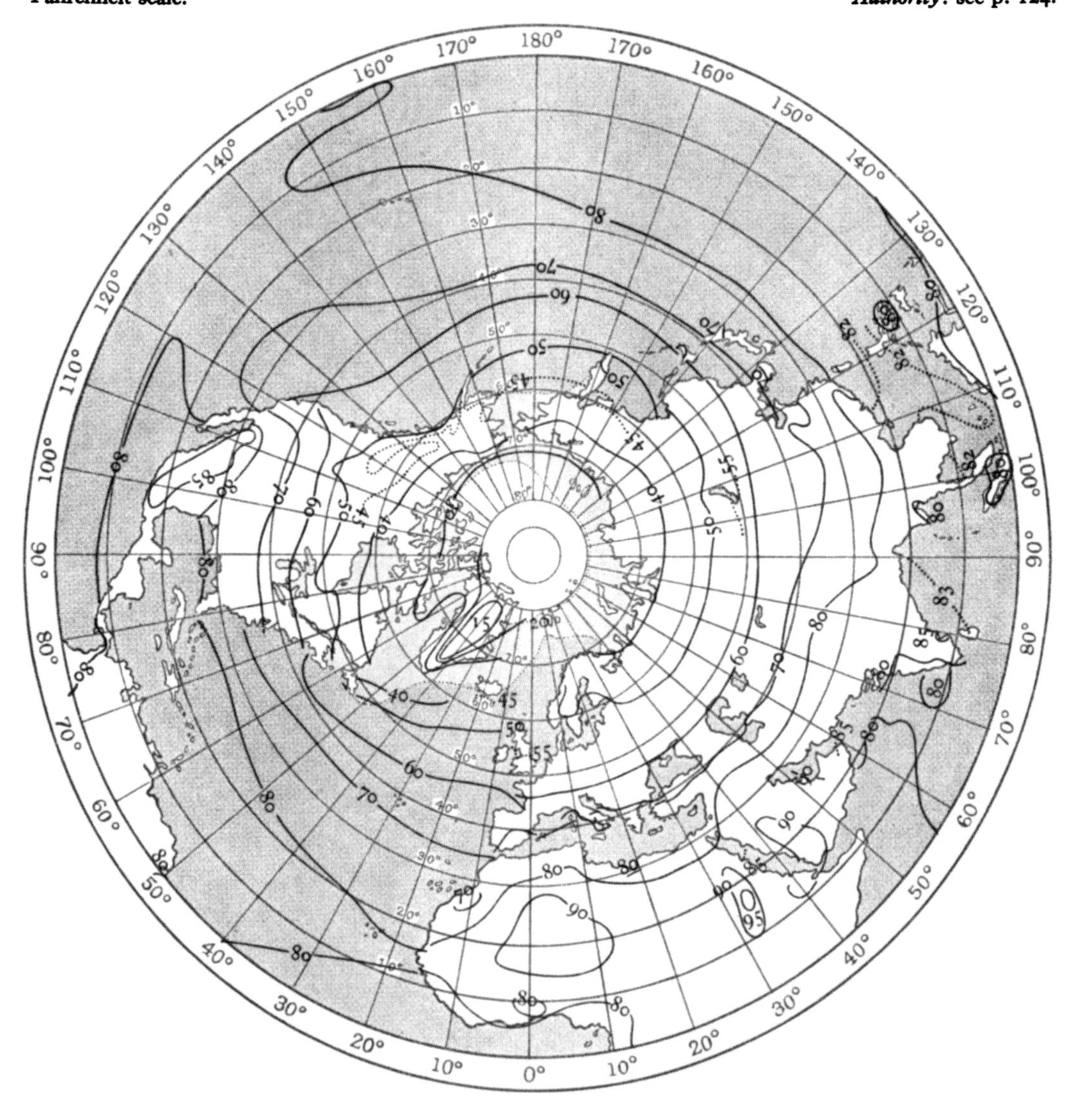

Diurnal variation of temperature (tercentesimal scale) at Parc St Maur, Paris, and at the Eiffel Tower.

Parc St Maur, 48° 49′ N, 2° 29′ E, 50 m, 1876–1900. The figures give the values *in excess of the minimum* for the respective months. Those underlined are below the mean value for the month. The column of minima refers to station level in the period 1851–1900.

Month	Min.	0	2	4	6	8	10	12	14	16	18	20	22	24	Mean
Jan.	274·04	·65	·41	·20	·04	·01	1·20	2·61	3·29	2·76	1·82	1·33	·95	·64	1·27
Feb.	274·56	·99	·64	·31	·07	·31	2·21	4·02	4·99	4·69	3·10	2·15	1·52	1·01	2·08
Mar.	275·63	1·47	·86	·38	·00	1·47	4·22	6·19	7·15	6·89	5·07	3·34	2·35	1·58	3·28
Apr.	278·68	1·85	·96	·22	·32	3·16	5·97	7·66	8·62	8·30	6·74	4·31	2·95	1·91	4·26
May	281·30	1·65	·73	·00	1·22	4·40	6·88	8·38	9·12	8·84	7·47	4·78	3·06	1·77	4·72
June	284·81	1·64	·65	·00	1·48	4·42	6·88	8·34	9·09	8·75	7·49	4·82	2·93	1·74	4·71
July	286·62	1·69	·71	·00	1·17	4·20	6·88	8·42	9·13	8·90	7·65	4·82	2·93	1·69	4·71
Aug.	286·08	1·77	·90	·16	·60	3·85	6·86	8·48	9·40	8·90	7·27	4·38	2·82	1·73	4·61
Sept.	283·79	1·52	·78	·20	·00	2·80	6·30	7·96	8·77	7·97	5·47	3·41	2·28	1·42	3·95
Oct.	280·36	1·08	·63	·25	·00	1·24	4·09	5·95	6·46	5·60	3·31	2·25	1·51	·91	2·69
Nov.	277·16	·77	·44	·15	·02	·29	2·15	3·78	4·30	3·36	2·13	1·52	1·04	·66	1·66
Dec.	274·57	·55	·31	·12	·03	·02	1·21	2·60	3·12	2·38	1·60	1·18	·83	·57	1·17

MEAN TEMPERATURE OF THE AIR AT SEA-LEVEL

FIGURE 36

Authority: see p. 124.

Fahrenheit scale.

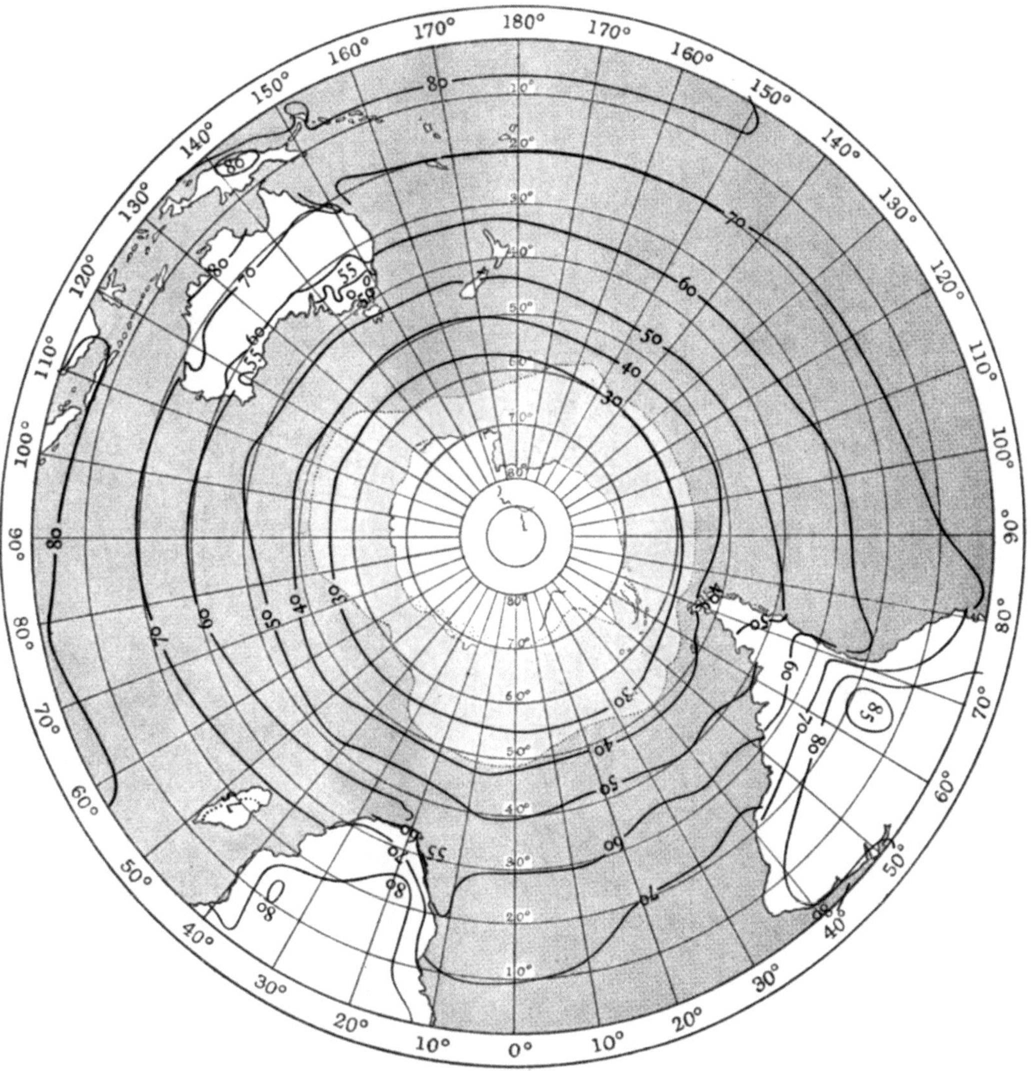

Diurnal variation of temperature (tercentesimal scale) at Parc St Maur, Paris, and at the Eiffel Tower.

Eiffel Tower, 48° 52′ N, 2° 17′ E, 335 m above M.S.L., 1890–1904. The figures give the values *in excess of the minimum* for the respective months. Those underlined are below the mean value for the month. Local time.

Month	Min.	0	2	4	6	8	10	12	14	16	18	20	22	24	Mean
Jan.	274·12	·49	·42	·19	·07	·00	·31	·78	1·13	1·06	·92	·80	·66	·46	·56
Feb.	274·53	·91	·63	·35	·11	·11	·60	1·51	2·17	2·32	1·95	1·64	1·35	·99	1·14
Mar.	276·35	1·14	·68	·30	·00	·15	·99	2·29	3·19	3·39	2·78	2·17	1·68	1·21	1·56
Apr.	279·11	1·32	·71	·24	·00	·39	1·96	3·50	4·45	4·61	3·98	3·02	2·20	1·41	2·20
May	282·58	1·31	·59	·03	·08	·83	2·45	3·92	4·82	4·99	4·41	3·22	2·34	1·46	2·42
June	285·38	1·37	·59	·00	·16	·97	2·58	4·09	5·07	5·26	4·70	3·47	2·44	1·46	2·57
July	287·05	1·50	·64	·02	·12	·94	2·63	4·16	5·07	5·29	4·76	3·47	2·43	1·46	2·59
Aug.	287·10	1·54	·74	·18	·03	·67	2·33	3·96	5·00	5·24	4·47	3·39	2·43	1·54	2·50
Sept.	285·19	1·38	·74	·32	·00	·40	1·61	3·18	4·06	4·09	3·31	2·60	1·94	1·32	1·97
Oct.	281·30	1·12	·73	·31	·01	·12	·85	2·09	2·82	2·71	2·11	1·78	1·39	·96	1·33
Nov.	277·09	·85	·58	·31	·10	·12	·55	1·32	1·73	1·55	1·39	1·21	·94	·69	·88
Dec.	274·66	·48	·35	·17	·06	·00	·38	·82	1·17	·96	·80	·76	·60	·47	·55

FIGURE 37

Fahrenheit scale.

MEAN TEMPERATURE OF THE AIR AT SEA-LEVEL

Authority: see p. 124.

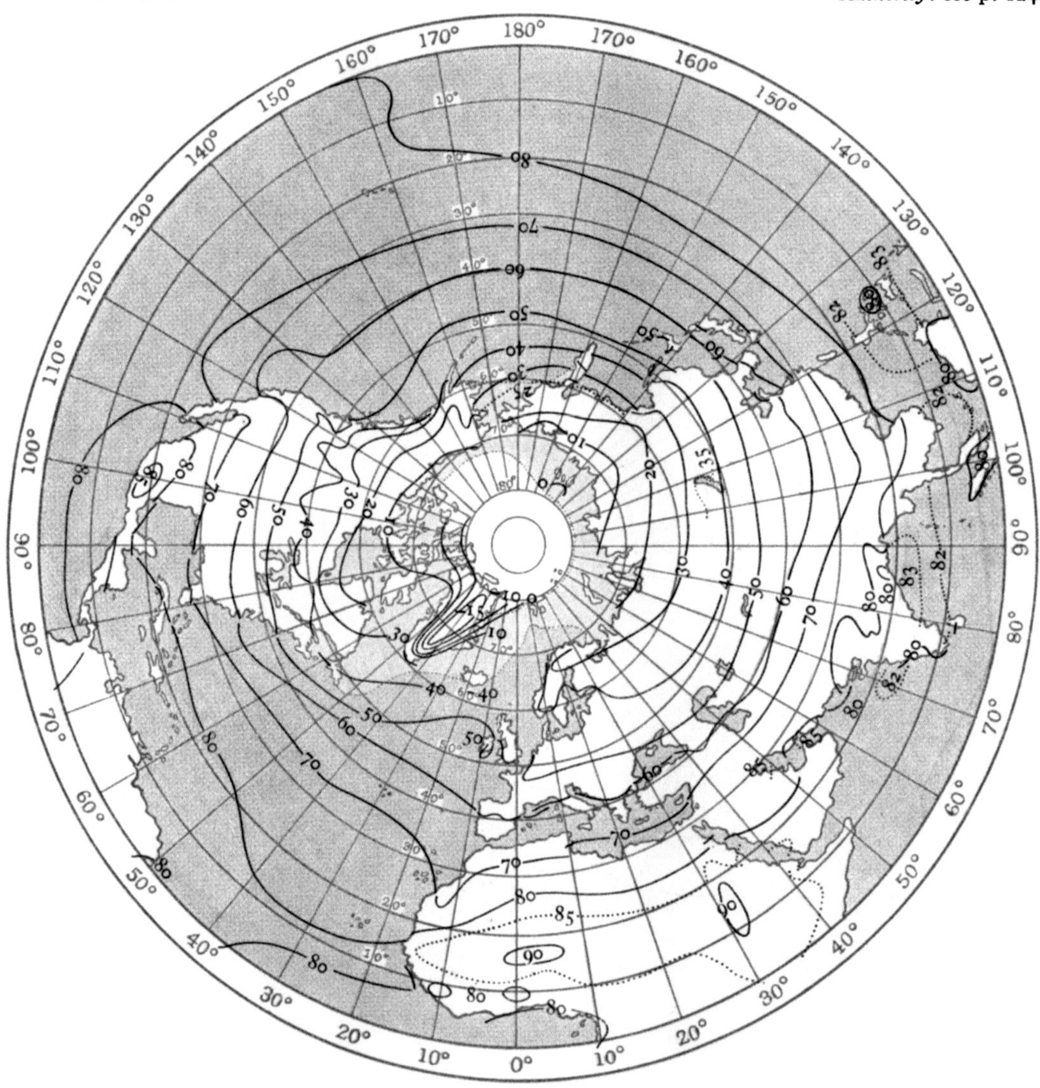

Diurnal variation of temperature (tercentesimal scale) at Clermont Ferrand and Puy de Dôme.

Clermont Ferrand, 45° 46′ N, 3° 05′ E, 388 m, 1881–1900. The figures give the values *in excess of the minimum* for the respective months. Those underlined are below the mean value for the month.

Month	Min.	0	2	4	6	8	10	12	14	16	18	20	22	24	Mean
Jan.	272·43	·89	·56	·35	·10	·18	2·49	4·77	5·47	4·41	2·37	1·82	1·34	·93	2·07
Feb.	273·42	1·43	·89	·52	·19	·82	4·14	6·66	7·65	6·87	4·17	3·06	2·17	1·43	3·23
Mar.	274·79	2·05	1·24	·49	·00	2·45	5·91	8·10	8·86	8·45	5·97	4·21	3·14	2·15	4·25
Apr.	277·70	2·35	1·20	·38	·60	4·28	7·49	9·11	9·70	9·07	7·08	4·90	3·65	2·48	5·00
May	280·92	1·88	·85	·00	2·15	5·13	7·95	9·54	10·03	9·41	7·64	4·96	3·21	1·96	5·22
June	283·97	2·19	·97	·00	3·08	5·90	8·57	10·27	10·69	10·06	8·47	5·65	3·58	2·28	5·77
July	285·76	2·22	1·03	·00	2·77	5·90	8·86	10·61	11·19	10·86	9·09	5·73	3·61	2·21	5·97
Aug.	284·84	2·41	1·17	·29	1·56	5·93	9·44	11·50	12·34	11·62	9·13	5·70	3·82	2·39	6·23
Sept.	282·53	2·06	1·03	·27	·24	4·64	8·56	10·77	11·33	10·52	6·86	4·55	3·23	1·96	5·35
Oct.	279·03	1·55	·91	·33	·00	2·38	6·39	8·35	8·85	7·67	4·08	3·01	2·11	1·35	3·79
Nov.	276·67	1·08	·66	·27	·05	·77	4·14	6·29	6·84	5·27	2·89	2·22	1·54	1·00	2·65
Dec.	273·45	·90	·57	·26	·11	·13	2·87	4·86	5·45	3·81	2·29	1·73	1·28	·88	2·02

MEAN TEMPERATURE OF THE AIR AT SEA-LEVEL

FIGURE 38

Authority: see p. 124.

Fahrenheit scale.

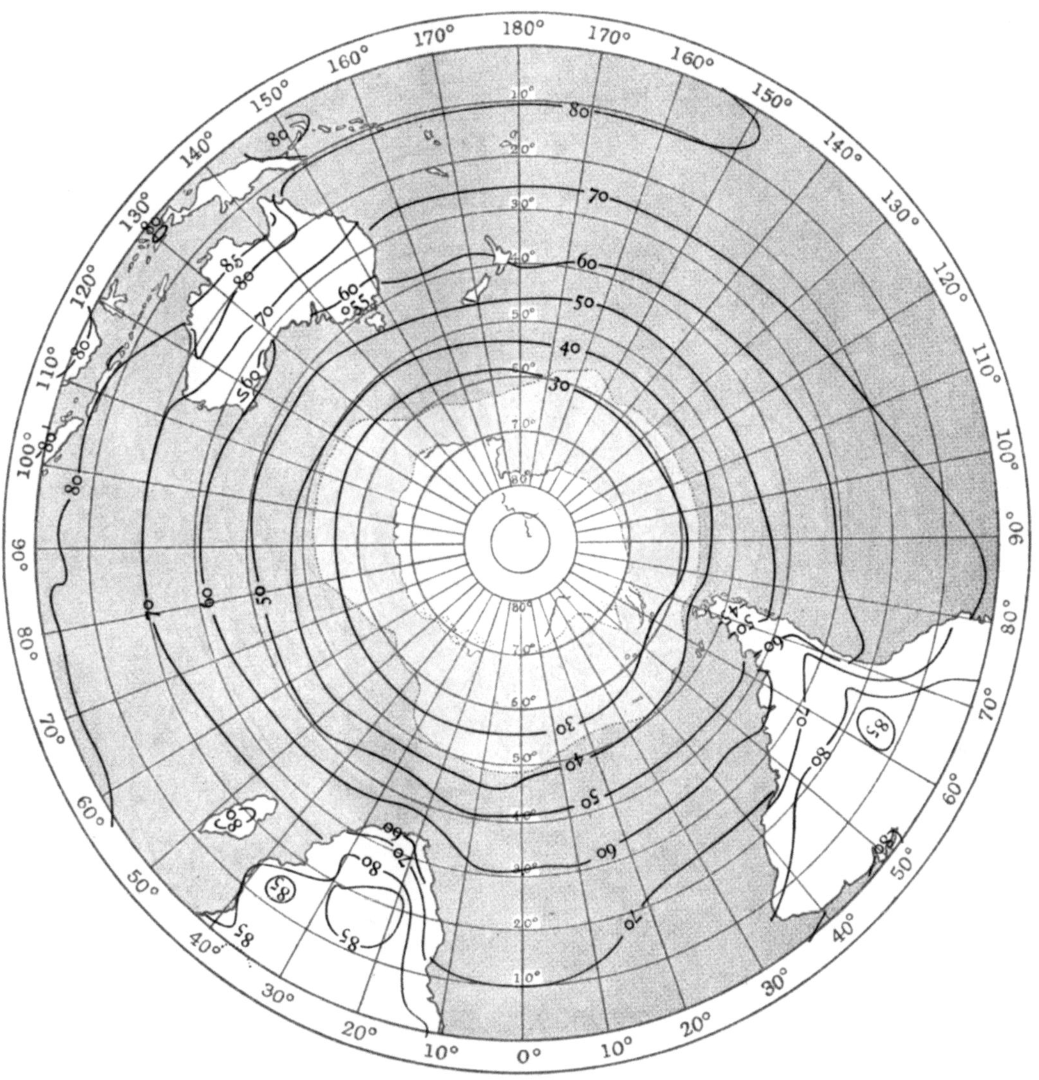

***Diurnal** variation of temperature (tercentesimal scale) at Clermont Ferrand and Puy de Dôme.*

Puy de Dôme, 45° 46′ N, 2° 58′ E, 1467 m, 1887–1902. The figures give the values *in excess of the minimum* for the respective months. Those underlined are below the mean value for the month.

Month	Min. 1881–1900	Local mean time. 0	2	4	6	8	10	12	14	16	18	20	22	24	Mean
Jan.	270·02	·26	·21	·15	·12	·00	·21	·59	·80	·70	·30	·24	·23	·22	·32
Feb.	270·76	·22	·18	·02	·01	·08	·32	·73	1·07	1·12	·56	·34	·27	·23	·41
Mar.	270·93	·49	·34	·14	·00	·14	·59	1·20	1·74	1·89	1·36	·98	·70	·56	·80
Apr.	273·36	·56	·31	·11	·00	·27	·95	1·74	2·40	2·44	1·91	1·30	·92	·67	1·08
May	276·42	·53	·27	·06	·18	·61	1·39	2·30	2·90	2·90	2·31	1·50	1·04	·67	1·34
June	280·34	·50	·23	·00	·19	·59	1·33	2·33	2·96	3·02	2·48	1·56	·99	·58	1·35
July	282·56	·63	·34	·05	·15	·53	1·20	2·17	3·01	3·21	2·53	1·60	1·05	·65	1·37
Aug.	282·64	·56	·28	·07	·07	·35	1·17	2·25	3·14	3·17	2·34	1·41	·94	·53	1·31
Sept.	280·75	·54	·40	·18	·00	·26	·94	1·91	2·65	2·57	1·58	·97	·63	·41	1·04
Oct.	276·31	·42	·30	·12	·00	·15	·65	1·36	1·77	1·47	·71	·53	·40	·31	·65
Nov.	274·18	·29	·19	·08	·03	·02	·35	·96	1·28	·86	·41	·24	·14	·13	·40
Dec.	271·07	·10	·05	·01	·03	·07	·27	·67	·87	·66	·30	·25	·19	·14	·29

FIGURE 39

Fahrenheit scale.

MEAN TEMPERATURE OF THE AIR AT SEA-LEVEL

Authority: see p. 124.

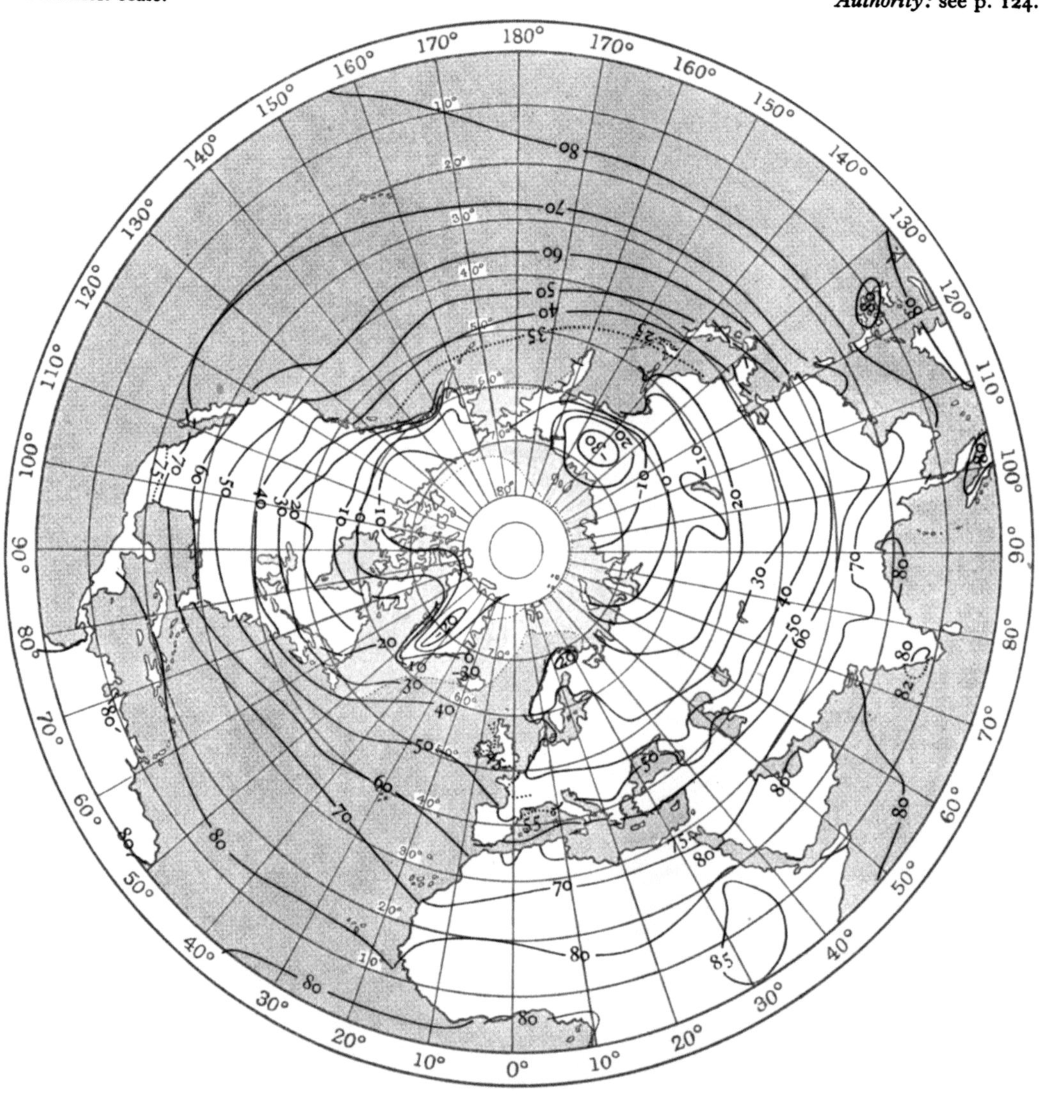

Diurnal variation of temperature (tercentesimal scale) at Lahore and Leh. (Local mean time.)

Lahore, 31° 34′ N, 74° 20′ E, 214 m, 1875–84 (36–40 observations per month). The figures give the values ***in excess of the minimum*** for the respective months. Those underlined are below the mean value for the month.

Month	Min.	0	2	4	6	8	10	12	14	16	18	20	22	Mean
Jan.	280·1	1·9	1·3	·8	·1	1·4	6·8	10·8	12·4	12·1	7·2	4·6	3·0	5·2
Feb.	282·6	2·3	1·3	·6	·0	1·7	6·9	10·3	11·4	11·2	7·9	5·0	3·3	5·2
Mar.	288·1	2·6	1·8	1·0	·0	3·2	8·2	11·6	13·2	12·9	9·7	6·0	3·7	6·2
Apr.	293·9	2·8	1·7	·4	·0	4·3	9·6	12·4	13·8	13·3	10·3	6·4	4·1	6·6
May	297·7	2·7	1·6	·5	·3	5·1	9·6	12·1	13·3	13·1	10·7	6·7	4·2	6·6
June	301·8	2·8	1·4	·4	·3	3·9	7·7	10·4	12·1	11·8	9·8	6·2	4·4	5·9
July	301·4	1·7	·9	·2	·1	2·4	4·7	7·0	7·6	7·5	5·9	3·8	2·6	3·7
Aug.	300·2	1·6	·9	·3	·0	2·2	4·6	6·6	7·4	7·1	5·4	3·4	2·2	3·5
Sept.	297·3	1·8	1·3	·6	·0	3·6	7·2	9·5	10·1	9·7	6·6	4·5	2·9	4·8
Oct.	290·2	2·4	1·7	·8	·0	4·6	10·2	13·7	14·8	14·2	8·8	5·8	3·9	6·7
Nov.	282·8	2·5	1·6	·9	·0	3·7	10·4	14·7	16·2	15·2	8·7	5·5	3·9	6·9
Dec.	279·6	2·4	1·8	1·1	·4	2·0	8·2	12·2	13·8	13·1	7·8	5·2	3·6	5·95

MEAN TEMPERATURE OF THE AIR AT SEA-LEVEL

FIGURE 40

Authority: see p. 124.

Fahrenheit scale.

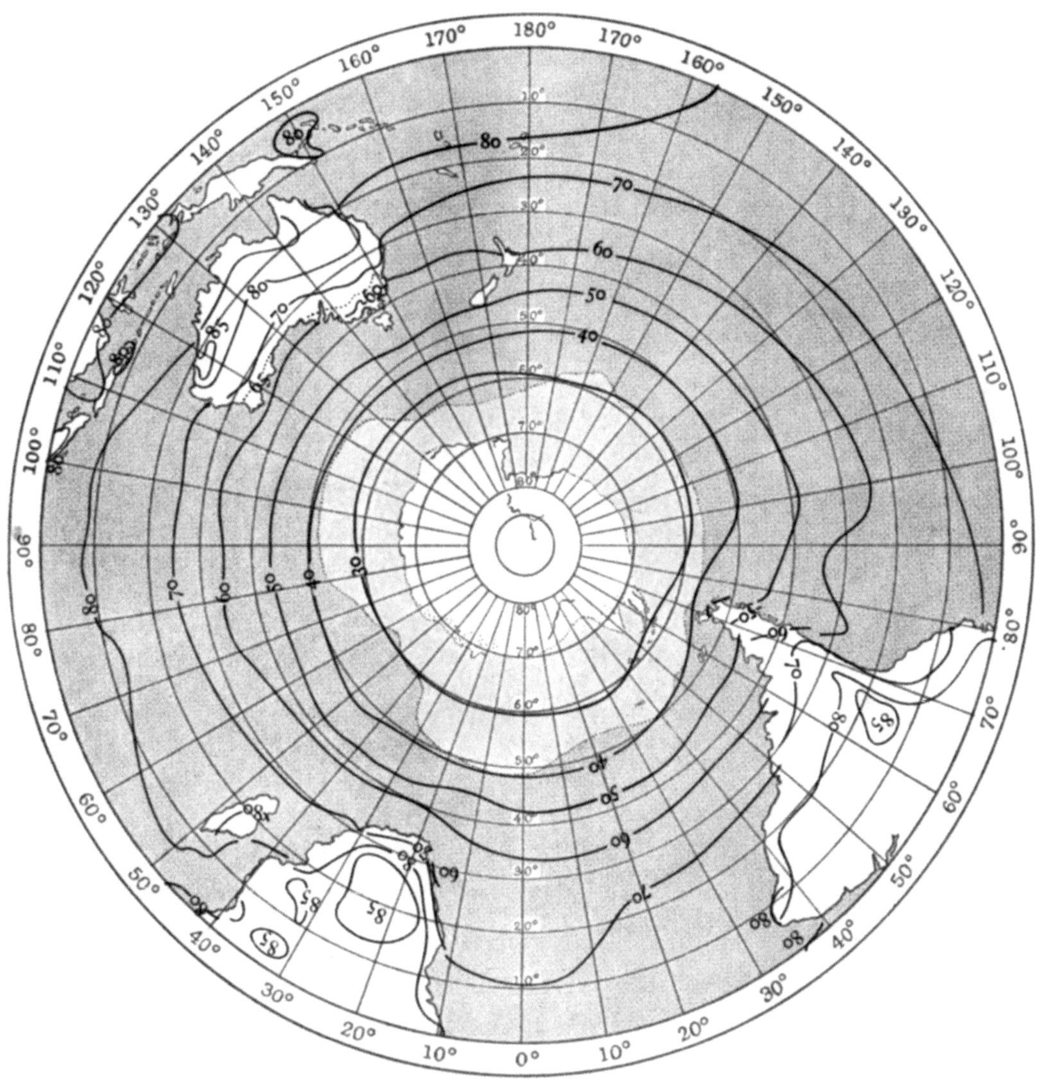

Diurnal variation of temperature (tercentesimal scale) at Lahore and Leh.

Leh, 34° 10′ N, 77° 42′ E, 3506 m, 1876–1891 (4 observations per month—60 observations). The figures give the values *in excess of the minimum* for the respective months. Those underlined are below the mean value for the month.

Month	Min.	0	2	4	6	8	10	12	14	16	18	20	22	Mean
Jan.	261·1	1·8	1·1	·4	·0	1·4	5·9	8·8	9·9	8·9	5·4	3·8	2·7	4·2
Feb.	262·6	2·1	1·2	·6	·0	2·3	6·7	8·9	9·7	8·8	5·8	3·9	2·9	4·4
Mar.	268·4	2·4	1·4	·6	·0	3·1	6·4	9·3	10·4	9·6	6·4	4·6	3·3	4·8
Apr.	273·4	2·8	1·6	·6	·2	4·5	8·1	10·7	11·4	10·6	7·7	5·4	3·9	5·6
May	276·2	2·6	1·4	·4	·7	4·9	8·4	10·8	11·9	10·8	8·1	5·1	3·6	5·7
June	280·5	2·8	1·5	·4	·7	5·2	8·7	11·7	13·0	12·7	9·7	5·9	4·1	6·4
July	284·1	2·7	1·7	·6	·3	4·1	7·8	10·9	12·2	11·8	8·4	5·6	3·8	5·8
Aug.	282·8	2·7	1·6	·4	·3	4·9	8·6	11·9	13·3	12·2	8·1	5·3	3·8	6·1
Sep.	278·5	2·6	1·5	·4	·1	4·7	9·1	12·3	13·6	13·0	8·0	5·5	3·8	6·2
Oct.	272·2	2·6	1·6	·6	·0	3·2	8·3	11·1	12·7	11·5	7·3	5·2	3·7	5·7
Nov.	266·8	2·7	1·8	·8	·0	2·9	7·8	10·8	12·3	11·2	6·7	4·7	3·1	5·4
Dec.	263·7	2·3	1·4	·6	·0	2·2	6·5	9·2	10·4	9·3	5·7	4·3	3·2	4·6

FIGURE 41

MEAN TEMPERATURE OF THE AIR AT SEA-LEVEL

Fahrenheit scale.

Authority: see p. 124.

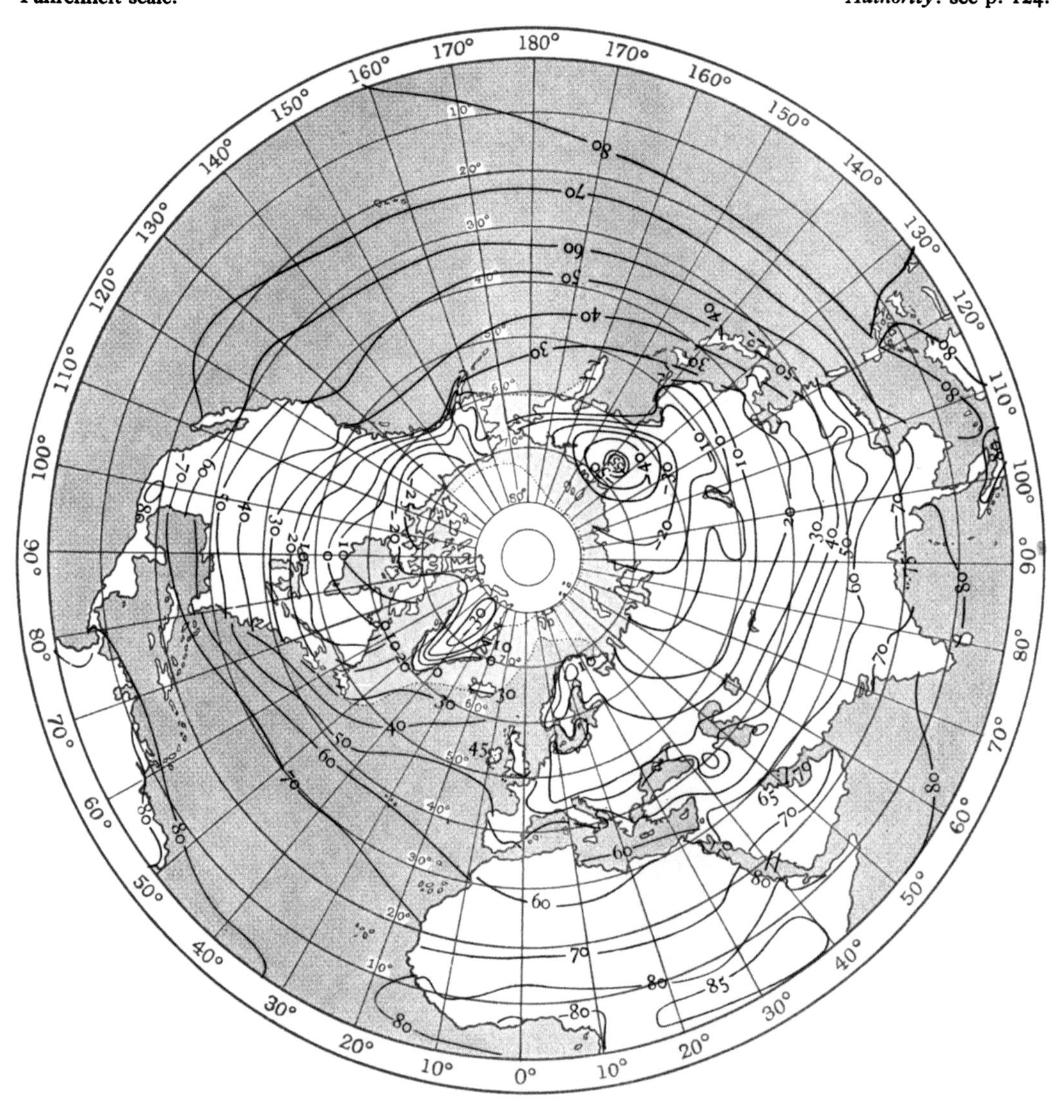

Lowest temperatures recorded in the "Réseau Mondial" at stations in the Northern Hemisphere, 1910–18.

	Jan.	Feb.	Mar.	Apr.	May	June	July	Aug.	Sept.	Oct.	Nov.	Dec.	Station of minimum
1910	217·4	217·4	227·4	237·4	255·0	266·5	270·5	266·3	260·8	243·4	226·7	223·6	Tanana
1911	209·2	214·8	223·8	234·6	247·3	267·4	270·0	269·6	256·9	239·3	223·5	213·2	Verkhoĭansk
1912	211·3	211·0	212·7	232·0	248·7	267·1	268·0	266·3	252·8	237·5	218·4	216·9	Verkhoĭansk
1913	213·7	212·7	222·7	234·7	249·5	259·9	268·4	265·2	257·6	239·8	221·4	217·4	Verkhoĭansk
1914	219·1	223·0	224·3	233·6	249·0	265·2	269·1	269·1	255·8	237·5	224·1	?217·6	? Iakoutsk
1915	211·3	214·4	217·9	235·2	248·5	267·4	269·1	268·6	261·3	240·7	223·1	214·5	Verkhoĭansk
1916	212·9	211·9	223·6	232·5	247·0	262·7	269·3	267·4	256·7	236·3	216·7	212·7	Iakoutsk
1917	210·4	219·5	218·4	230·2	246·9	261·0	270·0	269·3	256·7	233·1	219·5	217·4	Verkhoĭansk
1918	214·6	213·1	225·2	233·4	241·2	265·1	269·1	268·6	260·0	240·2	219·0	213·7	Verkhoĭansk
Station of minimum	Verkhoĭansk	Verkhoĭansk	Verkhoĭansk	Spitsbergen	Doudinka	Doudinka	Belle Isle	Tanana Eagle	Doudinka	Verkhoĭansk	Verkhoĭansk	Tourouk-hansk	Verkhoĭansk

MEAN TEMPERATURE OF THE AIR AT SEA-LEVEL

FIGURE 42

Authority: see p. 124.

Fahrenheit scale.

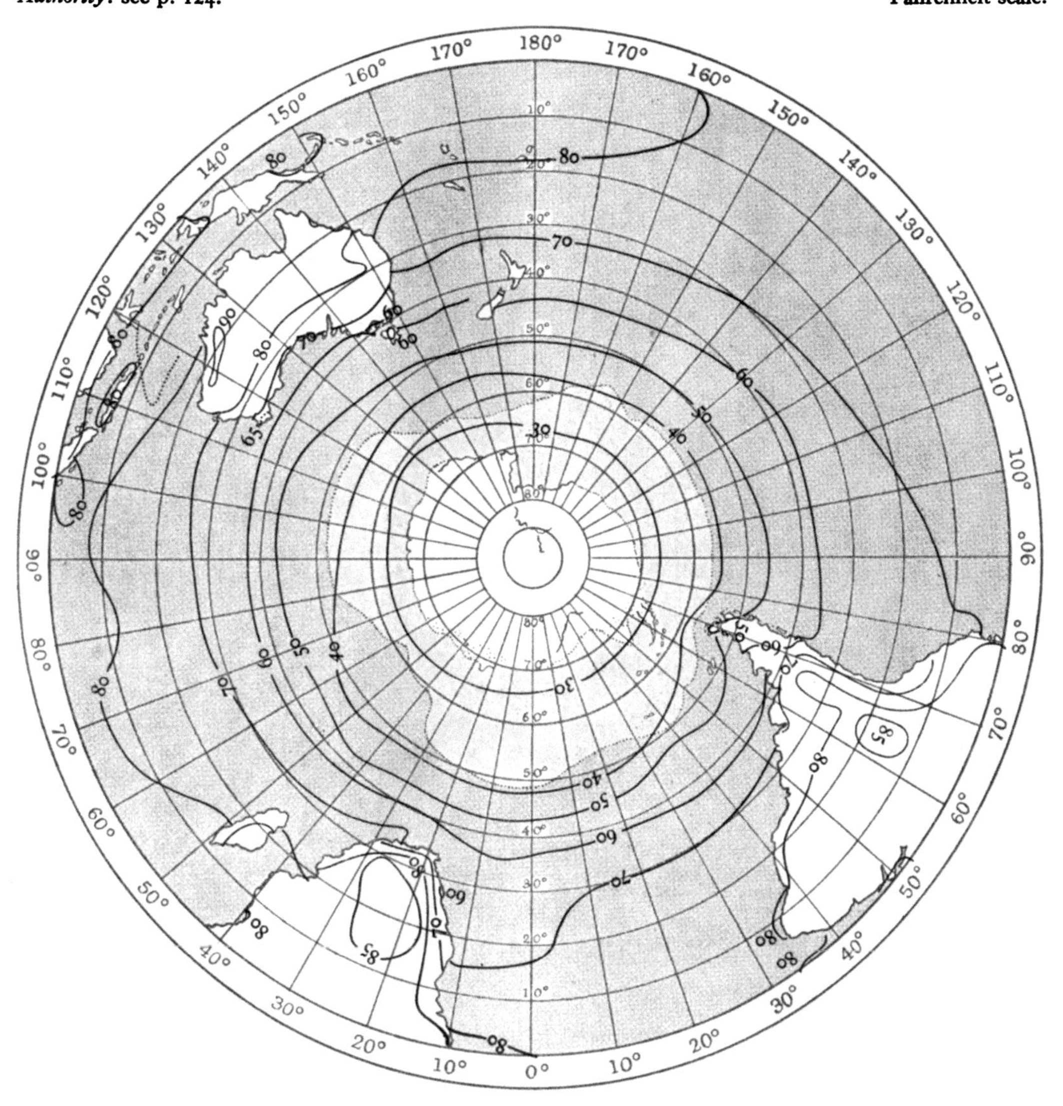

Highest temperatures recorded in the "Réseau Mondial" at stations in the Southern Hemisphere, 1910–18.

	Jan.	Feb.	Mar.	Apr.	May	June	July	Aug.	Sept.	Oct.	Nov.	Dec.	Station of maximum
1910	319·7	318·8	315·8	312·4	308·3	308·0	308·0	309·7	311·9	314·9	316·3	318·7	Eucla
1911	318·9	316·2	315·7	312·4	307·9	309·7	306·3	307·9	313·7	314·8	318·0	317·7	Onslow
1912	319·1	317·9	315·5	313·7	309·0	308·6	308·0	310·2	312·4	316·1	317·4	322·3	Eucla
1913	320·5	317·8	317·6	311·2	308·9	307·8	307·7	307·7	310·8	315·6	319·7	319·1	Bourke
1914	319·1	318·0	318·6	133·9	311·1	308·0	308·6	308·0	310·9	314·7	317·4	318·0	William Creek, [Bourke
1915	318·6	321·3	318·6	313·0	311·0	309·3	310·2	311·0	314·1	314·9	319·7	319·8	Boulia
1916	319·1	318·6	316·4	312·0	312·0	312·0	308·0	312·0	313·9	314·8	315·8	318·0	Onslow
1917	319·1	315·8	314·1	312·0	312·4	311·0	311·0	309·1	314·1	315·4	315·1	321·1	Catamarca
1918	316·9	316·3	314·7	313·0	308·7	308·3	307·9	309·8	312·0	316·3	317·4	319·1	Bourke
Station of maximum	Bourke	Boulia	Onslow, Boulia	Onslow	Corumba	Corumba	Corumba	Corumba	Daly Waters	William Creek	Eucla Boulia	Eucla	

FIGURE 43 SEASONAL VARIATION OF DAILY RANGE OF TEMPERATURE

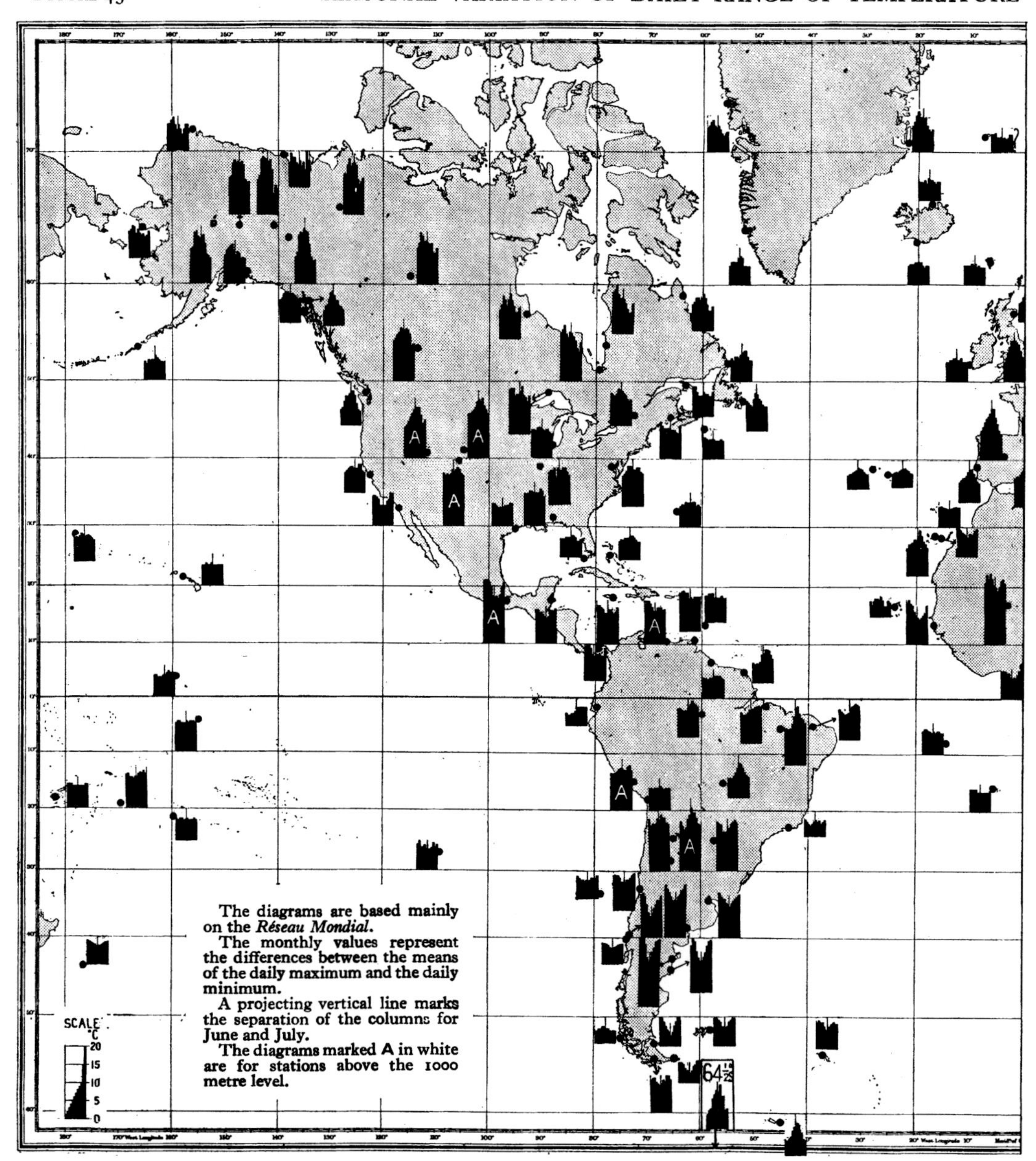

Isopleth diagram for Barnaoul.

Observations from H.M.S. *Challenger*

Diurnal range of temperature over the sea

Lat.	Long.	Range °F	Range t
37° N	168° W	3·1	1·7
30° N	42° W	3·2	1·8
0°	18° W	2·6	1·4
0°	145° E	2·1	1·2
36° S	36° W	2·5	1·4
36° S	100° W	4·0	2·2
63° S	85° E	0·8	0·4
54° S	73° E	1·8	1·0
52° S	118° E	1·5	0·8
47° S	56° E	1·9	1·1

Isopleth diagram for Agra.

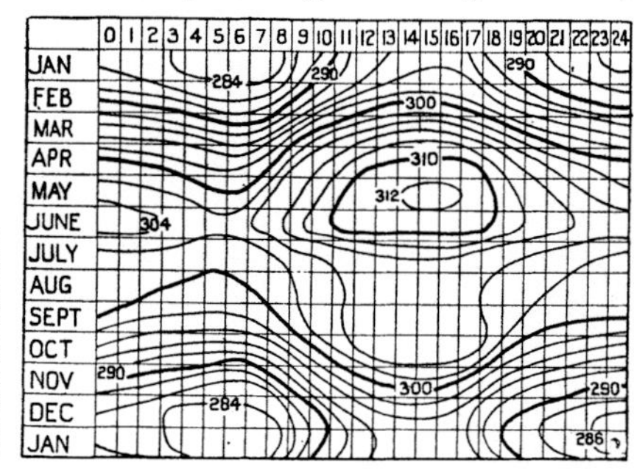

STEP-DIAGRAMS WITH A COLUMN FOR EACH MONTH. Scale 2° C. to 1 mm. FIGURE 44

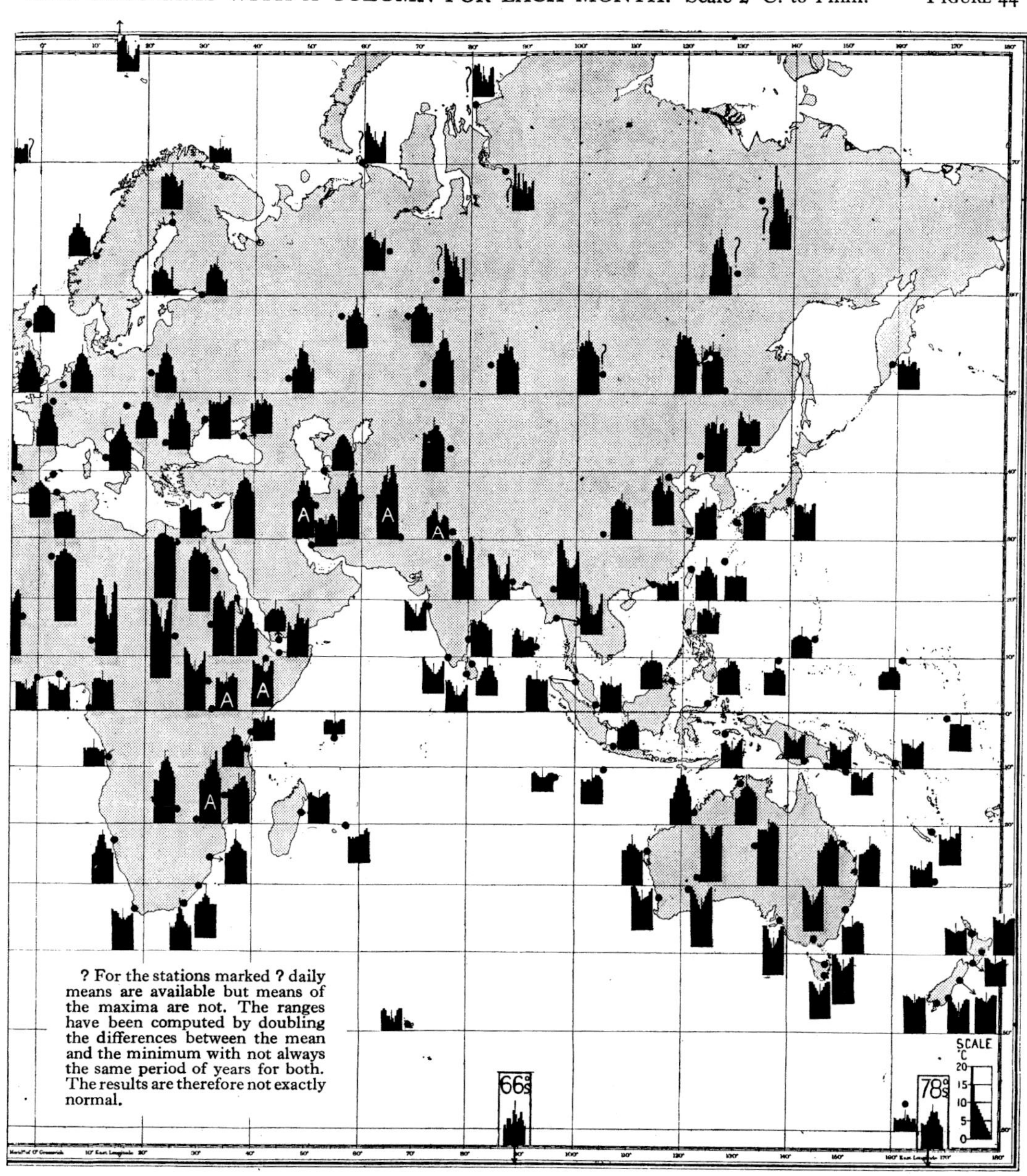

Isopleth diagram for St Helena.

Conversion table from range in Fahrenheit to range in degrees on the tercentesimal scale

°F	t
5	2·8
10	5·6
15	8·3
20	11·1
25	13·9
30	16·7

Isopleth diagram for the Antarctic.

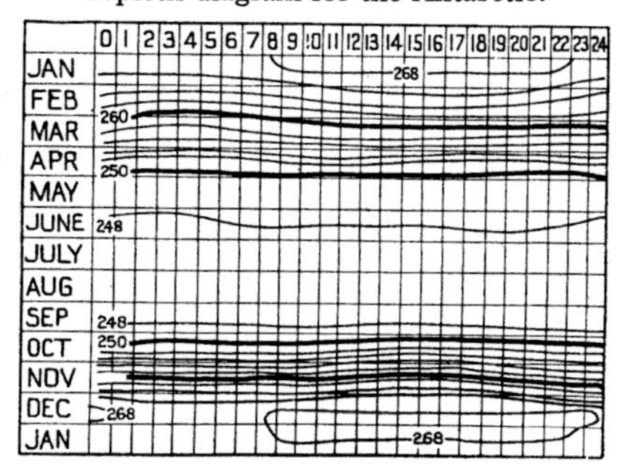

FIGURE 45

Authority: see p. 125.

MEAN SEASONAL RANGE OF TEMPERATURE

In tercentesimal or centigrade degrees.

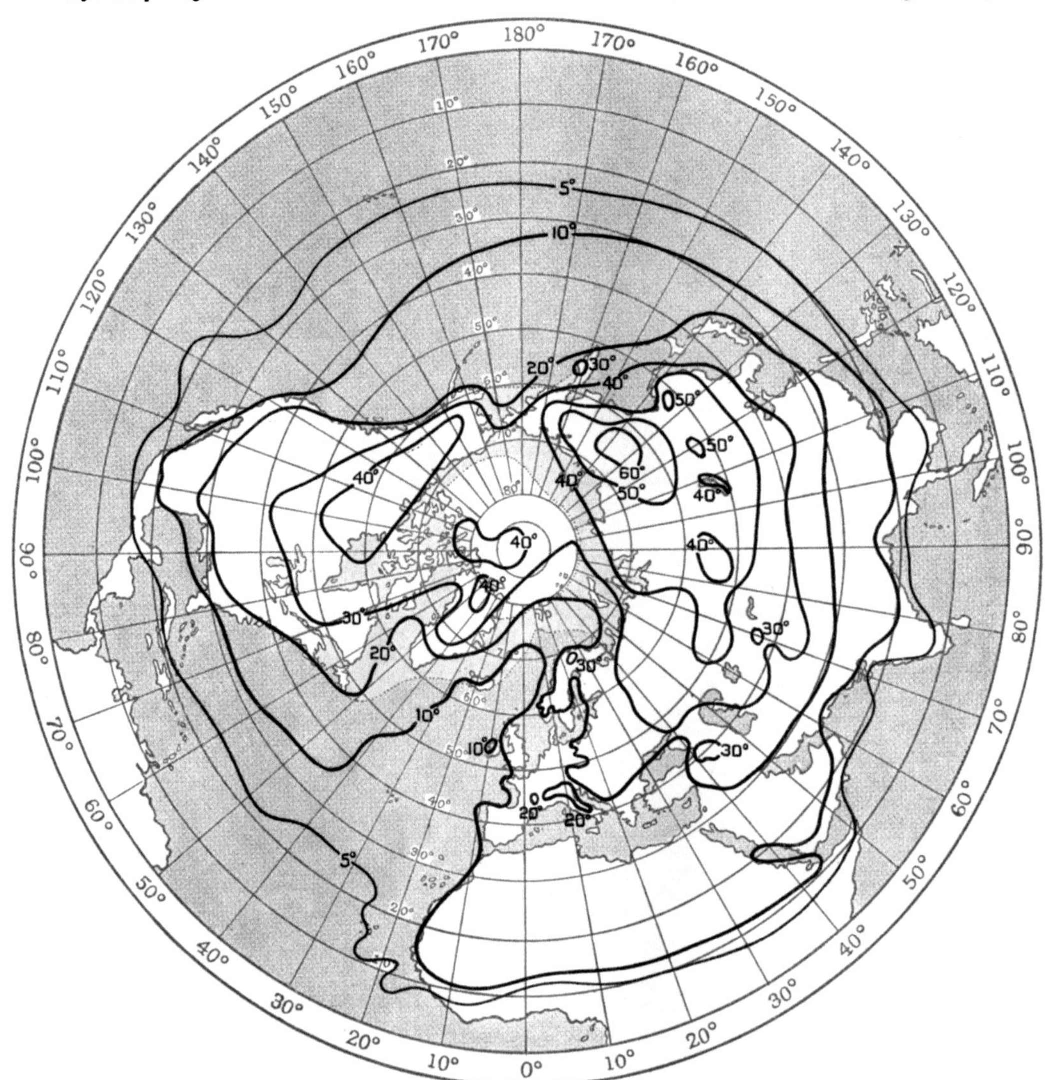

Soundings of the air of the Arctic—temperature and height—Ship *Maud*.

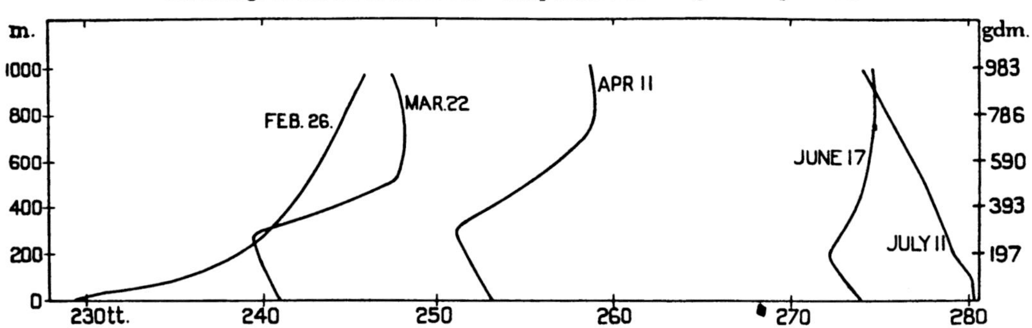

MEAN SEASONAL RANGE OF TEMPERATURE

FIGURE 46

In tercentesimal or centigrade degrees.

Authority: see p. 125.

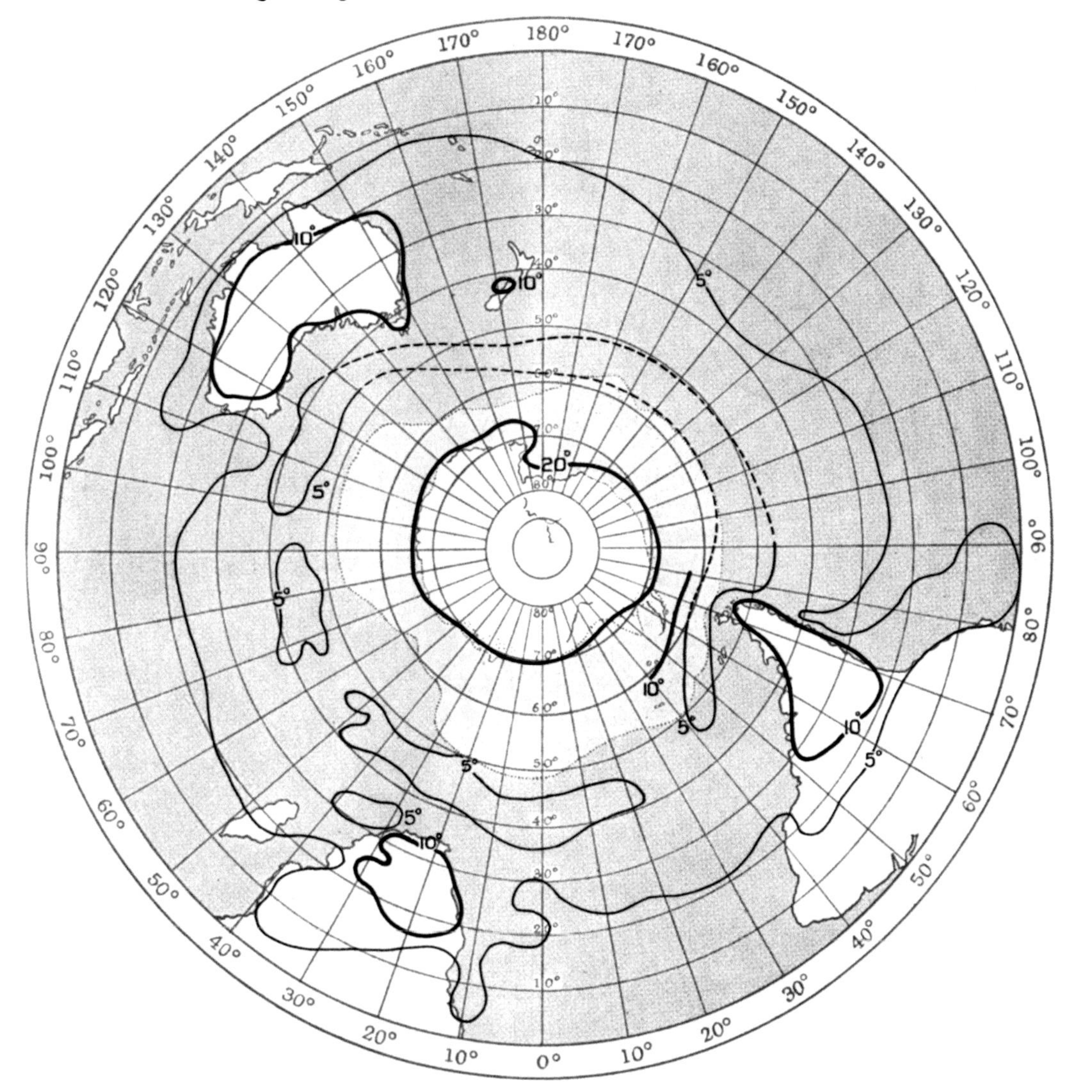

Soundings of the air of the Antarctic—temperature and height—McMurdo Sound.

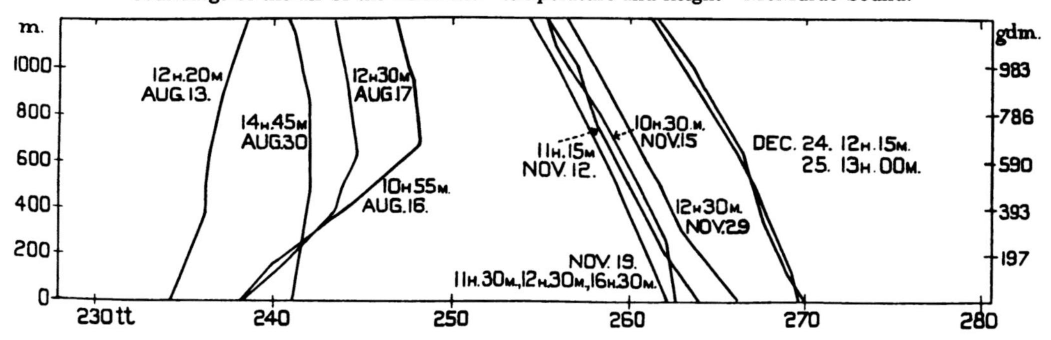

FIGURE 47

MEAN TEMPERATURE OF THE SEA-SURFACE

Fahrenheit scale with equivalents.

Authority: see p. 125.

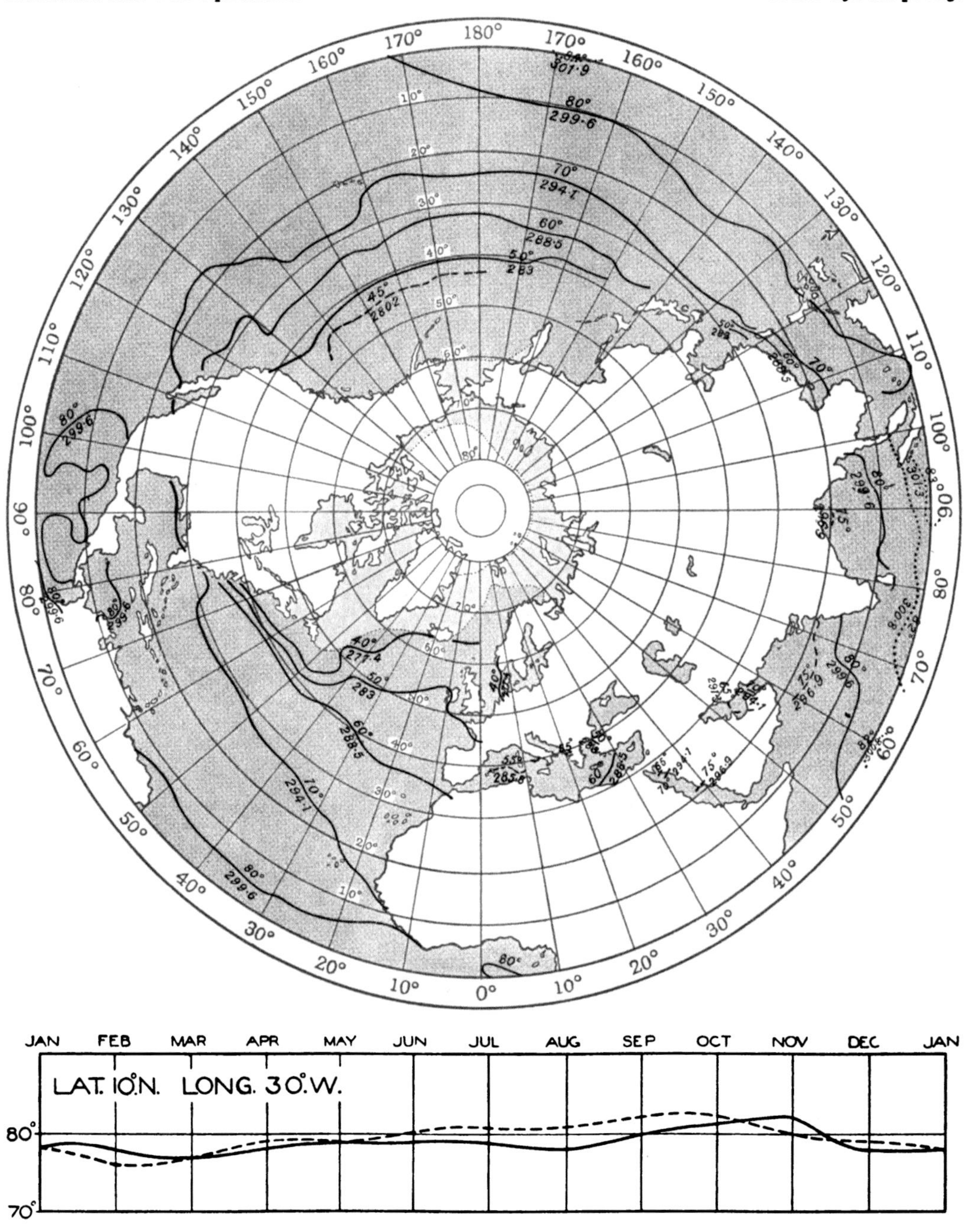

JAN FEB MAR APR MAY JUN JUL AUG SEP OCT NOV DEC JAN

LAT. 10°N. LONG. 30°W.

80°

70°

Normal temperature of the sea (broken line) in relation to that of the air (full line) in mid-Atlantic, 10° N.

MEAN TEMPERATURE OF THE SEA-SURFACE

FIGURE 48

Authority: see p. 125.

Fahrenheit scale with equivalents.

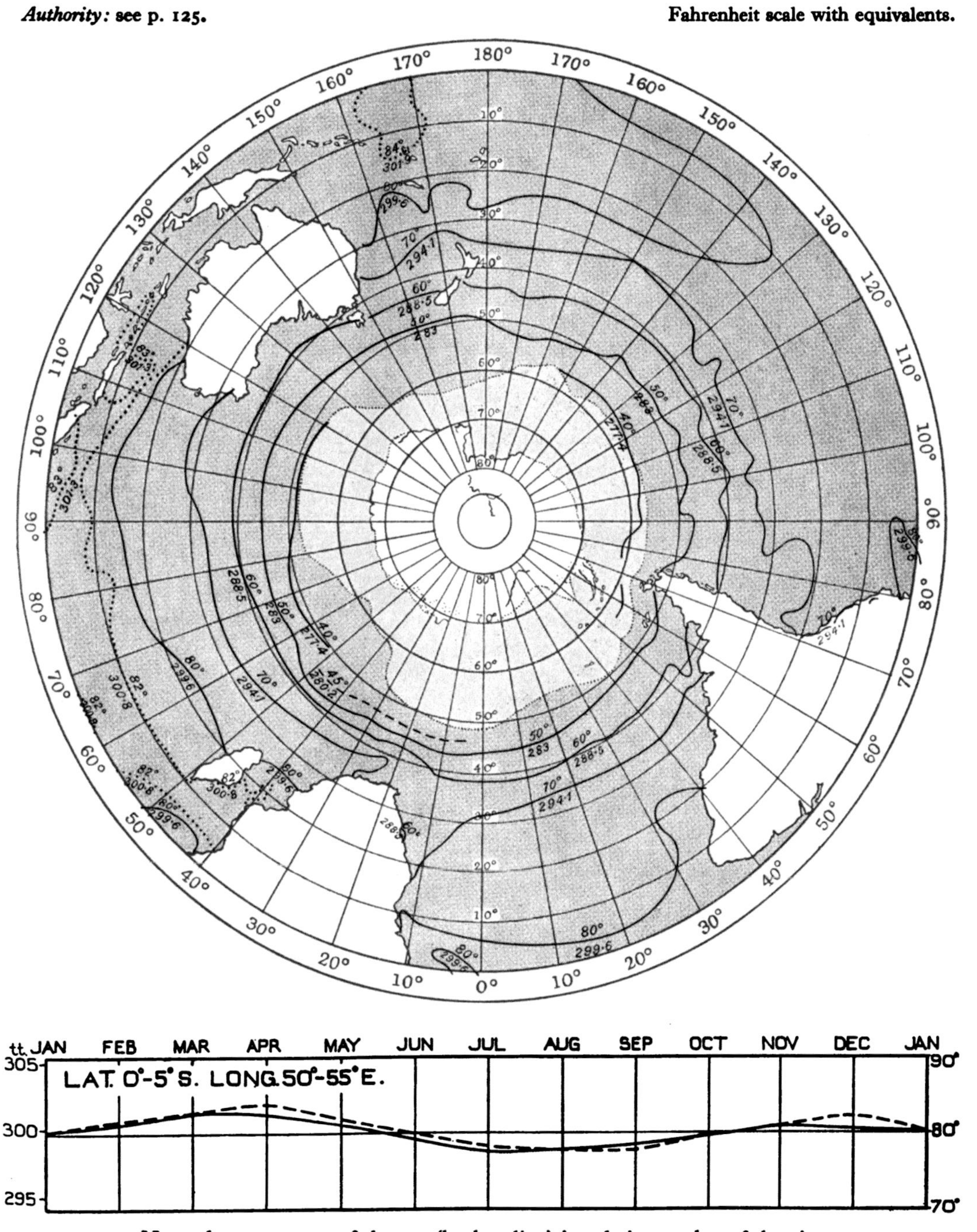

Normal temperature of the sea (broken line) in relation to that of the air (full line) in the equatorial Indian Ocean.

MAY

FIGURE 49

MEAN TEMPERATURE OF THE SEA-SURFACE

Fahrenheit scale with equivalents.

Authority: see p. 125.

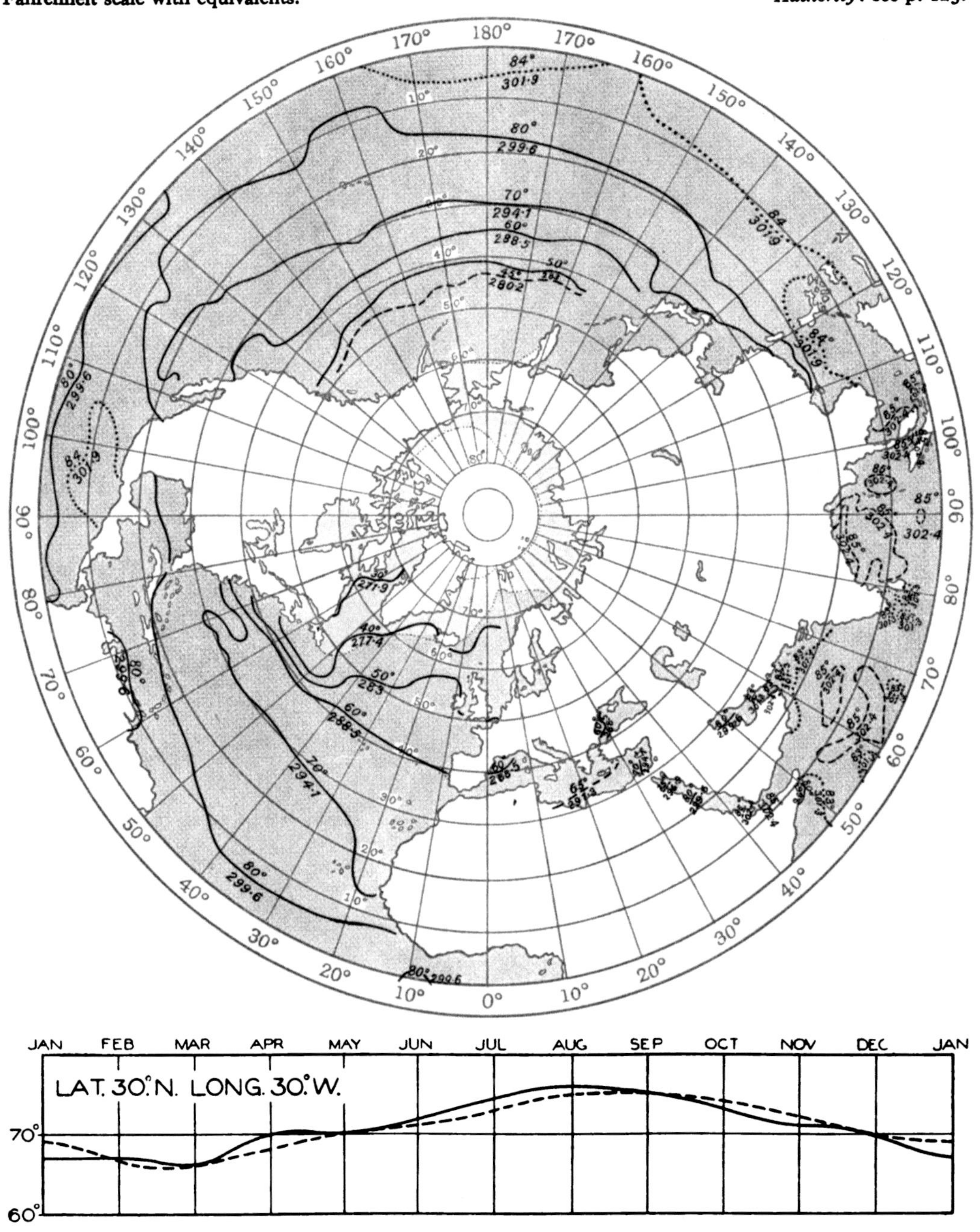

Normal temperature of the sea (broken line) in relation to that of the air (full line) in mid-Atlantic, 30° N.

MEAN TEMPERATURE OF THE SEA-SURFACE

FIGURE 50

Authority: see p. 125.

Fahrenheit scale with equivalents.

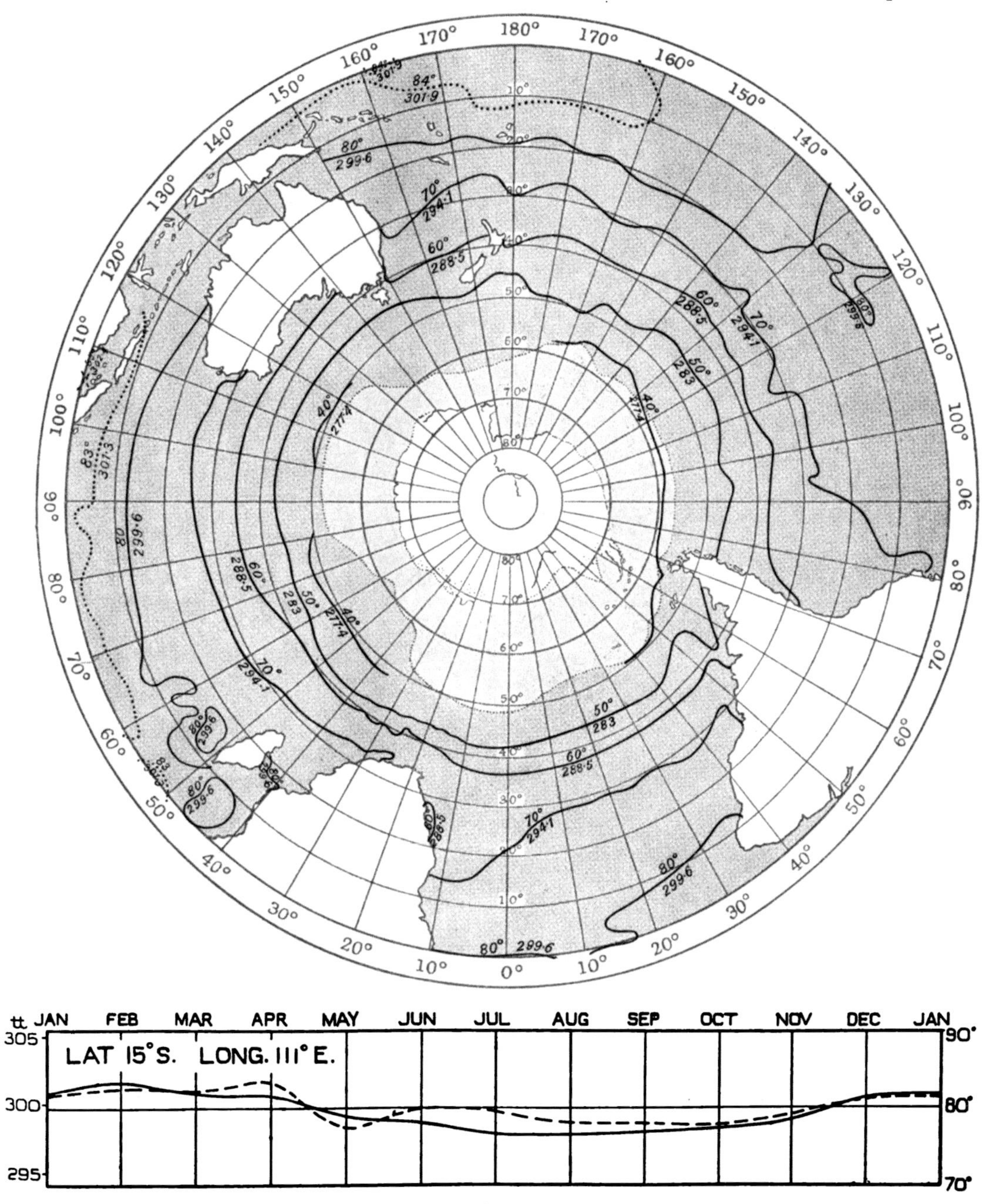

Normal temperature of the sea (broken line) in relation to that of the air (full line) in the Indian Ocean, 15° S.

FIGURE 51

MEAN TEMPERATURE OF THE SEA-SURFACE

Fahrenheit scale with equivalents.

Authority: see p. 125.

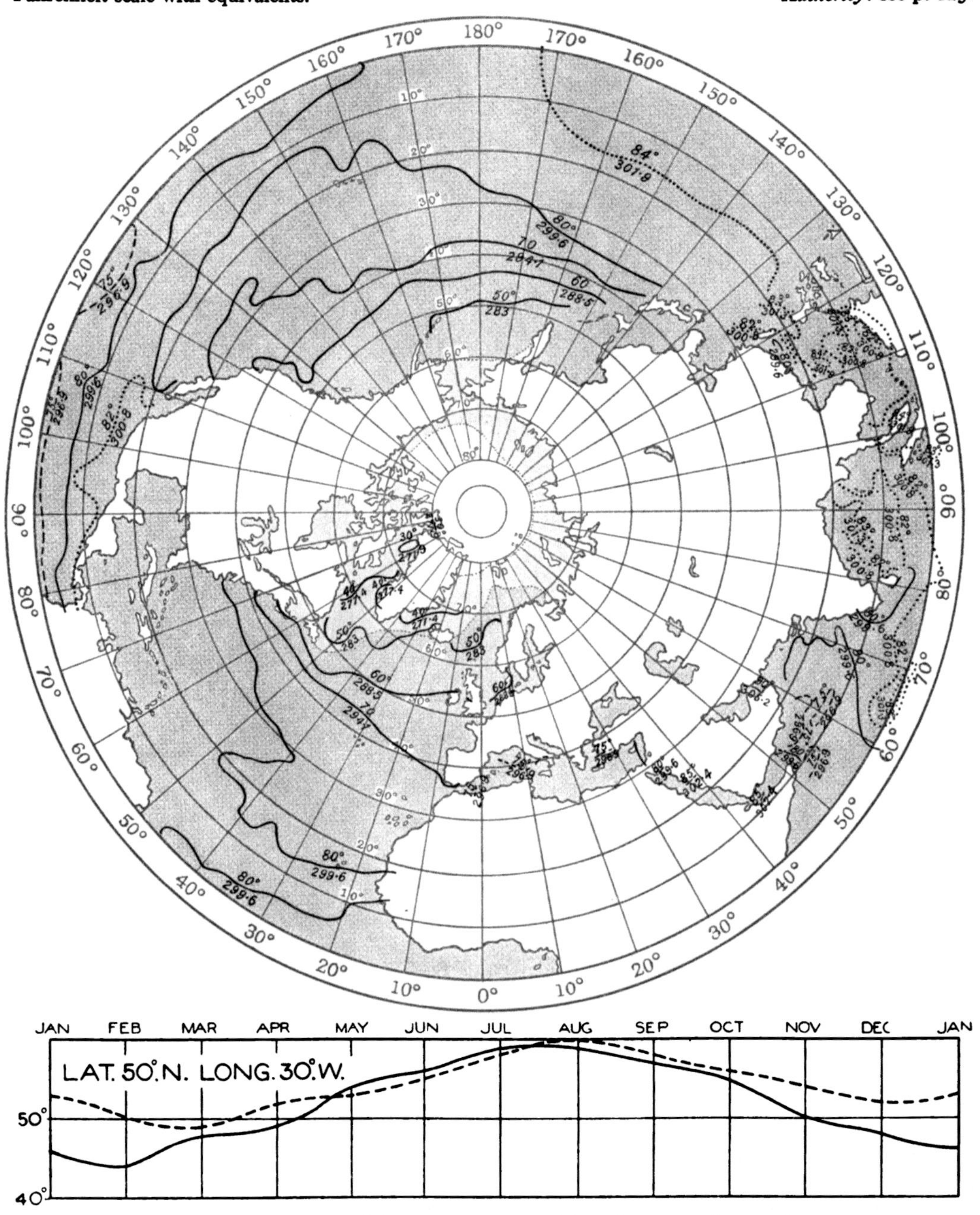

Normal temperature of the sea (broken line) in relation to that of the air (full line) in mid-Atlantic, 50° N.

MEAN TEMPERATURE OF THE SEA-SURFACE

FIGURE 52

Authority: see p. 125.

Fahrenheit scale with equivalents.

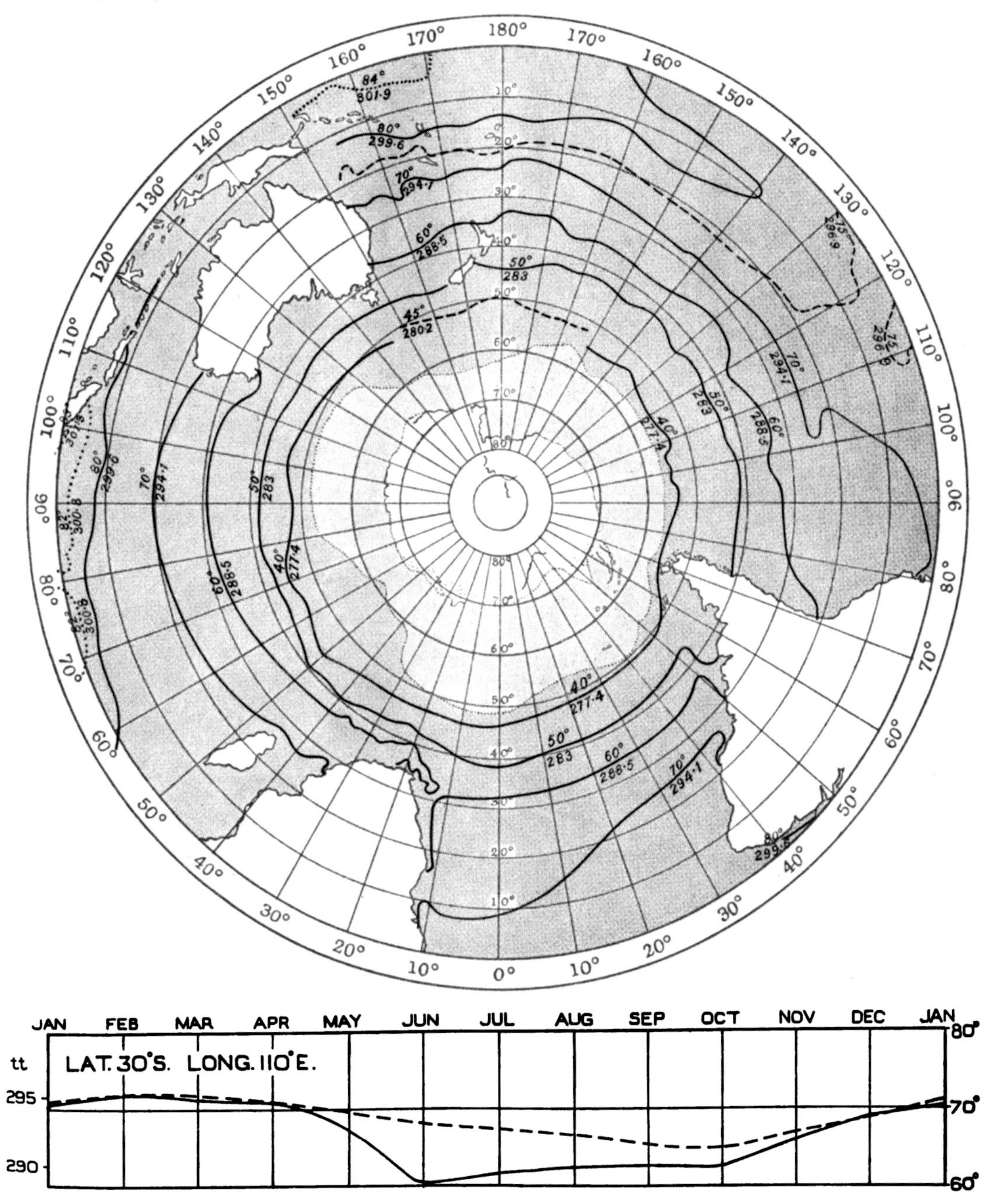

Normal temperature of the sea (broken line) in relation to that of the air (full line) in the Indian Ocean, 30° S.

FIGURE 53

MEAN TEMPERATURE OF THE SEA-SURFACE

Fahrenheit scale with equivalents.

Authority: see p. 125.

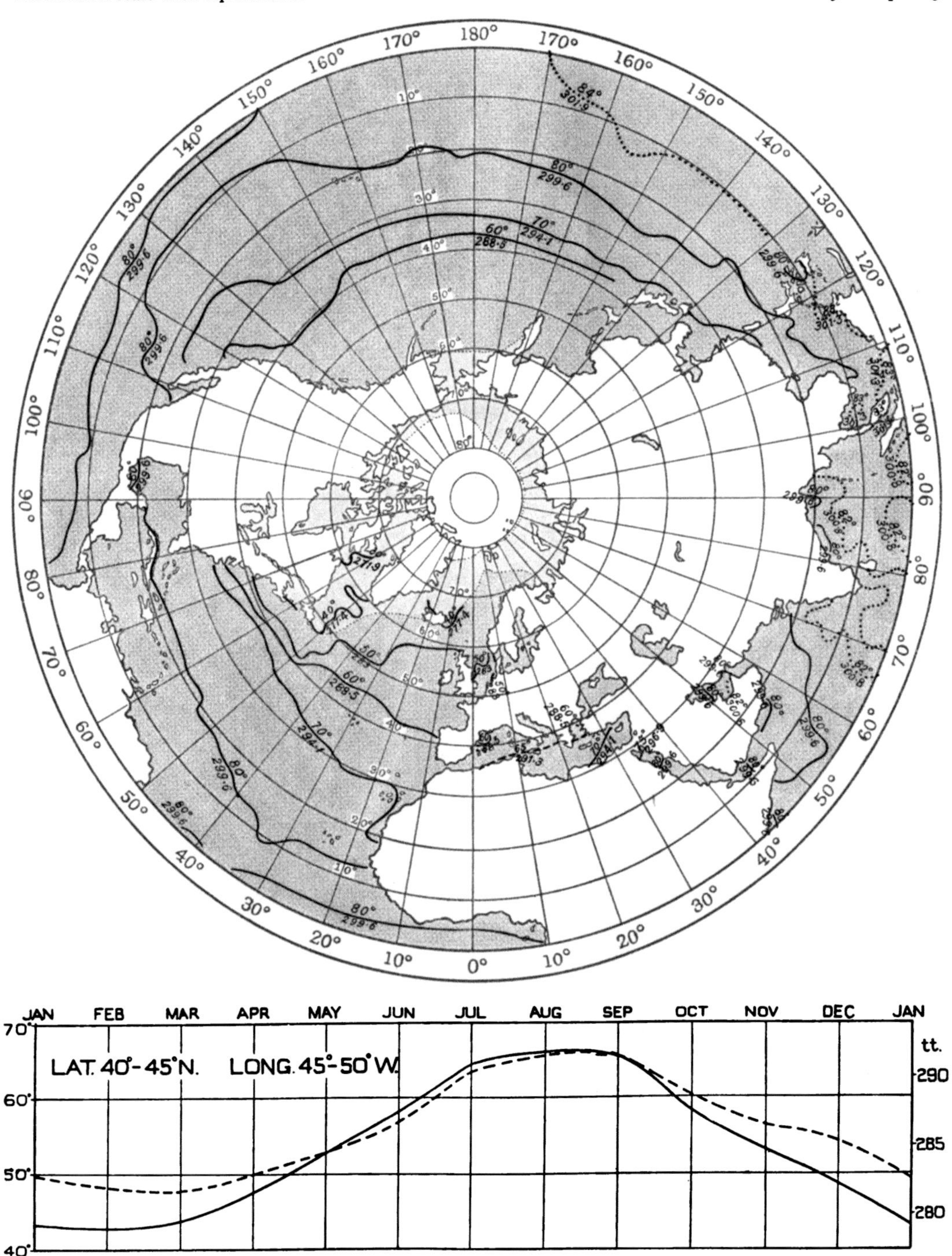

Normal temperature of the sea (broken line) in relation to that of the air (full line) in mid-Atlantic, 40–45° N.

MEAN TEMPERATURE OF THE SEA-SURFACE

FIGURE 54

Authority: see p. 125.

Fahrenheit scale with equivalents.

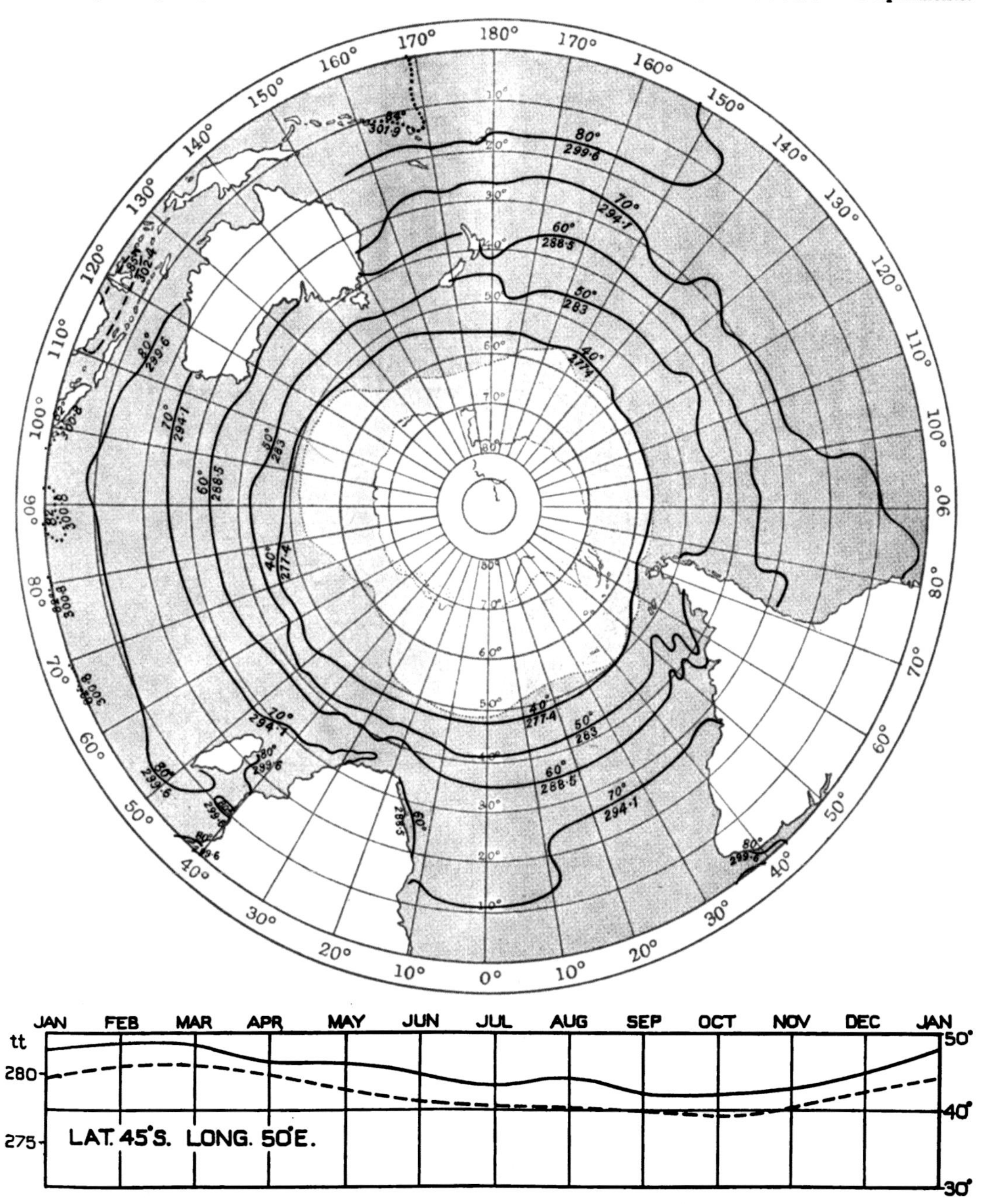

Normal temperature of the sea (broken line) in relation to that of the air (full line) in the Indian Ocean, 45° S.

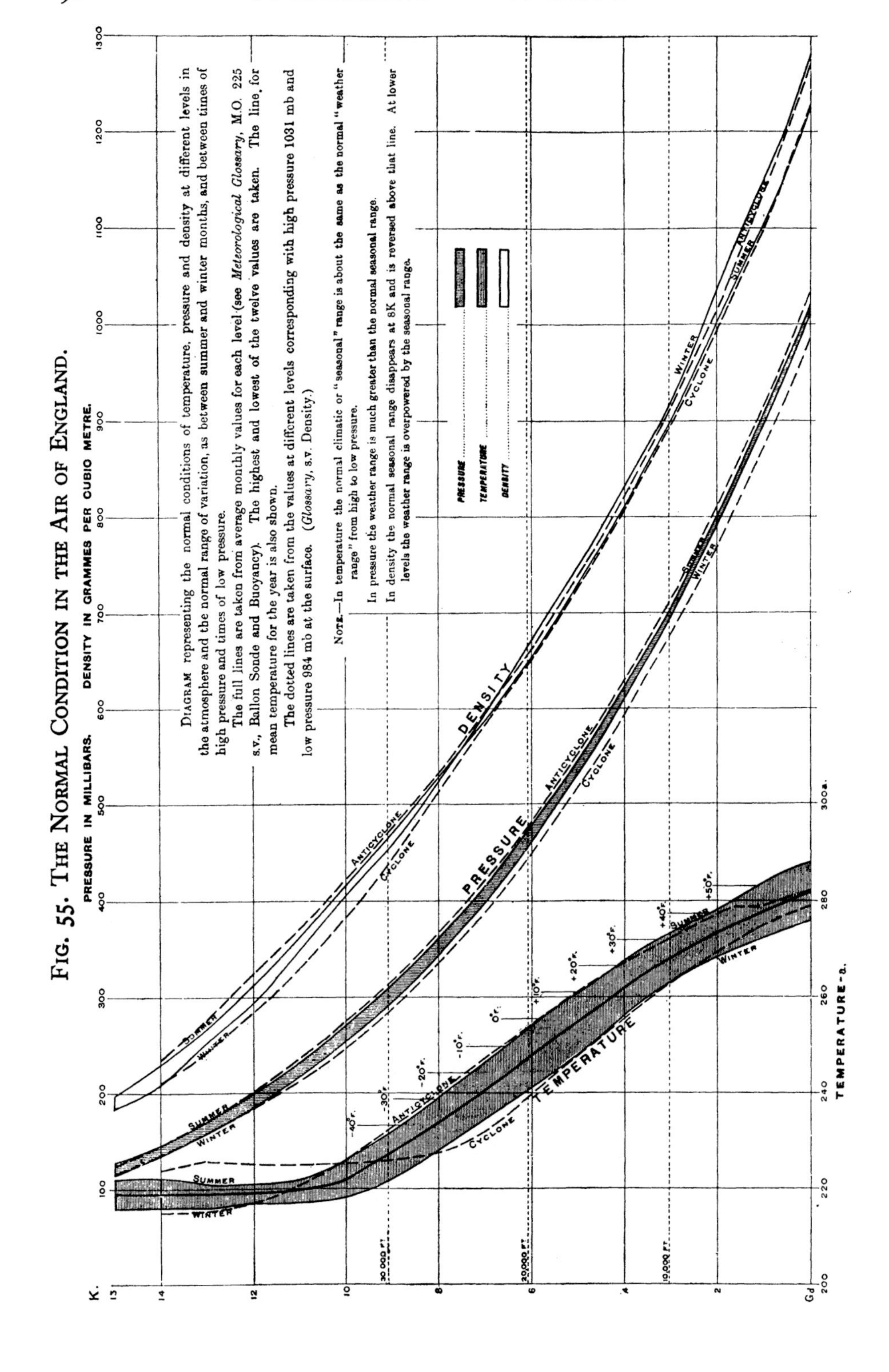

FIG. 55. THE NORMAL CONDITION IN THE AIR OF ENGLAND.

TEMPERATURES OF THE UPPER AIR

We pass now to display the information which is available to us for the representation of the results of observations of the upper air. We begin by representing the temperature referred to the ordinary or geometric height above sea-level because that was the practice in the earlier days of the systematic observation of the upper air, and a large amount of information is available on that basis. We have given already in volume I (fig. 104), as an example of the use of the polygraph, a composite diagram of the earlier observations of temperature of the air above a number of points in England. The diagram shows clearly the differentiation between the troposphere in which temperature falls off consistently with height, and the stratosphere in which the changes of temperature with height are at least notably smaller. We used to regard the stratosphere as everywhere isothermal in the vertical but now it seems to show a gradual recovery, a general counterlapse, of temperature up to and beyond 20 kilometres. The polygraph of the observations of Rotch and Teisserenc de Bort over the Atlantic (fig. 56) tells a similar story.

We reproduce in fig. 55 a diagram compiled in 1917 from figures given in W. H. Dines's memoir on *The Characteristics of the Free Atmosphere*, which exhibits the normal condition of the upper air over England as regards temperature, pressure and density, showing by shading the range between summer and winter, and, by separate lines, the range of conditions between anticyclones (high pressures at the surface) and cyclones (low pressures at the surface).

One of the first points that claims attention in the diagram of normal conditions, on comparing it with the polygraph, is the continuous change of lapse-rate of temperature between the troposphere and the stratosphere. The change in separate ascents is much more sudden than in the mean value. The result is reached by associating together for the purpose of a mean value occasions when the discontinuity is at different heights. Obviously, to take mean values for the several heights obtained in the ordinary way is not a satisfactory method of dealing with the results of observations which cover discontinuities. In fact the representation of the tropopause is necessarily obliterated in spite of the importance which is attached to it in meteorological works, as for example in the memoir on the *International Kite and Balloon Ascents*[1] by E. Gold.

But in spite of these disadvantages the diagram presents some interesting features. First, throughout the troposphere the fractional variation of temperature between summer and winter or between high pressure and low pressure is much greater than the fractional variation of density at the same height. At 6 kilometres it is as much as fifteen parts in two hundred and fifty, 6 per cent.; the change in pressure is about the same, while at the same level the change in density is less than 2 per cent. The graphs of density for summer and winter cross one another near to 8 kilometres. Conditions are different in the stratosphere; there the variation of temperature is fractionally the least important.

[1] Geophysical Memoir, No. 5. M.O. publication, No. 210 e, London, 1913.

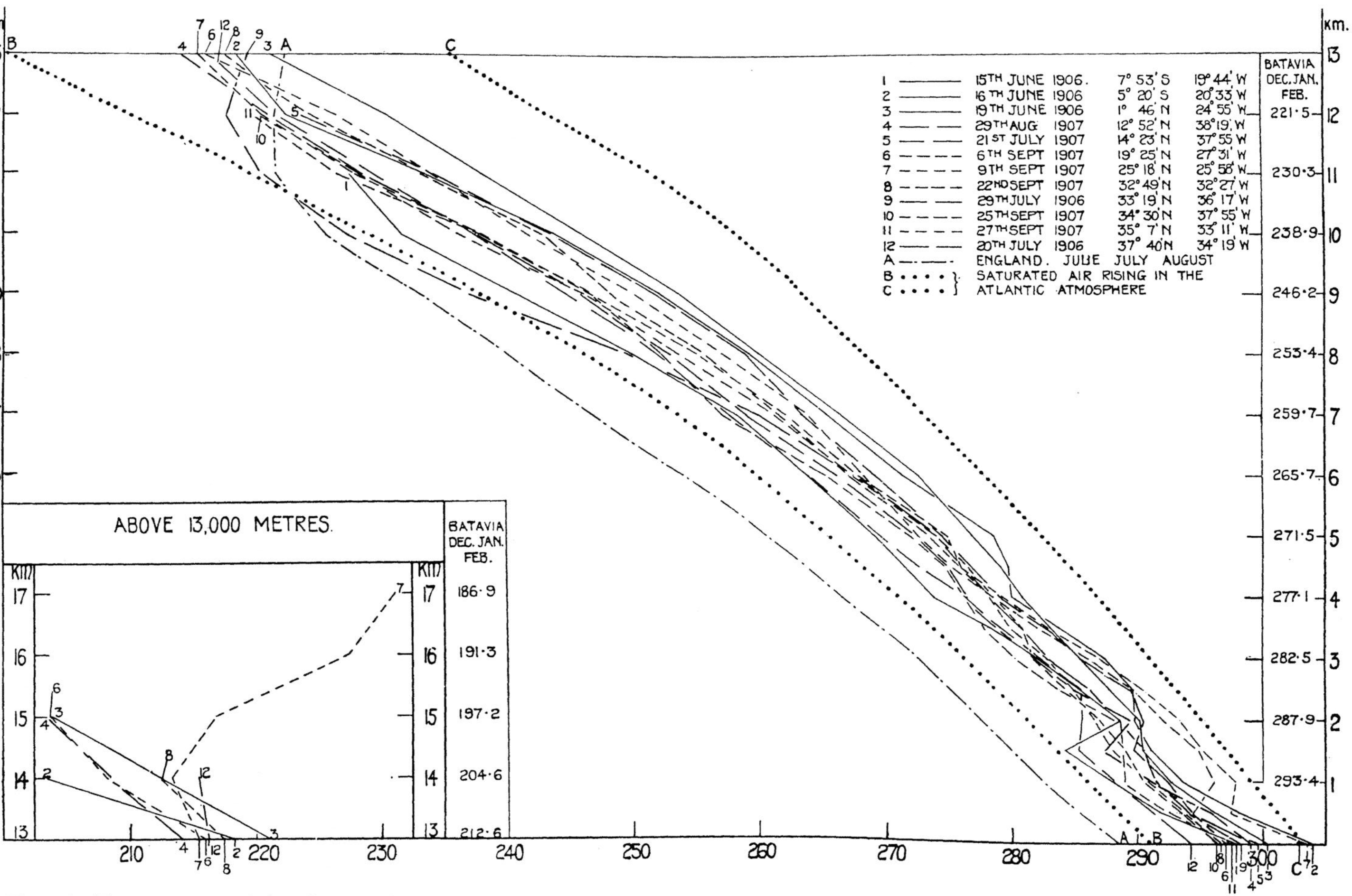

Fig. 56. Temperature of the air over the Atlantic Ocean in June, July, August and September (after Rotch and Teisserenc de Bort), with reference lines for (*A*) South East England, (*B* and *C*) saturated air rising in the Atlantic atmosphere from the surface with temperatures 291 tt and 303 tt. Figures for Batavia in Southern summer are also given at the side.

The change of temperature at the surface between cyclone and anticyclone is much less than the seasonal change, not more than a quarter of it, whereas the comparison is reversed in the differences of pressure at the surface. The conclusion is confirmed by the practical equality of the range of density at the surface between cyclone and anticyclone and between summer and winter.

Within the troposphere there is not much difference with height in the range of temperature; the seasonal change of pressure increases with height, that of density becomes zero and changes sign.

In contrast with the smooth curves of fig. 55, we notice the striking irregularities in the polygraph of Teisserenc de Bort's soundings over the Atlantic with which the smooth curve for England has been plotted, lower in temperature up to the stratosphere than any one of the graphs of the polygraph. We notice first that the tropopause is not always reached and that the air of the English stratosphere in summer is warmer than air at the same level over the tropical Atlantic, but the feature to which attention ought most to be directed is the large number of "inversions" below the level of 5 kilometres and particularly at about 2 kilometres. Initially the lapse-rate is very large, often as much as the adiabatic rate for dry air, but above that surface-layer is a different stage of the atmosphere where the whole scheme is on a warmer scale as though the upper layer was separate from the lower layer and distinguished from it by relative warmth. The temperature relation across the boundary marks a change almost sudden enough to be called a discontinuity. Two marked changes in lapse-rate are to be found at about 4 kilometres, but the majority are much lower, namely at 2 kilometres. We note these because soundings with that characteristic are found more or less frequently at all the stations. The origin of the double layer is one of the first problems in the study of the upper air. We notice that, for the steps which do not cross one of these discontinuities, the slope of the curve is of a single type to which nearly all the soundings conform. So much has that idea impressed itself upon explorers of the upper air that in obtaining mean values from a number of soundings some of them prefer to work from the mean temperature at the surface with a mean lapse-rate, rather than to take the means of the "brut" figures. But E. Gold recommends that means should be formed from the *actual temperatures recorded*[1].

Further interesting points in the results obtained at a single station may be found in the memoirs of W. H. Dines and E. Gold to which reference has been made already. We are for the present interested more closely in the combination of the observations from various places. Before the aeroplane age there were very few opportunities of soundings with ballons-sondes sufficiently near together to sketch out the relations of changes of pressure and temperature in the upper levels to those at the surface. On two occasions when observations had been made simultaneously in England, Scotland and Ireland, it has been possible to combine them and the combination leads to some interesting results.

[1] Geophysical Memoir, No. 5, p. 71.

TEMPERATURES AND PRESSURES IN A BLOCK OF ATMOSPHERE FIFTEEN MILES THICK OVER A PORTION OF THE BRITISH ISLES, JULY 27TH AND 29TH, 1908

The lower pair (numbered 3, 4) are for 27th, the upper pair (1, 2) for 29th July. On the left the view is from NE, on the right from SE. The models stand on a map, the back line in 2 and 4 is from Limerick to Crinan. The front line of the base in 3 is from Petersfield through Pyrton Hill (Oxon.) and Manchester to Crinan and in 1 from Pyrton Hill through Manchester to Crinan.

Isotherms are drawn for steps of 5 t from 280 tt to 215 tt on 27th and from 285 tt to 205 tt on the 29th when the surface was warmer, the freezing-point and the stratosphere higher.

The freezing-point 273 tt is indicated by the broad band filling the space between the isotherms of 270 tt and 275 tt. Elsewhere the width of an isotherm is adjusted to cover ½ t.

The change of temperature at the tropopause is marked by the development of the isotherm of 215 tt from a small closed curve over Pyrton Hill on the 27th to a protruding wedge covering the greater part of the area and including the projections of two additional wedges for 210 tt and 205 tt.

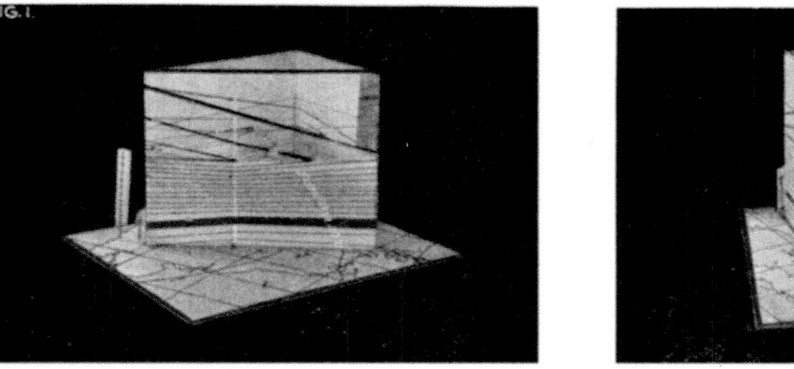

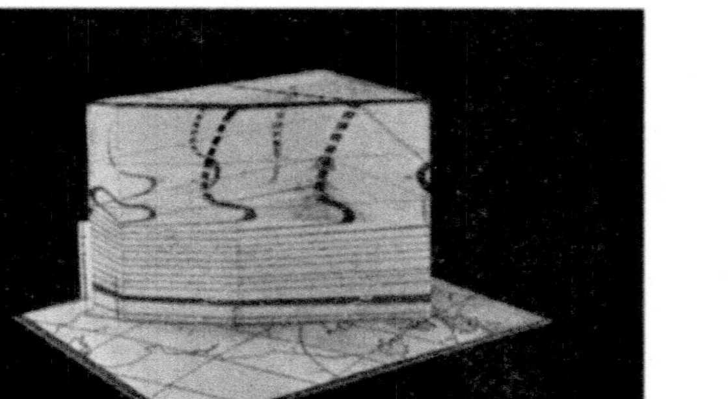

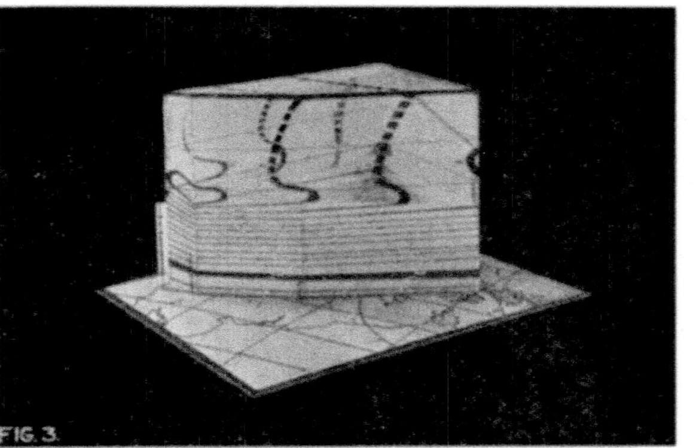

Fig. 57. Four photographs of two glass models of temperature in the upper air, July 27 and 29, 1908.

For the first occasion the initial pressures and temperatures with the subsequent changes are given in the following table[1]:

Simultaneous changes in pressure and temperature at different levels over the British Isles, 1908, *July* 27–29.

Height	Pyrton Hill				Limerick				Crinan			
	July 27		Change 29–27		July 27		Change 29–27		July 27		Change 29–27	
km	mb	tt	mb	t	mb	tt	mb	t	mb	tt	mb	t
10	268	220	+15	+7	279	234	+ 5	−2	272	230	+17	+4
9	311	27	+17	+8·5	321	34·5	+ 8	+4·5	314	34	+18	+8
8	360	36	+17	+7	371	41·5	+ 6	+6·5	363	40	+18	+9
7	417	42	+18	+8·5	427	49	+ 5	+7	417	46	+21	+9
6	478	50	+19	+7	487	57	+ 5	+5	478	53	+18	+7·5
5	548	55	+19	+8	554	64	+ 5	+4	545	60	+19	+6
4	624	63	+19	+5·5	628	70·5	+ 6	+2·5	621	65·5	+19	+6·5
3	707	70	+18	+4	711	78	+ 5	+0·5	705	68·7	+18	+7·8
2	800	76·5	+11	+3·3	803	82	+ 8	+2·5	798	74·5	+15	+8
1	905	80	+11	+1	907	87	+ 8	+1	901	78·5	+18	—
Ground	1003	86	+10	+1	1013	89	+11	+4	1016	89	+11	0

The results are striking; in the interval of 2 days pressure increased all over the area and at all levels but especially at the levels 6 km to 9 km, for which the average changes were 18 mb at Pyrton Hill, 6 mb at Limerick, 19 mb at Crinan. At the same time temperature increased for those levels, 8t at Pyrton Hill, 6t at Limerick and 8½t at Crinan. The change in the situation is illustrated by photographs of models (fig. 57) in which isotherms are seen as lines on a glass enclosure representing a block of atmosphere up to 24 kilometres. The models show an elevation of the tropopause and the spreading at that level of a layer of cold air, most marked at the Southern station but extending over all, and a corresponding increase of pressure.

The table for 1913, the second occasion, is shown below[2].

			Pyrton Hill: Changes						Limerick: Changes						Eskdalemuir: Changes			
	May 6		7–6		9–7		May 6		7–6		9–7		May 6		7–6		9–7	
Height	mb	tt	mb	t	mb	t	mb	tt	mb	t	mb	t	mb	tt	mb	t	mb	t
10	256	221	− 3	+7	+ 6	−9	251	226	−3	+2	0	+2	257	216	−1	+5	+1	0
9	300	22	− 5	+4	+ 5	−2	291	26	−3	0	−1	+2	300	21	−4	−2	+1	+1
8	348	30	− 7	−1	+ 8	+2	336	26	−1	−3	−4	+2	348	29	−4	−4	0	+3
7	403	40	− 8	−7	+ 8	+6	388	33	0	−4	−4	−5	401	38	−1	−4	−1	+3
6	463	48	−11	−8	+ 9	+6	448	41	+1	−6	−1	−1	464	47	−3	−6	0	+2
5	529	54	− 9	−7	+ 7	+6	515	45	+2	−1	0	0	529	55	+2	−5	−3	+1
4	605	61	−11	−8	+11	+7	592	50	+1	+2	−1	+1	604	57	0	0	1	0
3	689	67	− 9	−8	+ 7	+7	677	57	+2	+2	−2	+1	690	63	−1	0	−2	−1
2	781	72	− 5	−5	+ 4	+2	773	63	−3	+3	+2	0	783	70	0	−2	0	−1
1	884	78	− 1	−6	0	+6	883	70	−6	+2	0	0	885	77	+3	−5	0	+2
0	1000	—	+ 1	—	− 1	—	999	—	−8	—	−2	—	1004	—	+4	—	−4	—

Thereupon we note: (1) at Pyrton Hill, without any marked change of pressure at the surface from the initial 1000 mb, pressure showed decrease between May 6 and May 7 throughout the range from 1 km to 10 km and temperature also showed decrease except in the two highest layers. Between the 7th and the 9th, on the other hand, pressure increased throughout the range from 2 km to 10 km and temperature rose in sympathy except again for the two highest layers; (2) at Limerick with a lower and falling surface-pressure the changes during both intervals are much less regular; (3) at Eskdalemuir, with a higher pressure at sea-level, there are also few cases of agreement between changes in pressure and temperature.

[1] *The free atmosphere in the region of the British Isles*, M.O. 202, London, 1909, p. 7.

[2] *Journal Scot. Met. Soc.*, vol. XVI, 1914, p. 172. Two figures have been altered, as misprinted.

SOME NOTABLE OBSERVATIONS THAT CALL FOR EXPLANATION

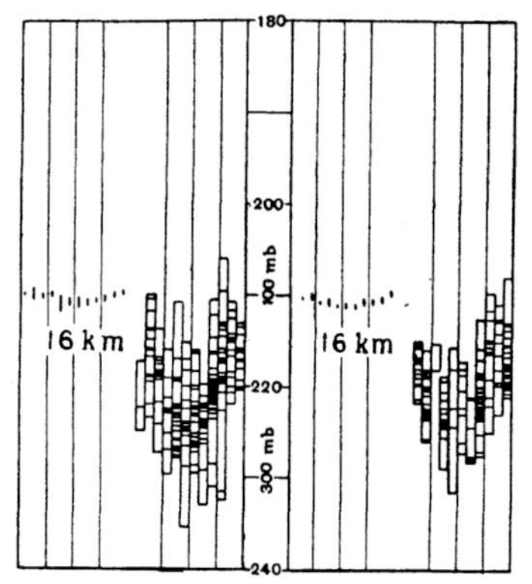

Observations of temperature and pressure at selected heights (km) or gravitational elevations (gkm). A vertical column is allotted for the observations recorded in each of the twelve months. An observation of temperature is indicated by a line across the vertical column. When the same value is to be repeated the depth of the cross-line is increased to correspond therewith. The extreme values for each month are also accentuated by a cross-line of double thickness.

Observations of pressure are not always available with the observations of temperature. The pressure-lines on the diagrams for Lindenberg are for the period 1905–27, and for Munich for a period of 9 years.

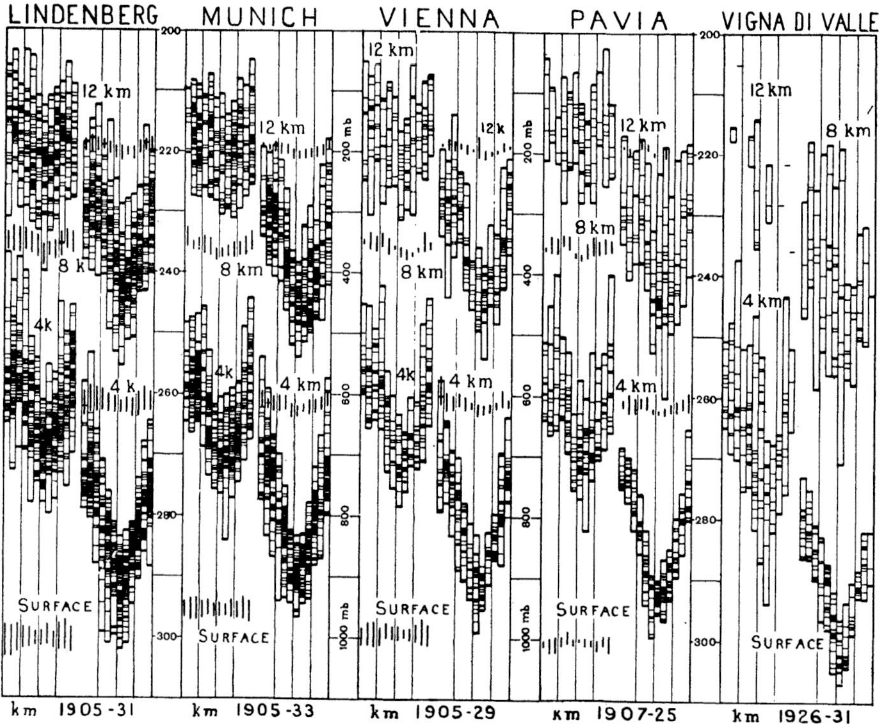

Fig. 58*a*. Temperature and pressure in the upper air at 12 km, 8 km, 4 km (geometric or geodynamic) and at the surface, at five stations in Europe with examples of observations at 16 km at two of the five stations.

The observations are grouped in columns for the calendar months January to December, vertical lines divide the calendar quarters. To avoid overlap of the records in the vertical the observations for 4 km and 12 km are set in separate columns on the left of the records for the surface and 8 km.

The ranges of pressure for the months are shown in the spaces left or right of the respective temperature records by vertical lines joining the extremes of pressure for each month as indicated on the scale 0 mb to 1000 mb.

Scales of temperature and pressure. In the main panel of the diagram the scale of temperature for each station runs downward from 200tt at the top to 310tt at the bottom and the scale of pressure likewise downward from 0 mb at the top to 1100 mb at the bottom. The steps of temperature are marked at successive intervals of 10t. Steps of pressure of 100 mb on the diagram correspond with successive intervals of 10t.

In the lateral columns for the 16 km level, graduation of temperature runs downward from 180tt at the top to 240tt at the bottom and steps of 10t are marked. For pressure zero corresponds with the mark of 200tt and steps of 100 mb range with those of 10t for temperature.

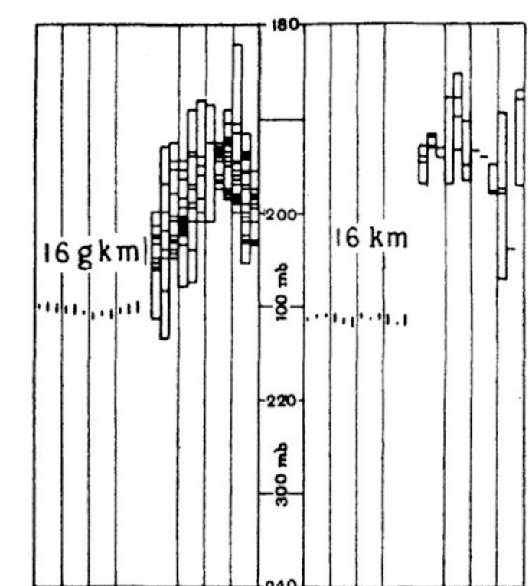

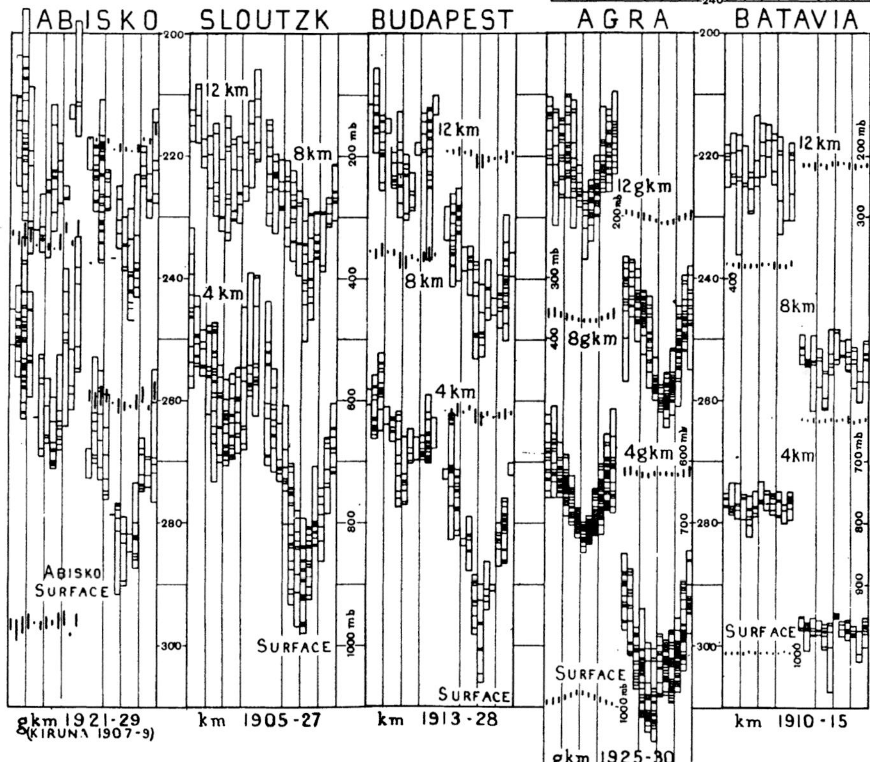

Fig. 58*b*. Temperature and pressure in the upper air at 12 km, 8 km, 4 km (geometric or geodynamic) and at the surface at five stations, from Abisko in the North to Batavia in the South, with examples for 16 km at two of the five stations to illustrate the characteristic variation with latitude of the temperature of the air at that level.

To avoid overlapping of pressure upon temperature in the diagram for Agra the pressure lines have been lowered on the page a distance representing 100 mb. Thus the horizontal line which represents 500 mb for the other diagrams represents 400 mb for Agra and so on for the other pressures. For temperature the scale is the same for all.

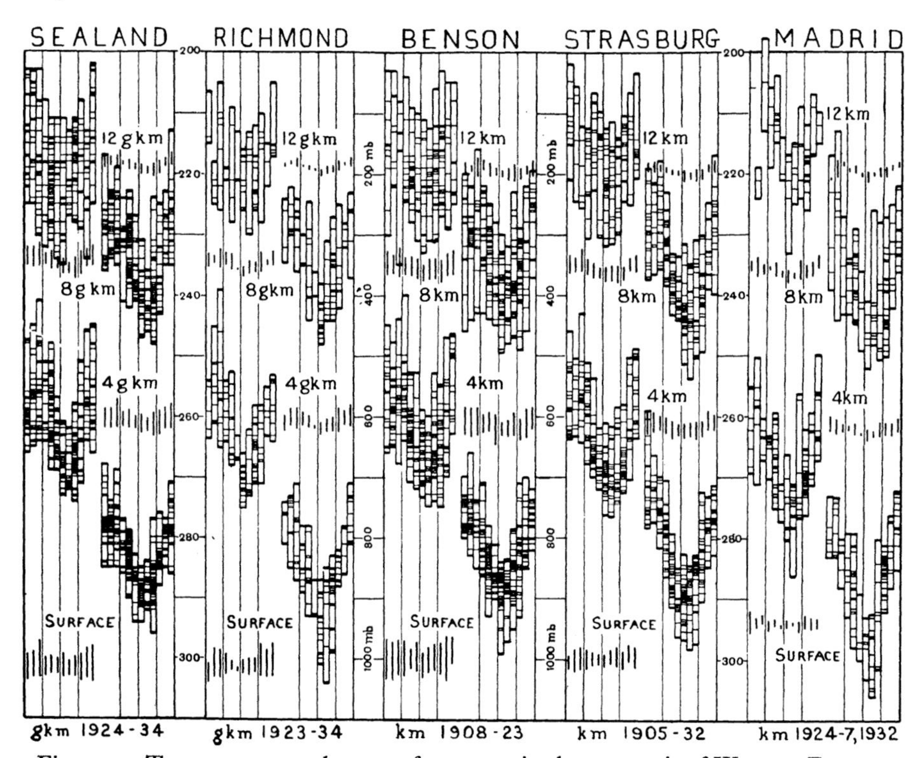

Fig. 59*a*. Temperatures and range of pressure in the upper air of Western Europe at the surface and at 4 km, 8 km, 12 km (see figs. 58*a* and *b*).

Temperatures in the upper air, with the ranges of pressure, at selected levels.

In the original edition of this volume the distribution of temperature of the upper air in a sector of the hemisphere, with the disturbance caused by a cyclone and an anticyclone between latitudes 40° and 60° and by a tropical hurricane in latitude 12° N, was suggested by a view, from above, of a model in cardboard and wax prepared for the British Empire Exhibition of 1924–5. There were also vertical sections in longitude and in latitude through the centres of the cyclone and anticyclone.

Since that time attention has been directed more especially to the outlying parts of the cyclone than to its centre, and the modern reader might regard the absence of any representation of "fronts", polar or other, as a serious omission. The actual situation is suggested by a comparison of the two stereoscopic diagrams of chap. IX, in one of which the movement of the air round the centre is elaborated while in the other it is left to the reader's imagination.

For reasons which are referred to in that chapter meteorologists may come once more to regard the centre of a cyclonic depression as a focus of dynamic interest, but in the mean time it is perhaps desirable to set out for the reader's information some actual facts about the temperature and pressure of the upper air, in place of the model suggested by ideas of what might be,

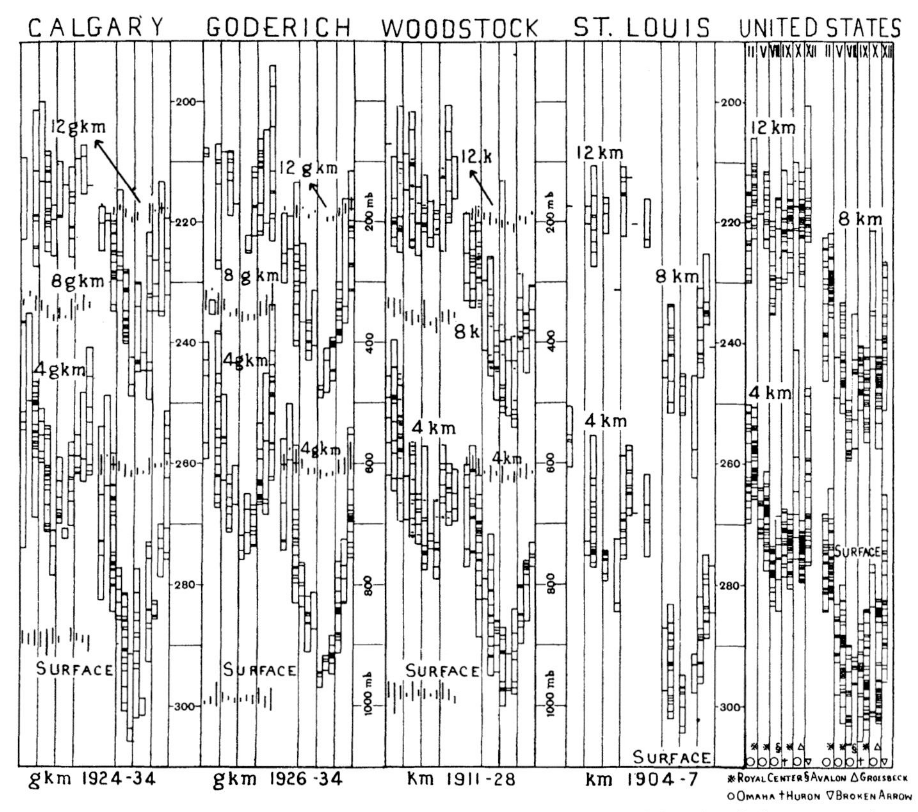

Fig. 59*b*. Temperatures and range of pressure in selected levels of the upper air of Canada and the United States.

before drawing general conclusions from the combination of observations into mean values. Such facts are represented in the panels of figures 58 and 59.

The columns of the diagrams exhibit in comparable form records of the observations of temperature at 4 km, 8 km, 12 km (geometric or geodynamic) as well as at the surface for stations ranging from Prussia through Bavaria and Austria to Italy (fig. 58*a*), from Abisko in the North through Russia, Hungary and India to Java (fig. 58*b*), from North-western England through France to Spain (fig. 59*a*) with in addition corresponding information from Canada and the United States (fig. 59*b*). For four of the stations Lindenberg, Munich, Agra and Batavia corresponding information is given for the level of 16 kilometres in order to exhibit notable variations with latitude.

The observations are as a rule transcribed from the published data, but for Canada the information has been communicated from the Canadian meteorological service through Mr D. C. Archibald, and those for the United States are compiled from the *Monthly Weather Review* with different stations for different months except for St Louis. A diagram giving some additional information about the tropopause and its variations is given as fig. 223 in chapter x, note 14.

TABLES OF MEAN VALUES OF THE TEMPERATURE

grouped according to the

From records in twenty-two localities numbered 1–22

SUMMER

(1) McMurdo Sound. (2) Kiruna. (3) Pavlovsk. (4) Ekaterinbourg. (5) Kutchino. (6) Hamburg. (7) Lindenberg. (8) S England. (9) Uccle. (10) Paris. (11) Nijni-Oltchédaeff.

Station ...		1	2	3	4	5	6	7	8	9	10	11
Latitude ...		78° S	68° N	60° N	57° N	56° N	54° N	52° N	52° N	51° N	49° N	49° N
Longitude ...		166° E	20° E	30° E	61° E	38° E	10° E	14° E	1° W	4° E	2° E	28° E
Months ...		Nov.–Dec.	Aug.	May–July	"Été"	May–July	May–July	May–July	May–July	May–July	May–July	"Été"
gdkm at 90°												
1960	20 km	—	—	—	—	—	—	—	—	—	—	—
1862	19	—	—	—	—	—	—	—	—	—	—	—
1765	18	—	—	—	—	—	—	—	—	—	—	—
1667	17	—	—	—	—	—	—	—	—	—	—	—
1569	16	—	—	—	—	—	—	—	—	—	—	—
1471	15	—	—	—	—	—	**224·7**	—	—	**218·7**	**225·8**	**223·3**
1373	14	—	**224·9**	—	**224·7**	—	**224·1**	**223·7**	**222·3**	**218·5**	**226·2**	**224·3**
1276	13	—	**221·9**	—	**224·7**	—	**221·3**	**222·5**	**222·7**	**216·9**	**225·9**	**224·9**
1178	12	—	**222·0**	**225·8**	**224·8**	**222·3**	**218·1**	**221·5**	**221·7**	**216·2**	**225·0**	**223·2**
1080	11	—	**223·9**	**225·2**	**223·8**	**222·5**	**218·1**	**221·5**	**221·0**	**217·3**	**223·5**	**223·7**
982	10	—	227·8	**226·5**	**223·1**	225·3	221·8	224·0	225·0	221·6	226·1	226·5
884	9	—	234·9	228·9	229·4	230·2	228·5	229·9	231·7	228·6	232·1	231·4
786	8	—	242·6	234·0	238·3	237·0	236·0	236·9	238·3	237·0	239·3	239·7
687	7	—	249·6	240·8	245·5	244·6	243·9	244·2	244·7	245·3	246·7	247·2
589	6	228·2	257·3	247·6	253·5	251·8	251·0	251·5	252·0	252·7	254·3	254·8
491	5	235·8	263·5	254·4	258·7	258·3	257·0	258·4	258·7	259·5	261·0	261·5
393	4	241·2	269·4	260·8	264·4	264·7	262·9	264·6	264·7	266·4	266·7	267·6
295	3	246·4	272·9	266·6	270·3	271·2	267·9	270·2	270·7	270·4	272·1	272·8
197	2	252·2	278·1	272·1	277·5	276·7	272·9	275·0	275·7	276·6	276·7	277·6
98	1	259·0	282·4	278·2	284·5	283·1	278·5	281·0	281·3	281·1	281·7	284·4
	0	265·7	283·9	285·3	293·6	290·0	284·6	287·2	287·3	285·6	284·4	289·0
		—	0·5 km	·03	·268	·14	·017	·116	—	·100	·171	—
Number of records	Top	1	?	8	—	?20	5	42	?	8	6	—
	Base	8	7	31	3	20	14	68	?	11	24	10

WINTER

Months ...		Aug.	Feb.–Mar.	Nov.–Jan.	"Hiver"	Nov.–Jan.	Nov.–Jan.	Nov.–Jan.	Nov.–Jan.	Nov.–Jan.	Nov.–Jan.	"Hiver"
gdkm at 60°												
1958	20 km	—	—	—	—	—	—	—	—	—	—	—
1860	19	—	—	—	—	—	—	—	—	—	—	—
1762	18	—	—	—	—	—	—	—	—	—	—	—
1665	17	—	—	—	—	—	—	—	—	—	—	—
1567	16	—	—	—	—	—	—	—	—	—	—	—
1469	15	—	—	—	—	—	**216·0**	—	—	**209·6**	**214·3**	**219·6**
1372	14	—	**213·6**	—	—	—	**216·0**	**215·6**	**215·7**	**210·6**	**214·7**	**220·2**
1274	13	—	**217·7**	—	—	—	**217·5**	**215·8**	**216·3**	**213·3**	**215·7**	**221·3**
1176	12	—	**216·3**	**218·4**	—	**213·9**	**217·8**	**215·4**	**217·3**	**212·6**	**216·5**	**219·9**
1078	11	—	**214·4**	**218·2**	209·7	**213·1**	**219·3**	**215·9**	**218·0**	**214·1**	**216·8**	**219·0**
980	10	—	**214·1**	**218·7**	212·2	**213·7**	222·1	218·4	221·3	216·8	220·6	**219·1**
882	9	—	221·5	220·8	216·4	219·5	225·1	223·1	225·7	222·5	227·2	221·6
785	8	—	225·9	224·4	223·3	226·7	231·1	229·3	232·3	229·6	234·7	227·1
687	7	—	232·5	230·1	231·8	234·4	237·0	236·6	238·7	236·5	242·3	234·3
589	6	—	239·1	236·8	239·1	241·6	243·7	244·2	245·7	243·7	249·2	241·5
491	5	—	243·7	243·3	246·2	248·1	250·9	251·5	252·3	251·3	255·7	248·9
393	4	—	250·8	248·8	252·7	254·2	257·9	257·8	258·7	258·5	262·2	255·1
294	3	235·8	256·2	254·8	259·1	260·0	263·0	264·1	264·7	265·4	268·0	259·9
196	2	239·0	259·2	260·3	265·5	264·0	269·0	269·0	269·3	270·6	272·6	265·5
98	1	**242·8**	**263·7**	264·0	**268·9**	**266·0**	273·4	273·1	272·7	**273·8**	275·6	268·2
	0	**238·0**	**254·0**	264·4	**265·4**	**265·2**	274·9	274·3	277·7	**273·5**	277·2	270·3
		—	0·5 km	·03	·268	·14	·017	·116	—	·100	·171	—
Number of records	Top	1	?	15	—	5	1	20	?	2	8	—
	Base	4	23	31	5	12	10	44	?	8	28	10

The values set against the height marked 0 are intended to refer to mean sea-level, but in some cases the temperature at ground-level is given and the height of the ground, in kilometres, is then marked beneath the temperature.

OF THE UPPER AIR AT STATIONS ON LAND

conditions of summer and winter.

arranged according to latitude irrespective of hemisphere.

SUMMER

(12) Strassburg. (13) Vienna. (14) Munich. (15) Zurich. (16) Pavia. (17) Woodstock (Canada). (18) United States. (19) St Louis. (20) Agra. (21) Batavia. (22) Victoria Nyanza.

12	13	14	15	16	17	18	19	20	21	22	Station	
49° N	48° N	48° N	47° N	45° N	43° N	40° N	39° N	27° N	6° S	1° S	Latitude	
8° E	16° E	12° E	9° E	9° E	81° W	100° W	90° W	78° E	107° E	33° E	Longitude	
May–July	May–July	May–July	May–July	May–July	May–July	"Summer"	"Summer"	May–July	Nov.–Jan.	Aug.–Sept.	Months	
											at 0°	gdkm
—	—	—	—	**222·3**	**214·0**	**221·9**	—	—	—	—	20 km	1950
—	—	—	—	**220·7**	**214·5**	**219·8**	—	—	—	**189·1**	19	1852
—	—	—	—	**219·4**	**212·7**	**217·5**	—	—	—	**190·5**	18	1755
—	—	—	—	**219·2**	**211·1**	**215·7**	—	—	189·5	197·1	17	1658
—	—	**222·5**	—	**219·1**	**211·5**	**215·0**	—	—	194·2	202·6	16	1561
222·3	**221·6**	**223·2**	—	**218·7**	**210·4**	216·9	**221·7**	—	200·3	206·8	15	1463
221·3	**221·6**	**222·2**	—	**216·9**	**210·7**	219·4	**217·1**	—	205·1	210·8	14	1366
221·0	**218·0**	**220·7**	**219·0**	**215·7**	**211·8**	223·0	**214·1**	—	213·1	216·0	13	1269
219·2	**215·6**	**219·6**	**220·0**	**217·0**	215·0	226·2	217·0	232·9	222·6	222·6	12	1171
219·1	**217·5**	222·0	**219·7**	220·0	219·7	232·5	222·5	235·3	230·4	231·4	11	1074
224·0	222·4	227·1	223·0	224·2	225·9	239·1	230·1	241·5	238·7	238·9	10	976
230·5	229·6	234·0	228·8	231·6	233·2	245·8	238·6	248·1	246·3	246·1	9	879
238·4	237·1	241·3	237·1	239·7	240·4	252·1	246·7	254·8	253·4	250·7	8	781
245·8	244·4	248·5	245·2	246·9	247·8	258·7	255·5	259·9	260·1	258·0	7	684
252·8	252·5	255·4	252·8	254·7	255·3	264·5	263·1	267·2	265·7	263·4	6	586
259·4	259·4	261·8	258·9	261·9	261·6	270·5	269·6	272·4	271·1	269·2	5	489
265·8	265·7	267·6	266·1	268·3	268·1	276·9	274·7	278·3	276·9	274·7	4	391
271·6	271·0	272·8	271·1	274·3	273·6	283·3	279·6	284·4	282·3	280·8	3	293
276·5	277·2	278·6	276·3	279·9	278·9	289·6	284·1	291·2	287·8	288·4	2	195
281·7	282·9	283·9	280·6	285·9	284·8	294·6	291·7	298·0	293·6	—	1	98
286·5	287·4	—	285·7	291·8	289·3	297·9	297·9	302·5	299·2	296·2	0	
·140	·190	·516	·48	—	·3	—	·167	—	—	1·140		
9	6	14	1	—	1	?	?	4	7	1	Top	Number of records
25	15	59	6	69	16	?	?	30	16	26	Base	

WINTER

12	13	14	15	16	17	18	19	20	21	22		
Nov.–Jan.	Nov.–Jan.	Nov.–Jan.	Nov.–Dec.	Nov.–Jan.	Nov.–Jan.	"Winter"	"Winter"	Nov.–Jan.	May–July	—	Months	
											at 30°	gdkm
—	—	—	—	**219·5**	—	**218·7**	—	—	—	—	20 km	1952
—	—	—	—	**219·4**	—	**217·8**	—	—	—	—	19	1855
—	—	—	—	**222·9**	**211·5**	**218·4**	—	—	—	—	18	1758
—	—	—	—	**218·6**	**212·0**	**217·8**	—	—	189·4	—	17	1660
—	—	**214·0**	—	**217·2**	**212·0**	**217·6**	—	—	192·3	—	16	1563
212·9	**217·3**	**215·5**	214·0	**216·4**	**207·0**	**218·2**	**210·6**	—	197·0	—	15	1465
212·6	**216·9**	**215·4**	215·0	**216·7**	**210·0**	**218·7**	**211·3**	—	203·7	—	14	1368
211·7	**216·8**	**216·0**	216·0	**215·7**	**209·8**	**220·4**	**214·5**	—	213·3	—	13	1270
212·5	**216·4**	**215·7**	209·7	**214·1**	**211·1**	**221·1**	**215·5**	221·4	222·0	—	12	1173
214·1	**216·3**	**216·5**	213·5	**214·3**	215·3	**222·4**	**216·7**	227·0	230·7	—	11	1075
218·3	**217·8**	220·0	221·5	218·6	220·0	224·4	221·4	232·9	239·0	—	10	978
224·6	222·1	224·9	227·4	223·6	226·1	227·9	225·8	238·2	247·2	—	9	880
231·7	228·1	231·4	233·9	230·6	233·2	234·0	230·9	244·0	254·2	—	8	783
239·3	235·1	238·3	241·1	237·8	241·7	240·0	240·2	251·1	260·7	—	7	685
246·9	243·1	245·6	247·9	245·2	249·4	246·9	248·8	257·5	266·3	—	6	587
253·6	250·3	252·6	255·1	253·4	256·2	253·6	254·7	264·1	271·9	—	5	489
260·2	257·4	259·1	261·1	260·0	262·6	260·1	261·5	270·6	277·4	—	4	391
266·4	263·6	265·1	267·2	266·2	268·4	265·2	266·4	277·4	283·1	—	3	294
271·7	269·2	270·5	272·4	271·2	272·9	269·7	270·2	282·7	288·6	—	2	196
274·6	272·1	274·4	276·0	275·0	274·9	271·5	272·3	286·9	293·7	—	1	98
275·6	273·6	—	278·1	276·6	275·5	279·0	273·9	289·4	299·5	—	0	
·140	·190	·516	·48	—	·3	—	·167	—	—	—		
11	8	5	1	—	1	?	?	3	4	—	Top	Number of records
23	21	49	5	44	12	?	?	13	15	—	Base	

Black type signifies a region in which the lapse-rate of temperature is zero (approximately) or negative.

FIG. 60. UPPER AIR TEMPERATURES IN NORTHERN SUMMER.

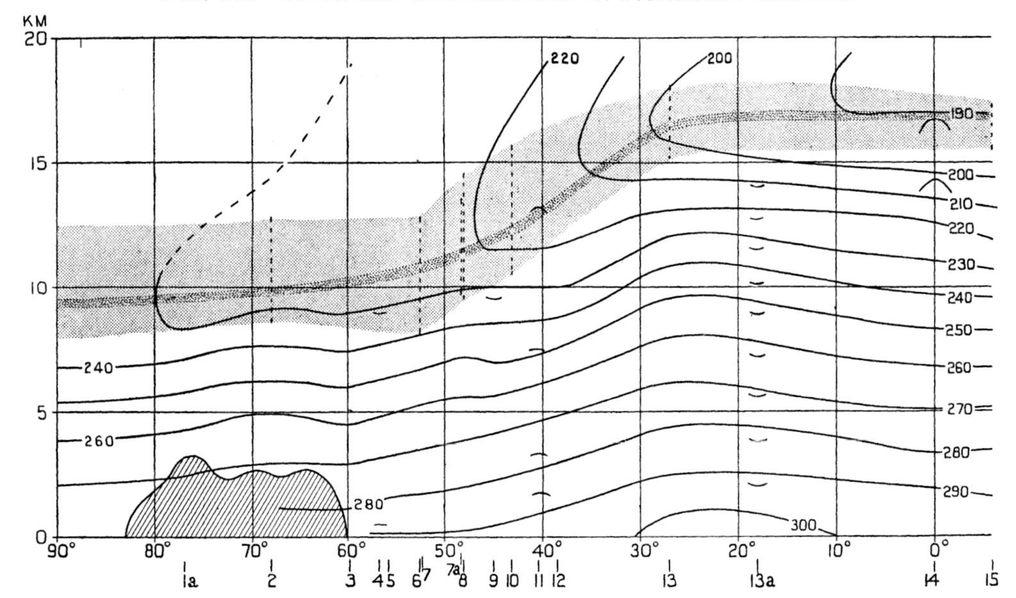

The latitudes of the stations for which observations of temperature have been used in constructing the diagram are indicated by numbers at the foot of the diagram with the following significance:

1 *a*, Arctic Sea; 1 *b*, Antarctic; 2, Abisko; 3, Pavlovsk; 4, Ekaterinbourg; 5, Kutchino; 6, England; 7, Lindenberg; 7*a*, Vienna; 8, Munich; 9, Pavia; 10, Canada; 11, Madrid; 12, St Louis; 13, Agra; 13*a*, Hyderabad; 14, Victoria Nyanza; 15, Batavia.

Mean temperature in the upper air in relation to latitude.

In the absence of co-ordination of the data for stations within the same meteorological system such as the models of fig. 57 may suggest it is perhaps natural to rely upon monthly means though they carry some uncertainty because the total number for any month in any locality is, as a rule, very small and the several ascents may differ considerably in respect of the height reached and the meteorological conditions at the time.

We have accordingly set out on pp. 106–7 tables of mean values grouped according to the vague distinction of summer and winter for a number of localities and have co-ordinated the results in a diagram, fig. 60, in order to show the variation with latitude of the height of the tropopause and other points of interest. The information supplied by the tables for land-stations is supplemented by observations from the sea as set out in the tables of mean values opposite.

The diagrams are arranged to give some idea of the distribution of temperature between the North pole and 6° S in summer and in winter with a little inset of the Antarctic to fill the vacancy caused by the lack of suitable observations in the Arctic winter. Fig. 60 is continued accross the opposite page to suggest a distribution from pole to pole.

We have restricted our information to that obtained from soundings which

FIG. 60. UPPER AIR TEMPERATURES IN NORTHERN WINTER.

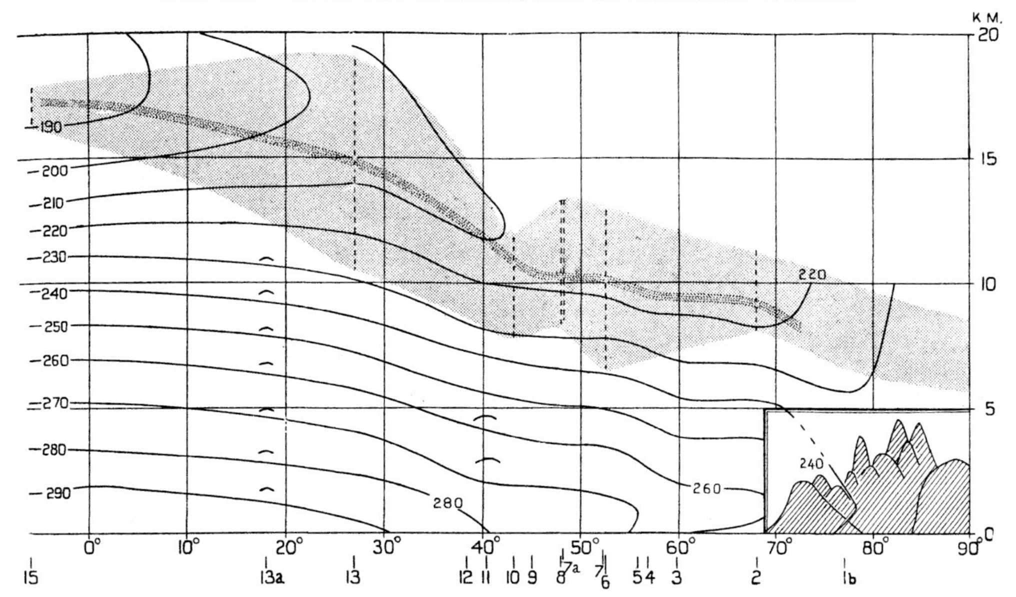

The shaded band indicates the probable position of the tropopause and the range of height at some of the stations is given by cross-lines along the vertical of the station.

Localities where there are exceptional results which cannot properly be incorporated into the run of the diagram without further research are indicated by detached peaks or bends.

Mean values of temperature of the upper air over the sea.

Authority	Hergesell	Peppler	Hildebrandsson		Peppler			Palazzo	
Locality	Arctic	Azores	North Atlantic					Zanzibar	
Latitude	ca 78° N	35–36° N	20–40° N	10° S–20° N	22½–37½° N	2½° S–22½° N	10–14° N	7° S	6° S
Longitude	ca 12·5° E	—	—	—	—	—	38–39° W	39° E	39° E
								1908	
Months	July–Aug.	July	May to September				July	July 30	Aug. 1
km									
15	—	216·9	—	—	208·5	206·9	—	—	—
14	—	221·3	208·7	212·8	209·9	212·5	—	—	—
13	—	224·9	216·3	221·9	213·3	219·4	—	—	—
12	—	230·3	219·5	226·3	217·5	227·6	223·7	—	—
11	224·0	236·7	227·0	232·6	224·9	234·1	234·1	—	—
10	219·5	242·1	235·3	240·7	233·2	242·4	243·4	—	—
9	225·2	248·0	243·8	248·8	241·1	250·3	251·6	—	—
8	234·2	253·5	252·0	256·4	248·4	257·3	258·6	—	—
7	240·4	260·2	259·1	263·6	255·5	263·7	264·6	—	—
6	247·8	268·0	265·9	268·7	261·9	269·4	270·2	—	261·9
5	254·7	271·9	271·9	273·9	268·3	274·6	270·6	—	269·7
4	261·4	280·5	276·7	278·1	273·9	279·0	279·6	276·6	275·0
3	266·7	285·5	282·4	284·7	279·6	284·3	285·8	278·2	278·8
2	271·9	289·2	287·9	288·9	284·6	288·3	290·1	285·4	284·7
1	274·5	291·5	291·7	291·9	287·6	292·2	293·4	290·4	290·7
0	278·3	296·8	297·9	300·0	295·3	299·0	299·8	297·6	297·4
Number of ascents:									
Top	2	1	11	8	4	4	1	1	1
Base	27	3	11	8	22	15	2	1	1

reached the stratosphere. There must now be many observations up to four or five kilometres obtained from aeroplanes which are available over sea and land. Their co-ordination is not yet effective for our purpose.

Diurnal variation of temperature in the upper air.

One of the most remarkable features of the data of temperature for the upper air is the occurrence of conspicuous changes from day to day extending up even to the greatest heights. The tables which we have quoted on p. 101 express changes of considerable magnitude and the ideas suggested by the figures are confirmed by the example of the soundings of the atmosphere on consecutive days as represented in fig. 108 of chapter XIII of volume I. Many diagrams of this kind have been published, conspicuous among them are the results of daily soundings throughout the whole year published by the Observatory of Berlin[1]. We may conclude therefrom that the surface has no monopoly of changes of temperature. For England, W. H. Dines gives the standard deviations of the meteorological elements, which are reproduced in the table on p. 113. That of temperature is least at the surface.

HOURLY SOUNDINGS OF THE UPPER AIR

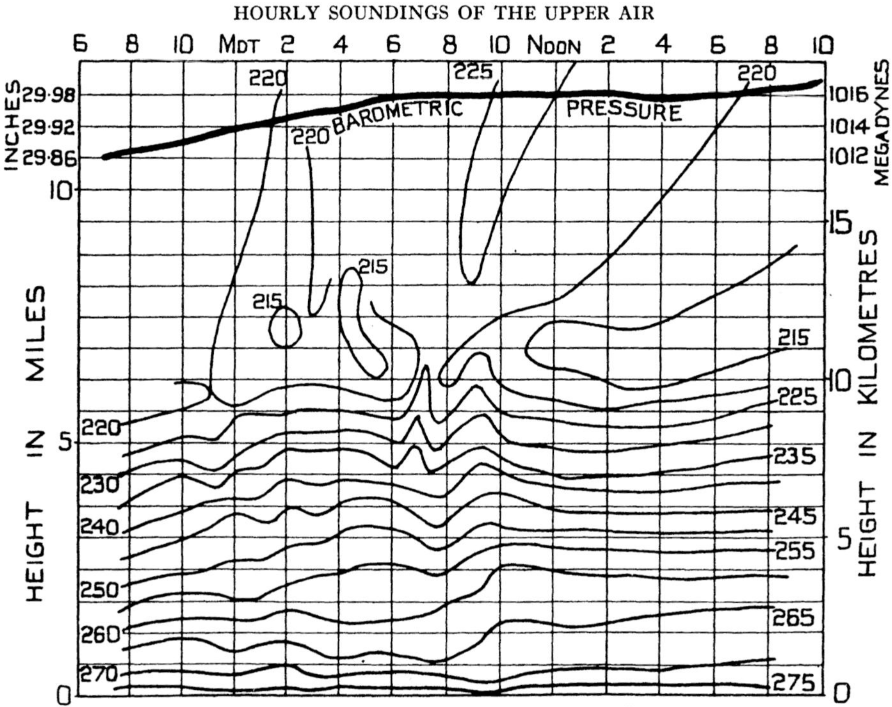

Fig. 61. Temperature of the upper air over Manchester from hourly soundings with balloons, March 18–19, 1910. The thick line shows the pressure at the surface.

Hence we are led to inquire whether the changes in the upper layers arrange themselves in an order of sequence from hour to hour, or from month to month, in a manner similar to the variations of the corresponding elements at the surface. We set out here therefore some data relating to the diurnal and seasonal variation of temperature in the upper air.

[1] *The temperature of the air above Berlin from October 1st 1902 until December 31st 1903*, by Richard Assmann, Berlin, 1904.

For variations within the day we give the diagrams (figs. 61–2) for the two series, each of 25 soundings, with ballons-sonde in 24 hours, carried out by the University of Manchester. They show some relation between change of pressure and change of temperature; but nothing, above a kilometre, that can be called diurnal variation in the sense used about the surface.

J. Bjerknes has made use of consecutive soundings such as these for a more definite purpose. In a memoir published in *Geofysiske Publikasjoner*, vol. IX, 1932 he has made the results of a series of consecutive soundings at Uccle 25 in number on 26th, 27th and 28th December, 1928, the basis of a graphic exposition in the language of fronts of the structure of the atmosphere which may be supposed to have passed over Belgium during the period mainly from

HOURLY SOUNDINGS OF THE UPPER AIR

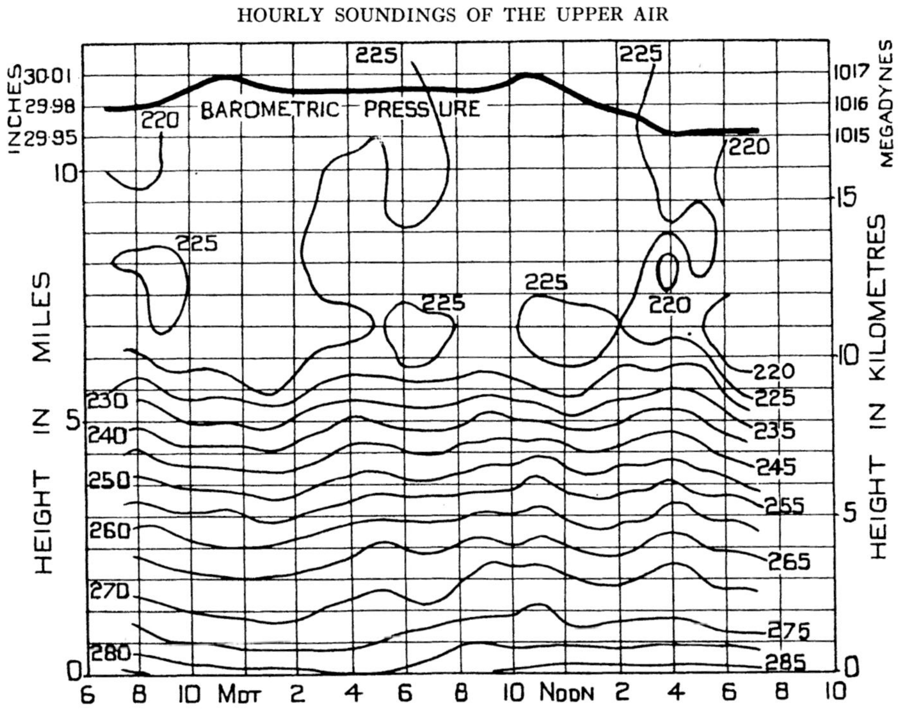

Fig. 62. Temperature of the upper air over Manchester, June 2–3, 1909. (*Q. Journ. Roy. Meteor. Soc.* 1910 and 1911.)

some point north of West. He marks the distinction between upper warm air, lower warm air and the advancing corner of aggressive cold air of the 26th and 27th and analyses in like manner a secondary cyclone of 28th December. The memoir includes a chapter on corresponding perturbations of the tropopause and another on the variations of pressure at different levels. It is supplemented by a discussion in similar terms of the results of seven soundings within 41 hours of 29–30 March of the same year. A more recent series 30–31 December 1930 is discussed in volume XI, no. 4, of the same publication.

We add also tables of the diurnal variation of temperature in the lower strata expressed as harmonic terms of the formula

$$\text{temperature} = A_0 + A_1 \sin(15t + \alpha_1) + A_2 \sin(30t + \alpha_2).$$

Diurnal variation of temperature in the upper air[1].

Height	Lindenberg, Germany				Drexel, Nebraska, U.S.A.				Batavia, Java			
	A_1	α_1	A_2	α_2	A_1	α_1	A_2	α_2	A_1	α_1	A_2	α_2
km	tt	°	tt	°	tt	°	tt	°	tt	°	tt	°
3·0	—	—	—	—	0·7	203	—	—	—	—	—	—
2·5	—	—	—	—	0·6	165	0·2	210	—	—	—	—
2·0	0·21	230	0·50	38	0·6	165	0·1	240	0·24	147	0·50	147
1·5	0·47	198	0·29	82	0·8	143	0·1	285	0·38	207	0·44	155
1·0	0·71	200	0·28	51	1·1	180	0·4	240	0·24	140	0·46	156
0·5	1·11	201	0·31	32	—	—	—	—	0·39	188	0·31	92
Ground	2·97	225	0·50	61	4·9	218	1·3	45	2·79	232	0·88	63

In regard to diurnal variation, we are concerned first with the amplitude of the first component, which is really negligible at 2 kilometres, and the phase angle of the same component which diminishes (i.e. the time of maximum is delayed) as the height of 2 kilometres is approached from the surface; at or beyond that height the figures show no regular relation to time of day.

We notice even here a relation between the change of pressure and corresponding change of temperature in the upper air. The chief component of the variation of temperature at and above the level of 1 kilometre at Batavia is semi-diurnal with an amplitude ranging from ·46t to ·50t, and we find also that the corresponding components for the diurnal variation of pressure at Mount Pangerango (3025 m) have an amplitude of ·80 mb for the 12-hour term as compared with ·09 mb for the 24-hour term. The phase angle of the second term in the variation of pressure is 124°, and indicates a lag of about an hour in the time of maximum of pressure relative to that of temperature.

Seasonal variation.

In the groundwork of the form used for exhibiting the results of soundings by aeroplane in the Upper Air Supplement of the Daily Weather Report of the Meteorological Office, we find a series of reference-graphs of the upper air. The co-ordinates are log(tercentesimal temperature) and log(pressure); the latter gives an indication of height. A shaded area is bounded by graphs marked February and August respectively, presumably mean values for those months. Away on the high temperature side, 5t to 10t beyond the graph for

[1] A_1, A_2 are the amplitudes of the 24- and 12-hour terms respectively, α_1, α_2 are the phase angles, that is the angles measured backwards from the origin, which define the points where the components cross the mean lines upwards.

Authorities: Lindenberg. 'Über den täglichen Gang der Lufttemperatur in höheren Luftschichten,' by E. Barkow, *Beiträge zur Physik der freien Atmosphäre*, Band VII, München, Leipzig, 1917, p. 30. For a later paper on the same subject by H. Hergesell *see* the Lindenberg volume XIV, 1922.

Drexel. W. H. Dines, *Q.J. Roy. Meteor. Soc.*, vol. XLV, 1919, p. 41; for later papers see *M.W. Rev.*, vol. LXI, Washington, 1933, p. 61, and *Physics of the Earth*, vol. III, chap. IV, Washington, 1931.

Batavia. 'Results of Registering Balloon Ascents at Batavia,' by W. van Bemmelen, *K. Mag. en Meteor. Obs. te Batavia*, Verhand. No. 4, Batavia, 1916, p. xxxviii.

August, is an outlying graph marked October 2nd, 1908; and away beyond the mean graph for February, about 6t below at the surface but nearly 25t below at 700 mb (about 10,000 ft.), is another outlying graph marked April 5th, 1911. The graphs are 25t apart at 1000 mb, 37t at 750 mb, 30t at 500 mb; hence the range is greater in the free air than near the surface. The frame of the diagram provides only for heights up to 450 mb, about 20,000 ft. We hazard the guess that these represent the extremes of the collection of records of British soundings, and call attention to the noteworthy fact that the minimum and maximum thereby indicated occur in April and October respectively. We may gather therefrom that seasonal changes of temperature in the upper air, though they are probably caused by seasonal changes at the surface, are not necessarily simultaneous therewith, and that the transitional months of April and October show conspicuous features; we may confirm that inference by the large figures for the standard deviations of temperature and pressure in the quarters April to June and October to December quoted in Dines's table below.

Standard deviations in England SE of temperature, pressure and density at different levels.

Height km	Temperature in tt				Pressure in mb				Density in g/m³			
	Jan.–Mar.	Apr.–June	July–Sept.	Oct.–Dec.	Jan.–Mar.	Apr.–June	July–Sept.	Oct.–Dec.	Jan.–Mar.	Apr.–June	July–Sept.	Oct.–Dec.
	t	t	t	t	mb	mb	mb	mb	g/m³	g/m³	g/m³	g/m³
13	5·5	5·0	5·1	6·9	4·7	6·0	5·2	6·5	12	11	11	16
12	5·2	5·0	5·7	5·9	5·9	6·8	5·3	5·2	14	14	12	18
11	4·5	4·5	4·0	4·5	7·1	8·0	6·1	10·8	15	15	12	20
10	3·3	5·1	3·5	3·9	9·0	9·6	6·7	13·1	13	17	10	20
9	3·2	5·7	4·2	5·2	10·0	10·7	6·8	14·2	11	15	6	15
8	4·1	6·7	4·9	7·0	11·5	10·6	7·1	14·9	9	9	7	12
7	4·6	7·0	4·7	8·4	13·1	10·3	7·2	15·0	11	7	7	11
6	5·5	6·9	4·7	8·7	13·1	10·3	7·5	14·9	10	8	7	12
5	5·5	6·8	4·4	7·8	13·5	10·1	6·9	14·9	10	10	8	11
4	5·4	6·6	3·9	6·8	12·7	10·1	6·3	14·5	9	11	8	15
3	5·2	6·4	3·7	6·6	12·6	9·4	6·6	14·5	11	14	9	15
2	5·3	6·3	4·2	6·0	12·2	9·2	6·1	14·5	12	20	14	13
1	4·3	6·3	4·3	4·3	11·5	10·3	5·7	14·3	16	21	17	18
0	3·0	4·7	3·6	4·5	11·1	10·9	5·8	15·2	19	23	17	23

In Geophysical Memoir, No. 5 (pp. 81–98), E. Gold has brought together the monthly values for stations in Europe. The months of minimum and maximum temperature are very various, but April and October are conspicuous as months of maxima and minima at various levels in the troposphere. In the present state of knowledge we cannot venture upon any satisfactory generalisation, except that the range of temperature in the upper air of any locality is certainly as large and possibly several times as large as the range at the surface; the conditions in the upper layers clearly depend not only upon the recent conditions of the surface but also on the lapse-rate which is probably a survival of past conditions. These controlling features are clearly indicated by the graphs of the upper air of India which are represented in fig. 66. The values in those graphs are referred to temperature and log (potential temperature) but heights may be taken as indicated roughly by the pressure-lines. [Indian curves have now been extended to reach the stratosphere.]

For further particulars of the results of the exploration of the upper air we may refer the reader to Gold's comprehensive memoir. Since its compilation in 1912 the material for the study has been increased, particularly by the Observatory at Lindenberg which has been for many years the most active centre of the investigation of the upper air. Discussions of many interesting subjects connected with it will be found in the *Beiträge zur Physik der freien Atmosphäre*, a publication instituted for the purpose by R. Assmann and H. Hergesell, successive directors of the Observatory.

The height of the tropopause, where the troposphere meets the stratosphere, and the temperature of the air there, have been attractive subjects of investigation; we quote the monthly and annual values for Lindenberg. March, April and October are conspicuous months in this case. Some individual values are shown in a diagram in note 14 of chapter x.

	Jan.	Feb.	Mar.	Apr.	May	June	
Height	10·09	10·19	9·55	9·61	10·09	10·43	
Tempr.	213·7	213·1	216·0	215·9	216·5	220·7	
	July	**Aug.**	**Sept.**	**Oct.**	**Nov.**	**Dec.**	**Year**
Height	10·57	10·65	10·59	10·89	10·42	9·28	10·19 km
Tempr.	221·0	218·9	219·8	216·5	215·1	218·0	217·1 tt

The uniformity of lapse-rate.

We are unable to enter into further details of the observation of the upper air, which refer for the most part to the results obtained at the separate observatories. The chief general results of the investigation are first the increase of the height of the tropopause and the decrease of temperature of the stratosphere with increase of pressure at the surface; secondly the decrease of the height of the tropopause as the latitude changes from the equator towards the poles. It implies a lower temperature of the stratosphere at the equator and consequent reversal of the gradient of temperature in the stratosphere as compared with the troposphere. There is thirdly the close approach to equality of lapse-rate in different localities and at different seasons for those parts of the troposphere which are above the surface-layer of inversion such as is found over the large continents in winter.

This most important and somewhat surprising generalisation is expressed by the near approach to parallelism of the graphs which we have reproduced. It is approximately true for mean values of the temperature of the air above the first two kilometres at all the stations at which observations have been obtained whether on land or sea.

As evidence for the generalisation we may refer to the tables of pp. 106–109 and to the uniformity of the distance apart of the isothermal curves in fig. 60.

Potential temperature and megadynetemperature.

The mode of presentation of the structure of the upper air which we have used hitherto in this chapter has relied upon the direct cartesian relation of temperature and height, or upon isopleths of temperature in a vertical section of the atmosphere. There are other modes of representation which in some ways are more instructive. One of them is that which is based upon the use of what is called potential temperature: the utility of this mode of presentation depends upon the fact that if the layers of the atmosphere were thoroughly mixed by stirring, the resulting mixture would not show the uniformity of temperature that is obtained by stirring a liquid; but would be notably colder at the top than at the bottom. Leaving out of account for the present the effect of water-vapour and therefore treating only of the properties of unsaturated air, and understanding that no heat is allowed to reach the air or to escape from it, the perfect mixture would be arranged in layers colder by nearly 10t for every kilometre of height. When that state of things has been reached the air can be stirred like a liquid without any alteration of the distribution of temperature and without expending any energy upon it except what is consumed in friction. It is then said to be in convective equilibrium, or the equilibrium is said to be neutral or labile. If one portion is exchanged with any other equal portion the final state is indistinguishable from the initial state; if one portion of the air is lifted it will have been cooled and a corresponding portion if depressed will have been warmed; but in the end if we imagine a vertical circulation in an endless tube nothing will have been changed, though the actual temperature of any portion passing along the circulation will have altered continuously on its way round. The continuous change will be due to the variation of pressure which the portion experiences on its route. In those portions of the circulation where the pressure remains unchanged the temperature will also remain unchanged and when the pressure changes the corresponding change of temperature can be calculated. What remains unchanged in an up and down circulation is not the ordinary temperature but the temperature as corrected for difference of pressure, or, to use a technical expression, the temperature reduced to a standard pressure. Calling the standard pressure p_0 and the reduced temperature t_0 the changes of temperature will be according to the formula for the change of temperature with change of pressure of unsaturated air under what are known as adiabatic conditions. The equation may be written

$$\log\left(\frac{p}{p_0}\right) = \frac{\gamma}{\gamma - 1}\log\left(\frac{t}{t_0}\right),$$

where t is the actual temperature at any point of the circulation where the pressure is p, and γ is the ratio of the specific heats of dry air (c_p/c_v).

The temperature of the free air corrected for difference of pressure from a chosen standard p_0 is called the *potential temperature* of the air. We have written as if the circulation were bounded by a closed tube but no boundary will be necessary; the air will remain in equilibrium with its surroundings, provided that the path chosen follows a line of equal potential temperature, not a line of equal pressure nor one of equal temperature. Hence the surfaces in the atmosphere along which air can circulate "of its own

Fig. 63. Variation of Realised Entropy in a Section of the Upper Air from North to South.
Realised Entropy = $C_p \log_e$ (Potential Temperature) + Constant.

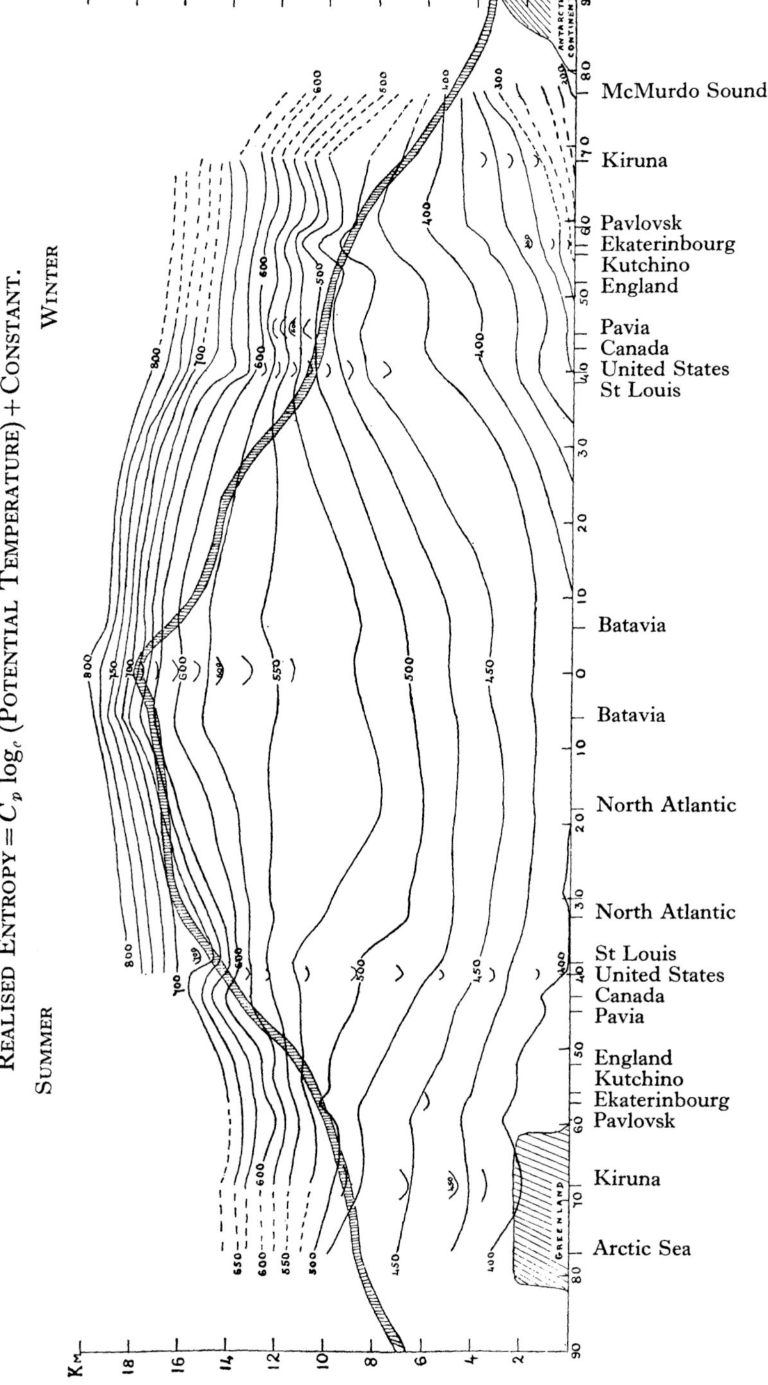

The shaded band with vertical hatching indicates the probable position of the tropopause.
The shaded areas with oblique hatching indicate the great land masses near the poles, namely Greenland and the Antarctic Continent.
◡ indicates observations which give exceptionally high realised entropy. ◠ indicates observations which give exceptionally low realised entropy.

accord," without requiring to have heat supplied or withdrawn, are the surfaces of equal potential temperature and no others.

In order to make use of this important idea in meteorology we require to choose a standard pressure to which the temperature is to be reduced. In that case potential temperature has a definite meaning. The standard pressure which we propose to use is that of 1000 mb, but in practice other standards are used; the standard of 760 mm of mercury at the freezing-point in latitude 45° is frequently employed. In these circumstances potential temperature of a mass of gas without further specification is insufficient as a name for the temperature which a gas would take up if its pressure were reduced to a *standard pressure* by compression (or in certain cases by expansion), without gain or loss of heat or moisture. Potential temperature may be any temperature whatever according to the standard chosen. We should like a name for the temperature which will be reached by adiabatic expansion or compression to the particular standard of 1000 millibars, and we think the name **megadynetemperature** would describe it effectively, because the standard pressure, 1000 millibars, is a megadyne per square centimetre.

We go a step further. For the purpose of thermodynamical reasoning the *entropy* of a mass of gas is often employed. We will not define it at present further than to say that it determines the conditions of the transformation of the intrinsic heat or energy of air into mechanical work. We have a simple conception of the entropy of a mass of air because (subject to a small correction on account of the water which the air contains) it is found to be proportional to the logarithm of the potential temperature, so for our purposes entropy and potential temperature are easily convertible terms. Surfaces of equal megadynetemperature are also surfaces of equal entropy.

We obtain therefore a new representation of the condition of the atmosphere as regards temperature from pole to pole by means of a diagram (fig. 63) of surfaces of equal entropy in a vertical section of the atmosphere. It indicates to us the surfaces along which air can move "of its own accord." It will be seen that the surfaces as represented in the diagram are not by any means horizontal in the troposphere. For free movement in the normal atmosphere without communication of heat or moisture air must descend as it approaches the equator from the poles and *vice versa*. There will of course be changes in the actual temperature in consequence of the increase in pressure as lower levels are reached and *vice versa*.

In the stratosphere the surfaces are much more nearly horizontal and it will be noted that they are very close together there. The distribution represents the normal stability of the atmosphere. The figures show that the entropy increases with height comparatively little in the troposphere but very rapidly in the stratosphere. In order that air may get upwards from one surface to the next above it must obtain the increase of entropy indicated on the diagram, 25 units for each step. In the troposphere 100 units will lift it through 10 kilometres, in the stratosphere the same increase would hardly lift it a kilometre.

We can in like manner represent the relative stability of the atmosphere under conditions of high pressure and low pressure (apart from any consideration of the condensation of water-vapour) by the section of the isentropic surfaces represented in fig. 64, in which the height in kilometres is marked on the edges, which represent the vertical through the points of highest pressure; the figures for entropy are entered at the points where the surfaces cross the central line of low pressure. The results disclosed are curious. Near the surface there is little difference between anticyclonic and cyclonic conditions as regards stability; in the middle layers the cyclonic air has less stability than the anticyclonic but it reaches the very stable condition of the stratosphere at a lower level. In the higher region of the troposphere there is not much to choose between the two until the cyclone begins to approach the stratosphere.

The effect of water-vapour with the possibility of the release of entropy by condensation has to be considered in the application of these conclusions to any particular case.

ENTROPY IN CYCLONES AND ANTICYCLONES

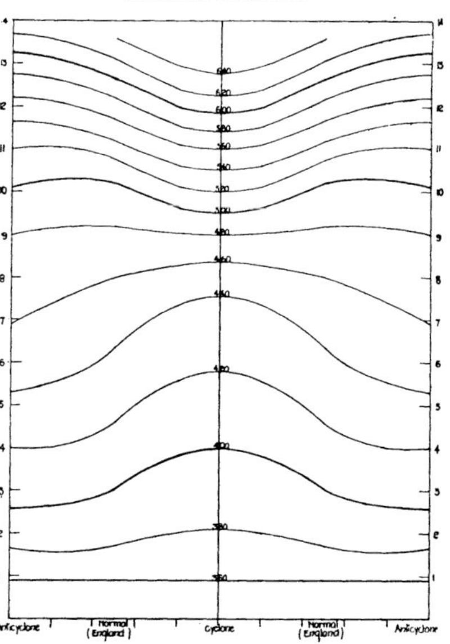

Fig. 64. The variation of level in lines of equal entropy in the atmosphere, according to changes of pressure from cyclone through normal to anticyclone. Entropy expressed in joules per degree, measured from a zero at 1000 mb and 200tt, is marked upon the several lines.

Entropy-temperature diagrams.

Finally we come to the new kind of representation to which we have referred in fig. 106 of vol. I. For the diagrams of figs. 63 and 64 we have used height as one of the co-ordinates and the other elements have been the horizontal distance in a vertical section, with lines of equal entropy for the isograms. But height is really dependent upon the pressure and temperature and is in consequence accounted for in the entropy. We can represent the structure of the atmosphere by using the entropy or log(megadynetemperature) as ordinate and temperature as abscissa. We can then make a graph representing the relation of these two co-ordinates in the course of a sounding by balloon or otherwise. We may regard such a sounding as supplying material for a vertical section of the atmosphere because the changes in the vertical are generally applicable to a horizontal area, at least as wide as that over which the balloon drifts.

We have already described these diagrams in chapter XIII of volume I. Upon the ground-work we have set out for a saturated atmosphere lines of equal pressure, lines of equal ratio of water-vapour to dry air from which the humidity can be obtained, and the adiabatic lines for "saturated" air which tell us how the temperature of air and its megadynetemperature change with continued reduction of pressure after the point of saturation is passed. On such a diagram the area of any closed curve represents the energy that would be required for the curve to be traced by a mass of atmosphere containing a kilogramme of dry air.

In this form, as fig. 65, we present once more the comparison and contrast of anticyclonic and cyclonic conditions which is partially represented in fig. 64. It disposes of a theory which was at one time generally accepted that air goes up automatically in a cyclone and comes down in a neighbouring anticyclone. From the law of equivalence of work and area for this diagram it will be seen that to get air up in a normal low pressure and bring it down again in a normal high pressure would require an aeroplane and a pilot to carry it. It could not perform the journey without an expenditure of energy represented by the area between the lines so marked. If it went round the other way, up in high pressure and down in low pressure, some work would be got out of the cycle but the local conditions of stability prevent that.

ENTROPY-TEMPERATURE DIAGRAM

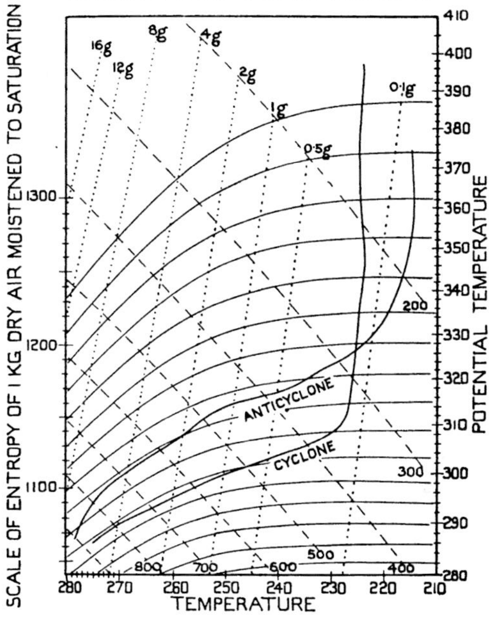

Fig. 65. Curves showing the relation of entropy (or potential temperature with 1000 mb standard) to temperature at successive levels in the atmosphere: in high pressure (ANTICYCLONE) and low pressure (CYCLONE). Entropy is measured from zero at 1000 mb, 100 tt.

The auxiliary lines are lines of equal pressure 200 mb...900 mb; lines of equal ratio of water-vapour to dry gas in saturated air, 0·1 g to 16 g per kg; and adiabatic lines (irreversible) for saturated air.

With diagrams of the same kind we are able to represent the normal conditions of the air in any locality for which sufficient observations have been accumulated. We give as an example the conditions at Agra in British India for each month of the year. The stability or instability of the normal state of the atmosphere is marked by the relation of the saturation adiabatic through any point of the graph to the graph itself. If the adiabatic runs forward above the graph, saturated air in that position would be unstable or "unsistible" as we prefer to call it; on the other hand, if the adiabatic passes to the right of the point where it cuts the graph even saturated air would be sistible and stable.

These diagrams are also furnished with a dotted line that indicates by the

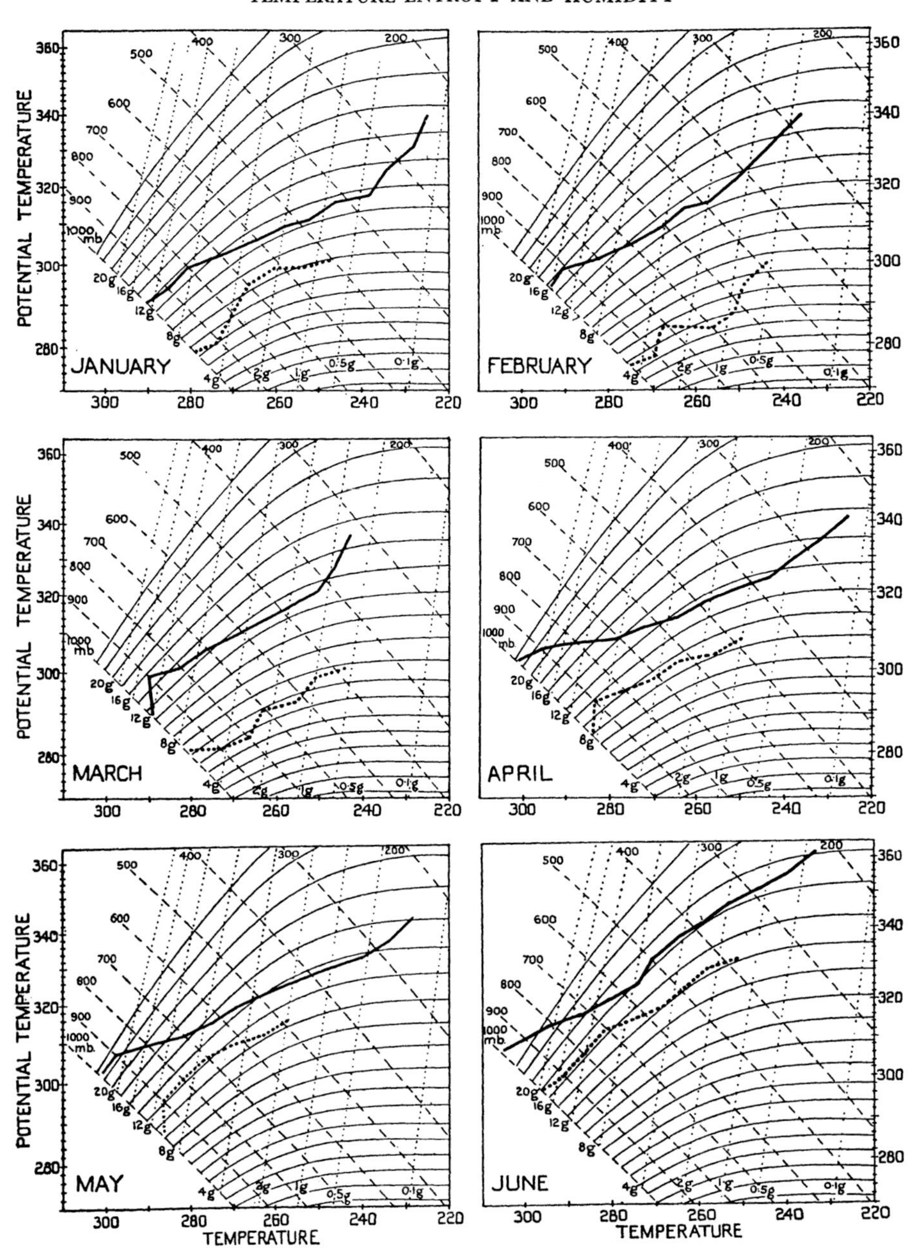

Fig. 66. Monthly mean results of soundings of the upper air of India (Agra)

(1) *tephigrams* represented by thick continuous lines which show the *temperature* at consecutive points of the ascent on the tercentesimal scale marked along the base, and the corresponding *megadynetemperature* (potential temperature for a standard pressure of 1000 millibars) on a logarithmic scale marked at the side; and

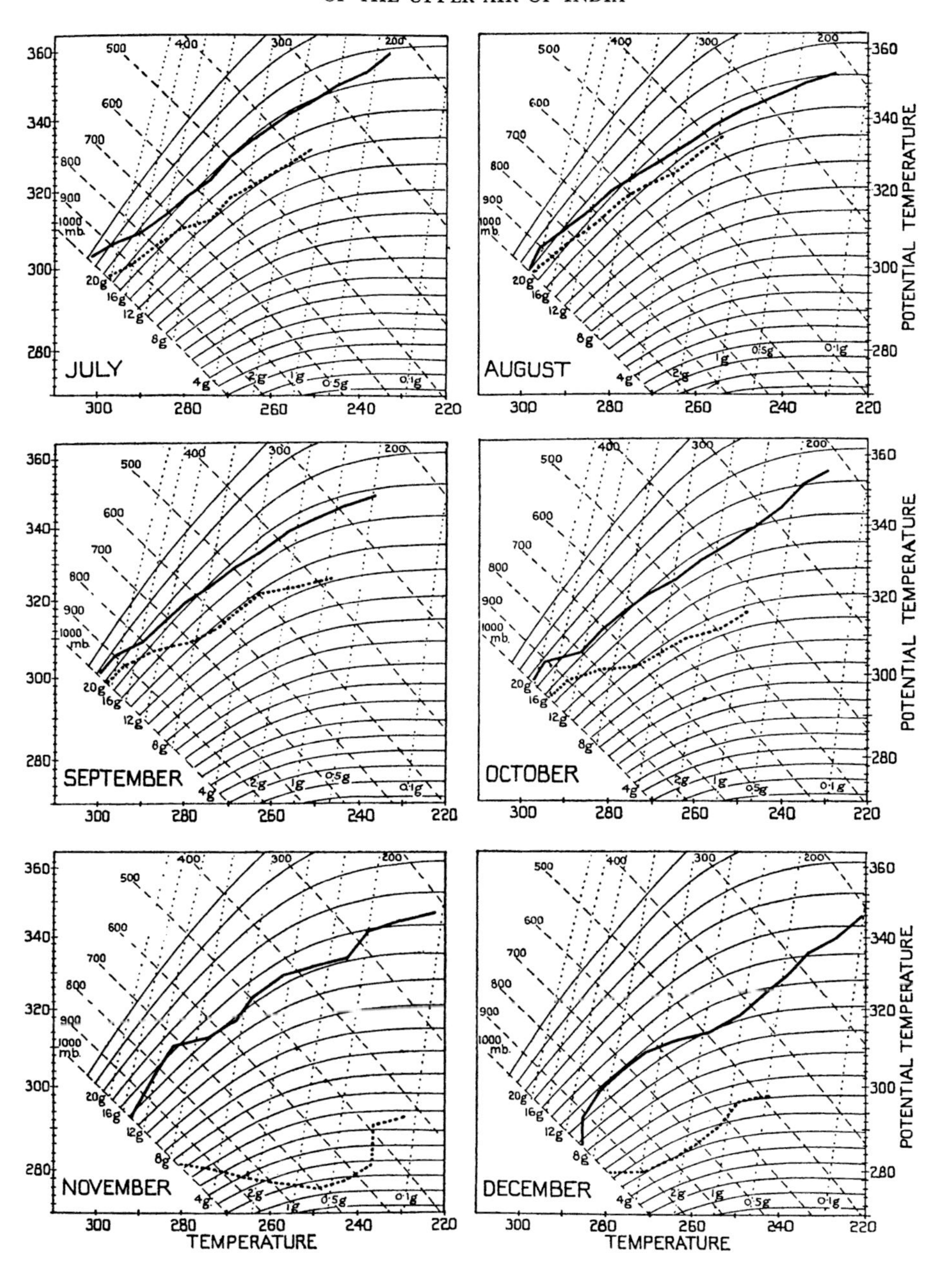

expressed by graphs with entropy and temperature as co-ordinates, viz.

(2) *depegrams*—dotted lines which represent the state of the air at consecutive points of the ascent as regards humidity by a line of temperatures of the "dew-point," arrived at from successive points of the tephigram by following a line of pressure until the temperature of the dew-point is reached.

temperature co-ordinate of any point on it the dew-point of the air at the point of the tephigram that bears the same pressure.

Whether the air is normally near saturation or far from it, is shown by the distance which has to be travelled from a point on the graph along the isobar to a corresponding point of the dotted line, the "depegram," or rather by the difference of the temperatures of the two points shown by the horizontal distance between them. It is easy to see from the tephigrams and depegrams of the charts for Agra that in November below the level of 700 mb the air would be sistible even if it were saturated; but it is in fact very dry and gets extraordinarily dry at higher levels, its dew-point being about 35t below its own temperature, whereas in August the air is nearly saturated the whole way up, having a dew-point not more than 2t below its own temperature and it is unsistible from a height of 900 mb upwards.

This method of representation of the structure of the upper air is summarised in the legend of fig. 66:

The lines which form the ground-work of the series of diagrams represent the thermal condition of a kilogramme of dry air charged with water-vapour to saturation under various conditions of pressure and temperature.

The thin continuous lines are the adiabatics of saturated air drawn for successive steps of 2tt at 1013 mb, the interrupted lines are the isobaric lines for 1000 mb, 900 mb, 800 mb and so on. The finer dotted lines are lines of equal quantity of water-vapour; they show the temperature and megadynetemperature of saturated air which contains the number of grammes of water-vapour indicated by the figures along the isobar of 1000 mb.

Geopotential, level surfaces and height.

In dealing with levels in our representation of the temperature of the upper air we have generally referred to height as commonly understood. That plan does not commend itself to all explorers of the upper air and so lately as April, 1925, the members of the International Commission for the Upper Air assembled in London assented to the expression of height in what Professor V. Bjerknes has called *dynamic metres*. They are not metres, and they are only dynamic in the sense that they are useful in gravitational calculations. They are intended to express the product gh, the gravitational potential or *geopotential*, the gravitational energy of unit mass at the geometric height h, assuming for the moment that g is constant in the vertical. If units are so chosen that g at the surface is represented by ·981, gh in dynamic metres will differ by less than 2 per cent. from height in metres, and for another value of g within the range of meteorological observations the difference will be of the same order. Hence for many practical purposes of meteorology a value in dynamic metres would be, within the limits of error of observation, the same as ordinary metres but better adapted to dynamical calculation. We can get over the difficulty of the name by calling the units geodynamic metres but the numerical similarity between heights in metres and geodynamic metres raises another question.

In an edition of the *Elements* of Euclid which has not yet been discovered, there may be found an axiom, "Things which can be mistaken for the same thing can be mistaken for one another." Deep respect for Euclid's logic based on long experience disposes us to think that the warning should be heeded, and that we should be careful to avoid rather than to encourage the use of quantities so alike numerically that a sort of judgment of Solomon is necessary to tell them apart. Perhaps the warning might be disregarded if it were not for the projection of a square root that trips one up if one expresses the geopotential at 1000 metres by a number with three digits.

We can arrive at three digits for geopotential at 1000 metres if h is expressed in metres and g in dekametres per second per second. But that means using two different units of length in one expression, dekametre in one part and metre in another, so that if we express the product, as we are entitled to do, as the square of a velocity, the unit of time remaining a second, the unit of length is $\sqrt{10}$ m. It seems better to avoid the square root and use the dekametre consistently as the unit of length and express the geopotential in geodekametres. Then the measure of g would be ·981, h at 1000 metres would be 100 dekametres, and the geopotential would be 98·1 geodekametres in place of 981 geodynamic metres. This would avoid the danger of metres being mistaken for geodynamic metres and the dynamical computation would not be impaired.

This is a digression. We have made it because it gives us the opportunity of pointing out to the reader that the diagrams which are represented in figs. 65 and 66 lend themselves very easily to the computation in C.G.S. units of the geopotential at any point of the curve which represents a sounding of the upper air. To find the difference of geopotential between the selected point on the curve and the start we have first to measure the area between the horizontal lines at start and finish and between the curve and the line of zero temperature. That can be expressed in C.G.S. units by the graduation of the diagram. For air which rises automatically this would be the dynamical equivalent of the heat obtained by condensation of vapour. To it must be added the dynamical equivalent of the heat derived from the cooling of the ascending unit, that is to say the product of the specific heat at constant pressure and the change of temperature between the start and finish.

From the geopotential thus obtained we can get the geometric height by a table. The derivation of this method of evaluating height is given in the *Avant Propos* of the *Comptes Rendus des Jours Internationaux* 1923, Commission Internationale de la haute atmosphère, and in greater detail in a memoir of the Royal Meteorological Society, 1927.

AUTHORITIES FOR THE DATA CONTAINED IN THE MAPS, DIAGRAMS AND TABLES OF CHAPTER IV

(See also chap. x, notes 24 and 25)

MONTHLY CHARTS OF MEAN ISOTHERMS OVER THE GLOBE

Temperature of the air over the land.

Alaska. Pacific Coast Pilot, coasts and islands of Alaska. Washington, 1879.

Canada. Atlas of Canada. Prepared under the direction of James White. Department of the Interior, 1906.

United States. Isotherms prepared by U.S. Weather Bureau in the Atlas of Meteorology. By J. G. Bartholomew, A. J. Herbertson and A. Buchan.

Scandinavia. Medeltal och extremer af Lufttemperaturen i Sverige 1856–1907. Bihang till meteor. iaktt. i Sverige, vol. XLIX, 1907.

Russia and Siberia. Atlas climatologique de l'Empire de Russie, 1849–1899. St Pétersbourg, 1900.

China. Zikawei, Observatoire Météorologique. La température en Chine et à quelques stations voisines, d'après des observations quotidiennes compilées par H. Gauthier, S.J., Changhai, 1918.

France. A. Angot. Études sur le Climat de France. Température. Ann. du Bur. Cent. Météor. Paris, 1897–1904.

Italy. F. Eredia. La Temperatura in Italia. Ann. Uff. Cent. di Meteor. e Geodin. Roma, vol. XXXI, parte I, 1909.

India. Calcutta, Meteorological Department. Climatological Atlas of India. 1906.

Africa. Maps based mainly on data given by A. Knox, The Climate of the Continent of Africa. Cambridge, 1911.

Brazil. Data from D. de Carvalho, Météorologie du Brésil. London, 1917.

Argentine. Data from Walter G. Davis, Climate of the Argentine Republic. Buenos Aires, 1910.

Australia. H. A. Hunt, Griffith Taylor and E. T. Quayle. The climate and weather of Australia. Melbourne, 1913.

General. J. G. Bartholomew, A. J. Herbertson and A. Buchan. Atlas of Meteorology. Bartholomew's Physical Atlas, vol. III. Westminster, 1899.

Report on the scientific results of the voyage of H.M.S. *Challenger* during the years 1873–1876. Physics and Chemistry, vol. II, part 5. Report on Atmospheric Circulation by Alexander Buchan. London, 1889.

Data and normals prepared for the British Meteorological and Magnetic Year-Book, Part 5, Réseau Mondial.

MS data compiled in the Meteorological Office for publication in the Admiralty Pilots and Sailing Directions.

Temperature of the air over the sea.

North Atlantic. Monthly Meteorological Charts of the North Atlantic Ocean. M.O. publication, No. 149.

North Pacific. Pilot Charts published by the U.S. Hydrographic Office, Washington.

South Atlantic. Monthly Wind Charts of the South Atlantic. M.O. publication, No. 168. 1903.

Red Sea. Meteorological Charts of the Red Sea. M.O. publication, No. 106. 1895.

Indian Ocean. Monthly Meteorological Charts of the East Indian Seas. M.O. publication, No. 181. 1911.

Southern Ocean. Meteorological Charts of the Southern Ocean between the Cape of Good Hope and New Zealand. M.O. publication, No. 123. 1917.

Mediterranean. Monthly Meteorological Charts of the Mediterranean Basin. M.O. publication, No. 224. 1917.

South Pacific and oceans in high latitudes. Report of the voyage of H.M.S. *Challenger.*

SEASONAL RANGE OF MEAN MONTHLY TEMPERATURE

North Polar Regions. The Norwegian North Polar Expedition, 1893–1896. Scientific results edited by Fridtjof Nansen, vol. VI, Meteorology by H. Mohn. Plate X. London, 1905.

Russia. Atlas climatologique de l'Empire de Russie, 1849–1899. St Pétersbourg, 1900.

Norway. Atlas de Climat de Norvège. Par H. Mohn. Nouvelle édition, par A. Graarud et K. Irgens. Geofysiske Publikationer, vol. II, No. 7. Kristiania, 1921.

France. A. Angot. Études sur le Climat de France. Température. Troisième partie. Ann. du Bur. Cent. Météor., 1903, Part I. Plate X. Paris, 1907.

Germany. Klima-Atlas von Deutschland. Berlin, 1921.

Italy. La Temperatura in Italia. Per F. Eredia. Ann. Uff. Cent. di Meteor. e Geodin. Roma, vol. XXXI, parte I, 1909. Rome, 1912.

Mexico. The Temperature of Mexico, by Jesus Hernandez. Monthly Weather Review, Supplement No. 23. Washington, 1923.

North Atlantic Ocean. Resultate Meteorologischer Beobachtungen von Deutschen und Holländischen Schiffen für Eingradfelder des Nordatlantischen Ozeans. Deutsche Seewarte. Hamburg.

Monthly Meteorological Charts of the North Atlantic Ocean. M.O. publication, No. 149. 1920.

South Atlantic and Indian Oceans. Mean values for five-degree squares supplied by the Marine Division of the Meteorological Office.

Pacific Ocean. U.S. Pilot Charts of the N. Pacific published by the Hydrographic Office, Washington.

Deutsche Seewarte. Stiller Ozean. Atlas von 31 Karten. Hamburg, 1896.

South Indian Ocean. K. Nederlandsch Meteorologisch Instituut De Bilt. Oceanographische en meteorologische Waarnemingen in den Indischen Oceaan. Data for two-degree squares from the charts.

General. Normals for all countries prepared for the British Meteorological and Magnetic Year-Book, Part 5, Réseau Mondial.

W. Köppen. Die Klimate der Erde. Berlin and Leipzig, 1923.

TEMPERATURE OF THE SEA-SURFACE

North Atlantic and Mediterranean. MS means prepared by the Marine Division of the Meteorological Office: for the North Atlantic means in two-degree squares to 1918; for the Mediterranean means in one-degree squares from observations 1900–14.

North Pacific and South Pacific. Charts of the Surface Temperature of the Atlantic, Indian and Pacific Oceans. M.O. publication, No. 59. 2nd edition. 1903.

South Atlantic (90° W–20° E), *Red Sea, Indian Ocean and Southern Ocean.* The authorities are the same as those set out above for the temperature of the air over the sea.

TABLES OF THE DIURNAL VARIATION OF PRESSURE AND TEMPERATURE

Arctic. The Norwegian North Polar Expedition, 1893–1896. Scientific results edited by Fridtjof Nansen. Vol. VI. Meteorology by H. Mohn. London, 1905. pp. 391, 395, 471, 483.

The data for diurnal variation in the Arctic regions given in the entablature of figs. 19 and 20 refer to the voyage of the *Fram* which changed its latitude and longitude continuously. To avoid misunderstanding it is therefore necessary to give the following list of degrees of latitude and longitude. We have rounded them off to the nearest degree from the table in Mohn's discussion of the data:

		Jan.	Feb.	Mar.	Apr.	May	June	July	Aug.	Sept.	Oct.	Nov.	Dec.
Lat. N	1893	—	—	—	—	—	—	—	—	—	78	78	79
Long. E	1893	—	—	—	—	—	—	—	—	—	136	138	137
Lat. N	1894	79	80	80	80	81	82	81	81	81	82	82	83
Long. E	1894	137	134	135	133	127	122	125	128	123	117	111	106
Lat. N	1895	84	84	84	84	85	85	85	84	85	85	86	85
Long. E	1895	103	103	101	96	87	81	74	77	79	77	65	51
Lat. N	1896	85	84	84	84	84	83	83	—	—	—	—	—
Long. E	1896	40	25	24	16	12	12	13	—	—	—	—	—

Aberdeen. British Meteorological and Magnetic Year-Book, 1917. Part IV. Hourly Values from Autographic Records. M.O. publication, No. 229 f. London, 1920. pp. 8–15.

Calcutta. Indian Meteorological Memoirs. Vol. IX, part 8. Calcutta, 1897. pp. 502, 529.

Batavia. Observations made at the Royal Magnetical and Meteorological Observatory at Batavia. Vol. XXXVIII, 1915. Batavia, 1920. pp. 74, 78.

Cape of Good Hope. Results of Meteorological Observations made at the Royal Observatory, Cape of Good Hope, discussed by E. J. Stone. Cape Town, 1871. pp. 2, 5, 10, 13.

Antarctic. Hut Point and Cape Evans. British Antarctic Expedition, 1910–1913. Meteorology, vol. III, Tables, by G. C. Simpson. London, Harrison and Sons, Ltd., 1923.

The data for the Antarctic are for Hut Point and Cape Evans; data are also available for Cape Adare, Borchgrevink expedition, which were discussed by L. Bernacchi; for the Scottish National Expedition to the Weddell Sea discussed by R. Mossman, and for the German expedition to Kerguelen Isle discussed by W. Meinardus.

Pairs of high and low level stations.

Fort William and Ben Nevis. The Meteorology of the Ben Nevis Observatories. Part III. Trans. Roy. Soc. Edin. Vol. XLIII. Edinburgh, 1905. pp. 489, 491.

Parc St Maur and Eiffel Tower. Études sur le Climat de France. Par A. Angot. Ann. du Bur. Cent. Météor., 1902, I, pp. 63, 90; 1903, I, p. 133, and 1894, I, p. B. 165.

Clermont Ferrand and Puy de Dôme. Études sur le Climat de France. Ann. du Bur. Cent. Météor., 1902, I, pp. 74, 96; 1903, I, pp. 155, 156.

Lahore and Leh. Indian Meteorological Memoirs. Vol. V. pp. 331, 488, 320, 476.

For additional data of diurnal variation of temperature the reader may consult:

J. Hann. Der tägliche Gang der Temperatur in der Inneren Tropenzone. Wien, 1905.

—— Der tägliche Gang der Temperatur in der Äusseren Tropenzone. Wien, 1907.

—— Die ganztätige (24 stündige) Luftdruckschwankung in ihren Abhängigkeit von der Unterlage (Ozean, Bodengestalt). Sitz. Akad. Wiss. Wien, Mathem.-naturw. Kl., Abt. II a, 128. Bd., 3. Heft. 1919.

SEASONAL AND DIURNAL VARIATION OF TEMPERATURE

Barnaoul. Die Temperatur-Verhältnisse des Russischen Reiches. Von H. Wild. Tabellen. St Petersburg, 1881. S. VII.

Agra. Indian Meteorological Memoirs. Vol. V, p. 278.

St Helena. The Trade-Winds of the Atlantic Ocean comprising Climatological Tables for St Helena. By J. S. Dines. M.O. publication, No. 203. London, 1910.

Antarctic. Hut Point and Cape Evans. British Antarctic Expedition, 1910–1913. Meteorology, vol. I. Discussion by G. C. Simpson. Calcutta, 1919. p. 53.

Arctic. Maud-ekspeditionens videnskabelige arbeide 1918–19 og nogen av dets resultater. Av H. U. Sverdrup. Sœrtryk av "Naturen," 1922, Bergen.

TEMPERATURE OF THE UPPER AIR FROM OBSERVATIONS OF BALLONS-SONDES

McMurdo Sound. British Antarctic Expedition, 1910–1913. Meteorology, vol. I. Discussion, pp. 275 and 284.

Kiruna. Étude préliminaire sur les vitesses du vent et les températures dans l'air libre à des hauteurs différentes. Par H. H. Hildebrandsson. Geografiska Annaler. Upsala, 1920. The original data are in L'expédition franco-suédoise de sondages aériens à Kiruna, 1907, 1908 et 1909, par H. Maurice, Nova Acta Soc. Reg. Scient. Ups., Sér. IV, tome III, No. 7, 1913.

Pavlovsk and Kutchino. Einige Ergebnisse der Registrierballonaufstiege in Russland. Von M. Rykatchew [junior]. Met. Zeitschr., Band XXVIII, 1911. Ss. 1–16.

Data for Moscow for February and March 1901 (not included in the table) are given in a report by A. de Quervain published in Travaux Scientifiques de l'Observatoire de Météorologie Dynamique de Trappes, tome III, Paris, 1908.

Ekaterinbourg and Nijni-Oltchédaeff. Geografiska Annaler. 1920.

Hamburg, Uccle, Paris, Strassburg, Vienna and Zurich. International Kite and Balloon Ascents. By E. Gold. Geophysical Memoir, No. 5; M.O. publication, No. 210 e. London, 1913.

Data for Uccle, grouped in two periods of six months, are given in Ann. mét. de l'Institut Royal Météorologique de Belgique, 1914, p. (53). Bruxelles, 1913.

Lindenberg. Die Lindenberger Registrierballonaufstiege in den Jahren 1906 bis 1916. Von J. Reger. Die Arbeiten des Preuss. Aeronaut. Obs. bei Lindenberg. Band XIII. Braunschweig, 1919. S. 41.

England SE. The Characteristics of the Free Atmosphere. By W. H. Dines. Geophysical Memoir, No. 13; M.O. publication, No. 220 c, London, 1919.

Munich. Die Temperatur der freien Atmosphäre über München nach den Registrierballonfahrten 1906–1914. Münchener Aerologische Studien, No. 8. Von A. Schmauss. Deutsches Meteorologisches Jahrbuch für 1918. München, 1920. Ss. D 1–10.

Pavia. Le caratteristiche dell' atmosfera libera sulla Valle Padana. Per Pericle Gamba. Venezia, 1923.

Canada. Upper Air Investigation in Canada. Part I. Observations by Registering Balloons. By J. Patterson. Ottawa, 1915.

United States. Mean values of free air barometric and vapour pressures, temperatures and densities over the United States. By W. R. Gregg. Monthly Weather Review, January, 1918, vol. XLVI, pp. 11–20. The values are means for Fort Omaha, Nebr., 41° N, 96° W; Indianapolis, Ind., 40° N, 86° W; Huron, Dak., 44° N, 98° W; Avalon, Cal., 33° N, 118° W.

St Louis. Exploration of the Air with ballons-sondes at St Louis and with kites at Blue Hill. By H. H. Clayton and S. P. Fergusson. Annals of the Astronomical Observatory of Harvard College, vol. LXVIII, part I. Cambridge, Mass., 1909.

Agra. The Free Atmosphere in India. Observations with kites and sounding-ballons up to 1918. By W. A. Harwood. Memoirs of the Indian Meteorological Department, vol. XXIV, part 6. Calcutta, 1924. [See also Ramanathan, vol. XXV, 1930.]

Batavia. Results of Registering Balloon Ascents at Batavia. By W. van Bemmelen. K. Mag. en Met. Obs. te Batavia, Verhand. No. 4. Batavia, 1916.

Victoria Nyanza. Bericht über die aerologische Expedition des K. Aeronaut. Obs. nach Ostafrika im Jahre 1908. Von A. Berson. Ergeb. der Arbeit. des K. Preuss. Aeronaut. Obs. bei Lindenberg. Braunschweig, 1910. S. 82.

Arctic. Aerologische Studien im arktischen Sommer von H. Hergesell. Beitr. Physik. Atmosph. Leipzig, VI, Heft 4, 1914, pp. 224–261.

Azores and North Atlantic (Peppler). Beitr. Physik. Atmosph. Leipzig, IV, 1912, pp. 17 and 224.

A compilation of results of individual ascents over the North Atlantic is given by H. U. Sverdrup, 'Der nordatlantische Passat,' Veröff. Geophys. Inst. der Universität, Leipzig, II, Heft 1, 1917.

North Atlantic (Hildebrandsson). Geografiska Annaler, 1920.

Zanzibar. 'La spedizione aerologica italiana a Zanzibar nel Luglio, 1908,' by L. Palazzo. Annali dell' Ufficio Centrale di Meteorologia e Geodinamica, vol. XXX, 1908, Parte I, Roma, 1910.

SUPPLEMENTARY DATA

Seasonal variation of temperature of the upper air at Pavia (P) lat. 45° N and Woodstock (W) 43° N (first figure omitted).

Height km.		Jan. tt	Feb. tt	Mar. tt	Apr. tt	May tt	June tt	July tt	Aug. tt	Sept. tt	Oct. tt	Nov. tt	Dec. tt
	P	215	215	220	222	220	219	218	222	218	216	219	217
16	W	—	—	—	16	15	—	08	13	08	—	12	—
	P	215	216	221	222	220	218	218	218	218	215	216	218
15	W	—	18	12	15	14	10	07	12	07	13	07	—
	P	216	215	221	223	219	214	218	219	218	218	215	219
14	W	—	19	14	16	16	08	09	12	09	15	10	—
	P	213	215	219	219	216	213	218	221	216	214	216	218
13	W	—	20	17	15	12	12	12	18	13	18	10	—
	P	210	215	218	218	217	215	219	223	218	217	215	217
12	W	—	20	18	14	12	14	19	21	18	17	10	13
	P	*211*	217	218	218	218	221	221	**224**	222	219	215	216
11	W	18	17	17	*14*	15	19	25	25	**27**	21	*14*	*14*
	P	*216*	219	220	221	221	226	226	**227**	**227**	224	220	220
10	W	20	18	*16*	18	19	26	33	31	**35**	23	21	19
	P	*222*	226	223	226	227	234	233	**235**	234	229	226	223
9	W	25	20	*18*	27	25	35	40	38	**41**	31	28	25
	P	231	233	*228*	233	235	**243**	241	242	241	236	232	*228*
8	W	31	*24*	*24*	35	33	41	47	45	**49**	39	37	33
	P	238	240	*236*	240	242	**250**	**250**	**250**	246	244	240	*236*
7	W	39	*31*	*31*	42	42	48	53	53	**58**	46	44	42
	P	246	246	*243*	245	250	**258**	257	257	256	252	247	*243*
6	W	46	*38*	39	51	50	56	60	60	**64**	53	53	50
	P	253	255	*251*	252	257	**264**	**264**	**264**	263	258	254	253
5	W	52	*45*	47	58	55	63	67	67	**70**	60	59	57
	P	260	261	259	259	264	**272**	269	269	269	265	262	*258*
4	W	59	*52*	53	64	62	69	73	73	**75**	66	64	65
	P	*265*	267	266	*265*	270	**277**	276	275	275	271	269	*265*
3	W	66	*58*	59	70	67	74	79	78	**80**	72	70	70
	P	*270*	272	272	271	276	**282**	**282**	281	281	277	274	*270*
2	W	71	*63*	*63*	77	73	78	**86**	84	**86**	78	73	75
	P	*272*	276	276	276	282	**288**	**288**	287	287	283	278	275
1	W	73	*65*	*65*	78	79	84	**91**	89	90	84	74	77
0	P	*273*	*273*	277	283	289	**293**	**293**	**293**	**293**	287	281	277
0·3	W	74	*67*	70	81	84	88	**96**	95	93	87	78	75

CHAPTER V

COMPARATIVE METEOROLOGY: AQUEOUS VAPOUR

The energy involved in rainfall and sunshine.

1 mm of rainfall gives 100,000 cc of water upon 1 square dekametre and 1 kg per square metre, 1000 metric tons per square kilometre.

In the condensation of 100,000 cc of water from the air, the energy which has to be disposed of is $2 \cdot 5 \times 10^{15}$ ergs: that is equivalent to 70 kilowatt-hours, or 93 horse-power-hours.

Sunshine through a perfectly transparent atmosphere would transmit 135 kilowatts per square dekametre.

Sun-power at the earth's surface varies from zero to about 100 kilowatts per square dekametre.

The energy which has to be disposed of to provide 1 mm of rainfall represents $\frac{1}{8}$ of a day's harvest of sunshine in rural England (Rothamsted) in May, June, July; $\frac{1}{3}$ day (Rothamsted) in Aug., Sept., Oct. or Feb., March, April; $1\frac{1}{4}$ days (Rothamsted) in Nov., Dec., Jan.

OUR next step in the representation of the structure of the atmosphere is to show the distribution of the aqueous vapour upon which depend the phenomena of cloud, rain, snow, hail; lightning and thunder are among the incidental accompaniments of the fall of heavy rain. In virtue of the large amount of energy which is rendered latent on the evaporation of water and thereby stored in the atmosphere, water-vapour is a most powerful dynamic agent. The study of its properties and distribution is therefore fundamental for meteorological science.

We show first the normal distribution of water-vapour by two pairs of maps with isograms of the normal mean temperature of the dew-point for January and July in the air at sea-level in the two hemispheres. The data for the maps have been compiled for this work from the published records of temperature and humidity at the earth's surface; the values of vapour-pressure have been reduced to sea-level by the formula[1] adopted by A. Kaminsky in the *Climatological Atlas of the Russian Empire*,

$$e_0 = e_h (1 + \cdot 0004h),$$

where h is the height in metres.

We note in passing that the dew-point, which is the temperature at which the air becomes saturated with water-vapour if it be cooled *without alteration of pressure*, is not quite the same as the temperature at which air becomes saturated when it is cooled *adiabatically*; the difference between the two rests upon thermodynamic considerations which are represented in fig. 106 of volume I.

Below the first pair of maps we have given in two portions a table of the pressures of water-vapour which correspond with the dew-points, and the density of the vapour at saturation. Below the second pair we have given tables of the pressure of water-vapour corresponding with different degrees of relative humidity, and some thermal constants for water, air and water-vapour.

[1] For heights up to 1000 m the formula gives results not very different from those obtained from Hann's formula $\log_{10} e_0 = \log_{10} e_h + \frac{h}{6300}$.

FIGURE 67

MEAN DEW-POINT OF THE AIR AT SEA-LEVEL

Saturation-temperatures on the tercentesimal scale.

Authority: see p. 208.

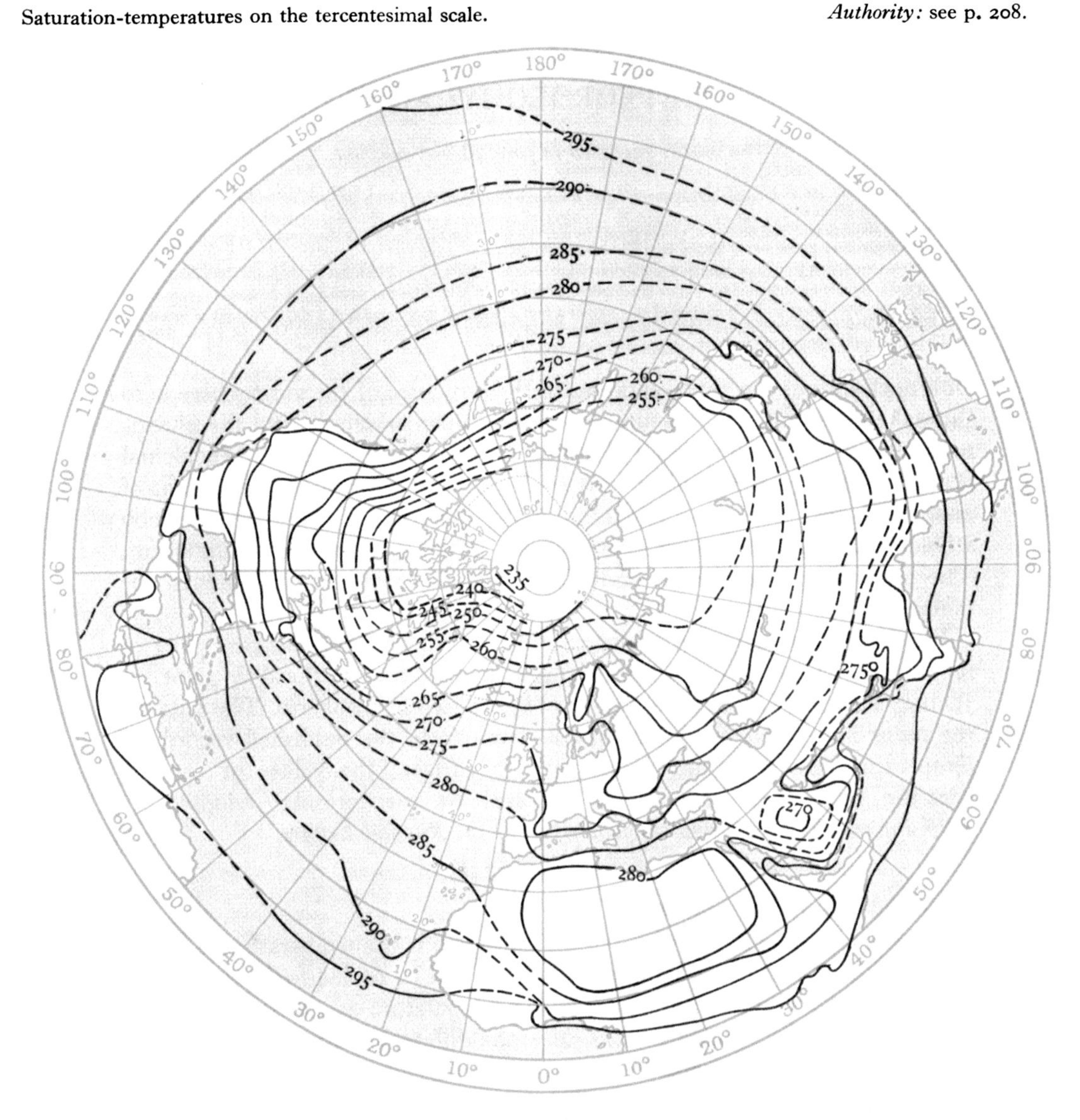

Table of dew-points with equivalent pressure of aqueous vapour in millibars and its density at saturation in grammes per cubic metre.

Dew-point	215	220	225	230	235	240	245	250	255	260
Corresponding vapour-pressure	—	—	·052	·096	·16	·28	·47	·78	1·27	1·99
Density at saturation	—	—	·050	·091	·15	·25	·42	·68	1·08	1·66
Dew-point	261	262	263	264	265	266	267	268	269	270
Corresponding vapour-pressure	2·18	2·39	2·61	2·85	3·10	3·38	3·68	4·01	4·37	4·75
Density at saturation	1·81	1·98	2·15	2·34	2·54	2·76	2·99	3·25	3·52	3·82
Dew-point	271	272	273	273	274	275	276	277	278	279
Corresponding vapour-pressure	5·16	5·61	6·13	6·09	6·54	7·03	7·55	8·09	8·68	9·29
Density at saturation	4·13	4·48	4·87	4·84	5·18	5·54	5·92	6·33	6·76	7·22

MEAN DEW-POINT OF THE AIR AT SEA-LEVEL — FIGURE 68

Authority: see p. 208.

Saturation-temperatures on the tercentesimal scale.

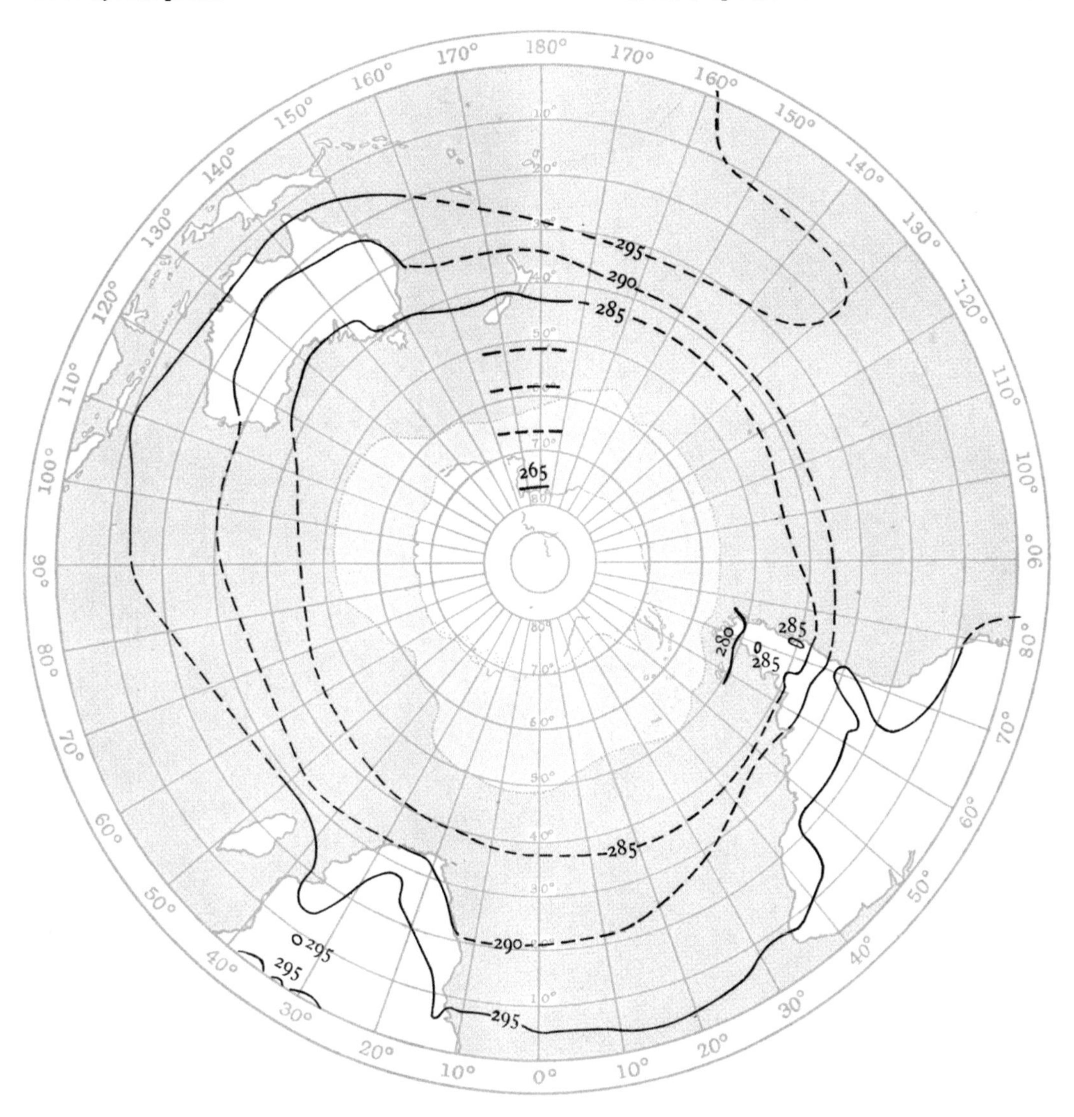

Table of dew-points with equivalent pressure of aqueous vapour in millibars and its density at saturation in grammes per cubic metre.

Dew-point	280	281	282	283	284	285	286	287	288	289
Corresponding vapour-pressure	9·96	10·65	11·40	12·19	13·02	13·91	14·85	15·84	16·89	18·01
Density at saturation	7·70	8·22	8·76	9·33	9·93	10·57	11·25	11·96	12·71	13·50
Dew-point	290	291	292	293	294	295	296	297	298	299
Corresponding vapour-pressure	19·20	20·44	21·76	23·14	24·62	26·17	27·81	29·53	31·36	33·28
Density at saturation	14·34	15·22	16·14	17·12	18·14	19·22	20·35	21·54	22·80	24·11
Dew-point	300	301	302	303	304	305	306	307	308	309
Corresponding vapour-pressure	35·29	37·42	39·65	42·01	44·49	47·09	49·83	52·69	55·71	58·88
Density at saturation	25·49	26·93	28·45	30·04	31·70	33·45	35·27	37·18	39·18	41·27

FIGURE 69 MEAN DEW-POINT OF THE AIR AT SEA-LEVEL

Saturation-temperatures on the tercentesimal scale. *Authority:* see p. 208.

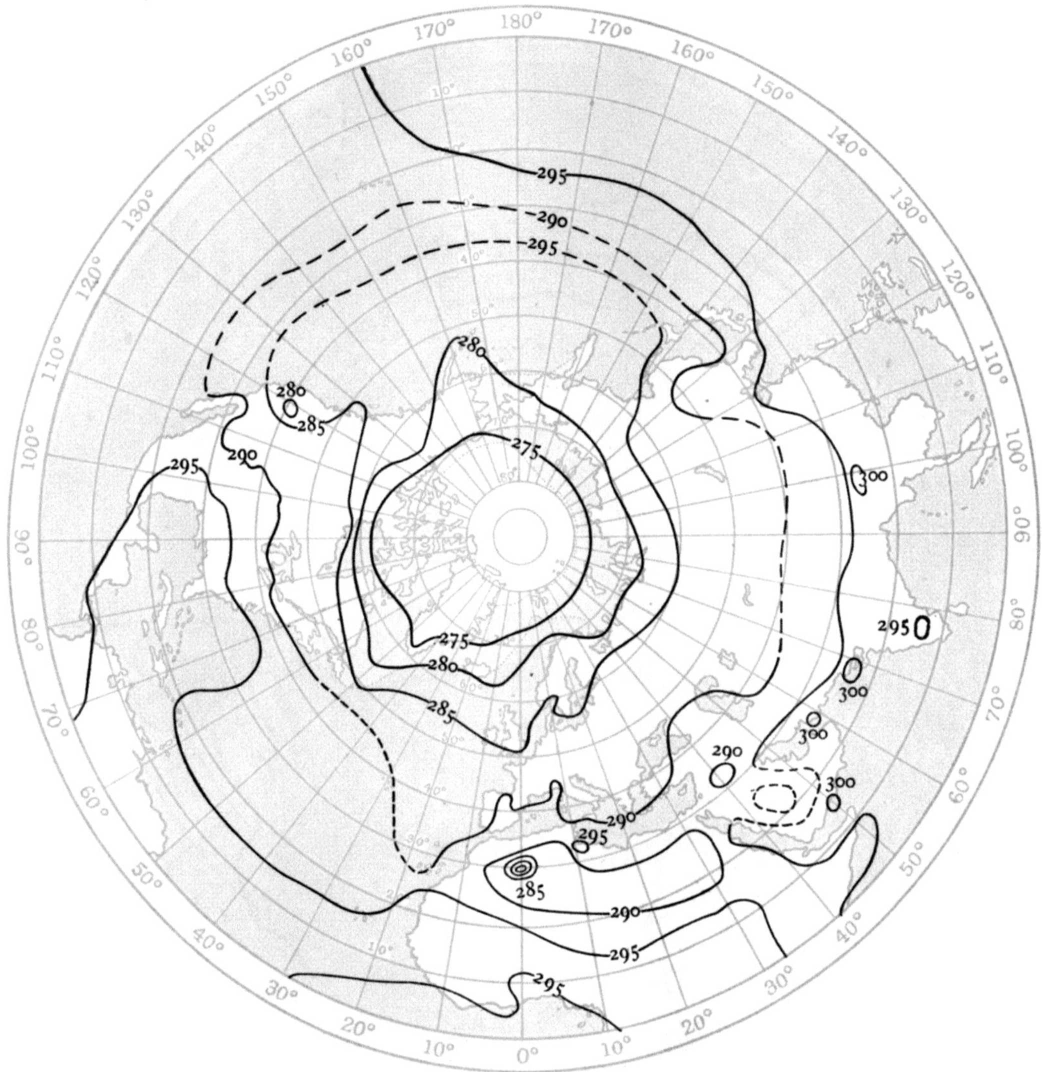

Vapour-pressure in millibars for different degrees of humidity at different temperatures on the tercentesimal scale.

Relative humidity	270	275	280	285	290	295	300	305	310
100	4·75	7·03	9·96	13·91	19·20	26·17	35·29	47·09	62·20
90	4·27	6·33	8·96	12·52	17·28	23·55	31·76	42·38	55·98
80	3·80	5·62	7·97	11·13	15·36	20·94	28·23	37·67	49·76
70	3·32	4·92	6·97	9·74	13·44	18·32	24·70	32·96	43·54
60	2·85	4·22	5·98	8·35	11·52	15·70	21·17	28·25	37·32
50	2·37	3·52	4·98	6·96	9·60	13·09	17·65	23·55	31·10
40	1·90	2·81	3·98	5·56	7·68	10·47	14·12	18·84	24·88
30	1·42	2·11	2·99	4·17	5·76	7·85	10·59	14·13	18·66
20	0·95	1·41	1·99	2·78	3·84	5·23	7·06	9·42	12·44
10	0·47	0·70	1·00	1·39	1·92	2·62	3·53	4·71	6·22

MEAN DEW-POINT OF THE AIR AT SEA-LEVEL FIGURE 70

Authority: see p. 208.

Saturation-temperatures on the tercentesimal scale.

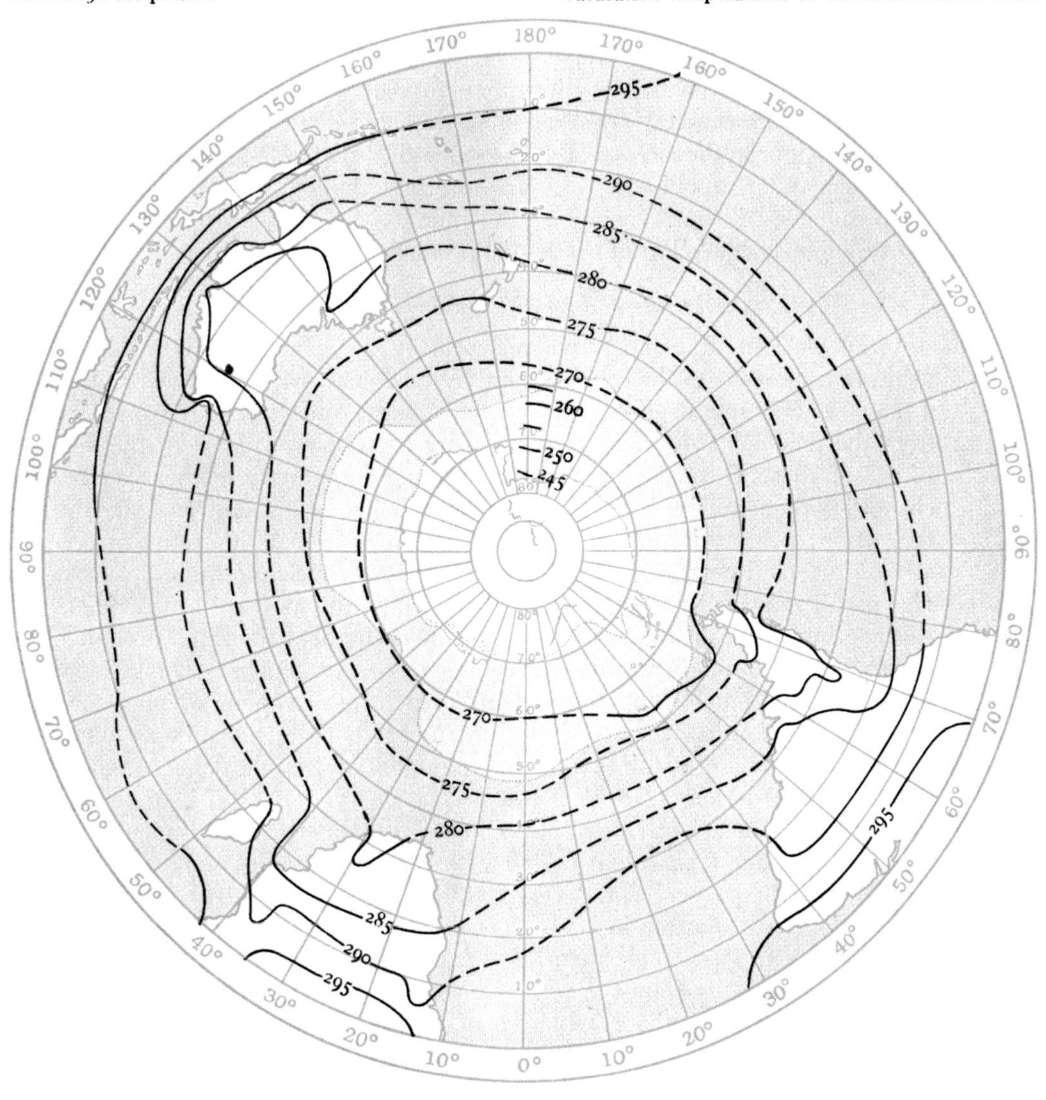

Some thermal constants of water and its vapour, and of air.

Latent heat of 1 gramme of water, 79·77 calories = 333·4 joules = 3·334 × 10⁹ ergs = 9·26 × 10⁻⁵ kilowatt-hours.

" 1 gramme of steam at 273tt, 597 calories = 2495 joules; at 373tt, 539 calories = 2252 joules.

Specific heat: sea-water at 290tt, 0·94; ice at 260tt, 0·502.

Dry air at 293tt, at constant pressure (c_p), 0·2417 (Swann, 1909); at constant volume (c_v), 0·1715 (Joly, 1891).

Water-vapour at 373tt, at constant pressure 0·4652 (Holborn and Henning, 1907); at constant volume 0·340 (Pier, 1909).

Ratio of specific heats of dry air c_p/c_v 1·40; of water-vapour 1·37.

Capacity for heat of one gramme of water and its dynamical equivalent:

tt	270	273	275	280	285	290	293	295	300	305
Calories	1·0130	1·0094	1·0076	1·0042	1·0021	1·0007	1·0000	0·9997	0·9990	0·9985
Joules	4·234	4·219	4·212	4·198	4·189	4·183	4·180	4·179	4·176	4·174

With the aid of these tables the physical properties of moist air can be computed. For this purpose the water-vapour and the so-called "dry air" are considered as existing side by side in the atmosphere quite independently one of the other, so that the pressure of either is the same as if the other were not present. Each contributes its quota to the resultant pressure of the atmosphere as measured with the aid of a barometer. Thus, if e is the pressure of water-vapour in the air, d that of the "dry air," that is to say, of the gases other than water-vapour, then the pressure p as derived from a reading of the barometer will be given by the formula $p=d+e$. This is the familiar expression of Dalton's law of the partial pressures of gases and vapours.

Absolute and relative humidity for maps.

As an element of comparative meteorology we have chosen the dew-point, in preference to the relative humidity of the air which perhaps makes a more general appeal. The table of pp. 130–131 makes it easy to derive from the dew-point the pressure or density of aqueous vapour in the air. Either of those elements is called the absolute humidity[1]; neither can be changed without adding water to the air or taking some away. Consequently they afford definite information about the composition of the atmosphere at the time; and changes of condition in respect of absolute humidity may give useful indications of the sources from which the air has been derived. Relative humidity, on the other hand, the measurement of which occupied the reader's attention in chapter X of volume I, represents the percentage ratio of the absolute humidity to the pressure or density of vapour which would saturate the air at its own temperature. It depends therefore quite as much on the temperature of the air as upon the amount of moisture which it contains.

Saturated air becomes "dry" if it is heated, or "very dry" if its temperature is raised sufficiently. In our climate, judging by the isopleths of fig. 107 of volume I, we might call air with relative humidity of 75 per cent. "dry," or with 50 per cent. "very dry," and the pressure or density of water might be the same in both. "Dry air" has so many meanings that it is not easy to disentangle them. The information necessary for obtaining the relative humidity from the vapour-pressure and the temperature, or the converse, is given in the entablature of p. 132. Lines of equal vapour-pressure could be drawn across the table: they obviously slope downwards steeply to the right. Thus an absolute humidity of 10 mb gives saturation at 280 tt, but relative humidity of less than 20 per cent. at a temperature only 30 t higher.

For this reason relative humidity wanders over a wide range wherever the range of temperature on any day is large, and a large diurnal variation is very characteristic of records of relative humidity at any land-station. It is on that account not quite suitable for comparative meteorological mapping, the maps would require a specification of the time of day. Mean values for

[1] For a reason given on p. 139 absolute humidity is perhaps best given as the amount of water associated with a kilogramme of dry air.

the twenty-four hours might of course be used, but nature does not operate by the mean for twenty-four hours, which may indeed be regarded as the most transient condition of the whole day. Extremes are of greater potency. At all stations where there are four hours of darkness the extreme in the dark period is generally not far from saturation. Our maps may therefore bear some resemblance to the distribution of 95 per cent. humidity in the night hours. The approximation to a condition of saturation during the night is shown in practice by the approach of the reading of the dry bulb to that of the wet bulb at night. The suggestion which is made here can be tested by examining the records of the seven observatories of the Meteorological Office in the Quarterly Weather Report. A specimen is given on p. 375.

Seasonal and diurnal variation of the amount of water-vapour in the surface-layer of air.

Among the most remarkable results of the co-ordination of observations of humidity are the differences which are disclosed in the amounts of moisture in the surface-layer of the atmosphere during the course of the year on the one hand, and the course of the day on the other. The first point to which we will call attention is the extraordinary similarity between the figures shown on p. 41 of volume I in the column for normal vapour-pressure at Richmond (Kew Observatory) and those in the corresponding column on p. 30 for Babylon, in spite of the difference of latitude and the great difference of climate which is shown by the figures for relative humidity in the adjacent columns of the two tables. Another remarkable feature is indicated by a comparison between the figures for moisture at Babylon on p. 30 and at Helwan on the previous page. The feature to which we wish to direct attention is the considerable loss of moisture in the surface-layer at Helwan between the morning and the mid-day observations, compared with the relative uniformity throughout the day at Babylon or at Richmond.

Some uncertainty may attach to the comparison on account of differences of practice with regard to the methods of reduction of the observations of humidity or the tables employed, but the loss of moisture at Helwan is too great to be accounted for in that way. It is possible that the difference is significant of heavy dews at night which are evaporated early on the following morning. Such evaporation might make the ground-surface assume the temperature of the wet bulb rather than the dry bulb, and the condition of the atmosphere at the surface during the process of drying would resemble that of the air in the near neighbourhood of a vast wet bulb, until the evaporation was completed and the moisture shared with the upper layers by the action of the wind.

So considerable is the field of inquiry into the variation of moisture in the air and the insight which may be obtained thereby into the physical processes operative in the atmosphere, that we have thought it well to supplement the information about moisture contained in the two pairs of maps, and in the notes about the climate of the "world known to the ancients" in chapter II of the first volume, by tables to show the diurnal and seasonal variation of moisture in a number of representative localities. These include (1) Barnaoul—a remote continental station, (2) Paris—a near and ordinary station of Western Europe, not very dissimilar from Richmond, (3) Calcutta—a tropical moist climate, (4) Dar es Salam—an equatorial coastal station, (5) Alice Springs—an arid and continental climate, (6) Naha—a very insular climate, and finally, some information for the Arctic and Antarctic regions.

Diurnal and seasonal variation of vapour-pressure in millibars.

Hour	Jan.	Feb.	Mar.	Apr.	May	June	July	Aug.	Sept.	Oct.	Nov.	Dec.
Barnaoul, 53° 20′ N, 83° 47′ E, ? 147 m, 1841–45, 1850, 1852–62.												
1	1·49	1·72	2·49	4·76	7·19	11·11	14·08	12·24	8·31	5·25	3·04	2·11
7	1·48	1·61	2·39	5·09	7·81	11·99	15·12	12·73	8·23	5·09	2·93	2·07
13	1·96	2·47	3·79	5·87	7·95	12·48	15·64	13·91	9·56	6·16	3·61	2·44
18	1·72	2·04	3·29	5·45	7·63	11·83	15·39	13·63	9·33	5·77	3·23	2·23
Paris (Parc St Maur), 48° 49′ N, 2° 29′ E, 50·3 m, 1891–1910.												
7	6·3	6·4	6·9	8·3	10·5	13·5	14·9	14·7	12·8	10·3	7·9	7·2
13	6·7	6·7	7·1	7·7	10·1	13·1	14·5	14·1	13·1	11·2	8·7	7·3
18	6·7	6·8	7·1	7·9	10·3	13·3	14·7	14·5	13·7	11·5	8·7	7·3
Calcutta, 22° 32′ N, 88° 20′ E, 6·5 m, 1881–93.												
1	15·2	17·1	24·5	29·1	30·4	32·8	32·8	32·6	32·5	29·1	20·5	15·2
7	14·1	16·1	24·0	29·4	31·5	33·5	33·0	32·8	32·8	29·1	19·7	14·3
13	13·8	14·3	19·9	25·8	31·0	33·7	33·7	33·4	32·7	27·4	18·8	14·1
18	15·9	16·6	22·1	27·6	30·6	33·2	33·4	33·0	32·9	29·3	21·5	16·5
Dar es Salam, 6° 49′ S, 39° 18′ E, 7·6 m, 1901–12.												
7	28·5	27·9	27·9	27·2	25·1	22·3	21·6	21·9	22·5	24·4	27·9	28·3
14	29·3	29·2	29·6	28·5	25·5	22·0	21·2	22·4	24·1	25·9	27·9	29·1
21	29·2	28·9	29·3	28·3	26·0	23·1	22·9	22·8	23·4	24·9	27·4	28·9
Alice Springs, 23° 38′ S, 133° 37′ E, 587 m, 1881–90.												
0	13·9	13·4	10·9	10·9	9·8	6·8	6·3	7·7	8·0	8·6	10·7	13·1
6	14·1	13·5	10·6	10·2	8·8	6·9	5·7	6·6	7·5	8·4	10·4	13·0
12	13·8	13·0	11·1	12·2	10·9	8·8	8·1	8·7	9·0	9·2	11·3	13·5
18	13·0	12·2	10·3	11·4	10·2	9·1	7·7	8·4	8·7	8·8	10·6	12·6
Naha (South of Japan), 26° 13′ N, 127° 41′ E, 10·5 m, 1906–10.												
1	14·1	14·1	15·7	19·4	22·1	28·9	30·3	30·3	28·6	23·6	18·5	14·6
7	14·1	14·2	15·7	19·7	22·8	29·3	30·7	30·5	28·5	23·5	18·5	14·5
13	14·9	15·1	16·3	20·4	23·1	29·4	30·5	30·8	28·7	24·4	19·1	15·4
18	14·5	14·4	15·9	19·8	22·4	29·0	30·2	30·1	28·3	23·9	18·7	15·2
Arctic, *Fram*, 1898–1902. (For the position of the *Fram* see p. 210.)												
0	·27	·63	·41	·83	2·27	5·29	6·23	5·85	2·65	1·32	·64	·47
4	·28	·63	·40	·76	2·12	5·13	6·29	5·77	2·68	1·33	·64	·47
8	·28	·60	·40	·83	2·28	5·23	6·36	5·83	2·59	1·32	·67	·45
12	·28	·64	·44	·93	2·44	5·35	6·49	5·81	2·67	1·35	·64	·47
16	·28	·65	·43	·96	2·56	5·35	6·53	5·88	2·73	1·33	·64	·47
20	·28	·63	·41	·91	2·45	5·36	6·45	5·87	2·71	1·32	·64	·47
*Antarctic, *Gauss*, 66° 2′ S, 89° 38′ E, Mar. 1902–Feb. 1903.												
0	4·88	4·33	2·87	1·55	1·75	1·55	1·52	1·04	1·47	1·93	3·15	4·57
4	4·77	4·36	2·80	1·51	1·85	1·53	1·53	1·01	1·39	1·81	2·91	4·56
8	5·19	4·40	2·85	1·37	1·88	1·55	1·59	1·04	1·39	2·16	3·31	4·93
12	5·31	4·67	2·97	1·59	1·85	1·61	1·59	1·11	1·72	2·56	3·71	5·21
16	5·33	4·64	3·03	1·65	1·76	1·57	1·60	1·07	1·68	2·49	3·73	5·37
20	5·20	4·49	2·83	1·56	1·79	1·57	1·55	1·04	1·45	2·01	3·41	5·03

* The figures for February represent the mean of 21 days' observations, March 25 days, April 18 days, May 16 days, December 29 days.

Among other important consequences of moisture in the atmosphere we will note the fact that its presence in sufficient quantity is held responsible for much of the discomfort which humanity has to suffer from atmospheric conditions. The human body is apparently susceptible to the discomfort of relatively moist conditions even in the Arctic or Antarctic region, where the absolute amount of moisture is very small; wood-work, cordage and other fibrous materials and structures are affected in like manner.

Within a certain narrow range of temperature from, say, 285tt to 295tt, human beings are not particularly sensitive to the influence of humidity but outside these limits of temperature, either on the higher or lower side, moist air feels disagreeably warm in the one case and disagreeably cold in the other. The impression caused by moist, warm air is the more pernicious. Consequently the temperature of the wet bulb has come to be regarded as a useful index of discomfort or otherwise of climatic conditions. So much so that with a wet bulb beyond 299tt (78° F) continuous hard labour is regarded as impracticable[1]. We give accordingly some information of high temperatures of the wet bulb extracted from the monthly tables of the *Meteorological Magazine* for the years 1921 to 1925.

High wet bulb temperatures.

Station	Months with mean temperature of wet bulb above 298·6tt (78° F)	Highest monthly maximum	
Lagos	February (3), March (4), April (4), May (3), June (1), Nov. (3), Dec. (1)	300·2	March, 1924
Calcutta	April (2), May (5), June (5), July (5), August (5), September (5)	300·7	May, 1921
Bombay	May (4), June (5)	299·1	June, 1923
Madras (1922–5)	April (4), May (3)	299·1	April, 1924
Colombo	April (5), May (5), June (3), Oct. (1)	299·6	May, 1921
Hong Kong	July (2), August (1)	298·8	August, 1922
Suva, Fiji	February (2)	299·1	February, 1921

The numbers in brackets indicate the number of years (out of the five) in which the mean temperature of the wet bulb for the month named exceeded the limit of 78° F. The observations are chiefly for hours between 8 h and 10 h local time, or the mean of observations in the morning and evening; for details see *Meteorological Magazine*, 1922, p. 174; 1924, p. 142; 1925, p. 155.

The vertical distribution of water-vapour.

The most notable feature about the distribution of water-vapour is that the amount diminishes very rapidly counting from the surface upwards. Attention was called to this rule by Sir Richard Strachey in six lectures on Geography[2].

The falling off is easily explained as the necessary consequence of the limitation of the pressure of water-vapour at any point to the saturation pressure which corresponds with the temperature of the environment. We can therefore assign a limit to the possible amount of vapour in the atmosphere

[1] Prof. J. W. Gregory, quoted by Griffith Taylor in *Australian Meteorology*, Clarendon Press, Oxford, 1920, p. 291. See also W. F. Tyler, 'The Psycho-physical Aspect of Climate,' *Journ. Trop. Med. and Hyg.*, April, 1907.

[2] *Lectures on Geography*, Macmillan and Co., London, 1888, p. 134.

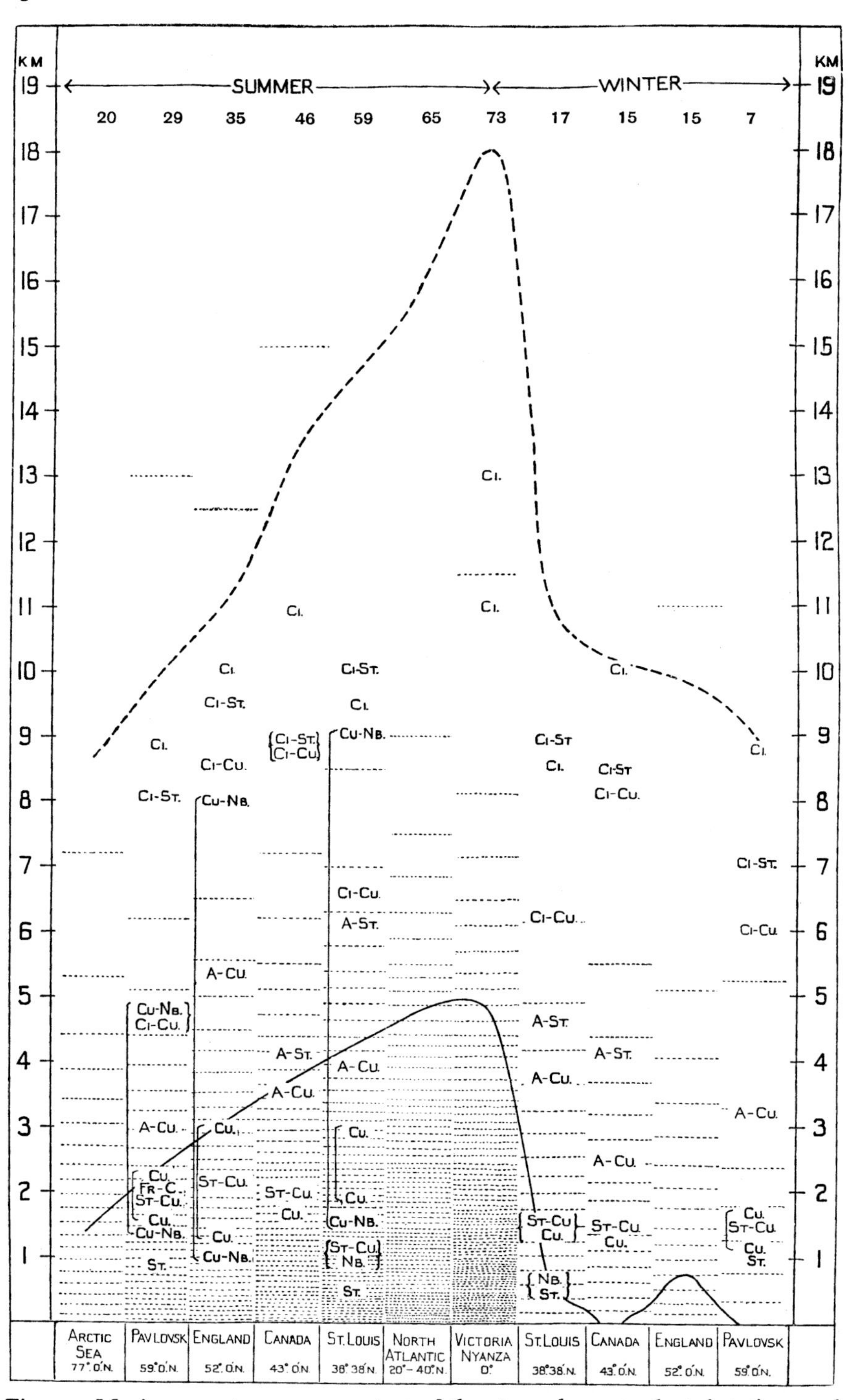

Fig. 71. Maximum water-vapour content of the atmosphere at selected stations and its vertical distribution calculated from the observed temperatures of the upper air, assuming saturation. The distribution is extended to sea-level by extrapolation.

in its normal condition by the pressure of saturation at the normal temperatures as indicated in the tables under figs. 67 and 68. We reproduce here a diagram on that basis (fig. 71) prepared originally for *The Air and its Ways*.

The number of lines in any portion of a vertical column gives the number of kilogrammes of water-vapour in that portion of a saturated column of cross-section 1 square metre, or the amount of rain in millimetres which would fall if the atmosphere were completely desiccated. The total rain-equivalent of the whole column in millimetres is given in figures at the head of it.

We have supplemented the information contained in the original diagram by lines representing the approximate elevation of the tropopause (an interrupted line) and of the freezing-point of water (a continuous line) and also by inserting the names of the different forms[1] of cloud at the height quoted in the entablatures of figs. 73–76 or elsewhere as being appropriate to the locality indicated in the diagram.

In this diagram we can easily see that the "falling-off" of water-vapour with height is in fact represented by the falling down of rain. If we suppose any one of the columns of the diagram pushed upwards bodily (without deranging the distribution of temperature) the number of horizontal lines crossed would represent the rainfall in millimetres.

It is pertinent here to remark that the saturation pressure of aqueous vapour, and consequently the dew-point and the density expressed in grammes per cubic metre at saturation, remain rigorously controlled by the temperature and nothing else. They do not vary with the pressure of the environment, but the density of the dry air does. It follows that saturated air at high levels in the atmosphere contains by weight a larger proportion of water-vapour than it would at the same temperature at the surface. At the freezing-point, for example, with a pressure of 1000 mb it takes 3·8 g of water to saturate a kilogramme of dry air, but at a level of 4 kilometres where the air-pressure is only 600 mb, the proportion of water to air is 3·8 g $\times$ 10/6, or 6·35 g per kilogramme of dry air.

The condensation of vapour.

From the consideration of air saturated with moisture we pass to the removal of the moisture by condensation of the vapour. This will take place in the form of dew or hoar-frost upon solid surfaces when the temperature of the surface falls below the dew-point of the air which is in contact with it; and that condition may be reached either by the travel of warm, moist air over surfaces which have been previously chilled or by the reduction of the temperature of a surface in consequence of radiation of its heat through the transparent atmosphere above it into space.

Considerable amounts of water may be collected by this process of condensation upon solid surfaces, and similar condensation may take place upon the surface-water of the sea or of lakes when the temperature of the water is below the dew-point of the superincumbent air; in these cases however, owing to the mobility of the water, we must suppose that condensation upon

[1] The definitions of cloud-forms have been re-edited in the new issue of the *International Cloud Atlas*, 1932.

a water-surface is more frequently attributable to the travel of warm air over colder water than to the cooling of the surface-water in consequence of radiation in calm air. What we wish to note here is that the travel of the moist air over cold sea or cold ground causes the condensation of water within the air itself, and in that way a special form of cloud is formed which lies on the surface. The conditions of its formation have been the subject of special study[1]. To this form of cloud the name *fog* is given. Other clouds are formed by the elevation of a general mass of air by some process of convection. It will be our duty in due course to call attention to such evidence as exists for ascribing the formation of the different forms of cloud to one or other of these two causes, and to consider any other physical process by which condensation of the vapour of the free air into water-drops may occur. In this chapter we are concerned only with the prevalence and distribution of clouds and mainly with those in the free air, but fog, regarded as cloud on the surface of water or in the hollows of land, though not essentially different in physical conditions from clouds in the free air, is so distinct for the purposes of practical meteorology that we shall preface our representation of the distribution of cloud with some preliminary information about fog.

CLOUD AT SEA-LEVEL. FOG

In its influence upon navigation fog is an important meteorological element, and it is by no means without interest from the scientific point of view; but it is not at all easy to give an effective representation of its relation to the general circulation of the atmosphere. The difficulty arises partly from the very capricious character of the formation of fog and also partly from the uncertainty which attaches to the use of the word fog and the related word mist, in meteorological observation at sea and elsewhere. Both are concerned with the impression produced upon the observer as regards visibility, for which a scale is now used, discontinuous indeed, because it is defined by the invisibility of selected objects at fixed distances, but continuous in the sense that it ranges from the excellent visibility of more than 50,000 metres to the extremely limited visibility of less than 50 metres, with eight intermediate stages. The visibilities 0, 1 and 2, ranging to 500 metres are called fog in the English version of the code; when the range of vision is extended beyond 500 metres the word mist might appear in the observer's "log" unless the atmosphere were so clear that the impression produced was more precisely absence of mist. [Visibility between 500 m and 1000 m is now *moderate fog*.]

In this connexion we may refer to the illustrations of nebula in the cloud-photographs of volume I, chapter XI, from which we form the impression that there is a real distinction between a cloud, *nubes*, which has a boundary of definite shape, and the nebula or milky appearance, the visible shape of which indicates the boundary of the illuminating beam that makes the nebula visible and not that of the material of which the visible cloudiness is evidence. So we may draw a distinction between fog, as a cloud, *nubes*, with a visible out-

[1] Shaw, *Forecasting Weather*, chaps. XIII and XIV, Constable and Co., 1923.

line when seen from a distance in suitable illumination, sometimes spoken of as a "wall of fog," and the mist or nebulous cloudiness which interferes with clear vision but has no definite boundary. With that distinction in mind we can speak of fog as filling a valley or capping a hill, though to a person within the cloud there may be continuous range of obstruction to visibility from slight obscuration of a very distant view to the inability to "see one's hand before one." Equal interference with vision may be caused by a small thickness of *nubes* or a greater thickness of *nebula*. The boundary of a cloud which looks so definite at a distance is generally quite indefinite as one passes through it; yet wreaths and wisps of real cloud are often distinctly visible and, in fact, it is quite possible that its power of reflecting light is an important means of protection for a fog against the consuming power of the sun.

In like manner a cloud of dust or smoke is often called a fog by persons who are within it though it would not be so called by observers from without. Nor does the definition of visibility dispose of the difficulty. The most convincing obstacle to visibility is a heavy snowstorm; it can bring railway traffic to a standstill, but it is not technically speaking fog; so also heavy rain can obstruct vision very effectively and the drizzling rain of a Scotch mist still more so, but neither of these is rightly called fog. A wisp of drifting cloud shuts out the vision of distant objects with a completeness that the rainstorm cannot emulate. A cloud can make even the sun invisible but a nebula cannot.

We do not here enter further into the implications of the physical difference between the fog-cloud with a clearly marked boundary and the nebula; that question belongs to a subsequent volume on the physics of the atmosphere, but we wish to point out that in regard to its relation to the general circulation it is the cloud with definite outline, formed at the surface and represented by the word fog, which we have in mind.

Another characteristic that makes it difficult to treat fog in a general manner is that being a surface-phenomenon it is subject to the exceptional conditions, particularly of heat, cold and moisture, that are peculiar to the surface. In consequence, conditions for the formation of fog may occur locally on a river-plain or in a sheltered valley which are altogether unlikely even a few kilometres away. There are frequent occasions of hill-tops in sunshine while the neighbouring valleys are in fog, and *vice versa* occasions when the valleys are clear but clouds cap the hills. Yet, while this must be recognised as true, there are occasions when fog may be reported from nearly every station in Western Europe north of the Alps, all over Western France and the British Isles from Lyons to Shetland. Fog on such occasions must be an indication of general conditions which are favourable for its development. There can, of course, be no question that on land the loss of heat by radiation to the sky is a primary condition for fog, as it is for dew or hoar-frost, and in discussing questions of the incidence of fog the lack of quantitative information about solar and terrestrial radiation is a serious disadvantage.

In order to illustrate the prevalence of this phenomenon we refer particularly to the best-known home of fog, the North Western Atlantic in the neighbour-

NORMAL FREQUENCY OF FOG

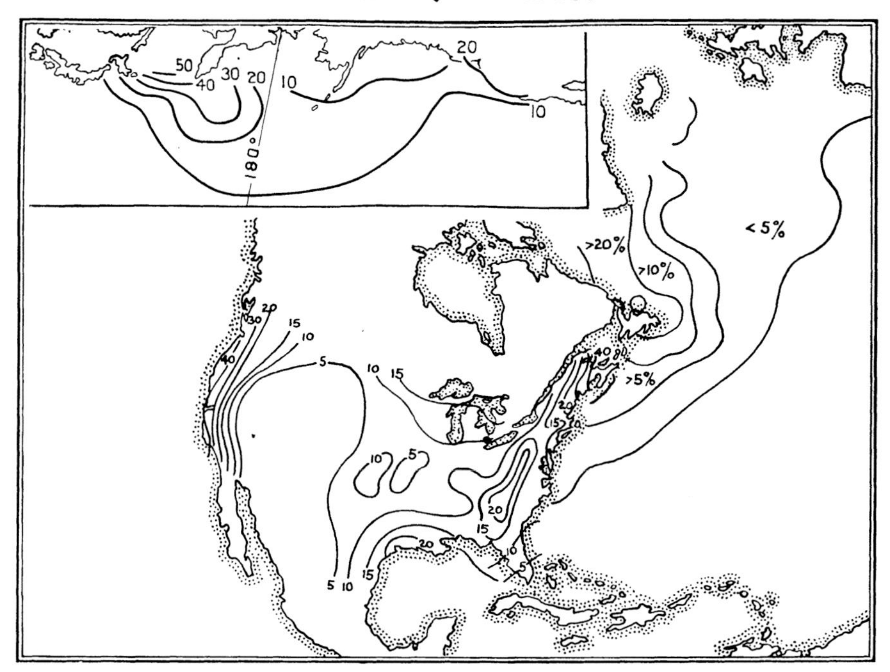

Fig. 72. Composite map: annual number of *days of fog* in the United States (de Courcy Ward) and annual number of fog-observations marked as percentage of the number of observations of weather over the ocean on the Eastern side *every four hours*. (Monthly Meteorological Charts of the North Atlantic.) With inset of corresponding percentage for July over the North Pacific (K. Tsukuda).

Seasonal variation of fog at land-stations in the Northern Hemisphere.

From M.O. normals: days per month and per year.

Station	Lat.	Long.	Jan.	Feb.	Mar.	Apr.	May	June	July	Aug.	Sept.	Oct.	Nov.	Dec.	Year
Dutch Harbour	54° N	167° W	0	0	0·2	0	0·8	0·7	2	2	0	0·2	0·2	0	6
Eureka	41	124	3	4	5	1	1	3	8	7	11	10	8	4	65
Eastport, Maine*	45	67	2	2	3	3	6	7	11	11	6	4	2	2	59
Belle Isle†	52	55	4	5	7	6	9	13	20	15	11	12	10	4	116
St John's, N.F.	48	53	2	2	3	5	6	5	5	4	3	3	3	2	43
Lerwick	60	1	0·4	1	1	2	4	6	7	5	3	2	1	1	33
Valencia	52	10	0·1	0·2	0·2	0·4	0·3	0·4	0·3	0·6	0·5	0·1	0·4	0·2	3·7
Aberdeen	57	2° W	0·5	0·3	1	0·9	3	3	1·5	2	2	2	1	0·7	18
Helsingfors	60	25° E	8	8	8	8	6	3	3	4	7	8	8	6	77
Vardo	70	31	—	—	—	0·2	2	4	6	5	1	0·4	—	—	19
Lyons	46	5	9	4	0·4	0·2	0·2	1·0	0·5	0·7	1·8	5	9	11	43
Alexandria	31	30	2	0·3	0·3	—	—	—	—	—	—	0·3	1	0	4
Baghdad	33	44	2·8	0·6	0·1	—	—	—	—	—	—	0·1	0·7	2·0	6·3
Barnaoul	53	84	8	10	8	4	1	1	1	5	6	6	5	8	63
Petropavlovsk	53	159	0·1	0·3	0·4	4	11	14	17	14	7	2	0·7	0·2	71
Vladivostok	43	132	2	2	4	7	12	15	17	12	2	3	2	2	80
Hong Kong	22	114	4	5	8	7	2	1	1	3	3	1	1	3	39
Osaka	35	135	1·3	0·8	0·9	0·5	0·3	0·5	0·2	0·2	0·5	1·1	1·6	1·8	9·7
Nemuro	43	146	0·7	1·4	3·1	7·6	11·0	16·0	17·4	16·4	6·5	2·8	1·4	1·1	85·4
Naha	26	128	0·1	0·1	0·1	0·1	0·2	—	0·1	—	0·0	0·0	—	—	0·7

* Number of dense fogs.

† Data of fog in the Gulf of St Lawrence are in course of publication as Memoir No. 1 of the Canadian Meteorological Service.

hood of Newfoundland. The map (fig. 72) shows a large area with a number of entries of fog exceeding 20 per cent. of the whole number of observations of which there are six daily. The distribution through the year is by no means uniform. At Belle Isle there is a maximum of 20 days in July, 15 in August, diminishing to 4 days in December and January.

Thus the Atlantic fogs are principally summer fogs. On the other hand, the fogs of the East coast of England are chiefly winter fogs. Yarmouth, for example, has recorded fog at 7 h on 25 days in the winter half-year and 5 in the summer. Indeed, a table of the prevalence of fog in the British Isles is interesting as showing a summer maximum for the more exposed parts of the coast in the North and West, whereas inland and on the shores of the narrow seas the maximum is in the winter. We should find the same in the case of inland stations on the continent. Land-fog, or East coast fog, is a creature of calm weather; sea-fog also has that characteristic but less markedly than the land—there may be fogs in the Newfoundland region even with considerable winds.

Fogs are prevalent in other parts of the world, notably so over the North Pacific Ocean about the Kurile Islands from May to September and on the West coast of the United States. Those who are at all familiar with meteorological literature will recall the number of beautiful photographs of fog-billows pouring from the sea over the land which are characteristic of the neighbourhood of San Francisco (vol. I, fig. 39). The particulars of fog in the United States have been put together by R. de Courcy Ward, and we have taken the liberty of combining his map, showing the number of days of fog in the year in different parts of the United States, with a map of the percentage frequency of fog in a year's observations in the neighbourhood of Newfoundland derived from the Monthly Meteorological Charts of the North Atlantic published by the Meteorological Office. The reader will notice that 40 days of fog at San Francisco correspond with the same number of days over Maine, while Nova Scotia in a different system just shows a line of 20 per cent. of the year's observations as fog. That figure is comparable with the figure for the greater part of Newfoundland and Belle Isle.

The prevalence of fog in the neighbourhood of Newfoundland suggests the juxtaposition of warm air and cold water as the dominant cause of the phenomenon, because the coast of Newfoundland marks a region where the warm South West air of the Gulf Stream often covers the so-called Labrador Current coming down from the cold North and bringing with it vast amounts of accumulated cold in the form of icebergs.

We have compiled two tables of data respecting fogs in coastal areas of the Northern and Southern Hemispheres. Observations of fog at sea are generally available and others are supplied in records of journeys by air. A bibliography is given in chap. X, note 15. We add a special table for the Arctic. The tables are not so eloquent as we could wish them to be. With the introduction of visibility as an international element we may expect an improvement in the situation.

Seasonal variation of fog in the Arctic.

Days per month and per year.

Station	Lat.	Long.	Jan.	Feb.	Mar.	Apr.	May	June	July	Aug.	Sept.	Oct.	Nov.	Dec.	Year
Fram	See p. 126		0	0	2	1	2	10	20	16	10	5	1	0	67
Maud, 1923	75°	158–167° E	—	—	—	—	—	19	27	30	13	5	2	0	—
1924	70–77°	135–165°	2	0	2	0	8	20	23	23	16	3	1	2	—
1925	71	162½	6	4	7	4	17	15	22	—	—	—	—	—	—
Inglefield Bay	77	67½ W	1	0	3	2	?	5	7	11	5	2	1	1	—
Upernivik	73	56	1	1	1	3	5	10	11½	7	2	1	<½	1	45
Angmagssalik	66	38° W	0·8	1	1	3	7	9	8	7	4	3	1	0·4	45
Greenland ice*	67°	42°	24	18	7	—	—	—	—	—	11	14	21	21	—
Gjesvaer	71	25½ E	0	0	0	<½	1	3	4	2½	1	<½	0	<½	11
Waigatz	70½	59	7	8	9	10	9	15	21	23	7	5	4	5	124
St Phoka Bay	76	60	6	2	6	13	16	27	21	20	9	5	1	0	126
Taimyr. Penin.	76	91–95	5½	8	9	8½	1½	7	15	12	4	4	2½	7½	85

* Visibility less than 880 yards. The chief cause of low visibility on the ice-cap is drifting snow.

Seasonal variation of fog in the Southern Hemisphere.

From M.O. normals: days per month and per year.

Station	Lat.	Long.	Jan.	Feb.	Mar.	Apr.	May	June	July	Aug.	Sept.	Oct.	Nov.	Dec.	Year
South America															
Pernambuco	8° S	35° W	5	6	8	9	8	6	4	6	8	5	4	5	74
Ondina	13	39	1	1	2	3	2	2	2	4	4	2	2	2	27
Rio de Janeiro[1]	23	43	10	11	14	16	19	19	21	21	18	15	11	8	183
Santos	24	46	0·6	1	3	5	5	8	8	8	6	4	2	1	52
Asuncion	25	58	2	0·6	2	4	6	7	4	5	2	1	0·6	0·6	35
Parana	32	61	0·0	0·3	0·1	0·4	1	2	2	2	0·3	0·1	0·1	0·0	8
Buenos Aires	35	58	0	0·1	0·3	0·4	1	2	1	1	0·5	0·2	0·1	0	7
Monte Video	35	56	0·7	0·5	2	4	7	10	11	7	5	3	1	0·2	51
Bahia Blanca	39	62	0	1	1	1	0	1	0	0·7	0·3	0·7	0	0	6
Falkland Is.*	52	58	6	3	5	4	3	5	6	4	5	5	3	5	54
Australasia															
Thursday Is.	11° S	142° E	0	0	0	0	0	0	0	0·1	0	0	0	0	0·1
Brisbane	27	153	0·4	0·6	1	3	4	4	4	4	2	1	0·4	0·3	26
Broome	18	122	0	0	0·3	0·2	0·8	1·6	2·6	1·8	2·3	0·4	0	0	10
Carnarvon	25	114	0	0	0	0·4	0·3	0	0·4	0·2	0	0·1	0	0	1·4
Perth	32	116	0·1	0·3	0·5	0·6	0·5	1·1	1·2	0·8	0·2	0·1	0·1	0·1	6
C. Leeuwin	34	115	1·5	0·8	0·3	0·3	0·3	0·4	0	0·2	0·1	0·2	0·9	0·3	5·3
Port Darwin	12	131	0	0	0	0	0·2	0·6	1·0	1·6	0·7	0·1	0	0	4
Adelaide	35	139	0	0	0	0·1	1	3	4	1	0·2	0	0	0	9
Norfolk Is.	29	168	1	1	1·5	0·3	0	0	0	0·5	3	2	1	2	13
Port Macquarie	31	153	0·2	0·2	0·7	0·8	0·1	0·1	0·1	0·4	0·2	0·2	0·2	0·2	3·4
Sydney	34	151	0	0	2	6	6	5	5	4	3	1	1	0	33
Auckland	37	175	0·1	0·2	0·2	0·2	0·7	0·8	1·1	0·4	0·4	0·1	0·1	0·1	4·4
Wellington	41	175	0	0	0	1	1	2	3	1	0	0	0	0	8
Christchurch	44	173	0·1	0·0	0·4	0·3	1·2	1·7	0·8	1·1	0·4	0·1	0·1	0·0	6·2
Chatham Is.	44	177	1·7	1·2	1·3	0·3	0·8	0·3	0·3	0·4	1·5	2·4	2·0	2·2	14·4
Dunedin	46	171	0·4	0·5	0·7	0·9	0·7	0·8	0·5	0·7	0·5	0·5	0·5	0·5	7·2
Campbell Is.	53	169	6	4	17	10	11	11	7	7	7	13	8	12	113
New Plymouth	39	174	0·5	0·2	0·2	0·0	0·0	0·2	0·3	0·0	0·1	0·3	0·3	0·6	2·7
Ocean Islands															
Easter Is.	27° S	109° W	0	0	0	0·5	3	1	1	0·3	0·3	0·3	0	0	6·4
Tonga Is.	19	174	2	1·5	1·5	0	0	0	0	0	0	0	0	0	5
St Paul's Is.†	39	78° E	11	0	2	4	6	4	1	0	·1	2	3	11	4%
Kerguelen	49	70	0	0	—	—	0	1	2	—	—	—	0	0	—
S. Georgia	54	37° W	3	3	4	3	3	1·5	2	0·9	2	2	1	3	30
Laurie Is.‡	61	45	13	16	27	17	24	26	26	26	21	22	23	14	255
S. Orkneys§	61	45	184	139	50	100	197	192	202	228	174	156	97	155	1874
New Year Is.‖	55	64	5	2	2	4	2	5	1	0	2	2	2	4	31

* From observations at Cape Pembroke lighthouse every 4 hours.
† Mist and fog, percentage of observations from ships every 4 hours.
‡ Mist and fog. § Hours. ‖ *Q.J. Roy. Meteor. Soc.*, vol. XLVI, 1920, p. 92.

[1] The figures for Rio de Janeiro are quoted as they appear under the heading "Nevoeiro" in the *Boletim Meteorológico* of the Brazilian Meteorological Service. In the *South America Pilot*, Part I, 1911, no note is made of fog at Rio.

At the surface of the earth, particularly at mountain-stations which are often immersed in cloud, or at stations at lower levels when fog is present, there is visibly more water in the atmosphere than is required to saturate the air according to our tables. Observations in the Austrian Alps have shown that a thick cloud may on occasions contain nearly 5 grammes per cubic metre of water-drops. Water suspended as mist, fog or cloud may be taken as ranging from 0·1 g to 5 g per cubic metre[1]. The higher limit corresponds with the amount of water in saturated air at about the temperature of the freezing-point. Vast quantities of water also may be contained in the atmosphere in the form of rain or hail, but such conditions are transitory and come rather within the scope of the physics and dynamics of the atmosphere than the normal general circulation. Cloud, on the other hand, is characteristic of the general circulation and must have an important influence upon it on account of the shadow which cloud casts and the associated reflexion and scattering of the sun's light and heat. The normal amount of cloud is very different in different localities on the earth's surface and is indeed a notable element of climate. We interpose here, therefore, a series of maps of the annual and monthly distribution of cloud over the globe which were compiled for us in the Meteorological Office by C. E. P. Brooks in 1919 and revised in 1921. They are based upon the recorded estimates of the number of tenths of the sky covered by cloud at the term hours of regular meteorological observations on sea and on land. In this universal extension of observations the science is as fortunate as it is with air-temperature and barometric pressure. The observations which are utilised for these maps, however, give only the total fraction of the sky covered irrespective of the form of the clouds.

Though, with hardly any exceptions, cloud-forms of similar character can be recognised in any part of the world the frequency of occurrence of the various forms is different in different localities, and a very interesting map, or a spherical model, could be made representing the types of cloud specially prevalent in different parts of the world, alto-cumulus and strato-cumulus in the countries bordering the North-East Atlantic, cumulus in the trade-winds, cumulus with castellated extensions in the intertropical oceans, cumulo-nimbus in the special thunder regions, and so on. But the material is not yet sorted out. [Particulars for the Arctic are given in chap. x, note 16.]

We use the space below the maps of cloudiness for some information about the height of clouds, etc. The motion of cloud as being associated with the currents of air by which the clouds are carried belongs to another chapter, namely that on pressure and winds.

[1] *Meteorological Glossary*, p. 67, quoting from Wegener's *Thermodynamik der Atmosphäre*, p. 262. It must not be understood that the condensation of moisture in the air to form water-drops is necessarily a sign that the air in which the condensation takes place is "saturated" in the sense that the vapour-pressure has reached the limit of saturation corresponding with the temperature. The conditions of condensation are very complicated. With suitable nuclei condensation may take place at 80 per cent. of saturation or on the other hand not until the "saturation" point has been passed.

"As much as 0·05 inch, 1¼ mm of water—equivalent to a moderate shower of rain—has been deposited from a California fog in a single night.... In all such cases the moisture from a drifting fog is caught by trees and bushes and then shed upon the ground. The process is called 'fog-drip.'" (C. F. Talman.)

FIGURE 73 NORMAL DISTRIBUTION OF CLOUD

Tenths of sky covered: mean for the day. *Authority:* see p. 208.

Mean height of upper clouds in summer (kilometres).

	Bosse-kop	Storlien	Upsala	Pavlovsk	Potsdam	Trappes	Toronto	Blue Hill	Wash-ington	Allaha-bad	Manila	Batavia*
Lat. ...	70° N	63° N	60° N	60° N	52° N	49° N	44° N	42° N	39° N	25° N	15° N	6° S
Long. ...	23° E	12° E	18° E	30° E	13° E	2° E	79° W	71° W	77° W	82° E	121° E	107° E
Ci. ...	8·32	8·27	8·18	8·81	9·03	8·94	10·90	9·52	10·36	10·76	11·13	11·49
Ci. St. ...	6·61	—	6·36	8·09	8·96	7·85	8·94	10·10	10·62	—	12·97	10·59
Ci. Cu. ...	5·35	6·34	6·46	4·60	6·37	5·83	8·88	6·67	8·83	11·28	6·82	6·30
A. St. ...	4·65	—	2·77	—	4·21	3·79	4·24	6·25	5·77	—	4·30	—
A. Cu. (sup.)	—	4·56	5·22	—	—	—	—	—	—	—	—	—
A. Cu. (inf.)	—	2·74	2·68	—	—	—	—	—	—	—	—	—
A. Cu. ...	3·42	—	3·43	3·05	3·97	3·68	3·52	3·76	5·03	4·50	5·71	5·40
St. Cu. ...	1·34	1·79	1·77	1·85	2·16	1·81	2·00	1·16	2·87	—	1·90	2·39

* Year.

NORMAL DISTRIBUTION OF CLOUD

FIGURE 74

Authority: see p. 208.

Tenths of sky covered: mean for the day.

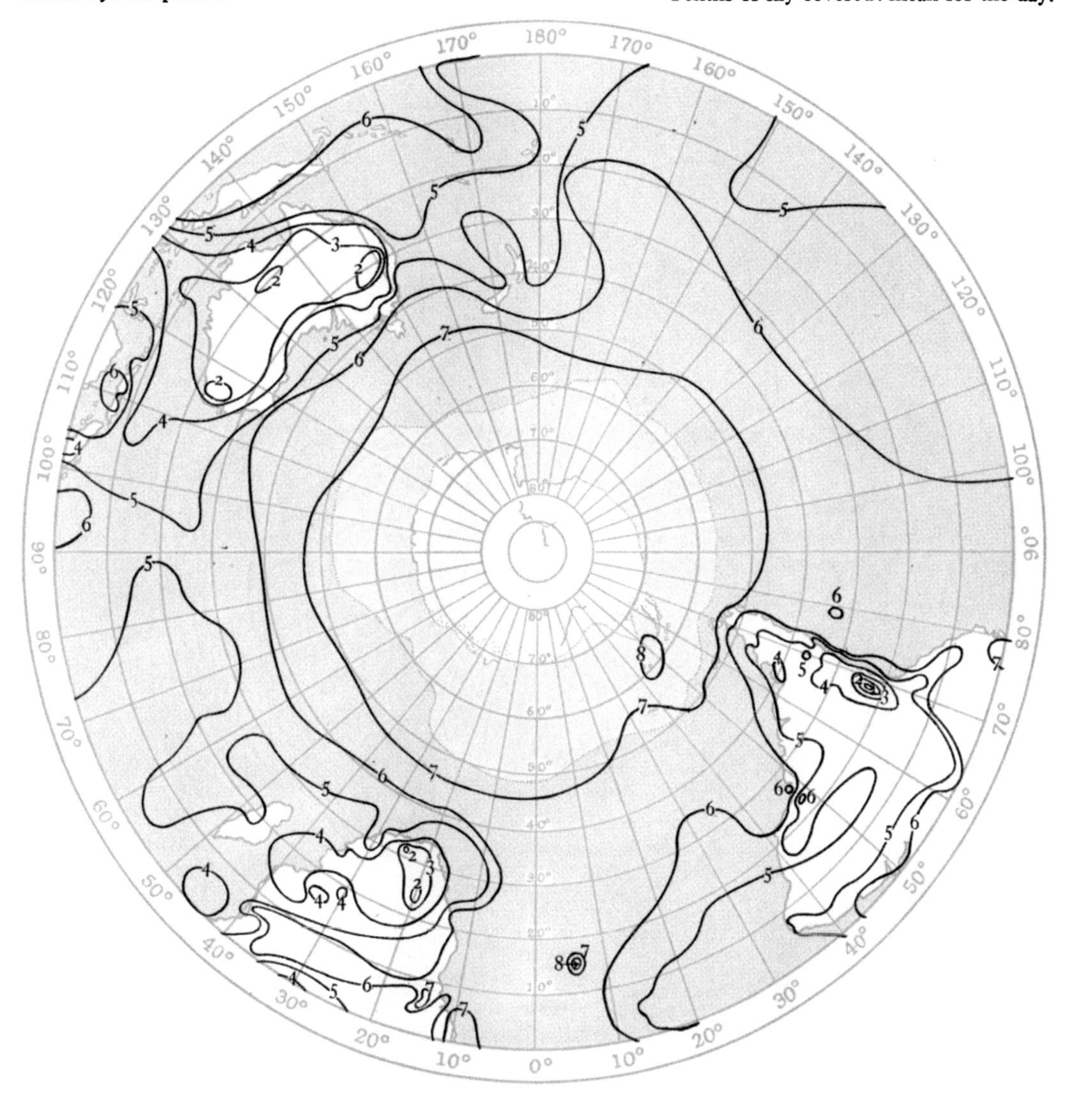

Mean height of lower clouds in summer (kilometres).

	Bossekop	Storlien	Upsala	Pavlovsk	Potsdam	Trappes	Toronto	Blue Hill	Washington	Allahabad	Manila	Batavia*
Lat. ...	70° N	63° N	60° N	60° N	52° N	49° N	44° N	42° N	39° N	25° N	15° N	6° S
Long. ...	23° E	12° E	18° E	30° E	13° E	2° E	79° W	71° W	77° W	82° E	121° E	107° E
Nb. ...	0·98	1·66	1·20	—	1·79	1·08	—	1·19	1·93	0·84	1·38	—
Cu. ...	—	1·68	1·68	1·76	1·88	1·57	1·70	—	—	1·76	1·83	1·74
Cu. (top)	2·15	2·18	2·00	2·41	2·10	2·16	—	2·90	3·07	—	—	—
Cu. (base)	1·32	1·40	1·45	1·63	1·44	—	—	1·78	1·18	—	—	—
Fr. Cu. ...	—	—	1·83	2·15	1·71	1·40	—	—	—	—	—	—
Cu. Nb. ...	—	—	—	—	—	—	—	—	—	1·71	6·45	2·00
Cu. Nb. (top)	3·95	—	3·97	4·68	3·99	5·48	—	9·03	4·96	—	—	—
Cu. Nb. (base)	2·04	—	—	1·61	2·06	2·52	—	1·60	1·75	—	—	—
St. ...	0·66	1·00	—	0·84	0·67	0·94	—	0·51	0·84	—	1·06	—

* Year.

FIGURE 75

NORMAL DISTRIBUTION OF CLOUD

Tenths of sky covered: mean for the day.

Authority: see p. 208.

	Upsala	Pavlovsk	Potsdam	Trappes	Toronto	Blue Hill	Washington	Allahabad	Manila
Lat.	60° N	60° N	52° N	49° N	44° N	42° N	39° N	25° N	15° N
Long.	18° E	30° E	13° E	2° E	79° W	71° W	77° W	82° E	121° E
Ci.	6·98	8·74	8·31	8·51	9·98	8·61	9·51	12·88	10·63
Ci. St.	5·45	7·09	8·05	5·85	8·53	8·89	9·53	13·34	11·64
Ci. Cu.	6·13	5·98	6·20	5·63	8·25	6·15	7·41	11·55	6·42
A. St.	4·09	—	2·99	3·82	4·18	4·57	4·80	—	3·90*
A. Cu.	4·11	3·17	3·96	4·27	2·49	3·66	3·82	6·26	4·64
St. Cu.	1·96	1·50	1·42	1·61	1·54	1·60	2·40	3·55	2·32

Mean height of upper clouds in winter (kilometres).

* From Hann-Süring, p. 96.

NORMAL DISTRIBUTION OF CLOUD

FIGURE 76

Authority: see p. 208.

Tenths of sky covered: mean for the day.

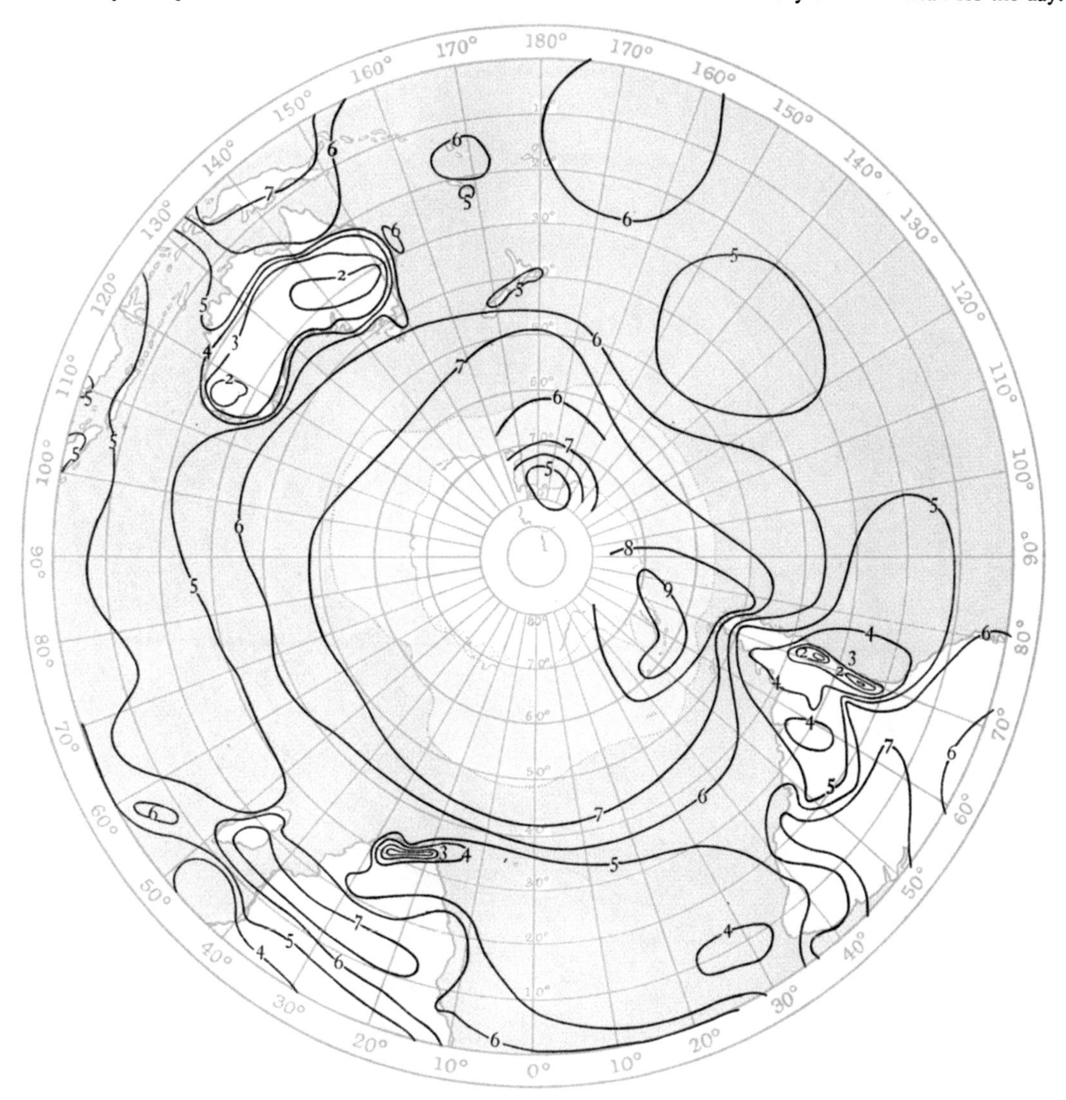

Mean height of lower clouds in winter (kilometres).

	Upsala	Pavlovsk	Potsdam	Trappes	Toronto	Blue Hill	Washington	Allahabad	Manila
Lat.	60° N	60° N	52° N	49° N	44° N	42° N	39° N	25° N	15° N
Long.	18° E	30° E	13° E	2° E	79° W	71° W	77° W	82° E	121° E
Nb.	0·99	—	1·28	1·05	—	0·65	1·80	5·00	1·49
Cu.	1·52	—	1·50	1·11	1·32	—	—	1·34	1·82
Cu. (top)	1·65	1·60	1·74	2·37	—	1·62	2·85	—	—
Cu. (base)	0·71	1·12	0·99	—	—	1·54	1·20	—	—
Fr. Cu.	1·22	—	1·02	1·43	—	—	—	2·62	—
Cu. Nb.	—	—	—	—	—	—	—	2·52	3·14
Cu. Nb. (top)	5·17	—	4·73	3·85	—	—	3·73	—	—
Cu. Nb. (base)	1·38	—	3·82	—	—	—	—	—	—
St.	0·51	1·00	0·61	—	—	0·61	1·13	—	—

FIGURE 77

NORMAL DISTRIBUTION OF CLOUD

Tenths of sky covered: mean for the day.

Authority: see p. 208.

Maximum and minimum altitudes of different types of cloud.

	Bossekop		Upsala		Pavlovsk		Exeter		Trappes		Toronto	
Lat.	70° N		60° N		60° N		51° N		49° N		44° N	
Long.	23° E		18° E		30° E		4° W		2° E		79° W	
	Max.	Min.	Max.	Min.	Max.	Min.	Max.	Min.	Max.	Min.	Max.	Min.
Ci.	11·79	—	11·34	3·61	11·69	4·67	27·41	4·11	12·07	6·35	11·78	8·22
Ci. St.	10·39	—	9·95	2·92	10·12	3·29	15·50	3·84	11·31	4·05	10·63	7·08
Ci. Cu.	8·39	—	10·63	2·46	7·92	2·20	11·68	3·66	10·72	2·83	11·50	5·42
A. St.	6·19	—	6·62	1·47	—	—	—	—	9·95	1·36	5·13	2·47
A. Cu.	6·66	—	(6·42)	—	7·80	1·42	9·39	1·83	7·69	0·83	4·39	2·30
Cu. Nb.	9·02•	—	9·02•	1·38†	6·63•	0·98†	6·41•	0·77†	10·35•	0·75†	—	—
Cu.	4·82•	—	4·40•	0·52†	5·71•	0·70†	4·58•	0·58†	5·03•	0·65†	3·86	0·76
St. Cu.	3·21	—	4·39	0·47	3·52	0·74	6·93	0·82	5·09	0·39	2·99	1·06

• Max. height of top. † Min. height of base.

NORMAL DISTRIBUTION OF CLOUD

FIGURE 78

Authority: see p. 208.

Tenths of sky covered: mean for the day.

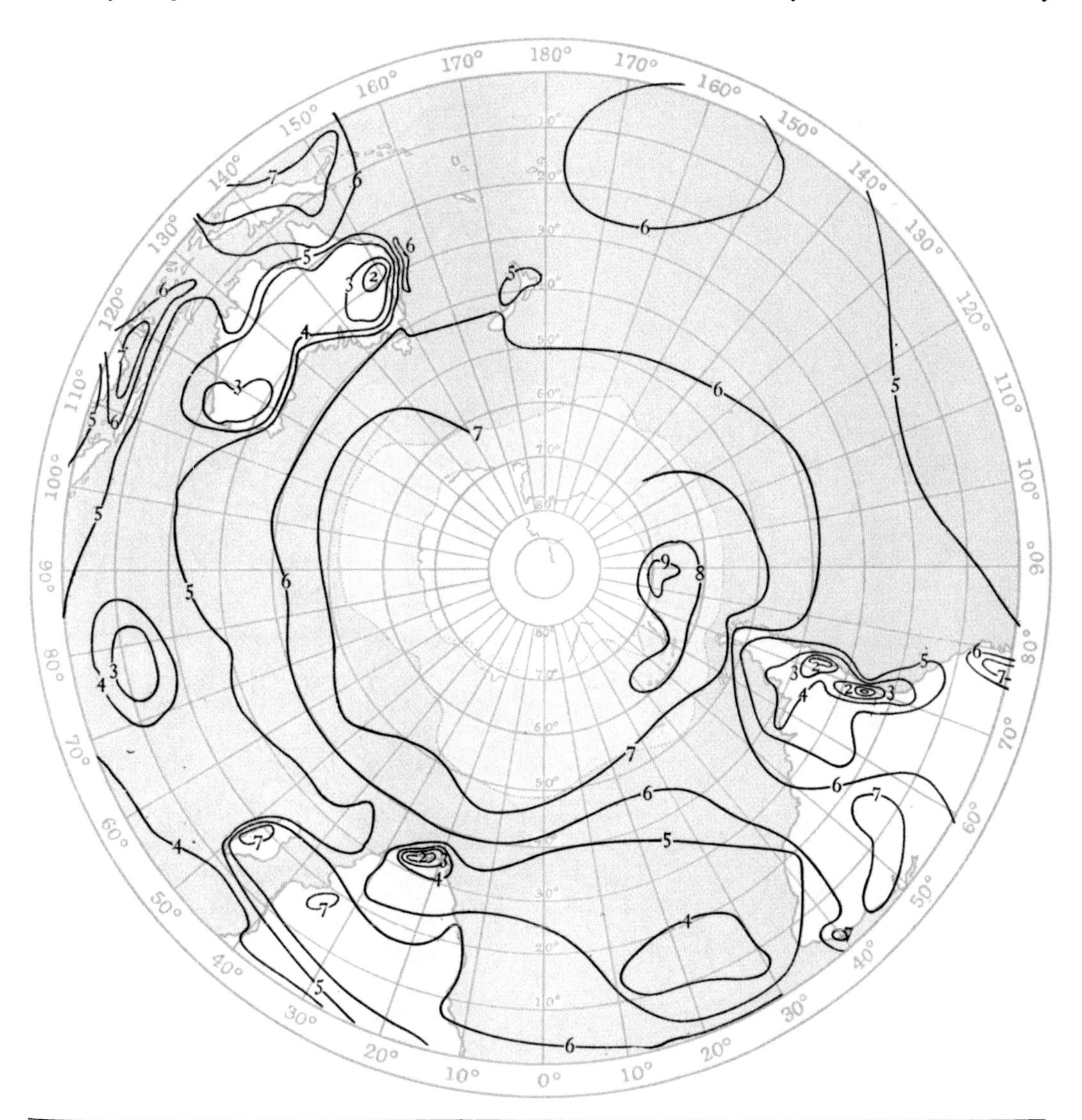

Maximum and minimum altitudes of different types of cloud.

	Blue Hill		Washington		Allahabad		Manila		Batavia		Melbourne	
Lat.	42° N		39° N		25° N		15° N		6° S		38° S	
Long.	71° W		77° W		82° E		121° E		107° E		145° E	
	Max.	Min.	Max.	Min.	Max.	Min.	Max.	Min.	Max.	Min.	Max.	Min.
Ci.	15·01	2·72	17·18	5·43	31·77	?·72	20·45	5·11	18·60	7·87	12·8	6·1
Ci. St.	13·60	4·04	16·14	5·14	27·34	3·51	17·14	6·88	14·21	4·80	11·9	6·1
Ci. Cu.	11·41	2·13	15·41	3·07	36·92	3·96	11·22	3·25	10·96	2·63	10·4	5·5
A. St.	9·69	1·23	15·55	1·61	—	—	7·14	3·21	—	—	9·4	4·3
A. Cu.	9·17	0·98	10·17	1·52	21·85	0·41	8·04	2·91	10·30	1·64	7·0	3·0
Cu. Nb.	13·88*	0·17†	15·90	1·25	7·70	0·47	12·86	0·89	—	—	—	—
Cu.	5·00*	0·54†	5·24	0·54	4·27	0·26	4·45	0·53	4·12	0·74	4·3	—
St. Cu.	4·60	0·31	7·28	1·37	3·46	3·34	3·83	1·34	3·26	1·73	3·7	—

For information as to cloud heights at Potsdam, Lindenberg, Friedrichshafen, Paris (Montsouris) and Melbourne, see pp. 152–3.

FIGURE 79

NORMAL DISTRIBUTION OF CLOUD

Tenths of sky covered: mean for the day.

Authority: see p. 208.

Seasonal variation in the height of cumulus (kilometres).

Station	Type	Jan.	Feb.	Mar.	Apr.	May	June	July	Aug.	Sept.	Oct.	Nov.	Dec.
Lindenberg	A. Cu. + A. St.	2·66	2·97	3·14	2·91	2·63	2·96	3·12	2·95	3·12	2·97	3·13	3·43
,,	Cu.	(0·95)	(1·13)	1·28	1·43	1·56	1·54	1·46	1·42	1·30	(0·78)	(0·98)	(1·60)
,,	St. Cu.	1·13	1·35	1·26	1·49	1·57	1·59	1·62	1·58	1·66	1·49	1·27	1·35
Friedrichshafen	A. Cu.	3·74	3·68	3·74	3·76	4·04	3·87	4·08	3·97	3·91	4·00	4·21	3·93
,,	Cu.	1·84	2·05	1·66	1·65	2·08	1·79	1·62	1·82	1·69	1·67	1·84	1·25
,,	St. Cu.	1·92	2·20	2·19	2·02	2·32	2·31	2·42	2·55	2·16	2·17	1·99	2·03
Potsdam	A. Cu.	3·82	4·19	3·85	3·08	3·92	3·73	3·88	4·13	4·06	4·28	4·16	4·51
Paris	A. Cu.*	1·96	2·23	2·18	2·26	2·46	2·36	2·41	2·35	2·34	2·42	2·09	2·14
,,	Cu.	1·09	1·15	1·35	1·34	1·49	1·42	1·47	1·38	1·28	1·17	1·13	0·96
,,	St. Cu.	1·07	1·01	1·03	1·13	0·97	1·24	1·24	1·13	1·06	0·89	0·93	1·06

* A. Cu. inférieure between 1375 and 3375 m.

NORMAL DISTRIBUTION OF CLOUD

FIGURE 80

Authority: see p. 208.

Tenths of sky covered: mean for the day.

Seasonal variation in the height of cirrus (kilometres).

Station	Type	Jan.	Feb.	Mar.	Apr.	May	June	July	Aug.	Sept.	Oct.	Nov.	Dec.
Pavlovsk	Ci.	7·9	8·3	8·1	8·4	8·2	9·0	9·4	9·2	8·8	9·4	8·5	8·4
Potsdam	Ci.	8·2	8·0	8·0	8·3	8·5	8·9	9·5	9·3	9·6	9·1	8·4	8·2
,,	Ci. St.	7·5	8·1	7·9	7·4	9·0	9·2	9·8	9·8	9·0	8·7	8·1	8·1
,,	Ci. Cu.	6·8	6·5	5·9	6·1	5·8	6·6	6·5	6·8	6·7	5·8	6·3	—
Melbourne	Ci.	10·0	9·4	9·9	8·5	8·9	8·0	8·9	8·5	8·2	8·7	9·4	10·0

Authorities for cloud heights. Les Bases de la Météorologie dynamique, by H. H. Hildebrandsson and L. Teisserenc de Bort. *Batavia*, R. Mag. and Met. Observatory Batavia, vol. XXX, App. II, Utrecht, 1910. *Potsdam*, Hann-Süring, Lehrbuch der Meteorologie, 4te Aus., S. 296; Veröff. des Preuss. Met. Inst. Nr. 317, Abhand. Bd. VII, Nr. 3, Berlin, 1922. *Lindenberg and Friedrichshafen*, Meteor. Zeitschr., XL, 1923, p. 152. *Paris*, Ann. des Services Techniques d'Hygiène, tome III, p. 232; tome IV, p. 255, Paris, 1922 and 1923. *Pavlovsk*, Meteor. Zeitschr., XXVIII, 1911, p. 11. *Melbourne*, Cloud-heights from Melbourne Observatory Photographs by E. Kidson, Report of the Australasian Ass. for the Advancement of Science, vol. XVI, 1923, pp. 153–192. *Exeter*, Cloud Studies by A. W. Clayden, London, 1905.

FIGURE 81

NORMAL DISTRIBUTION OF CLOUD

Tenths of sky covered: mean for the day.

Authority: see p. 208.

Diurnal and seasonal variation of cloudiness in the Arctic, 1898–1902.

Percentage of sky covered.

	Mdt.	2	4	6	8	10	12	14	16	18	20	22	Mean
Jan.	14	13	16	16	19	24	25	27	23	17	18	17	18·9
Feb.	31	34	30	32	44	44	41	41	43	39	34	31	37·0
Mar.	23	22	25	29	36	35	34	36	38	37	29	27	30·9
Apr.	45	45	44	45	47	46	48	46	45	48	50	49	46·7
May	57	58	57	56	57	57	57	56	54	54	56	59	56·5
June	63	60	60	62	60	61	58	59	61	62	64	62	61·0
July	64	62	64	65	63	64	64	59	61	62	62	63	62·8
Aug.	79	77	75	78	77	73	81	80	81	80	75	79	77·9
Sept.	73	72	75	76	72	65	64	65	65	71	69	71	69·4
Oct.	57	62	64	63	68	69	68	72	69	64	60	59	64·6
Nov.	25	26	28	27	35	41	42	40	34	27	30	29	32·3
Dec.	29	27	26	24	25	31	34	31	25	24	29	30	27·9

NORMAL DISTRIBUTION OF CLOUD

FIGURE 82

Authority: see p. 208.

Tenths of sky covered: mean for the day.

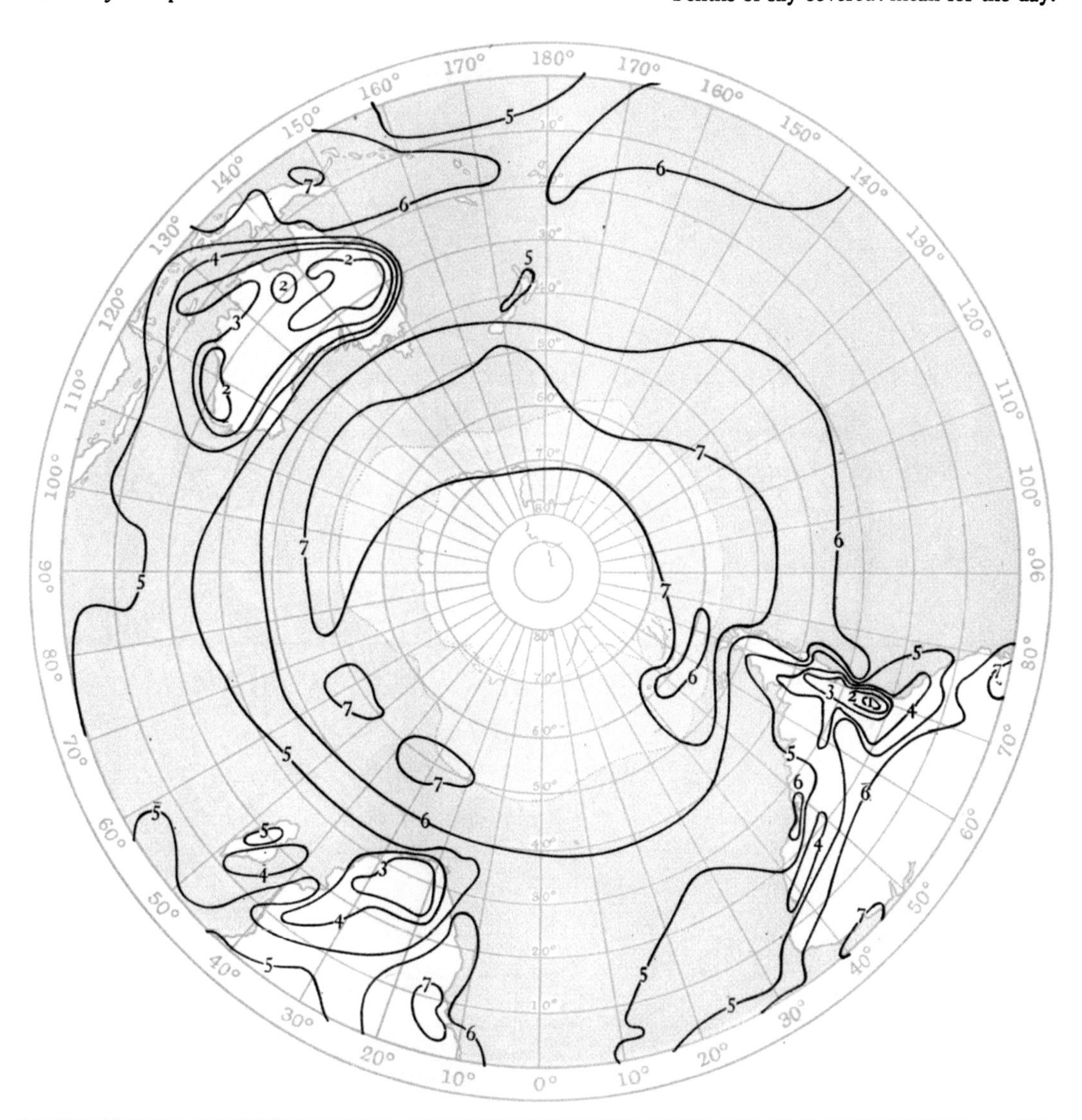

Diurnal variation of cloudiness in the Antarctic, 1911–12, *and at sea* ("*Challenger*").

	Mdt.	2	4	6	8	10	12	14	16	18	20	22	Mean
Spring	75	77	80	85	86	84	83	77	79	77	69	76	78·9
Summer	82	83	83	87	84	80	83	80	76	77	79	80	81·3
Autumn	64	75	78	76	78	82	83	84	78	72	68	61	75·0
Winter	48	50	49	50	50	59	62	55	54	46	44	48	52·2
Occasional observations at sea													
	57	59	59	62	62	58	56	58	59	57	57	57	

Seasonal variation at Helwan, Egypt, 1904–10.

Local time	Jan.	Feb.	Mar.	Apr.	May	June	July	Aug.	Sept.	Oct.	Nov.	Dec.
8 h	44	42	39	31	30	12	20	23	20	27	33	40
11	42	44	45	32	25	6	2	4	7	28	31	39
14	55	48	48	40	32	9	5	6	7	33	37	50
17	53	46	47	41	34	12	5	7	7	32	36	47
20	36	32	29	28	22	6	3	1	2	18	24	30

FIGURE 83

NORMAL DISTRIBUTION OF CLOUD

Tenths of sky covered: mean for the day.

Authority: see p. 208.

Diurnal and seasonal variation of cloudiness at Potsdam, 52° 23′ N, 13° 4′ E, 85 m. 1894–1900.

Percentage of sky covered.

	Mdt.	2	4	6	8	10	12	14	16	18	20	22	Mean
Jan.	74	71	73	76	**84**	82	81	80	80	75	72	71	77
Feb.	69	73	74	76	**83**	80	78	75	73	70	63	65	73
Mar.	59	60	63	73	73	**75**	73	73	73	70	60	59	68
Apr.	57	57	66	70	69	71	**73**	72	70	67	60	53	65
May	49	56	65	63	63	65	67	**69**	66	62	60	53	62
June	50	58	59	57	60	61	**65**	64	64	59	57	55	59
July	51	58	66	66	65	66	**68**	68	68	64	63	55	63
Aug.	44	46	60	58	57	60	**63**	63	60	55	51	43	55
Sept.	47	50	58	**66**	63	62	64	63	60	60	47	46	57
Oct.	62	62	66	**73**	71	70	70	72	68	62	59	61	66
Nov.	66	67	66	70	75	**75**	74	71	70	64	62	65	69
Dec.	71	72	71	70	**78**	78	76	75	76	65	67	72	73

NORMAL DISTRIBUTION OF CLOUD

FIGURE 84

Authority: see p. 208.

Tenths of sky covered: mean for the day.

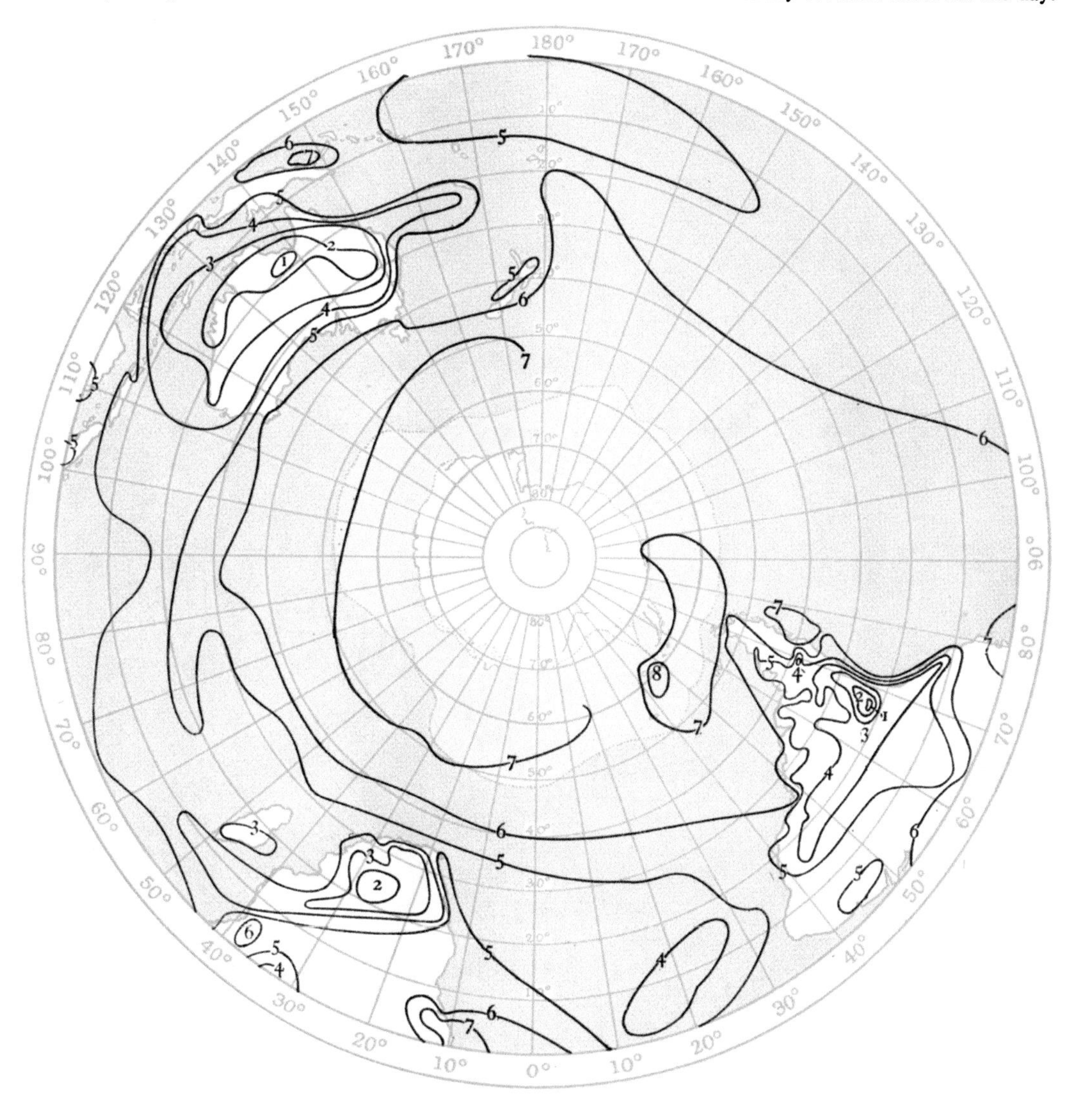

Frequency of snow in Newfoundland, days per month and per year.

	Jan.	Feb.	Mar.	Apr.	May	June	July	Aug.	Sept.	Oct.	Nov.	Dec.	Year
Port aux Basques	21	17	14	8	2	0·1	0	0	0·2	1	11	19	92
St George	19	14	13	6	0·7	0	0	0	0·4	1	8	14	76
Point Riche	6	6	6	4	1	0·2	0	0	0·1	0·5	5	5	32
Cape Norman	11	10	9	8	5	0·8	0	0	0·1	2	7	9	61
St John's	6	9	5	4	2	0·2	0	0	0	0	1	5	32

Normal chances of snow in the British Isles. The odds against one about snow on any day.

	Jan.	Feb.	Mar.	Apr.	May	June	July	Aug.	Sept.	Oct.	Nov.	Dec.	Snow days per year
Shetland	9	6	5	10	38	—	*	—	—	38	17	8	22
Aberdeen	4	3	3	11	33	—	—	—	300	21	11	5	34
N. Shields	5	4½	5	16	76	*	—	—	*	51	15	7	23
Valencia	33	27	30	97	*	—	—	—	—	309	150	38	4
Plymouth	30	27	30	100	309	—	—	—	—	—	100	43	4
Dungeness	9	11	9	32	*	—	—	—	—	154	74	17	12
Yarmouth	6½	6½	6	17	76	—	—	—	*	309	37	12	17

* An off-chance; not more than once in 20 years; about 1000 to 1 against.

FIGURE 85

NORMAL DISTRIBUTION OF CLOUD

Tenths of sky covered: mean for the day.

Authority: see p. 208.

Diurnal and seasonal variation of cloudiness at Trichinopoly, 10° 50′ N, 78° 44′ E, 78 m. 1881–9.

Percentage of sky covered.

	Mdt.	2	4	6	8	10	12	14	16	18	20	22	Mean
Jan.	21	17	18	34	31	42	45	50	49	43	23	16	33
Feb.	12	18	27	35	29	42	30	26	27	19	14	12	24
Mar.	18	22	25	38	37	42	38	37	36	32	24	19	30
Apr.	30	28	34	40	34	39	39	45	45	49	32	30	37
May	44	37	36	46	46	46	46	54	57	60	52	51	48
June	57	60	61	72	68	66	60	64	59	65	51	53	61
July	65	65	65	73	75	71	65	66	67	69	67	67	68
Aug.	70	68	68	76	76	70	60	63	66	72	71	69	70
Sept.	64	61	59	64	61	56	52	56	65	72	70	66	62
Oct.	64	68	72	77	77	75	74	78	84	83	63	61	73
Nov.	51	50	50	60	62	68	69	74	75	74	57	54	61
Dec.	53	53	57	71	69	75	72	77	72	69	54	52	64

NORMAL DISTRIBUTION OF CLOUD

FIGURE 86

Authority: see p. 208.

Tenths of sky covered: mean for the day.

Diurnal and seasonal variation of cloudiness at Batavia, 6° 11′ S, 106° 50′ E, 8·0 m. 1880–1915.

Percentage of sky covered.

	Mdt.	2	4	6	8	10	12	14	16	18	20	22	Mean
Jan.	72	70	69	76	73	76	79	75	75	80	77	75	75·0
Feb.	72	69	70	77	74	76	79	76	76	79	74	73	74·8
Mar.	66	62	59	67	60	64	71	67	71	77	74	70	67·2
Apr.	58	54	50	57	50	54	63	61	64	70	66	62	58·9
May	51	48	44	54	47	49	59	57	57	63	56	54	53·0
June	49	48	44	53	46	46	57	55	52	58	54	52	51·0
July	47	45	41	50	39	41	52	49	43	51	53	55	46·8
Aug.	46	41	36	46	36	40	55	45	37	47	52	51	44·2
Sept.	51	44	39	49	39	47	59	46	43	57	60	57	49·0
Oct.	51	48	46	56	46	56	65	56	58	69	63	58	56·0
Nov.	64	60	56	66	58	67	73	68	73	81	74	69	67·5
Dec.	69	64	62	73	67	72	78	77	80	82	76	71	72·8

FIGURE 87

Tenths of sky covered: mean for the day.

NORMAL DISTRIBUTION OF CLOUD

Authority: see p. 208.

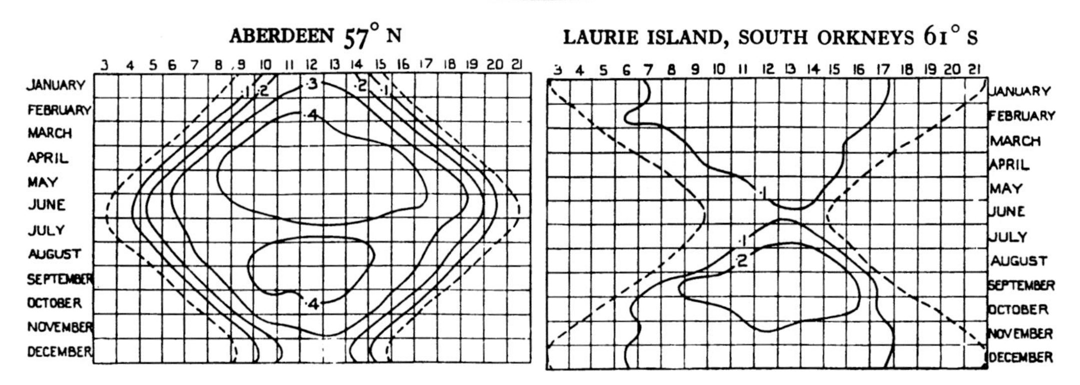

Isopleths of normal sunshine.

NORMAL DISTRIBUTION OF CLOUD

FIGURE 88

Authority: see p. 208.

Tenths of sky covered: mean for the day.

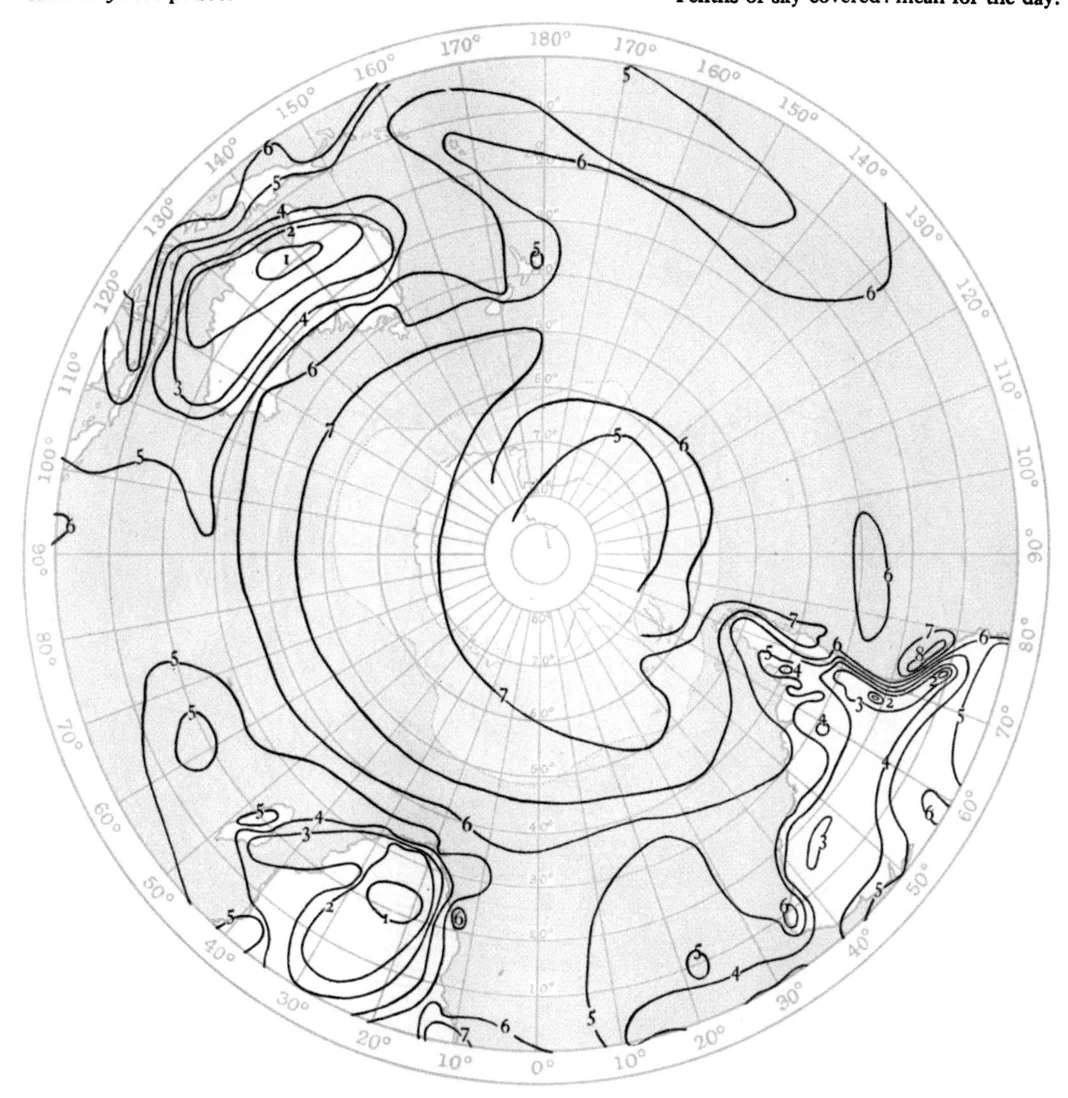

KEW OBSERVATORY, RICHMOND $51\frac{1}{2}°$ N

VALENCIA OBSERVATORY, CAHIRCIVEEN 52° N

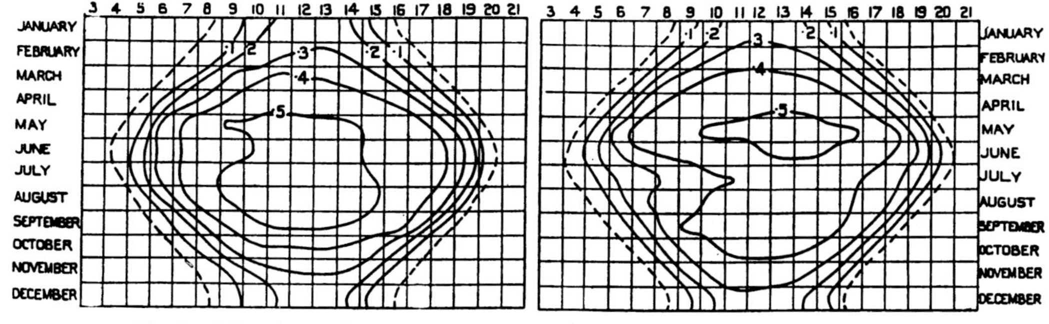

Decimal fractions of the several hours with the limits of sunrise and sunset.

FIGURE 89

NORMAL DISTRIBUTION OF CLOUD

Tenths of sky covered: mean for the day.

Authority: see p. 208.

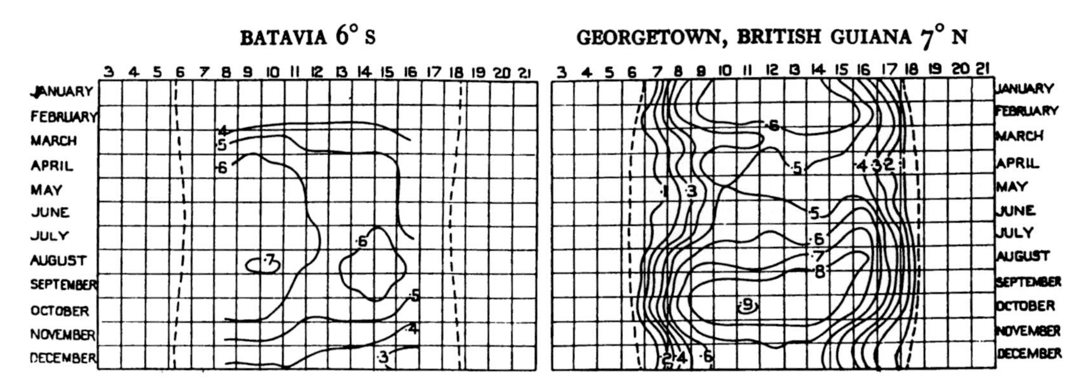

Decimal fractions of the several hours with the limits of sunrise and sunset.

NORMAL DISTRIBUTION OF CLOUD

FIGURE 90

Authority: see p. 208.

Tenths of sky covered: mean for the day.

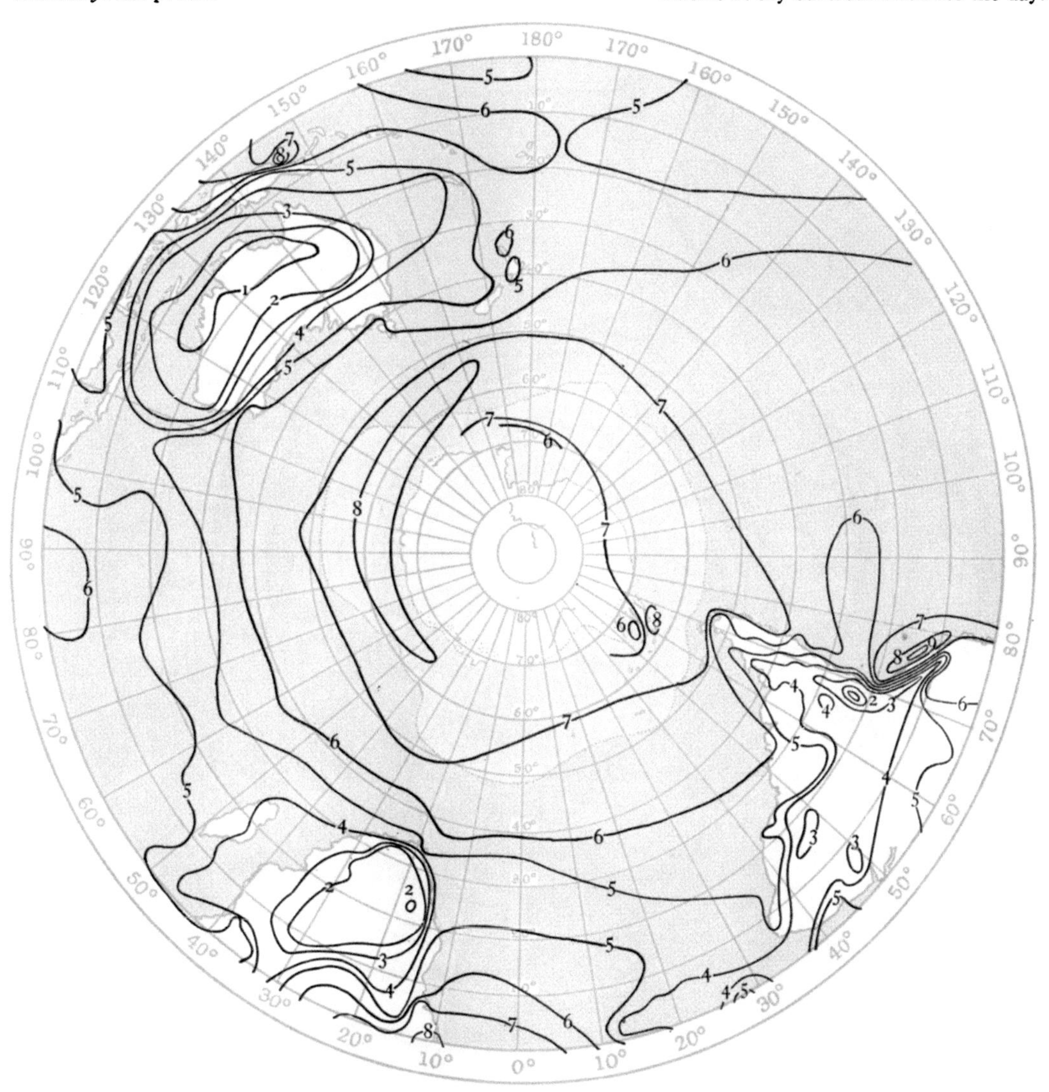

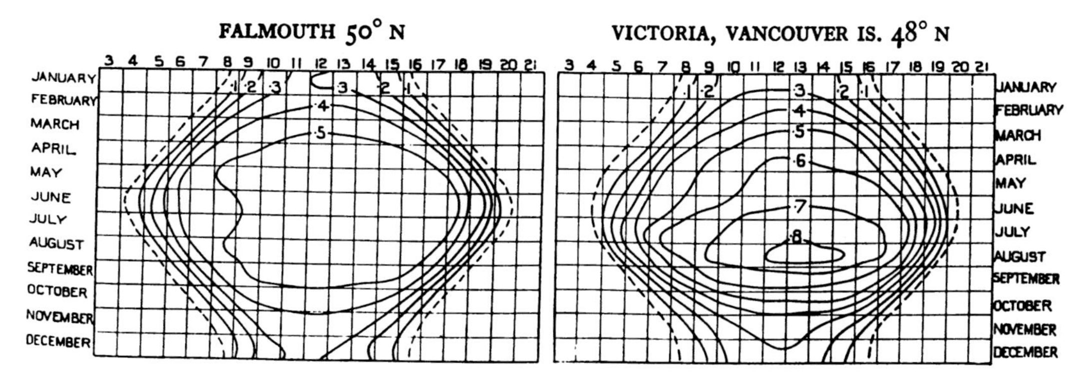

Isopleths of normal sunshine (*continued*).

FIGURE 91

NORMAL DISTRIBUTION OF CLOUD

Tenths of sky covered: mean for the day.

Authority: see p. 208.

Odds against a clear sky at Richmond (Kew Observatory) for given wind directions.

		Calm	N	NE	E	SE	S	SW	W	NW
10 h	Spring	∞•	7	7	3	3	12	10	13	20
	Summer	∞•	8	4	2	3	8	21	16	21
	Autumn	9	7	7	5	5	13	9	5	8
	Winter	2	4	6	5	15	18	12	5	3
16 h	Spring	—	13	8	3	5	19	28	18	19
	Summer	—	11	4	2	5	11	41	32	23
	Autumn	1	7	9	2	17	8	18	14	7
	Winter	—	6	6	6	10	41	15	9	5
22 h	Spring	∞•	2	2	1	1	3	2	2	1
	Summer	1	3	2	1	1	2	2	3	2
	Autumn	1	2	3	2	3	3	3	2	2
	Winter	∞•	2	3	4	4	11	4	2	1

Spring: March to May.
Summer: June to August.
Autumn: Sept. to Nov.
Winter: Dec. to Feb.

• One observation.

NORMAL DISTRIBUTION OF CLOUD

FIGURE 92

Authority: see p. 208.

Tenths of sky covered: mean for the day.

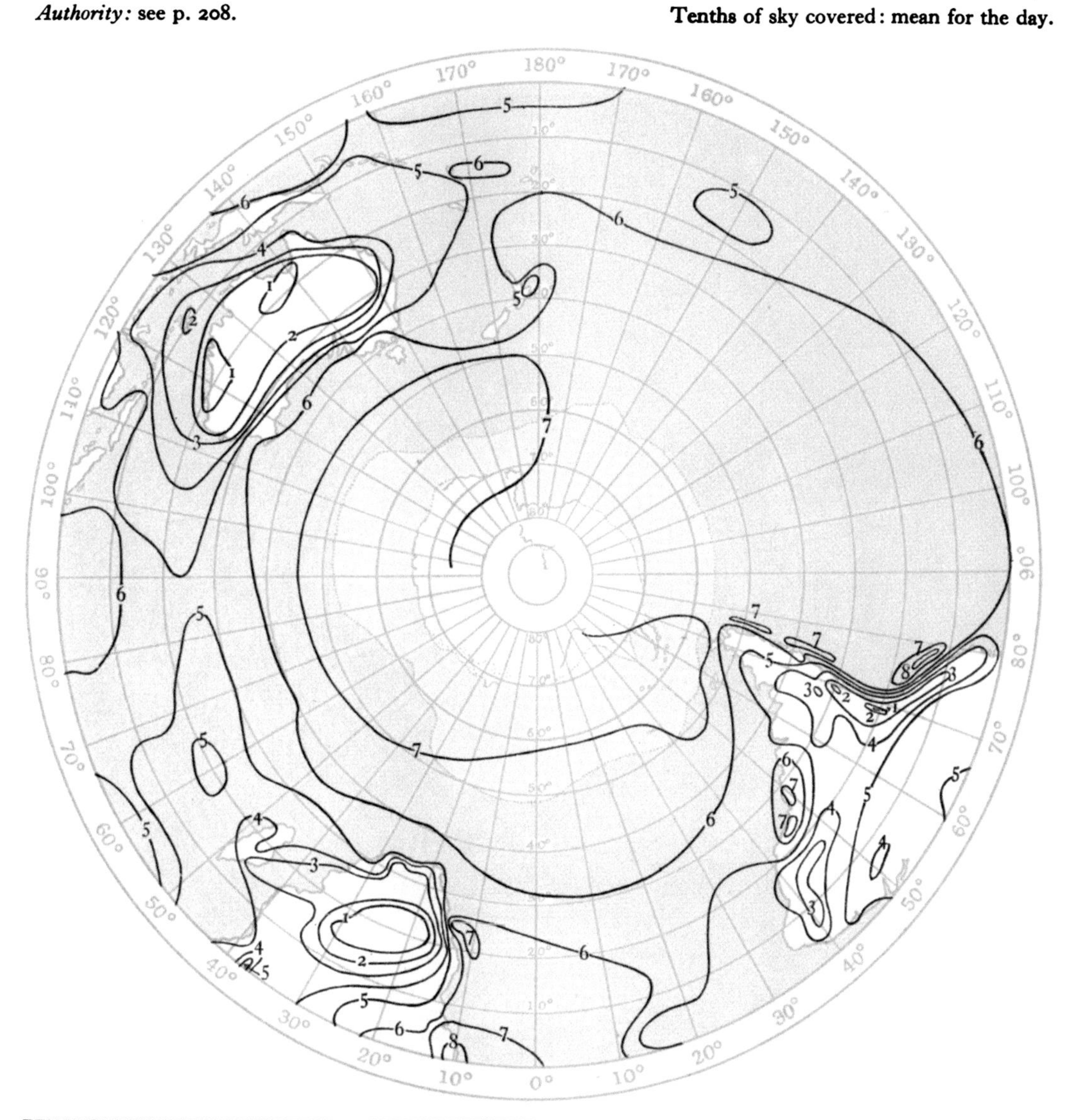

The stratification of the sky. Levels of predominant cloud-frequency.

As representing the result of the consideration of observations by Vettin (Berlin), Ekholm and Hagström (Upsala), Süring (Potsdam), Clayton (Blue Hill) and Bigelow (Washington) we quote the following from *Les Bases de la Météorologie Dynamique*, tome II, p. 324:

"Comme résultat définitif de ces travaux il [Süring] donne les hauteurs suivantes:

500, 2000, 4300, 6500, 8300 et 9900 m.

"Ces hauteurs ne sont que des moyennes générales. Elles varient d'une station à une autre et il y a aussi des variations diurnes et annuelles. On voit cependant que ces hauteurs correspondent sensiblement avec celles des (1) Stratus, (2) nuages inférieurs, (3) Al. st. et Al. cu., (4) Ci. cu., (5) et (6) deux étages dans la région des nuages supérieurs.

"Nous n'entrons pas ici dans une discussion sur les causes de ces couches différentes, ni sur les conditions météorologiques de ces étages séparés. Ces questions, encore peu étudiées, ne trouveront leurs solutions que par les travaux poursuivis avec tant de zèle dans les stations aéronautiques."

[The details of formation and behaviour of clouds at different levels have been discussed by W. Peppler in a number of papers in the *Meteorologische Zeitschrift* or in the *Beiträge zur Physik der freien Atmosphäre.*]

FIGURE 93

NORMAL DISTRIBUTION OF CLOUD

Tenths of sky covered: mean for the day.

Authority: see p. 208.

Weather in the equatorial Atlantic. Square 3, 0–10° N, 20–30° W.

The figures give the number of occurrences as percentage of the number of observations of wind.

Weather	Beaufort letter	Jan.	Feb.	Mar.	Apr.	May	June	July	Aug.	Sept.	Oct.	Nov.	Dec.
Blue sky ...	b	29	28	34·5	32	32	26	34	40	34	35·5	31	27
Cloudy	c	41	37	44	40	44	37	44	43	40	43	41	41
Gloom	g	10	6	4	4	7	3	5	6	5·5	4	6	4
Lightning ...	l	13	9	10	10	7	6	2	1	3	7	10	11
Mist	m	16	14	13	11	7	4	7	7	7	4·5	6	12
Overcast ...	o	14	8	9	9	12	9	11	10	10	10	13	9
Passing showers	p	11	7	8	11	13	10	10	8	10	13	15	10
Squalls	q	10	6	8	8	10	10	10	9	11	14	15	9
Rain	r	20	15	14	16	19	21	19	12	15	20	20	15
Thunder ...	t	5	4	5	4	2	2	<1	<1	1	2	3	5
Unusual visibility	v	1	1	2	1	2	<1	3	3	2	3	2	2
Dew	w	<1	1	1	1	2	1	1	3	1	<1	<1	1

f (fog) and u (ugly) always less than one observation per hundred, h (hail) no occurrence, d (drizzle) included with r never more than two occurrences in 100 observations.

NORMAL DISTRIBUTION OF CLOUD

FIGURE 94

Authority: see p. 208.

Tenths of sky covered: mean for the day.

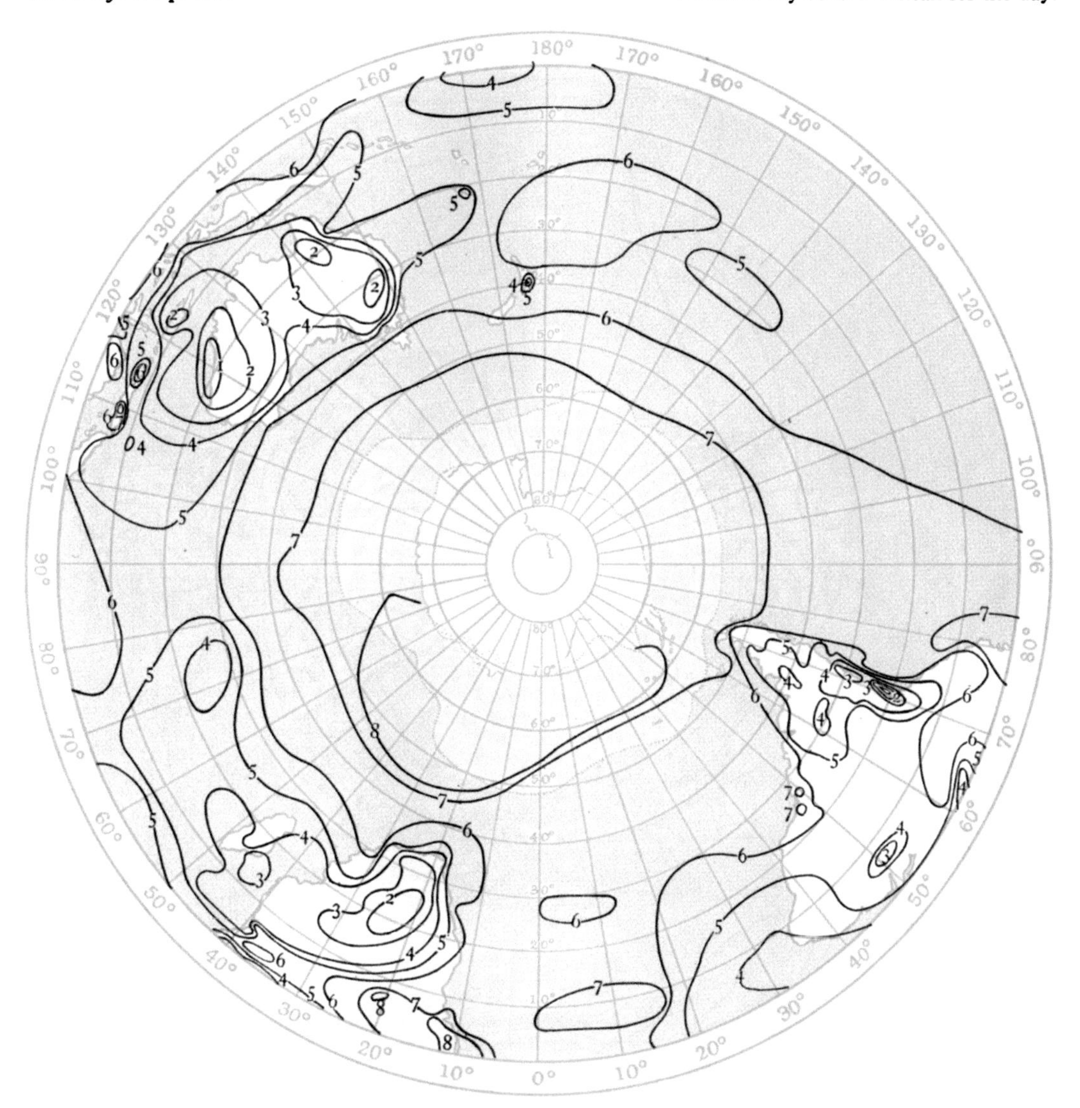

Weather in the equatorial Atlantic. Square 302, 0–10° S, 20–30° W.

The figures give the number of occurrences as percentage of the number of observations of wind.

Weather	Beaufort letter	Jan.	Feb.	Mar.	Apr.	May	June	July	Aug.	Sept.	Oct.	Nov.	Dec.
Blue sky ...	b	41	41	42	41	42	45	49	54	48	54	52	45
Cloudy	c	40	42	47	46	43	41	44	44	40·5	46	57	45
Gloom	g	1	2	<1	2	3	<1	2	<1	1	<1	<1	<1
Lightning ...	l	2	2	6	3	<1	<1	<1	—	—	<1	—	<1
Mist	m	5	5	5	3	4	7	8	8	2	4	4	6
Overcast ...	o	5	4	7	7	5	1	1	2	1	3	2	3
Passing showers	p	8	8	11	13	12	8	9	7	9	6	8	6
Squalls	q	4	5	9	9	11	8	7	7	7	5	4	4
Rain	r	5	7	9	10	9	3	3	1	2	2	2	2
Thunder ...	t	<1	<1	1	<1	<1	—	—	—	—	—	—	<1
Unusual visibility	v	5	3	2	2	5	4	6	3	4	7	2	2
Dew	w	2	<1	<1	<1	<1	3	4	4	2	1	2	1

f (fog) and u (ugly) always less than one observation in a hundred, h (hail) no occurrence, d (drizzle) included with r never greater than two occurrences in 100 observations.

FIGURE 95

NORMAL DISTRIBUTION OF CLOUD

Tenths of sky covered: mean for the day.

Authority: see p. 208.

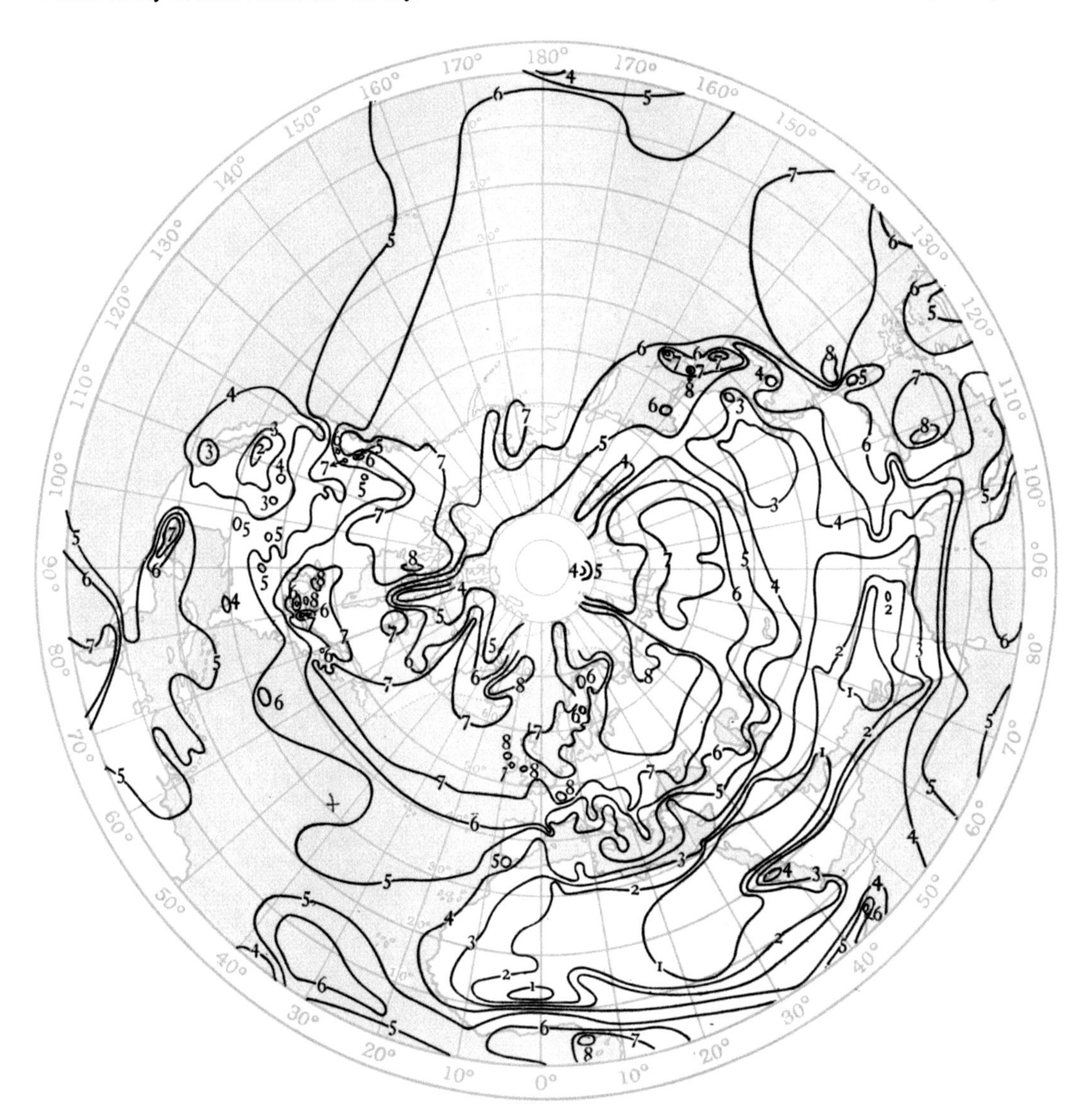

Weather in the equatorial Atlantic. Square 39, 10–20° N, 20–30° W.

The figures give the number of occurrences as percentage of the number of observations of wind.

Weather	Beaufort letter	Jan.	Feb.	Mar.	Apr.	May	June	July	Aug.	Sept.	Oct.	Nov.	Dec.
Blue sky ...	b	40	38	38	39	43	30	36	35	41	42	36	37
Cloudy ...	c	46	39	37	40	48	38	49	42	42	40·5	40·5	41
Gloom	g	4	3	3	2	4	3	5	3	3	3	4	2
Lightning ...	l	<1	<1	<1	<1	<1	<1	1	2	4	4	5	<1
Mist	m	30	25	24	21	22	18	16	15	14	20	16	22
Overcast ...	o	9	5	5	5	6	9	10	11	7	8	8	6·5
Passing showers	p	2	1	1	<1	1	2	8	11	8	5	4	3
Squalls	q	3	2·5	<1	<1	2	2	4	6	8	8	3	2
Rain	r	1	1	1	1	1	1	7	12	9	8	4	1
Thunder ...	t	—	<1	<1	<1	<1	<1	<1	<1	<1	1	<1	<1
Unusual visibility	v	3	2	2	2	2	<1	2	1	2	2	2	2
Dew	w	2	4	3	6	7	4	1	2	2	2	2	3

f (fog) and u (ugly) always less than one observation in a hundred, h (hail) no occurrence, d (drizzle) included with r never greater than two occurrences in 100 observations.

NORMAL DISTRIBUTION OF CLOUD

FIGURE 96

Authority: see p. 208.

Tenths of sky covered: mean for the day.

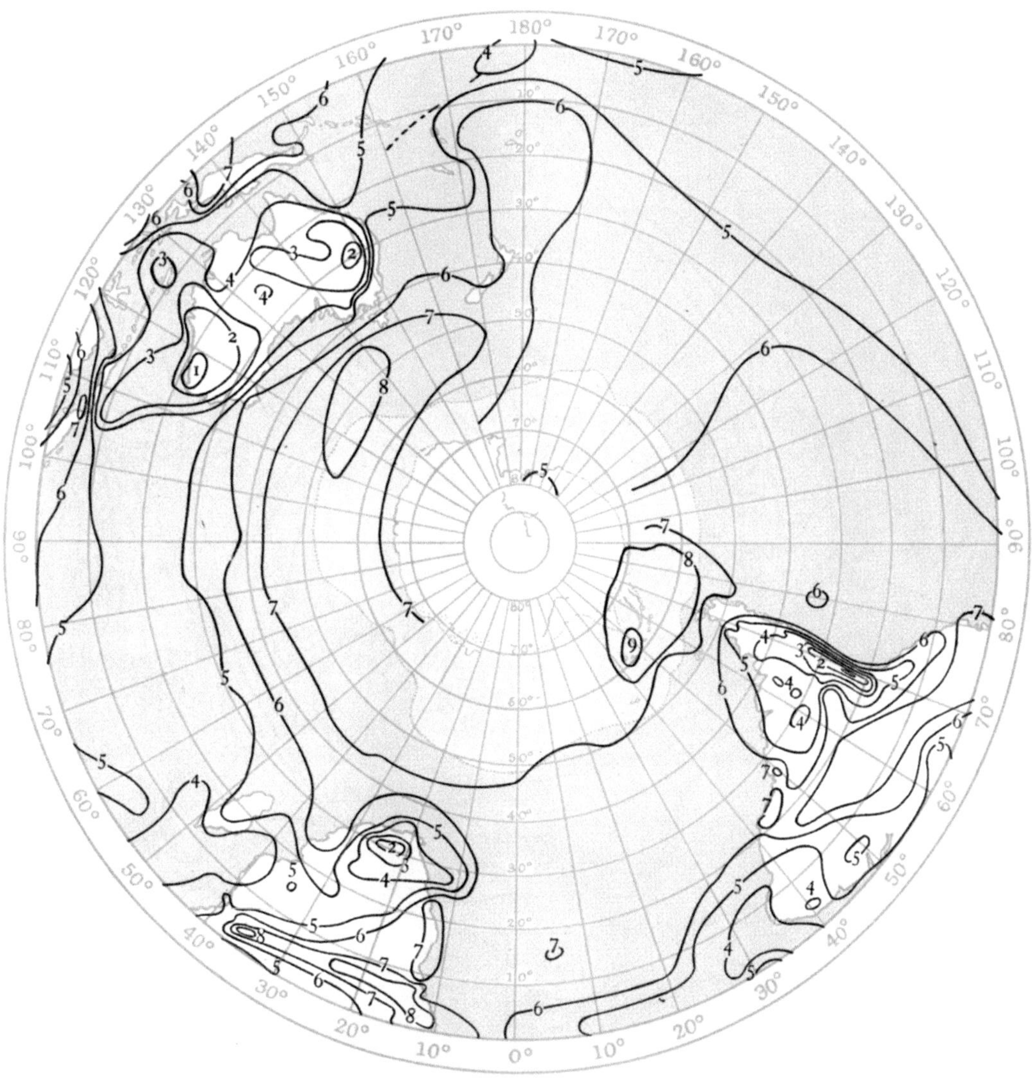

S 13 Feb. 1919 16 Nov. 1917 21 Dec. 1917 S

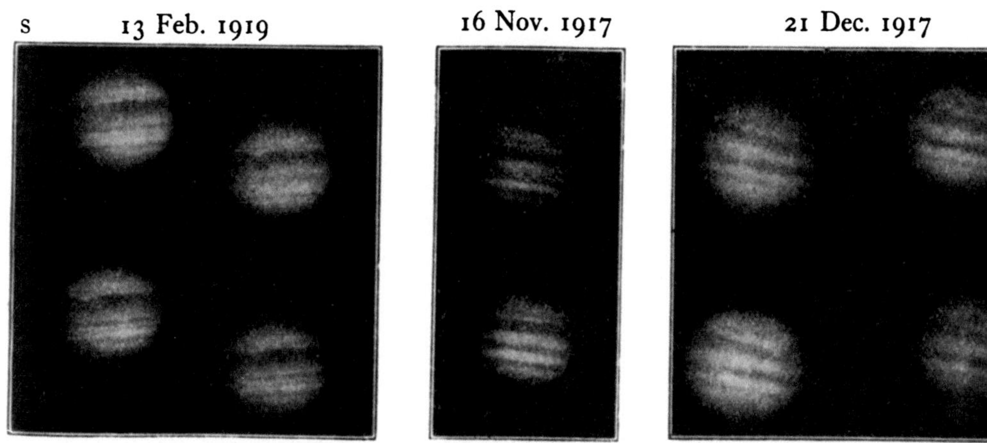

N Photographs of Jupiter by J. H. Reynolds, Birmingham. N

The length of a "day" on Jupiter is 9 h. 55 min. 21 sec.

FIGURE 97

NORMAL DISTRIBUTION OF CLOUD

Tenths of sky covered: mean for the day.

Authority: see p. 208.

19 December, 1917

S S

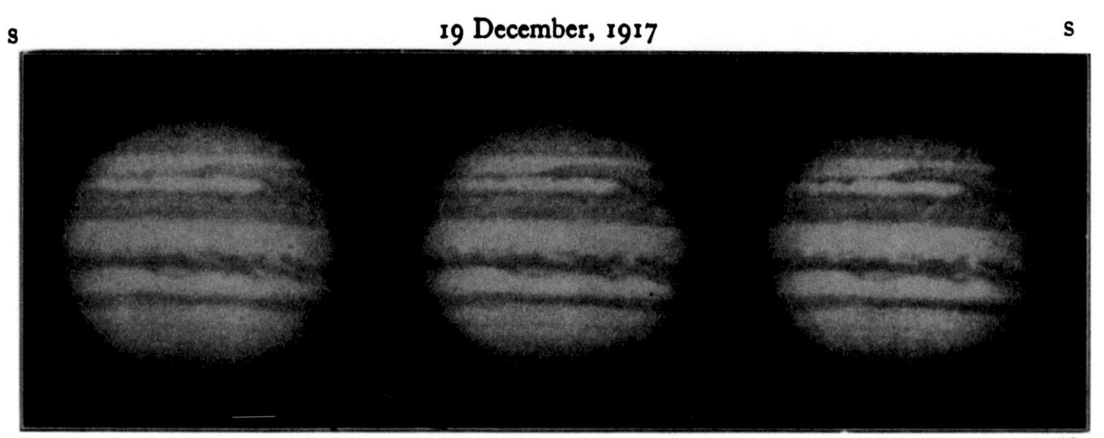

N N

Jupiter. Flagstaff Observatory, Arizona.

NORMAL DISTRIBUTION OF CLOUD

FIGURE 98

Authority: see p. 208.

Tenths of sky covered: mean for the day.

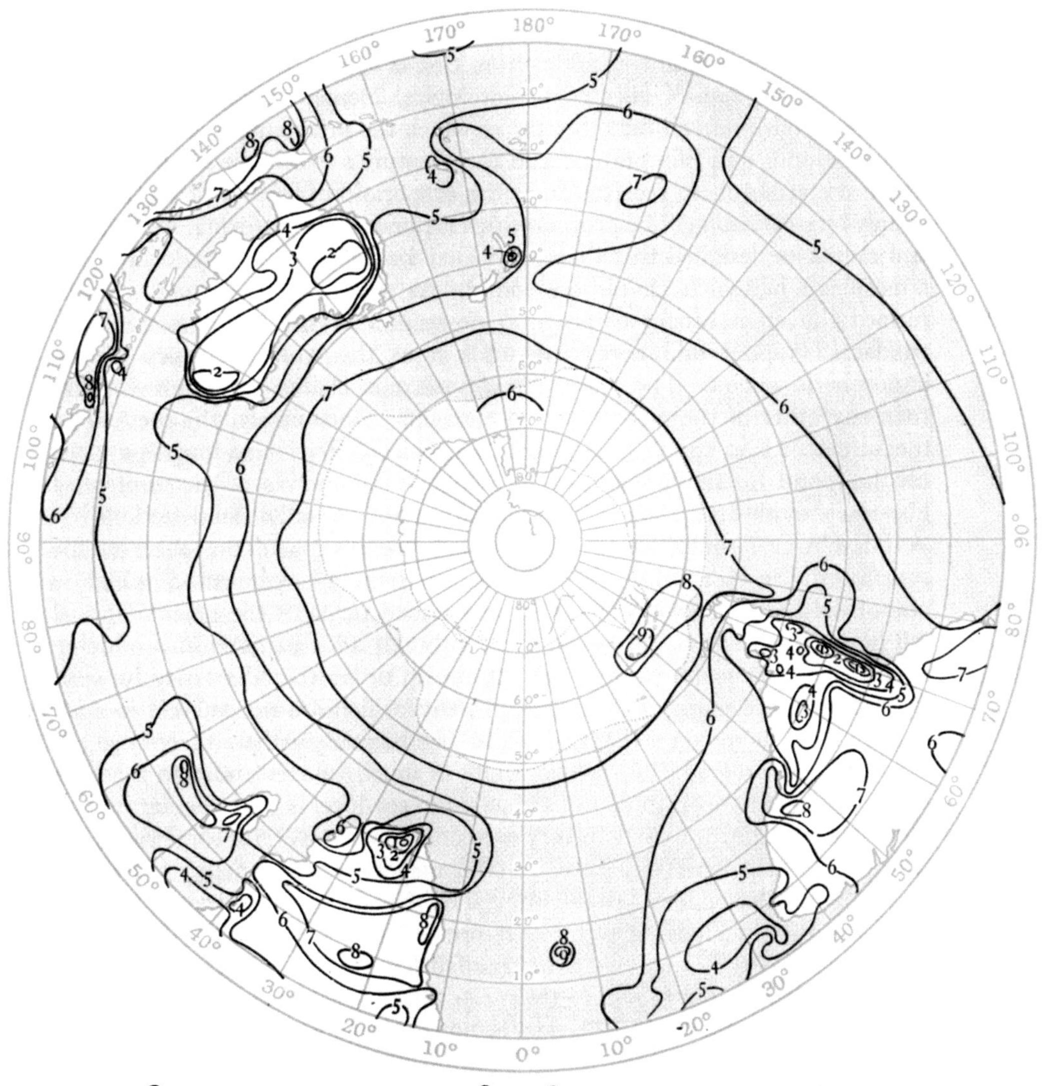

S 19 Oct. 1915 12 Oct. 1891 19 Oct. 1915 S

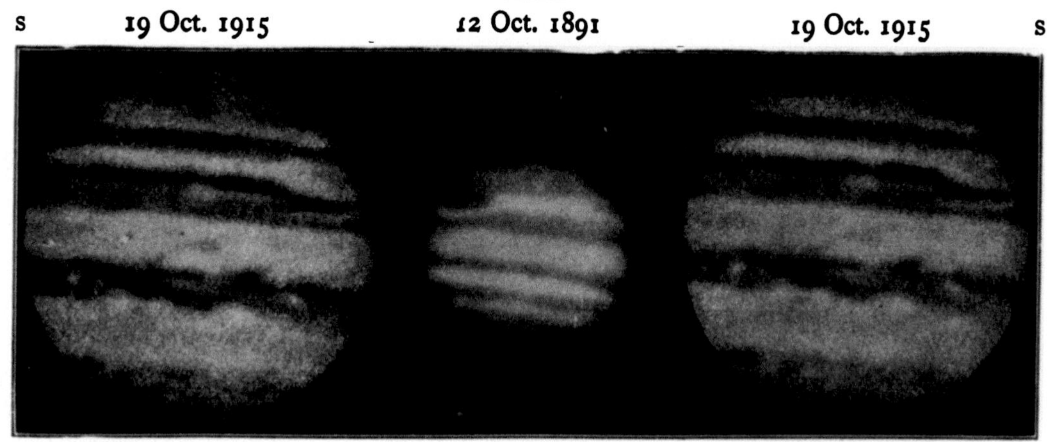

N Flagstaff Observatory. Lick Observatory, California. Flagstaff Observatory. N

RAINFALL

We pass, by a natural transition, from clouds to rainfall and thereby touch the most important of all the meteorological elements. Including in the convenient term rain all the other forms which the precipitation of water may assume, clouds and rain may be said to be nature's valid receipts for energy which the atmosphere receives from the evaporation of water at the surface mainly by solarisation. The process which is pursued in the formation of cloud and rain is so designed that the energy may be delivered at any level of the atmosphere in which clouds are found. As soon as the water-drops are formed the environment comes into possession of the latent heat of the condensed vapour and necessarily uses it as the working-agency of the atmospheric engine. The measurement of rain claims far more votaries than any other of the meteorological elements. Fortunately, the method of measurement is as simple and inexpensive as it is effective so long as a place can be found for fixing a rain-gauge. Hence the shelves of meteorological libraries provide an abundance of observations of rainfall at land-stations.

It is however much less easy to express the results of the observations satisfactorily in monthly maps. The British Rainfall Organization, which is one of the best known in the world, has issued maps of the normal annual fall in the British Isles but has only recently felt itself justified in producing normal monthly maps, such as we require in order that they may be confronted with the maps of the other elements which are not subject to such disabilities. The reason is that rainfall, as may be easily gathered from fig. 3, is influenced by orographical features quite as much or even more so than the observations of temperature; and reduction to sea-level is too irregular in time and locality to afford any satisfactory correction. Moreover, for some recondite reason of atmospheric structure, rainfall may be exceptionally heavy in a particular locality on level ground or even on the sea. A water-spout or cloud-burst or other incidents of a thunderstorm with prodigious local convection may give a perfectly unprecedented rainfall for a single hour. Hundreds of years may be required to reduce the mean values to what a rainfall enthusiast likes to regard as "normal." The fact is, however, that the idea of an invariable normal underlying the vagaries of weather is something of a chimera and no living meteorologist can afford to wait for its realisation. He must regard the normal values as being conventional within certain limits. The result of the exceptional local values due to orographic or atmospheric causes is to place upon the map a series of numbers which do not lend themselves to smooth lines like pressure. A third cause of difficulty arises from the inequality in the length of the period for which observations are available at different stations. This difficulty can, however, be got over in a satisfactory manner in the case of a station with short period in the neighbourhood of one with a normal period by comparing the figures of the two stations for the period common to both and thus obtaining a factor by which the average for the short period can be multiplied to give a substitute for the missing normal[1]. Thus, the land areas of the globe for the most part provide material enough to plot upon a map, and it is useful, in spite of the outstanding uncertainties, to attempt the division of the areas of monthly maps by *isohyets* or lines of equal rainfall.

Seasonal variation.

In the entablatures of the maps of annual rainfall we have placed a table for the conversion of measurements in inches to millimetres and a list of some heavy falls of rain. The available spaces on the monthly maps are occupied by a series of monthly values of rainfall at selected stations included within the volumes of the *Réseau Mondial*, in order that the reader may examine for himself the peculiarities of the seasonal variation of rainfall over the globe. We shall remark upon some of the features disclosed by these tables at the end of our sketch of the normal circulation in chapter VI.

Diurnal variation.

We add here isopleth diagrams (fig. 99) which illustrate both the seasonal and diurnal variation of rainfall; first, for Batavia, in which lines are drawn to show the normal hourly fall in tenths of a millimetre, secondly, for Richmond (Kew Observatory) and thirdly, for Cahirciveen (Valencia Observatory) in which the figures mark the number of hours with rain in ten years. We have already included isopleths for two British observatories in the specimen of these useful diagrams in fig. 107 of volume I. Many others could be added from the abundance of published data about rainfall at "self-recording" observatories.

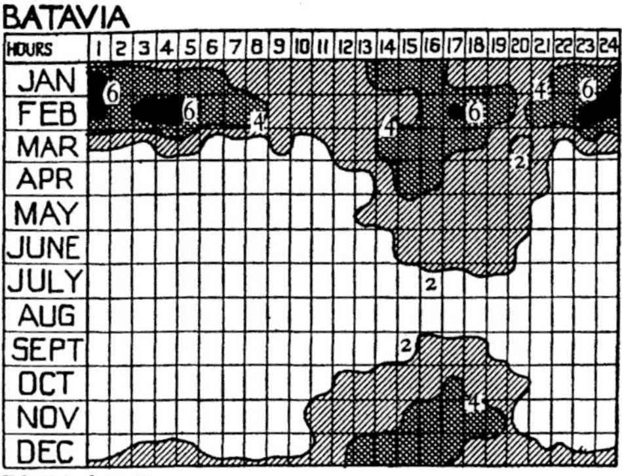

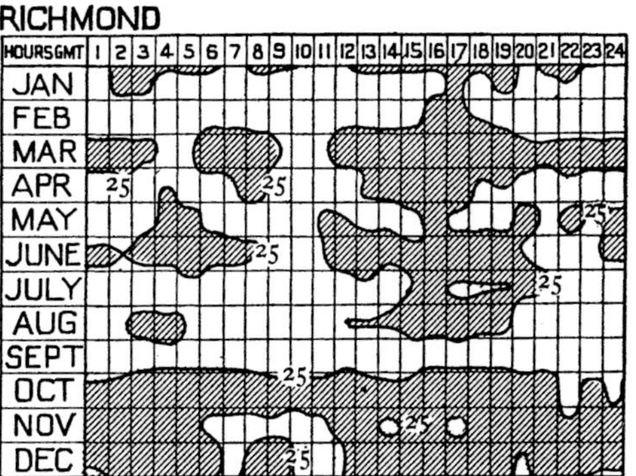

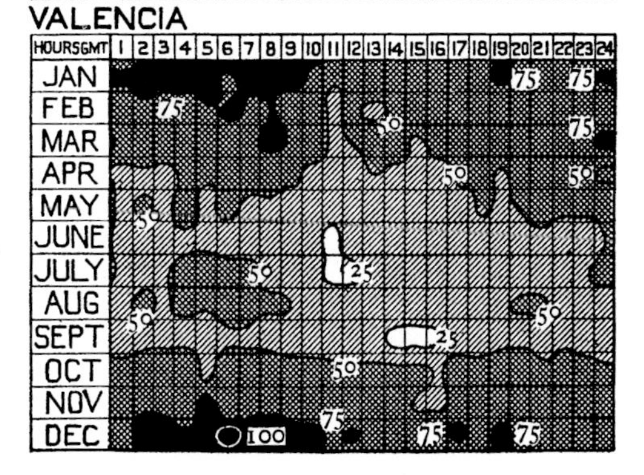

Fig. 99. (1) Batavia, rain amount: mean hourly fall in tenths of a millimetre, (2) Richmond, and (3) Valencia, frequency of rain, normal number of hours with rain in ten years.

[1] *Meteorological Magazine*, July, 1925.

We quote one example, that of Samoa[1], and with it also an example of the diurnal distribution of rainy weather over the sea taken from the Report of the Voyage of the *Challenger*. It introduces us to another aspect of rainfall.

Diurnal variation of rainfall.

Samoa, 13° 48′ S, 171° 46′ W.

Hour		0–3	3–6	6–9	9–12	12–15	15–18	18–21	21–24
Amount of rainfall as percentage of total	May–Oct.	12·6	6·8	9·9	12·5	13·8	17·3	13·9	12·2
	Nov.–Apr.	12·5	15·0	12·0	9·5	12·2	14·0	11·7	12·1
Duration of rainfall as percentage of total	May–Oct.	14·2	12·6	11·7	11·6	12·2	14·1	11·4	11·6
	Nov.–Apr.	11·7	14·1	12·2	11·4	12·5	14·2	12·5	11·4

The Sea: frequency of occurrence of observations of rainfall from the report of the "Challenger."

Hour ...	2	4	6	8	10	Noon	14	16	18	20	22	Mdt.
Open sea	**130**	118	117	115	113	110	103	95	101	113	114	112
Near land	87	**90**	75	75	82	79	75	71	74	82	79	83

Rainfall over the sea (see also Bibliography, chap. x, note 24).

In figs. 103 to 128 we have been able to present with some confidence the maps of the normal distribution of rainfall for the year, and monthly maps of the distribution over the land. For the area covered by sea, no such maps are available. Instead we can only offer the few isolated endeavours to obtain a working idea of the distribution of rain or rainy weather over the sea. The late Professor Supan[2] extended isohyets over the maps of annual rainfall to include the Atlantic, Indian and, with less certainty, the Southern Ocean between South America and New Zealand, but the monthly distribution has still to be sought. An estimate of the annual fall in different regions is subjoined.

Annual rainfall over the ocean estimated by W. G. Black[3].

Atlantic Ocean, 1864–84		Days	In.	Indian Ocean, '64–75	Days	In.
North temperate zone	(40–52° N)	67	27	—	—	—
North extratropical zone	(30–39° N)	87	9	—	—	—
North tropical zone	(12–29° N)	21	1	—	—	—
				(8–17° N)	18	5
North equatorial zone	(0–11° N)	133	129	(7–8° N)	157	81
South equatorial zone	(0–4° S)	149	58	(0–4° S)	134	64
				(5–12° S)	134	26
South tropical zone	(5–17° S)	74	7	(13–19° S)	190	116
South extratropical zone	(18–42° S)	94	20	(20–30° S)	104	12
South temperate zone	(43–51° S)	64	11	(31–50° S)	106	21
South pressure trough	(50–60° S)	137	21	—	—	—
Austro-Chinese Seas, 1864–75				**Pacific Ocean, '75–80**		
North temperate zone	(24–40° N)	108	92	(41–50° N)	207	28
North extratropical zone	(19–23° N)	73	38	(31–40° N)	152	30
North tropical zone	(10–18° N)	135	129	(20–30° N)	175	76
North intertropical zone	(8–9° N)	73	.16			
North equatorial zone	(0–7° N)	210	108			
South equatorial zone	(0–5° S)	188	92	No observations		
South intertropical zone	(6–13° S)	260	37			
South tropical zone	(14–24° S)	116	60	(13–29° S)	100	11
South extratropical zone	(25–40° S)	35	10	(30–39° S)	60	71
South temperate zone	—	—	—	(40–53° S)	105	16

[1] Apia Observatory, Samoa, *A Summary of the meteorological observations of the Samoa Observatory* (1890–1920), by G. Angenheister, Wellington, N.Z., 1924, p. 34.

[2] 'Die jährlichen Niederschlagsmengen auf den Meeren,' *Geographische Mitteilungen*, 1898.

[3] 'Rain-Gauge...at Sea 1864–75–81,' *Manchester Geog. Soc.*, Edinburgh, 1898.

A collection of data from trade-routes and elsewhere sufficient to give us the distribution of rainfall over the sea is in reality an indispensable requirement for tracing the structure of the atmosphere and its general circulation; but it does not exist, by reason mainly of course of the difficulty of finding a satisfactory exposure for a rain-gauge on board ship. The late Captain Hepworth, Marine Superintendent of the Meteorological Office in London, devoted himself with much energy to the endeavour to remedy this signal omission from our libraries but his effort has not yet borne much fruit[1]. We are therefore dependent mainly upon inference from the frequency of observation of rain in ships' logs. Of these there are large numbers, but few compilations have been made. One of the earliest is contained in one of the publications of the Meteorological Office, *Charts of Meteorological Data for the Nine* 10° *squares of the Atlantic* (M.O. No. 27), which was intended to aid the passages of sailing ships crossing the equatorial region of the Atlantic. We represent the frequency of observations of rain and the normal relative humidity in the two months January and July.

FREQUENCY OF RAINFALL OVER THE SEA

January July

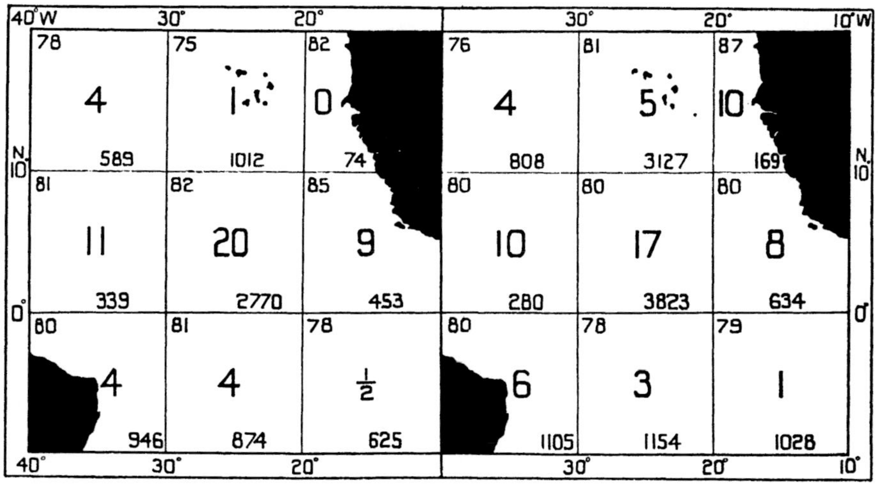

Fig. 100. Sixteen years' observations from the equatorial Atlantic. The central numbers show frequency of entries of rainfall as percentage of the whole number of observations marked at the bottom of each square.

The numbers at the top give the normal relative humidity derived from the whole number of observations.

We have already given some figures for the weather of three squares, which form the middle column of the nine squares, as entablatures to the maps of figs. 93, 94, 95. The curious reader will find some discrepancies between the diagram and the table because in the latter we have included observations

[1] The data are published on the M.O. Charts of the North Atlantic, June 1918–April 1919. C. S. Durst quotes 17 mm per day for the Atlantic doldrums. M.O. 254 h, 1926.

of drizzle "d" (which are not more than 2 per cent. of the whole number of observations in any month) with the observations of rain "r" but in the diagram the observations of rain only are counted. The position of square 3, the central square, as the region of most frequent rainfall in both the months is noteworthy as indicating that the conditions of weather along the equatorial belt are not uniform all round the globe as some theoretical representations of the general circulation would lead us to expect.

Our next illustration of the frequency of rainfall over the sea (fig. 101) is taken from the meteorological charts of the region near the Cape of Good Hope (M.O. No. 43) and shows the seasonal variation. The figures represent the frequency of rainfall as a percentage of the whole number of observations, which are taken every 4 hours. For the purposes of comparison a small table is inset which shows the seasonal frequency of hours of rain at four British observatories arranged in order of Westerly longitude. If it is fair to regard 10 per cent. of rain-hours as the same as 10 per cent. of observations every fourth hour, the approximate parallelism of the tables means that rain is about as frequent at our observatories as in the stretch of sea for 1500 miles Southward from the Cape; and the comparative wetness going Southward is not unlike that which we find going Westward in the British Isles.

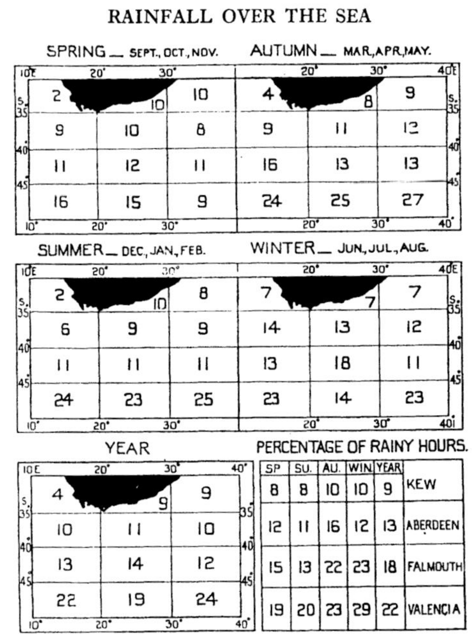

Fig. 101. Number of entries of rainfall expressed as percentage of the number of observations in certain squares near the Cape of Good Hope: together with the number of rainy hours as percentage of the whole number of hours at certain British and Irish observatories in corresponding seasons.

We touch here a problem which must be regarded as a meteorological conundrum rather than an ordinary question; that is, the possibility of evaluating the monthly and annual rainfall over any part of the sea from the frequency of observations of rainfall in ships' logs. It is a conundrum of this kind to which the isohyets drawn by Supan over various oceans must be regarded as an attempted answer.

For stations on land there would be no difficulty in arriving at an empirical solution; but the influence of orographic features upon the rainfall of any land-station makes the use of land-observations to interpret the meaning of the frequency of rain at sea too hazardous for normal practice.

Nevertheless the reward for a successful solution is so great that, while we are waiting for effective measurements of rainfall at sea, we are justified in trying to interpret in terms of rainfall such observations bearing upon the subject as we possess. Very much depends upon the exposure. Hourly observations from Sable Island or Willis Island could be tried with confidence; observations from the station at St Helena, 632 m. above the sea, could not. We have neither time nor space to work out the subject, but we give on p. 204 a table which shows factors of conversion from percentage frequency of rain-hours to mean daily rainfall for each month at Falmouth in South-West England, Valencia in South-West Ireland and at Batavia. It will be seen that the divisor 5 is good enough to convert the percentage-frequency of rain-hours into mean daily rainfall in millimetres at the English stations but a divisor not much different from 1·7 is required in the case of Batavia.

RAINFALL OVER THE SEA

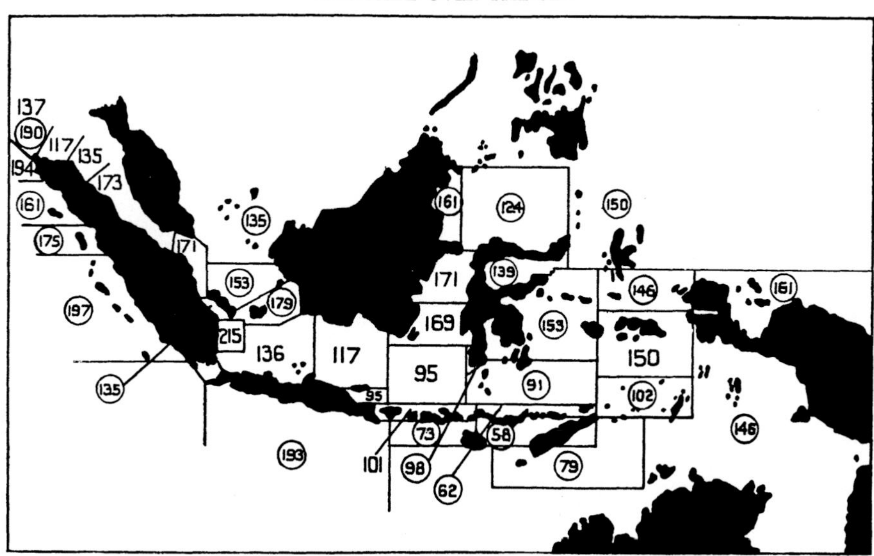

Fig. 102. East Indian Seas. Mean number of days of rain per year in various districts. The figures are actual averages, except those enclosed by circles which are computed from the "probability" or percentage frequency. (*Het Klimaat van Nederlandsch-Indië*, Deel 1, Afl. 3, by Dr C. Braak. *K. Mag. en Met. Obs. te Batavia,* Verh. No. 8, p. 216, Batavia 1923.)

A third illustration of the frequency of rain over the sea shows the number of rain-days in different parts of the East Indian Seas. The areas referred to by the several figures are the enclosures in which the figures are placed. The map extends from 10° N to 10° S and the figures should be compared with those for the nine 10-degree squares of fig. 100, if we could find a divisor to relate the percentage of rain-days per annum with the percentage of rainfall frequency in four-hourly observations. It may be that there are some regions in the East Indian Seas close to the equator in which rain is more frequent than in the equatorial Atlantic.

FIGURE 103

Inches.

NORMAL RAINFALL

Authority: see p. 211.

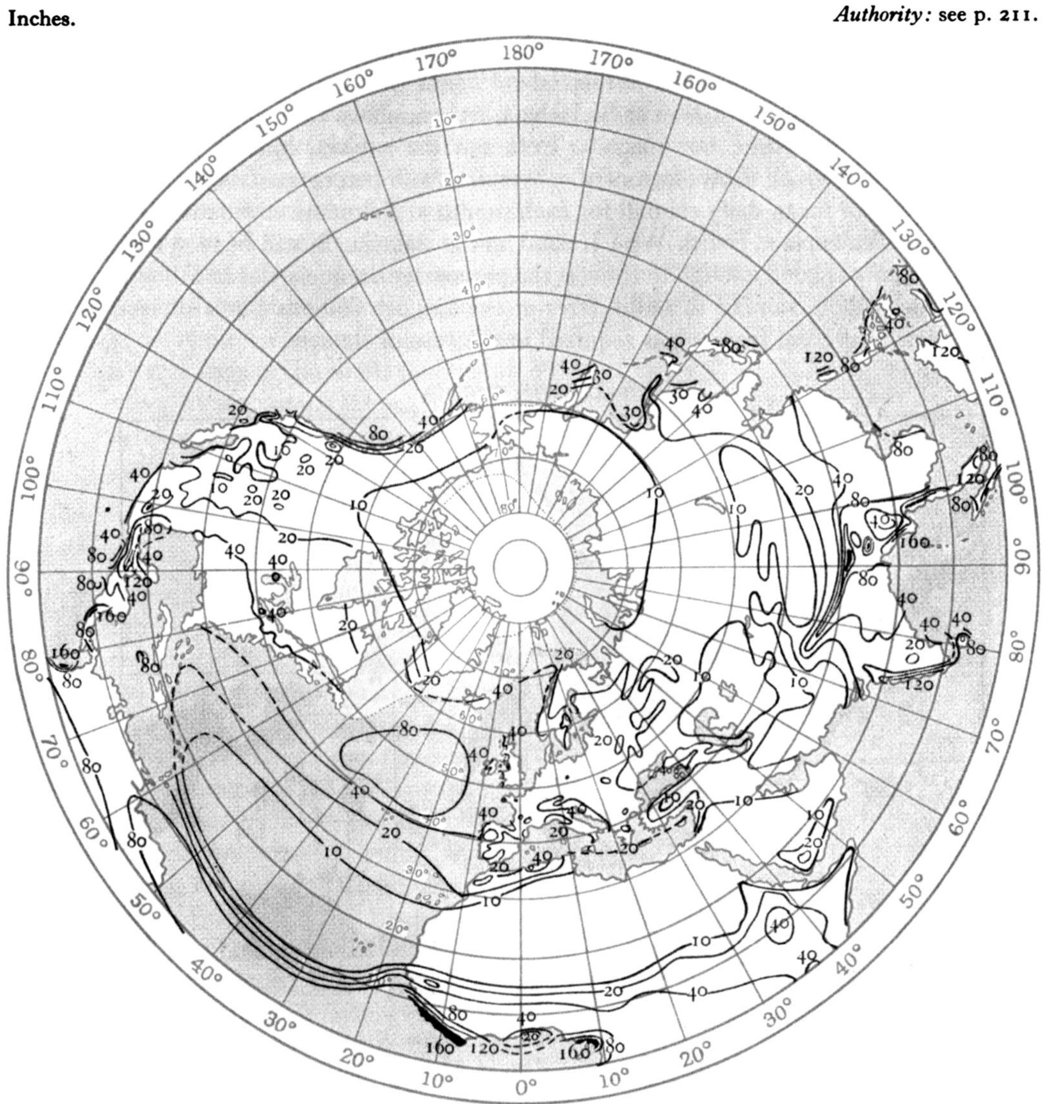

Table of equivalents of inches and tenths, in millimetres.

1 inch = 25·400 mm.

In.	0·0	0·1	0·2	0·3	0·4	0·5	0·6	0·7	0·8	0·9
	mm	mm	mm	mm	mm	mm	mm	mm	mm	mm
0·0	0	2·5	5·1	7·6	10·2	12·7	15·2	17·8	20·3	22·9
1·0	25·4	27·9	30·5	33·0	35·6	38·1	40·6	43·2	45·7	48·3
2·0	50·8	53·3	55·9	58·4	61·0	63·5	66·0	68·6	71·1	73·7
3·0	76·2	78·7	81·3	83·8	86·4	88·9	91·4	94·0	96·5	99·1
4·0	101·6	104·1	106·7	109·2	111·8	114·3	116·8	119·4	121·9	124·5
5·0	127·0	129·5	132·1	134·6	137·2	139·7	142·2	144·8	147·3	149·9
6·0	152·4	154·9	157·5	160·0	162·6	165·1	167·6	170·2	172·7	175·3
7·0	177·8	180·3	182·9	185·4	188·0	190·5	193·0	195·6	198·1	200·7
8·0	203·2	205·7	208·3	210·8	213·4	215·9	218·4	221·0	223·5	226·1
9·0	228·6	231·1	233·7	236·2	238·8	241·3	243·8	246·4	248·9	251·5

For amounts between 10 and 20 inches, add 254
,, ,, 20 ,, 30 ,, 508

For amounts between 30 and 40 inches, add 762
,, ,, 40 ,, 50 ,, 1016

NORMAL RAINFALL

Authority: see p. 211.

FIGURE 104

Inches.

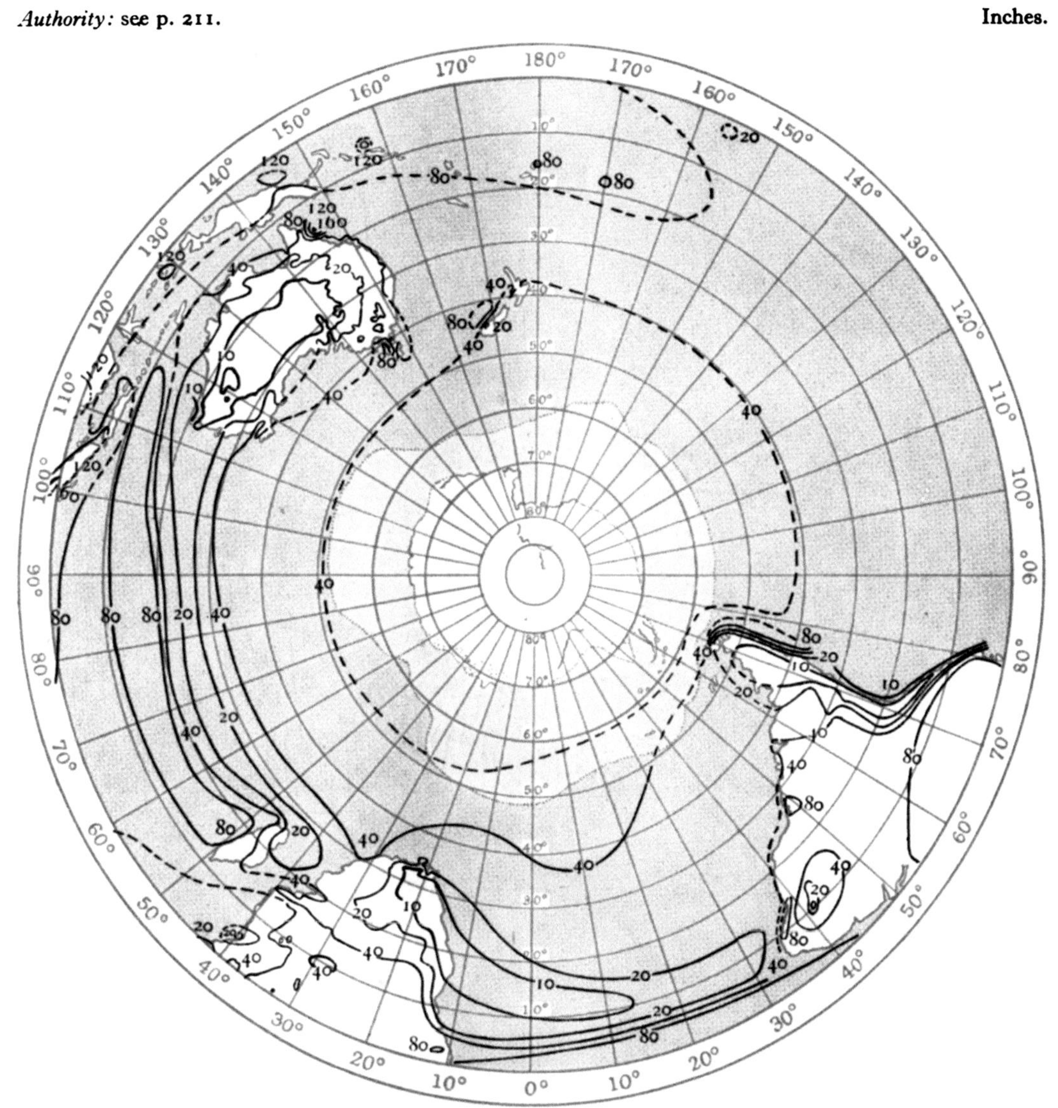

Heavy falls of rain in short periods.

	mm/min.
U.S.A., Campo, Cal., Aug. 12, 1891, 292 mm in 80 minutes ...	3·65
U.S.A., Palmetto, Nev., Aug. 1890, 224 mm in 60 minutes ...	3·73
U.S.A., Virginia, Aug. 24, 1906, 235 mm in 30 minutes	7·83
Roumania, July 7, 1889, 204·6 mm in 20 minutes	10·23
Porto Bello, Panama, Nov. 29, 1911, 62·5 mm in 3 min.	20·83
Gt Britain, Norwich, July 14, 1917, 12 mm in 2½ minutes ...	4·8
U.S.A., Opid's Camp, Calif., April 5, 1926, 25·9 mm in 1 minute	25·9
Gt Britain, Kew Observatory, July 18, 1934, 5·0 mm in 1 minute	5·0

Heavy falls of rain in a month.

	mm
Assam, Manoyuram, Aug. 1841	6710
Europe, Hermsburg, Oct. 1889	1451
Wales, Snowdon, Oct. 1909	1435
America, Helen Mine, Cal., Jan. 1909	1817

Maximum falls recorded in 24 hours.

	mm
Philippine Is., Baguio, July 14–15, 1911 ...	1168
India, Cherrapunji, June 14, 1876	1036
Japan, Funkiko, Aug. 31, 1911 and July 20, 1913	1034
Australia, Crohamhurst, Feb. 2, 1893 ...	907
Hawaii, Honomu, Feb. 20, 1918	812
Ceylon, Nedunkem, Dec. 15–16, 1897 ...	807
Jamaica, Silver Hill, Nov. 6, 1909	775
Mauritius, 1865	610
United States, Altapass, Mitchell Co., N.C., July 15–16, 1916	564
Hong Kong, July 19, 1926 (9 hours) ...	519
Dutch East Indies, Besokor, Jan. 31–Feb. 1, 1901	511
England, Bruton, June 28, 1917	243

FIGURE 105 NORMAL RAINFALL

Inches. *Authority:* see p. 211

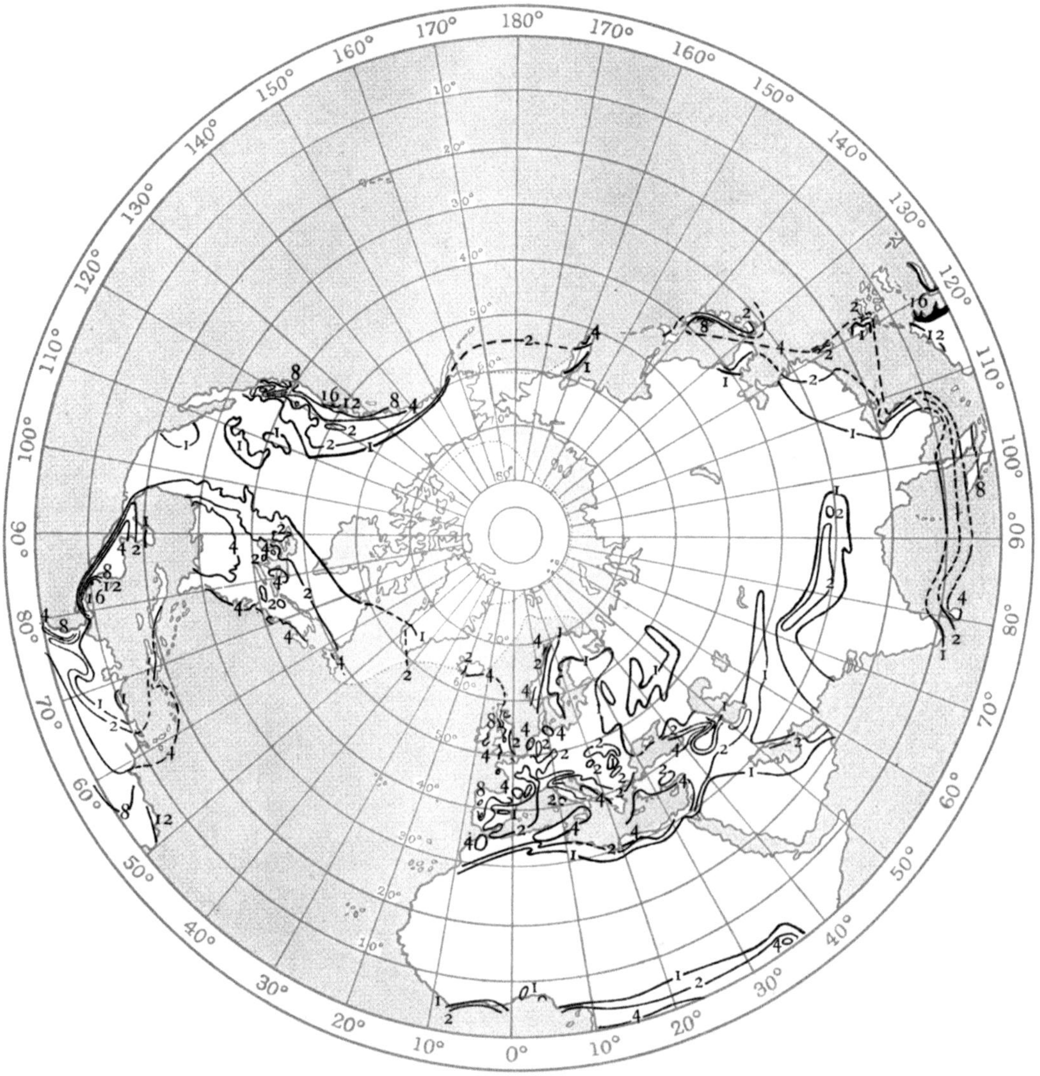

Seasonal variation of rainfall in millimetres, 70–80° N.

Station	Lat.	Long.	Jan.	Feb.	Mar.	Apr.	May	June	July	Aug.	Sept.	Oct.	Nov.	Dec.	Year
Upernivik ...	73° N	56° W	12	13	19	14	14	13	23	**29**	26	26	24	13	226
Spitsbergen ...	78	14° E	31	33	24	23	12	13	19	21	19	29	30	**39**	293
Treurenberg Bay	80	17	14	5	10	7	4	19	20	**38**	20	15	14	11	177
Cap Thordsen ...	78	16	10	37	3	7	10	4	7	14	**36**	31	22	8	189
Gjesvaer ...	71	25	50	54	47	37	36	37	51	44	69	**78**	71	67	641
Mehavn ...	71	28	71	59	54	46	43	43	56	52	**83**	81	76	78	742
Vardö	70	31	70	**71**	51	39	37	38	47	54	62	64	62	64	659
Malye Karmakouly	72	53	7	6	7	7	13	16	31	**43**	40	30	13	10	223
Sagastyr ...	73	124	3	2	0	1	5	11	7	**36**	11	2	3	5	86

The figures for the month of greatest rainfall are indicated by black type.

[A table of mean monthly and yearly rainfall at stations in the Arctic north of 70° is given by Franz Baur, *Arktis*, Heft 4, 2 Jahrg/1929.]

NORMAL RAINFALL

FIGURE 106

Authority: see p. 211.

Inches.

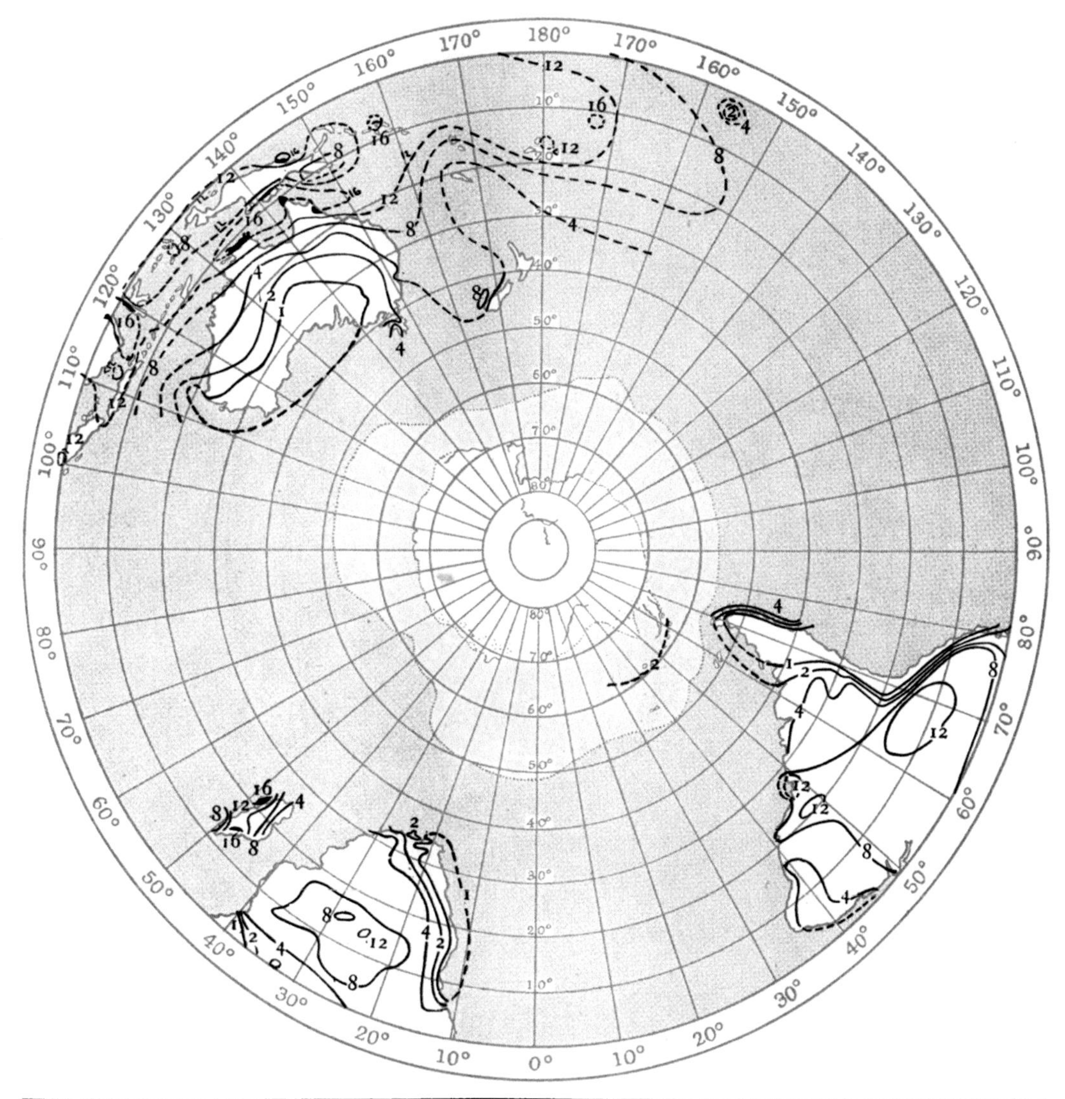

Seasonal variation of rainfall in millimetres, 60–70° N.

Station	Lat.	Long.	Jan.	Feb.	Mar.	Apr.	May	June	July	Aug.	Sept.	Oct.	Nov.	Dec.	Year
Nome	64½° N	165° W	21	25	23	14	23	30	**69**	68	51	37	24	27	412
Valdez	61	146	90	127	80	82	65	51	72	147	**212**	158	102	159	1345
Carcross	60	135	13	25	15	8	10	15	**29**	22	25	24	25	18	229
Godthaab	64	52	38	48	43	29	44	34	59	78	**83**	63	47	39	605
Angmagsalik	66	38	87	43	52	66	78	53	56	63	115	**164**	91	76	944
Vestmanno	63	20	146	114	109	99	82	81	81	75	145	**149**	132	140	1352
Thorshavn	62	7° W	**175**	139	128	95	85	65	79	95	119	160	166	168	1474
Christiansund	63	8° E	112	85	80	55	62	51	78	103	133	**148**	111	113	1131
Helsingfors	60	25	42	37	35	35	45	44	59	**73**	62	66	62	51	611
Berezov	64	65	15	10	14	16	34	52	57	**58**	40	24	19	14	353
Iakoutsk	62	130	9	5	5	9	17	38	41	**46**	27	19	10	11	237
Novo-Mari-inskii Post	65	178	9	7	6	5	10	18	30	**41**	35	14	7	10	192

FIGURE 107 NORMAL RAINFALL

Inches. *Authority:* see p. 211.

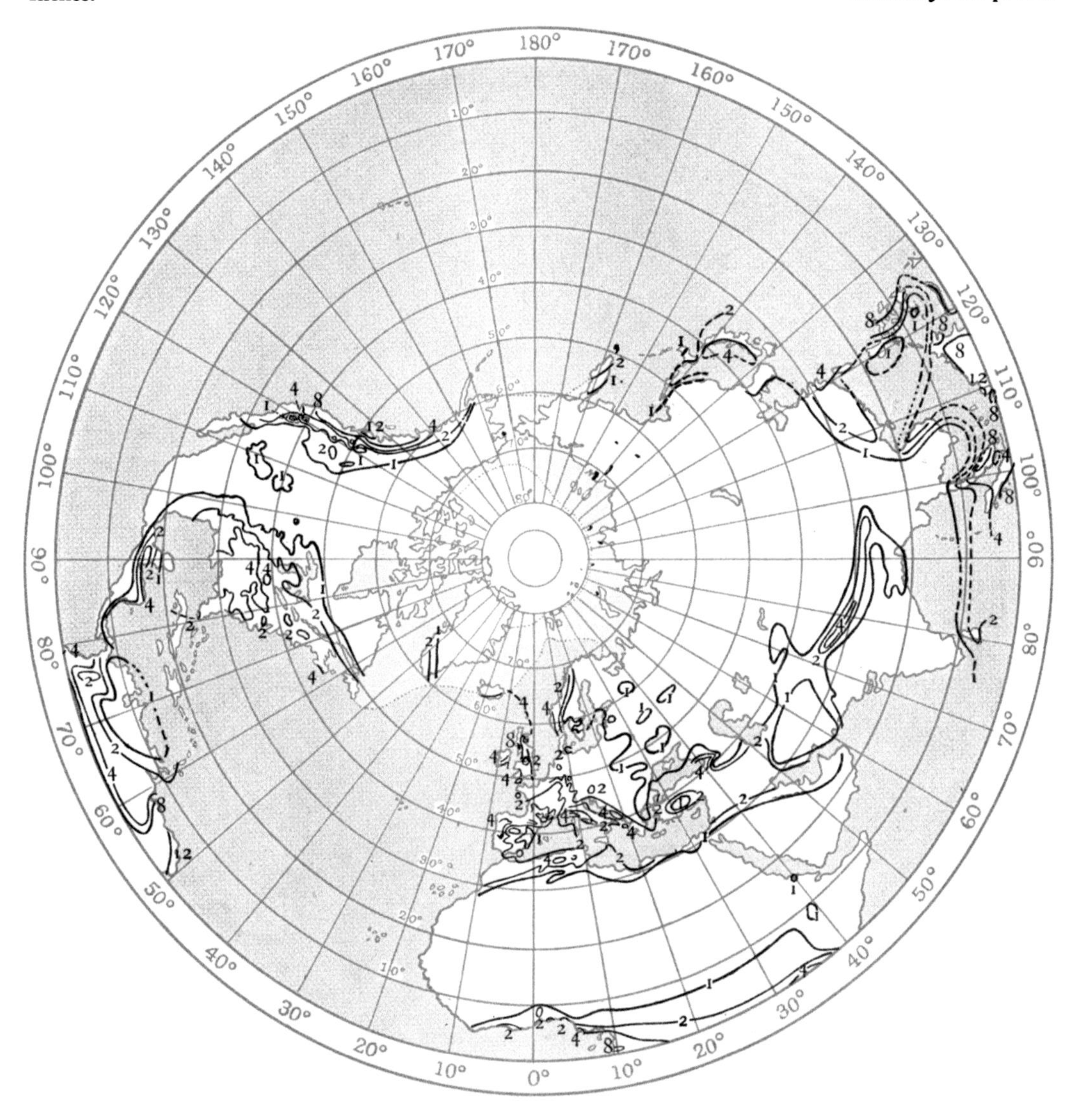

Station	Lat.	Long.	Jan.	Feb.	Mar.	Apr.	May	June	July	Aug.	Sept.	Oct.	Nov.	Dec.	Year
Seasonal variation of rainfall in millimetres, 50–60° N.															
Dutch Harbour	54° N	166½° W	135	157	129	84	124	74	59	82	154	**216**	173	173	1560
Sitka	57	135	169	124	132	160	96	73	102	181	262	**298**	237	235	2069
Kamloops	51	120	23	20	8	9	24	31	32	27	24	15	27	**39**	279
Qu'Appelle	51	104	18	20	25	28	58	**86**	69	51	40	26	24	19	464
Port Nelson	57	93	16	13	19	24	23	**66**	50	61	56	29	33	30	420
Mistassini Post	50	74	22	55	49	16	43	42	63	**73**	40	38	27	17	485
Belle Isle	52	55	55	52	89	69	65	136	126	152	144	**153**	98	69	1208
Valencia	52	10	139	132	115	93	81	81	96	122	105	142	139	**169**	1414
Aberdeen	57	2° W	55	52	61	48	59	43	71	70	56	76	75	**82**	748
Greenwich	51	0	43	40	44	37	44	51	57	56	45	**64**	58	57	596
Uccle	51	4° E	64	58	58	53	62	69	78	**80**	70	78	70	69	809
Potsdam	52	13	40	33	42	33	53	59	**82**	58	46	45	39	44	574

NORMAL RAINFALL

FIGURE 108

Authority: see p. 211.

Inches.

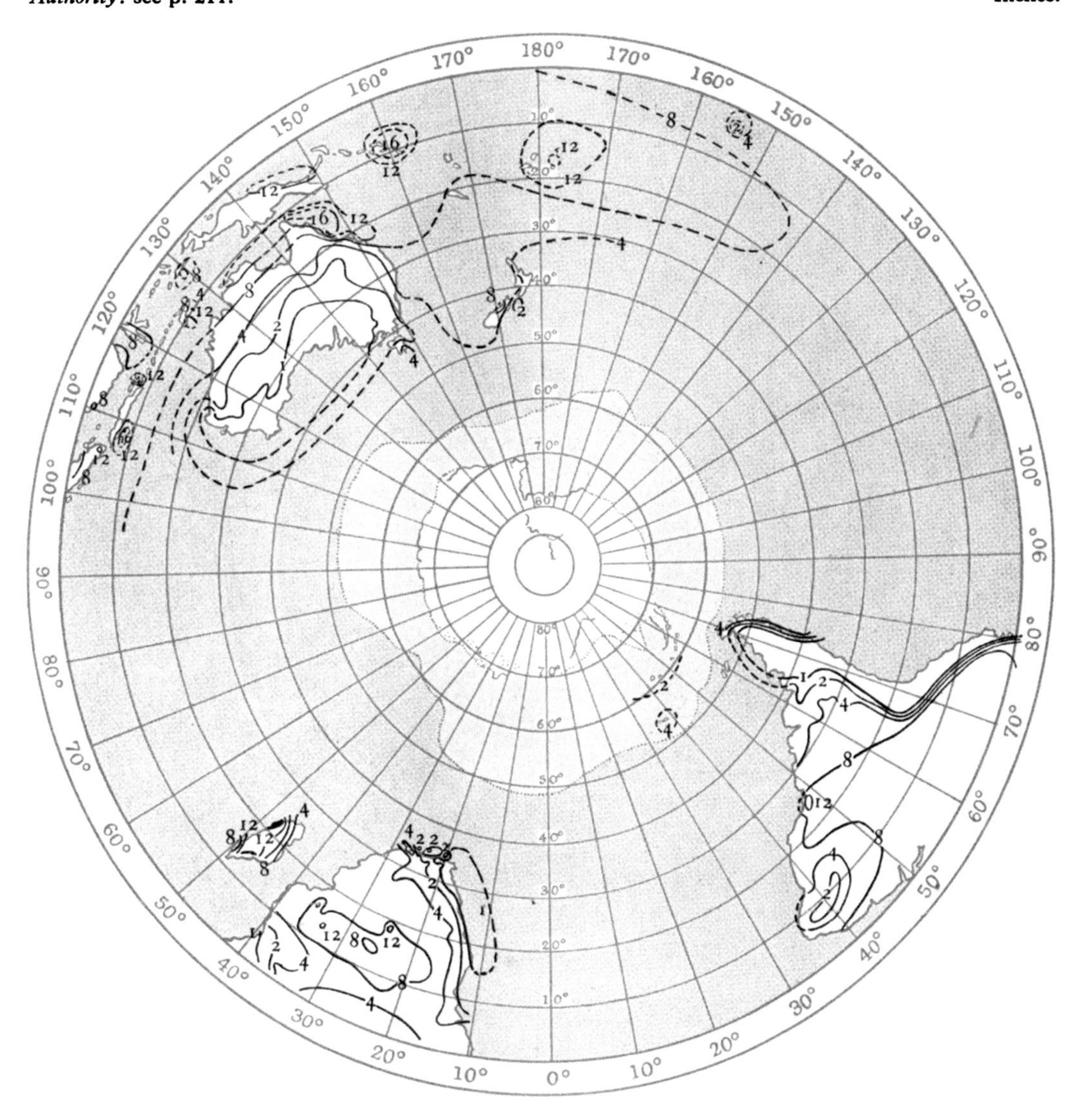

Station	Lat.	Long.	Jan.	Feb.	Mar.	Apr.	May	June	July	Aug.	Sept.	Oct.	Nov.	Dec.	Year
Upsala	60° N	18° E	32	30	31	28	44	48	69	**75**	49	53	42	39	540
Petrograd	60	30	24	23	23	26	42	50	64	**71**	53	45	36	32	489
Moscow	56	38	32	27	31	37	49	58	**77**	73	55	44	43	40	566
Kazan	56	49	18	15	18	25	37	58	**62**	54	41	33	30	22	413
Ekaterinbourg	57	61	10	8	10	15	43	71	**73**	64	36	23	19	14	386
Omsk	55	73	14	10	9	13	31	52	**53**	49	27	24	17	20	319
Tomsk	56° 30′	85	30	21	23	24	41	**71**	70	67	40	55	45	44	531
Irkutsk	52	104	13	9	8	15	33	55	**80**	71	43	18	16	18	378
Nertchinskii Zavod	51	120	2	2	5	14	29	60	**111**	109	45	13	7	4	401
Blagoveshchensk	50	127° 30′	2	1	10	23	42	74	123	**133**	74	19	5	1	507
Nikolaevsk-sur-Amour	53	141	15	12	17	29	33	39	51	**79**	68	51	32	21	447
Okhotsk	59	143	2	2	5	9	22	45	**52**	47	47	24	6	3	264

Seasonal variation of rainfall in millimetres, 50–60° N.

FIGURE 109 NORMAL RAINFALL

Inches.

Authority: see p. 211.

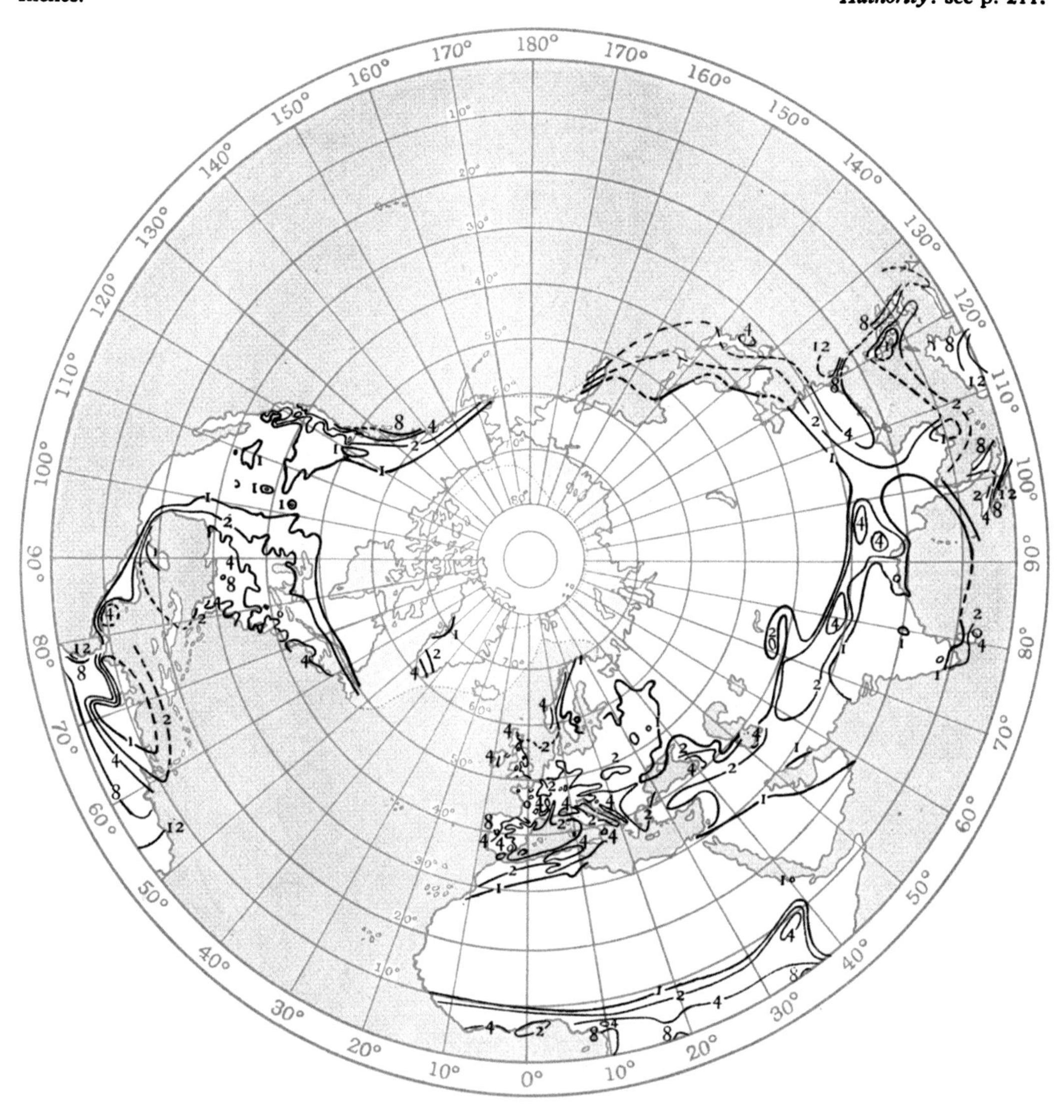

Seasonal variation of rainfall in millimetres, 40–50° N.															
Station	Lat.	Long.	Jan.	Feb.	Mar.	Apr.	May	June	July	Aug.	Sept.	Oct.	Nov.	Dec.	Year
Victoria	48° N	123° W	115	90	65	44	33	24	9	17	51	65	164	150	827
Salt Lake City	41	112	34	37	53	51	52	22	13	20	23	38	35	35	413
Winnipeg	50	97	21	21	28	38	58	83	80	62	58	36	24	25	534
Port Arthur	48	89	19	18	23	39	52	68	96	70	83	63	37	21	589
Chicago	42	88	53	55	62	70	93	88	88	79	77	63	60	51	839
St John, N.B.	45	66	122	99	115	89	94	83	92	98	95	115	112	106	1220
SW Pt. Anti-costi	49	64	55	41	47	34	51	59	83	87	70	83	71	55	736
St John's, Newfoundland	48	53	146	129	124	105	90	90	90	90	92	144	145	143	1388
Madrid	40	4	33	32	40	43	43	35	10	12	36	45	50	42	421
Nantes	47	2° W	58	55	58	47	56	56	57	49	47	62	80	85	730
Paris	49	2° E	39	33	38	44	51	55	52	50	53	55	49	42	562
Zürich	47	9	53	56	73	91	110	135	129	132	105	94	69	73	1120

NORMAL RAINFALL

Authority: see p. 211.

FIGURE 110

Inches.

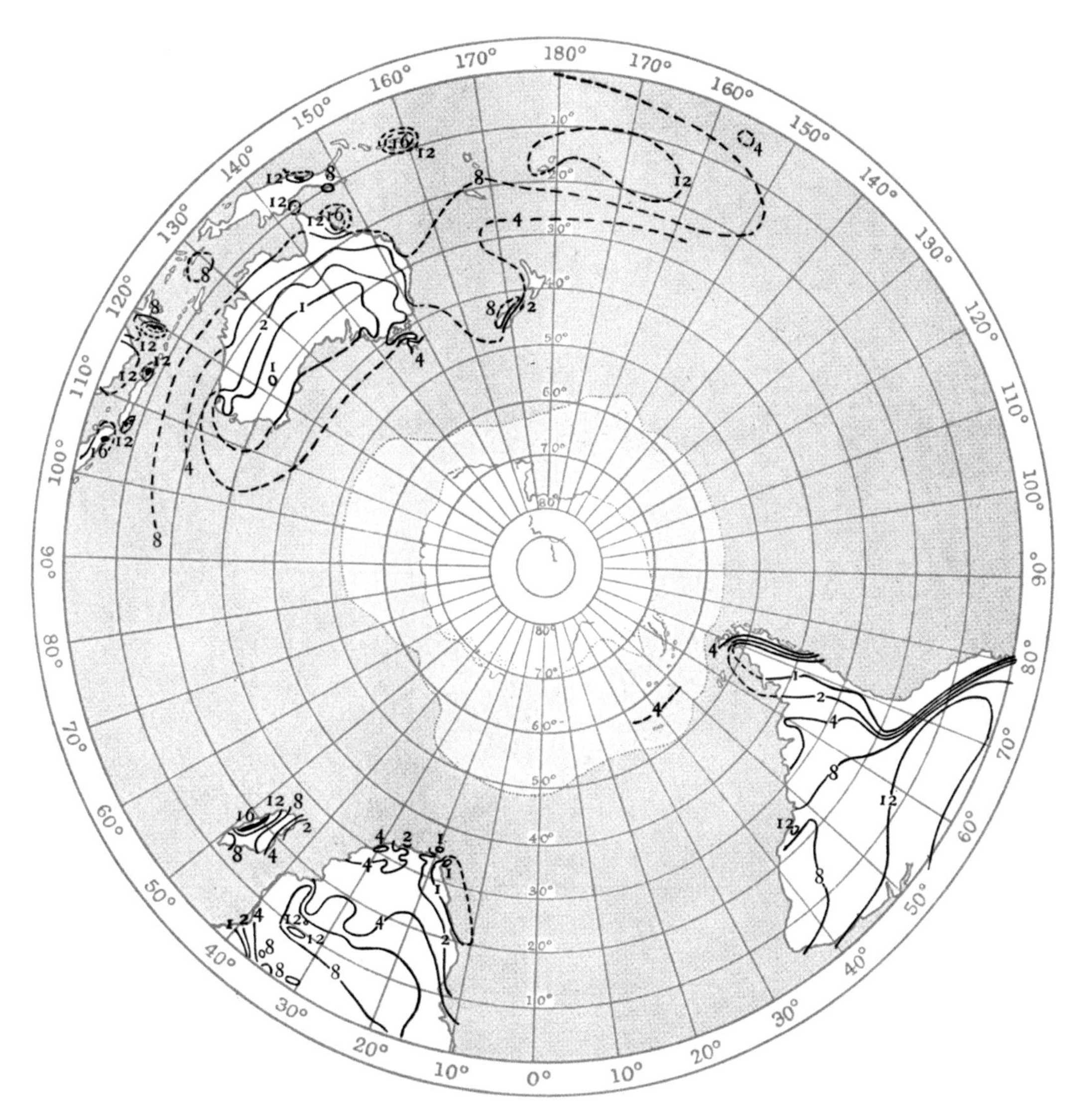

Seasonal variation of rainfall in millimetres, 50–70° S.

Station	Lat.	Long.	Jan.	Feb.	Mar.	Apr.	May	June	July	Aug.	Sept.	Oct.	Nov.	Dec.	Year
Islota de los Evangelistas	52° S	75° W	319	253	306	292	244	215	232	204	223	246	251	258	3043
Punta Arenas	53	71	30	33	43	55	53	57	44	58	44	30	34	37	518
Punta Dungeness	52	68	31	19	23	31	17	20	24	18	13	10	9	27	242
Santa Cruz	50	68	13	7	6	15	14	13	17	13	5	8	11	20	142
Ushuaia	55	68	62	62	51	69	56	53	36	37	32	40	38	50	586
Año Nuevo	55	64	72	69	58	69	53	55	45	41	31	28	44	49	614
Stanley	52	58	71	63	59	59	60	47	52	45	32	40	47	67	642
S. Georgia	54	37	75	118	138	134	151	124	139	123	82	72	99	90	1345
S. Orkneys	61	45	38	39	46	44	33	29	30	35	25	26	35	23	403
Port Charcot	65	63	46	33	44	24	16	28	10	39	27	22	58	31	376

FIGURE 111 NORMAL RAINFALL

Inches. *Authority:* see p. 211.

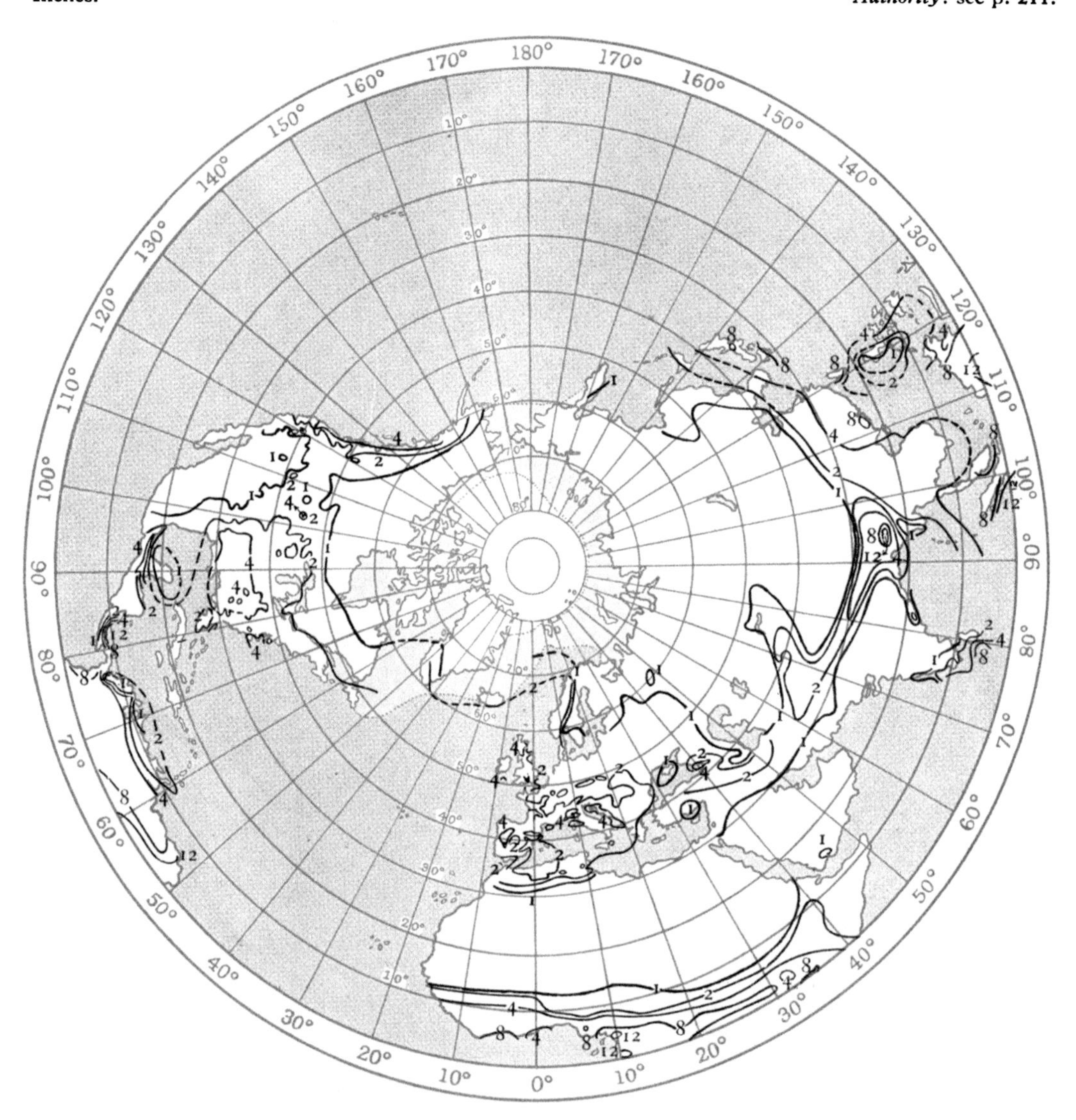

Seasonal variation of rainfall in millimetres, 40–50° N.

Station	Lat.	Long.	Jan.	Feb.	Mar.	Apr.	May	June	July	Aug.	Sept.	Oct.	Nov.	Dec.	Year
Rome	42° N	12° E	79	64	69	68	58	41	19	27	71	**118**	116	90	820
Vienna	48	16	37	33	46	50	70	**71**	70	70	44	49	41	42	623
Sofia	43	23	35	29	41	54	**86**	73	67	56	55	61	51	31	639
Odessa	46	31	26	21	27	25	31	**56**	49	32	32	31	34	30	394
Novorossiisk	45	38	70	69	58	49	42	62	56	29	55	40	69	**90**	689
Astrakhan	46	48	13	9	11	12	**18**	**18**	13	12	14	11	11	14	156
Krasnovodsk	40	53	14	15	19	**24**	12	11	4	5	7	10	16	15	152
Tachkent	41	69	51	30	**67**	56	33	12	4	1	4	28	34	38	358
Mukden	42	123	4	6	19	26	61	83	**157**	137	91	37	23	6	650
Vladivostok	43	132	7	9	16	31	50	70	77	110	**112**	46	29	13	570
Ochiai	47	143	42	40	37	50	61	88	106	93	**114**	106	79	70	886
Syana	45	148	98	50	68	65	80	61	80	70	86	103	**118**	112	991

NORMAL RAINFALL

Authority: see p. 211.

FIGURE 112

Inches.

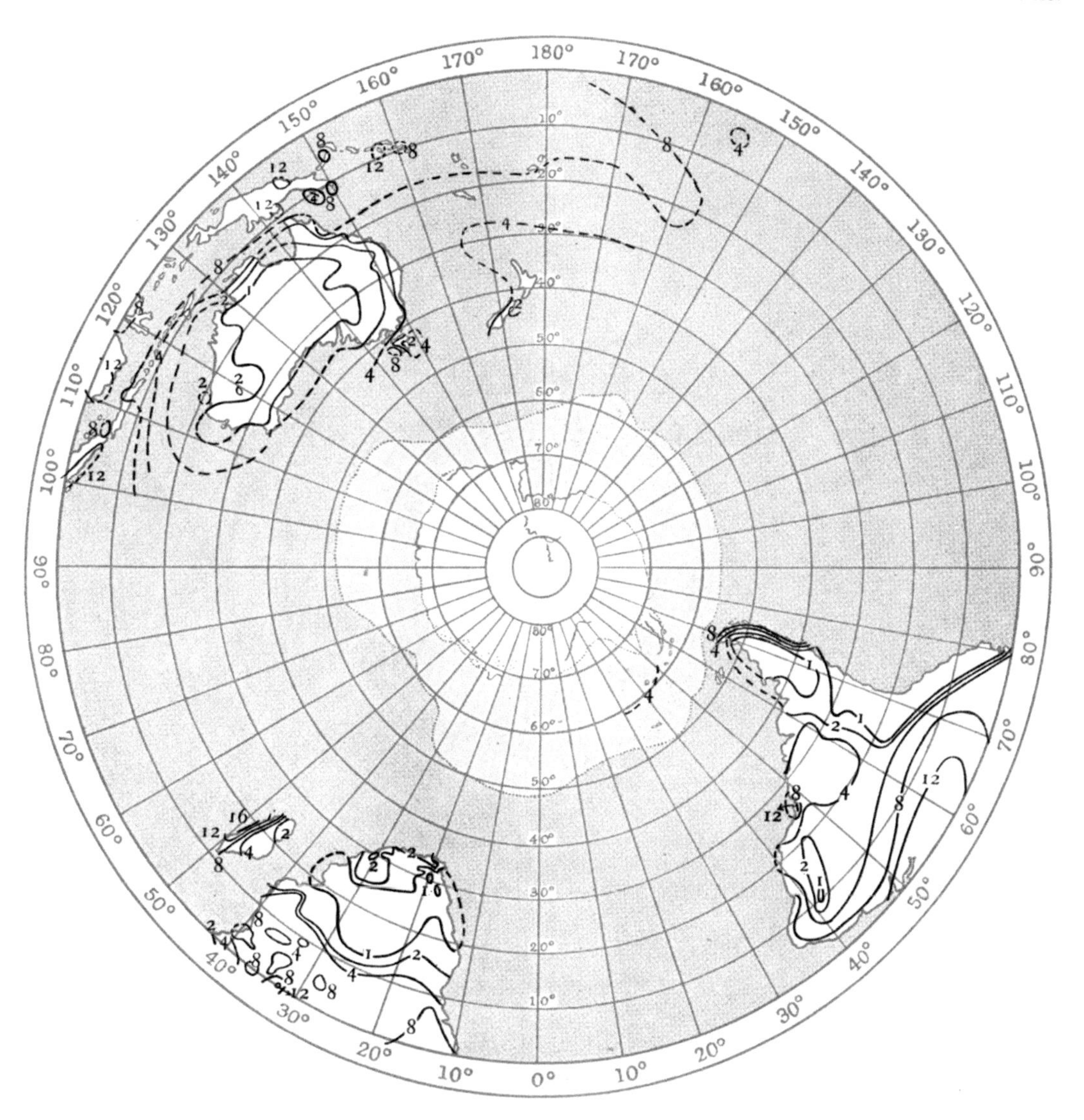

Seasonal variation of rainfall in millimetres, 40–50° S.

Station	Lat.	Long.	Jan.	Feb.	Mar.	Apr.	May	June	July	Aug.	Sept.	Oct.	Nov.	Dec.	Year
Chatham Is.	44° S	177° W	61	67	81	82	98	83	**107**	89	67	58	73	61	927
Isla Guafo	44	75	65	60	62	103	136	145	119	**148**	73	55	70	56	1092
Punta Galera	40	74	60	76	105	223	323	**359**	329	269	177	107	111	78	2217
Sarmiento	45	69	8	11	16	13	22	9	**26**	13	8	5	10	5	146
Puerto Madryn	43	65° W	10	11	9	19	**25**	20	13	8	10	16	8	4	153
Launceston	41	147° E	46	28	45	54	68	**89**	77	70	75	68	46	52	718
Hobart	43	147	45	37	43	48	47	56	54	47	54	57	**64**	50	602
Invercargill	46	168	109	73	91	111	117	88	87	86	78	**121**	113	110	1184
Dunedin	46	171	86	71	75	69	84	79	78	78	69	77	82	**89**	937
Christchurch	44	173	52	47	56	49	60	**71**	69	53	46	42	50	48	643
Wellington	41	175	84	83	83	101	121	126	**146**	114	104	105	88	81	1236

FIGURE 113 NORMAL RAINFALL

Inches. *Authority:* see p. 211.

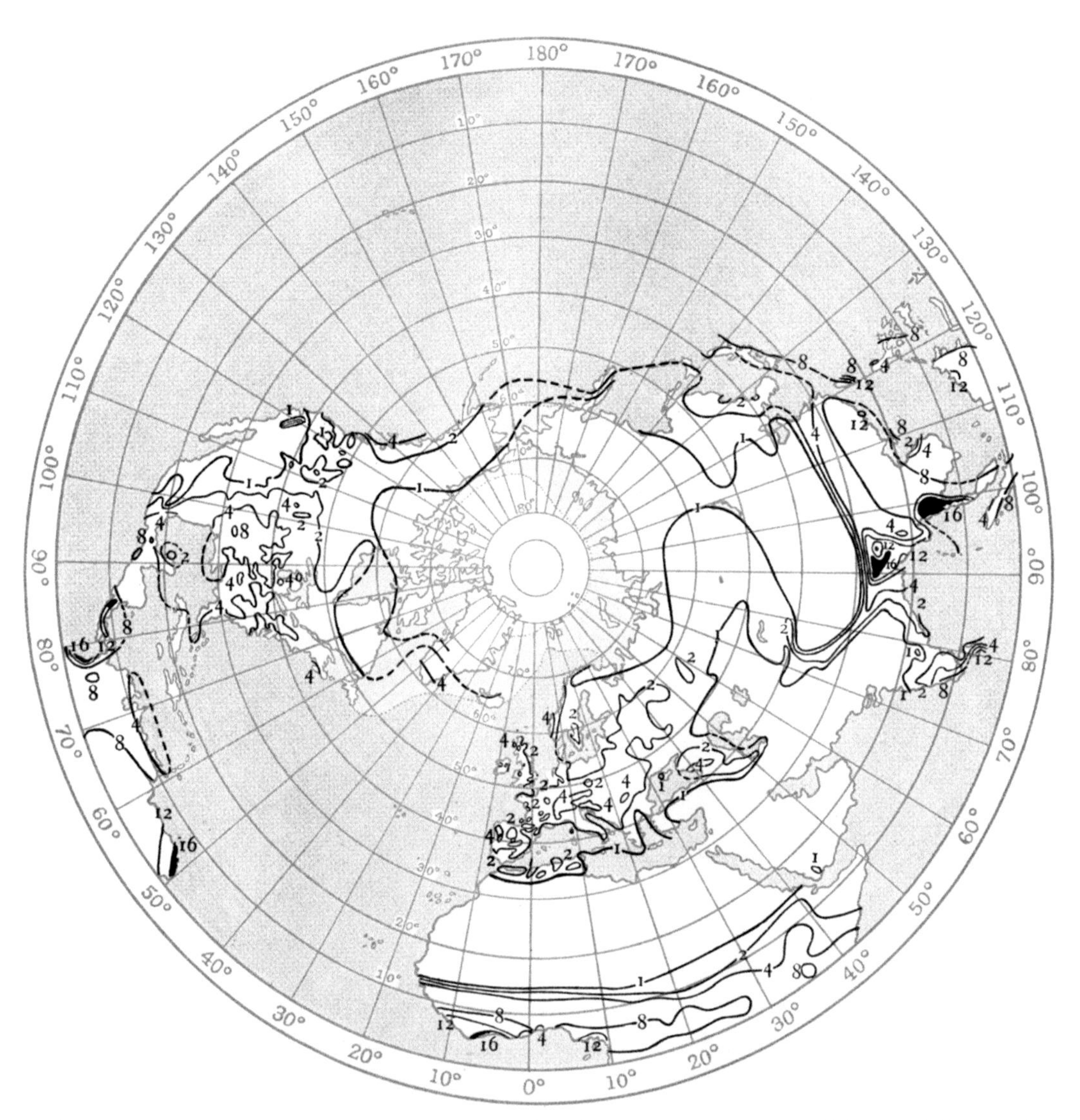

Seasonal variation of rainfall in millimetres, 30–40° N.

Station	Lat.	Long.	Jan.	Feb.	Mar.	Apr.	May	June	July	Aug.	Sept.	Oct.	Nov.	Dec.	Year
San Francisco	38° N	122° W	**125**	91	82	37	20	5	1	1	10	27	60	104	563
Modena	38	114	26	28	29	23	21	8	**37**	**37**	26	22	13	13	283
St Louis	39	90	65	70	84	90	**109**	107	90	73	81	63	69	54	955
Mobile	31	88	116	135	167	123	109	141	**181**	173	136	85	93	120	1579
Washington	39	77	86	84	98	81	92	108	**116**	109	86	77	63	82	1082
Bermuda	32	65	110	116	120	103	111	111	116	140	132	**151**	124	120	1454
Flores	39	31	**220**	180	154	109	105	73	61	98	102	144	169	197	1612
Horta	39	29	119	100	85	75	84	62	54	81	76	113	121	**122**	1092
Madeira	33	17	91	79	80	48	26	11	3	3	29	85	**130**	104	689
Lisbon	39	9° W	94	91	88	69	47	20	5	6	37	82	**111**	103	753
Palma	40	3° E	43	35	39	37	40	19	11	18	55	**77**	55	55	484
Algiers	37	3	**100**	62	81	53	46	17	4	4	28	79	87	95	656

NORMAL RAINFALL

Authority: see p. 211.

FIGURE 114

Inches.

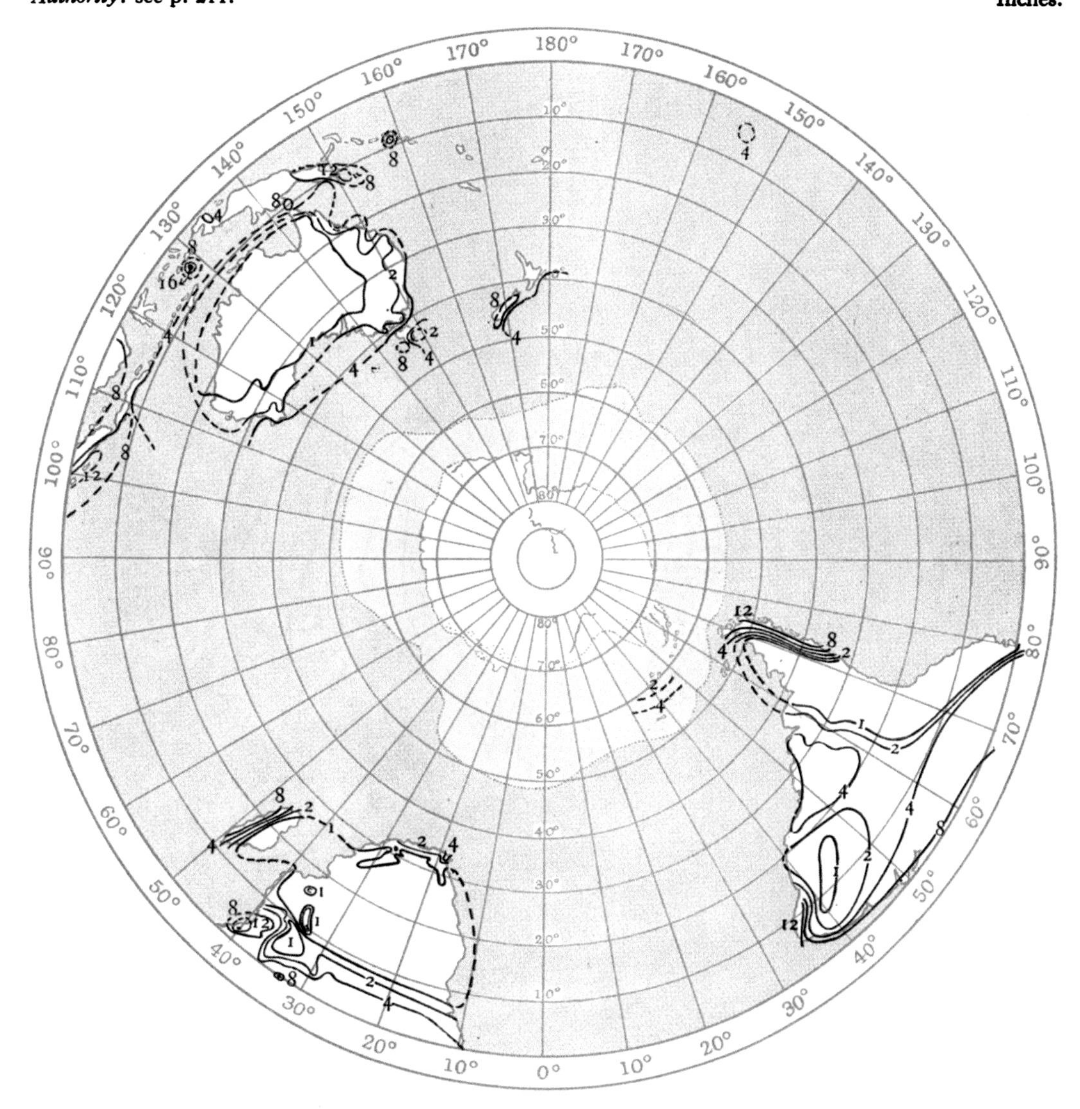

Seasonal variation of rainfall in millimetres, 30–40° S.

Station	Lat.	Long.	Jan.	Feb.	Mar.	Apr.	May	June	July	Aug.	Sept.	Oct.	Nov.	Dec.	Year
Juan Fernandez	34° S	79° W	18	30	43	85	147	154	153	95	70	43	52	14	904
Valdivia	40	73	69	79	154	233	380	480	443	370	214	155	124	113	2814
Valparaiso (Punta Angeles)	33	72	0	0	10	11	92	144	107	69	31	11	8	5	488
Santiago	33	71	1	2	5	14	59	83	87	58	30	14	6	5	364
Cordoba	31	64	106	110	88	43	25	7	8	13	23	61	102	118	704
Bahia Blanca	39	62	44	56	66	65	31	25	27	25	41	57	53	48	538
Buenos Ayres	35	58° W	80	62	112	86	77	68	56	61	79	86	77	98	942
Cape Town	34	18° E	18	13	22	51	93	109	96	83	61	43	26	22	637
Heidelberg	34	21	25	41	41	47	44	33	25	36	41	49	33	38	453
E. London	33	28	80	81	87	78	54	42	25	58	76	103	85	82	851

FIGURE 115 NORMAL RAINFALL

Inches. *Authority:* see p. 211.

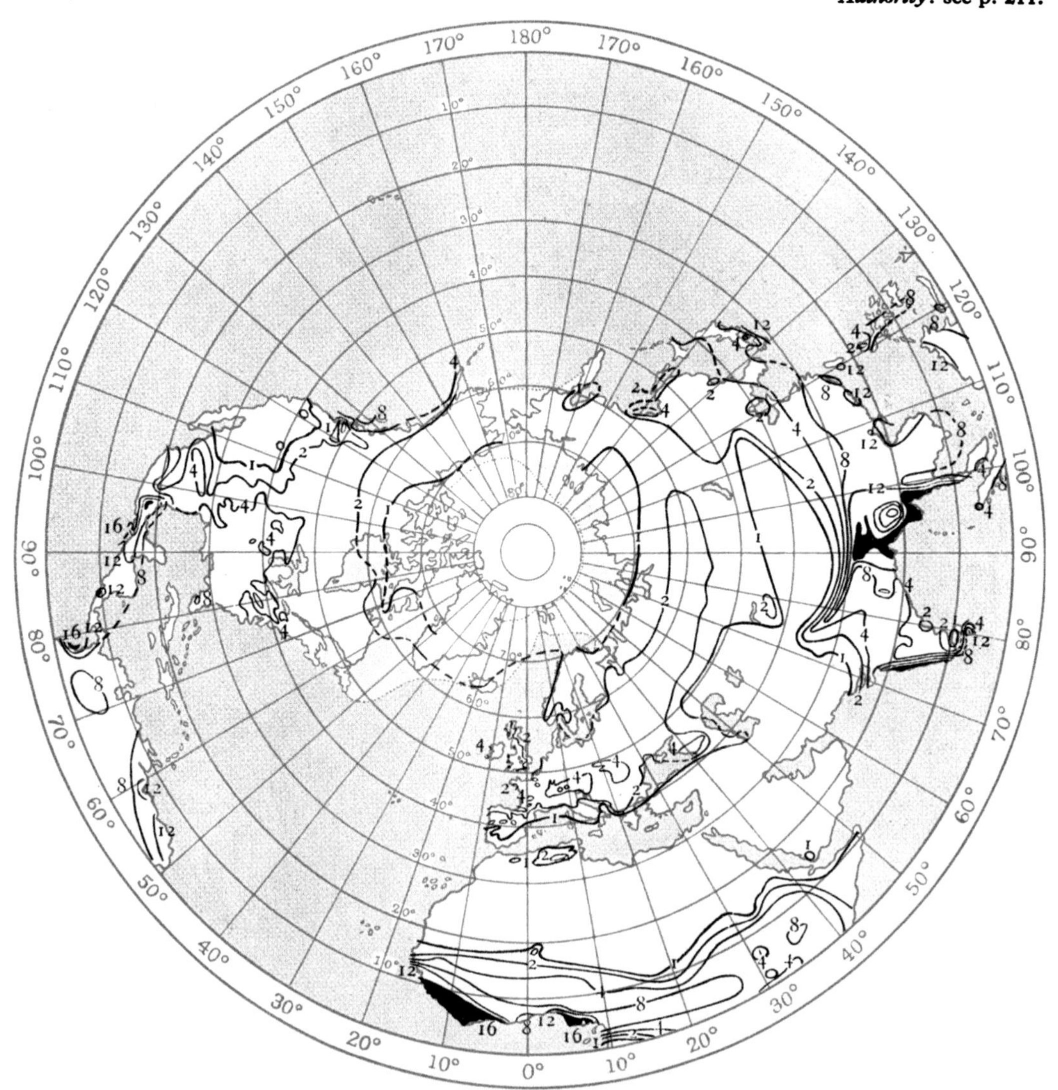

Seasonal variation of rainfall in millimetres, 30–40° N.

Station	Lat.	Long.	Jan.	Feb.	Mar.	Apr.	May	June	July	Aug.	Sept.	Oct.	Nov.	Dec.	Year
Alexandria	31° N	30° E	53	24	13	4	1	0	0	0	1	7	34	67	204
Cairo	30	31	10	5	6	6	1	0	0	0	0	0	2	4	34
Nicosia	35	33	66	43	40	22	30	13	2	2	5	22	55	77	376
Lenkoran	39	49	95	70	81	53	33	23	21	47	182	199	154	99	1057
Teheran	36	51	43	26	49	29	11	2	6	1	2	8	27	32	236
Meshed	36	60	19	24	59	45	35	8	2	1	0·5	10	15	20	238
Quetta	30	67	49	51	49	26	10	4	19	11	2	2	7	24	254
Simla	31	77	67	84	62	44	68	175	431	461	145	22	11	32	1603
Tientsin	39	117	5	3	13	13	29	70	174	142	48	20	13	3	533
Shanghai	31	121	51	58	89	94	89	185	153	144	113	82	53	35	1146
Nagasaki	33	130	77	83	133	191	169	326	238	172	218	123	89	86	1903
Tokio	36	140	56	67	112	131	153	163	140	163	226	191	104	55	1561

NORMAL RAINFALL

Authority: see p. 211.

FIGURE 116

Inches.

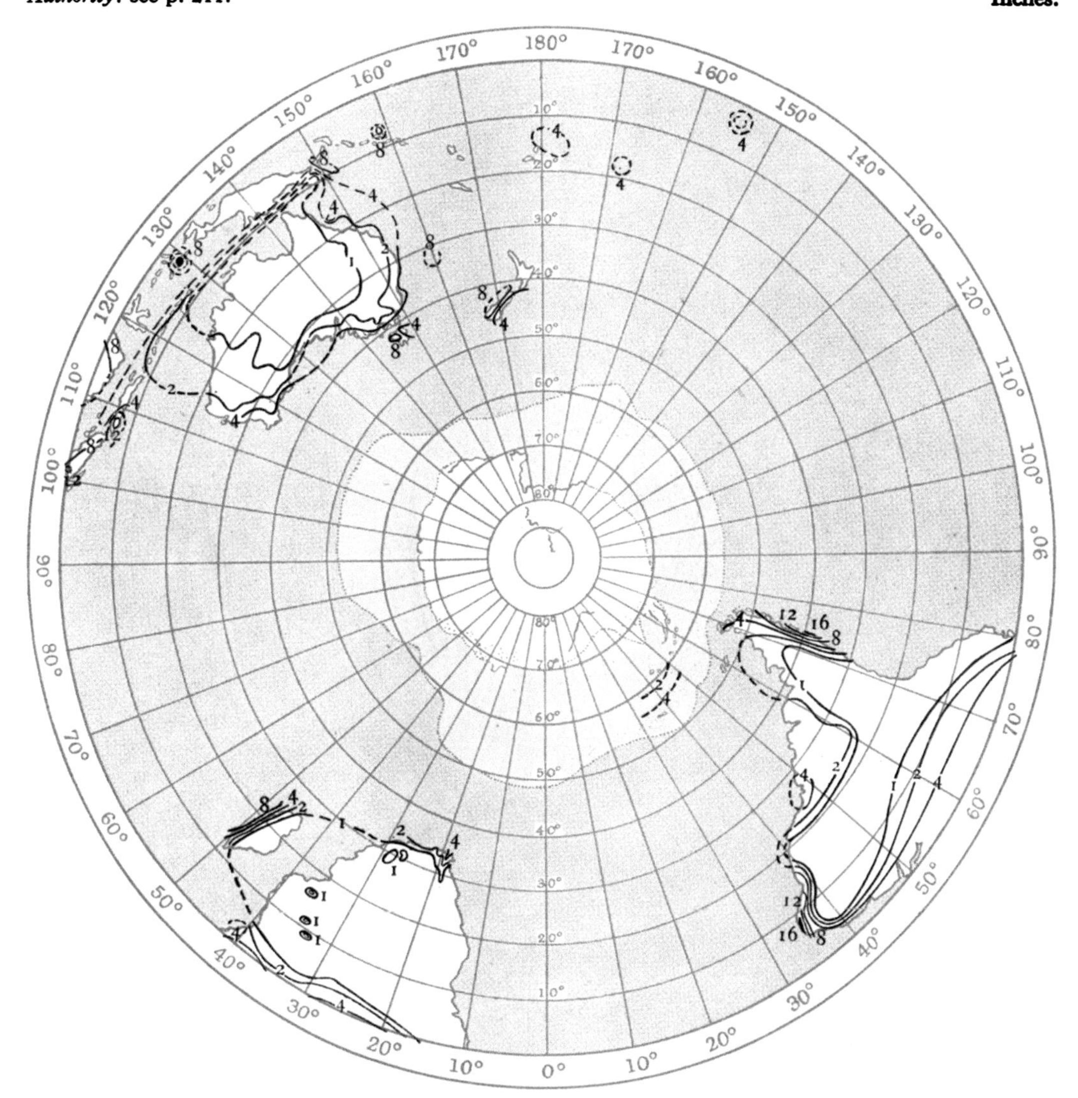

Seasonal variation of rainfall in millimetres, 30–40° S.

Station	Lat.	Long.	Jan.	Feb.	Mar.	Apr.	May	June	July	Aug.	Sept.	Oct.	Nov.	Dec.	Year
Perth	32° S	116° E	8	12	19	40	124	174	166	144	85	54	20	15	861
Katanning	34	118	10	15	24	30	52	71	78	63	50	40	17	14	464
Coolgardie	31	121	12	19	19	24	35	31	23	26	15	19	17	18	258
Eucla	32	129	15	16	23	28	32	27	23	24	19	18	18	11	254
Streaky Bay	33	134	11	12	15	26	49	71	60	49	35	24	17	10	379
Adelaide	35	139	18	17	26	46	69	78	67	64	50	44	29	24	532
Melbourne	38	145	47	44	57	57	55	53	46	46	61	68	56	59	647
Bourke	30	146	38	43	32	30	27	28	23	22	20	26	32	31	352
Sydney	34	151	93	112	126	135	131	123	126	76	74	75	72	72	1215
Lord Howe Is.	31	159	125	127	138	145	138	190	202	133	134	139	106	124	1701
Auckland	37	175	65	76	77	85	113	122	130	108	92	91	83	71	1113
Napier	40	177	42	61	94	64	108	67	101	79	47	64	53	52	832

FIGURE 117 NORMAL RAINFALL

Inches. *Authority:* see p. 211.

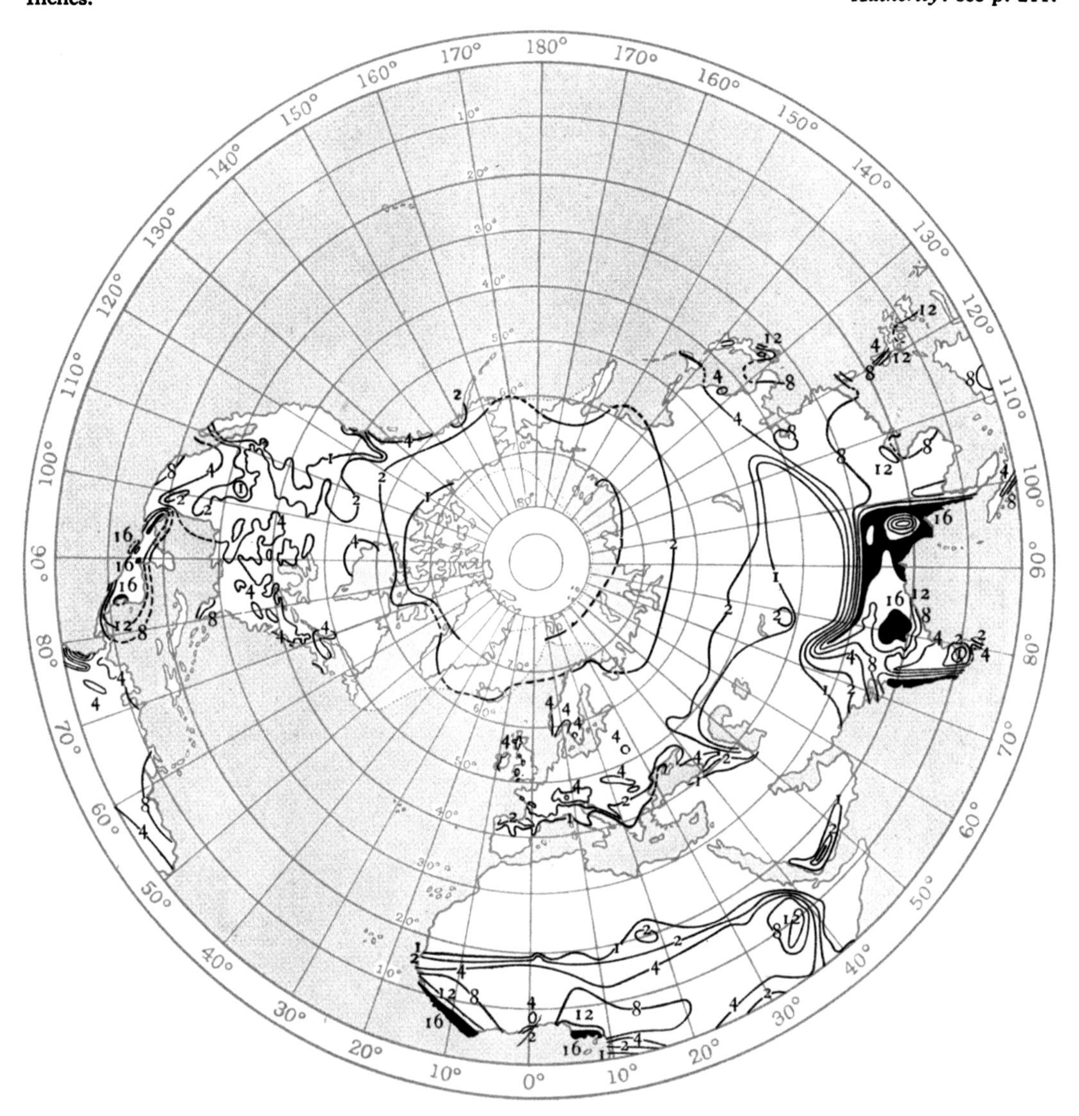

Seasonal variation of rainfall in millimetres, 20–30° N.

Station	Lat.	Long.	Jan.	Feb.	Mar.	Apr.	May	June	July	Aug.	Sept.	Oct.	Nov.	Dec.	Year
Midway Is.	28° N	177° W	86	95	102	65	97	79	83	84	124	**147**	42	93	1097
Honolulu	21	158	100	104	104	49	40	29	30	36	46	50	99	**114**	801
Leon	21	102	8	8	7	6	27	109	**154**	141	125	38	12	11	646
Tulancingo	20	98	17	8	28	37	58	70	**95**	59	86	57	18	14	547
Galveston	29	95	81	73	70	81	91	99	96	125	**149**	114	94	98	1171
Havana	23	82	65	36	47	43	118	134	113	119	130	**163**	76	61	1105
Nassau	25	77	55	41	39	64	150	169	148	167	**192**	161	72	38	1296
La Laguna	28	16° W	110	75	84	59	18	5	4	2	12	61	88	**126**	644
Insalah and Aswan. No rain measured.															
Bushire	29	51° E	69	49	25	13	1	0	0	0	0	2	36	**76**	271
Robat	30	61	16	**30**	18	13	1	0	0	0	0	3	17	17	105
Jaipur	27	76	12	7	8	5	13	52	200	**211**	87	7	4	6	612

NORMAL RAINFALL

Authority: see p. 211.

FIGURE 118

Inches.

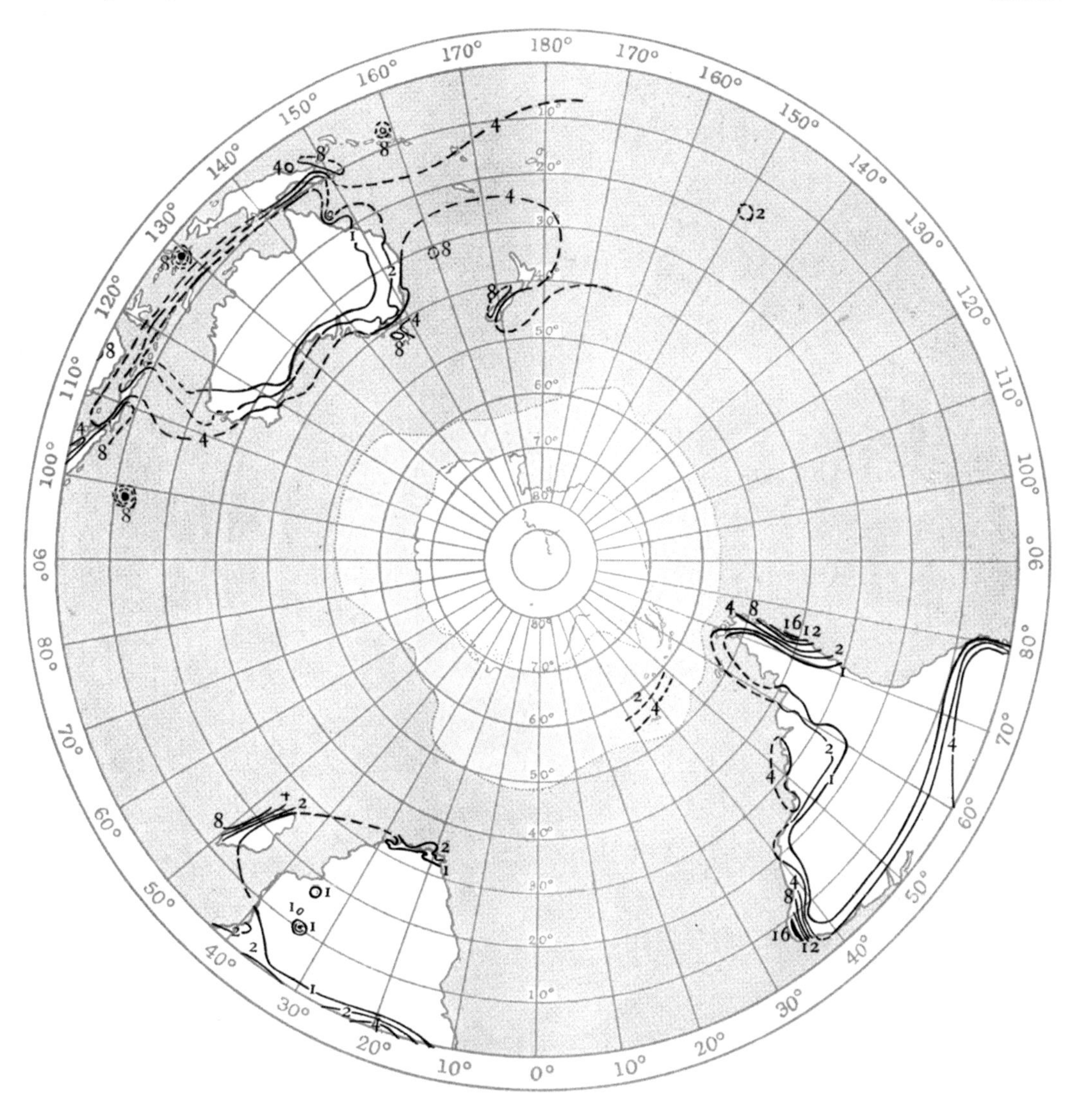

Seasonal variation of rainfall in millimetres, 20–30° S.

Station	Lat.	Long.	Jan.	Feb.	Mar.	Apr.	May	June	July	Aug.	Sept.	Oct.	Nov.	Dec.	Year
Rarotonga	21° S	160° W	242	254	**269**	196	163	109	78	113	111	103	153	178	1969
Easter Is.*	27	109	133	41	229	131	99	**241**	81	65	90	55	128	74	1367
Punta Tortuga	30	71	0	0	2	0	34	**47**	34	13	3	0	2	0	135
Iquique	20	70	0	0	0	0	0	0	**1**	0	0	0	0	0	1
Salta	25	65	**145**	125	103	31	9	1	1	3	6	20	60	107	611
Asuncion	25	58	146	117	103	130	97	61	57	41	68	134	**154**	151	1259
Rio de Janeiro	23	43° W	128	115	**139**	108	84	58	45	48	67	87	105	**139**	1123
Windhoek	23	17° E	**93**	74	78	40	6	1	1	2	1	9	21	50	376
Kimberley	29	25	71	**78**	**78**	36	22	8	9	7	20	25	40	54	448
Bulawayo	20	29	**150**	102	80	16	7	1	1	1	3	23	83	131	598
Durban	30	31	113	119	**136**	84	49	29	24	39	83	118	128	127	1049
Lourenço Marques	26	33	**138**	125	79	38	30	14	13	16	29	76	81	108	747

* Very short period 1911–13.

FIGURE 119 NORMAL RAINFALL

Inches. *Authority:* see p. 211.

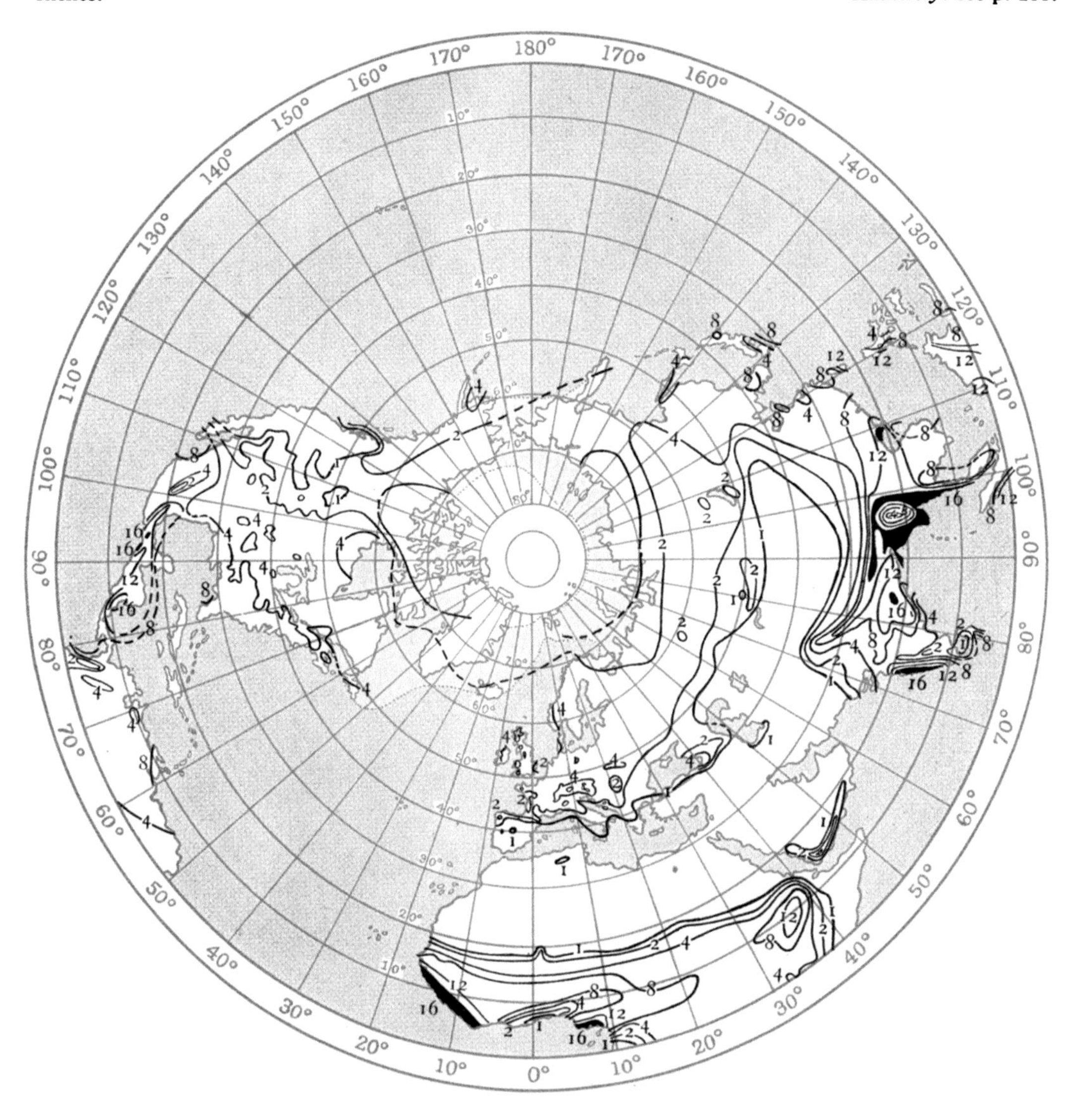

Seasonal variation of rainfall in millimetres, 20–30° N.

Station	Lat.	Long.	Jan.	Feb.	Mar.	Apr.	May	June	July	Aug.	Sept.	Oct.	Nov.	Dec.	Year
Nagpur	21° N	79° E	9	11	13	14	19	221	**365**	299	218	49	17	14	1249
Darjiling	27	88	16	25	45	93	222	578	**822**	676	469	114	7	6	3073
Calcutta	23	88	10	29	32	44	146	290	**327**	309	263	99	14	5	1568
Gauhati	26	92	14	22	64	161	249	**324**	307	269	189	75	14	6	1694
Cherrapunji	25	92	13	75	250	807	1210	2357	**2542**	2096	893	555	74	10	10882
Mandalay	22	96	1	2	5	30	145	148	82	115	**151**	119	43	10	851
Phu Lien	21	107	27	27	52	76	213	221	264	265	**292**	132	64	29	1662
Hong Kong	22	114	35	41	71	135	295	**409**	340	357	254	123	43	29	2132
Taihoku	25	122	93	124	178	134	229	278	225	**303**	257	133	78	82	2114
Santo Domingo	20	122	236	118	121	109	245	154	283	376	347	355	356	**404**	3104
Naha	26	128	136	135	152	160	254	**264**	181	253	181	171	144	105	2134
Titizima, Bonin Is.	27	142	124	88	113	120	194	97	114	**202**	122	124	157	130	1585

Waialeale, Kauai, 22° N 159° W. In five years 216 feet of rain fell (*Science*, 23 Nov. 1917).

NORMAL RAINFALL

Authority: see p. 211.

FIGURE 120

Inches.

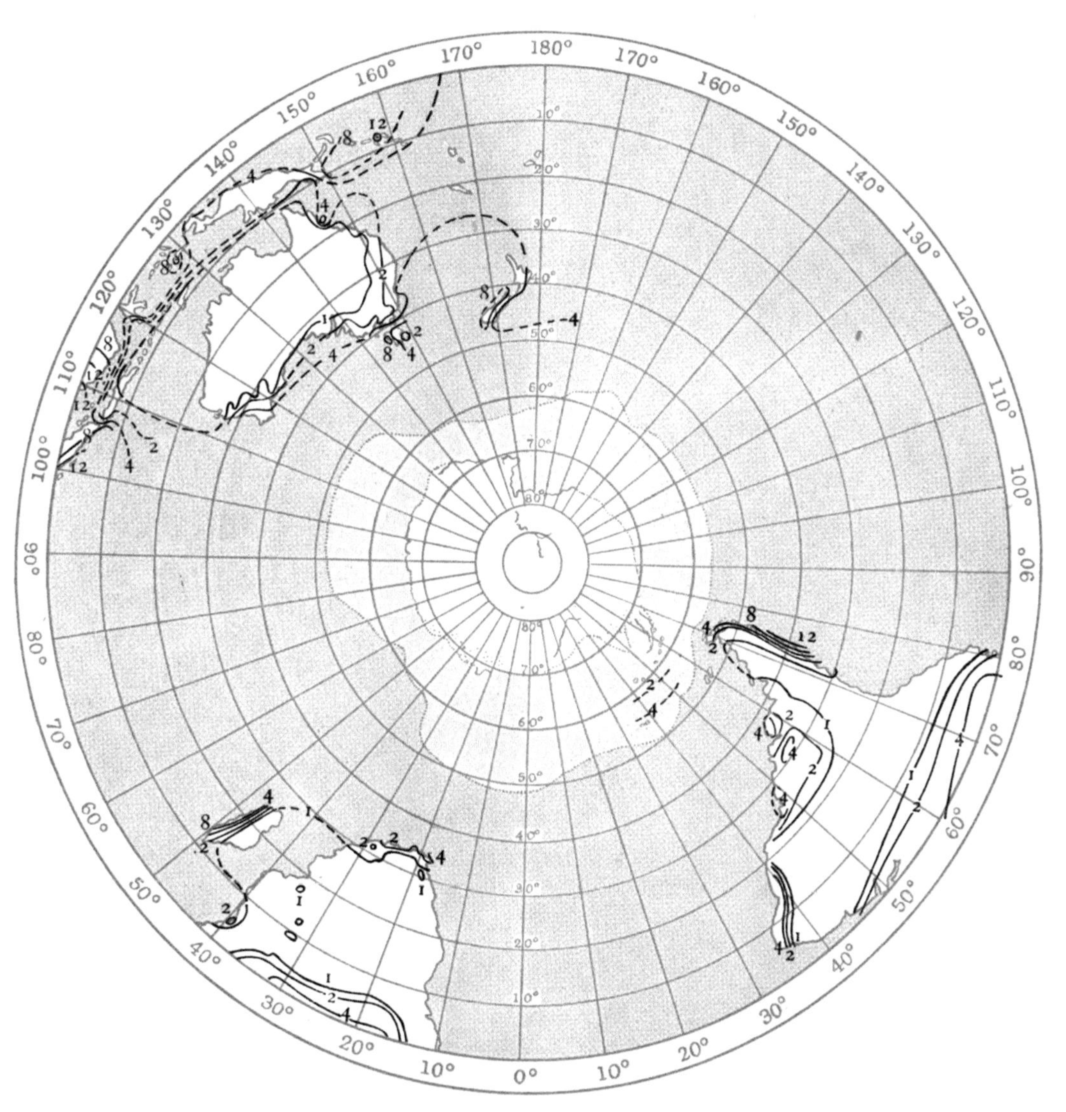

Station	Lat.	Long.	Jan.	Feb.	Mar.	Apr.	May	June	July	Aug.	Sept.	Oct.	Nov.	Dec.	Year
Mauritius	20° S	58° E	197	213	**238**	114	77	71	63	60	33	35	40	120	1261
Onslow	22	115	24	25	23	23	37	**45**	24	9	1	1	1	3	216
Peak Hill	26	119	**42**	31	33	31	23	36	18	13	6	5	10	15	263
Nullagine	22	120	**80**	55	60	22	13	27	14	10	0	6	13	34	334
Laverton	29	122	26	29	**31**	26	25	24	17	16	8	8	21	18	249
William Creek	29	136	14	11	**17**	10	10	16	7	7	11	10	11	10	134
Thargomindah	28	144	36	**42**	20	21	21	22	13	14	13	21	24	36	283
Mitchell	27	148	**86**	**86**	81	37	39	43	35	25	34	35	50	64	615
Rockhampton	23	150° 30′	**233**	196	132	60	41	50	35	25	34	46	54	109	1015
Brisbane	27	153	**163**	161	148	91	73	67	59	54	53	67	93	126	1155
Ouitchambo	22	166	170	**270**	174	105	98	156	83	73	93	67	93	91	1473
Norfolk Is.	29	168	92	137	101	105	141	150	**161**	150	97	104	71	87	1396

Seasonal variation of rainfall in millimetres, 20–30° S.

FIGURE 121

Inches.

NORMAL RAINFALL

Authority: see p. 211.

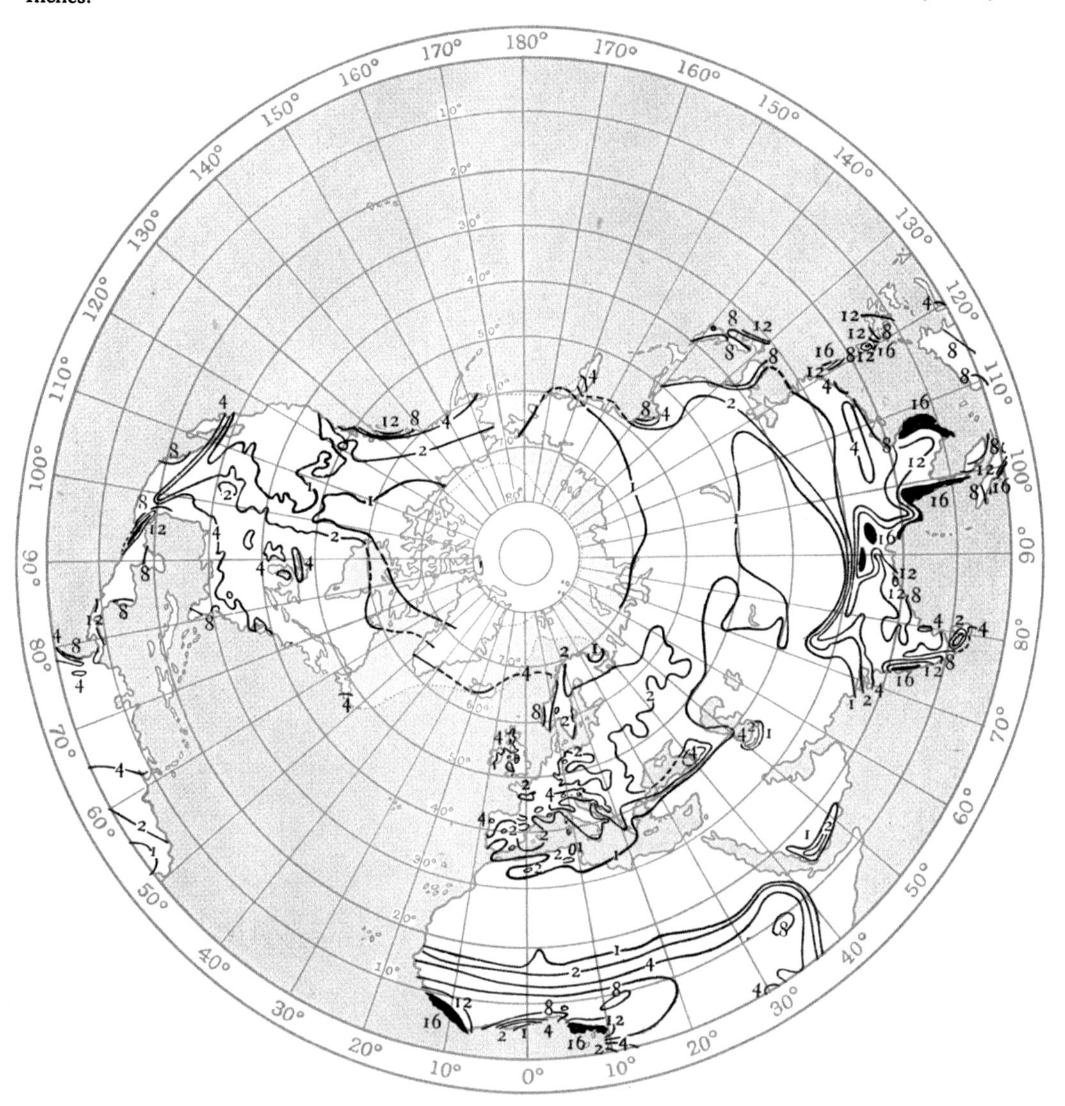

Seasonal variation of rainfall in millimetres, 10–20° N.

Station	Lat.	Long.	Jan.	Feb.	Mar.	Apr.	May	June	July	Aug.	Sept.	Oct.	Nov.	Dec.	Year
Tacubaya	19° N	99° W	9	7	11	19	51	104	**128**	122	104	53	15	6	629
S. Salvador	14	89	2	4	11	39	167	265	**302**	292	284	266	51	12	1695
Belize	17	88	164	75	43	67	146	288	253	240	235	**345**	302	193	2351
Jamaica	18	78	34	39	59	77	149	159	150	172	191	**213**	82	54	1379
Port au Prince	19	72	31	57	95	165	**244**	88	62	135	201	182	78	41	1379
Caracas	10° 30′	67	23	11	17	42	71	106	**113**	108	93	97	84	48	813
Barbados	13	60	68	35	38	39	71	128	142	179	**187**	157	155	95	1294
St Vincent	17	25	4	5	1	0	0	0	6	42	**51**	31	8	7	155
Bathurst	13	17° W	0	0	0	1	5	75	284	**499**	256	99	7	3	1229
Sokoto	13	5° E	0	0	3	4	50	103	163	**204**	107	11	0	0	645
Kaduna Capital	11	7	0	1	11	84	151	199	208	246	**292**	55	3	0	1250
Khartoum	16	33	0	0	0	0	3	8	40	**56**	18	5	0	0	130

NORMAL RAINFALL

Authority: see p. 211.

FIGURE 122

Inches.

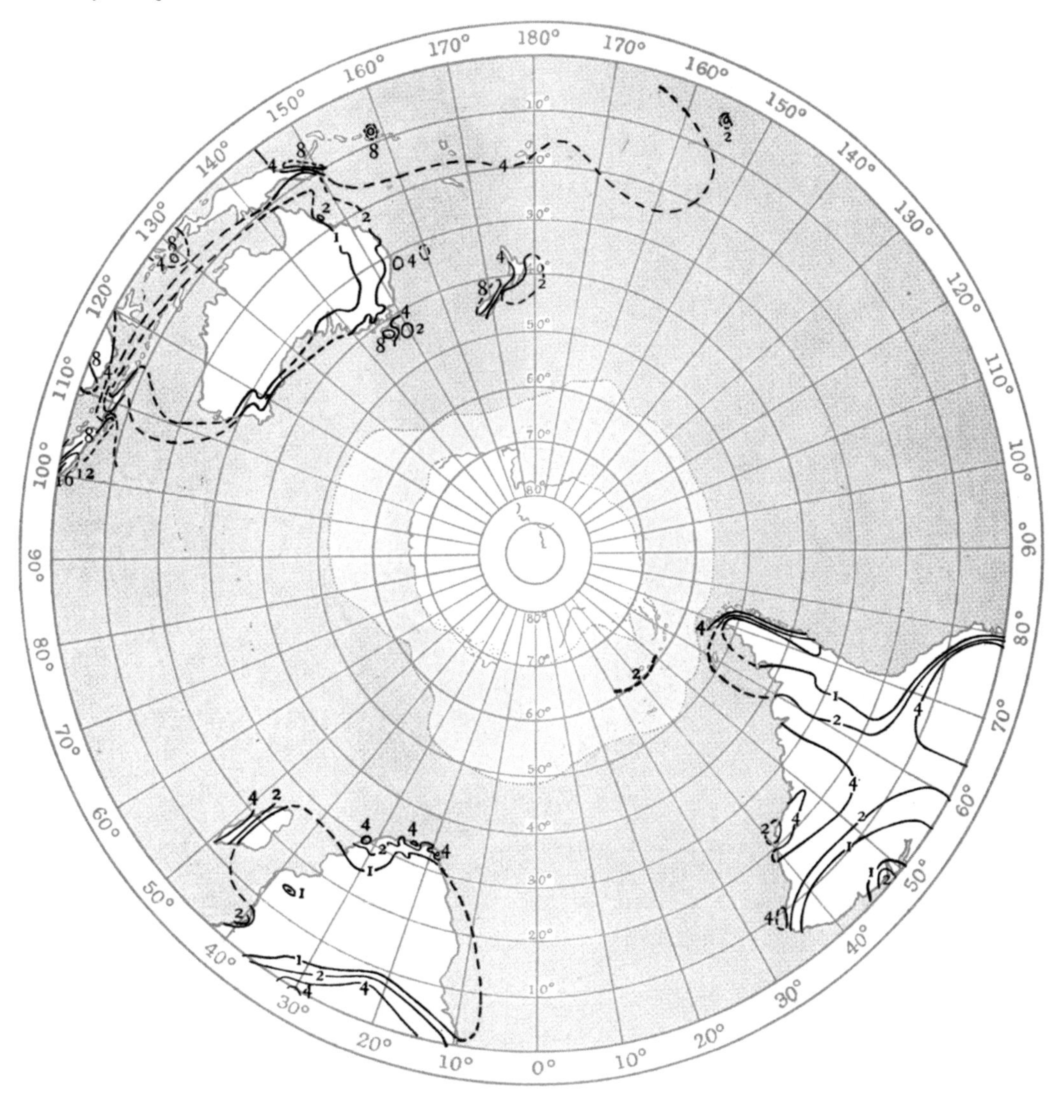

Seasonal variation of rainfall in millimetres, 10–20° S.

Station	Lat.	Long.	Jan.	Feb.	Mar.	Apr.	May	June	July	Aug.	Sept.	Oct.	Nov.	Dec.	Year
Samoa	14° S	172° W	**427**	399	344	260	140	131	67	80	130	154	236	346	2714
Niue Is.	19	170	250	265	**304**	188	139	94	61	118	124	90	127	267	2027
Makatea	16	148	218	340	215	130	120	61	58	78	92	182	175	**366**	2035
Arequipa	16	72	31	**44**	10	4	0	1	2	0	0	0	1	3	96
Puerto de Arica	18	70	**1**	0	0	0	0	0	0	0	0	0	0	0	1
Sucre	19° 30′	64	**161**	119	95	45	7	2	4	4	20	36	61	111	665
Cuyaba	16	56	**253**	216	219	98	49	7	4	28	51	116	156	198	1395
Bello Horizonte	20	44	**386**	220	165	67	14	10	9	19	36	118	213	284	1541
Caetite	14	43	139	85	80	67	14	11	9	6	24	71	130	**150**	786
Ondina	13	39	81	106	164	280	**293**	260	194	122	78	132	141	114	1965
Aracaju	11	37	39	43	96	81	**159**	113	110	65	38	57	21	71	893
St Helena	16	6	77	92	**119**	104	105	99	99	101	73	49	44	54	1016

FIGURE 123

Inches.

NORMAL RAINFALL

Authority: see p. 211.

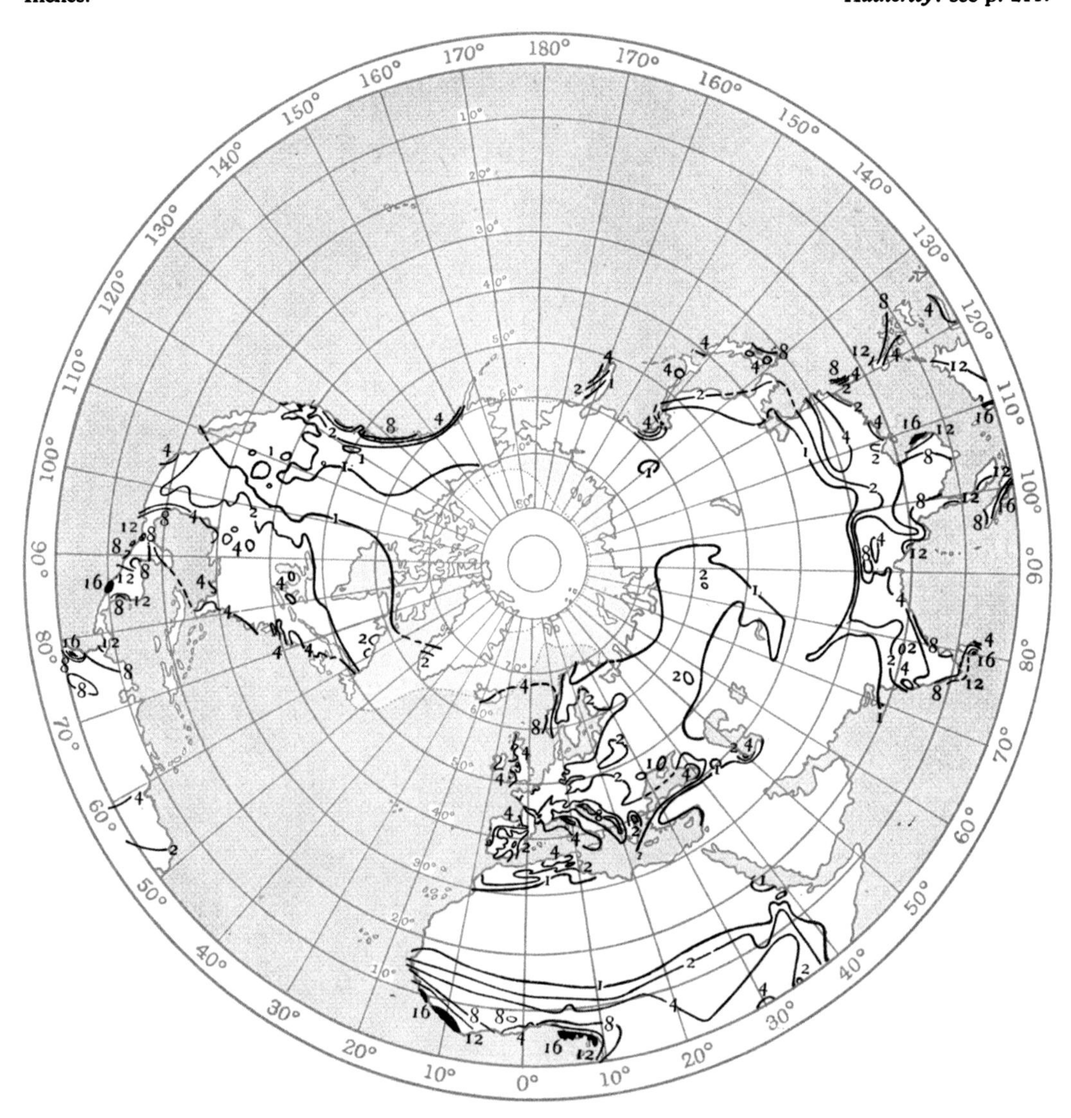

Seasonal variation of rainfall in millimetres, 10–20° N.

Station	Lat.	Long.	Jan.	Feb.	Mar.	Apr.	May	June	July	Aug.	Sept.	Oct.	Nov.	Dec.	Year
Berbera	10° N	45° E	3	8	**23**	11	8	0	2	1	1	1	5	2	65
Bombay	19	73	2	1	1	2	18	469	**639**	360	275	47	10	2	1826
Cochin	11	76	16	19	53	132	289	**715**	581	332	221	332	160	41	3021
Bangalore	13	78	7	5	12	35	110	73	106	139	**172**	160	72	12	903
Madras	13	80	24	8	9	17	49	50	98	118	124	278	**331**	138	1244
Waltair	18	83	18	26	9	22	59	116	102	142	**182**	164	81	27	948
Pt. Blair	12	93	26	26	20	73	429	**489**	399	376	471	277	252	173	3011
Rangoon	17	96	2	7	9	40	311	446	**532**	500	397	171	60	4	2479
Nhatrang	12	109	62	17	20	27	64	56	53	37	179	310	**378**	192	1395
Bolinao	16	120	11	11	15	27	179	381	**655**	560	552	184	46	12	2633
Manila	15	121	26	10	19	34	106	235	**411**	393	366	193	128	61	1981
Guam	13	145	62	76	80	46	104	138	378	**403**	400	340	184	129	2340

NORMAL RAINFALL

Authority: see p. 211.

FIGURE 124

Inches.

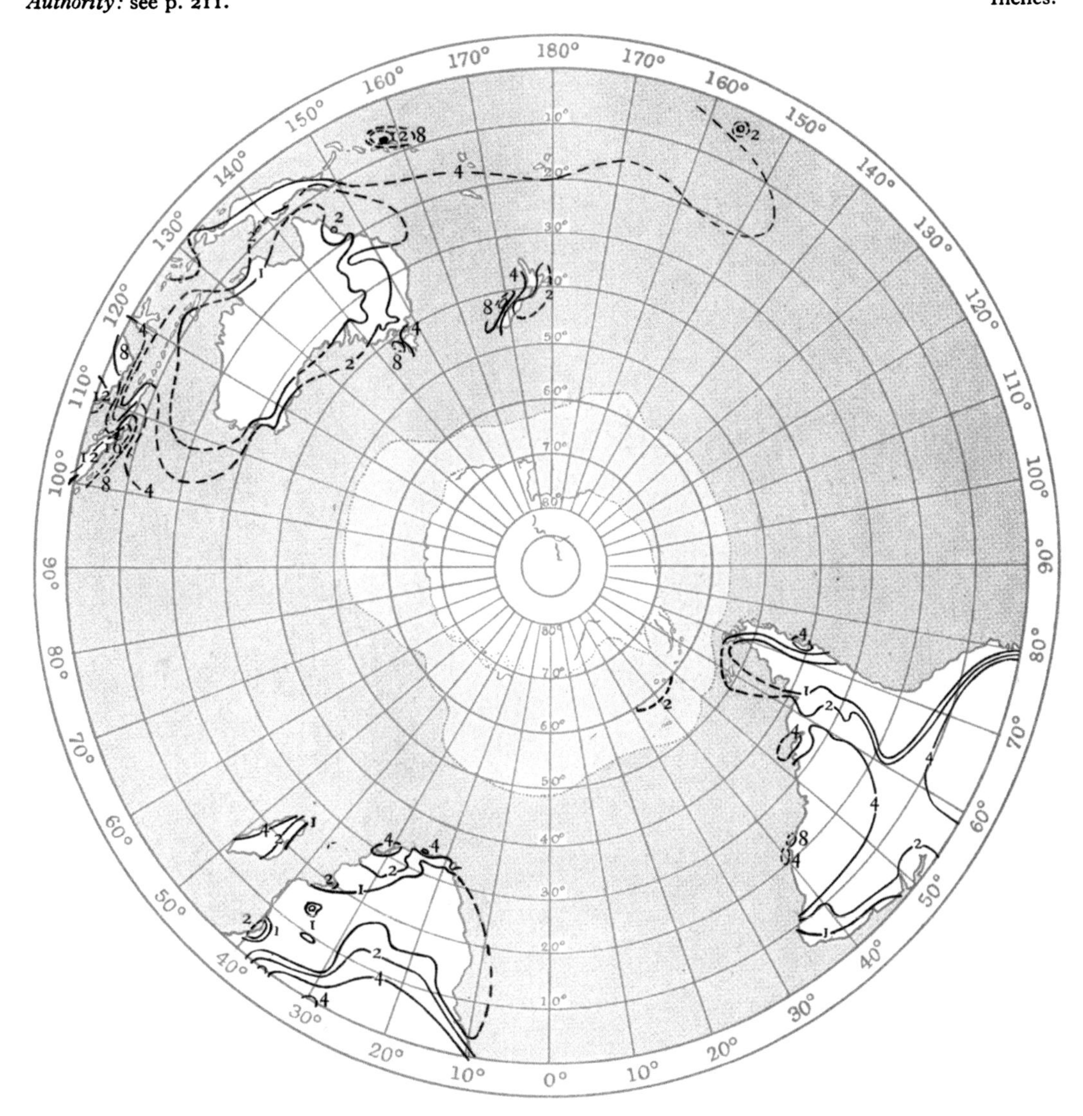

Seasonal variation of rainfall in millimetres, 10–20° S.

Station	Lat.	Long.	Jan.	Feb.	Mar.	Apr.	May	June	July	Aug.	Sept.	Oct.	Nov.	Dec.	Year
Gwelo	19° S	30° E	**165**	138	71	16	10	2	1	3	5	17	98	148	674
Zomba	15	35	**279**	274	229	97	27	14	8	9	9	39	140	272	1397
Tananarivo	19	48	**305**	292	188	51	14	7	6	8	15	64	129	286	1365
Cocos Is.	12	97	134	187	237	229	208	241	**262**	146	117	115	120	112	2108
Christmas Is.	10	106	250	**292**	254	201	242	147	137	62	88	90	252	203	2217
Derby	17	124	**199**	154	109	37	21	15	5	3	0	1	30	114	688
Darwin	12	131	**402**	327	251	106	17	4	2	3	13	55	124	264	1568
Daly Waters	16	133	**162**	**162**	116	25	4	7	2	4	6	21	56	101	666
Mein	13	143	**330**	278	232	80	12	9	3	3	2	15	67	180	1211
Harvey Creek	17	146	814	576	**815**	536	336	216	110	139	97	97	218	299	4252
Samarai	11	151	140	214	273	239	**309**	272	258	251	300	168	177	116	2717
Suva, Fiji	18	178	272	257	**373**	287	258	156	117	209	177	198	242	308	2854

FIGURE 125 NORMAL RAINFALL

Inches. *Authority:* see p. 211.

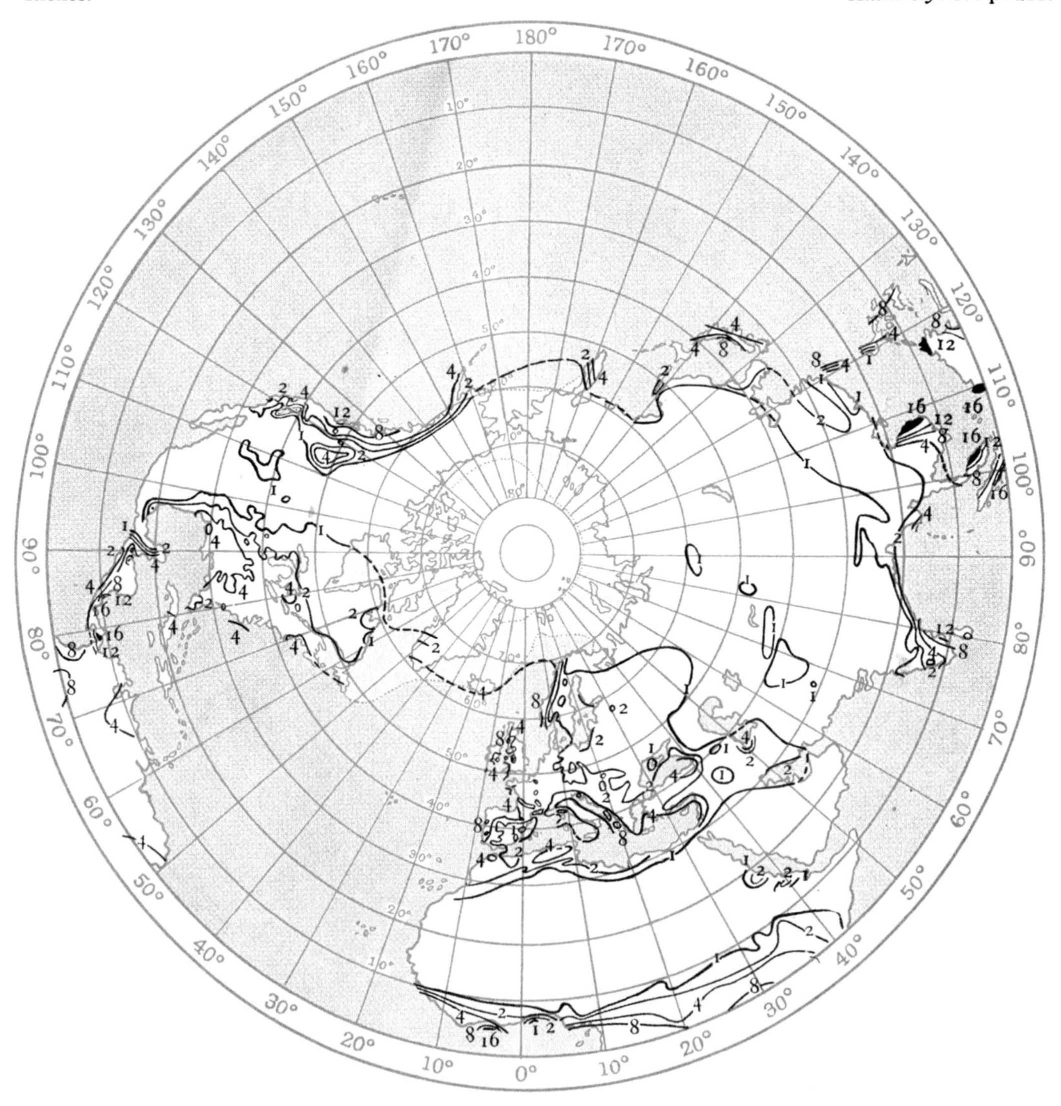

Seasonal variation of rainfall in millimetres, 0–10° N.

Station	Lat.	Long.	Jan.	Feb.	Mar.	Apr.	May	June	July	Aug.	Sept.	Oct.	Nov.	Dec.	Year
Fanning Is.	4° N	159° W	284	280	320	**434**	373	255	215	102	77	97	73	224	2734
Colon	9	79	98	42	41	107	323	340	407	381	322	378	**542**	297	3278
Bogota	5	74	57	58	100	144	115	59	51	56	59	**163**	114	64	1040
El Peru	7	62	100	75	55	78	150	**199**	178	181	115	89	90	106	1416
Paramaribo	6	55	224	179	212	250	**316**	298	205	149	69	73	141	222	2338
Cayenne	5	52	365	312	402	481	**557**	394	176	70	31	34	117	272	3211
Conakry	10	14	0	0	2	24	187	535	**1424**	1110	748	427	116	8	4581
Accra	6	0° W	17	25	46	94	144	**178**	43	15	25	49	38	18	692
Zungeru	10	6° E	1	1	14	58	123	163	195	230	**262**	80	4	2	1133
Libreville	0	9	263	237	348	341	236	13	2	19	105	341	**352**	237	2494
Yola	9	12° 30′	0	0	9	53	113	138	161	**209**	186	78	2	0	949
Wau	8	28	1	6	22	53	139	169	185	**207**	162	127	17	0	1088

NORMAL RAINFALL

FIGURE 126

Authority: see p. 211.

Inches.

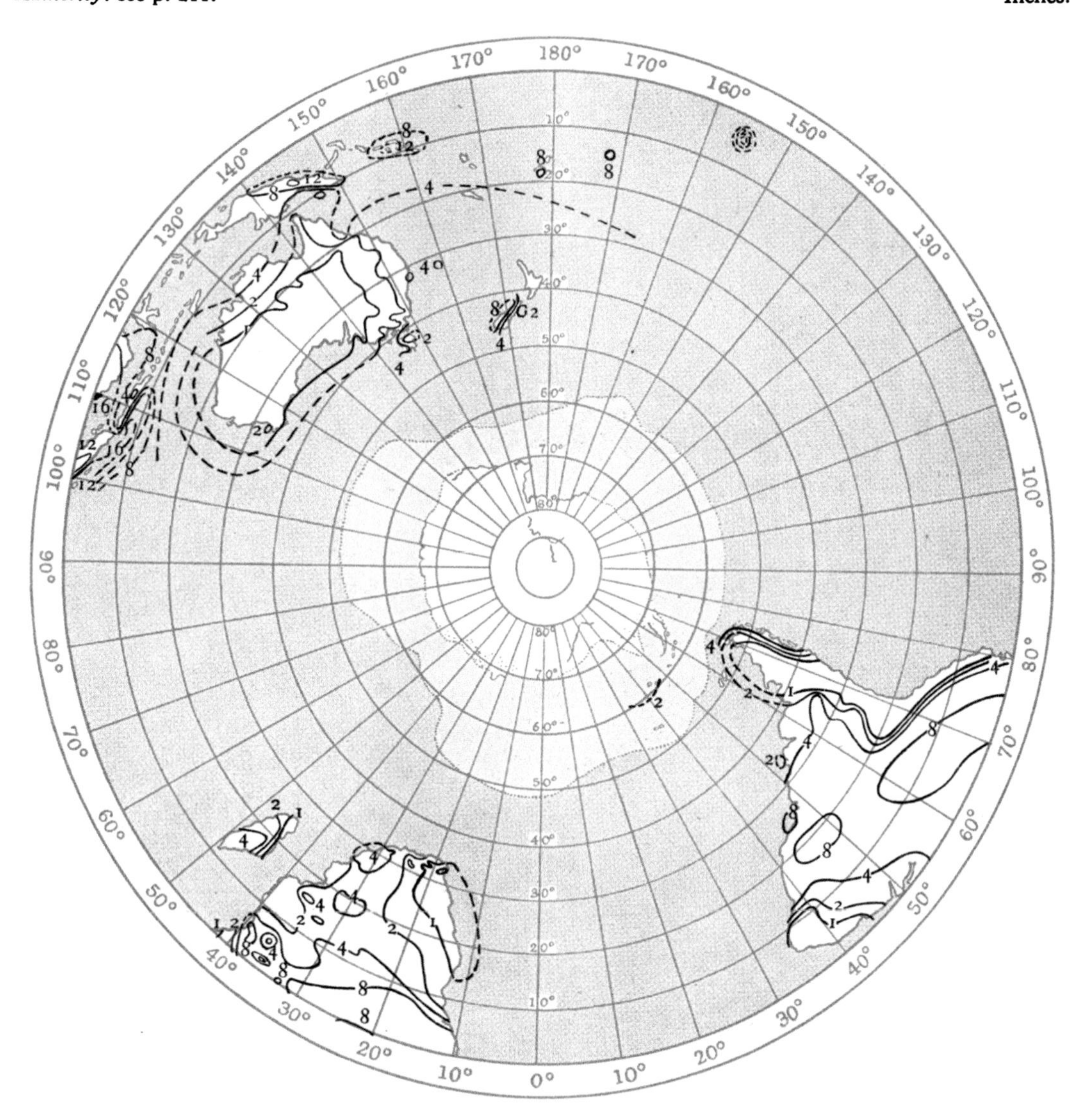

Seasonal variation of rainfall in millimetres, 0–10° S.

Station	Lat.	Long.	Jan.	Feb.	Mar.	Apr.	May	June	July	Aug.	Sept.	Oct.	Nov.	Dec.	Year
Malden Is.	4° S	155° W	103	51	**120**	118	108	49	51	41	22	25	19	20	727
Manaos	3	60	**216**	207	202	203	158	73	38	40	41	101	127	205	1651
Turyassu	2	45	228	309	**425**	418	274	211	132	69	11	12	11	58	2158
Barra do Corda	5° 30′	45	**192**	179	162	99	61	19	8	12	27	43	84	121	1007
Quixeramobim	5	39	76	101	**151**	129	93	41	20	12	1	1	3	30	658
Pão dé Assucar	10	37	45	19	37	49	55	86	**101**	61	22	22	34	73	594
Fernando Noronha	4	32° W	64	111	193	**248**	221	113	64	34	7	6	7	15	1083
Loanda	9	13° E	27	30	65	**117**	13	0	0	1	2	7	31	23	316
Brazzaville	4	15	107	132	157	**227**	139	7	1	6	23	128	179	147	1253
Nairobi	1	37	44	69	118	**230**	150	49	14	27	30	47	158	65	1001
Lamu	2	41	1	3	23	182	**434**	147	80	43	32	32	43	24	1044
Seychelles	5	55	**411**	331	241	186	154	117	65	59	140	128	251	347	2430

FIGURE 127 NORMAL RAINFALL

Inches. *Authority:* see p. 211.

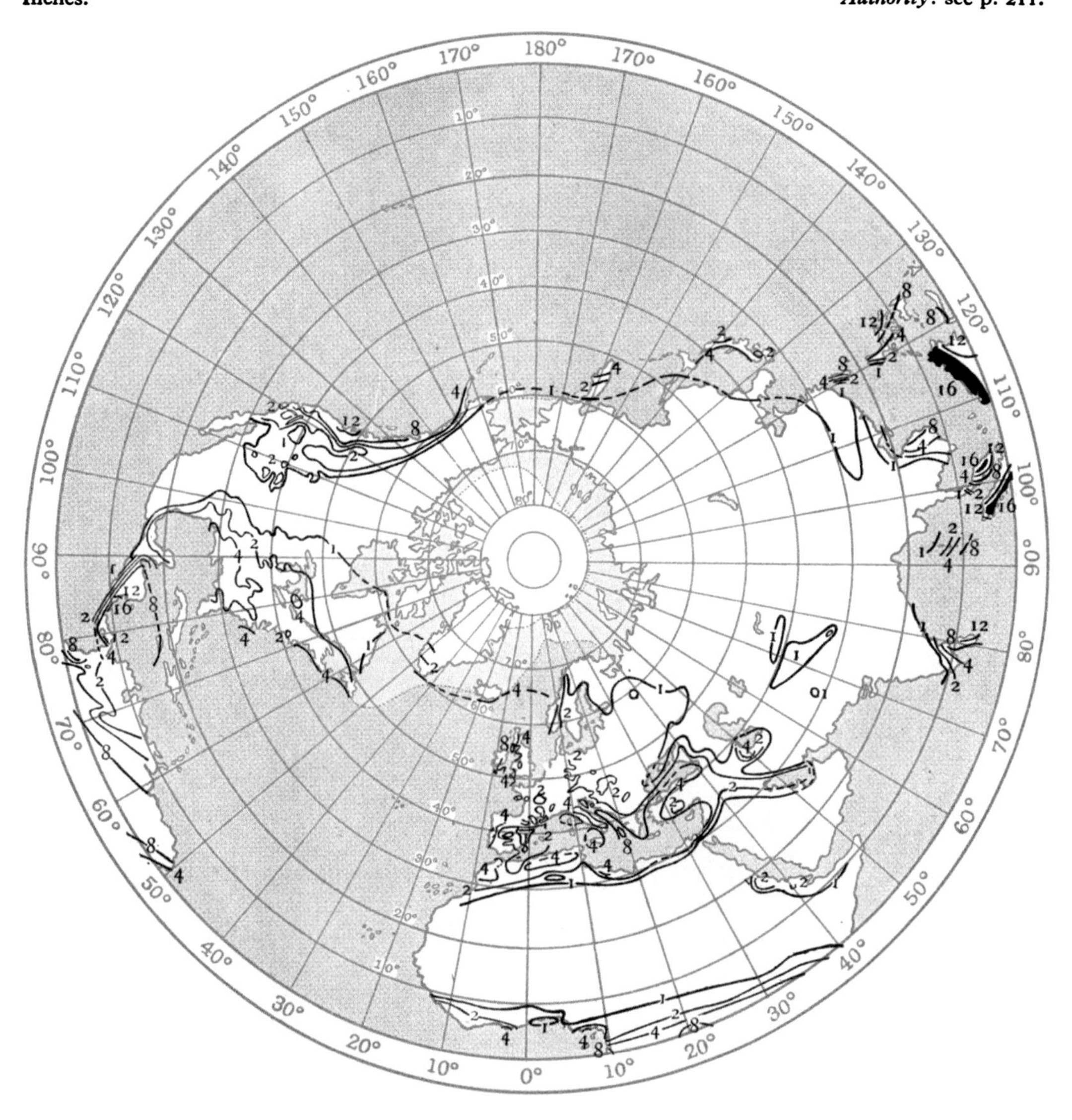

Seasonal variation of rainfall in millimetres, 0–10° N.

Station	Lat.	Long.	Jan.	Feb.	Mar.	Apr.	May	June	July	Aug.	Sept.	Oct.	Nov.	Dec.	Year
Mongalla	5° N	32° E	2	19	39	106	137	117	132	**148**	124	109	46	8	987
Entebbe	0	32	66	91	148	**249**	216	129	76	77	78	89	125	129	1473
Minicoy	8	73	41	16	24	70	144	**289**	230	186	162	215	136	88	1601
Colombo	7	80	89	45	115	199	329	196	164	77	149	**332**	279	132	2107
Trincomalee	9	81	148	52	39	52	62	33	55	110	117	208	355	**357**	1588
Kota Radja	6	95	139	90	86	103	155	95	111	119	167	185	192	**224**	1666
Penang	6	100	97	74	119	176	274	195	212	326	410	**422**	300	126	2731
Singapore	1	104	251	168	188	194	169	174	172	202	172	205	252	**268**	2415
Sandakan	6	118	**469**	244	206	106	152	185	166	205	239	254	373	449	3048
Iwahig	10	119	88	62	57	49	155	220	222	149	230	260	298	**337**	2127
Menado	1° 30′	125	**467**	361	285	206	155	167	125	102	83	112	225	368	2654
Yap	9	138	157	171	146	125	277	246	**423**	394	308	259	223	200	2929

NORMAL RAINFALL

FIGURE 128

Authority: see p. 211.

Inches.

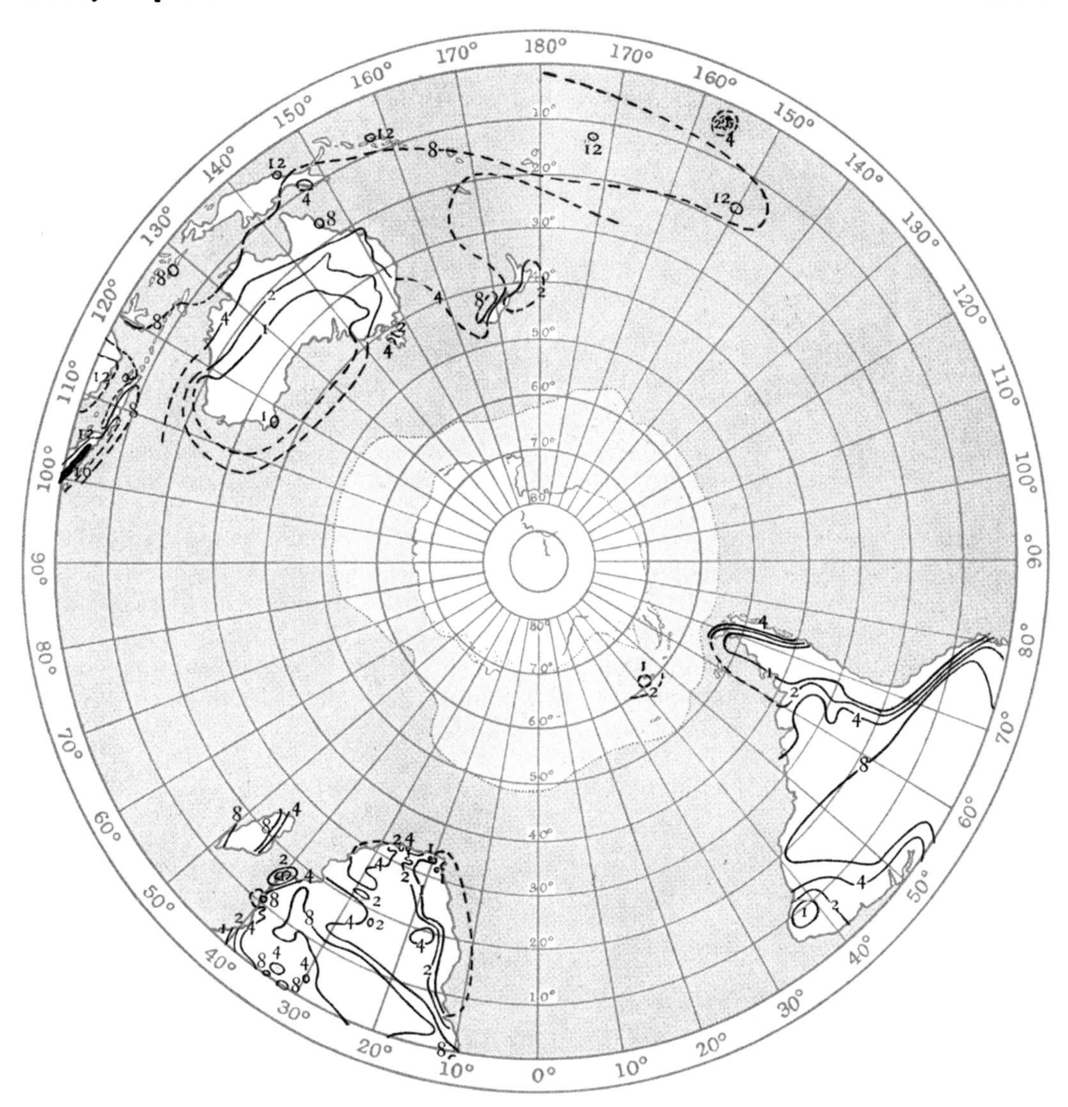

Seasonal variation of rainfall in millimetres, 0–10° S.

Station	Lat.	Long.	Jan.	Feb.	Mar.	Apr.	May	June	July	Aug.	Sept.	Oct.	Nov.	Dec.	Year
Padang	1° S	100° E	349	261	291	365	321	340	293	361	415	511	**525**	494	4526
Batavia	6	107	316	**323**	192	122	102	92	66	36	74	112	139	210	1784
Pontianak	0	109	273	195	243	284	270	232	163	224	204	376	**395**	355	3214
Pasoeroean	8	113	241	**270**	206	125	91	65	32	6	8	15	62	161	1282
Kajoemas	8	114	**454**	447	**454**	216	154	73	37	15	14	39	191	388	2482
Ambon	4	128	132	112	134	269	516	**624**	589	395	219	160	116	140	3406
Manokwari	1	134	273	275	**323**	274	237	215	168	150	115	91	157	289	2567
Daru	9	143	**358**	286	329	338	272	105	101	56	45	55	97	194	2236
Pt. Moresby	9	147	204	**236**	172	81	68	20	29	21	33	21	50	94	1029
Rendova	8° 30′	157	473	475	**531**	286	324	350	321	370	315	367	326	301	4439
Tulagi	9	160	344	408	**419**	212	184	144	162	176	191	200	217	273	2930
Ocean Is.	1	170	**327**	229	214	216	154	130	144	68	99	122	146	204	2053

Our last illustration of this part of the subject is given in a pair of charts of the number of rain-days in the North Atlantic in two special months, August 1882 and February 1883, taken from the daily charts compiled by the Meteorological Office.

RAINFALL OVER THE SEA

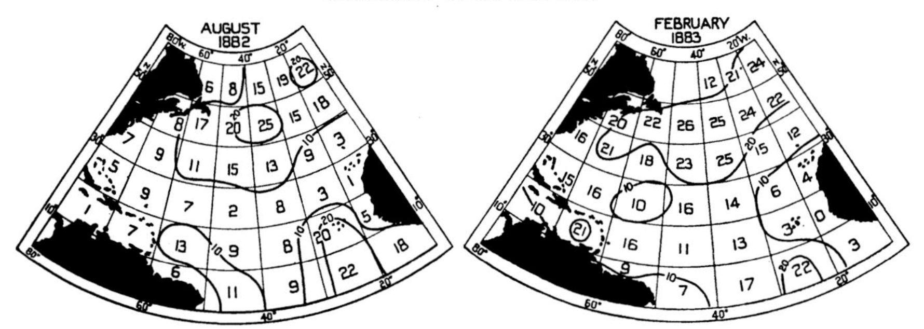

Fig. 129. Number of days of rain in 10° squares of the North Atlantic during August 1882 and February 1883.

Square 3 appears on this map with a frequency of 22 out of 31 days in August and of 28 in February and thus maintains the character which it displays in fig. 100. An equally or even more rainy area is indicated by the familiar region of cyclonic depressions in the more Northern part of the map. The extension of the area of maximum frequency between August and February is suggestive of the difference of conditions over that part of the Atlantic between a summer and a winter month.

Returning to the question raised on p. 177 we have for—

Percentage frequency of rain-hours.

	Jan.	Feb.	Mar.	Apr.	May	June	July	Aug.	Sept.	Oct.	Nov.	Dec.
Falmouth	17	18	14	13	11	8	10	11	8	18	17	22
Valencia	23	22	20	17	15	13	13	15	12	20	20	25
Batavia	16	15	10	7	6	5	4	3	4	6	10	11

Divisors for converting percentage-frequency of rain-hours into daily rainfall in millimetres.

Falmouth	5·0	4·9	4·7	5·7	5·2	4·2	5·0	4·4	4·7	4·2	4·3	4·7
Valencia	5·1	5·1	4·8	5·2	4·8	4·2	4·8	4·4	3·5	4·3	5·3	4·9
Batavia	1·6	1·7	2·0	1·6	1·3	1·6	1·8	1·7	1·3	1·3	1·8	2·1

Hence taking the mean value for Batavia as applying equally to the three squares 39, 3 and 302 of the equatorial Atlantic we get an answer to the conundrum for the mid-Atlantic which is not far from Durst's estimate (p. 175).

Computed monthly rainfall in mm for three squares of the equatorial Atlantic

Square 39	18	18	18	18	18	18	126	**216**	162	144	72	18
,, 3	360	270	252	288	342	**378**	342	216	270	360	360	270
,, 302	90	126	162	**182**	162	54	54	18	36	36	36	36

EVAPORATION in the ordinary sense means the passing of water from the liquid to the gaseous condition presumably by the passage of a number of the constituent molecules of the liquid across the surface outward in excess of the number which pass in the opposite direction from the gas to the liquid. Evaporation will cease when the two streams carry equal amounts of the material, and that condition must be regarded as reached when the space above the liquid is "saturated" with the vapour. But evaporation takes place from various surfaces other than those of water. All animals evaporate water, and in torrid climates they have to keep themselves cool by sufficient evaporation from the skin and lungs. Foliage of every kind is also a means of transfer of water-vapour to the air. "A tree of average dimensions exhales into the air some two and a half gallons of water daily; while a single plant of the ordinary sunflower will exhale in twelve hours in dry warm weather about two pounds weight of water. To attain a weight of 1000 lb. an elm tree must absorb from the soil and evaporate into the air 35,000 gallons of water...admitting that in the six or seven months during which the leaf is on the tree the weight of an elm increases by 50 lb., the water absorbed and then evaporated in order to obtain this annual increase would fill a tank 30 ft. long by 3 ft. wide and 3 ft. deep."

Irregularities and the difficulty of applying data for stations to large areas.

If the particulars available for the representation of rainfall are regarded as unsatisfactory, still more unsatisfactory is the information which we have been able to compile about the important subject of evaporation. It is important not only from the economic point of view in respect of the water-supply of those countries where the sufficiency of rainfall is a matter of vital concern, but also from the scientific point of view because, as we have already pointed out, it is nature's chief method of supplying thermal energy to the upper levels of the atmosphere. Having regard to the uncertainties which arise from differences in the instruments employed and their exposure, we have not found it possible to present the compilation of the results of observations in various countries in the form of maps with isograms of evaporation; and the limited size of the page is an obstacle to the reproduction of the figures for annual evaporation as set out on a map which was exhibited at the meeting of the International Union for Geodesy and Geophysics at Madrid in 1924. We give tables however of annual evaporation which display the results of the observations at land-stations. The instruments used are indicated by signs. For the far more important part of the subject, the evaporation from sea and ice, we are able to give only the results of one memoir[1]. Our limited experience hardly enables us to say that the method of approaching a satisfactory knowledge has been adequately formulated.

Rainfall and evaporation are generally quoted according to the amounts recorded in the months and the year of our kalendar. Neither provides so satisfactory a unit of time for the purpose as the day, or the week of seven days. Accordingly, in bringing to a close our representation of rainfall and evaporation we give on p. 212 a table reducing the accumulation in months or years of various lengths to the comparable expression of the amount in millimetres per day.

[1] 'Die Verdunstung auf dem Meere' von Dr G. Wüst, *Veröff. d. Inst. für Meereskunde*, Berlin, 1920.

Country and Station	Period	mm
Hawaii		
Hoaeae Upper[4]	'20–23	1630
Maunawili[4]	'20–23	1134
United States and West Indies		
Chula Vista[4]	'19–23	1565
Oakdale[4]	4 years	2115
Mesa[4]	'19–23	2077
Yuma[4] (Citrus)	'21–23	3039
New Mexico[4]	'19–23	2346
Elephant Butte Dam[4]	'20–23	2606
Santa Fé[4]	'19–23	1578
Hill's Ranch, Tex.[4]	'21–22	1572
Silverhill[4]	'21–23	1472
San Juan, P. R.[4]	'19–23	2104
Havana	'11–15	1726
Azores		
Angra do Heroismo[1]	'16–20	1370
Horta[1]	'16–20	967
Ponta Delgada[1]	'16–20	1315
Western Europe		
Talla Water[3]	'08–25	415
Harrogate[3]	'08–25	481
Ardsley[3]	'08–25	441
Revesby[3]	'08–25	371
Otterbourne[3]	'08–25	498
Kennick[3]	'08–25	474
Ghent[1]	'11–12	842
St Genis Laval	'20–22	1212
Spain and Portugal		
Coimbra[4]	'66–90	2189
Lisbon[1]	'80–93	1222
Barcelona	'66–90	1056
Oviedo	'66–75	694
,,	'76–90	1182
Santiago	'66–80	631
,,	'81–90	822
S. Fernando[4]	'68–78	2197
,,	'79–90	1447
Valencia	'66–82	2863
,,	'83–90	2290
Alicante	'66–76	533
,,	'77–90	1135
Madrid[4]	'66–90	1610
Burgos	'66–80	1431
,,	'81–90	1060
Ciudad-Réal	'66–76	1566
,,	'78–90	1144
Badajoz	'66–70	3016
,,	'71–90	1926
Granada	'66–70	803
,,	'71–82	716

Country and Station	Period	mm
Northern Europe		
Dorpat[2]	'66–'15	340
Stockholm[2]	'19–22	692
Central Europe		
Bremen	'00–09	322
Potsdam[2]	'16–20	327
Dresden	—	300
N. Germany	—	399
Vienna	5 years	343
Klagenfurt	'97–'07	303
Roumania[2]	—	325
Menes	'06–10	837
Tarczal	'06–10	732
Temesvar	'06–10	362
Budapest	—	650
Nagytagyos	5 years	318
Tapolcza	'07–10	839
Mediterranean		
Balearic Is.	'14–17	950
Trieste[2]	'05–09	496
Venice	'20–25	605
Pola	'08–12	579
Modena	'06–11	698
Bologna	'03–12	1217
Florence	'19–21	1007
Naples	'86–'00	730
,,	'04–13	909
Catania	'92–'10	1226
Athens[1]	'07–11	1597
Haifa[1]	'22–25	1720
Jenin[1]	,,	2040
Jerusalem[1]	3 years	2680
Gaza[1]	'22–25	1610
Beersheba[1]	3 years	2682
Jericho	'24–25	3290
Atlantic Ocean		
70°–80° N	—	70
60°–70° N	—	110
50°–60° N	—	370
40°–50° N	—	660
30°–40° N	—	910
20°–30° N	—	1170
10°–20° N	—	1240
0°–10° N	—	910
0°–10° S	—	1200
10°–20° S	—	1170
20°–30° S	—	1060
30°–40° S	—	840
40°–50° S	—	550
50°–60° S	—	220

Country and Station	Period	mm
Russia[2]		
St Petersburg	'73–92	320
Moscow	'79–92	417
Skopin	'80–92	588
Pinsk	'79–92	460
Wassilewitschi	'78–92	664
Nikolaewskoe	'80–92	625
Urjupinskaja	'81–92	805
Malyi-Usen	'82–92	911
Kiew	'80–92	481
Elissawetgrad	'76–87	725
Lugan	'79–92	741
Chersson	'82–92	639
Boasta	'80–92	956
Petrowsk	'81–92	671
Tifliss	'79–91	537
Lenkoran	'83–92	428
Katherinenburg	'77–92	453
Perowsk	'82–92	1436
Wernyi	'82–92	498
Petro-Alexandrowsk	'74–92	1624
Ssamarkand	'82–92	710
Chodsent	'81–92	1096
Margelan	'81–92	1338
Nertschinsk-Hüttenwerk	12 years	420
China		
Kharbin[2], Manchuria	'99–'03	647
,,	'04–06	432
Zyosin	'11–20	1049
Heizyo	,,	1330
Zinsen	,,	1255
Fusan	,,	1405
Tsingtau	'16–20	1571
Japan		
Nemuro	'89–'01	872
Suttsu	'99–'01	1215
Yamagata	'98–'01	829
Kanazawa	'90–'01	1000
Gifu	'87–'01	1113
Osaka	'85–'01	1288
Hamamatsu	'91–'01	989
Numazu	'91–'01	1215
Matsuyama	'90–'01	975
Wakayama	'90–'01	1394
Kumamoto	'90–'01	1321
Taihoku	'97–'01	1266
Koshun	'00–01	1910
Philippine Is.		
Baguio	5 years	1113
Manila	—	1331

Further information is now available for Lake Superior. Information about the instrumental arrangement in Tokyo and Johannesburg is given in the bibliography of pp. 211–2. A new form of instrument is now available designed by Dr J. S. Owens.

Seasonal variation of evaporation in millimetres.

Station	Lat.	Long.	Jan.	Feb.	Mar.	Apr.	May	June	July	Aug.	Sept.	Oct.	Nov.	Dec.	Year
Toronto[3]	44° N	79° W	—	—	—	52	71	85	99	85	59	40	25	—	—
Dorpat[2]	58	27° E	4	5	12	29	57	64	60	46	30	17	9	5	340
Barnaoul[1]	53	84	3	4	13	40	95	106	106	84	63	39	9	3	565
Bremen	53	9	9	11	21	35	47	49	45	41	28	22	11	10	327
London[3]	51	0	3	6	17	38	62	74	75	59	35	16	6	2	394
Tachkent[2]	41	69	29	39	87	97	146	198	215	201	139	88	57	43	1339
Tokio*	36	140	59	70	78	92	109	111	127	144	95	70	60	58	[1088]
Candia[1]	35	25	107	83	112	113	130	149	168	160	140	116	111	101	1489
Helwan[2]	30	31	85	99	164	224	292	321	305	269	225	198	127	84	2387
Suakin[1]	19	37	136	132	151	157	210	311	396	360	234	145	141	143	2511
Atbara[1]	18	34	422	440	571	608	642	601	572	526	503	517	448	417	6267
Mongalla[1]	5	32	356	301	303	188	127	102	92	88	106	139	179	269	2250

* The monthly totals are reduced to months of 30 days.

[1] Piche. [2] Wild. [3] Tank. [4] Some other instrument.

Country and Station	Period	mm
Ecuador		
Quito[2]	'14	773
Quinta[1]	'14–17	1260
Ambato[2]	2 years	795
Peru		
Lima[5]	'18	666
,,	,,	1188
Chile		
Iquique	9 years	352
Isla de Pascua	'12–13	964
Punta Angeles (40 m)	'11–20	339
Los Andes (816 m)	,,	706
Valdivia	,,	553
Frutillar	'16–20	302
Isla Guafo	9 years	122
Evangelistas	4 years	182
Punta Arenas	'11–19	467
British Guiana		
Georgetown[3]	'17–21	1471
Brazil[1]		
Turyassu	'11–20	712
S. Luiz	'09–20	1178
Porangaba	'11–20	1226
Fernando Noronha	'10–20	1985
Guaramiranga	'11–20	640
Quixeramobim	'96–'20	1336
Barro do Corda	'12–20	1021
Natal	'04–20	1920
Jaboatão	'11–20	913
Pesqueira	'13–20	1409
Ondina	'97–'20	996
Caetite	'09–20	1421
Cuyaba	'00–20	946
Goyaz	'11–20	1553
Montes Claros	'05–20	870
Theophilo Ottoni	'11–20	580
Uberaba	'97–'20	1055
Campos	'11–20	1103
Friburgo	'01–20	422
Vassouras	'11–20	875
Rio de Janeiro	'90–'20	1172
Florianopolis	'09–20	555
Santa Maria	'12–20	1073
Uruguayana	'12–20	1394
Bolivia		
Sucre[1]	'17–21	1784
Uruguay		
Montevideo[1]	'01–20	1172
Argentine, Cordoba		
Copper dish	About 20 years	1723
Wild balance	,,	1625
Glass dish	,,	1259
Tank	,,	1104

Country and Station	Period	mm
Mexico		
Zacatecas[6]	'07–11	3089
Obs. de Leon[5]	"Mean"	1110
,,	,,	2820
Guanajuato[5]	3 years	1606
,,	,,	2899
Morelia	'09–10	2100
Tacubaya	'81–'20	2421
Puebla[5]	'08–10	1500
,,	,,	2538
Oaxaca[5]	'07–08	1379
,,	,,	3206
Central America		
San Salvador	'11–12	1244
Lake Nicaragua:		
Fort San Carlos	'98–'00	1080
Sapoa	1900	1251
Tipitapa	'99–'00	1546
Panama Canal		
Balboa Heights	—	1365
Gatun Lake	'18–22	1452
Colon	—	1342
Canary Islands		
Las Palmas[1]	4 years	1650
La Laguna[1]	'13–16	621
N. Africa		
Tangier[1]	'14–17	910
Sidi Barrani	'10–15	2280
Port Said[2]	'01–20	905
El Arish[1]	'07–14	1920
Qurashiya[1]	'07–20	1630
Zagazig[2]	'13–18	1030
Heliopolis[1]	'08–20	3070
Bene Suef[1]	10 years	1830
Tor[1]	'05–20	3000
Asyut[2]	'00–20	2570
Dakhla Oasis[1]	'05–15	3260
Esna[2]	'07–16	2460
Aswan[2]	'01–20	3700
Wadi Halfa[1]	'05–20	5920
Dongonab[1]	'08–19	4110
Merowe[1]	'05–20	5990
Tokar[1]	'13–20	4640
Khartoum[1]	'00–20	5400
Kassala[1]	'01–20	3730
El Obeid[1]	'01–20	4960
Roseires[1]	'04–20	4050
Kadugli[1]	'10–17	3980
Gambela[1]	'09–20	2410
Wau[1]	'02–20	2700

Country and Station	Period	mm
Southern Rhodesia		
Cleveland Dam	9 years	2327
Bulawayo Dam	4 years	2588
Mazoe Dam	'21–22	2325
Portuguese E. Africa[1]		
Lourenço Marques	4 years	1451
Quissico	'17–18	1608
Inhambane	'17–21	1282
Beira	'17–21	1693
Quelimane	'17–18	1490
South Africa[3]		
Johannesburg	'05–13	1869
Kimberley	'98–'03	1557
Cape Town	'10–21	1996
Dunbrody	—	1834
Dutch East Indies[2]		
Instruments under screen		
Buitenzorg	'13–18	910
Tjibodas	'12–18	620
Pekalongan	'12–18	695
Bandoeng	'12–18	1020
Pasoeroean	'14–18	1310
Tosari	'12–18	730
Djember	'13–18	1130
Instruments inside screen		
Tjipetir	'05–18	475
Sawahan	'05–18	550
Tjinjiroean	'05–16	330
Karanganjar	'09–18	475
Salatiga	'06–18	510
Modjowarno	'05–14	765
Wedi	'05–18	585
Bangelan	'05–18	510
Australia		
Cue[4], W.A.	—	3960
Wiluna[4], W.A.	—	3800
Laverton[4], W.A.	—	3590
Coolgardie[3], W.A.	—	2228
Perth[3], W.A.	23 years	1675
Alice Springs[3], S.A.	—	2394
Adelaide[3], S.A.	52 years	1386
Brisbane[3], Qu.	13 years	1125
Walgett[3], N.S.W.	—	1624
Wilcannia[3], N.S.W.	—	1597
Hay[3], N.S.W.	—	1169
Young[3], N.S.W.	—	1201
Sydney[3], N.S.W.	42 years	967
Lake George[3], N.S.W.	—	616
Prospect[3], N.S.W.	—	1168
Melbourne[3], Vic.	49 years	987
Hobart[3], Tas.	11 years	830

The evaporation from a wet soil surface is surprisingly high. "Fortier found that a wet soil surface evaporated 4·75 inches per week, while a free water surface evaporated only 1·88 inches per week." (H. D. Leppan, T.U.C. Bulletin No. 12, Pretoria, 1927.)

Seasonal variation of evaporation in millimetres.

Station	Lat.	Long.	Jan.	Feb.	Mar.	Apr.	May	June	July	Aug.	Sept.	Oct.	Nov.	Dec.	Year
Quixeramobim[1]	5° S	39° W	124	87	81	69	70	84	109	128	143	154	146	141	1336
Ondina[1]	13	38	89	83	88	73	81	70	88	88	84	88	82	81	996
Rio de Janeiro[1]	23	43	109	98	98	92	92	88	91	101	91	94	100	113	1172
Cuyaba[1]	16	56	58	47	50	53	63	78	99	120	121	106	82	69	946
Porto Alegre[1]	30	51	103	94	85	62	54	43	43	58	62	75	85	109	873
Punta Arenas	53	71° W	129	109	94	61	42	21	21	35	62	93	106	119	885
Johannesburg	26	28° E	165	144	129	130	130	110	125	168	201	214	194	188	1897
Batavia[2]	6	107	44	35	42	43	43	39	51	64	64	64	46	46	588
Alice Springs[3]	24	134	323	276	242	174	124	86	93	131	185	243	278	311	2466
Brisbane[3]	27	153	170	137	124	94	73	56	62	76	98	136	156	172	1356
Perth[3]	32	116	263	218	194	120	69	44	44	60	84	133	194	248	1671
Hobart[3]	43	147	130	96	76	51	35	23	23	32	51	79	103	116	815

[1] Piche. [2] Wild. [3] Tank. [4] Some other instrument.
[5] Upper figures screened, lower figures free exposure. [6] "Sistema Pastrana."

AUTHORITIES FOR THE DATA CONTAINED IN THE CHARTS, TABLES AND DIAGRAMS OF CHAPTER V

CHARTS OF THE DISTRIBUTION OF VAPOUR-PRESSURE AND CLOUDINESS

Charts.

North America. Bewölkungsverhältnisse und Sonnenscheindauer von Nordamerika. Von A. Gläser. Hamburg, Arch. D. Seewarte, Bd. XXXV, 1912, Nr. 1.

Scandinavia. Stockholm, Meteor. Centralanstalt. Observations météor. suédoises. Tome L, 1908, App. Nébulosité et soleil dans la péninsule scandinave. Par H. E. Hamberg. Uppsala, 1909.

Russia. Petrograd, Observatoire Physique Central. Atlas climatologique de l'Empire de Russie. St Pétersbourg, 1900.

Mediterranean. Bewölkung und Sonnenschein des Mittelmeergebietes. Von J. Friedemann. Hamburg, Arch. D. Seewarte, Bd. XXXV, 1912, Nr. 2.

India. Calcutta, Indian Meteorological Department. Climatological Atlas of India. Edinburgh, 1906.

Arabian Sea. Investigation into the mean temperature, humidity and vapour-tension conditions of the Arabian Sea and Persian Gulf. Indian Meteor. Memoirs, vol. VI, part 2, 1894.

Numerical data.

L. Teisserenc de Bort. Etudes sur la distribution moyenne de la nébulosité à la surface du globe. Paris, Ann. Bur. Cent. Météor., 1884, IV, seconde partie, p. 27.

Meteorologische Zeitschrift. Numerous small tables.

Copenhagen, Dansk Meteor. Inst. Éléments météorologiques des îles Féroé, de l'Islande et du Groenland. Aarbog, 1895. 2me partie, App. Copenhagen, 1899.

A. Schönrock. Die Bewölkung des Russischen Reiches. St Pétersbourg, Mém. Acad. Sci. (8), I, No. 9, 1895.

V. Kremser. Die Strombeschreibungen. 4 Bände. Berlin (Reimer), 1896–1901.

Utrecht, K. Nederl. Meteor. Inst. Maandelijksch Overzicht der Weergesteldheid in Nederland. 1914.

F. Bayard. English climatology, 1881–1900. Q. Journ. Roy. Meteor. Soc., vol. XXIX, 1903, p. 1.

J. Maurer u. a. Das Klima der Schweiz, 1864–1900. Band II, Tabellen. Frauenfeld, 1910.

Vienna, K. K. Zentralanstalt für Meteor. und Geodynamik. Klimatographie von Österreich. Teile I–IX. 1904–19.

St C. Hepites. Album climatologique de Roumanie. Bucureşci, 1900.

H. E. Hamberg. De l'influence des forêts sur le climat de la Suède. Pt. 3, Humidité de l'air. Bihang till Domänstyrelsens underdåniga berättelse rörande Skogsväsendet för År 1887. Stockholm, 1889.

H. Mohn. Klima-Tabeller for Norge. Kristiania, 1895–1906.

G. Hellmann. Feuchtigkeit und Bewölkung auf der iberischen Halbinsel. Utrecht, Nederl. Meteor. Jaarb., 1876, I Deel, p. 267.

Hamburg, Deutsche Seewarte. Ergebnisse der meteor. Beobachtungen im Systeme der D. Seewarte für 1876–1900, 1901–1905, 1906–1910. Hamburg, 1898–1912.

H. Fritsche. Über das Klima Ostasien's insbesondere des Amurlandes, China's und Japan's. L. v. Schrench's Reisen und Forschungen im Amurlande. Bd. 4, L. 2, s.l. (K. Akad. Wiss. [St Petersburg], 1877).

Sven Hedin. Scientific results of a journey in Central Asia, 1899–1902. Vol. v, part 1, a, b. Meteorologie, von N. Ekholm. 2 vols. Stockholm, 1905, 1907.

Tokio, Central Meteorological Observatory. Results of the meteorological observations in Japan for the...years 1876–1905, 1906–1910. 2 vols. Tokio, 1906, 1913.

H. Kondo. The climate, typhoons and earthquakes of the island of Formosa. The Government-General of Formosa. Taihoku, 1914.

Calcutta, Indian Meteorological Department. Monthly and annual normals of... vapour-tension and cloud. By G. T. Walker. Indian Meteorological Memoirs, vol. XXII, part 3. 1914.

Batavia, K. Magn. en Meteor. Observatorium. Observations made at secondary stations in Netherlands East India. Vols. I–VII. 1913–1919.

Cairo, Survey Department. Meteorological report. 1907.

F. Foureau. Documents scientifiques de la mission saharienne. Premier fascicule. Paris, 1903.

A. Thévenet. Essai de climatologie algérienne. Alger, 1896.

Brussels, Société Royale de Médecine. Congrès national d'hygiène et de climatologie médicale de la Belgique et du Congo...1897. 2me partie, Congo. Bruxelles, 1898.

Washington, U.S. Coast and Geodetic Survey. Pacific coast pilot, coasts and islands of Alaska. App. 1. Washington, 1879.

Toronto, Meteor. Service of Canada. Monthly record of meteorological observations.

Washington, D.C., U.S. Department of Agriculture, Weather Bureau, Bulletin S. Report on the temperatures and vapor-tensions of the United States...1873–1905. Washington, 1909.

*R. C. Mossman. Cloud amount in Brazil and Chile. Q. Journ. Roy. Meteor. Soc., vol. XLVI, 1920, p. 294.

W. G. Davis. Climate of the Argentine Republic. Buenos Aires, 1910.

Melbourne, Commonwealth Bureau of Census and Statistics. Official year-book, No 13. Melbourne, 1920: No. 17. Melbourne, 1924.

Perth, Govt. Astronomer. The climate of Western Australia...1876–99. By W. E. Cooke. Perth, 1901.

London, Meteorological Office. Contributions to our knowledge of the Meteorology of the Arctic regions. Vol. 1, parts 1–5. Official, No. 34. London, 1885, 1888.

London, Meteorological Office. Monthly charts of meteorological data for the nine 10° squares of the Atlantic...between 20° N and 10° S, and 10–40° W. London, 1876.

P. Schlee. Niederschlag, Gewitter und Bewölkung im südwestlichen und in einem Theile des tropischen Atlantischen Ozeans. Hamburg, Arch. D. Seewarte, XV, 1892, Nr. 3.

Utrecht, K. Nederl. Meteor. Inst. No. 107 A, Monthly meteorological data for ten-degree squares in the Atlantic and Indian oceans. Utrecht, 1914.

No. 104, Oceanographische en meteor. Waarnemingen in den Indischen Oceaan (1856–1914). Tabellen. 4 Bände.

Meteorological tables compiled in the Meteorological Office, London, for the Admiralty Pilots and Sailing Directions.

Reports of the several expeditions to the Arctic and Antarctic.

* Received after the original maps were drawn.

DIURNAL VARIATION OF VAPOUR-PRESSURE

Arctic. Report of the Second Norwegian Arctic Expedition in the *Fram*, 1898–1902. No. 4. Meteorology by H. Mohn. Kristiania, 1907. pp. 168–9.

The positions of the stations during the four years of observations are as follows:

Sept. 1898 to July 1899. Rice Strait, 78° 46′ N, 74° 57′ W.
Oct. 1899 to Aug. 1900. Havnefjord, 76° 29′ N, 84° 4′ W.
Sept. 1900 to Aug. 1901. Gaasefjord, 78° 49′ N, 88° 40′ W.
Sept. 1901 to July 1902. Gaasefjord, 76° 40′ N, 88° 38′ W.

Barnaoul. Über den täglichen und jährlichen Gang der Feuchtigkeit in Russland. Repertorium für Meteorologie. Von H. Wild. Band IV, Nr. 7. St Petersburg, 1875. S. 15.

Paris and Dar es Salam. The Weather Map. M.O. publication, No. 225 i, 4th issue. London, 1918.

Naha. Results of the Meteorological Observations in Japan for the Lustrum 1906–1910. Tokio, 1913.

Calcutta. Indian Meteorological Memoirs, vol. IX, part VIII. Calcutta, 1897.

Alice Springs. Zum Klima des Innern von Australien. Von J. Hann. Meteorologische Zeitschrift, Band XII, 1895. S. 398. The vapour-pressure has been calculated from the values of the relative humidity and the dry bulb temperature.

Antarctic. Deutsche Südpolar Expedition, 1901–1903. Band IV, Meteorologie. Band II, Tabellen. Berlin, 1913. Ss. 36 et seq.

HEIGHTS OF CLOUDS

See fig. 80, p. 153.

DIURNAL AND SEASONAL VARIATION OF CLOUDINESS

Arctic. Report of the Second Norwegian Arctic Expedition in the *Fram*, 1898–1902. No. 4. Meteorology by H. Mohn. Kristiania, 1907. p. 328. For the position of the *Fram* see above.

Antarctic. Veröff. des Preuss. Met. Inst. Abhand. Bd. VII, Nr. 6. Berlin, 1924.

Ocean. Report of the work of H.M.S. *Challenger*, vol. II, part V, p. 28.

Helwan. Survey Department. Meteorological Report, 1910. Cairo, 1913.

Potsdam. Bewölkung und Sonnenschein in Potsdam (1894 bis 1900). Von Otto Meissner. Meteorologische Zeitschrift, Band XXIV, 1907, S. 406. Observations for January and July for the years 1894–1920 are given in Hann-Süring, Lehrbuch der Meteorologie, S. 308.

Trichinopoly. Indian Meteorological Memoirs, vol. IX, part VI. Calcutta, 1896. p. 404.

Batavia. Observations made at the Royal Magnetical and Meteorological Observatory at Batavia. By Dr W. van Bemmelen. Vol. XXXVIII, 1915. Batavia, 1920. p. 95.

ISOPLETH DIAGRAMS OF THE DIURNAL AND SEASONAL VARIATION OF SUNSHINE

Laurie Island. Annals of the Argentine Meteorological Office. Vol. XVII. Climate of the S. Orkney Islands. 2nd part. Buenos Aires, 1913. p. 108.

Aberdeen, Falmouth and Valencia. British Meteorological and Magnetic Year-Book, 1913, part IV, section 2. Hourly Values from Autographic Records. M.O. publication, No. 214 f. Edinburgh, 1915. pp. 60–61.

Georgetown and Richmond (Kew Observatory). Fifth Annual Report of the Meteorological Committee. London, 1910.

Batavia. Observations made at the Royal Magnetical and Meteorological Observatory at Batavia. Vol. XXXVIII, 1915. Batavia, 1920. p. 97.

Vancouver Is. Canadian Monthly Weather Review, 1908–1915, Canadian Monthly Record, 1916–1917. Issued by the Meteorological Service of Canada.

CHARTS OF MEAN MONTHLY AND ANNUAL RAINFALL

North America. J. G. Bartholomew, A. J. Herbertson and A. Buchan. Atlas of Meteorology. Bartholomew's Physical Atlas, vol. III. Westminster, 1899.

India. Calcutta, Indian Meteorological Department. Climatological Atlas of India. Edinburgh, 1906.

Africa. The Climate of the Continent of Africa. By Alexander Knox. Cambridge, 1911.

Australia. Melbourne, Commonwealth Bureau of Meteorology. Revised average annual rainfall map of Australia and Tasmania, with maps of mean monthly rainfall. Sydney, 1912. (A further revision has been made to the annual map.)

New Zealand. Mean Annual Rainfall Map of New Zealand. Published by the Meteorological Office, New Zealand (D. C. Bates, 7. vi. 1911).

Oceans. Supan. Die jährlichen Niederschlagsmengen auf den Meeren. Geographische Mitteilungen, 1898, Heft VIII. The chart is reproduced in Hann's Lehrbuch der Meteorologie.

General. Distribution of Rainfall over the Land. By A. J. Herbertson. London, Murray, 1901.

Normals prepared for the British Meteorological and Magnetic Year-Book, part 5, Réseau Mondial.

SEASONAL VARIATION OF RAINFALL

Data from normals prepared for the British Meteorological and Magnetic Year-Book, part 5, Réseau Mondial.

SEASONAL VARIATION OF EVAPORATION

Toronto. MS data supplied by the Meteorological Office, Toronto (J. Patterson).

Dorpat. Fünfzigjährige Mittelwerte aus den meteorologischen Beobachtungen 1866–1915 für Dorpat. Tartus, 1919. The evaporation measurements are for 1886–1915.

Barnaoul and Taschkent. Ueber den jährlichen Gang der Verdunstung in Russland. Repert. für Meteorol., XVII, Nr. 10. St Petersburg, 1894. S. 5. (See also Meteorologische Zeitschrift, 1895, S. (76).) The values for Barnaoul are for the years 1876–92 and for Taschkent 1877–1892.

Bremen. Deutsches Meteorologisches Jahrbuch für 1909. Bremen, 1910. S. viii.

London (Camden Square). British Rainfall, 1924. M.O. publication, No. 275. London, 1925. p. 131.

Tokyo. The Bulletin of the Central Meteorological Observatory of Japan. Vol. I. Evaporation in Japan. By T. Okada. Tokyo, 1910. p. 12. The observations are for the period 1879–1901, and are for the evaporation from a "copper vessel of circular cylinder 2 decimetres in diameter and 1 decimetre in depth," with water of a depth of 2 cm.

Egypt. Ministry of Public Works, Egypt, Physical Department. Climatological Normals for Egypt.... Cairo, 1922. The periods of observation are as follows: Candia 1908–20, Helwan 1904–20, Suakin 1906–20, Atbara 1902–20, Mongalla 1903–20.

Brazil. Ministerio da Agricultura, Industria e Commercio. Directoria de Meteorologia. Boletim de Normaes. 1922. The date of the foundation of the stations is as follows: Quixeramobim 1896, Ondina 1897, Rio de Janeiro (evaporation) 1890, Cuyaba 1900, Porto Alegre 1910.

Punta Arenas. Boletín Meteorológico del Observatorio Salesiano M. José Fagnano, Año XXXV. Punta Arenas, 1923.

South Africa. Union of South Africa, Office of Census and Statistics. Official Year-Book of the Union, No. 7. Pretoria, 1925. p. 58. The observations are for the evaporation from a "free water surface."

Batavia. Observations made at the Royal Magnetical and Meteorological Observatory at Batavia. Vol. XXXVIII, 1915. Batavia, 1920. p. 103.

Australia. Official Year-Book of the Commonwealth of Australia, No. 17. 1924. (Perth 25 years, Brisbane 15 years, Hobart 13 years.)

Meteorology of Australia, Commonwealth Bureau of Meteorology. Monthly Weather Report. Vol. III, 1912. Melbourne. (Alice Springs.)

DESERT REGIONS

Rain falls occasionally in desert regions. On December 20, 1934, a Dutch air-liner was destroyed in a thunderstorm in the Syrian desert with so much water that safe landing was not possible.

In 1919 in the neighbourhood of Cairo, with a normal rainfall of 1·3 inches (33 mm), a single storm gave 1·77 inches of rain; and in the Libyan desert near Siwa oasis 28 mm. fell within a day, 28 December 1930, and on the same day heavy rainfall occurred at Sollum on the coast, and between Sollum and Alexandria (L. J. Sutton, *Meteorological Magazine*, 1931, p. 29).

Storage of water in plants. "A vine in the arid regions of Arizona and Sonora stores water in an expanded base in such quantity that it has been known to live on its reserve for 15 years. A tree-cactus may hold many hundreds of gallons of surplus water....An acre of cabbage will use 2 million quarts of water in a season, and 200 beech trees on an acre require nearly double that amount. One of these trees loses about 80 quarts of water as vapour daily from its leaves" (D. T. MacDougal quoted in *Bull. Amer. Met. Soc.* April 1932, p. 80).

Conversion of quantities of rainfall or evaporation in years and months of different lengths to their equivalents in millimetres per day.

mm	in years of 366	365·25	365	in months of 31	30	29	28·25	28	in a week of 7 days
100	·2732	·2738	·2740	3·226	3·333	3·448	3·540	3·571	14·287
90	·2459	·2464	·2466	2·903	3·000	3·103	3·186	3·214	12·857
80	·2186	·2190	·2192	2·580	2·667	2·759	2·832	2·857	11·429
70	·1912	·1917	·1918	2·258	2·333	2·414	2·478	2·500	10·000
60	·1639	·1643	·1644	1·935	2·000	2·069	2·124	2·143	8·571
50	·1366	·1369	·1370	1·612	1·667	1·724	1·770	1·786	7·143
40	·1093	·1095	·1096	1·290	1·333	1·379	1·416	1·429	5·714
30	·0820	·0821	·0822	·968	1·000	1·034	1·062	1·071	4·287
20	·0546	·0548	·0548	·645	·667	·690	·708	·714	2·857
10	·0273	·0274	·0274	·323	·333	·349	·354	·357	1·429

CHAPTER VI

COMPARATIVE METEOROLOGY: PRESSURE AND WIND

Graphic integration of monthly charts of normal pressure.

Subject to a correction of not more than 5 per cent. for air displaced by mountains, the mass of air on the Northern Hemisphere exceeds its mean value

by	0·1	1·7	4·2	5·1	4·4	1·7	0·3 billion tons
in	Oct.	Nov.	Dec.	Jan.	Feb.	Mar.	April

and falls short of its mean value by

by	1·6	4·2	5·1	4·1	2·5 billion tons
in	May	June	July	Aug.	Sept.

The values range themselves very nearly on a sine-curve which follows the curve of total power of solar radiation over the hemisphere with a lag of 27 days.

We now pass to the representation of barometric observations in all parts of the world by maps of normal isobaric lines or lines of equal pressure, and we associate therewith the normal conditions as to wind, or the normal resultant flow of air, because under the conditions represented by normal values there is an unmistakable relation between the wind and the distribution of pressure. The details of the relation form the subject of volume IV. Here it is sufficient for us to say that the development of the original idea of such a relation is associated with the name of Buys Ballot[1] in the formulation of a law which may be regarded as the first law of dynamical meteorology. It came to be regarded as being in some way connected with the mathematical relation between the rotation of the earth ω, the velocity of a mass of air moving freely with reference to the earth along a great circle without change of its velocity, v, and the gradient of pressure, bb, which is necessary to prevent the air deviating from the line of the great circle. The relation is expressed by the equation $bb = 2\omega v\rho \sin\phi$, where ρ is the density of the moving air and ϕ the latitude. The value of v deduced from this equation is called the *geostrophic wind*.

The equation has been extended to include the case in which the motion without change of velocity is along a small circle of angular radius r lying in a horizontal or level surface, when it takes the form

$$bb = 2\omega v\rho \sin\phi \pm v^2\rho \cot r/E,$$

where E is the earth's radius.

The value of v deduced from this equation is called the *gradient wind*, and if the latitude ϕ be so small that the first term is negligible compared with the second the corresponding value of v is called the *cyclostrophic wind*. The relation has been generalised further in equations of motion which do not include any limitations as to the direction or speed of the motion; but the consideration of these equations belongs to the section on the dynamics of the atmosphere in volume IV.

Here we only take into account the ideas associated with the first equation which would be properly applicable when the motion is along a great circle and without change of velocity. The uncertainties in the measurements or estimates of wind make any close numerical comparisons impracticable, but

[1] *Les Bases de la Météorologie dynamique*, tome I, p. 91.

we are justified in saying that at the earth's surface the relationship is very considerably impaired by the eddy-motion due to friction between the moving air and the surface. The interference shows itself quite definitely in a diminution of velocity and consequent flow across the isobars everywhere from high pressure to low pressure; the effect of the interference becomes less in the free air above the surface, and at the height of a kilometre the agreement between the observed measure of wind and the value calculated by the formula is so close that we cannot say certainly whether it is in excess or in defect[1]. Furthermore, we are constrained to allow first that observations indicate for the most part that with the time unit of an hour velocity generally shows very little change, and secondly that, although the dynamical conditions with which we are most familiar are on the side of retardation of the balancing velocity, we must nevertheless allow that pressure in a travelling mass of air can be increased by the sacrifice of velocity; and we find it best to assume that the relation indicated holds for wind-velocity and pressure-gradient in the free air, within the limits of accuracy of observation, when the isobar does not deviate much from a great circle; we can use the relation as a guide to the motion in the general circulation of the atmosphere in all but the lowest layer.

Thus the geostrophic wind may be looked upon as a primary and even vital consideration in the general circulation of the atmosphere.

Pressure-gradient is less easily measured on a map than its numerical reciprocal, the distance apart of the isobars; and for the purpose of rapid estimation a scale can be used on which, when it is properly adjusted for density and for latitude, the geostrophic wind can be read off by laying the scale across the isobars. We have already used this method of indicating the velocity of wind in fig. 4 of volume I. Our hemispherical charts are very ill-adapted to the use of a scale because the scale of distance is so different in different parts of the map; but we can still utilise the idea of close isobars upon a map being significant of strong wind. A pair of geostrophic scales for the separate components was given on p. 220 of the original edition.

With these remarks we introduce a pair of maps for the normal distribution of pressure at sea-level over the globe for the year, figs. 130–1, and twelve pairs of monthly normals, figs. 132–55.

MONTHLY MAPS OF THE DISTRIBUTION OF PRESSURE

It is unnecessary to enter into detailed description of the maps; they speak for themselves. Two points, however, may be noticed, first that the isobaric lines are closed curves which group themselves round centres of high pressure and low pressure. Of centres of high pressure there are several in either hemisphere, and for low pressure there are several centres in the Northern Hemisphere but in the Southern only a circumpolar belt. Second, that with the seasons the general distribution changes markedly in the Northern Hemisphere from high pressures over the Eastern and Western continents in winter towards the formation of a vast area of low pressure over the Southern portion

[1] *Barometric Gradient and Wind-Force*, by E. Gold, M.O., No. 190, London, 1908.

of the Asiatic continent in summer with a corresponding change over North America. It is the transition from one set of these conditions to the other and back again that the maps are intended to illustrate. The explanation of the transition is one of the principal problems in meteorology.

In order to furnish suggestions for the method of procedure a number of tables of seasonal variation of pressure combined with diurnal variation are given in the available spaces of the maps; and other tables give similar particulars for conjugate stations at high level and low level, so that the effect of the minor changes which difference of level can produce may not be overlooked. These tables are preceded by tables of conversion between millimetres and millibars, and in the case of the maps of isobaric lines in the free air a new kind of conversion table is introduced, namely, between height and the gravitational potential or geopotential corresponding therewith. This table is introduced here in order to keep the reader in touch with the system of dynamical meteorology which Professor V. Bjerknes has based upon the values of meteorological elements referred to the same horizontal level instead of the same "height." Reference has already been made to this method in chapter IV.

Geodynamic metres and geodekametres for computation of wind.

The result as exhibited in the publications of the Geophysical Institute at Leipzig is to express heights in "geodynamic metres" instead of metres. The difference between the two is hardly to be considered important for our present purpose but with greater heights the difference in the two methods of indicating successive layers of the atmosphere would be appreciable. There can be little doubt that what meteorologists set out as referring to the same height is really intended to represent conditions at the same horizontal level. It has indeed been customary to obtain heights by taking a fixed value of g, and it is fair to say that the meteorological practice has been to compute geopotential and transform to what has been called height by taking g constant, a process which though perhaps accurate enough for practical purposes is nevertheless illicit. We have accordingly given a conversion table from "heights" in different latitudes to corresponding "level" surfaces over which geopotential is the same and which are at every point normal to the resultant force of gravity at the point. We quote the geopotential of the level surfaces in units of the C.G.S. system based on the dekametre instead of the centimetre. The figures are the same as those in the tables prepared by Professor Bjerknes, but the unit is ten times as great as the 10-metre-metre units used by Professor Bjerknes for the expression of geopotential in "dynamic metres." We have already explained that we should prefer to call those units geodynamic metres. In our numbers for the geopotential of the level surfaces the decimal point is one place to the left as compared with Bjerknes's expression in dynamic metres. We express that fact by calling the units geodekametres. The reader can convert the figures to geodynamic metres by multiplication by ten.

FIGURE 130

Millibars.

NORMAL PRESSURE AT SEA-LEVEL

Authority: see p. 288.

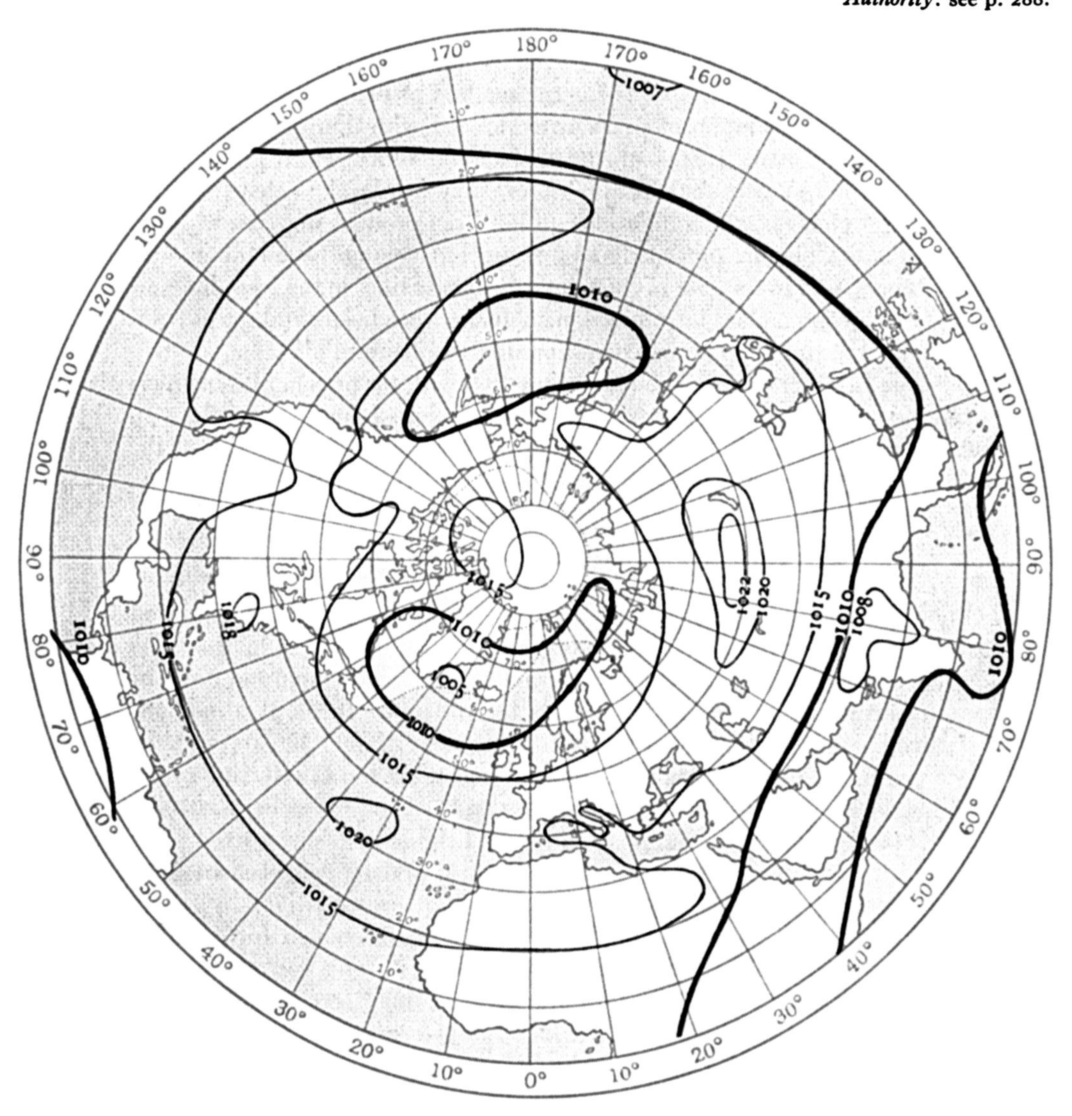

Pressure in millibars to millimetres of mercury at 0° C. in latitude 45°.

1 mb = ·750076 mm. 1 mm = 1·333200 mb.

mb	0	1	2	3	4	5	6	7	8	9
950	712·6	713·3	714·1	714·8	715·6	716·3	717·1	717·8	718·6	719·3
960	720·1	720·8	721·6	722·3	723·1	723·8	724·6	725·3	726·1	726·8
970	727·6	728·3	729·1	729·8	730·6	731·3	732·1	732·8	733·6	734·3
980	735·1	735·8	736·6	737·3	738·1	738·8	739·6	740·3	741·1	741·8
990	742·6	743·3	744·1	744·8	745·6	746·3	747·1	747·8	748·6	749·3
1000	750·1	750·8	751·6	752·3	753·1	753·8	754·6	755·3	756·1	756·8
1010	757·6	758·3	759·1	759·8	760·6	761·3	762·1	762·8	763·6	764·3
1020	765·1	765·8	766·6	767·3	768·1	768·8	769·6	770·3	771·1	771·8
1030	772·6	773·3	774·1	774·8	775·6	776·3	777·1	777·8	778·6	779·3
1040	780·1	780·8	781·6	782·3	783·1	783·8	784·6	785·3	786·1	786·8
1050	787·6	788·3	789·1	789·8	790·6	791·3	792·1	792·8	793·6	794·3

NORMAL PRESSURE AT SEA-LEVEL

FIGURE 131

Authority: see p. 288.

Millibars.

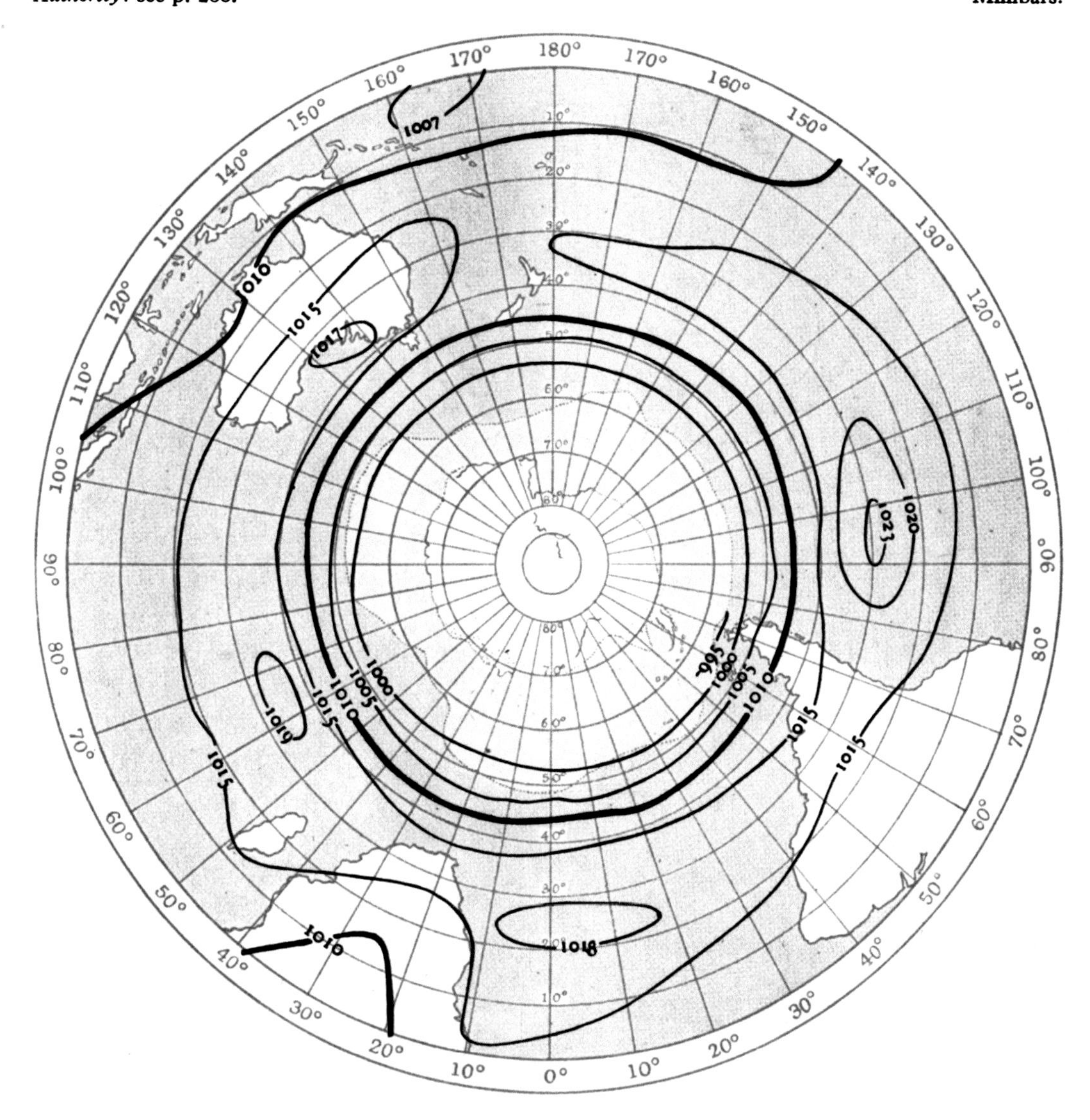

Pressure in millibars to inches of mercury at 32° F. in latitude 45°.

1 mb = ·0295306 in. 1 in = 33·8632 mb.

mb	0	1	2	3	4	5	6	7	8	9
950	28·05	28·08	28·11	28·14	28·17	28·20	28·23	28·26	28·29	28·32
960	28·35	28·38	28·41	28·44	28·47	28·50	28·53	28·56	28·59	28·62
970	28·64	28·67	28·70	28·73	28·76	28·79	28·82	28·85	28·88	28·91
980	28·94	28·97	29·00	29·03	29·06	29·09	29·12	29·15	29·18	29·21
990	29·24	29·26	29·29	29·32	29·35	29·38	29·41	29·44	29·47	29·50
1000	29·53	29·56	29·59	29·62	29·65	29·68	29·71	29·74	29·77	29·80
1010	29·83	29·86	29·88	29·91	29·94	29·97	30·00	20·03	30·06	30·09
1020	30·12	30·15	30·18	30·21	30·24	30·27	30·30	30·33	30·36	30·39
1030	30·42	30·45	30·48	30·51	30·53	30·56	30·59	30·62	30·65	30·68
1040	30·71	30·74	30·77	30·80	30·83	30·86	30·89	30·92	30·95	30·98
1050	31·01	31·04	31·07	31·10	31·13	31·15	31·18	31·21	31·24	31·27

FIGURE 132
Millibars.

NORMAL PRESSURE AT SEA-LEVEL

Authority: see p. 288.

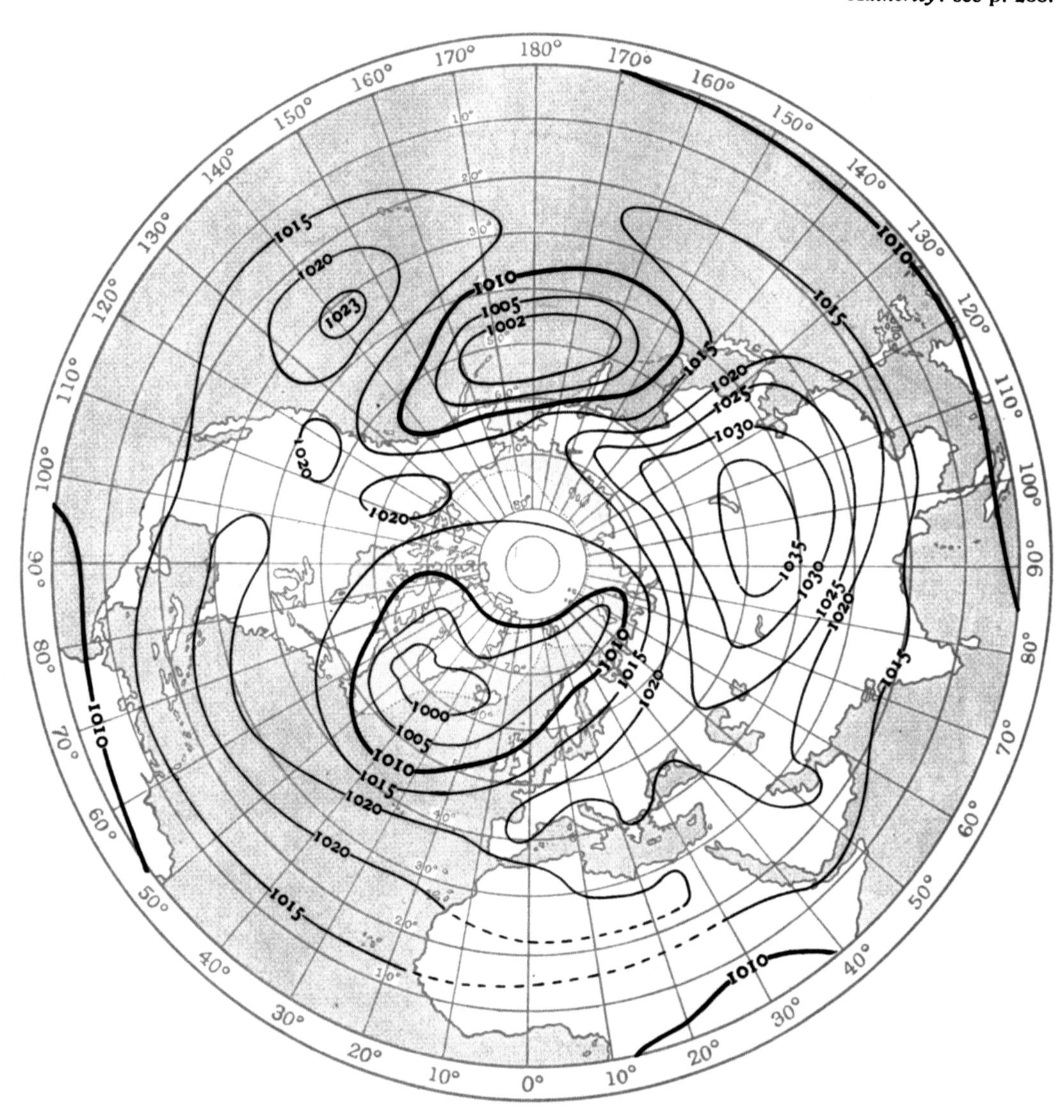

Seasonal variation of pressure in the Arctic regions.

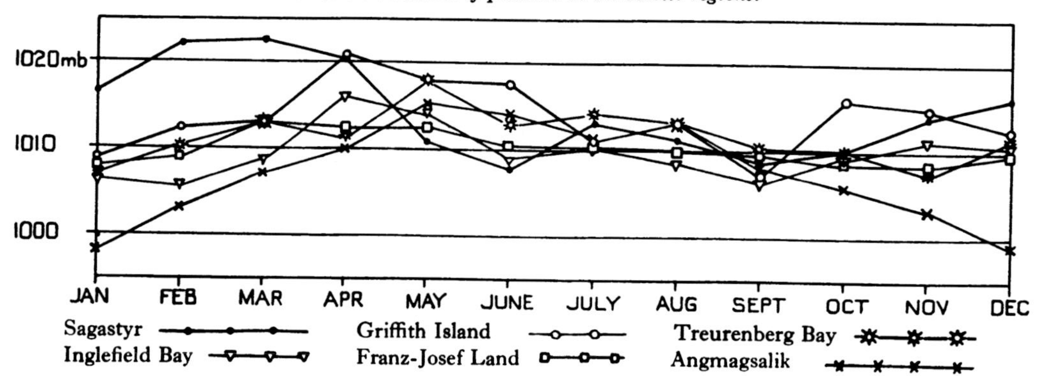

NORMAL PRESSURE AT SEA-LEVEL FIGURE 133

Authority: see p. 288. Millibars.

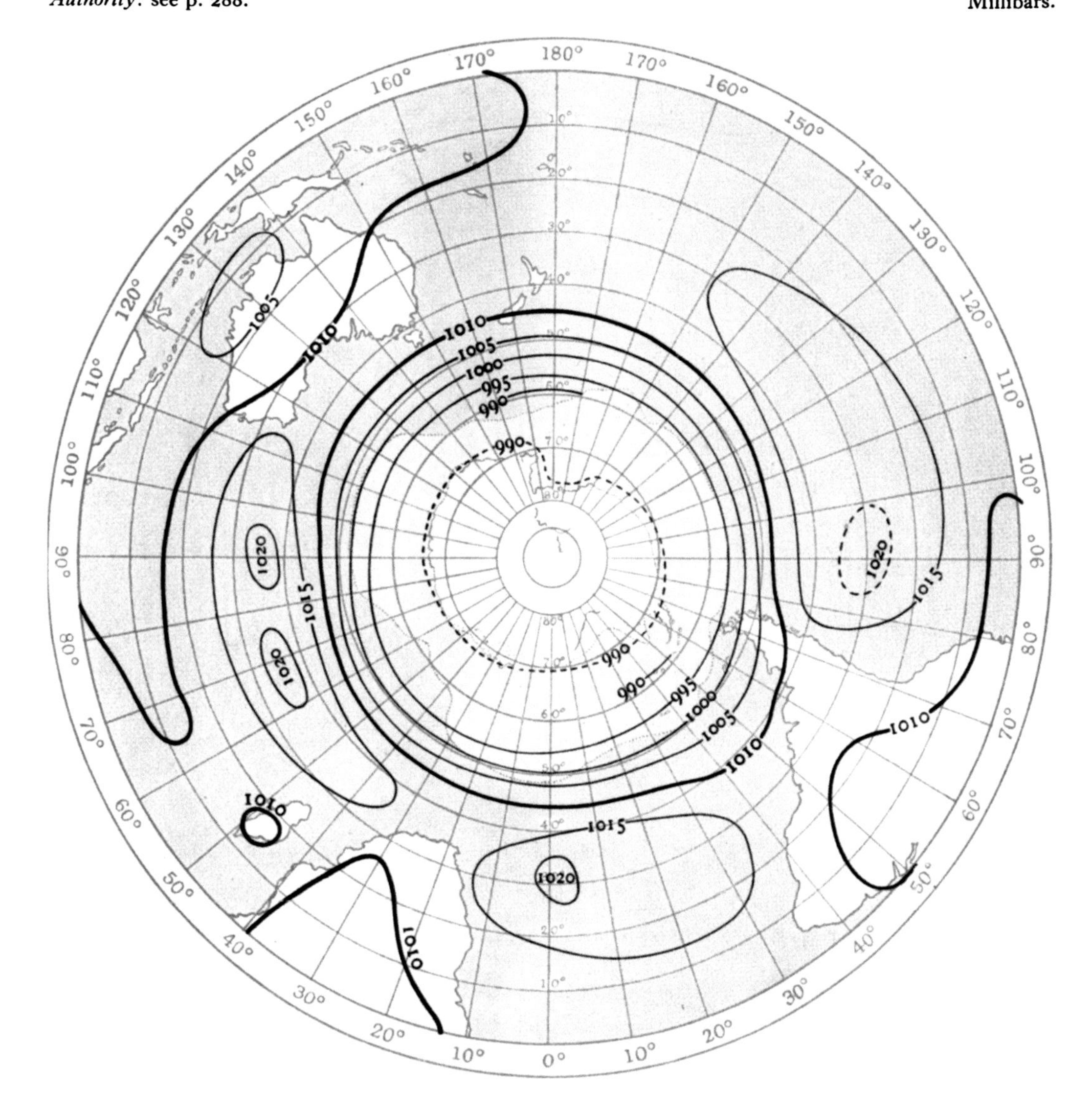

Station	Period	Lat.	Long.	Jan.	Feb.	Mar.	Apr.	May	June	July	Aug.	Sept.	Oct.	Nov.	Dec.	Year
				Excess of pressure over 900 millibars.												
Kerguelen*	1902–3	49° S	70° E	98	99	101	100	97	97	97	97	97	97	96	96	98
Cape Horn*	—	56	68° W	93	94	93	93	95	97	96	97	98	94	90	91	94
S. Georgia	1906–15	54	37	94	96	97	95	94	98	97	99	101	99	92	97	97
Port Charcot*	1904–5	65	63	98	92	85	85	93	97	96	95	93	92	93	97	93
Snow Hill*	1902–3	64	57	92	88	88	90	90	90	88	87	87	88	91	93	89
S. Orkneys	1903–23	61	45° W	91	91	91	91	93	96	95	95	94	93	89	93	93
Gauss*	1902–3	66	90° E	89	88	88	88	88	89	86	80	78	83	90	92	87
Belgica*	1898–9	70	86° W	93	86	84	87	94	99	97	94	92	93	94	96	92
Discovery*	1902–4	78	167° E	93	93	96	95	94	94	91	88	87	87	91	94	92
C. Adare	1899–1900	71	170	91	—	87	94	85	83	98	84	71	80	96	92	87
,,	1911–12	71	170° E	—	—	86	90	84	86	82	84	81	74	101	107	88
Framheim	1911–12	79	164° W	96	91	87	86	84	79	79	81	80	70	100	106	86

Seasonal variation of pressure in the Antarctic (see also pp. 261, 288).

* Smoothed.

FIGURE 134 NORMAL PRESSURE AT SEA-LEVEL

Millibars. *Authority:* see p. 288.

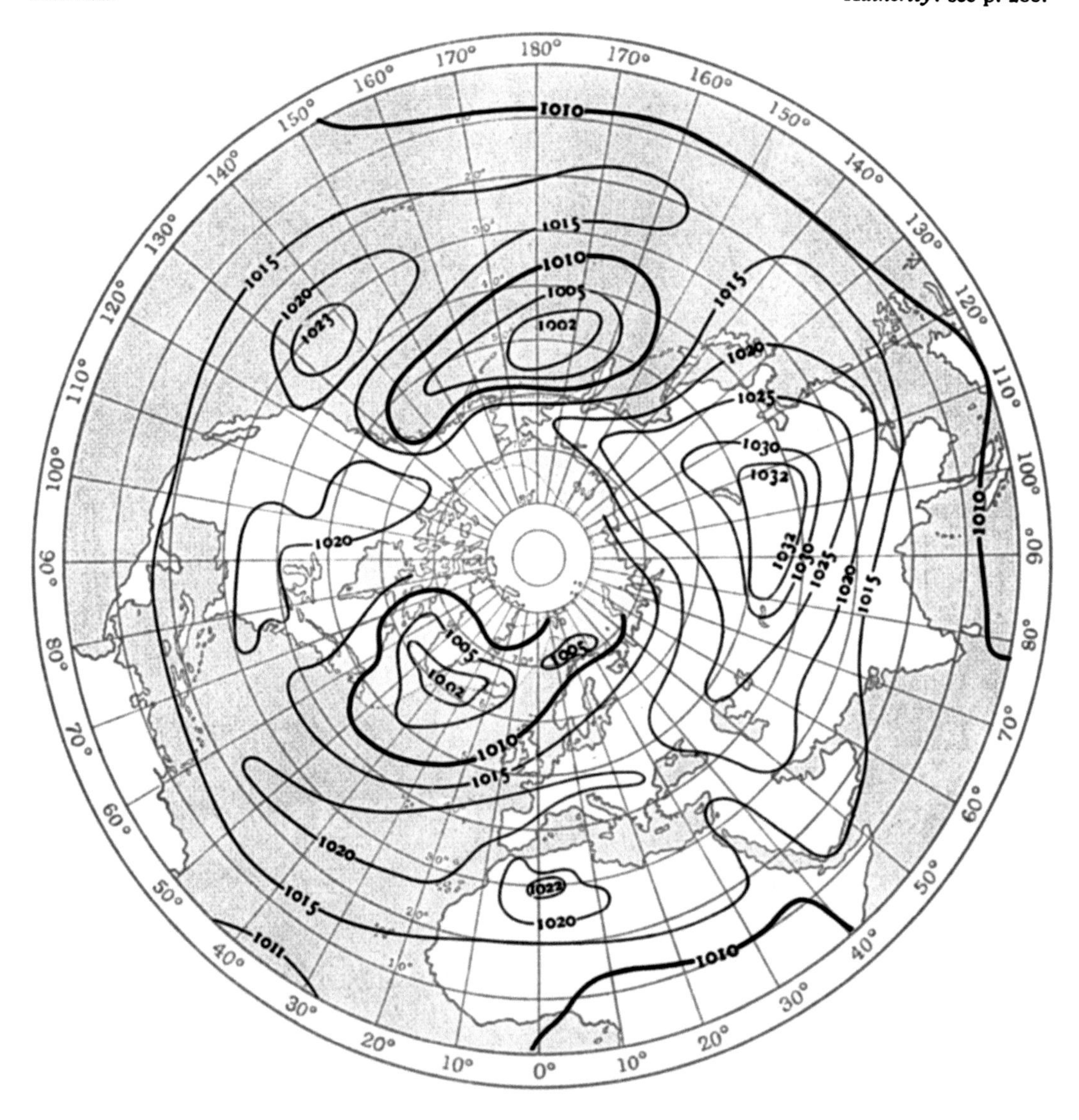

Seasonal variations of local pressure and their meteorological significance.

The study of the seasonal variations of pressure which has been briefly noted in the caption of this chapter has been carefully elaborated and discussed by H. Arctowski in vol. IV of *Communications de l'Institut de Géophysique et de Météorologie de l'Université de Lwów*, 1928–9. Among other results is a series of charts of the changes of normal pressure over the globe from month to month expressed as differences from the normals for the year based upon the data of the Réseau Mondial.

A single chart of the world with step-diagrams of changes of pressure representing imports and exports of air in tons per square dekametre for the months at a number of stations is given in *The Drama of Weather*. The implications are clearly apparent in Arctowski's discussion of the data. Among notable features we may mention the exchange of air in the course of the year between the two hemispheres and between continents and oceans as represented in fig. 185.

In the volume already cited Arctowski discusses the inferences to be drawn from graphs of the seasonal change of pressure and returns to the subject in *Communication* No. 83 of the same publication. Annual variation is the subject of a paper in *Gerlands Beiträge*, vol. XXXVII, 1929, and the variation from day to day in another paper of vol. IV of the *Communications*.

NORMAL PRESSURE AT SEA-LEVEL

FIGURE 135

Authority: see p. 288.

Millibars.

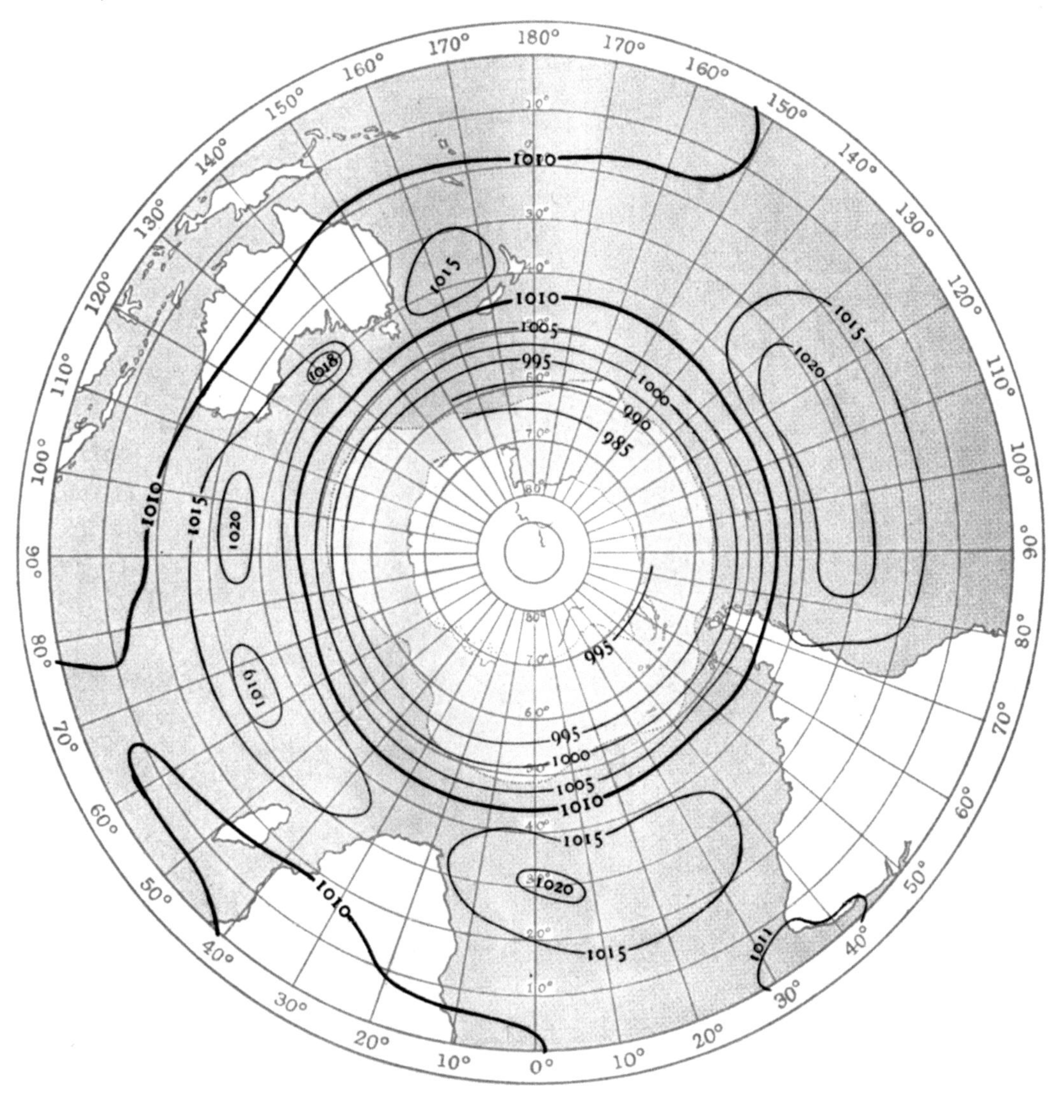

Wind-velocity in metres per second to miles per hour.

1 m/s = 2·23694 mi/hr; 1 mi/hr = 0·44704 m/s.

m/s	0	1	2	3	4	5	6	7	8	9	
0	0·00	2·24	4·47	6·71	8·95	11·18	13·42	15·66	17·90	20·13	mi/hr
10	22·37	24·61	26·84	29·08	31·32	33·55	35·79	38·03	40·27	42·50	
20	44·74	46·98	49·21	51·45	53·69	55·92	58·16	60·40	62·64	64·87	
30	67·11	69·35	71·58	73·82	76·06	78·29	80·53	82·77	85·01	87·24	
40	89·48	91·72	93·95	96·19	98·43	100·66	102·90	105·14	107·37	109·61	

Metres per second to kilometres per hour.

1 m/s = 3·6 km/hr; 1 km/hr = 0·278 m/s; 1 km/hr = ·6214 mi/hr.

m/s	0	1	2	3	4	5	6	7	8	9	
0	0·0	3·6	7·2	10·8	14·4	18·0	21·6	25·2	28·8	32·4	km/hr
10	36·0	39·6	43·2	46·8	50·4	54·0	57·6	61·2	64·8	68·4	
20	72·0	75·6	79·2	82·8	86·4	90·0	93·6	97·2	100·8	104·4	
30	108·0	111·6	115·2	118·8	122·4	126·0	129·6	133·2	136·8	140·4	
40	144·0	147·6	151·2	154·8	158·4	162·0	165·6	169·2	172·8	176·4	

FIGURE 136
Millibars.

NORMAL PRESSURE AT SEA-LEVEL

Authority: see p. 288.

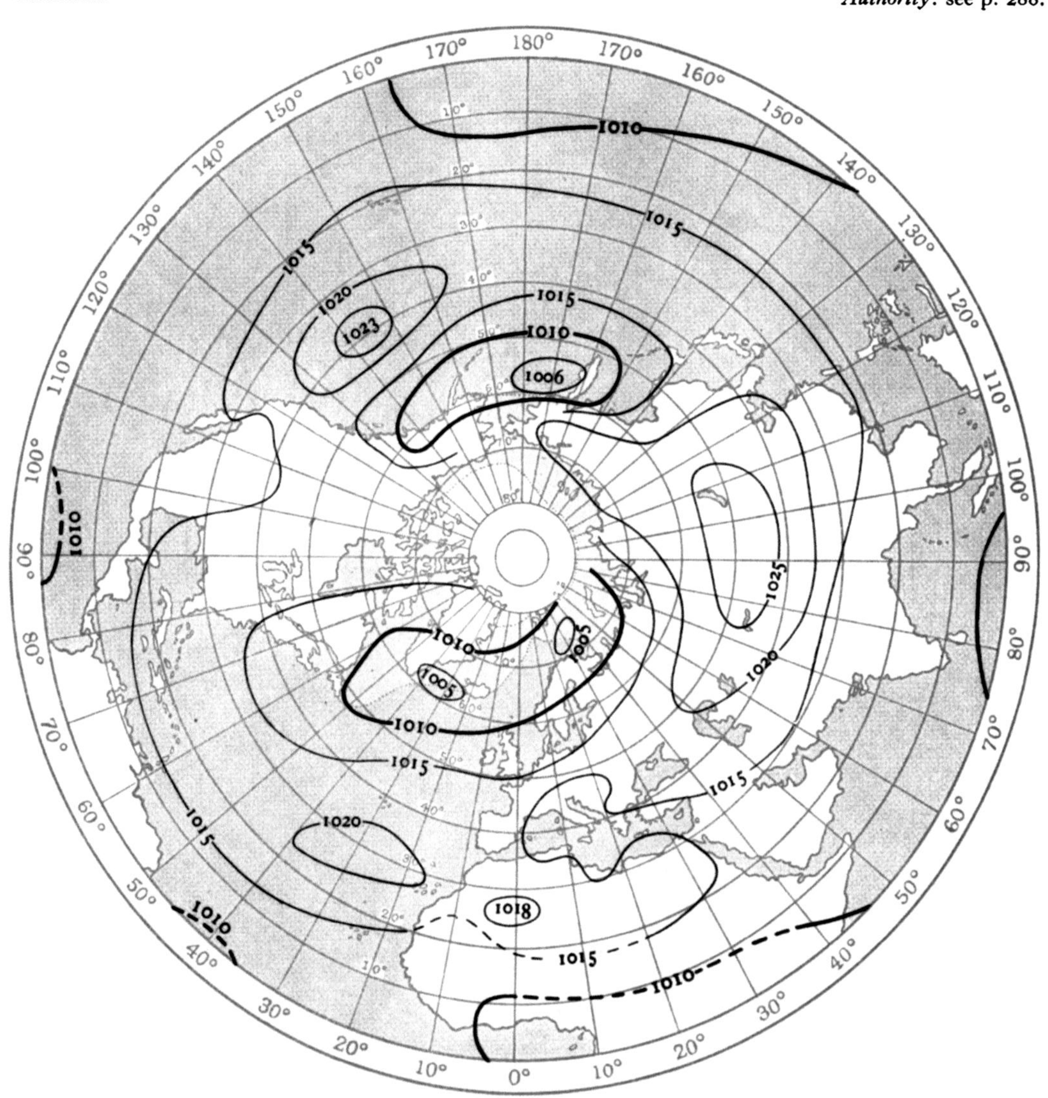

Month	Mean	0	1	2	3	4	5	6	7	8	9	10	11 G.M.T.
Jan.	1007·53	·15	·02	·01	·07	·20	·34	·36	·30	·07	·14	·32	·33
Feb.	1007·25	·24	·09	·01	·20	·33	·40	·39	·29	·05	·07	·21	·28
Mar.	1006·51	·26	·13	·00	·24	·36	·41	·34	·23	·04	·05	·16	·19
Apr.	1009·42	·18	·00	·14	·32	·43	·44	·26	·13	·01	·06	·12	·09
May	1011·83	·21	·04	·10	·26	·32	·29	·17	·08	·03	·05	·07	·07
June	1011·95	·25	·10	·04	·20	·22	·20	·10	·01	·08	·07	·09	·09
July	1009·60	·25	·09	·06	·24	·26	·24	·14	·05	·04	·03	·04	·06
Aug.	1008·48	·22	·05	·07	·25	·34	·34	·21	·11	·01	·06	·09	·11
Sept.	1010·57	·22	·10	·01	·21	·33	·37	·33	·09	·06	·14	·18	·12
Oct.	1007·39	·13	·01	·14	·35	·42	·47	·36	·20	·04	·13	·24	·23
Nov.	1006·62	·17	·01	·04	·20	·27	·32	·27	·16	·10	·19	·33	·30
Dec.	1004·19	·13	·01	·02	·14	·27	·39	·38	·32	·12	·07	·31	·27

Diurnal and seasonal variation of pressure in millibars at Aberdeen,

The pressures are for station-level, corrected for temperature and gravity.

NORMAL PRESSURE AT SEA-LEVEL

Authority: see p. 288.

FIGURE 137

Millibars.

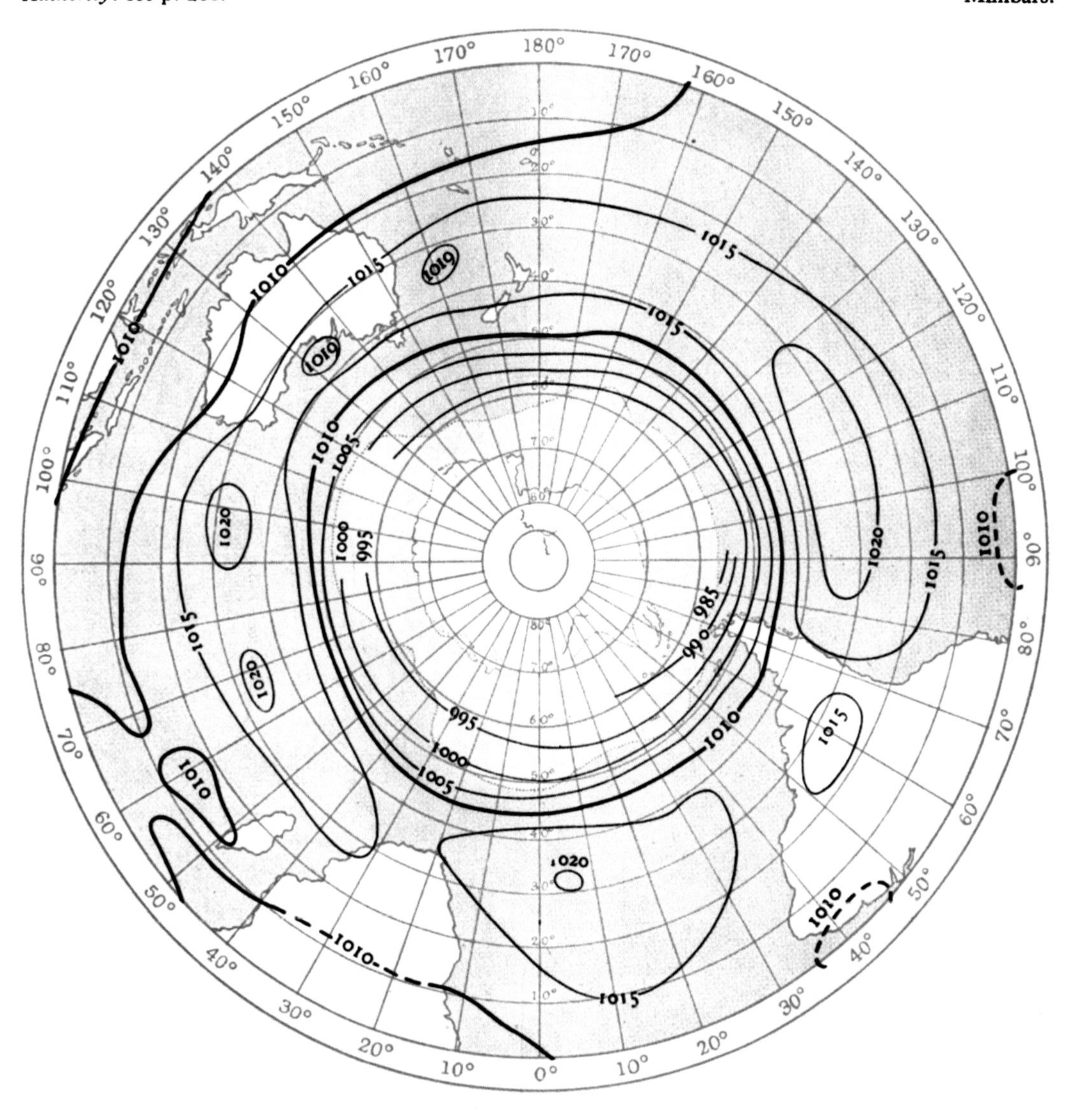

57° 10′ N, 2° 6′ W, 26·8 m. 1871–1915.														
The figures underlined represent departures from the mean on the negative side.														
G.M.T. 12	13	14	15	16	17	18	19	20	21	22	23	Mdt	Range	Month
·14	·12	·21	·24	·12	·07	·06	·10	·21	·21	·24	·17	·13	·69	Jan.
·22	·01	·14	·27	·20	·12	·11	·17	·25	·23	·26	·21	·20	·68	Feb.
·16	·00	·15	·25	·26	·20	·04	·20	·33	·33	·35	·30	·16	·76	Mar.
·09	·02	·06	·20	·21	·21	·05	·13	·39	·43	·43	·37	·30	·87	Apr.
·06	·01	·03	·14	·18	·23	·14	·02	·20	·35	·40	·34	·26	·72	May
·08	·00	·03	·13	·18	·25	·18	·10	·07	·26	·35	·31	·25	·60	June
·06	·01	·00	·06	·14	·20	·13	·04	·12	·27	·35	·29	·22	·61	July
·09	·05	·01	·08	·14	·18	·12	·00	·24	·30	·33	·26	·18	·67	Aug.
·09	·01	·11	·24	·27	·24	·07	·13	·32	·31	·32	·25	·17	·69	Sept.
·16	·01	·11	·21	·18	·08	·17	·23	·30	·30	·29	·20	·16	·77	Oct.
·12	·06	·15	·23	·13	·06	·10	·12	·15	·15	·12	·05	·04	·65	Nov.
·09	·13	·17	·19	·02	·01	·15	·18	·26	·25	·27	·22	·19	·70	Dec.

FIGURE 138

Millibars.

NORMAL PRESSURE AT SEA-LEVEL

Authority: see p. 288.

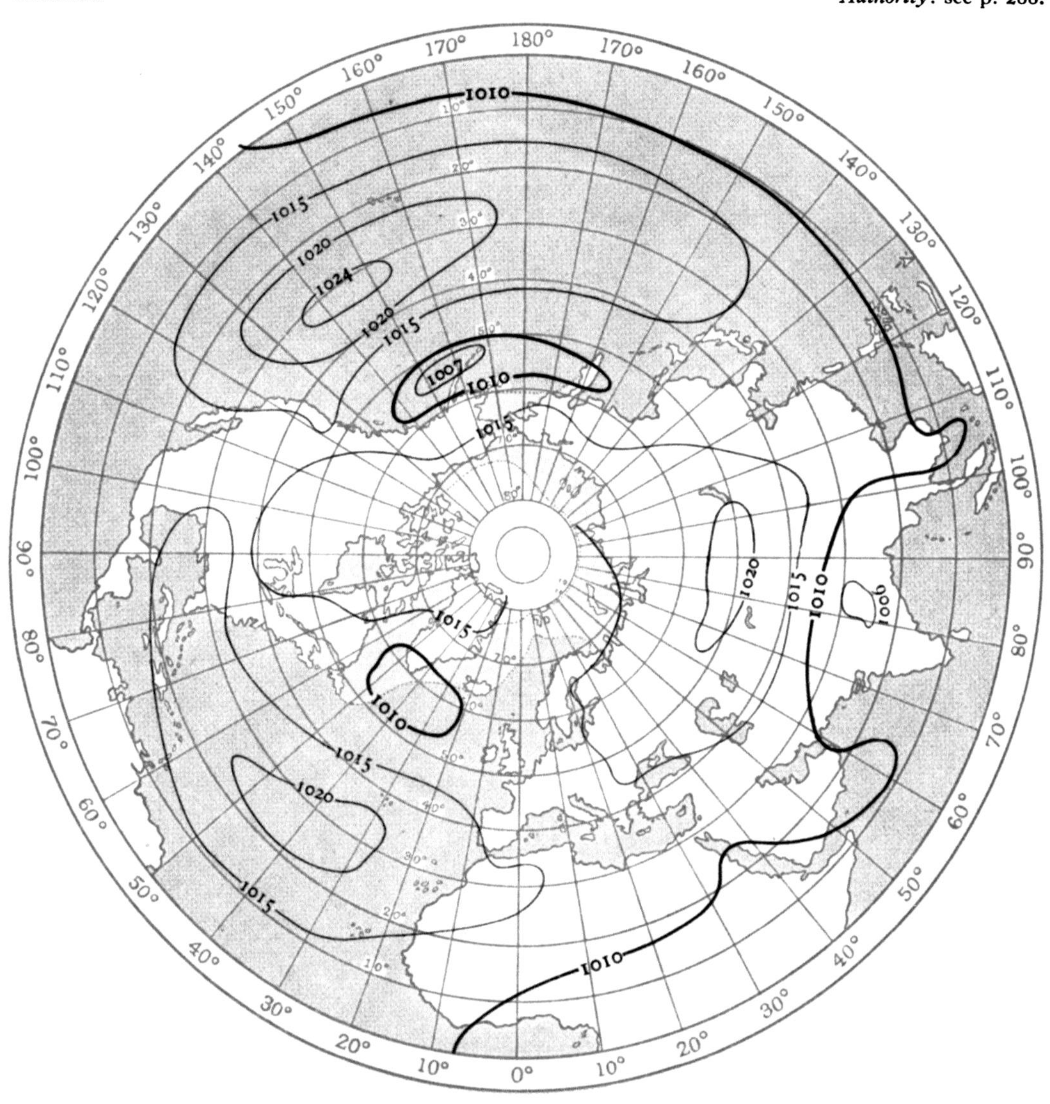

Diurnal and seasonal variation of pressure in millibars at Calcutta (Alipore),

The figures underlined represent departures from the mean on the negative side.

Month	Mean	0	1	2	3	4	5	6	7	8	9	10	11 L.M.T.
Jan.	1016·51	·20	·14	·37	·71	·81	·71	·17	·54	1·46	2·27	2·57	2·00
Feb.	1014·68	·34	·03	·30	·68	·81	·71	·10	·64	1·76	2·20	2·57	2·20
Mar.	1010·95	·51	·07	·34	·81	·85	·57	·10	·85	1·76	2·34	2·54	2·17
Apr.	1007·06	·44	·10	·37	·78	·68	·34	·30	·98	1·79	2·20	2·30	1·90
May	1004·48	·41	·14	·47	·64	·51	·30	·27	·88	1·49	1·86	1·96	1·56
June	1000·69	·81	·17	·14	·47	·47	·30	·17	·64	1·15	1·42	1·49	1·19
July	999·95	·91	·34	·03	·37	·41	·41	·00	·47	·98	1·29	1·42	1·15
Aug.	1002·22	·78	·27	·17	·54	·58	·54	·14	·44	1·02	1·39	2·03	1·25
Sept.	1005·36	·47	·00	·34	·71	·71	·58	·07	·51	1·25	1·69	1·86	1·46
Oct.	1010·55	·24	·20	·44	·71	·68	·41	·20	·85	1·59	2·07	2·17	1·52
Nov.	1014·17	·34	·07	·34	·61	·58	·44	·17	·85	1·63	2·23	2·23	1·52
Dec.	1016·81	·27	·03	·27	·61	·61	·51	·07	·74	1·59	2·34	2·44	1·73

NORMAL PRESSURE AT SEA-LEVEL

FIGURE 139

Authority: see p. 288.

Millibars.

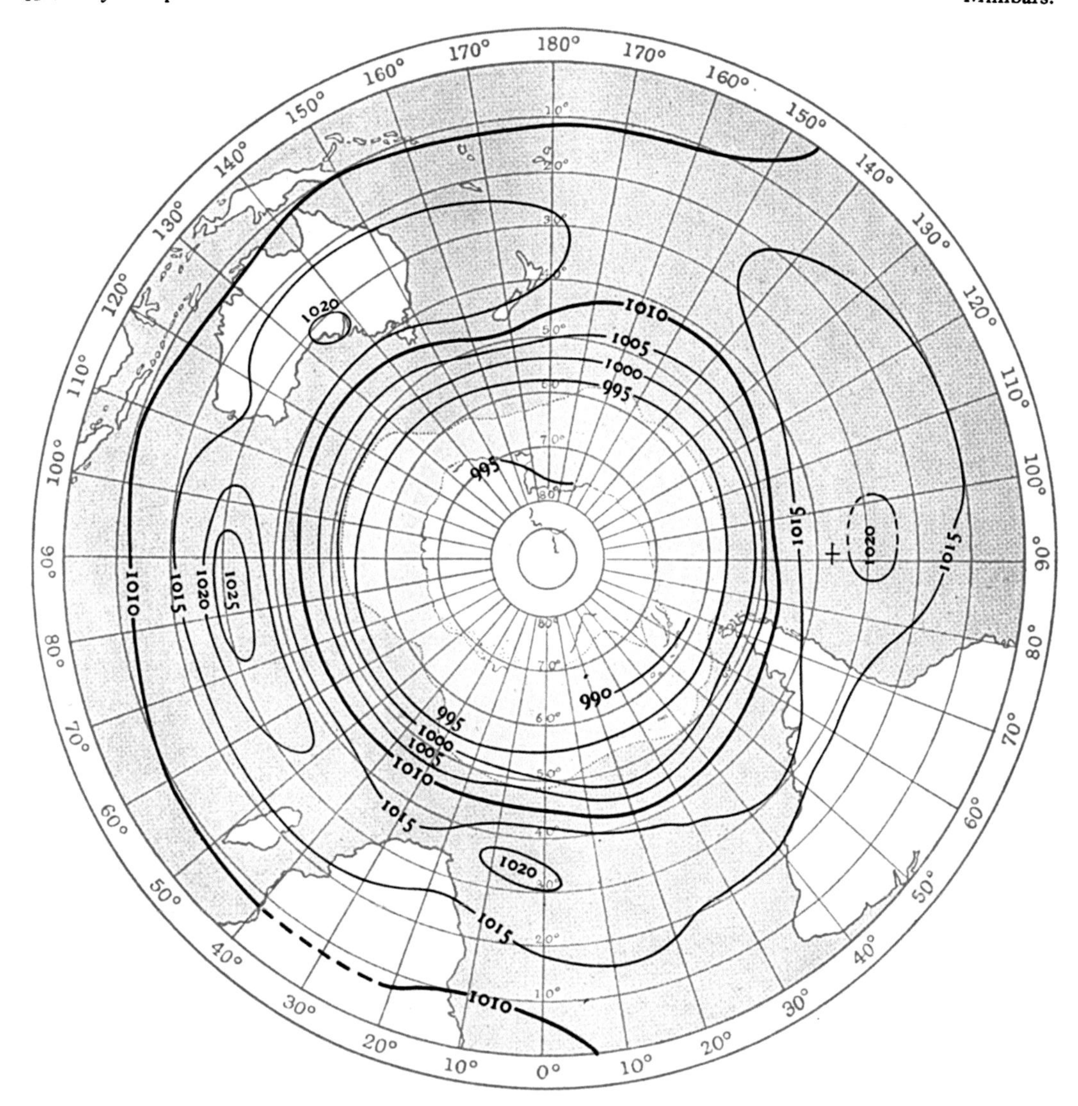

22° 32′ N, 88° 20′ E, 6·5 m. 1881–93.

The figures underlined represent departures from the mean on the negative side.

L.M.T. 12	13	14	15	16	17	18	19	20	21	22	23	Range	Month
·98	·24	1·12	1·76	1·86	1·73	1·29	·74	·10	·27	·58	·41	4·43	Jan.
1·39	·14	·91	1·73	2·00	1·93	1·73	1·19	·41	·14	·44	·41	4·57	Feb.
1·42	·20	·78	1·76	2·07	2·17	1·79	1·29	·51	·17	·64	·51	4·71	Mar.
1·32	·24	·71	1·63	2·23	2·47	2·03	1·35	·47	·17	·68	·54	4·77	Apr.
1·08	·24	·61	1·52	2·07	2·27	1·76	1·15	·30	·37	·88	·78	4·23	May
·74	·07	·71	1·46	1·86	2·00	1·56	·88	·10	·37	·91	·88	3·49	June
·74	·03	·58	1·35	1·76	1·93	1·52	·95	·17	·44	1·05	1·02	3·35	July
·78	·03	·74	1·49	1·86	1·96	1·63	·95	·07	·64	1·12	1·12	3·99	Aug.
·91	·17	·98	1·79	1·90	1·96	1·42	·81	·03	·71	1·12	·95	3·82	Sept.
·74	·34	1·12	1·73	1·76	1·63	1·25	·64	·14	·51	·68	·47	3·93	Oct.
·64	·58	1·35	1·90	1·83	1·63	1·19	·58	·10	·44	·64	·47	4·13	Nov.
·78	·58	1·39	1·93	1·93	1·76	1·25	·71	·03	·37	·61	·44	4·37	Dec.

FIGURE 140

NORMAL PRESSURE AT SEA-LEVEL

Millibars.

Authority: see p. 288.

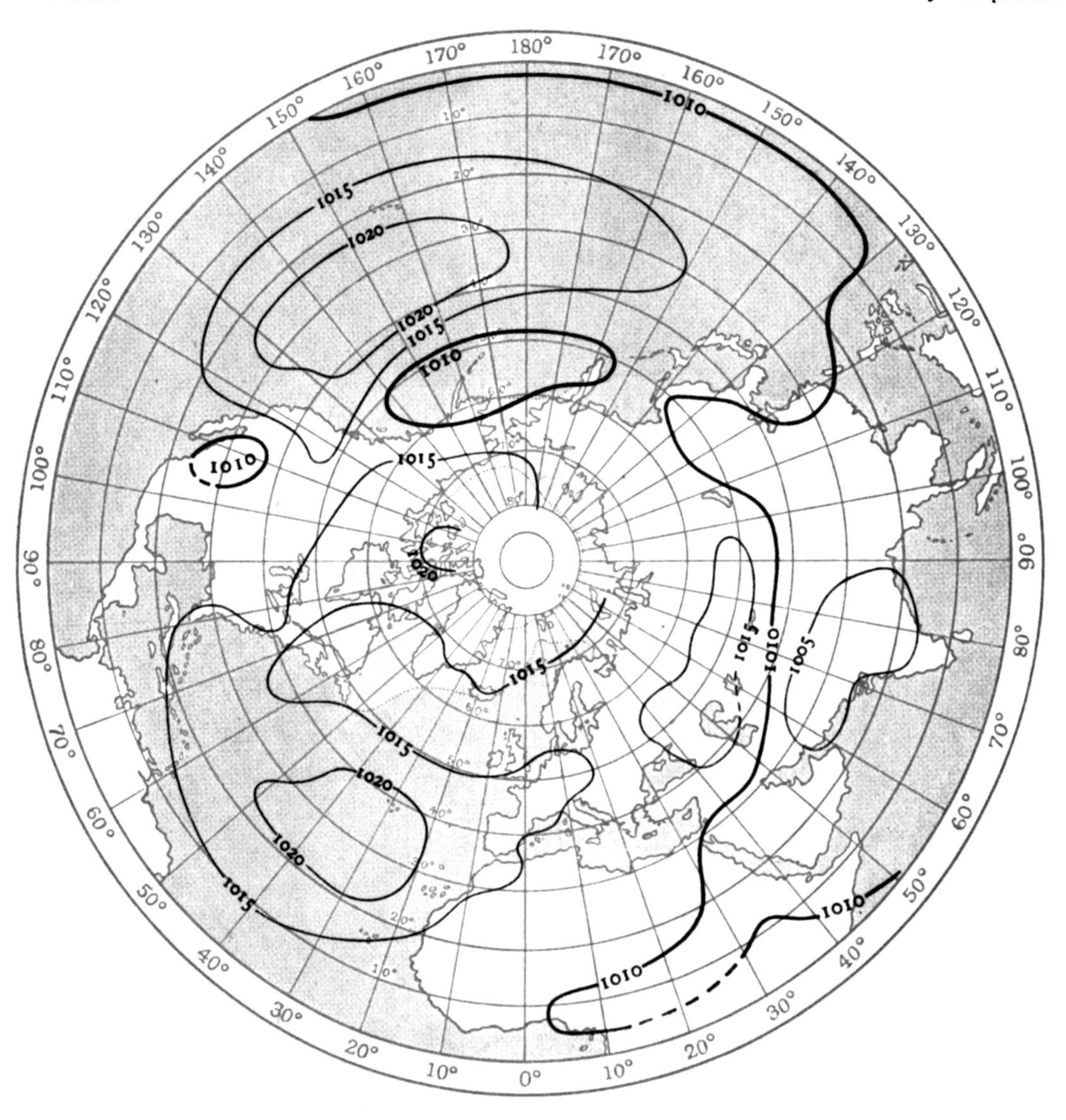

Diurnal and seasonal variation of pressure in millibars at Batavia,

Corrections for latitude (−2·57 mb) and for altitude (+·92 mb) not applied.

Month	Mean	0	1	2	3	4	5	6	7	8	9	10	11 L.M.T.
Jan.	1011·57	·77	·21	·27	·60	·65	·36	·19	·87	1·35	1·53	1·39	·97
Feb.	1011·66	·83	·28	·20	·52	·57	·33	·07	·67	1·33	1·59	1·47	1·08
Mar.	1011·47	·89	·36	·15	·45	·52	·29	·07	·68	1·27	1·60	1·52	1·05
Apr.	1010·98	·96	·40	·11	·41	·51	·35	·09	·72	1·24	1·56	1·48	·92
May	1011·09	·91	·43	·07	·40	·49	·27	·13	·72	1·28	1·53	1·35	·83
June	1011·50	·92	·49	·03	·33	·40	·19	·13	·68	1·25	1·44	1·24	·75
July	1011·86	·93	·55	·12	·25	·31	·11	·21	·75	1·32	1·53	1·29	·75
Aug.	1012·09	·96	·49	·07	·24	·29	·09	·31	·91	1·45	1·71	1·44	·83
Sept.	1012·27	·85	·33	·07	·29	·29	·03	·45	1·11	1·60	1·83	1·55	·80
Oct.	1011·87	·73	·23	·21	·45	·39	·07	·43	1·09	1·61	1·71	1·40	·79
Nov.	1011·45	·69	·16	·35	·60	·56	·21	·31	·99	1·45	1·57	1·31	·75
Dec.	1011·22	·76	·19	·29	·63	·68	·36	·20	·88	1·32	1·52	1·33	·83

NORMAL PRESSURE AT SEA-LEVEL

FIGURE 141

Authority: see p. 288.

Millibars.

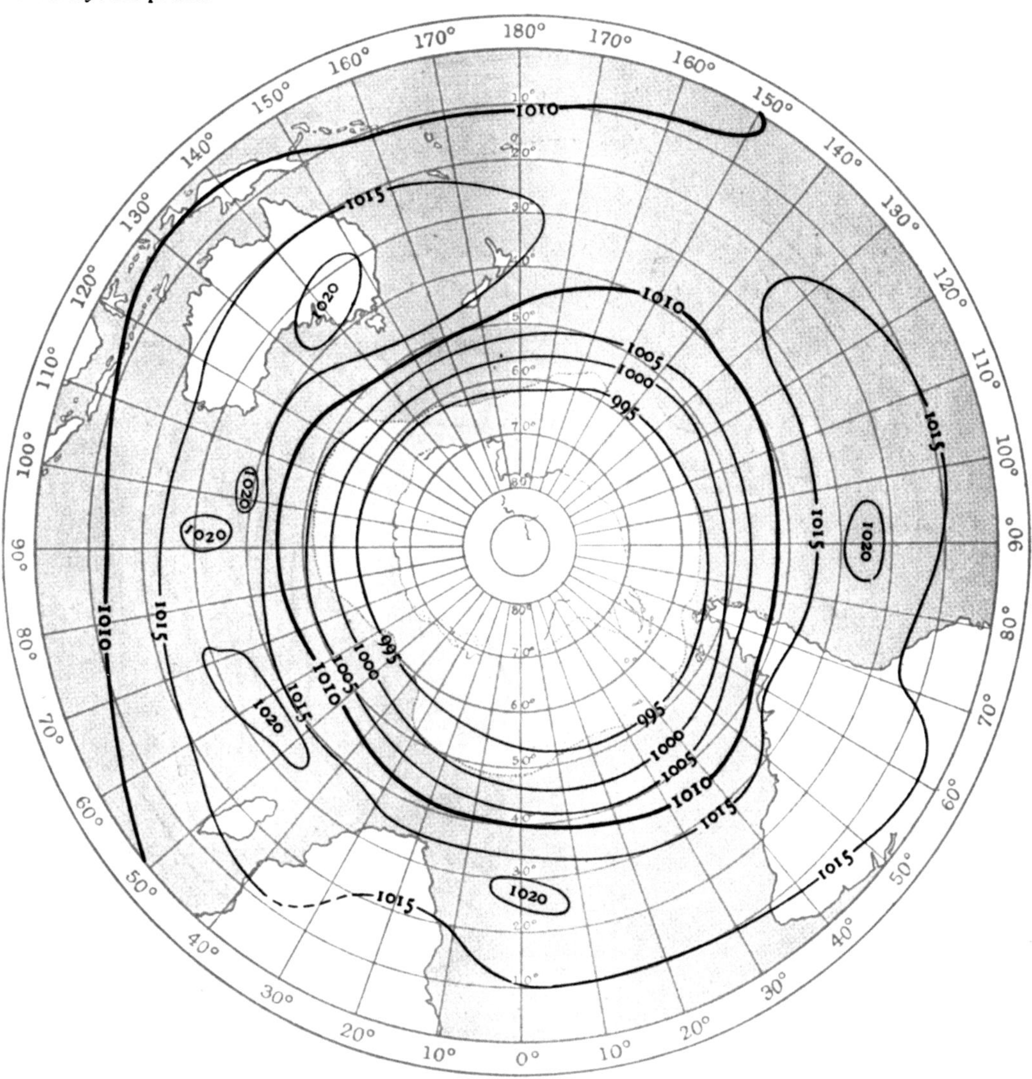

6° 11′ S, 106° 50′ E, 8 m. 1866–1915.

The figures underlined represent departures from the mean on the negative side.

L.M.T. 12	13	14	15	16	17	18	19	20	21	22	23	Range	Month
·39	·43	1·27	1·97	2·15	1·77	1·11	·39	·28	·84	1·15	1·13	3·68	Jan.
·40	·48	1·28	1·89	2·15	1·87	1·23	·51	·19	·79	1·17	1·17	3·74	Feb.
·29	·68	1·52	2·07	2·23	1·85	1·19	·45	·25	·84	1·23	1·24	3·83	Mar.
·08	·81	1·64	2·20	2·19	1·69	1·08	·35	·40	·97	1·24	1·25	3·76	Apr.
·03	·81	1·61	2·09	2·05	1·64	1·08	·35	·40	·96	1·21	1·15	3·62	May
·01	·79	1·55	1·97	1·97	1·59	1·05	·39	·32	·89	1·13	1·08	3·41	June
·00	·77	1·56	2·03	2·07	1·73	1·23	·57	·17	·77	1·05	1·07	3·60	July
·00	·88	1·67	2·15	2·21	1·88	1·32	·59	·17	·79	1·13	1·20	3·92	Aug.
·11	1·04	1·81	2·29	2·36	1·96	1·31	·52	·27	·84	1·20	1·21	4·19	Sept.
·08	1·07	1·83	2·24	2·29	1·80	1·04	·27	·41	·99	1·31	1·17	4·00	Oct.
·03	·93	1·71	2·20	2·16	1·63	·88	·11	·56	1·09	1·28	1·13	3·77	Nov.
·19	·61	1·43	2·01	2·07	1·65	·96	·20	·48	1·01	1·21	1·13	3·59	Dec.

FIGURE 142 NORMAL PRESSURE AT SEA-LEVEL

Millibars. *Authority:* see p. 288.

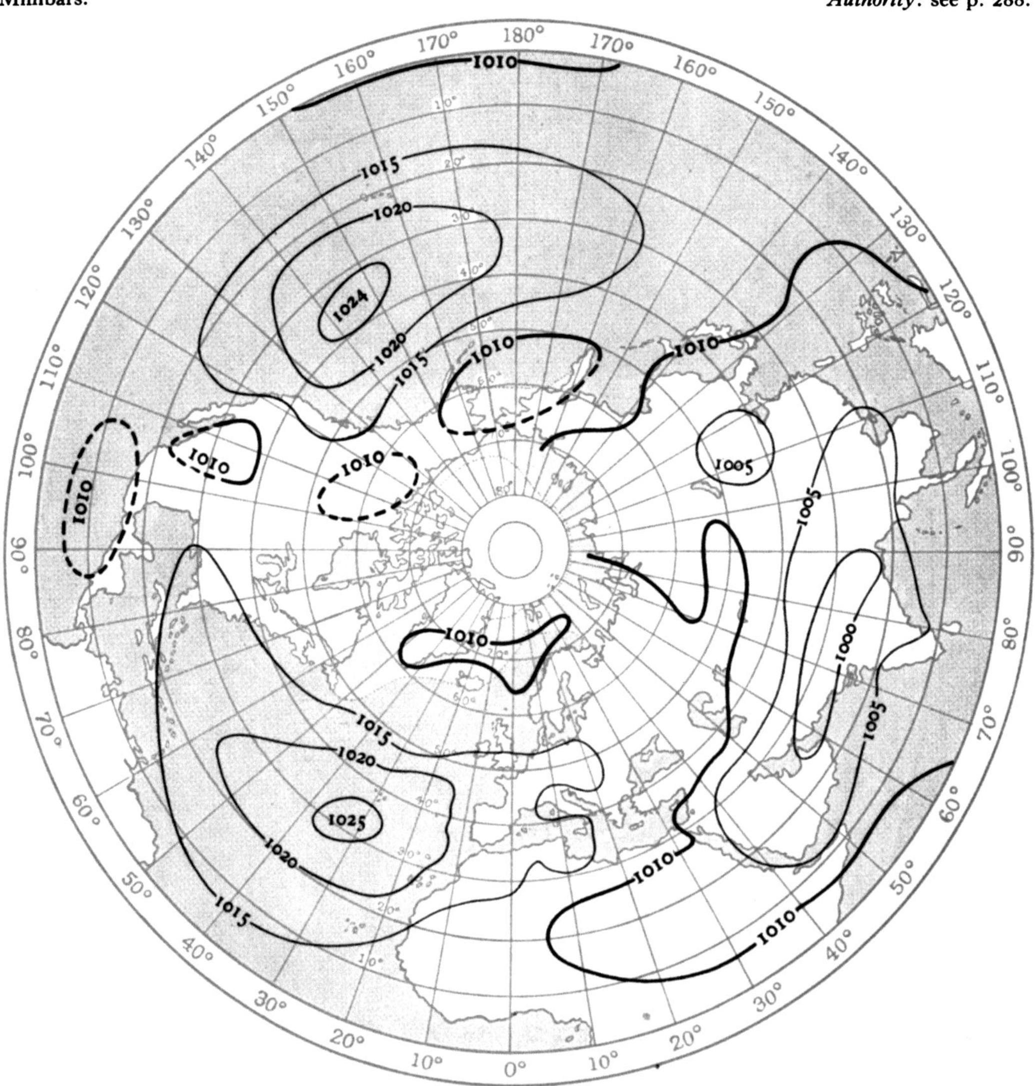

Diurnal and seasonal variation in millibars at the Cape of Good Hope,

The values in the column headed "Mean" are for the period 1841–70, corrected to sea-level by the addition of ·037 in. (1·25 mb).

Month	Mean	0	1	2	3	4	5	6	7	8	9	10	11 L.M.T.
Jan.	1014·74	·30	·10	·56	·80	·77	·46	·10	·30	·44	·46	·49	·43
Feb.	1014·64	·29	·03	·35	·64	·72	·49	·12	·29	·54	·66	·75	·55
Mar.	1015·96	·13	·25	·18	·41	·46	·30	·03	·41	·69	·86	·94	·71
Apr.	1017·69	·10	·10	·22	·59	·71	·46	·21	·14	·62	·88	·88	·71
May	1019·45	·13	·02	·03	·23	·39	·36	·12	·21	·65	·98	1·06	·85
June	1021·28	·27	·25	·27	·34	·59	·55	·39	·17	·44	·86	1·06	·97
July	1022·67	·07	·01	·15	·25	·44	·40	·14	·22	·54	·90	1·12	·85
Aug.	1021·72	·17	·00	·20	·42	·51	·46	·15	·26	·64	·89	·91	·73
Sept.	1020·40	·02	·01	·41	·62	·61	·33	·08	·47	·76	·92	·92	·70
Oct.	1018·50	·27	·17	·53	·69	·63	·34	·13	·47	·62	·73	·73	·35
Nov.	1016·84	·20	·24	·51	·68	·66	·34	·05	·37	·53	·54	·53	·37
Dec.	1015·46	·30	·16	·52	·69	·62	·32	·03	·41	·54	·50	·49	·47

NORMAL PRESSURE AT SEA-LEVEL

FIGURE 143

Authority: see p. 288.

Millibars.

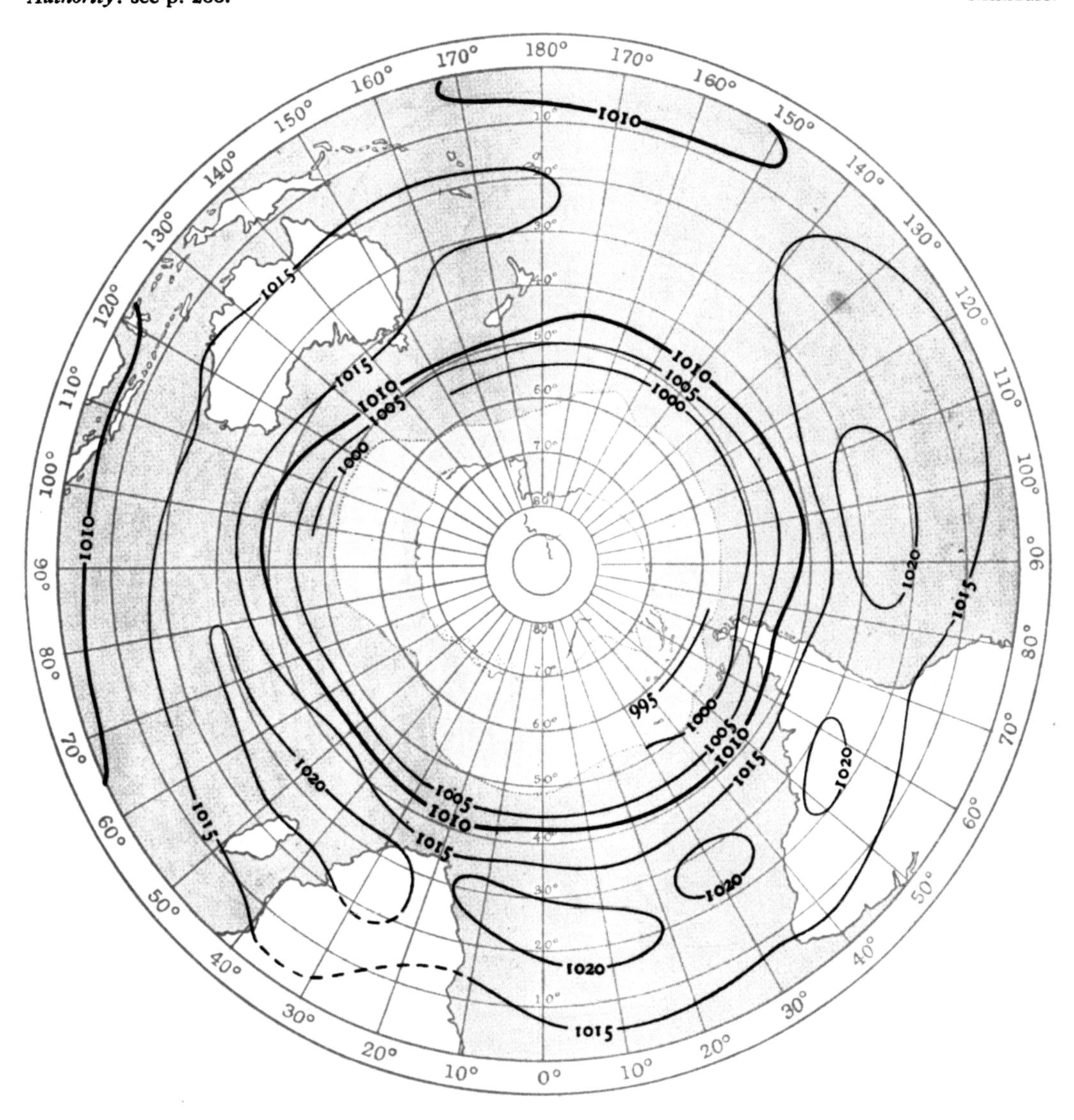

33° 56′ S, 18° 29′ E. 1841–6.

The figures underlined represent departures from the mean on the negative side.

L.M.T. 12	13	14	15	16	17	18	19	20	21	22	23	Range	Month
·30	·12	·14	·43	·69	·72	·37	·03	·35	·59	·70	·58	1·50	Jan.
·29	·04	·27	·54	·71	·75	·53	·14	·23	·55	·59	·42	1·50	Feb.
·33	·14	·51	·78	·80	·70	·52	·29	·14	·32	·28	·14	1·74	Mar.
·31	·24	·60	·63	·52	·51	·31	·01	·23	·30	·39	·27	1·59	Apr.
·20	·57	·90	·94	·78	·56	·32	·04	·21	·29	·34	·24	2·00	May
·19	·42	·76	·77	·60	·42	·14	·15	·30	·38	·46	·36	1·83	June
·26	·31	·73	·79	·71	·58	·29	·05	·13	·23	·26	·22	1·91	July
·19	·31	·74	·82	·82	·55	·24	·07	·28	·42	·41	·24	1·73	Aug.
·34	·29	·62	·81	·79	·58	·34	·07	·29	·41	·34	·15	1·73	Sept.
·09	·14	·52	·81	·78	·56	·31	·00	·56	·63	·51	·48	1·54	Oct.
·12	·14	·45	·64	·70	·60	·26	·12	·47	·68	·66	·54	1·38	Nov.
·25	·01	·31	·57	·78	·78	·46	·07	·27	·60	·77	·61	1·55	Dec.

FIGURE 144

NORMAL PRESSURE AT SEA-LEVEL

Millibars.

Authority: see p. 288.

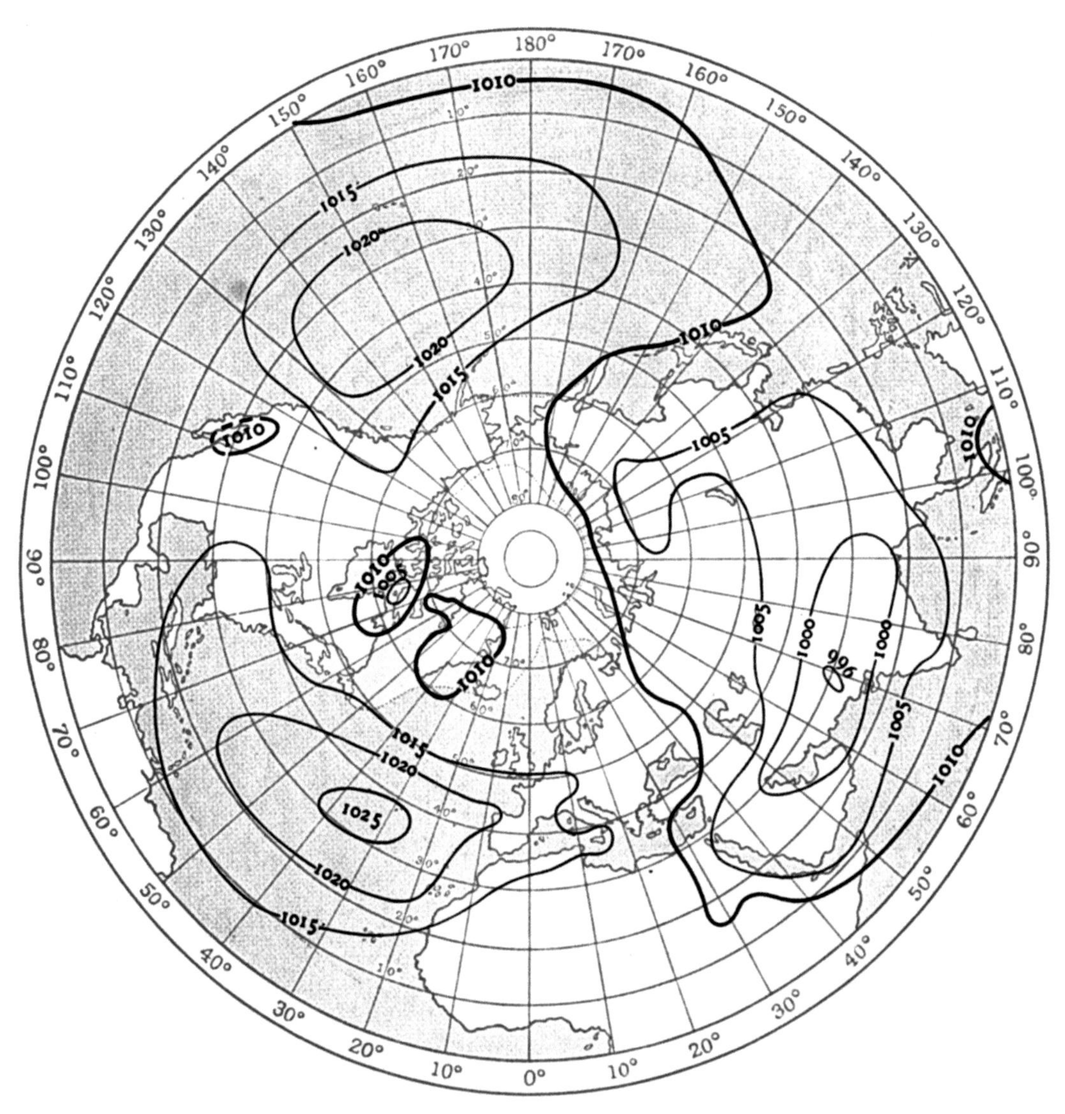

Diurnal and seasonal variation of pressure in millibars in the Arctic. 1893–6. (Local time.) See p. 126.

Corrected for seasonal variation and reduced to standard-gravity and sea-level.

Month	Mean	Mdt	2	4	6	8	10	12	14	16	18	20	22	Range
Jan.	1012·93	·01	·08	·09	·04	·07	·12	·03	·04	·12	·11	·13	·01	·26
Feb.	1010·89	·09	·07	·23	·20	·21	·17	·05	·11	·19	·19	·12	·16	·50
Mar.	1011·01	·04	·09	·15	·13	·16	·16	·04	·04	·16	·19	·12	·08	·46
Apr.	1017·35	·01	·01	·03	·00	·07	·09	·08	·03	·01	·01	·00	·00	·21
May	1015·19	·11	·09	·03	·05	·12	·13	·00	·04	·04	·08	·08	·12	·28
June	1011·31	·09	·07	·07	·09	·12	·13	·03	·03	·05	·11	·12	·11	·29
July	1008·26	·04	·01	·12	·15	·23	·24	·13	·03	·17	·24	·23	·13	·49
Aug.	1014·91	·03	·07	·01	·08	·15	·24	·13	·05	·11	·23	·21	·09	·47
Sept.	1007·74	·09	·04	·03	·19	·16	·01	·09	·09	·00	·01	·04	·08	·31
Oct.	1014·75	·04	·07	·13	·05	·15	·24	·17	·05	·16	·24	·23	·15	·48
Nov.	1011·98	·08	·13	·17	·15	·16	·19	·01	·13	·24	·23	·17	·04	·48
Dec.	1015·37	·03	·04	·01	·01	·05	·12	·03	·01	·07	·04	·07	·00	·22

NORMAL PRESSURE AT SEA-LEVEL

FIGURE 145

Authority: see p. 288.

Millibars.

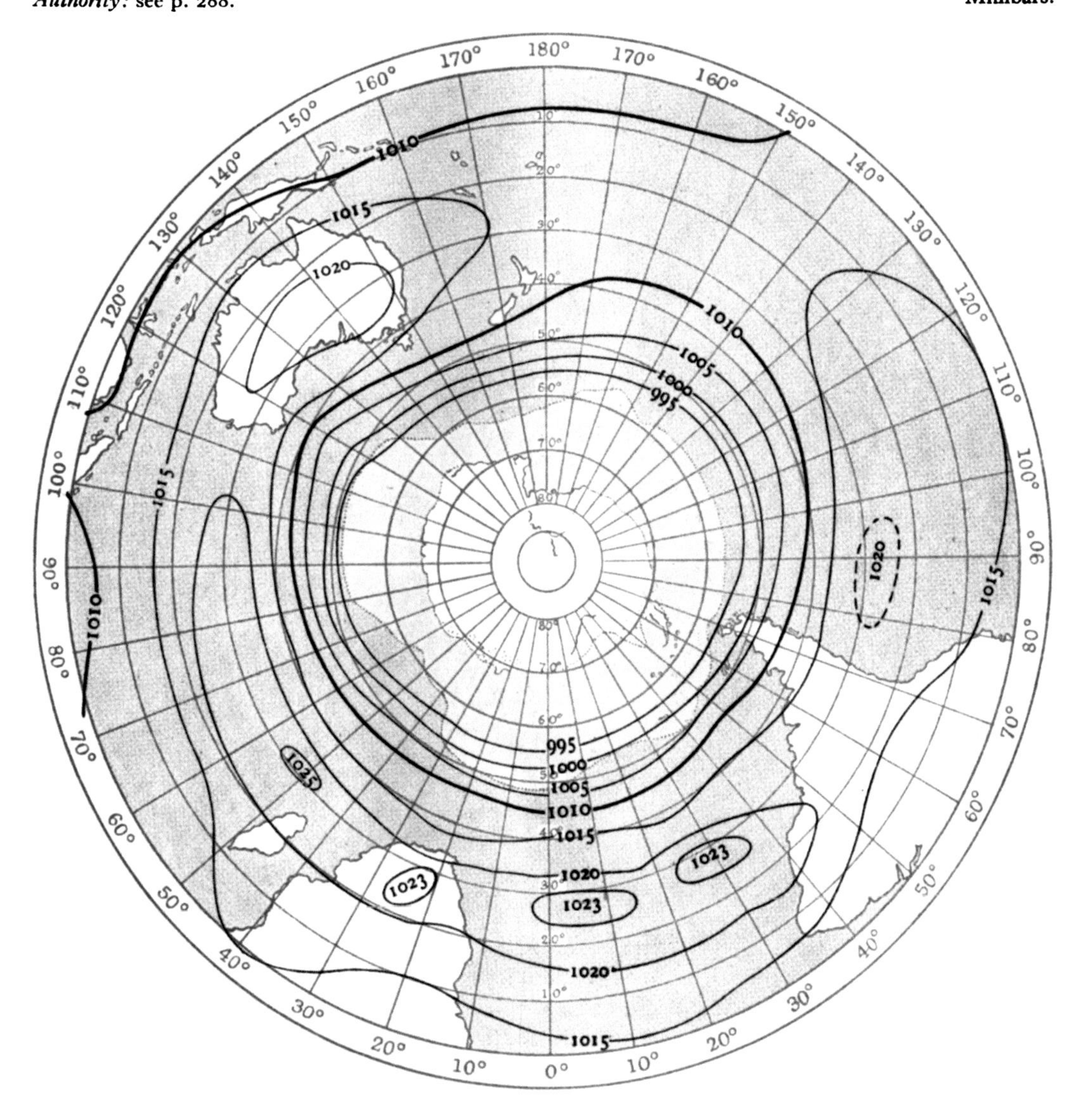

Hut Point and Cape Evans, 77½° S, 167° E. 1902, 1903, 1911 and 1912. (Local time.)

Reduced to 32° F, sea-level and gravity at 45°. The figures underlined represent departures from the mean on the negative side.

Month	Mean	0	2	4	6	8	10	12	14	16	18	20	22	Range
Jan.	993·21	·02	·11	·15	·08	·09	·23	·12	·08	·04	·08	·06	·00	·38
Feb.	994·39	·03	·08	·07	·03	·01	·17	·10	·03	·05	·07	·01	·03	·25
Mar.	992·77	·00	·16	·21	·10	·08	·19	·14	·03	·04	·02	·05	·01	·40
Apr.	994·80	·04	·16	·26	·29	·20	·06	·10	·06	·12	·17	·28	·21	·58
May	990·06	·03	·04	·03	·07	·02	·13	·05	·07	·09	·05	·04	·03	·22
June	989·38	·13	·11	·01	·19	·19	·11	·10	·02	·02	·02	·10	·20	·39
July	987·48	·03	·01	·01	·06	·01	·11	·02	·01	·09	·09	·05	·05	·20
Aug.	985·83	·01	·07	·24	·36	·25	·10	·08	·07	·20	·24	·31	·25	·68
Sept.	990·84	·06	·00	·04	·07	·02	·07	·00	·06	·05	·04	·10	·06	·17
Oct.	981·93	·05	·06	·09	·06	·08	·16	·11	·03	·02	·06	·02	·05	·25
Nov.	994·49	·01	·20	·21	·06	·01	·07	·08	·08	·04	·03	·09	·05	·30
Dec.	996·76	·09	·20	·23	·11	·05	·22	·20	·13	·02	·04	·02	·00	·45

FIGURE 146 NORMAL PRESSURE AT SEA-LEVEL

Millibars. *Authority:* see p. 288.

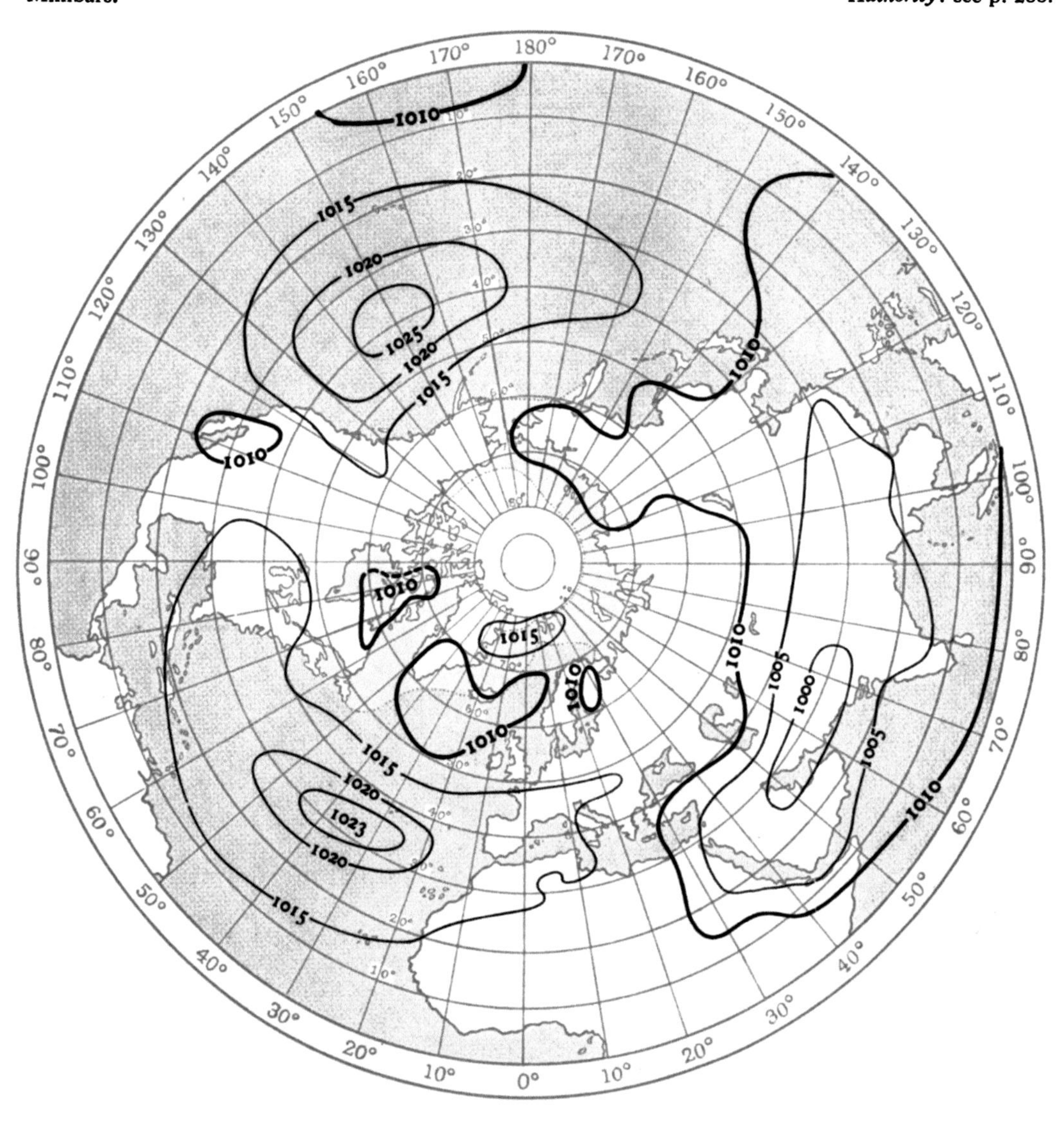

Diurnal and seasonal variation of pressure in millibars

Fort William, 56° 48′ N, 5° 5′ W. The values are corrected for seasonal change, reduced to 32° F and sea-level. The figures in each row are excesses above the minimum for the month. Values underlined are below the mean for the month.

Month	Min.	Mdt	2	4	6	8	10	12	14	16	18	20	22	Mean
Jan.	1010·61 +	·78	·51	·24	·00	·27	·71	·61	·37	·54	·74	·91	·91	·54
Feb.	1010·55 +	·64	·47	·27	·27	·58	·74	·71	·20	·07	·41	·61	·61	·44
Mar.	1007·77 +	·74	·54	·17	·27	·51	·58	·54	·24	·00	·24	·61	·71	·41
Apr.	1011·53 +	·74	·44	·24	·41	·68	·68	·47	·27	·03	·14	·58	·74	·44
May	1014·30 +	1·02	·78	·64	·78	·88	·71	·47	·24	·07	·17	·61	·98	·58
June	1014·37 +	·91	·74	·61	·78	·88	·71	·54	·30	·07	·10	·44	·85	·54
July	1012·81 +	·71	·51	·34	·51	·68	·58	·47	·34	·14	·10	·34	·71	·44
Aug.	1009·94 +	·61	·34	·10	·24	·51	·51	·41	·24	·07	·07	·44	·64	·34
Sept.	1011·39 +	·74	·54	·20	·24	·47	·54	·41	·20	·00	·17	·64	·78	·37
Oct.	1007·94 +	·54	·30	·07	·14	·54	·68	·51	·20	·07	·41	·54	·58	·34
Nov.	1011·02 +	·64	·37	·10	·07	·44	·68	·51	·10	·14	·44	·61	·64	·37
Dec.	1005·40 +	·64	·41	·14	·03	·24	·68	·51	·17	·30	·47	·71	·74	·41

NORMAL PRESSURE AT SEA-LEVEL

Authority: see p. 288.

FIGURE 147

Millibars.

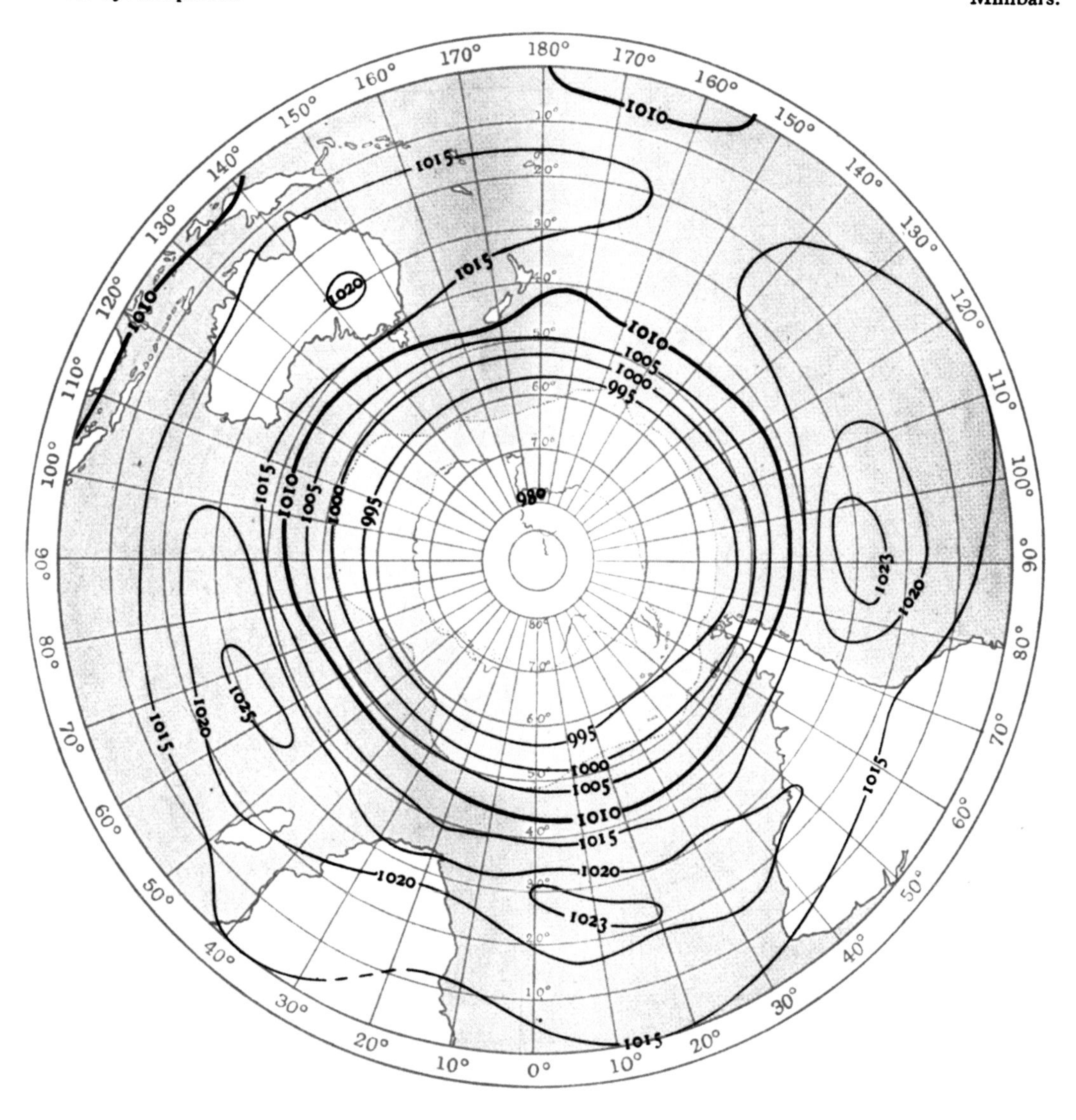

at Fort William and Ben Nevis. 1891–1903. (G.M.T.)

Ben Nevis, 56° 48′ N, 5° W, 1343 m. The values are corrected for seasonal change and reduced to 32° F. The figures in each row are excesses above the minimum for the month, the maximum figure gives the range.

Month	Min.	Mdt	2	4	6	8	10	12	14	16	18	20	22	Mean
Jan.	853·66 +	·64	·41	·20	·00	·24	·58	·58	·41	·58	·74	·85	·85	·51
Feb.	853·93 +	·47	·27	·03	·03	·24	·54	·64	·34	·24	·47	·58	·54	·37
Mar.	851·59 +	·64	·41	·00	·03	·27	·47	·74	·64	·44	·64	·78	·74	·51
Apr.	855·93 +	·64	·30	·03	·14	·44	·74	·95	1·02	·88	·88	·91	·85	·64
May	859·89 +	·68	·34	·03	·10	·37	·58	·85	·98	·85	·78	·88	·91	·61
June	861·85 +	·68	·27	·00	·10	·37	·58	·81	·91	·85	·78	·81	·91	·58
July	860·80 +	·64	·30	·03	·07	·37	·58	·78	·95	·85	·74	·74	·81	·58
Aug.	858·16 +	·68	·34	·00	·07	·41	·61	·85	·91	·81	·74	·88	·88	·61
Sept.	858·64 +	·74	·41	·07	·03	·30	·54	·71	·74	·61	·64	·91	·91	·54
Oct.	853·93 +	·51	·30	·00	·03	·41	·58	·71	·61	·51	·68	·71	·61	·47
Nov.	855·86 +	·54	·30	·07	·03	·37	·61	·51	·27	·27	·51	·61	·61	·41
Dec.	849·90 +	·58	·37	·14	·00	·17	·58	·51	·27	·34	·54	·68	·68	·41

FIGURE 148 NORMAL PRESSURE AT SEA-LEVEL

Millibars. *Authority:* see p. 288.

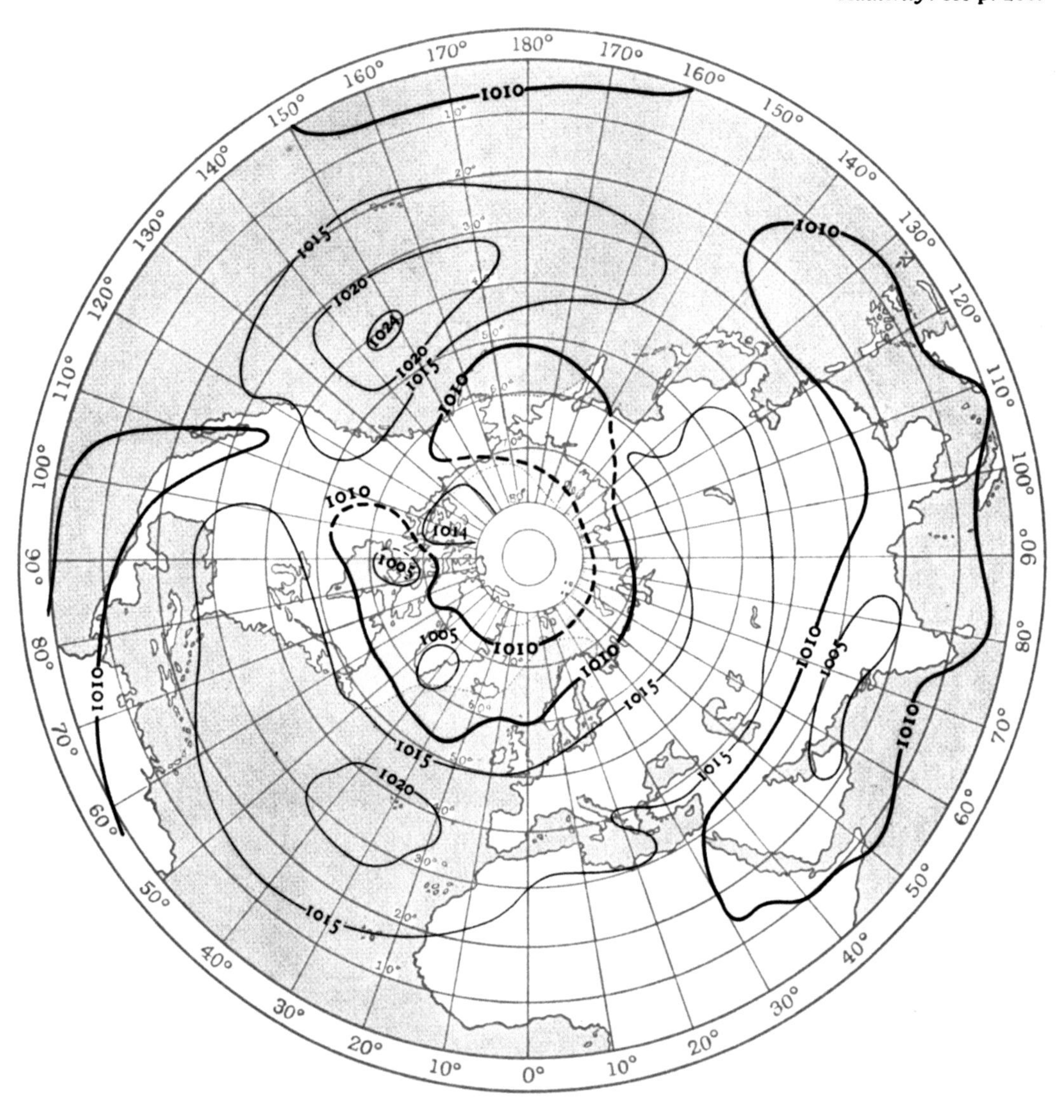

Diurnal and seasonal variation of pressure in millibars

Clermont Ferrand, 45° 46′ N, 3° 5′ E, 388 m, 1881–1900. The correction to normal gravity (Cg = −·01 mm = −·013 mb) and the reduction to sea-level have not been made. Values underlined are below the mean for the month.

Month	Min. 1876–1900	0	2	4	6	8	10	12	14	16	18	20	22	Mean
Jan.	972·52 +	·75	·57	·36	·23	·67	1·04	·48	·00	·16	·52	·76	·81	·53
Feb.	971·16 +	1·01	·79	·55	·55	·92	1·23	·87	·12	·05	·52	·93	1·07	·72
Mar.	967·95 +	1·16	·89	·60	·69	1·05	1·29	·88	·19	·00	·40	·93	1·19	·77
Apr.	966·10 +	1·33	1·08	·85	1·04	1·29	1·28	·80	·28	·00	·28	·92	1·39	·88
May	968·69 +	1·25	·92	·77	1·01	1·19	1·11	·71	·27	·00	·23	·88	1·41	·81
June	970·78 +	1·15	·81	·75	·99	1·17	1·05	·71	·29	·03	·15	·67	1·21	·75
July	971·64 +	1·24	·93	·79	1·05	1·24	1·11	·77	·39	·03	·11	·68	1·24	·80
Aug.	971·21 +	1·21	·96	·80	·99	1·29	1·27	·77	·32	·03	·04	·71	1·21	·80
Sept.	971·77 +	1·24	·96	·71	·93	1·33	1·40	·85	·32	·00	·32	·93	1·28	·85
Oct.	969·86 +	·87	·51	·31	·41	·88	1·13	·55	·04	·07	·51	·88	1·03	·60
Nov.	970·49 +	·88	·65	·41	·37	·79	1·17	·51	·00	·13	·57	·89	·97	·61
Dec.	971·43 +	·75	·55	·28	·31	·79	1·24	·56	·00	·20	·52	·80	·89	·57

NORMAL PRESSURE AT SEA-LEVEL

FIGURE 149

Authority: see p. 288.

Millibars.

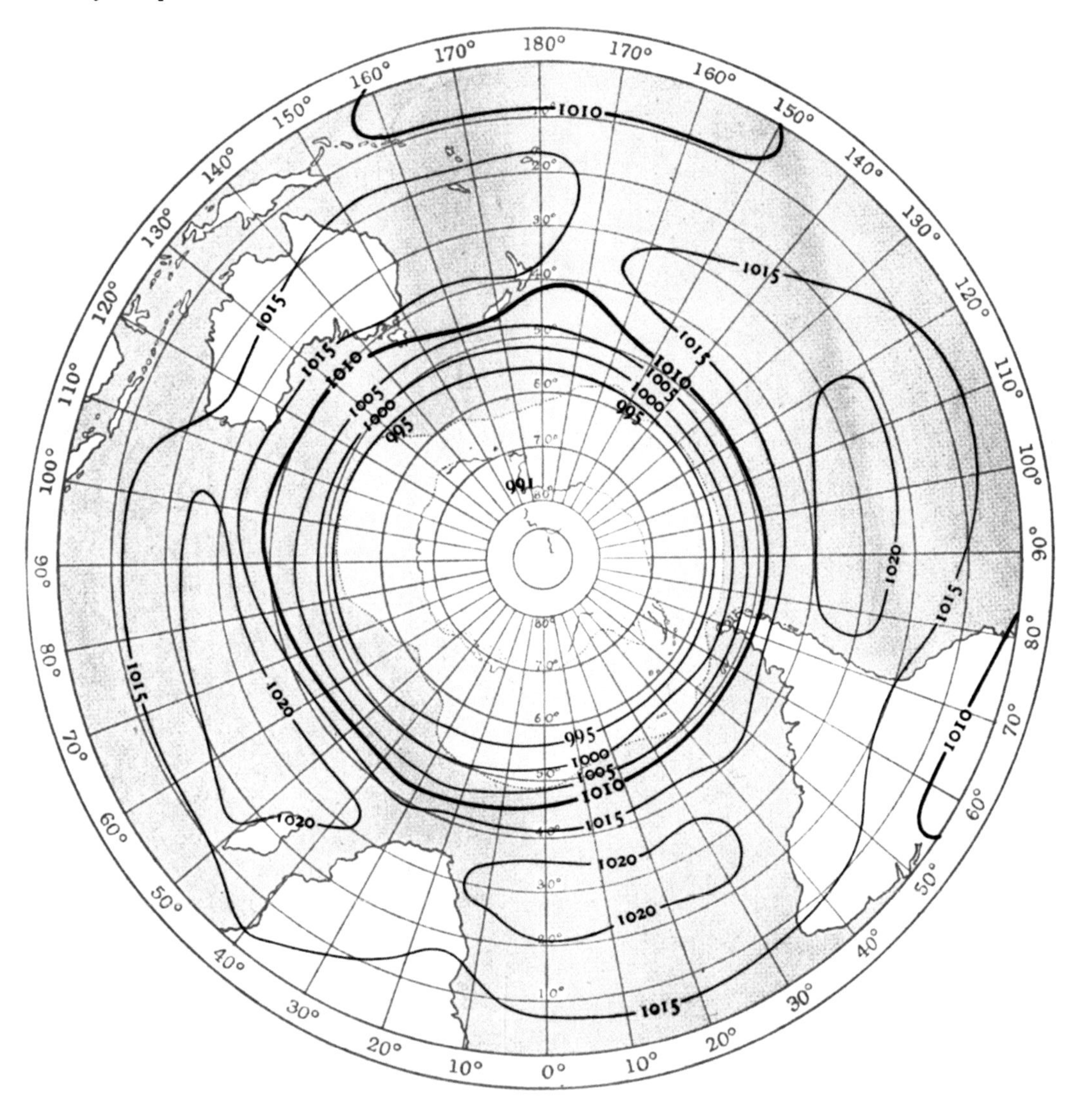

at Clermont Ferrand and Puy de Dôme. (L.M.T.)

Puy de Dôme, 45° 46′ N, 2° 58′ E, 1467 m, 1881–1900. The correction to normal gravity (Cg = −·14 mm = ·19 mb) and reduction to sea-level have not been made. The figures in each row are excesses above the minimum for the month.

Month	Min.	0	2	4	6	8	10	12	14	16	18	20	22	Mean
Jan.	849·09 +	·57	·36	·13	·04	·39	·83	·61	·33	·41	·59	·75	·75	·48
Feb.	849·25 +	·65	·39	·05	·03	·31	·73	·81	·44	·35	·49	·65	·69	·47
Mar.	846·69 +	·87	·45	·07	·04	·37	·84	1·00	·77	·65	·83	1·04	1·07	·67
Apr.	847·11 +	·87	·41	·04	·11	·49	·92	·97	·76	·65	·67	1·03	1·08	·67
May	850·36 +	·79	·37	·00	·15	·52	·87	·97	·87	·72	·71	·95	1·09	·67
June	853·94 +	·73	·25	·00	·13	·49	·85	·99	·87	·72	·63	·84	1·00	·63
July	855·45 +	·79	·29	·00	·12	·48	·81	1·00	·93	·77	·61	·84	1·00	·64
Aug.	855·42 +	·77	·35	·03	·12	·56	·95	1·09	1·01	·83	·68	·92	1·01	·69
Sept.	854·72 +	·80	·37	·04	·07	·56	1·00	1·00	·80	·61	·65	·91	1·00	·65
Oct.	850·49 +	·68	·28	·03	·03	·48	1·00	·88	·69	·61	·75	·97	·97	·61
Nov.	850·25 +	·64	·32	·07	·03	·40	·79	·61	·37	·37	·60	·77	·79	·48
Dec.	849·43 +	·51	·31	·12	·07	·43	·81	·56	·20	·28	·43	·59	·67	·41

FIGURE 150

Millibars.

NORMAL PRESSURE AT SEA-LEVEL

Authority: see p. 288.

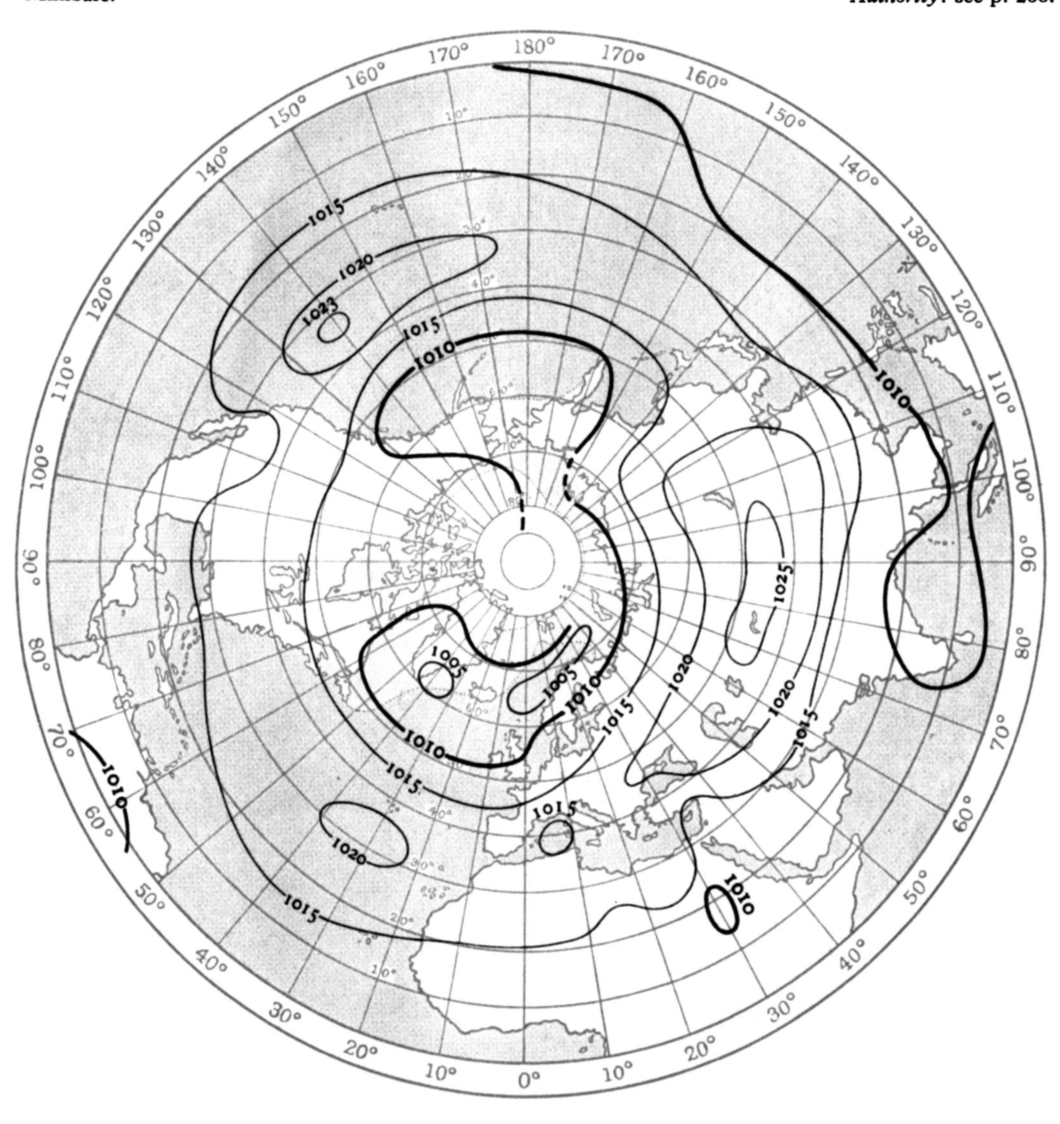

Diurnal and seasonal variation of pressure in millibars

Bureau Central Météorologique, 48° 52′ N, 2° 18′ E, 33 m. The correction to normal gravity (Cg = + ·26 mm = ·35 mb) and the reduction to sea-level have not been made. Values underlined are below the mean for the month.

Month	Min. 1851–1900	0	2	4	6	8	10	12	14	16	18	20	22	Mean
Jan.	1013·42 +	·47	·40	·20	·05	·48	·93	·57	·00	·11	·29	·51	·59	·39
Feb.	1013·18 +	·68	·49	·21	·24	·63	·99	·83	·16	·00	·31	·53	·65	·48
Mar.	1009·88 +	·76	·59	·33	·51	·93	1·20	·96	·35	·00	·28	·67	·80	·61
Apr.	1009·38 +	·96	·72	·56	·81	1·19	1·31	·92	·44	·01	·07	·64	·92	·71
May	1010·23 +	1·04	·84	·72	1·00	1·27	1·20	·85	·43	·07	·11	·60	1·01	·76
June	1011·78 +	·99	·80	·72	·99	1·24	1·19	·95	·51	·11	·08	·45	·95	·75
July	1011·82 +	·95	·72	·60	·84	1·09	1·05	·81	·47	·11	·04	·43	·89	·67
Aug.	1011·40 +	·96	·80	·68	·95	1·23	1·35	·96	·52	·12	·00	·53	·89	·75
Sept.	1012·22 +	·80	·67	·51	·75	1·13	1·33	·92	·39	·00	·09	·59	·83	·67
Oct.	1010·30 +	·71	·52	·28	·29	·89	1·07	·69	·17	·00	·36	·61	·77	·53
Nov.	1011·36 +	·51	·33	·15	·19	·69	1·01	·52	·03	·05	·37	·56	·61	·41
Dec.	1012·95 +	·64	·53	·36	·29	·65	1·11	·59	·00	·12	·39	·65	·73	·51

NORMAL PRESSURE AT SEA-LEVEL

FIGURE 151

Authority: see p. 288.

Millibars.

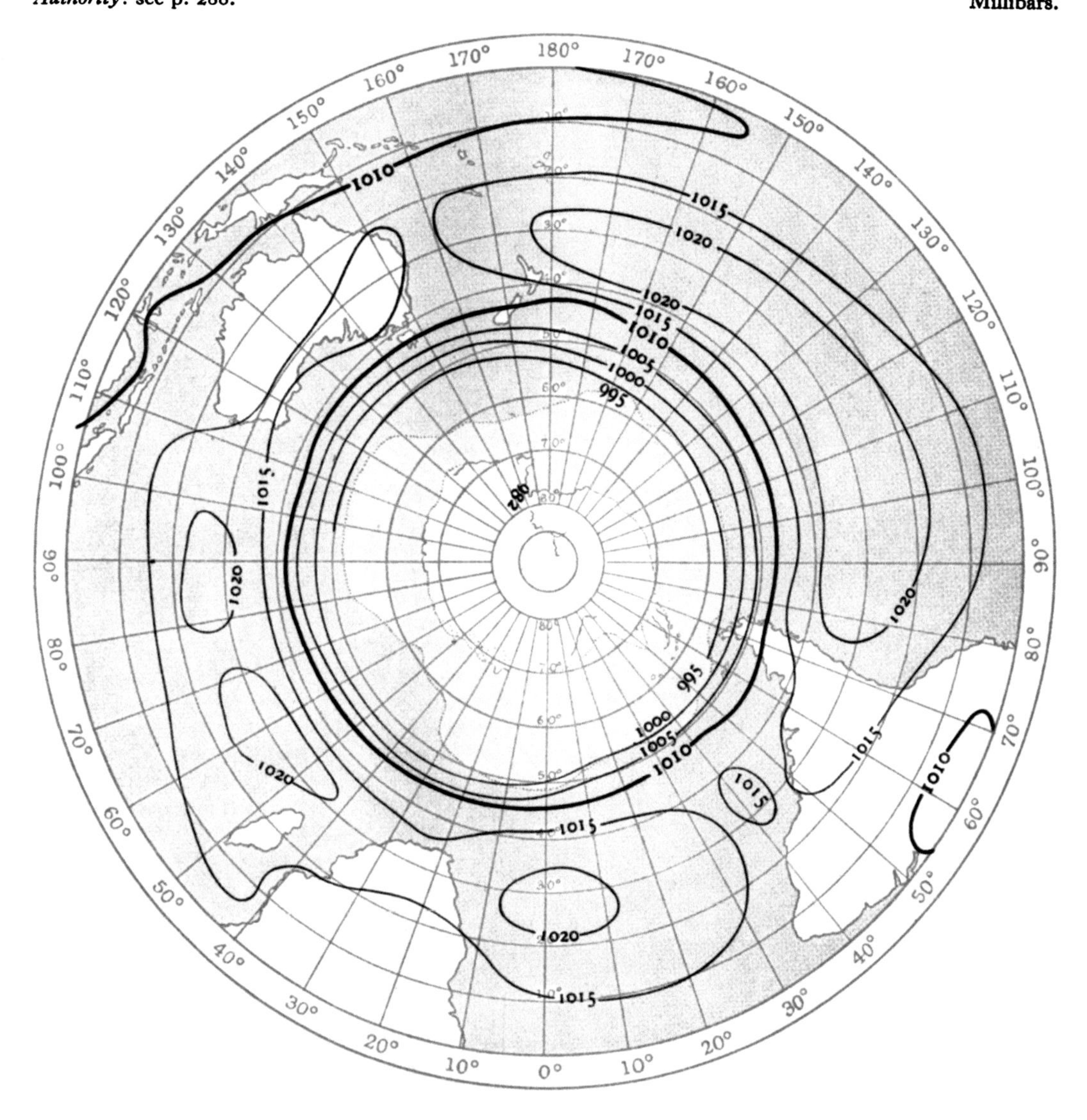

at the Bureau Central and the Eiffel Tower. 1891–1900. (L.M.T.)

Eiffel Tower, 48° 52′ N, 2° 17′ E, 313 m. The correction to standard gravity (Cg = ·18 mm = ·24 mb) and the reduction to sea-level have not been made. The figures in each row are excesses above the minimum for the month.

Month	Min. 1891–1900	0	2	4	6	8	10	12	14	16	18	20	22	Mean
Jan.	978·53 +	·51	·41	·15	·00	·41	·85	·60	·11	·24	·41	·60	·64	·41
Feb.	979·00 +	·57	·32	·00	·03	·40	·75	·69	·19	·07	·36	·53	·60	·37
Mar.	976·00 +	·57	·32	·00	·15	·60	·93	·83	·31	·00	·31	·61	·71	·44
Apr.	977·13 +	·71	·32	·08	·32	·75	1·03	·75	·35	·00	·03	·55	·79	·47
May	977·50 +	·81	·49	·27	·55	·92	·97	·73	·37	·08	·09	·55	·91	·56
June	979·25 +	·73	·40	·24	·48	·84	·92	·77	·43	·11	·08	·43	·84	·52
July	979·64 +	·72	·33	·12	·33	·71	·80	·67	·43	·09	·05	·40	·79	·45
Aug.	979·27 +	·69	·41	·17	·41	·79	1·03	·77	·43	·12	·03	·51	·76	·51
Sept.	980·21 +	·53	·28	·03	·24	·69	1·01	·77	·31	·00	·11	·53	·65	·43
Oct.	976·69 +	·53	·27	·01	·03	·59	·84	·61	·17	·07	·41	·61	·69	·40
Nov.	978·81 +	·48	·24	·01	·04	·51	·85	·53	·11	·17	·47	·61	·61	·39
Dec.	978·88 +	·53	·41	·19	·12	·45	·89	·48	·00	·15	·37	·57	·65	·40

FIGURE 152 — NORMAL PRESSURE AT SEA-LEVEL

Millibars.

Authority: see p. 288.

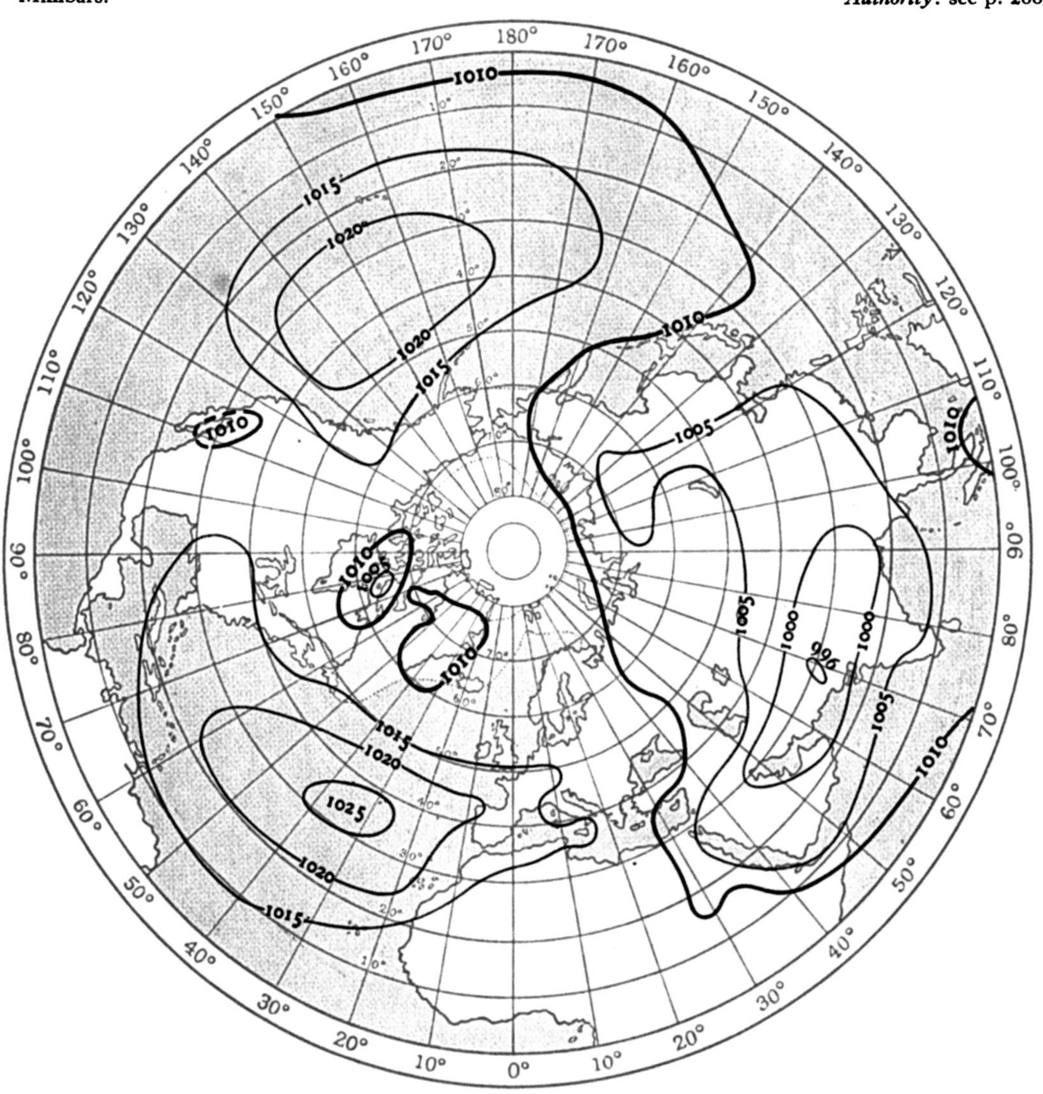

Diurnal and seasonal variation of pressure in millibars

Lahore, 74° 20′ E, 31° 34′ N, 214 m, 1880–8 (4 days per month). Corrected for seasonal change and reduced to 32° F. L.M.T. Values underlined are below the mean for the month.

Month	Min.	0	2	4	6	8	10	12	14	16	18	20	22	Mean
Jan.	991·21 +	1·19	·98	·47	·68	1·76	2·78	1·90	·54	·00	·24	·95	1·35	1·08
Feb.	989·75 +	1·25	·81	·30	·61	1·69	2·81	2·40	·91	·03	·20	·85	1·35	1·12
Mar.	985·93 +	1·49	1·19	·81	1·19	2·44	2·95	2·57	1·15	·07	·27	·98	1·59	1·39
Apr.	981·05 +	1·52	1·08	·98	1·59	2·64	3·25	2·68	1·32	·17	·17	1·02	1·66	1·52
May	976·41 +	1·46	·91	1·15	2·03	2·91	3·22	2·81	1·46	·24	·07	·81	1·56	1·56
June	971·81 +	1·35	1·22	1·42	2·40	3·39	3·52	3·12	1·73	·44	·00	·64	1·35	1·73
July	972·42 +	1·56	1·35	1·46	2·10	3·01	3·18	2·74	1·66	·34	·03	·68	1·46	1·63
Aug.	974·62 +	1·39	1·15	1·15	1·86	3·01	3·39	2·84	1·56	·27	·03	·78	1·46	1·59
Sept.	979·93 +	1·15	·91	·85	1·69	2·71	3·22	2·57	1·02	·07	·14	·88	1·12	1·35
Oct.	986·03 +	·98	·78	·71	1·46	2·57	3·18	2·30	·78	·07	·17	·85	1·12	1·25
Nov.	990·67 +	1·15	·81	·58	1·05	2·13	2·78	1·76	·44	·00	·30	·95	1·29	1·12
Dec.	992·02 +	1·12	·98	·61	·85	2·00	2·98	2·03	·61	·00	·24	·85	1·29	1·15

NORMAL PRESSURE AT SEA-LEVEL

FIGURE 153

Authority: see p. 288.

Millibars.

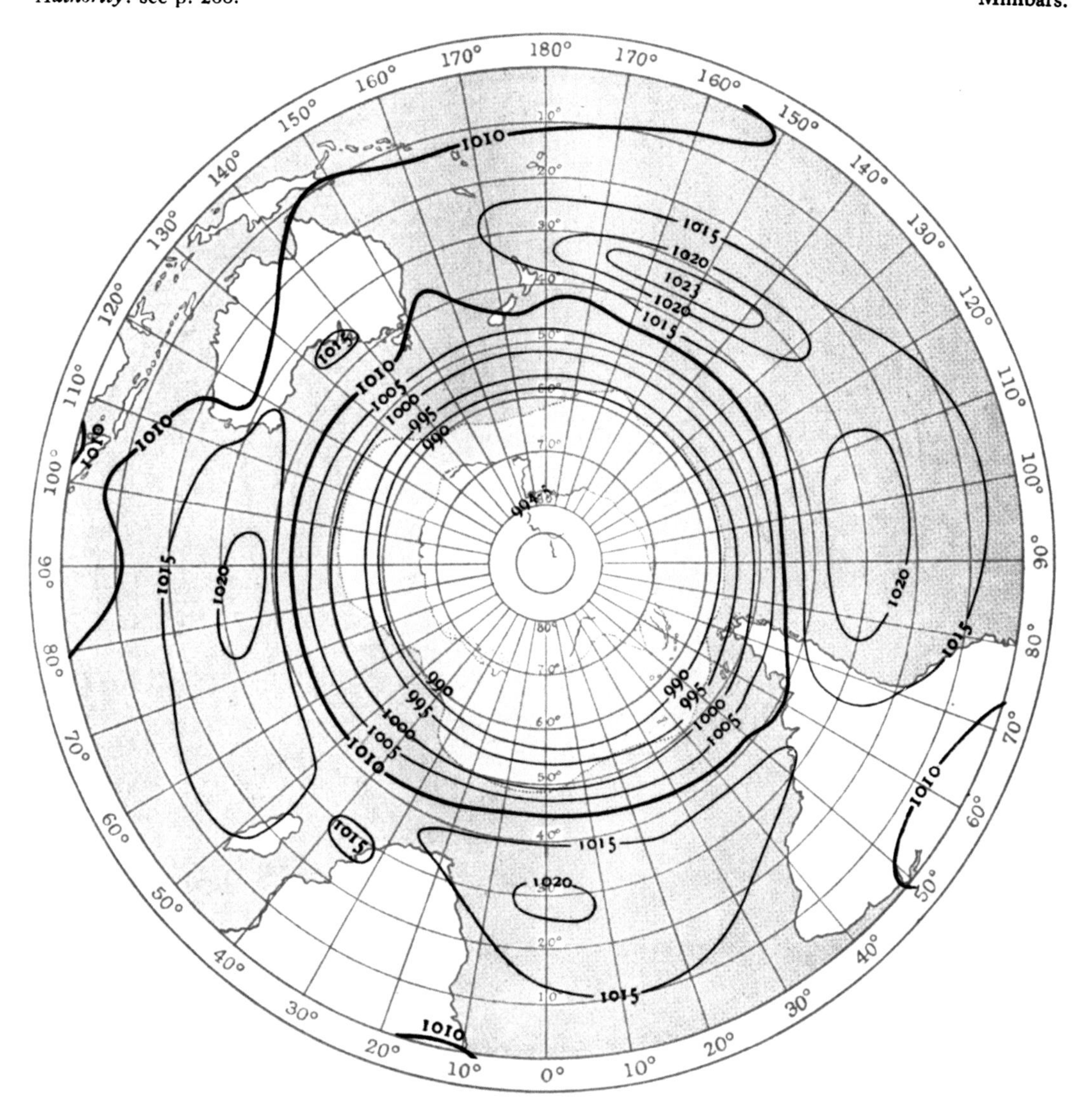

at Lahore and Leh.

Leh, 77° 42′ E, 34° 10′ N, 3506 m, 1876–91 (4 days per month). Corrected for seasonal change and reduced to 32° F. L.M.T. The figures in each row are excesses above the minimum for the month.

Month	Min.	0	2	4	6	8	10	12	14	16	18	20	22	Mean
Jan.	663·68 +	1·05	·98	·88	1·22	1·96	2·61	1·46	·37	·00	·37	·81	1·08	1·05
Feb.	661·79 +	1·52	1·32	1·15	1·46	2·20	2·57	1·86	·47	·00	·34	·95	1·35	1·25
Mar.	664·16 +	1·76	1·66	1·59	1·96	3·01	3·32	2·23	·88	·00	·37	·91	1·49	1·59
Apr.	664·97 +	1·63	1·59	1·66	2·07	2·81	2·84	1·79	·61	·00	·07	·78	1·39	1·42
May	664·66 +	2·07	1·93	1·86	2·34	2·91	2·74	1·79	·71	·00	·37	1·08	1·76	1·63
June	663·44 +	1·73	1·83	1·96	2·57	3·25	2·98	2·03	·81	·17	·20	·71	1·32	1·63
July	661·62 +	2·10	2·34	2·57	3·15	3·79	3·56	2·40	1·05	·17	·14	·88	1·63	1·96
Aug.	661·08 +	2·20	2·47	2·64	3·01	3·66	3·59	2·37	1·05	·00	·17	1·19	1·83	2·03
Sept.	664·70 +	2·00	2·27	2·51	3·12	3·93	3·93	2·64	·95	·00	·20	1·12	1·59	2·03
Oct.	666·22 +	1·90	2·03	2·27	2·68	3·49	3·59	2·30	·81	·00	·20	1·08	1·66	1·83
Nov.	666·60 +	1·73	1·76	1·83	2·20	3·05	3·28	2·00	·47	·00	·34	·98	1·35	1·59
Dec.	665·38 +	1·25	1·29	1·35	1·69	2·44	3·08	1·86	·51	·00	·30	·81	1·22	1·32

FIGURE 154 NORMAL PRESSURE AT SEA-LEVEL

Millibars. *Authority:* see p. 288.

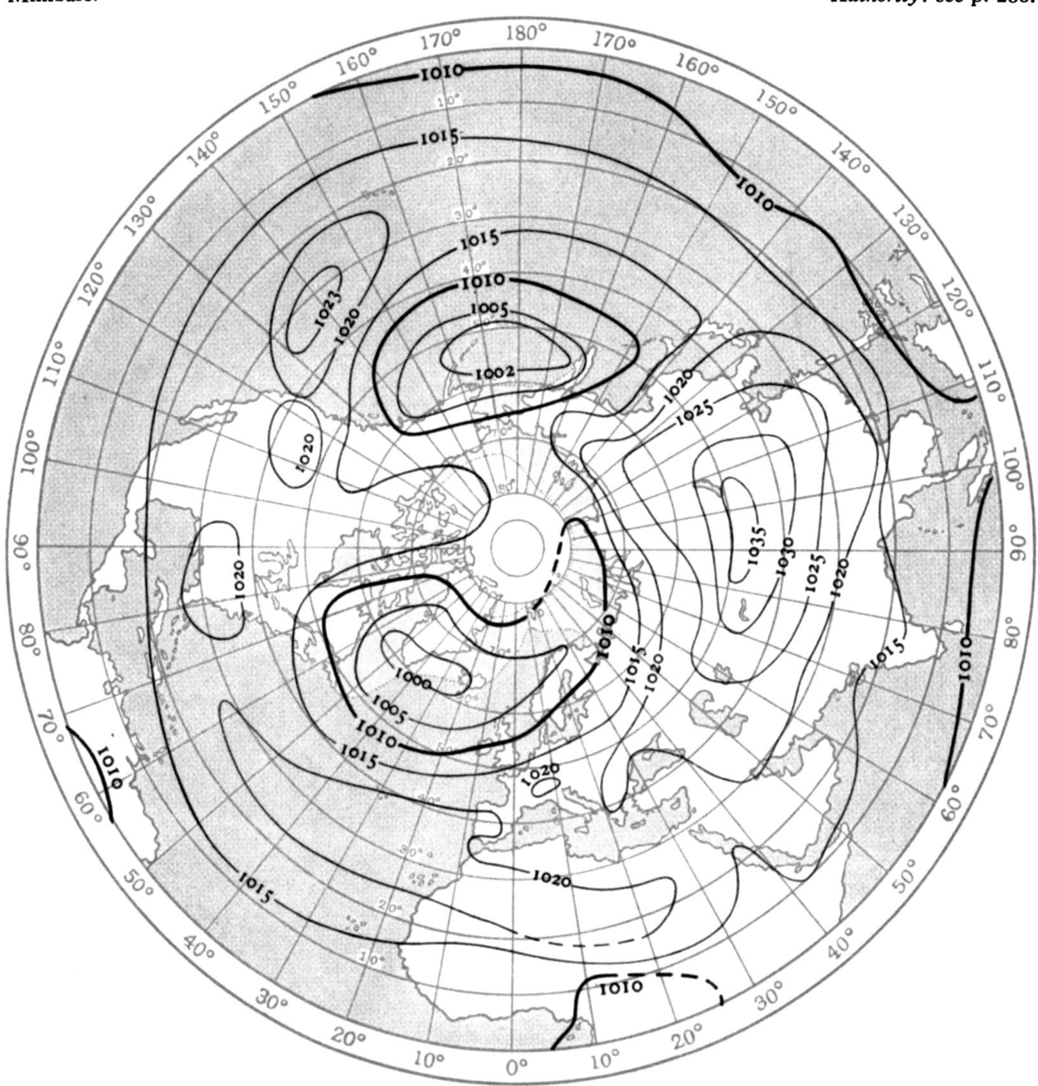

Diurnal and seasonal variation of pressure in millibars at Pike's Peak, 38° 50′ N, 105° 2′ W, 4308 m.

Nov. 1892 to Sept. 1894. The figures in each row are excesses above the minimum for the month. Local time.

Month	Min.	0	2	4	6	8	10	12	14	16	18	20	22	Mean
Jan.	592·27 +	·85	·41	·10	·27	1·08	1·83	1·42	1·15	1·22	1·32	1·32	1·15	1·02
Feb.	590·24 +	1·02	·54	·00	·30	·88	1·42	1·46	1·05	1·15	1·15	1·42	1·32	·98
Mar.	591·86 +	1·15	·54	·00	·47	1·29	2·07	2·34	2·30	2·37	2·51	2·30	2·03	1·63
Apr.	595·38 +	1·29	·61	·27	·20	1·15	1·59	1·76	1·96	1·96	2·10	2·17	2·00	1·42
May	600·90 +	·98	·37	·00	·37	1·08	1·59	1·93	1·86	1·66	1·69	1·76	1·76	1·25
June	606·15 +	·74	·24	·00	·20	·88	1·39	1·56	1·56	1·42	1·19	1·35	1·56	1·02
July	612·18 +	·41	·07	·00	·14	·54	·95	1·15	·95	·61	·41	·54	·74	·54
Aug.	611·57 +	·68	·27	·07	·03	·47	·95	1·08	·95	·68	·41	·61	·74	·58
Sept.	605·95 +	·88	·44	·10	·07	·51	1·22	1·56	1·49	1·15	1·08	1·02	·88	·88
Oct.	601·65 +	·85	·37	·00	·24	·91	1·59	1·96	1·73	1·69	1·73	1·59	1·32	1·15
Nov.	596·53 +	·61	·27	·00	·14	·85	1·35	·95	·88	1·02	1·22	1·35	1·02	·81
Dec.	593·62 +	·81	·27	·00	·30	·81	1·46	1·08	·81	·95	1·22	1·42	1·22	·88

NORMAL PRESSURE AT SEA-LEVEL

FIGURE 155

Authority: see p. 288.

Millibars.

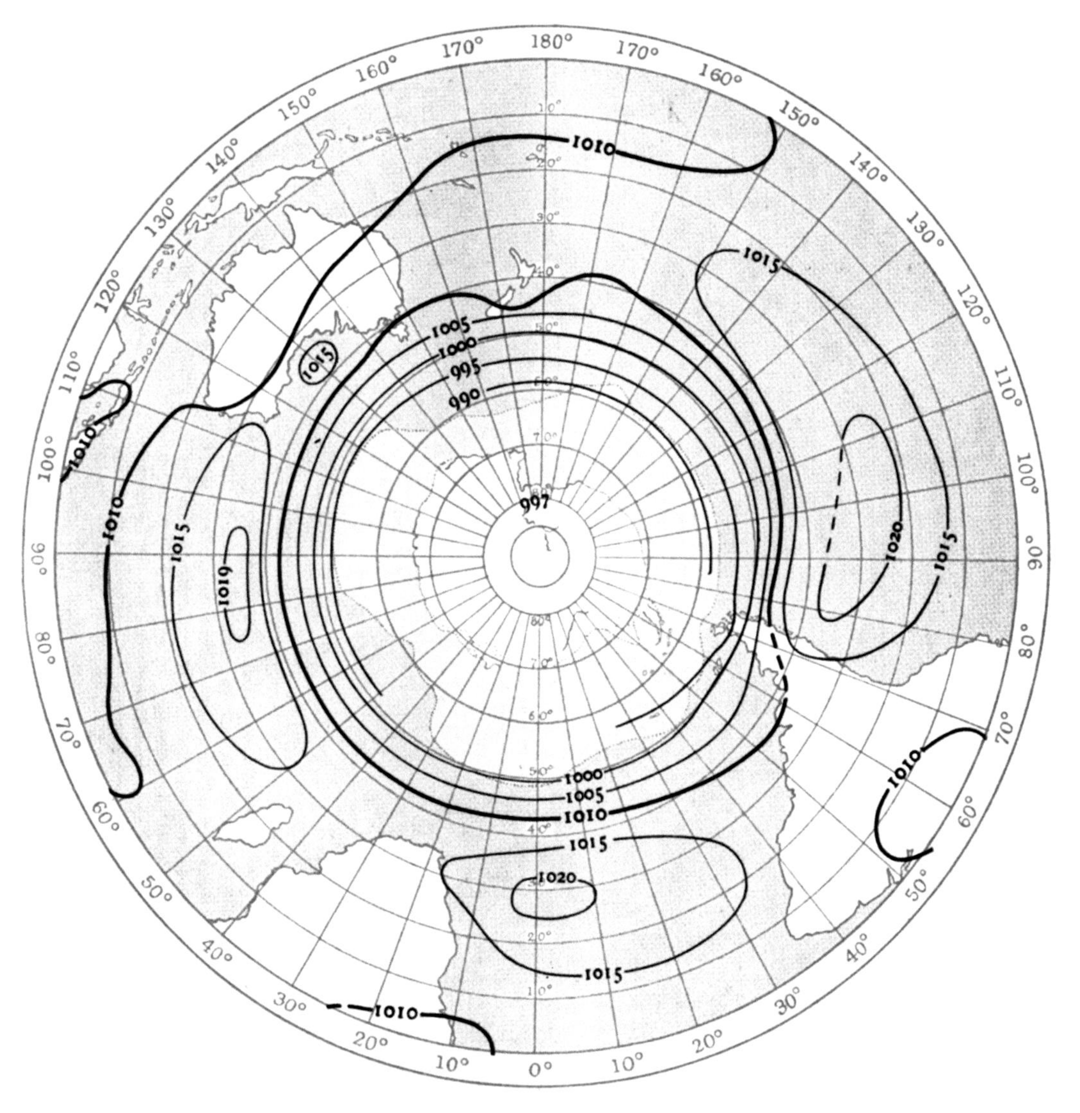

Seasonal and diurnal variation of pressure in millibars at Sonnblick, 47° 3′ N, 12° 57′ E, 3106·5 m, 1909–18.

The figures in each row are excesses above the minimum for the month. Local time.

Month	Min.	0	2	4	6	8	10	12	14	16	18	20	22	24	Mean
Jan.	685·96 +	·53	·39	·18	0	·17	·44	·44	·19	·26	·32	·48	·49	·41	·32
Feb.	686·18 +	·64	·50	·14	0	·18	·43	·62	·44	·42	·51	·62	·67	·61	·43
Mar.	684·55 +	·56	·38	·06	0	·14	·42	·65	·59	·55	·59	·78	·83	·72	·47
Apr.	687·91 +	·75	·38	·07	0	·19	·51	·72	·82	·77	·70	·88	·91	·79	·56
May	693·11 +	·62	·28	0	·04	·27	·56	·87	·96	·92	·83	·90	·97	·80	·61
June	696·07 +	·65	·32	·00	0	·19	·47	·74	·82	·79	·68	·68	·78	·61	·51
July	697·50 +	·66	·33	·01	0	·18	·46	·75	·86	·85	·73	·81	·91	·78	·55
Aug.	698·14 +	·74	·39	·07	0	·14	·46	·71	·82	·79	·63	·75	·81	·68	·52
Sept.	696·41 +	·69	·39	·08	0	·20	·56	·75	·75	·71	·65	·79	·78	·60	·52
Oct.	693·39 +	·55	·34	·08	0	·21	·46	·49	·33	·36	·44	·55	·60	·46	·36
Nov.	687·89 +	·56	·35	·09	0	·22	·49	·42	·19	·25	·42	·58	·65	·59	·35
Dec.	687·13 +	·50	·34	·15	0	·17	·48	·33	·06	·15	·22	·36	·41	·37	·26

CHARTS OF NORMAL PRESSURE, AND WIND-ROSES IN THE INTERTROPICAL BELT.

Scale: 1 cm to 2000 km at the equator

JUNE

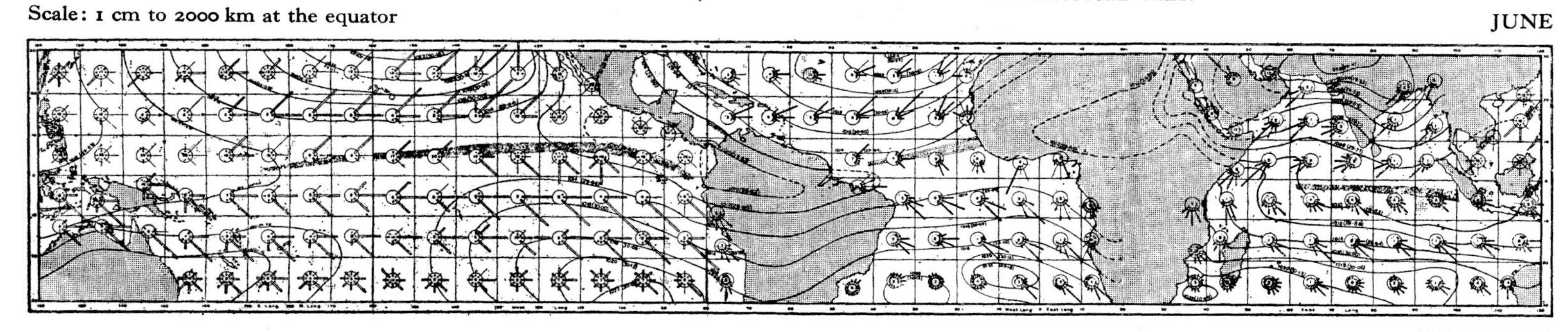

DECEMBER

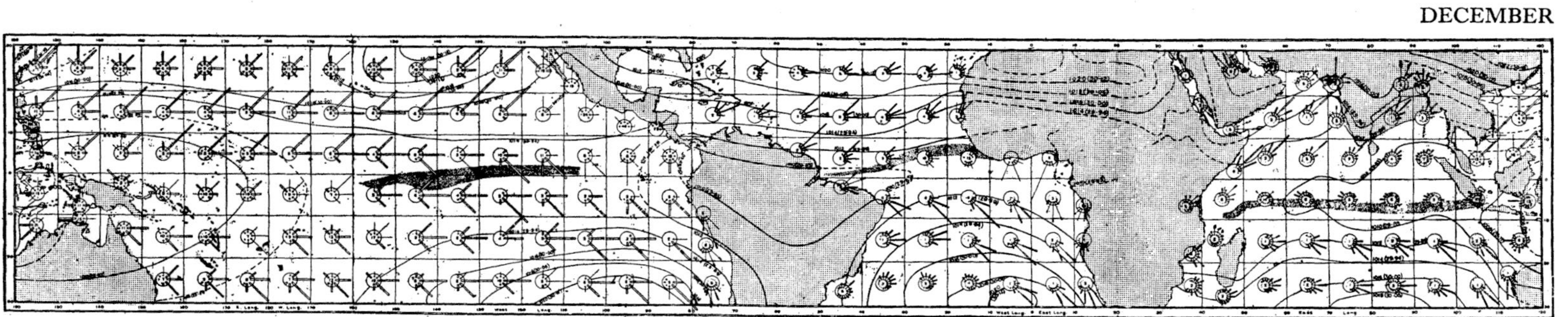

FIG. 156. WIND-ROSES FOR EACH 10° SQUARE; the lengths of the arrows represent the frequency of winds from each of the sixteen even points of the compass, viz. 0°, 22½°, 45° etc. from true North. The scale for the arrows is approximately 1 cm to 100 per cent. of frequency. For the **South Atlantic and Indian Oceans** the length of the single fine line from the point of the arrow within the circle represents the number of light winds (less than 5·5 m/s) as a percentage of the whole number of observations including calms, the double line moderate winds (5·5 to 17 m/s) and the "blocked in" outer end (if any) represents gales (greater than 17 m/s). The figure in the circle gives the percentage of calms. For **other oceans** the mean force of the winds from the several directions is represented. A single line represents a mean force of light winds and a double line a mean force of moderate wind. Where there is not space enough for an arrow of full length the lines are broken and the percentage given in figures in the break. For details see figs. 161, 162; also the *Barometer Manual for the Use of Seamen*, Plates XXXVI–XLIII, from which these charts are compiled. Isobars are drawn for even millibars (1010 to 1020 between the equator and the tropics over the Atlantic). Shaded areas show the "doldrums" of the equatorial seas.

WINDS ON THE EARTH'S SURFACE [see chap. x, note 17]

Our next step is to display what information we have about the distribution of winds beginning with two pairs of maps, figs. 157–160, transformed from the original maps of Köppen for January-February and July-August, which are reproduced in Bartholomew's *Atlas*. These are surface-winds and for the ocean areas only. The local influence of radiation and elevation upon the winds on land is so disturbing that no one has yet given us a satisfactory generalisation. With judicious reserve in the circumstances the reader may utilise the isobars as an indication of the régime to which the winds conform to some extent[1].

General relation of wind to pressure.

From an inspection of the maps of the winds over the sea in winter and summer amplified by the isobars over the land we may derive some general ideas of the circulation of air over the surface of the globe.

We remark first that a comparison of the winds with the distribution of pressure over the sea for the corresponding months (figs. 132–5 and 144–7) shows very good agreement, and there is, as a rule, fair continuity between the run of the isobars over the land and the general direction of the wind at the coast-lines where both wind and isobars are indicated. The notable exceptions are over the Indian peninsula in winter, over the Guinea coast and the corresponding coast of Central America in both summer and winter. The flow in these cases, though not very strong, is indicated as being roughly speaking at right angles to the isobars. The western margins of the tropical anticyclones are much less clearly marked by the winds than the eastern margins are. There is also a curious closed isobar in the North of Madagascar in the Southern summer which is only vaguely indicated in the winds.

With this brief indication of some interesting details of the question of the general relation of wind to the distribution of pressure, we may abstract or summarise the results by calling attention to some salient features of the general scheme of winds.

The equatorial flow westward.

There is first a general flow in the equatorial region from East to West fed by air from both North and South and, between the two, narrow belts of what Halley called "calms and tornados" and are now called the equatorial doldrums. Ships which depend on sails are "in the doldrums" when the winds are so fitful and irregular that no proper course can be kept, and progress is in consequence very slow, but the name is now practically restricted to those equatorial belts which lie between the wind-systems of the Northern and Southern Hemispheres. They fluctuate somewhat in position with the seasons. In the Atlantic with a range from 1° N to 12½° N the belt is at its furthest North in August and furthest South in March. The belts of doldrums in the months of June and December are indicated on the maps of pressure and winds over the intertropical region (fig. 156).

[1] For a detailed comparison of observed and calculated winds, see S. N. Sen, *Surface and Geostrophic Wind-Components at Deerness, Holyhead, Great Yarmouth and Scilly*, Geophysical Memoir, No. 25, M.O. publication, No. 254 e, London, 1925; also vol. IV of this *Manual*.

FIGURE 157

PREVAILING WINDS OVER THE OCEANS

After W. Köppen.

Beaufort scale.

Length of arrow, steadiness of wind

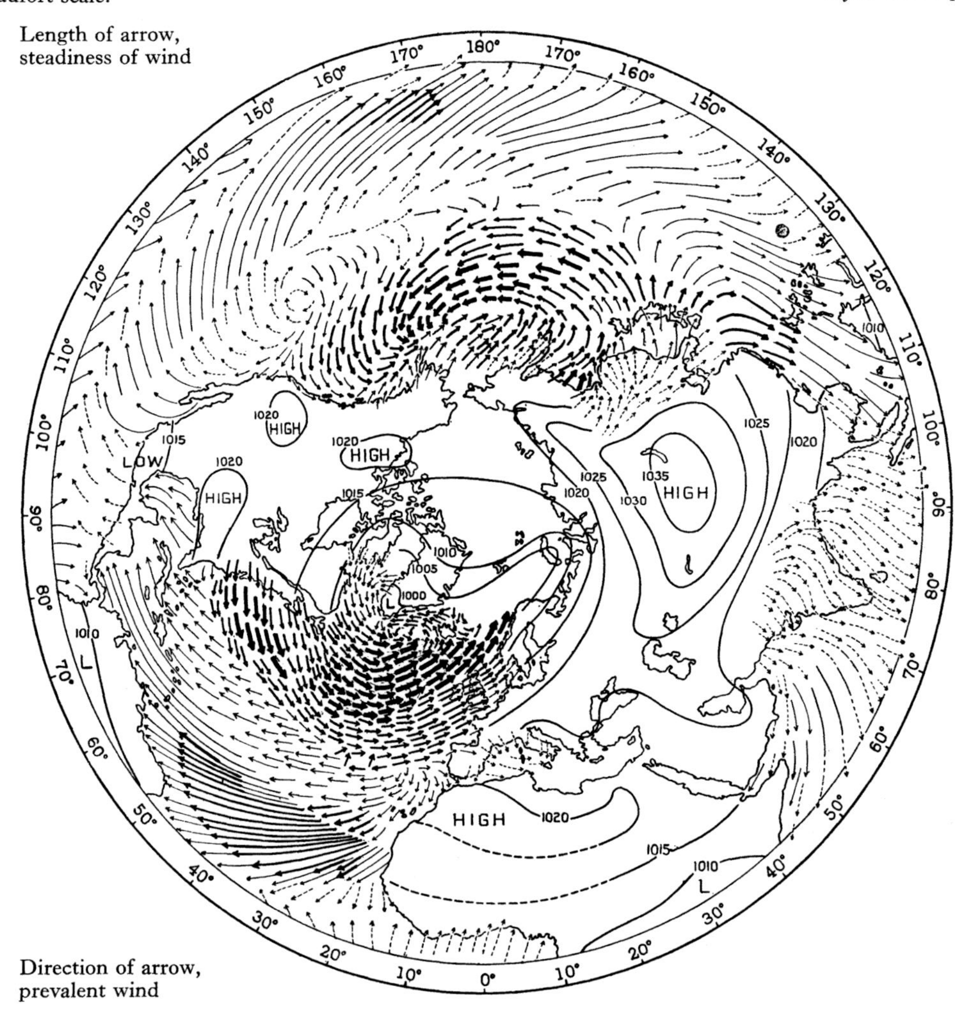

Direction of arrow, prevalent wind

TRAJECTORIES OF AIR OVER THE ATLANTIC REGION, 1882–3

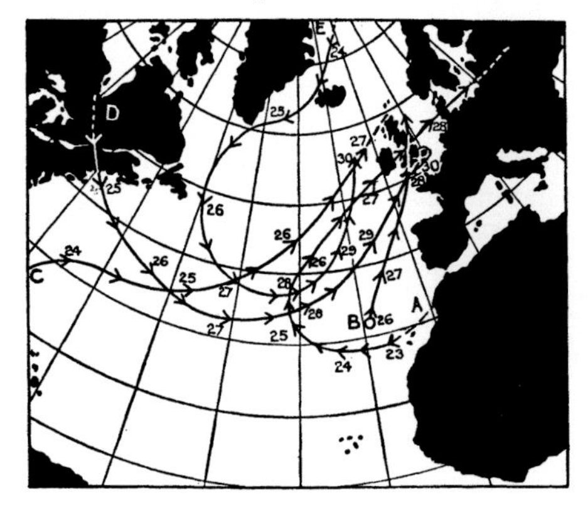

December 23–30, 1882

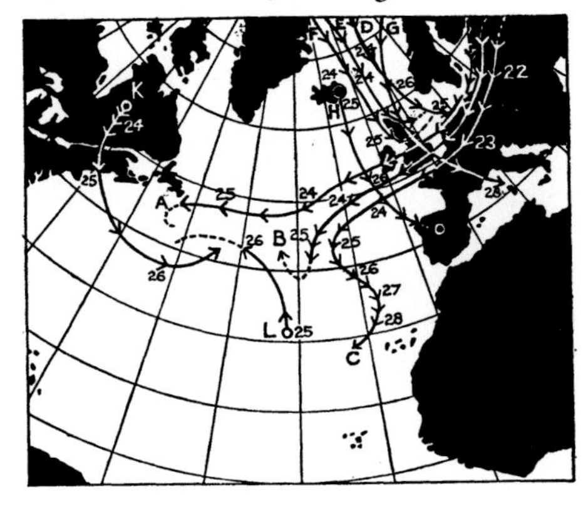

March 22–28, 1883

PREVAILING WINDS OVER THE OCEANS

After W. Köppen.

FIGURE 158

Beaufort scale.

Width of arrow, strength of wind

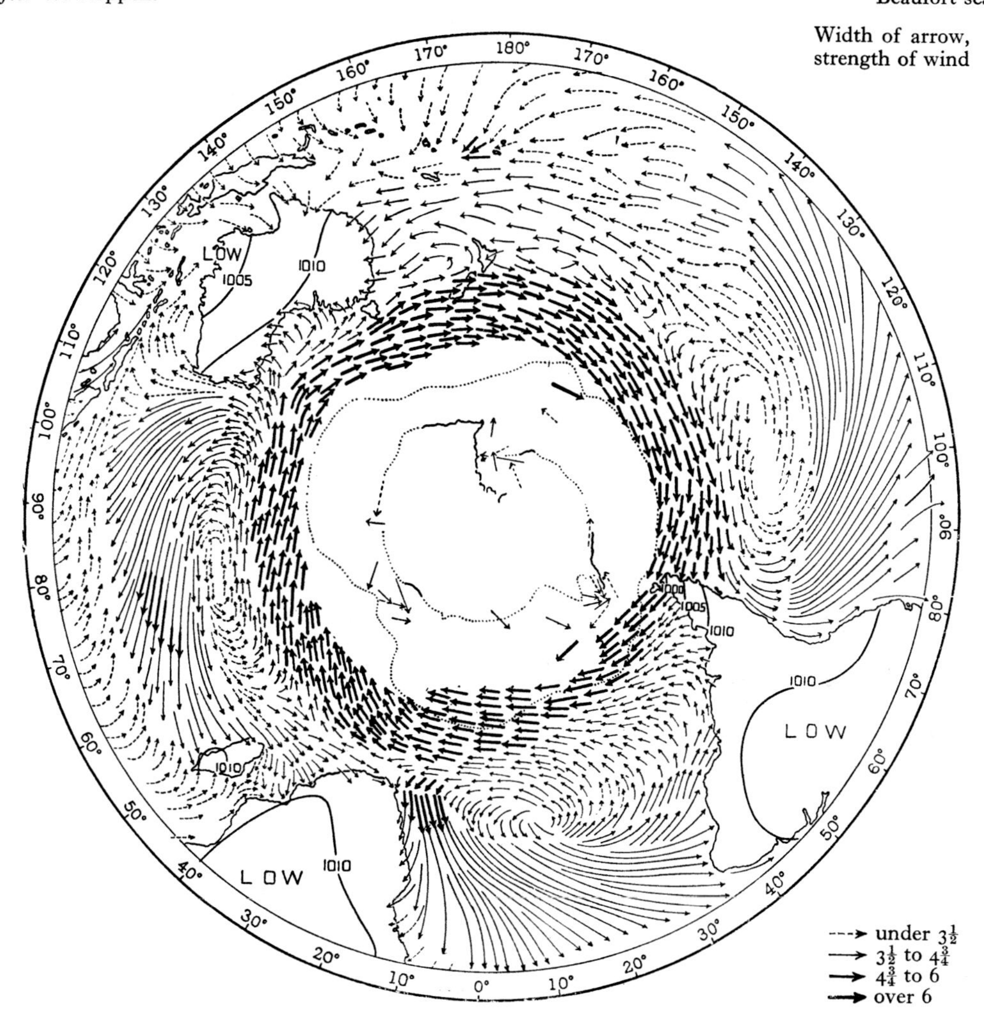

WINDS AND WEATHER ON THE SOUTH POLAR PLATEAU, 87° TO 90° S

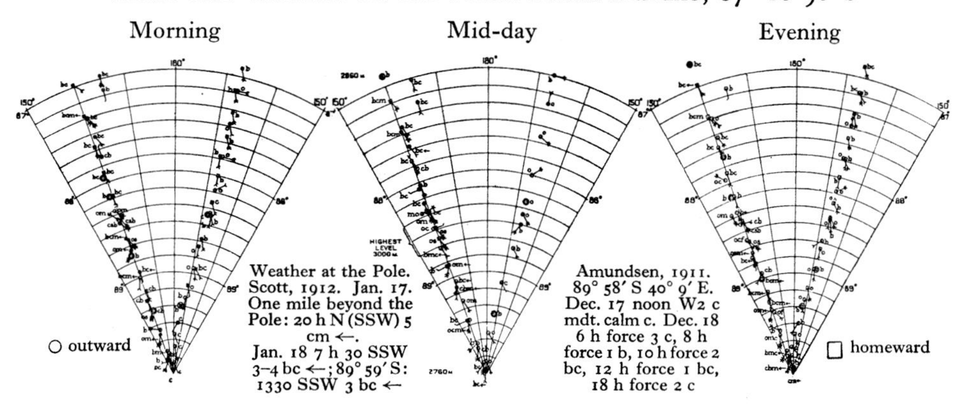

FIGURE 159 PREVAILING WINDS OVER THE OCEANS

Beaufort scale. *After* W. Köppen.

Length of arrow, steadiness of wind

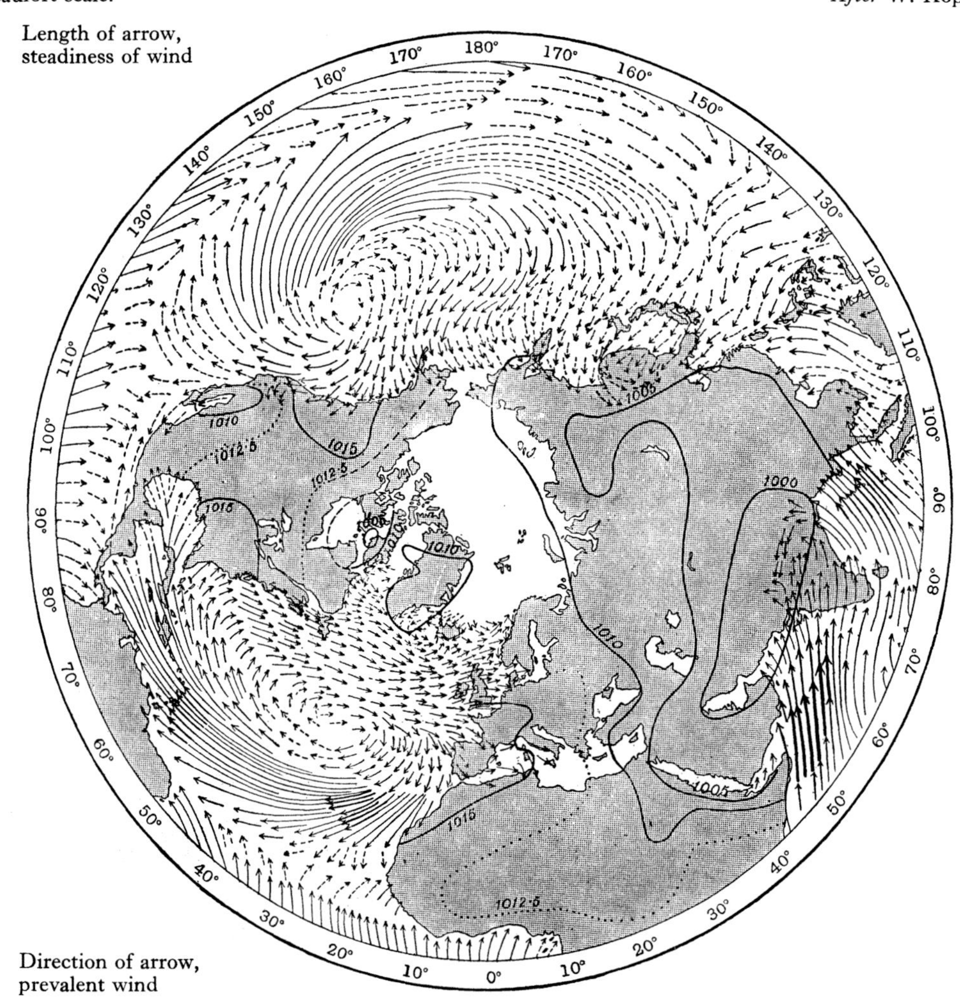

Direction of arrow, prevalent wind

TRAJECTORIES OF AIR OVER THE ATLANTIC REGION, 1882–3

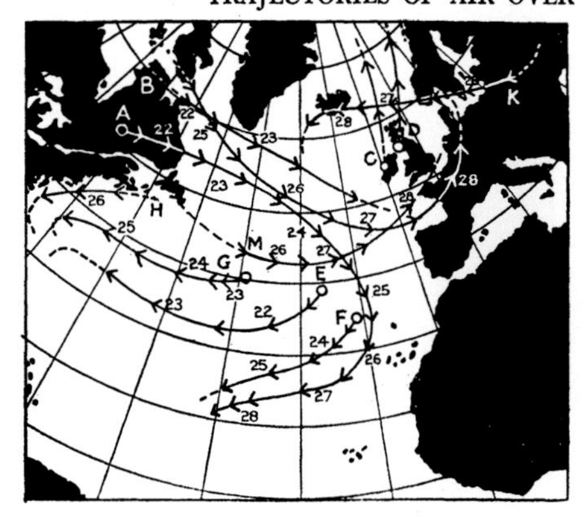

September 22–28, 1882

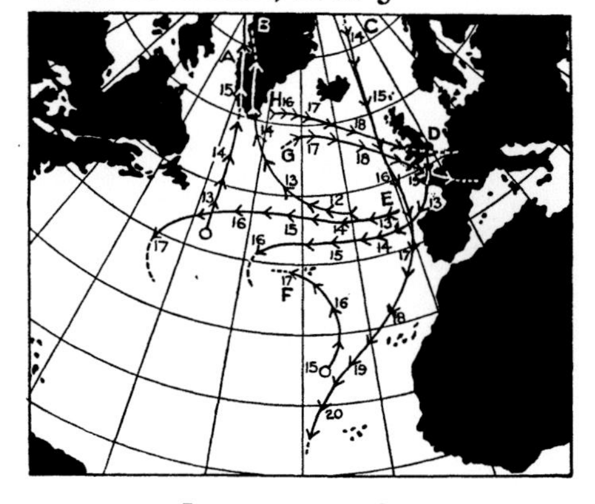

June 13–20, 1883

PREVAILING WINDS OVER THE OCEANS

After W. Köppen.

FIGURE 160

Beaufort scale.

Width of arrow, strength of wind

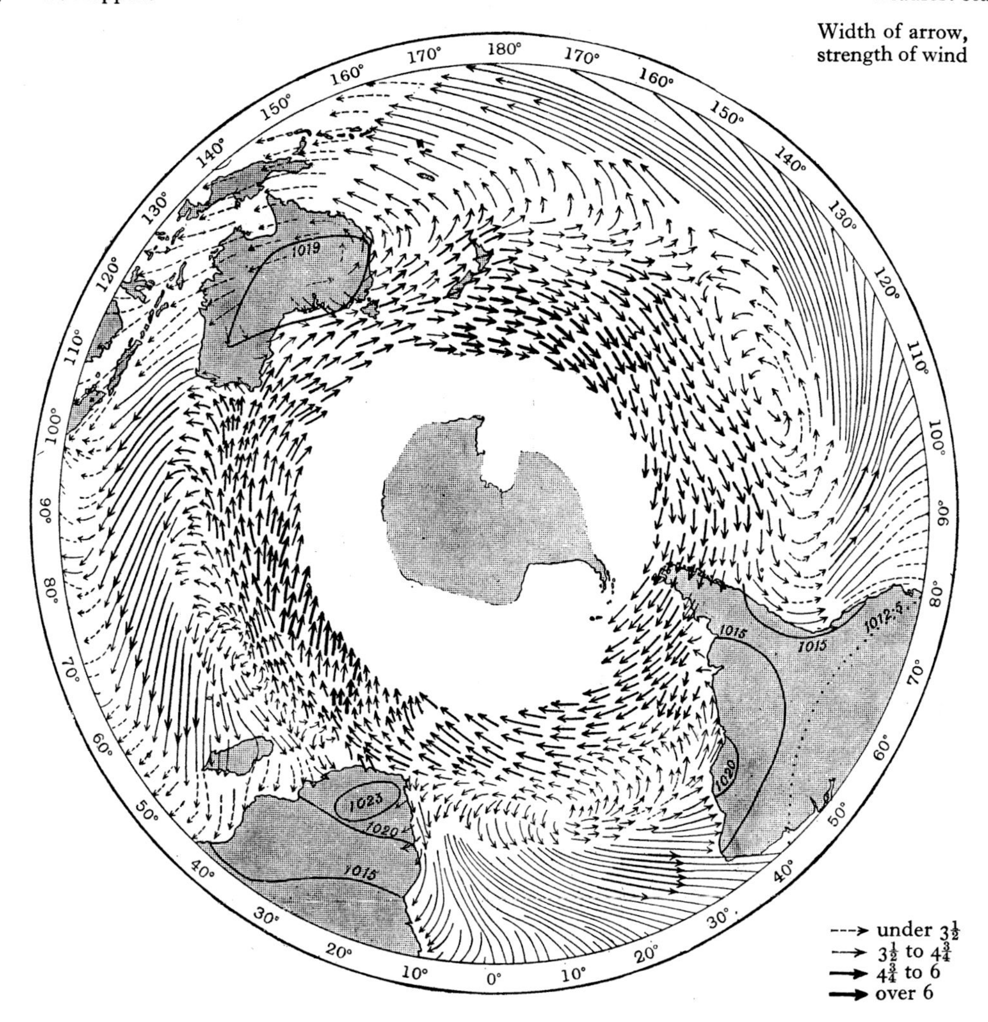

---→ under $3\frac{1}{2}$

→ $3\frac{1}{2}$ to $4\frac{3}{4}$

→ $4\frac{3}{4}$ to 6

→ over 6

WINDS OF THE ANTARCTIC

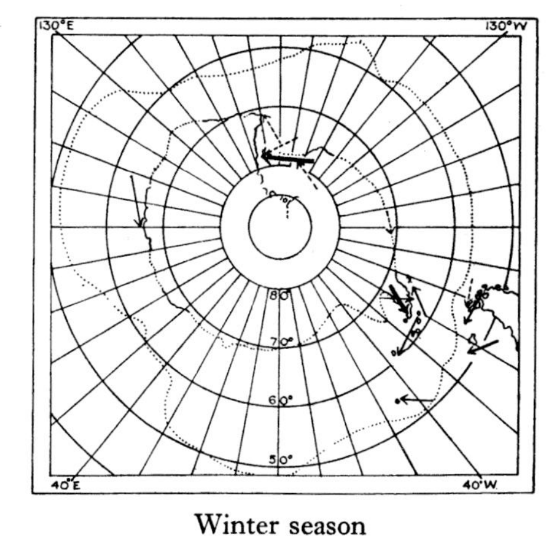

Winter season

WIND-ROSES OF THE ANTARCTIC

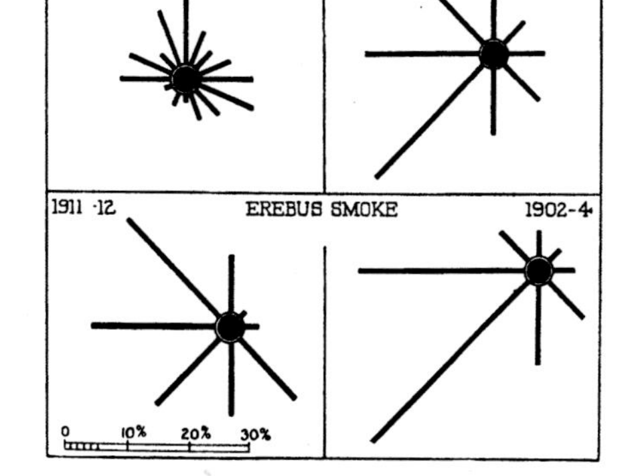

Year

SURFACE-AIR OF THE EQUATORIAL BELT IN JULY

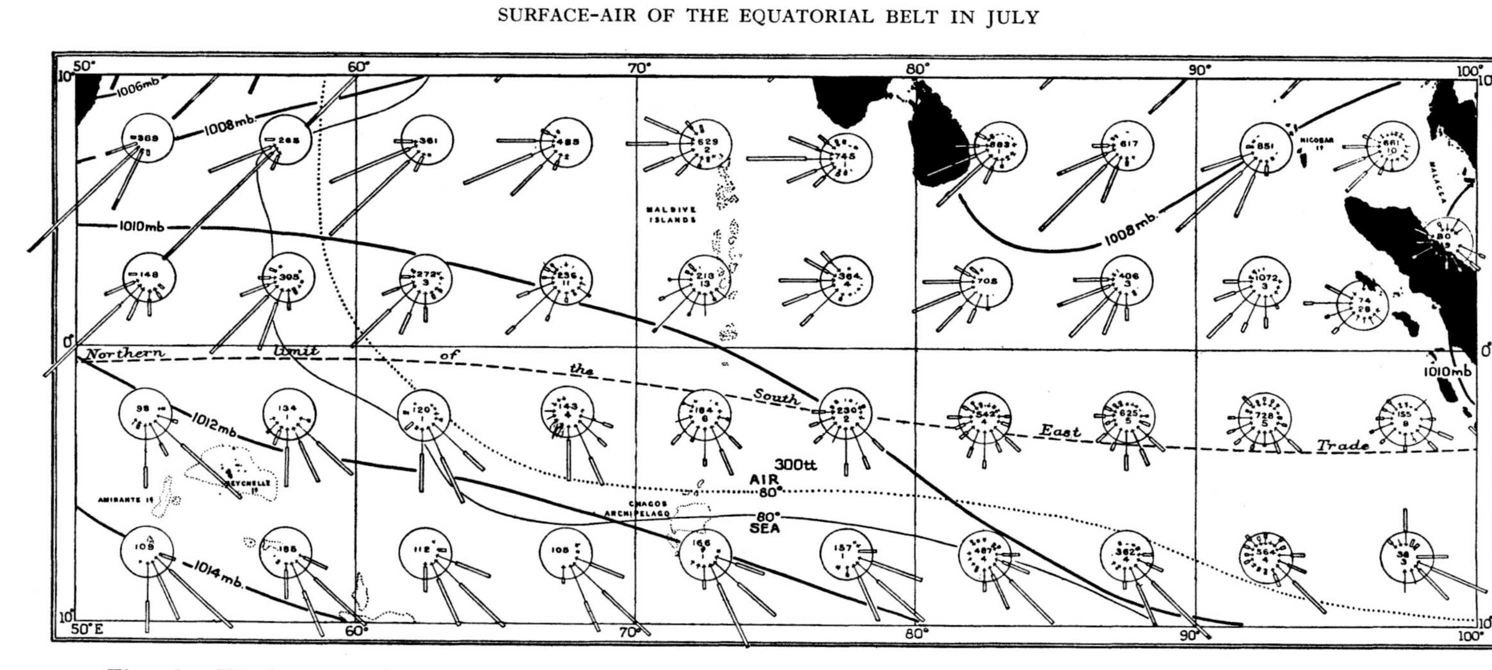

Fig. 161. Wind-roses of the equatorial belt of the Indian Ocean with isobars and isotherms for sea and air. From Monthly Meteorological Charts of the Indian Ocean, M.O. publication, No. 181, July.

A wind-rose represents the normal frequency of light winds (Beaufort scale 1–3) from sixteen points of the compass as a percentage of the whole number of observations by the lengths of the thin lines, of moderate winds (forces 4–7) by the lengths of the doubled lines, and of gales (forces 8–12), if any, by the lengths of the black outer ends of the arrows. The distance from the point of an arrow to the circumference of the circle represents 5 per cent. of the whole number of observed winds, a length of 10 mm represents $33\frac{1}{3}$ per cent. The upper figures within the wind-roses give the total number of observations, the percentage logged as calms being given underneath.

The maps on p. 242 show peculiarities of the region. The winds are not subject to the control of the distribution of pressure in the same way as the other parts of the earth's surface because the influence of the rotation of the earth becomes feebler as the equator is approached and at the equator itself is represented as merely a slight reduction in the weight of the air.

North of the equator an air-current keeps low pressure on its left, South of the equator on its right; at the equator itself the winds are only controlled by the pressure in the case of revolving storms with more or less circular isobars, which are local and transient and could not be indicated on monthly average maps. It follows that if there should be a continuous slope of pressure extending across the equator, the flow of air along the isobars must be in opposite directions on the two sides of the equator, and this we find to be the case in the Indian Ocean in its summer with its "monsoon" (fig. 161). With a region of lower pressure to the Northward along the line of the Himalayas in that season, the isobars range from Northern India over the equator to a high-pressure region about 30° S. There is a Westerly current North of the equator and an Easterly current South of it with a line of doldrums between the two.

The peculiar weather of the doldrums, which is showery and often electrical, may fairly be attributed to the fact of their being at the line of junction of the air of the two hemispheres with completely different life-histories. There is no reason to suppose that their condition is the same in respect of any of the meteorological elements except pressure. Temperature, humidity, electrical condition may all be different, and the line of junction is a sort of physical laboratory in which the effects of differences of physical state are disclosed.

The trade-winds.

The main supplies of air for the equatorial flow are found in streams of air from the North or North East in the Northern Hemisphere, and from South or South East in the Southern, near the Western shores of the continents. These can be identified in the maps quite easily and form the second noteworthy characteristic of the flow of air over the globe. They are the real trade-winds; their most pronounced feature is that they form the Eastern flanks of the high-pressure areas of the oceans. They are shown as pouring a great stream of air from extratropical regions towards the equatorial belt, and to the same stream the anticyclonic regions of the tropics make contributions also, from their equatorial sides, by currents that are marked on maps as from the North East and South East respectively. Hence the name trade-wind has come to be applied to any wind of the intertropical regions between the equator, or more properly the doldrums, and the tropics. Maps of the globe have been put forward showing continuous belts of high pressure about 30° N and S making uniform contributions from NE and SE to a belt of calms at or near the equator. So it comes about that the name trade-wind is often given to any wind outside the doldrums within the tropics. That is not a sufficiently accurate description of the conditions represented in our maps. If it were,

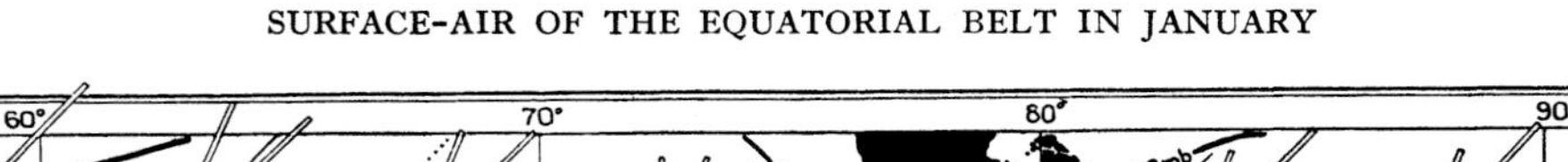
SURFACE-AIR OF THE EQUATORIAL BELT IN JANUARY

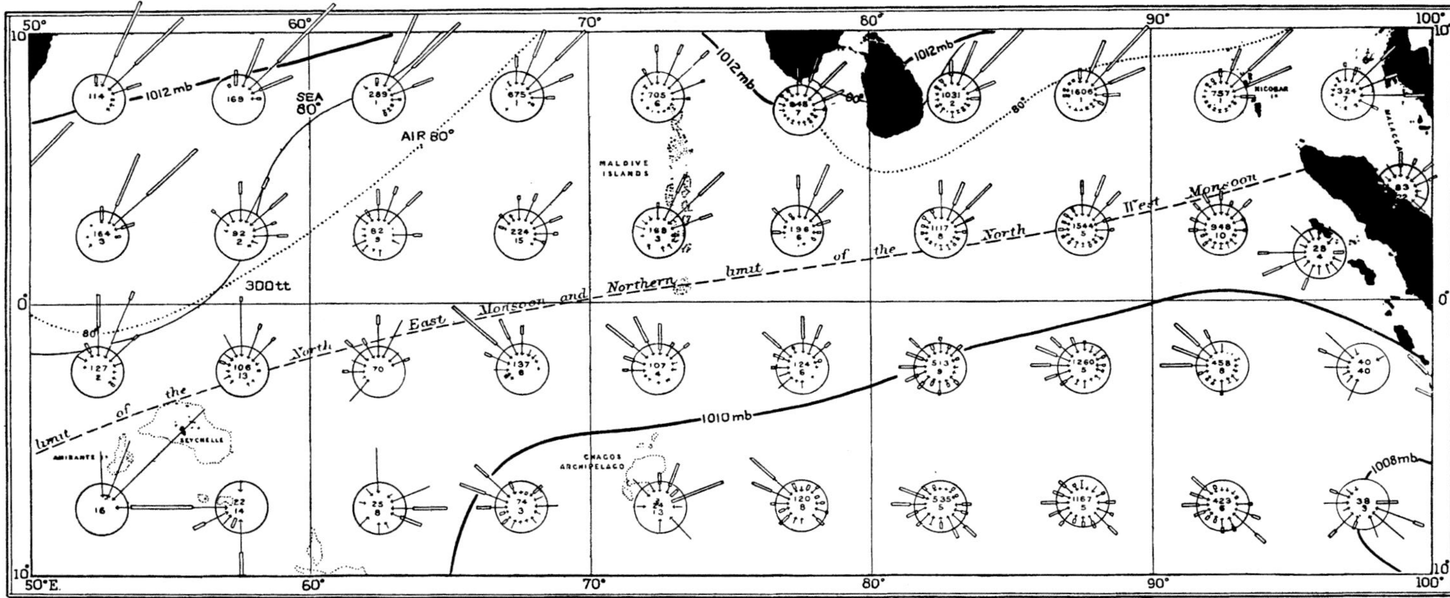

Fig. 162. Wind-roses of the equatorial belt of the Indian Ocean with isobars and isotherms for sea and air. From Monthly Meteorological Charts of the Indian Ocean, M.O. publication, No. 181, January.

A wind-rose represents the normal frequency of light winds (Beaufort scale 1–3) from sixteen points of the compass as a percentage of the whole number of observations by the lengths of the thin lines, of moderate winds (forces 4–7) by the lengths of the doubled lines, and of gales (forces 8–12), if any, by the lengths of the black outer ends of the arrows. The distance from the point of an arrow to the circumference of the circle represents 5 per cent. of the whole number of observed winds, a length of 10 mm represents $33\frac{1}{3}$ per cent. The upper figures within the wind-roses give the total number of observations, the percentage logged as calms being given underneath.

those who investigate "the trade-winds" might take any part of the belt as the immediate subject of investigation; but that is not done. If we take, for example, H. U. Sverdrup's memoir, *Der nordatlantische Passat*[1], we find on p. 9 the area of observation of temperature used for the study of the trade-wind to extend from 3° S to 43° N and from 12° W to 47° W, the great majority of the observations being between 20° W and 40° W. The same is true of the observations of wind-velocity. No meteorologist could assume that the results given in that memoir are applicable to any other part of the circles of latitude between 3° S and 43° N over the Atlantic; and clearly the results obtained, which are beautifully represented by stereoscopic photographs of a model, could not be repeated in the next section either East or West; the assumed uniformity of conditions throughout a whole belt of latitude has indeed no foundation in fact. If the model really represents "Der nordatlantische Passat" we want another name for the winds of any other part of the North Atlantic. We prefer the name of "intertropical flow" for the winds of the West Indies. It leaves us at liberty to explore the actual conditions with the necessary freedom from prejudice which cannot be secured so long as we are nominally dealing with the North East trade. Judging by the isobars our maps would indicate that North East or South East trades are to be found over the following regions:

	Lat.	Long.		Lat.	Long.
Pacific	10–45° N	120–140° W	Atlantic	10–40° N	20–40° W
	10–20° S	70–110° W		0–40° S	10° E–15° W
		150° W–150° E			
Indian and Southern Ocean (Monsoon area)				20° N–25° S	40–140° E

Monsoons. The seasonal variation of surface-winds.

In the last-mentioned area we find the régime which is typical of trade-winds influenced by the orographical distribution to such an extent that in the Northern summer there is no North East wind in the Indian Ocean, North of the equator, but a South West wind instead which is called the South West monsoon, or simply the monsoon, on account of the economic importance of the rainfall which is associated with it in India. The alternation of the wind-circulation between that of the South West monsoon of India and its correlative the North East monsoon which becomes North West over Australia is shown in the wind-maps, pp. 244–7, and the process of transition is indicated by the pressure-maps, pp. 218–41.

The word monsoon really means season, and it is associated with the winds and rains of India because they are seasonal. From that circumstance there has developed a habit among meteorologists of calling any marked seasonal variation in wind-direction with the corresponding rainfall "monsoonal." It is an unnecessary and an unfortunate habit, not only because the name monsoon has been attributed from the very earliest times to part of a certain definite cyclonic circulation, but also because practically everything meteorological is seasonal but is not on that account monsoonal. We give in tables an example of the seasonal character of winds over the Atlantic Ocean. The

[1] *Veröff. d. Geophys. Inst. der Univ. Leipzig*, Leipzig, 1917.

information is derived from an investigation of the geostrophic winds of the Atlantic carried out at the Meteorological Office by Mr R. Nahon, who offered his assistance in meteorological investigation and, after a long analysis of the winds of the British telegraphic reporting stations, tabulated the geostrophic winds at two points of the Atlantic, 40° W and 25° W, latitude 50° N, using the daily charts of the Seewarte and Danish Meteorological Office. There is clearly a very marked seasonal variation, but we forbear to head the table the monsoonal character of Atlantic winds[1].

Geostrophic wind-velocities over the North Atlantic Ocean (1881–1908) *reduced to a standard of* 1000 *observations per month.*

50° N, 40° W. *Seasonal variation in the frequency of geostrophic velocities greater than* 20 m/s.

	Jan.	Feb.	Mar.	Apr.	May	June	July	Aug.	Sept.	Oct.	Nov.	Dec.
N	20	20	16	11	7	7	3	—	11	15	20	—
NE	20	13	14	2	10	1	0	—	4	14	17	—
E	3	11	13	8	3	4	0	—	2	6	4	—
SE	6	15	6	5	2	4	0	—	1	9	8	—
S	28	14	20	11	9	1	0	—	10	16	20	—
SW	76	35	28	7	6	1	3	—	14	26	37	—
W	45	29	18	5	7	4	1	—	18	16	26	—
NW	60	41	26	12	3	5	1	—	23	29	44	—
Totals	258	178	141	61	47	27	8	—	83	131	176	—

Seasonal variation in the frequency of winds of different direction.

	Jan.	Feb.	Mar.	Apr.	May	June	July	Aug.	Sept.	Oct.	Nov.	Dec.
N	52	96	58	54	81	75	45	—	56	85	61	—
NE	53	39	53	26	51	27	15	—	17	37	32	—
E	23	59	38	36	33	27	8	—	15	20	10	—
SE	39	65	52	64	48	35	28	—	12	51	37	—
S	70	89	89	75	89	60	71	—	65	67	85	—
SW	228	175	176	151	154	171	262	—	193	160	179	—
W	199	165	159	145	154	179	206	—	225	160	189	—
NW	199	154	149	139	124	136	120	—	179	180	218	—
Indeterminate	136	158	227	310	265	290	245	—	238	241	190	—

50° N, 25° W. *Seasonal variation in the frequency of geostrophic velocities greater than* 20 m/s.

	Jan.	Feb.	Mar.	Apr.	May	June	July	Aug.	Sept.	Oct.	Nov.	Dec.
N	10	11	12	10	9	2	1	—	8	10	13	—
NE	9	5	1	2	0	0	0	—	1	10	4	—
E	3	1	0	1	0	0	0	—	0	5	1	—
SE	10	14	2	4	2	0	0	—	4	10	6	—
S	32	28	15	6	7	1	2	—	14	23	24	—
SW	88	77	56	18	12	1	10	—	29	39	75	—
W	73	28	43	13	8	2	7	—	24	23	25	—
NW	30	20	16	21	8	1	5	—	18	29	36	—
Totals	255	184	145	75	46	7	25	—	98	149	184	—

Seasonal variation in the frequency of winds of different direction.

	Jan.	Feb.	Mar.	Apr.	May	June	July	Aug.	Sept.	Oct.	Nov.	Dec.
N	51	52	68	61	103	57	58	—	56	82	71	—
NE	29	23	41	30	50	30	13	—	17	41	29	—
E	20	28	38	37	33	19	6	—	12	28	19	—
SE	32	61	40	50	47	32	9	—	19	46	35	—
S	94	125	98	77	99	89	45	—	88	91	92	—
SW	278	237	199	190	162	213	182	—	189	157	220	—
W	256	205	217	174	153	183	265	—	230	197	224	—
NW	122	106	122	156	109	89	168	—	148	147	151	—
Indeterminate	119	163	176	225	243	287	255	—	242	211 –	160	—

[1] Geostrophic winds of other latitudes of the Atlantic have been investigated by C. S. Durst.

The prevailing Westerly winds of temperate latitudes.

The tables which we have just cited show a marked preponderance of Westerly winds and that may serve as a stepping-stone to the next type of winds in the circulation over the earth's surface, the Westerlies, the brave West winds or South West winds which are to be found from 40° Northwards over the North Atlantic and North Pacific, and as an uninterrupted belt of some 10° or 20° wide in the great Southern Ocean, from 40°, the "roaring forties," Southward as far as our information extends towards the Antarctic circle. These are the regions in the two hemispheres in which cyclonic depressions travel, and in which they are often formed and as often disappear. Over the North Atlantic the Westerly or South Westerly winds follow the extension of the Gulf Stream and carry warmth to the Norwegian coast and to Spitsbergen in latitude 80°. The warmth keeps the sea navigable throughout the year up to Murmansk on the North West of Russia, and provides thereby a remarkable contrast with the Eastern coast of America where the St Lawrence, which is below 50° N, is ice-bound in the winter.

These winds which form the Northern flanks of the oceanic anticyclones are sometimes called the counter-trades and have been supposed to represent air lifted in the doldrums and dropped in the high-pressure belt of the tropics. We are not yet clear as to where they come from, the important thing is that they are on the map. But they are represented by short interrupted lines, very different from the long continuous lines used to indicate the steadiness of the trade-winds. The shortness means that the winds are fluctuating as they naturally would be if they formed part of travelling cyclonic depressions.

It is important therefore to understand that the intermittent lines of the Northern Hemisphere, or the interrupted circles of the Southern Hemisphere, do not indicate the actual paths of the air of the Westerly winds. In fact the actual path of air over the earth's surface is hardly ever to be guessed from the single indication of a variable prevailing wind. So far as they could be ascertained from consecutive daily maps of the Atlantic, the tracks of air have been drawn for a number of occasions in the year 1882–3 for which specially detailed maps exist. They are represented in the entablatures below figs. 157 and 159. It will be seen that the horizontal travel of the air from any starting point may be extensive; its termination is either near the centre of an area of low pressure or in the equatorial belt. It is to these characteristics that the fluctuations of the wind and weather of temperate climates are due.

The circulations of the polar zones.

Our maps give us little information about the circulation of Southern regions beyond the "roaring forties." We know enough about the prevalence of an Easterly wind on the margin of the Antarctic continent to be practically certain that the circulation would correspond with a permanent anticyclone covering the continent if the space which the anticyclone should occupy were not already nearly filled by a huge mass of land. There are some indications also in what is known of the pressures (see table, p. 219) that a circumpolar trough of low pressure is passed somewhere about the Antarctic circle and that beyond that circle pressure begins to rise in conformity with the change in the direction of the wind. The vast extent of the area thus occupied is little realised and we may therefore reproduce an illustration (fig. 163) by which the late Dr W. S. Bruce impressed the point.

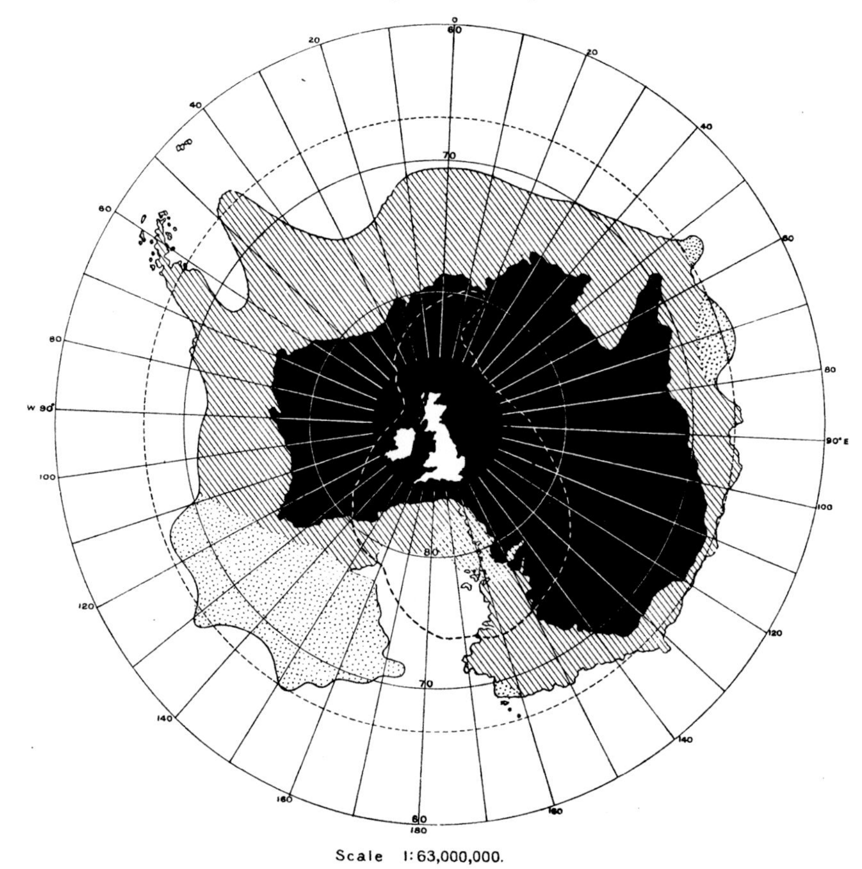

Fig. 163. The Antarctic continent with the area of Australia and of the British Isles superposed (Scottish Geographical Society, July 1906).

The broken outline represents the unknown North Polar regions, longitude for longitude the same as the area upon which it is superposed.

We might in like manner suppose that the region round the North Pole, which is mainly sea or sea-ice, is also an anticyclonic region in conformity with the general distribution of temperature which falls with increasing latitude. But there are too many irregularities for the inference to be justified. The centres of extreme of winter-cold, for example, are on the two facing continents and not apparently at the pole itself, and such observations of pressure in the polar circle as are available (see the diagram on p. 218) do not indicate pressure above the average for the hemisphere.

And as a matter of fact there is an insuperable obstacle to any regularity of circulation in the polar regions of the Northern Hemisphere in the shape of Greenland which is a mass extending over some 22° of latitude, about 1500 miles, and some 3000 metres high. Such a wall forms an absolute barrier to any orderly circulation of cold air. We have seen from our maps of ice that icebergs are carried down the East side of Greenland and, rounding Cape Farewell in latitude 60°, pass up the West coast. The air is less subject to restriction than the water, yet nevertheless something of the same kind must happen. The air cannot flow over a wall 3000 metres high.

The maps of trajectories show that air might come down from Greenland along the West coast or the East coast to take part in the circulation of the temperate zone, and we may conclude that the dominant factor of the circulation of air in the North Polar regions is not a centralised continent, as in the South Polar regions, but the vast dome formed by Greenland covered by perpetual ice and therefore cold in winter and summer alike.

Katabatic winds.

This leads to the last type of wind to which we shall call attention as forming part of the general circulation of the atmosphere, that is the winds that represent air which has come down the cold mountain-sides in consequence of the cooling of the air in contact with the ice-covered land. Air so cooled must find its way downward and there are a large number of examples in the shape of valley-winds in India, the Norwegian fjords and elsewhere. A very good example on small scale but in full detail is represented in E. V. Newnham's paper on a nocturnal wind in the valley of the Upper Thames[1]. Such winds may be very strong. The boras of the Adriatic and of the Black Sea belong to that class. Sir Douglas Mawson has described his experiences of the Antarctic continent in *The Home of the Blizzard*[2], and we understand that Norwegians wishing to make their way up a fjord in winter prefer to do their climbing first on a flanking ridge and move along at the higher level rather than face the descending stream at its lowest.

Such winds are gravity-winds which pay no attention to isobars until they get into the open where they have time to adjust themselves to the requirements of the earth's rotation.

Professor Hobbs[3] has given much attention to the influence of glaciated mountains in this connexion and describes the final effect as a glacial anticyclone; but it must always be remembered that the main bulk of the so-called anticyclone, either in the North or South polar regions, is a great land-mass. We should not expect to find evidence of the so-called anticyclone on the top of the land-mass, the conditions there may be the reverse of those which are indicated by the margins at sea-level.

[1] *Notes on Examples of Katabatic Wind in the Valley of the Upper Thames*, Professional Notes, No. 2, M.O. publications, No. 232 b, London, 1918.

[2] Heinemann, 1915.

[3] *The Glacial Anticyclones: the Poles of the Atmospheric Circulation*, by W. H. Hobbs. University of Michigan Studies, Scientific Series, vol. IV. New York: the Macmillan Co., 1926.

In a review[1] of *Die Ergebnisse der meteorologischen Beobachtungen der Deutschen Antarktischen Expedition*, 1911–1912, Dr H. R. Mill writes as follows:

Herr Barkow thus summarises his view of Antarctic atmospheric circulation—"The circulation over the Antarctic continent is dominated by an anticyclonic cap of air which probably flows outward in a series of waves moving in a south to north direction. Above this cap of cold air there is a cyclonic stratum connected with the circulation of temperate latitudes and, in this, wandering depressions travel in a west to east direction. The two air-systems mutually influence each other; on the whole the lower system dominates the upper at least so far as atmospheric pressure is concerned. The surface-winds are dominated on the whole by the lower system for the most part on the continent and in less degree as the distance from land increases. In the cirrus region, and especially in the stratosphere, cloud movements appear to show that there is a current of air moving right across the continent from the Indian Ocean to the region of the West Antarctic as indicated by the northern component of cirrus-movement in the former region and the southern component in the latter." Herr Barkow says that he concurs in Professor Hobbs' conception of the formation of a south polar glacial anticyclone.

Katabatic winds have been less studied than they deserve to be and either Greenland or the Antarctic continent offers special attractions from that point of view, but they are not the only examples. Sir Henry McMahon has described in impressive language the extraordinary winds of Sistan which register day after day in early summer something like 100 miles per hour on an anemometer.

We reached our main camp at Robat on May 1, [1896]...and by May 14 our final agreements and maps were prepared and signed, and we were able at last to start homewards. The Afghans returned to the Helmand en route for Kandahar, while we followed, as far as Nushki, much the same route as that by which we had come. It was a trying journey, as the heat was very severe, registering 116° Fahr. in our tents. We marched as before, always at night, and now were able to get little or no rest by day, for the "Bad-i-sad-o-bistroz," i.e. the wind of 120 days, had now sprung up, and blew with hurricane violence day after day the whole day long, blowing down our tents, and smothering us in sand. This charming wind gets up every year about May, and blows without ceasing from the north-west for four months. While it lasts, it makes life along the Helmand Valley and the deserts on either side a perfect purgatory. Right glad were we to at last reach the edge of the desert at Nushki, and ascend out of the hot, wind-swept plain into the cool, refreshing air of the high mountains west of Quetta.

'The Southern Borderlands of Afghanistan,' by Capt. A. H. McMahon, C.I.E., *Geographical Journal*, vol. IX, 1897, p. 414.

There is no pressure-gradient that can account for such winds, and they occur in spring when the Asiatic heights are covered with snow and the sun on the surface is very powerful; it is therefore desirable in the absence of any other explanation to inquire further into the possibilities of katabatic winds.

The land and sea breezes of physical geography are examples of katabatic and anabatic winds. On p. xxxi a pertinent illustration is cited. Facile generalisation brings the East Indian monsoons into the same category. The subject is however not so simple as these explanations suggest. On 31 August, 1906, the Western Hebrides furnished a neat demonstration of the next stage in the reasoning. (*Barometric Gradient and Wind Force*, M.O. 190, p. 9, 1907.)

[1] *Meteorological Magazine*, July, 1925.

Following the representation of pressure and winds at sea-level we present for the reader's attention a number of maps of the distribution of pressure in the upper levels computed from the distribution at the surface with the assumption that the lapse-rate of temperature is the same all over the world. Maps of the distribution of pressure in the upper air may be said to have begun with Teisserenc de Bort's construction of maps for 1467, 2859 and 4000 metres[1], and the same idea has been used by many authors either in the form of the distribution of pressure at selected heights which we shall give in this volume or, after the method of Bjerknes, in the form of lines of equal geopotential, i.e. lines of equal heights in geodynamic metres, of points on the same isobaric surfaces.

Without any special search we have found the practice of drawing maps for upper levels in C. J. P. Cave, *The Structure of the Atmosphere in Clear Weather*, Cambridge, 1912; S. Fujiwhara, *Monthly Weather Review*, 1921; T. Kobayasi, *Q. Journ. Roy. Met. Soc.*, 1922. The subject has also been discussed by F. H. Bigelow, 'Report on the barometry of the United States, Canada, and the West Indies,' *Report of the Chief of the Weather Bureau*, 1900–1, vol. II, and by J. W. Sandström in *Trans. Amer. Phil. Soc.*, N.S., vol. XXI, part I, Philadelphia, 1906, pp. 31–95, who refers to Köppen, *Met. Zeits.* 1888, p. 476. It also forms the subject of Supplement No. 21 to the *Monthly Weather Review*, 1922, by the late C. Le Roy Meisinger. The use of daily charts for 2500 metres is discussed by Köppen in the report of the meeting of the Upper Air Commission in Vienna, 1912. Charts for the 1000 and 2000 metre levels are now included in the British *Daily Weather Report*.

We reproduce the maps by Teisserenc de Bort for both hemispheres at 4000 m in January and for the Southern Hemisphere in July, and also maps compiled for this volume representing the distribution of pressure in the Northern Hemisphere at 2000, 4000, 6000 and 8000 metres in July and at 2000 metres in January.

The process which has been followed in the construction of the maps of pressure in the upper air for the purpose of this work may be briefly indicated as follows.

The surface-pressure is taken from the maps of the *Barometer Manual for the Use of Seamen* supplemented by the information which is incorporated in the pressure-charts of this chapter.

Figures were obtained for the temperature at the points of intersection of successive lines of 10° of latitude and longitude from a paper by C. E. P. Brooks on *Continentality and Temperature*[2]. Figures for lapse-rate in July were tabulated for 17 land-stations and localities at sea. From these a general mean for the Northern Hemisphere was derived as follows:

Kilometres	0–2	2–4	4–6	6–8	8–10	10–12
Lapse-rate tt per km	5·2	5·5	6·0	7·0	7·3	7·35

[1] The maps which are intended to give the normal pressure at the levels of Puy de Dôme, Pic du Midi and Pike's Peak are given in a *Report on the Present State of our Knowledge respecting the general Circulation of the Atmosphere.* Presented to the Meteorological Congress at Chicago, 1893, London, Edward Stanford.

[2] *Q. Journ. Roy. Meteor. Soc.*, vol. XLIV, 1918, pp. 253–69.

The values adopted for the purposes of computation were 5·0, 5·5, 6·0, 7·0 respectively. The value 5·0 was adopted in place of 5·2 for the first stage because it had been used by C. E. P. Brooks for the reduction of temperature at land-stations to sea-level.

Laplace's original equation for the variation of pressure with height is $dp = -g\rho\, dz$. The formula for computation of pressures in the upper air derived from it thus becomes $\log p = \log p_0 + g/(R\beta)\ (\log tt - \log tt_0)$, where β is the lapse-rate.

The numerical equations for successive layers are obtained by substituting values for $g/(R\beta)$ appropriate to the different levels; the values are:

	0–2 km	2–4 km	4–6 km	6–8 km
$g/(R\beta)$	6·83	6·21	5·69	4·88

With these formulae the changes of pressure from the surface at the points of intersection of the lines of latitude and longitude were computed for successive steps of 2 kilometres. Corrections were made for the geographical variation in the value of gravity.

The values corresponding with the intersections were then plotted on charts and the isobars drawn accordingly. The particulars of these and other computations of like character are preserved in a MS volume compiled during the preparation of the illustrations of *The Air and its Ways*.

If the process thus sketched has given a satisfactory representation of the distribution of pressure, the mean values of pressure at the several levels which have been determined by direct observation should fit into the maps. The available values have accordingly been inserted in the maps for the levels of 2, 4 and 6 kilometres, which now appear for the first time; for the level of 8 kilometres, for which a map was published in 1922, the available values have been put into a table for the panel at the foot of the map.

It will be remarked that the maps are drawn only for the month of July. The reason is that in the winter months the lowest layer of the air is frequently represented by an inversion of lapse-rate of temperature. It is not permissible in such case to proceed by the method of surface-temperature and assumed lapse-rate. The same objection does not apply however to the air over the sea which is not liable to such counterlapses and their consequences. We have accordingly computed a map for the level at 2 kilometres over the sea in January. The temptation to complete the isobars by hypothetical lines over the land is hardly resistible—we have indicated the appropriate hesitation by broken lines. The result of the extrapolation is not otherwise than satisfactory.

References to charts of normal isobars at high levels for special localities are given in chap. x, note 18.

Entablatures of the maps of pressure in the upper air.

We have utilised the entablatures of the maps to give a conversion table from millibars to millimetres for pressures down to 100 mb; and the height in metres of the level surfaces of 100, 200, ... 1000 geodekametres. Information is also added of pressure on the South Polar plateau and of the seasonal variation of density in the upper air computed from observations of ballons-sonde. A table of densities for England is given in chap. x, note 19.

NORMAL PRESSURE AT 4000 METRES

FIGURE 164

Authority: Teisserenc de Bort.

Millibars.

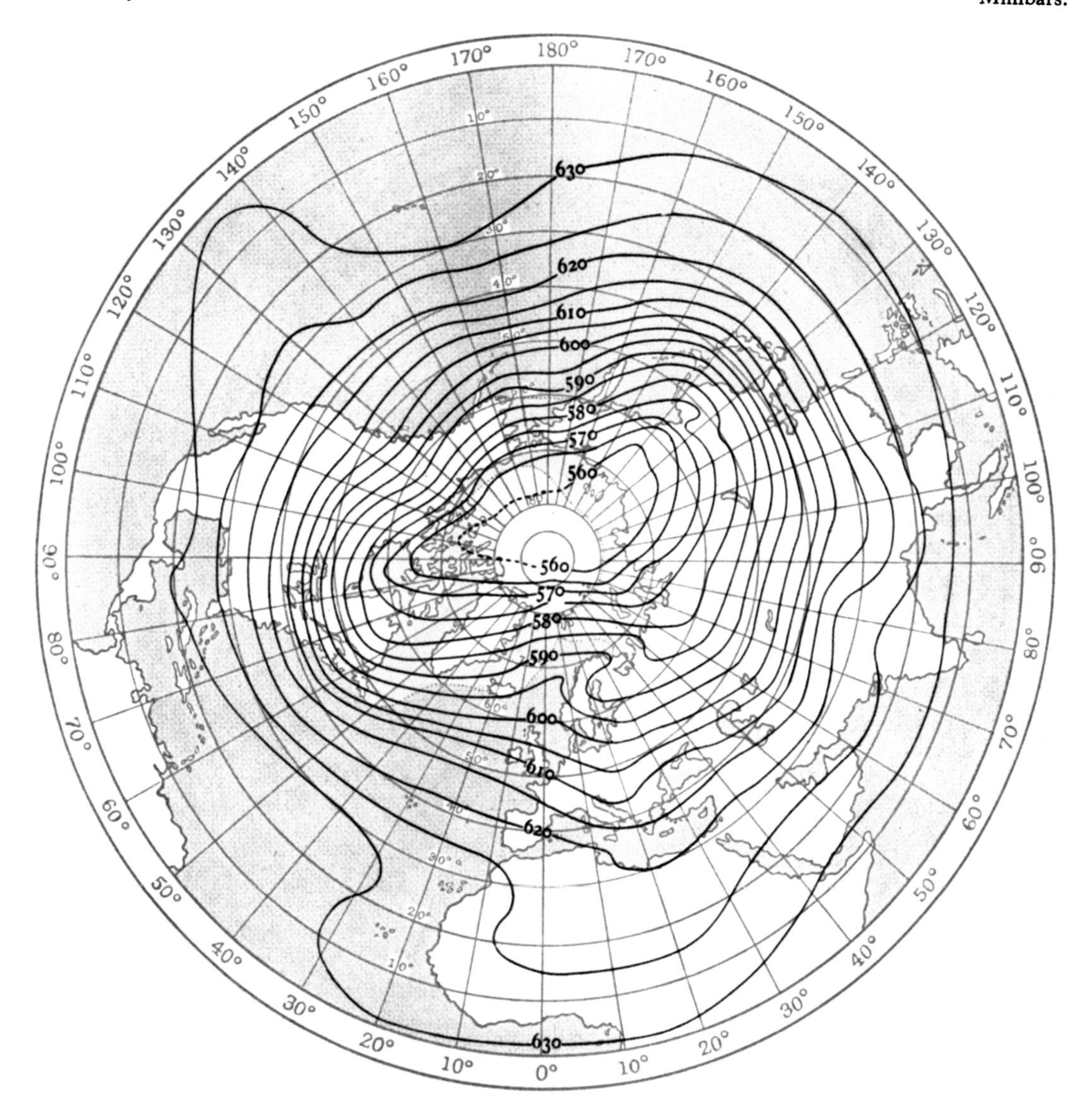

Pressure in millibars to millimetres of mercury at 0° C in latitude 45°.

mb	0 mm	10 mm	20 mm	30 mm	40 mm	50 mm	60 mm	70 mm	80 mm	90 mm
100	75·0	82·5	90·0	97·5	105·0	112·5	120·0	127·5	135·0	142·5
200	150·0	157·5	165·0	172·5	180·0	187·5	195·0	202·5	210·0	217·5
300	225·0	232·5	240·0	247·5	255·0	262·5	270·0	277·5	285·0	292·5
400	300·0	307·5	315·0	322·5	330·0	337·5	345·0	352·5	360·0	367·5
500	375·0	382·5	390·0	397·5	405·0	412·5	420·0	427·5	435·0	442·5
600	450·0	457·5	465·0	472·5	480·0	487·5	495·1	502·6	510·1	517·6
700	525·1	532·6	540·1	547·6	555·1	562·6	570·1	577·6	585·1	592·6
800	600·1	607·6	615·1	622·6	630·1	637·6	645·1	652·6	660·1	667·6
900	675·1	682·6	690·1	697·6	705·1	712·6	720·1	727·6	735·1	742·6
1000	750·1	757·6	765·1	772·6	780·1	787·6	795·1	802·6	810·1	817·6

FIGURE 165 NORMAL PRESSURE AT 4000 METRES

Millibars. *Authority:* Teisserenc de Bort.

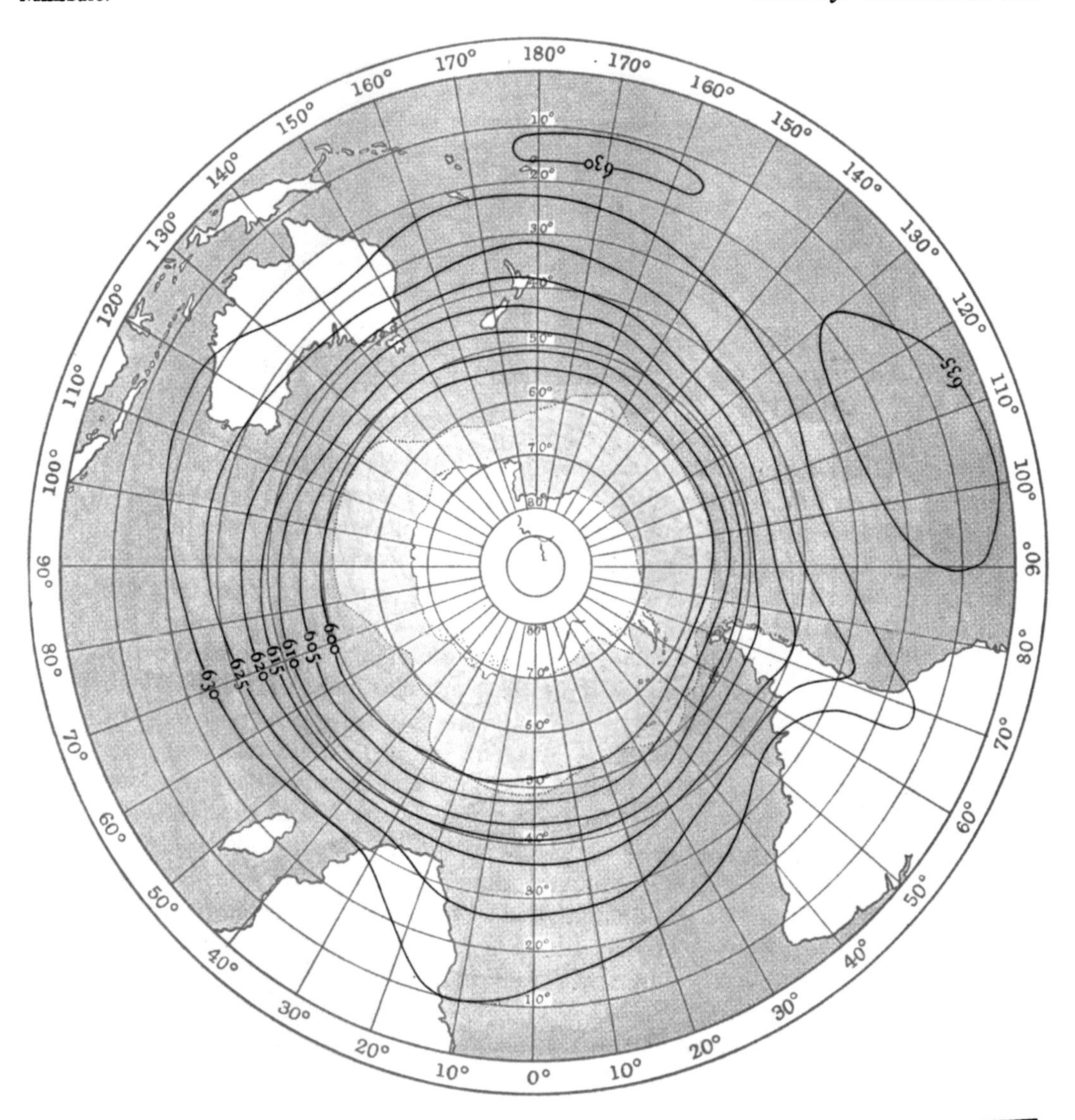

Height in metres of the level surfaces of 100, 200, ... 1000 (10 m)²/sec² *for latitudes* 0–90°.

	Latitude	0	10	20	30	40	50	60	70	80	90
	g 900 +	78·03	78·19	78·64	79·32	80·17	81·07	81·91	82·61	83·06	83·22 cm/sec²
		m	m	m	m	m	m	m	m	m	m
Geopotential gdkm (10 m)²/sec²	1000	10241	10239	10234	10227	10218	10209	10200	10193	10188	10187
	900	9216	9214	9210	9203	9195	9187	9179	9173	9168	9167
	800	8191	8189	8185	8179	8173	8165	8158	8152	8149	8147
	700	7166	7164	7161	7156	7150	7143	7137	7132	7129	7127
	600	6140	6139	6136	6132	6127	6122	6117	6112	6109	6108
	500	5116	5115	5113	5109	5105	5101	5096	5092	5090	5089
	400	4092	4091	4089	4087	4083	4080	4076	4074	4072	4071
	300	3069	3068	3067	3065	3062	3060	3057	3055	3054	3053
	200	2046	2046	2045	2043	2042	2040	2038	2037	2036	2036
	100	1023	1023	1022	1022	1021	1020	1019	1018	1018	1018

NORMAL PRESSURE AT 4000 METRES

Authority: Teisserenc de Bort.

FIGURE 166

Millibars.

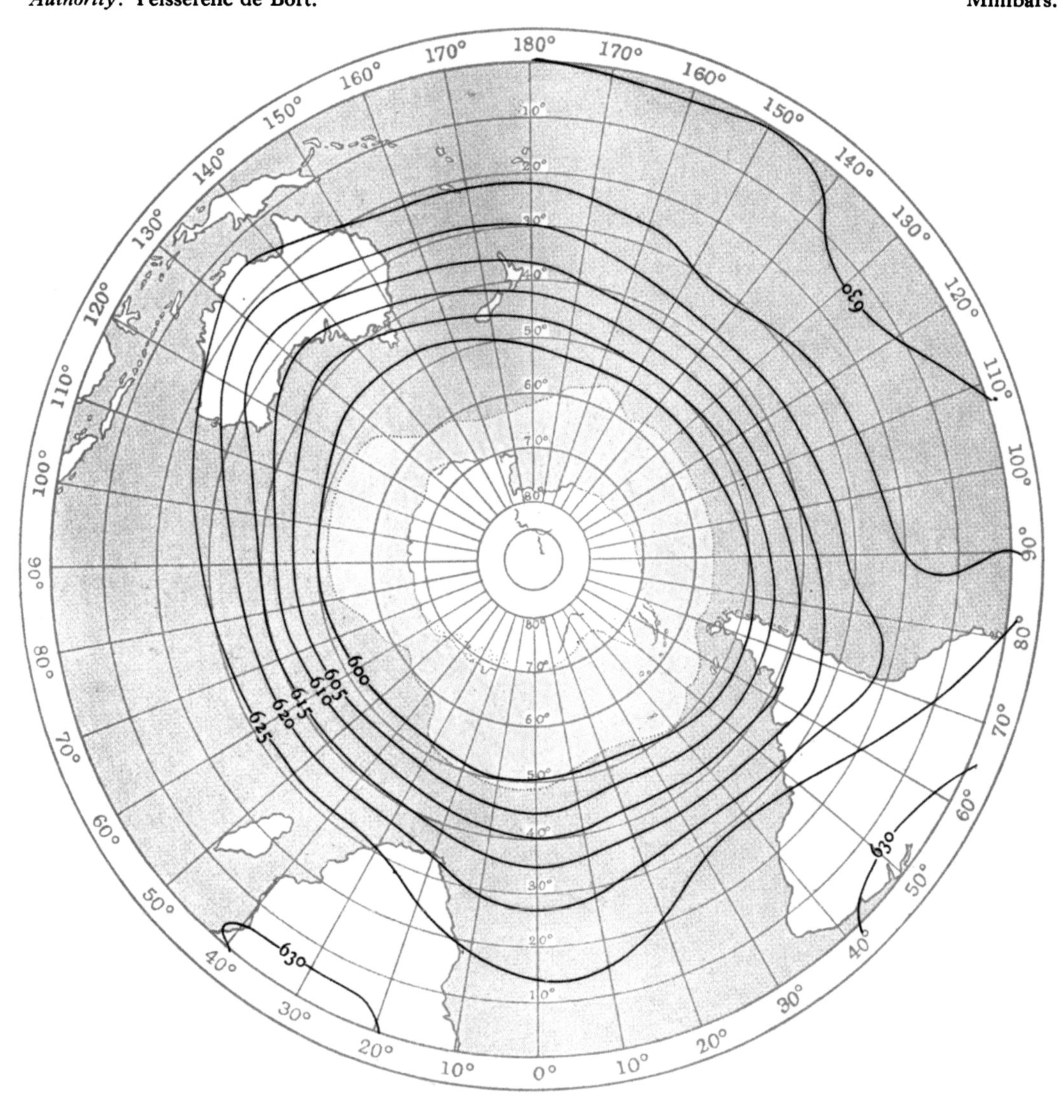

Pressure in the Antarctic. See also pp. 219, 288.

South Polar Plateau. The mean pressure recorded by Scott's party on the plateau, January 1–31, 1912, reduced to the level of 3005 metres, the highest traversed by the party, was 663 mb. It varied from the extreme of 655 mb on January 12 to 681 mb on January 26.

The height of the pole works out at 2765 m. The pressure recorded there by Scott's party on January 17 was 684 mb and the mean of the observations recorded there by Amundsen's party December 16–18, 1911, was 693 mb. Allowing a difference of 18 mb for height, the reduced pressures for the level of 3005 metres would be 666 mb (Scott) and 675 mb (Amundsen).

McMurdo Sound. The mean pressures observed in McMurdo Sound 77° 38′ S, 166° 24′ E in excess of 900 mb are:

	Jan.	Feb.	Mar.	Apr.	May	June	July	Aug.	Sept.	Oct.	Nov.	Dec.
1911	92	92	89	93	90	86	85	88	87	76	103	107
1912	97	100	88	94	87	78	82	86	97	85	91	88

FIGURE 167 NORMAL PRESSURE AT 8000 METRES

Millibars. *Authority:* Original.

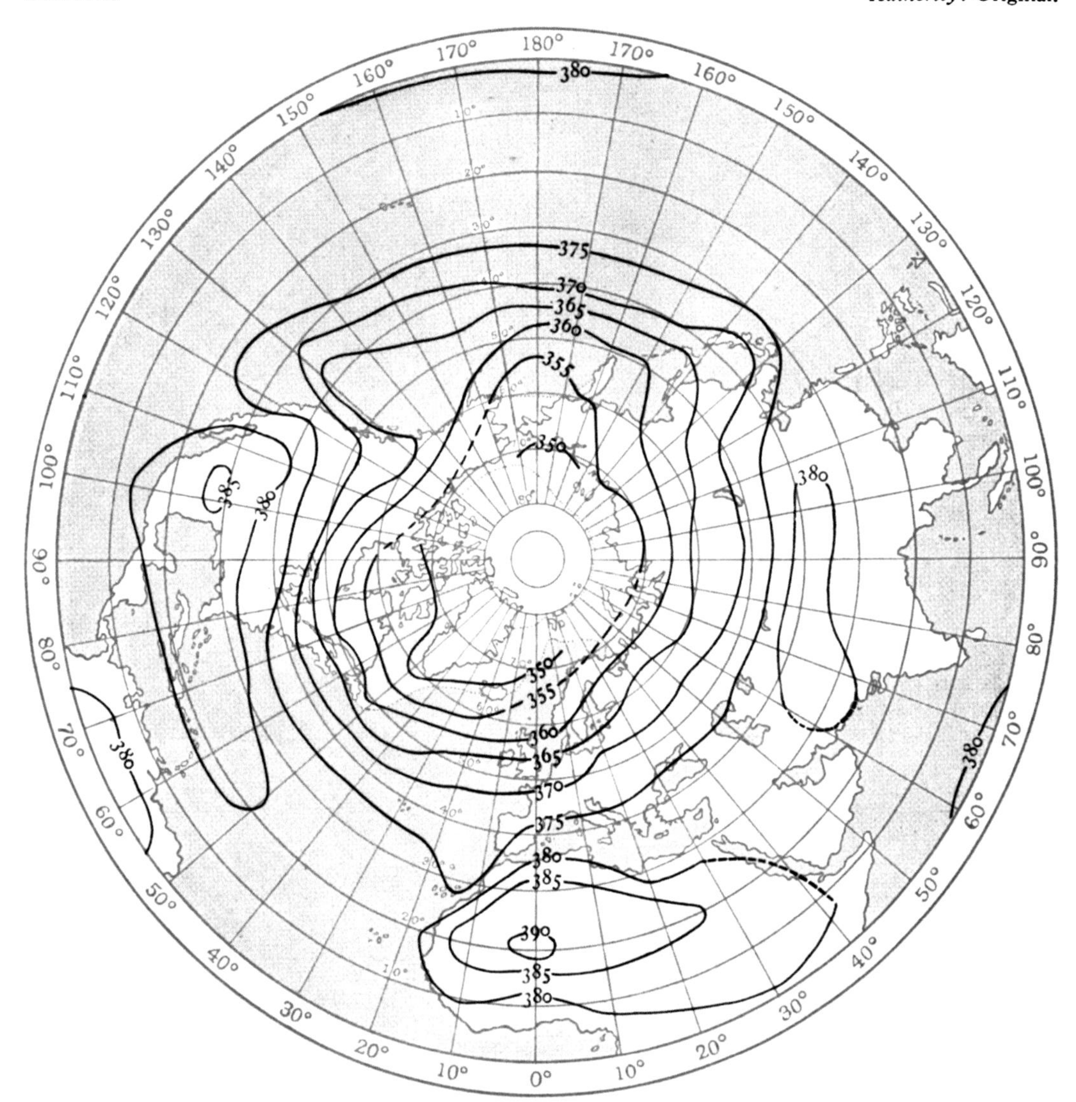

Pressure at 8 kilometres in summer from observations by ballons-sonde.

Station	Lat.	Long.	Month	Pressure mb
Arctic Sea	78° N	12° E	July–Sept.	351
Kiruna	68° N	20° E	Aug.	360
Pavlovsk	60° N	30° E	"Summer"	356
Ekaterinbourg	57° N	61° E	,,	360
Kutchino	56° N	38° E	,,	361
England, SE	51° N	1° W	July	362
Europe	48–50° N	2–12° E	"Summer"	367
Pavia	45° N	9° E	July	365
Canada	43° N	81° W	June–Aug.	369
U.S.A.	33–44° N	86–118° W	—	379
St Louis	39° N	90° W	—	374
Agra	27° N	78° E	July	377
N. Atlantic	30–35° N	—	—	378
,,	20–40° N	—	—	378
,,	10–26° N	—	—	381
Batavia	6° S	107° E	Dec.–Feb.	377
Victoria Nyanza	1° S	34° E	July–Oct.	377

NORMAL PRESSURE OF THE LAYER BETWEEN SEA-LEVEL AND 8000 METRES FIGURE 168

Authority: Original. Millibars.

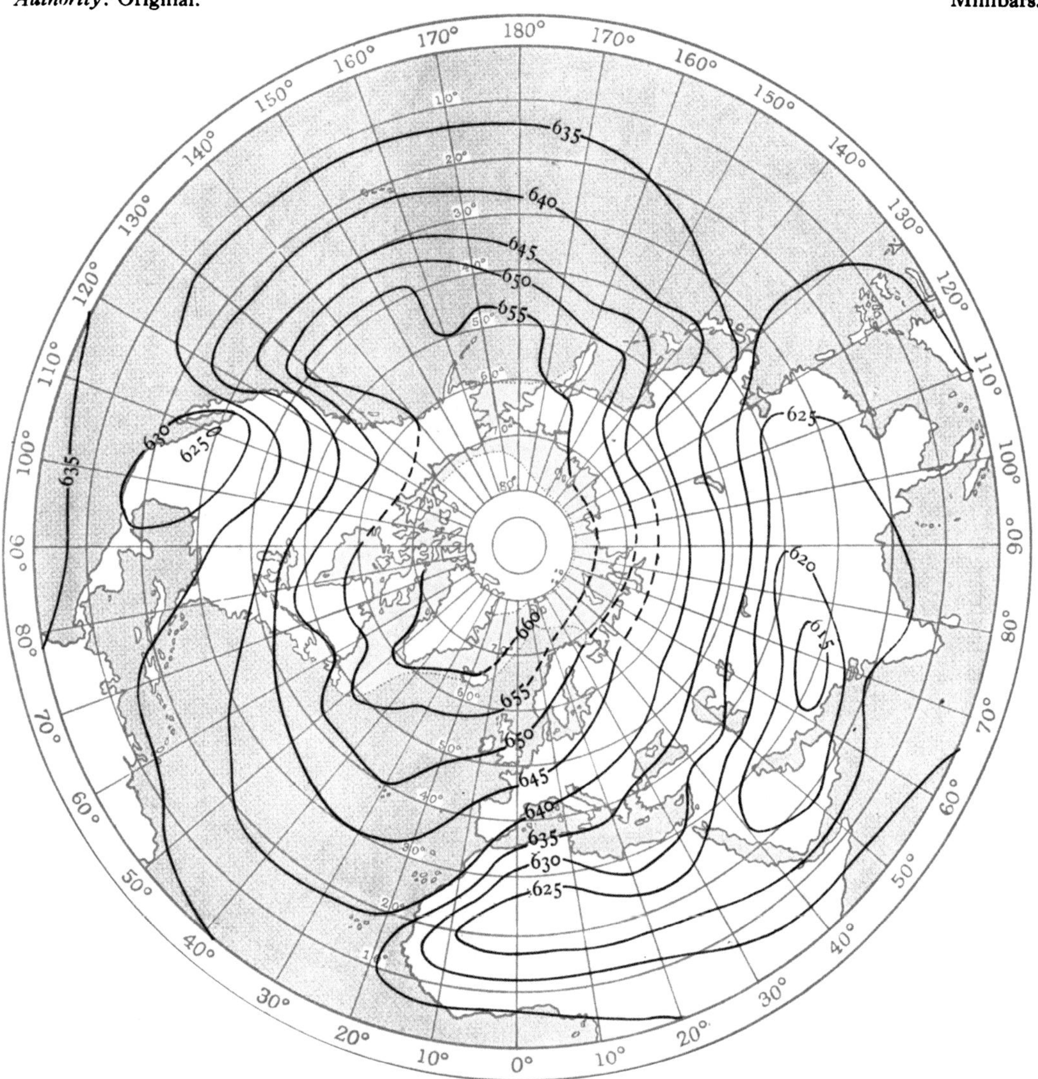

Seasonal variation of density in the upper air (grammes per cubic metre).

UNITED STATES	Winter	Spring	Summer	Autumn	km	Dec.–Jan.	Feb.–Mar.	Apr.–May	June–July	Aug.–Sept.	Oct.–Nov.	MUNICH * 516 m.
	408	419	418	419	10	419	418	416	421	422	414	
	526	527	523	529	8	532	533	529	527	526	524	
	662	648	647	657	6	668	668	664	653	655	657	
	825	805	795	804	4	830	830	830	812	816	818	
	1027	995	965	986	2	1026	1027	1019	997	1001	1010	
	1270	1207	1177	1182	0	1232*	1220*	1195*	1163*	1164*	1188*	

PAVIA	Dec.–Jan.	Feb.–Mar.	Apr.–May	June–July	Aug.–Sept.	Oct.–Nov.	km	Feb.	Apr.	June	Aug.	Oct.	Dec.	AGRA † 170 m.
	413	419	418	421	419	416	10	405	418	411	408	413	421	
	532	536	526	529	528	529	8	526	523	509	505	518	526	
	665	669	662	652	652	659	6	648	648	633	626	640	652	
	824	826	818	806	807	816	4	800	795	775	770	787	795	
	1025	1019	1012	989	990	1004	2	973	957	929	942	962	979	
	1293	1292	1234	1207	1209	1245	0	1173†	1129†	1089†	1115†	1139†	1175†	

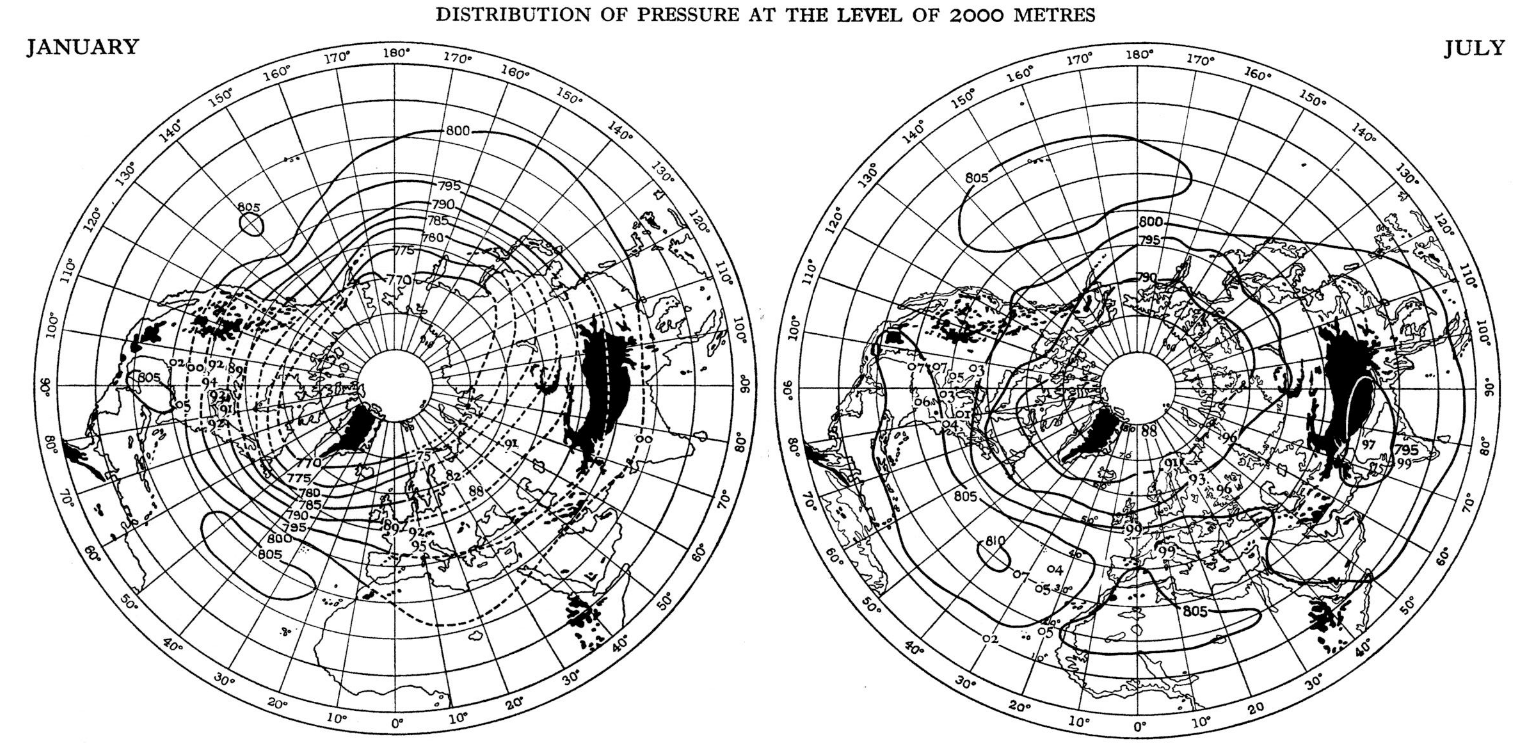

Fig. 169. January. Isobars computed from the distribution of mean pressure and temperature over the sea, connected by dotted lines across the land.

Fig. 170. July. Isobars computed from the distribution of mean pressure and temperature over land and sea.

(The mean value for July by observation for Central Europe is 800 mb and agrees with the isobar. The figures 96 on long. 60° E. are too far North by 5° of latitude.)

Isobars for steps of 5 mb are marked with three figures (770 to 805 or 790 to 810) indicating the pressure in millibars.

The mean results, also in millibars, of the observations by means of registering balloons at 2000 metres are shown by figures for the tens and units: the figures for the hundreds are omitted.

The contour of land at the height of 2000 metres is shaded in black. The range of geopotential at that height is from 1965 geodynamic metres at the pole to 1955 gdm at the equator.

DISTRIBUTION OF PRESSURE IN JULY AT 6000 METRES AND 4000 METRES

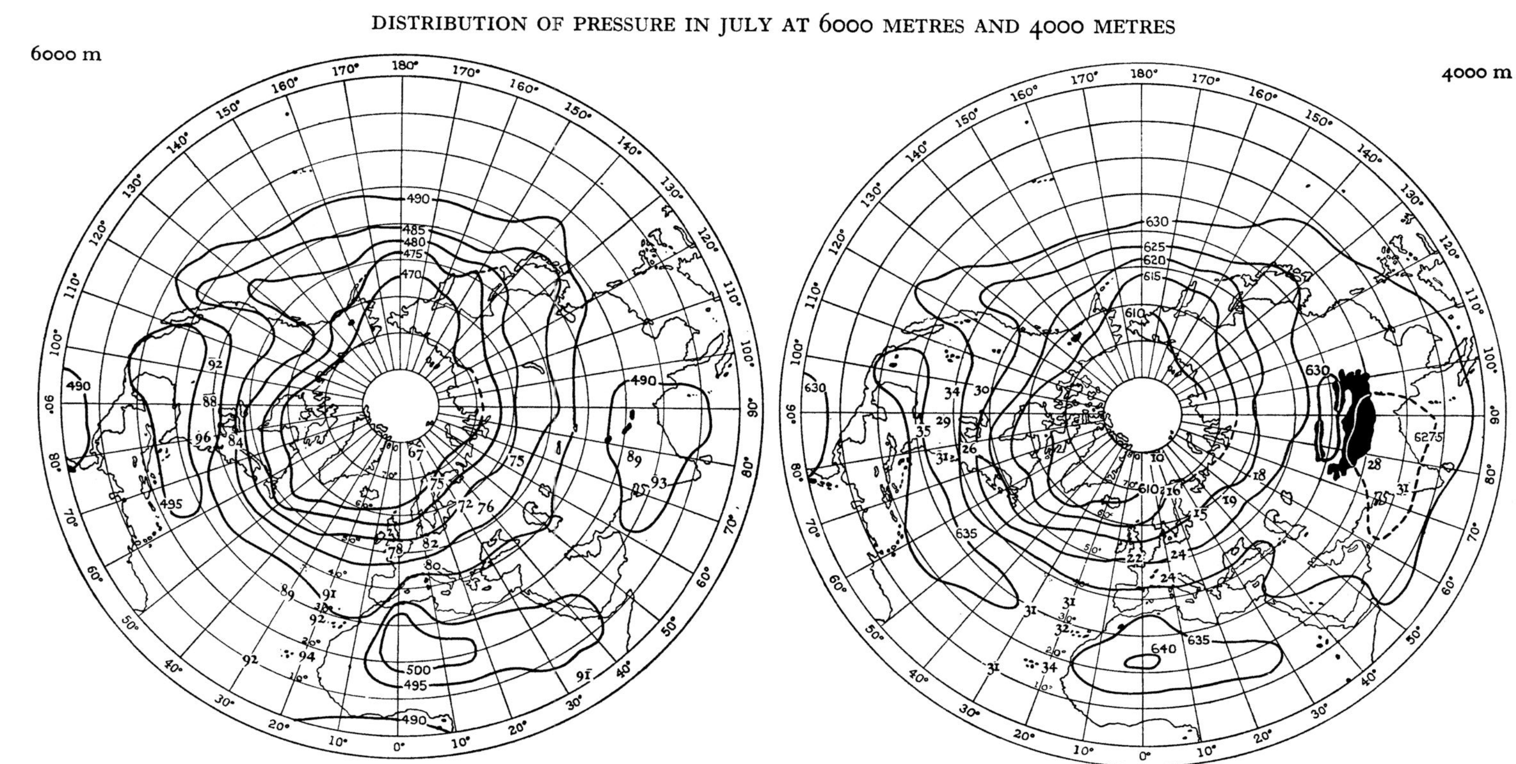

Fig. 171. Height 6000 metres. Isobars for steps of 5 mb are marked with three figures, 470 mb to 500 mb.

The mean results, also in millibars, of the observations by means of registering balloons at 6000 metres are shown by figures for the tens and units: the figures for the hundreds are omitted.

The small areas of the contour of land at 6000 metres are shaded black. The range of geopotential at that height is from 5893 gdm at the pole to 5862 gdm at the equator.

Fig. 172. Height 4000 metres. Isobars for steps of 5 mb are marked with three figures, 610 mb to 640 mb.

The mean results, also in millibars, by means of registering balloons at 4000 metres are shown by figures for the tens and units: the figures for the hundreds are omitted.

The contours of land at the 4000-metre height are shaded black. The range of geopotential at that height is from 3931 gdm at the pole to 3910 gdm at the equator.

WINDS IN THE UPPER AIR

Our knowledge of the normal circulation in the upper layers of the atmosphere is very imperfect. We get the best idea of the whole from the maps of the distribution of pressure in the month of July which we have given for sea-level, as base, and for 2000, 4000, 6000 and 8000 metres, and in January for sea-level, 2000 and 4000 metres. These maps, as tested by such direct observations of mean pressure as are available, are generally trustworthy. From them we conclude that pressure is so arranged in the upper layers as to give a circulation round the pole from West to East, which becomes stronger with increasing height up to the limit of the maps, and at 8000 metres actually overrides all the distinctive features of the South West monsoon.

From the maps we can calculate the direction and velocity of the wind, but the calculation cannot be extended satisfactorily below latitude 40° N because the gradient becomes indefinite and the directing force of the earth's rotation is so much weakened that we can no longer assume that the motion is along isobars for the equatorial region; it is controlled by some other consideration, and we do not know how far the new form of control may extend North and South of the equatorial belt. The general agreement between the air-motion and the direction of the isobars can be tested to some extent by the motion of clouds, and outside the equatorial region it was shown by Hildebrandsson to be satisfactory for upper clouds, taking Teisserenc de Bort's maps of the pressure at 4000 metres as a guide.

The reader may examine this question for himself by comparing the direction of the isobars with the resultant directions of motion of upper clouds which are here reproduced from Hildebrandsson's tables.

DIRECTION OF MOTION OF CIRRUS CLOUDS OR "UPPER" CLOUDS

The values are given on the scale 01 to 36; 9 East, 18 South, 27 West, 36 North.

70° *to* 90° N.

Annual.

Behring St. to Spitsbergen 17. Karajac 70° N, 50° W, 22 (prevailing direction). Bossekop 70° N, 23° E, 26.

Half-Yearly.

			Oct.–Mar.	Apr.–Sept.				Oct.–Mar.	Apr.–Sept.
Upernivik	73° N	56° W	9	18	Treurenberg Bay	80° N	17° E	27	19
Jan Mayen	71° N	9° W	29	23	Cap Thordsen	78° N	18° E	32	26

Quarterly.

Franz-Josef Land 80–81° N, 50–58° E. Winter 9. Spring 32. Summer 2. Autumn 4.

45° *to* 90° S.

Annual.

Ushuaia	55° S	68° W	28	New Year's Is.	55° S	64° W	29
Snow Hill	64° S	57° W	27 (prevailing direction)	"Discovery"	78° S	167° E	25

Half-Yearly.

			Oct.–Mar.	Apr.–Sept.				Oct.–Mar.
Cape Pembroke	52° S	58° W	29	30	Cape Adare	71° S	170° E	34
Cape Evans	78° S	166° E	3					

Quarterly.

Belgica	71° S	81–99° W	Dec.–Feb. 19	Mar.–May 24	June–Aug. 25	Sept.–Nov. 27
C. Horn and S. Georgia	55° S	36–67° W	,, 28	,, 28	,, 27	,, 28
S. Orkneys	61° S	45° W	,, 26	,, 27	,, 25	,, 26
S. Georgia	55° S	36° W	,, 27	,, 22	,, 25	,, 31

50° *to* 70° N.

Half-Yearly.

Station	Lat.	Long.	Oct.–Mar.	Apr.–Sept.
Stykkisholm	65° N	23° W	31	27
Reykjavik	64° N	22° W	4	2
Teigarhorni	65° N	14° W	34	30
Thorshavn	62° N	7° W	25	21
Drontheim	63° N	10° E	27	24
Kristiania	60° N	11° E	30	26
Aasnes	61° N	12° E	30	25
Lödingen	68° N	16° E	22	21

Quarterly.

Canada, Dec.–Feb. 28, Mar.–May 28, June–Aug. 27, Sept.–Nov. 28

Monthly.

Station	Lat.	Long.	Jan.	Feb.	Mar.	Apr.	May	June	July	Aug.	Sept.	Oct.	Nov.	Dec.
England	—	—	27	29	29	28	26	27	28	27	27	28	28	33
Bergen	60° N	5° E	27	22	27	20	25	24	—	—	21	23	24	23
Aarhus and Hinnerup	56	10	31	31	30	25	25	26	28	26	27	26	30	27
Potsdam	52	13	26	27	26	26	26	26	26	26	28	24	24	23
Germany	—	—	26	31	28	28	23	26	26	26	28	23	26	27
Nora	60	15	32	30	30	28	27	27	26	27	28	28	30	30
Sundsvall	62	17	29	30	30	28	26	27	24	26	28	28	30	31
Qvickjock and Arjeploug	67	18	32	31	31	32	31	31	30	31	33	31	32	35
Upsala	60	18	31	31	29	28	28	28	25	25	29	28	31	31
Tomsk	56	84	31	22	24	28	34	23	25	25	27	27	17	25
Irkoutsk	52	104	31	31	31	29	31	28	30	28	29	30	31	31

40° *to* 50° N.

Half-Yearly.

Barcelona, 41° N, 2° E. Oct.–Apr. 28, Apr.–Oct. 29

Monthly.

Station	Lat.	Long.	Jan.	Feb.	Mar.	Apr.	May	June	July	Aug.	Sept.	Oct.	Nov.	Dec.
St Paul	45° N	93° W	26	24	20	21	23	25	25	26	26	27	27	27
Detroit	42	83	27	27	27	28	29	30	30	30	29	28	27	27
Cleveland	41	82	27	27	27	27	28	29	29	29	28	28	28	27
Buffalo	43	79	28	29	29	28	27	26	26	26	26	26	27	28
Blue Hill	42	71	27	28	28	28	27	27	26	26	26	26	26	27
Madrid	40	4° W	32	29	31	29	26	25	30	27	27	31	31	30
Paris	49	2° E	29	28	28	25	26	25	27	24	27	25	27	26
Perpignan	43	3	30	31	30	30	30	29	29	29	28	28	30	30
Pavia	45	9	29	32	30	30	25	27	32	28	26	27	31	29
Pola	45	14	35	29	29	30	28	29	28	29	28	28	27	28
Tiflis	41	45	32	34	32	29	?17	30	33	?12	25	29	32	32
Zyosin	41	129	28	27	28	27	27	27	27	27	27	27	27	27
Hakodate	42	141	27	29	29	27	28	28	28	29	28	27	28	28

30° *to* 40° N.

Half-Yearly.

Azores 38° N, 29° W. Winter 25, Summer 31

Monthly.

Station	Lat.	Long.	Jan.	Feb.	Mar.	Apr.	May	June	July	Aug.	Sept.	Oct.	Nov.	Dec.
Abilene	32° N	100° W	23	23	23	24	23	22	18	20	19	24	24	24
Kansas	39	95	26	27	27	26	24	23	22	22	22	23	25	26
Vicksburg	32	91	27	27	27	27	27	25	18	15	18	27	29	29
Louisville	38	86	27	27	26	27	28	29	29	28	28	26	26	26
Waynesville ... ... Ocean City	35 38	88 } 75 }	27	27	27	27	27	27	28	28	27	27	27	27
Washington	39	77	27	27	27	27	27	27	27	27	27	27	27	27
Lisbon	39	9	24	26	24	28	25	26	25	25	26	26	26	26
San Fernando ...	37	6° W	6	29	26	28	27	27	27	27	27	27	29	30
Kojak	31	67° E	26	27	26	26	26	—	—	19	—	27	27	26
Lahore	32	74	27	27	25	27	27	22	22	16	25	25	25	27
Simla	31	77	27	27	27	27	27	29	25	22	25	25	27	27
Tsingtau	36	120	26	27	26	25	27	27	27	26	27	26	26	28
Zi-ka-Wei	31	121	27	27	27	27	27	27	31	36	28	27	26	27
Hiroshima	34	132	27	27	27	28	26	27	26	26	26	26	27	26

20° *to* 30° N.

Annual.

Tenerife 28° N, 17° W, 25 (Winter); Persian Gulf, 27° N, 51° E, 31
Atlantic Ocean 25–30° N, 30° W, Apr.–Sept. 22

Quarterly.

Arabian Sea, 28–32° N, Dec.–Feb. 31, Mar.–May 29
,, 24–28° N, Mar.–May 30, Sept.–Nov. 32

Arabian Sea, 20–24° N, Mar.–May 25, Sept.–Nov. 25
Assam, 26° N, 92° E. Each quarter 26

Monthly.

Station	Lat.	Long.	Jan.	Feb.	Mar.	Apr.	May	June	July	Aug.	Sept.	Oct.	Nov.	Dec.
Mexico	19–23°	—	21	22	23	24	23	13	15	10	14	15	21	21
Key West	25	82° W	27	27	27	29	31	35	4	5	13	24	25	26
Havana	23	82	26	26	27	27	28	31	3	5	3	27	26	26
Cienfuegos	22	81° W	24	24	25	25	26	27	var.	var.	var.	25	25	25
N. India*	23–31	74–88° E	27	26	26	25	14	15	13	11	10	8	var.	26
Jaipur	27	76	27	27	27	27	29	29	31	31	31	25	27	27
Agra	27	78	26	27	26	26	26	9	10	9	7	24	25	27
Allahabad	25	82	27	27	27	25	27	4‡	13‡	9‡	18‡	27	25	25
Darjiling	27	88	27	—	26	27	31	—	10	11	18	27	28	—
Calcutta	23	88	27	27	25	27	25	7	9	9	9	20	25	25
Central India†	16–25	67–86	27	27	25	25	22	22	25	9	9	27	27	23
Taihoku	25	122	26	27	26	28	27	31	6	9	30	30	30	28
Naha	26	128	24	27	26	26	27	31	1	2	34	26	26	25

* Lahore, Roorkee, Agra, Lucknow, Allahabad, Patna, Hazaribagh, Calcutta.
† Kurachee, Deesa, Bombay, Poona, Belgaum, Nipur, Jubbulpore, Cuttack.
‡ Very few observations.

10° *to* 20° N.

Annual.

Hawaii, 19° N, 155° W, 27

Half-Yearly.

Square 38. 10–20° N, 10–20° W, "Winter" 22, "Summer" 21. Cape Verde, 15° N, 25° W, "Summer" 13

Quarterly.

Arabian Sea	16–20° N	Dec.–Feb. 29	Mar.–May 31	June–Aug. 12	Sept.–Nov. 34
,,	12–16° N	,, 3	,, 13	,, 6	,, 16

Monthly.

Station	Lat.	Long.	Jan.	Feb.	Mar.	Apr.	May	June	July	Aug.	Sept.	Oct.	Nov.	Dec.
Santiago	20°	76° W	25	26	26	27	28	28	30	30	25	24	24	25
Kingston	18	77	25	24	26	26	27	29	29	35	33	29	27	25
Santo Domingo	18	70	29	29	29	29	28	26	24	20	18	29	29	29
Curaçao	12	69	24	24	24	24	24	var.	var.	var.	var.	27	25	25
San Juan	18	66	27	28	28	28	28	26	26	27	33	33	30	28
St Kitts	17	63	27	27	28	28	28	26	var.	var.	28	31	29	28
Dominica	15	61	26	26	26	26	26	26	22	var.	var.	33	30	29
Trinidad	11	61	26	26	27	27	26	var.	var.	9	7	var.	var.	25
Barbados	13	60	26	26	26	26	26	26	23	var.	var.	33	27	25
Square 39	15	25° W	24	22	22	5	?35	12	11	8	15	15	21	21
Bangalore	13	78° E	19	21	22	22	14	8	10	10	9	11	12	15
Kodaikanal	10	78	17	15	21	20	12	8	8	9	9	9	12	13
Madras	13	80	22	25	22	22	16	9	9	9	9	11	11	20
Manila	15	121	17	17	16	26	7	7	8	8	6	13	11	12

THE UPPER AIR OF THE INTERTROPICAL BELT. DECEMBER TO FEBRUARY

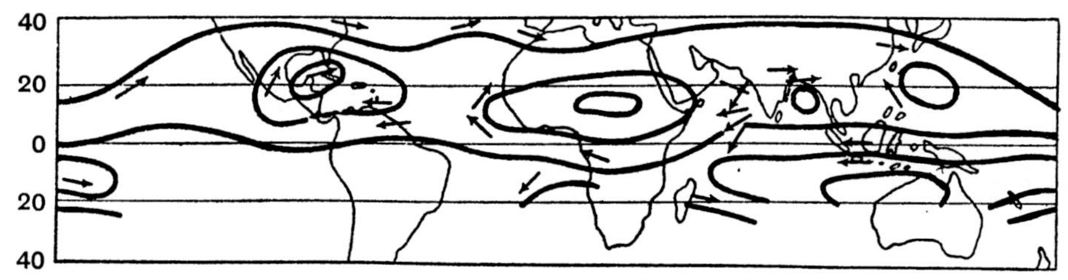

Fig. 173. Lines of flow of cirrus cloud compiled by W. van Bemmelen.

0° *to* 10° N.

Half-Yearly.

Square 4	0–10° N	30–40° W	"Winter" 8	"Summer" 6	Bay of Bengal	5° N	Nov.–Feb. 22
" 2	"	10–20° W	" 10	" 9			

Quarterly.

Paramaribo	6° N, 55° W	Dec.–Feb. 9	Mar.–May 9	June–Aug. 9	Sept.–Nov. 9
Arabian Sea	8–12° N	" 8	" 26	" 23	" 27
"	4–8° N	" 6	" 15	" 22	" 34
"	0–4° N	" 3	" 26	" 26	" 32

Monthly.

Station	Lat.	Long.	Jan.	Feb.	Mar.	Apr.	May	June	July	Aug.	Sept.	Oct.	Nov.	Dec.
San José	10°	84° W	8	11	4	7	2	6	9	6	7	8	7	7
Square 3	0–10	20–30	17	15	11	11	12	10	10	6	7	11	15	10

SOUTHERN HEMISPHERE.

0° *to* 45° S.

Annual.

Mercedes	33° S	58° W	28	Mukimbungu	5° S	14° E	13
Kaikoura	42° S	174° E	26	Ascension	14° S	8° W	4

Half-Yearly.

Square 303, 0–10° S, 30–40° W, "N. Winter" 9, "N. Summer" 9
Square 302, 0–10° S, 20–30° W, " 10, " 11
Square 301, 0–10° S, 10–20° W, " 4, " 4
Johannesburg, 27° S, 28° E, July–Sept. 26
Bay of Bengal, 5° S, Nov.–Feb. 15

Quarterly.

Pontianak	0° S	109° E	Dec.–Feb. 9	July–Sept. 6	Oct.–Nov. 12
Batavia	6° S	107° E	" 9	" 6	" 7

Monthly.

Station	Lat.	Long.	Jan.	Feb.	Mar.	Apr.	May	June	July	Aug.	Sept.	Oct.	Nov.	Dec.
Samoa	14°	172° W	28	26	27	29	34	30	33	28	30	27	29	30
Mauritius	20	58° E	28	28	28	29	30	30	32	34	34	31	29	28
Melbourne	38	145	27	26	27	28	28	28	27	29	28	27	27	26

W. van Bemmelen has extended this form of inquiry to find the general circulation over the intertropical regions by observations of cirrus. We reproduce (figs. 173–4) the maps which he prepared. They are quite reasonably well in accord with our pressure-map for the Northern Hemisphere at 8000 metres for July. W. A. Harwood has given corresponding results for India, which suggest that the elongation over India from the super-African high pressure at high levels ought to have a separate centre over India, but otherwise the agreement is good.

THE UPPER AIR OF THE INTERTROPICAL BELT. JUNE TO AUGUST

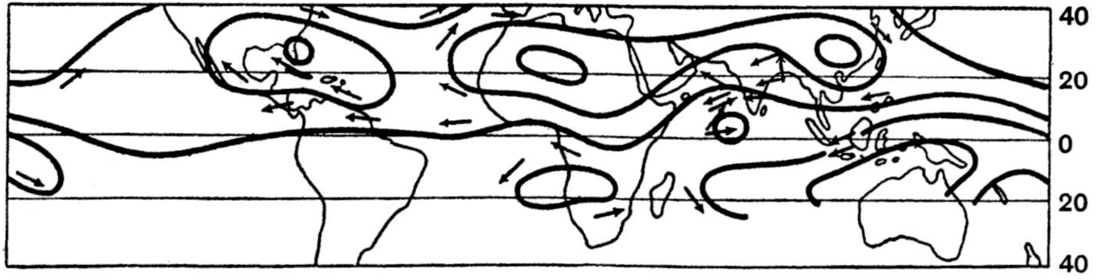

Fig. 174. Lines of flow of cirrus cloud compiled by W. van Bemmelen.

Summaries of direct observations of wind by means of kites are sufficiently numerous to check the results of the first 5 kilometres over Lindenberg and also over the United States. We quote here summaries for six stations in the United States. A summary for the station at Lansing (Mich.) is also available in the *Monthly Weather Review* 1925, p. 17.

Circulation in the upper air over the United States[1].

Direction in degrees from North, velocity in metres per second.

Height over m.s.l.	Broken Arrow Okla.		Drexel Nebr.		Due West S. C.		Ellendale N. Dak.		Groesbeck Tex.		Royal Center Ind.	
	233 m	4–7 yrs	396 m	9–10 yrs	217 m	4–5 yrs	444 m	7–8 yrs	141 m	7 yrs	225 m	7–8 yrs
km	°	m/s	°	m/s	°	m/s	°	m/s	°	m/s	°	m/s
						JANUARY						
5	265	10·1	282	17·8	—	—	—	—	268	14·7	—	—
4	267	10·9	274	17·0	262	16·1	300	16·9	253	12·6	256	15·2
3	270	9·6	280	14·2	268	16·4	291	14·1	260	10·1	266	14·2
2	259	7·0	285	10·4	269	12·0	294	10·5	258	7·5	262	11·9
1	230	3·5	281	5·5	261	5·1	296	6·3	243	3·4	254	7·3
0·5	211	2·1	275	2·4	270	2·3	297	3·2	236	1·4	238	4·7
Surface	224	0·8	270	1·4	289	1·0	301	2·8	326	0·6	229	2·0
						APRIL						
5	275	13·7	278	14·9	—	—	286	15·3	—	—	—	—
4	263	12·4	274	11·1	283	13·6	281	8·8	265	11·9	267	13·5
3	259	8·3	268	8·7	263	10·1	278	6·2	244	10·5	274	9·8
2	240	7·6	266	4·3	261	7·7	280	3·0	228	8·0	264	8·3
1	205	5·6	235	1·2	243	3·7	307	1·0	201	6·1	235	5·8
0·5	189	4·2	170	0·6	255	2·4	349	1·5	184	4·7	222	4·0
Surface	181	2·8	154	0·4	262	1·4	349	1·6	175	2·5	232	1·9
						JULY						
5	210	4·5	235	8·4	—	—	287	15·3	—	—	—	—
4	254	6·7	260	7·6	279	8·4	292	11·2	277	1·1	290	9·5
3	242	4·3	254	6·2	266	7·9	278	7·6	202	3·7	269	11·2
2	221	3·6	238	4·9	268	5·6	267	4·2	210	4·0	268	7·3
1	207	4·8	206	4·3	268	2·7	230	1·7	210	6·0	260	4·8
0·5	191	4·6	184	2·9	258	2·0	187	0·2	210	6·2	256	3·3
Surface	178	3·0	183	2·0	247	1·3	341	0·1	201	3·6	260	1·7
						OCTOBER						
5	251	9·8	275	11·8	326	6·9	—	—	303	4·0	—	—
4	242	8·4	266	11·3	267	6·5	271	11·4	229	2·7	263	12·2
3	243	6·0	261	9·5	272	3·2	274	8·8	237	3·4	267	10·7
2	226	4·7	252	7·0	20	0·4	271	6·1	221	2·4	260	9·0
1	198	4·4	226	4·2	61	3·6	266	3·1	172	2·9	248	6·3
0·5	185	3·8	206	2·1	54	4·4	270	1·7	153	2·9	231	4·5
Surface	178	2·5	201	1·6	50	2·8	281	1·6	109	0·9	222	2·2

For observations extending to the higher levels we have results from Lindenberg (fig. 175), Batavia (fig. 176) and Samoa (fig. 177), India (fig. 178) and also a month of observations in Egypt (fig. 179).

Prevalence and average speed of the wind in the four quadrants.

From observations at the Aeronautical Observatory, Lindenberg.

Height	N–E		E–S		S–W		W–N	
km	Number	Speed	Number	Speed	Number	Speed	Number	Speed
15	20	7·6	7	5·4	22	9·7	18	7·7
14	24	9·1	12	4·7	22	11·6	28	9·9
13	32	11·6	13	6·9	30	12·6	34	12·3
12	30	15·3	15	10·3	34	14·9	42	14·5
11	35	18·6	13	10·0	40	17·1	39	15·4
10	33	18·3	13	11·3	41	16·4	44	17·4
9	39	16·4	14	10·9	40	16·5	39	17·0
8	39	15·5	16	10·7	36	15·2	41	13·7
7	41	12·7	15	12·7	40	12·8	36	12·7

[1] *Monthly Weather Review*, vols. LII–LIII, October 1924 to July 1925.

There are now probably sufficient observations of pilot-balloons at many other stations both in Europe and America to make a comparison with the distribution of pressure in the lower layers; we give references to compendious summaries by W. J. Humphreys for the United States[1], L. Marini for Genoa[2] and A. de Quervain for Greenland[3].

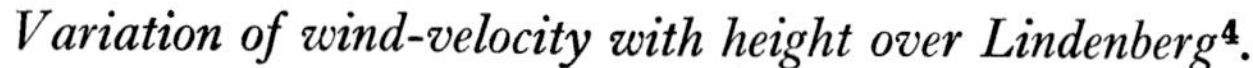

Variation of wind-velocity with height over Lindenberg[4].

Height km	Mean Velocity m/s Summer	Mean Velocity m/s Winter	Direction in degrees from N Summer	Direction in degrees from N Winter
25	9·2	9·4	—	—
24	8·1	10·5	—	—
23	7·4	11·7	—	—
22	8·8	12·9	—	—
21	9·2	13·2	—	—
20	8·3	14·1	—	—
19	8·7	12·5	—	—
18	8·7	12·9	297	286
17	10·1	13·9	300	288
16	10·0	14·6	302	288
15	10·9	15·4	303	286
14	12·6	15·5	300	279
13	14·2	15·8	296	274
12	16·1	17·4	296	271
11	17·5	18·5	296	266
10	16·8	19·0	296	264
9	15·7	18·6	297	263
8	14·0	17·6	297	264
7	12·6	15·7	297	265
6	11·2	14·0	298	264
5	9·8	12·2	297	262
4	8·6	10·3	297	260
3	7·5	8·7	297	260
2	6·3	7·3	295	259
1	5·6	5·8	287	260

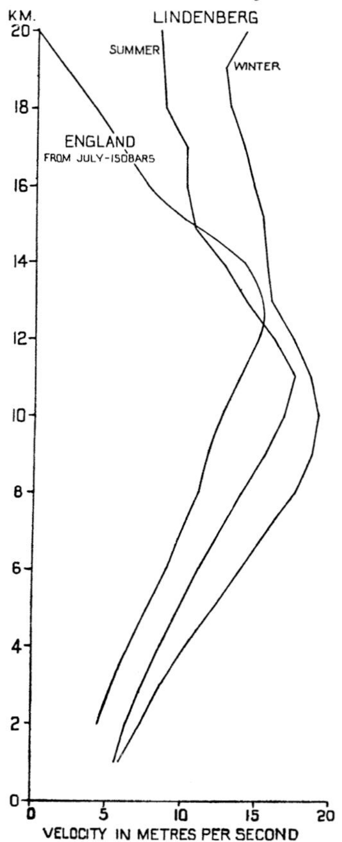

Fig. 175. Average winds for England between 2 km and 20 km computed from the estimated distribution of pressure, with mean values of observations of balloons at Lindenberg Observatory.

The summary of observations at Lindenberg for the winds of thc stratosphere is particularly instructive, and gives the mean of a number of observations which are included in those plotted separately by Dobson[5]. We give the table summarising the results and note the close parallelism between the number of North East winds and South West winds at different heights as shown by the second and sixth columns of the table on p. 270. The frequency of winds from the East in a region where on the average the flow of air is predominantly from the West raises a question to which we are not able to return a satisfactory answer.

[1] 'The Way of the Wind,' *Journal of the Franklin Institute*, September, 1925.
[2] *Mem. d. R. Uff. Cent. Met. e Geof.*, vol. I, Serie 3, Roma, 1924.
[3] Schweizerische-Grönlandexpedition, 1912–3, Zürich, 1920.
[4] W. Peppler, see p. 291.
[5] *Q. Journ. Roy. Meteor. Soc.*, vol. XLVI, 1920, p. 54. (See also *M.*, vol. IV, p. 184.)

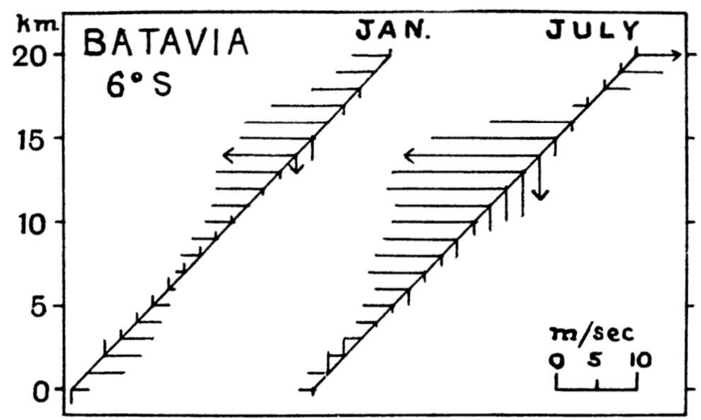

Fig. 176. Clothes-line diagrams according to the examples cited in vol. IV, p. 186, of means of the cardinal components of winds in the upper air of Batavia, based on van Bemmelen's observations numbering more than twenty up to 6 km in January and 11 km in July.

The results of the observations at Batavia in January and July are set out in fig. 176. Direction is represented as well as velocity by the use of the device of "clothes-line" diagrams for the mean values of the West-East component and the South-North component at the different levels. The method is quite effective in this case because when everything is taken into account the dominant component is that from the East. At the same time the remarkable change in July from an East wind of 16 m/s at 14 km to a West wind of 5 m/s at 19 km is a striking phenomenon. It asks for further investigation by comparison with results which might be obtained at other stations within the intertropical belt. The variations in the S-N component are comparatively slight except for a development of the Northerly component to a velocity of 6 m/s at 14 km in July when the Easterly component reaches its maximum.

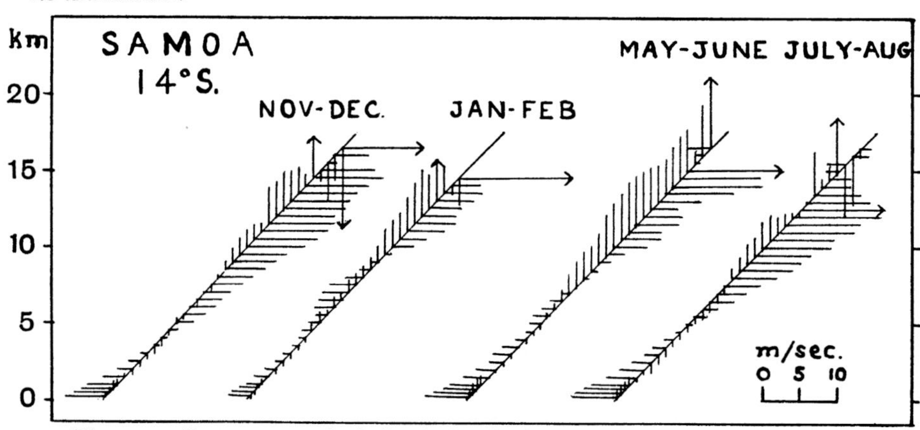

Fig. 177. Components of wind-velocity for Samoa in summer (Nov.—Feb.) and in winter (May—Aug.).

In fig. 177 corresponding diagrams are given for the means of the cardinal components of winds in the upper air of Samoa. They are based upon the results given in a memoir of *Observations of upper air-currents at Apia, Samoa* (second series) by Andrew Thomson, Director of the Observatory, 1929. The diagrams show the means of the results for two pairs of months in the summer half of the year and for two other pairs in the winter half.

In all four diagrams winds from the East (with a touch of South in it) are shown at the surface diminishing almost to zero under 5 km with not much more than a suspicion of winds from the North, and beyond the five kilometre level components from the West begin to show in two of the four pairs, one summer and one winter, and become notably stronger with increased elevation; in the other two pairs the Westerly wind begins only at 10 km. In three of the groups considerable velocity is attained within the range of 20 km.

In all of them South components show themselves with more or less vivacity over a considerable range of height above seven or eight kilometres and somewhat irregular conditions are indicated in the diagrams above 15 km including the largest of the components from each of the cardinal points except the East.

In his report on the observations of the international cloud-year 1896–7 Hildebrandsson showed the prevailing wind for each month by a line representing the direction; the sequence from month to month was shown by placing the lines for each month to follow on, and the lines for every month had the same length. In fig. 178, overleaf, for the upper winds of India we have adopted a similar plan; but have made allowance for representing the percentage frequency of the prevalent wind and also the frequency of the winds of other directions.

Fig. 178. The percentage frequency of the most prevalent direction for each month is shown by a thick line with a scale of 1 mm to 6 per cent., the less prevalent directions follow in order of frequency as thin lines on the same scale. The initial of the month is at the angle of thin and thick. The winds of the months succeed each other as shown by the arrow for January.

The directions of the prevailing winds in the monsoon months are marked by arrows on the corresponding lines.

* * *

Pilot-balloons are now sent up daily in many countries in the interests of aerial navigation.

Within the past seven years many thousands have been observed and their records noted. We give, p. 436, a list of publications in which some of the results can be found. A conspectus of the inferences to be drawn from them to correspond with the maps of the distribution of pressure at different levels is very desirable but beyond the capacity of this work.

Since 1 January, 1933, the Meteorological Office of the British Air Ministry has utilised the observations of pressure-temperature given by aeroplanes, and those of wind by pilot-balloons and clouds, in maps for the previous day of the distribution of pressure and wind at 1000 and 2000 metres over Europe, and the extension of the map North, South, East and especially West is very desirable.

One of the points at issue in any enterprise of this kind is to decide to what degree of accuracy the motion of a balloon or of a cloud can be accepted as an indication of the general drift of air in which the balloon is carried or the cloud formed and for which the distribution of pressure may be some guide. Any layer of air is liable to carry local disturbances of the drift due to eddy-motion and the apparent motion of cloud will naturally be affected by the formation of cloud on the boundary which is apparently defined by the cloud itself. Nearly every view of cloud-groups in the sky asks awkward questions of this kind about their motion. It is a difficult problem and synoptic charts day by day seem to be the most direct method of approach to its solution.

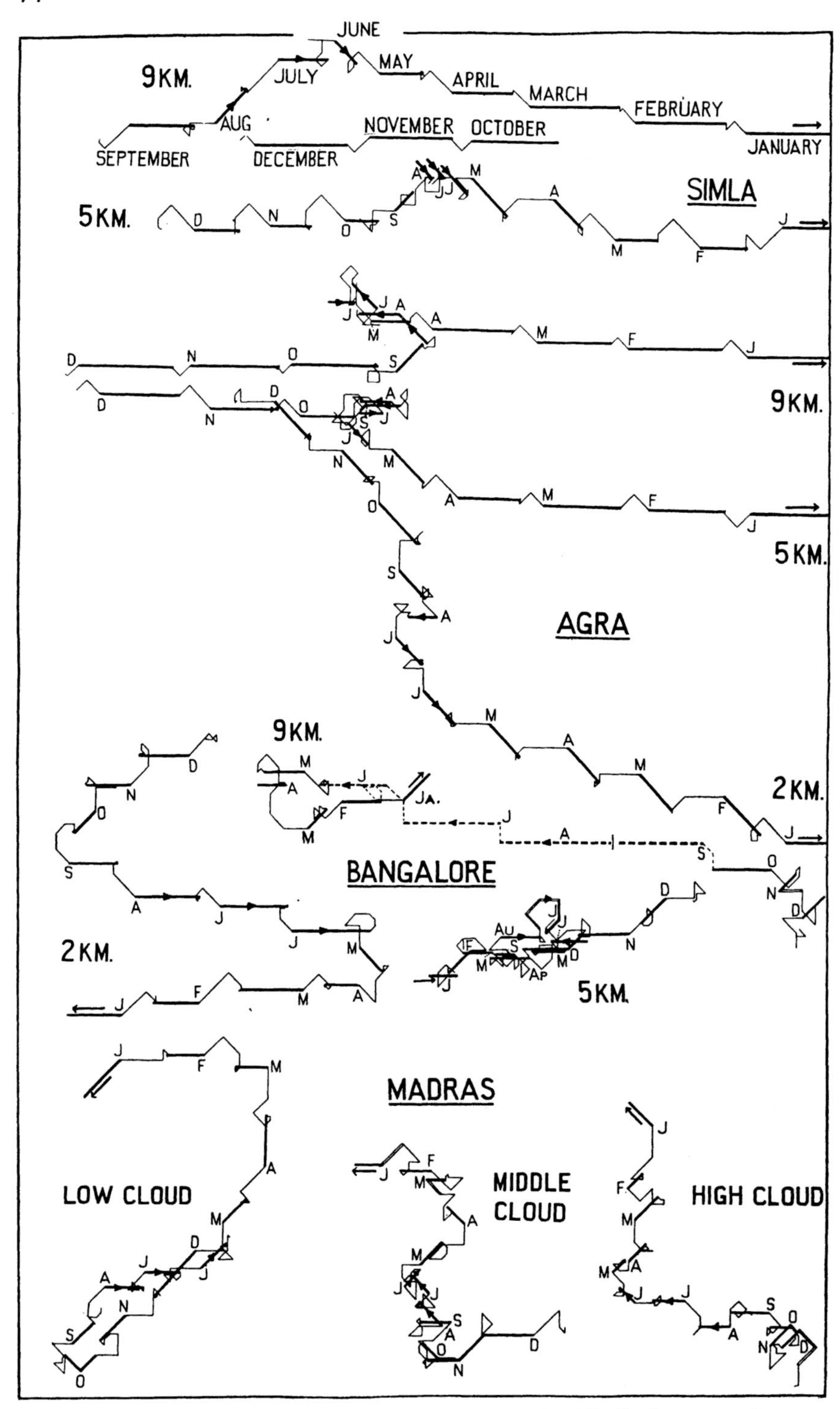

Fig. 178. The normal movement of the upper air of India (see p. 273).

RAINFALL AND WINDS AT THE SOURCES OF THE NILE

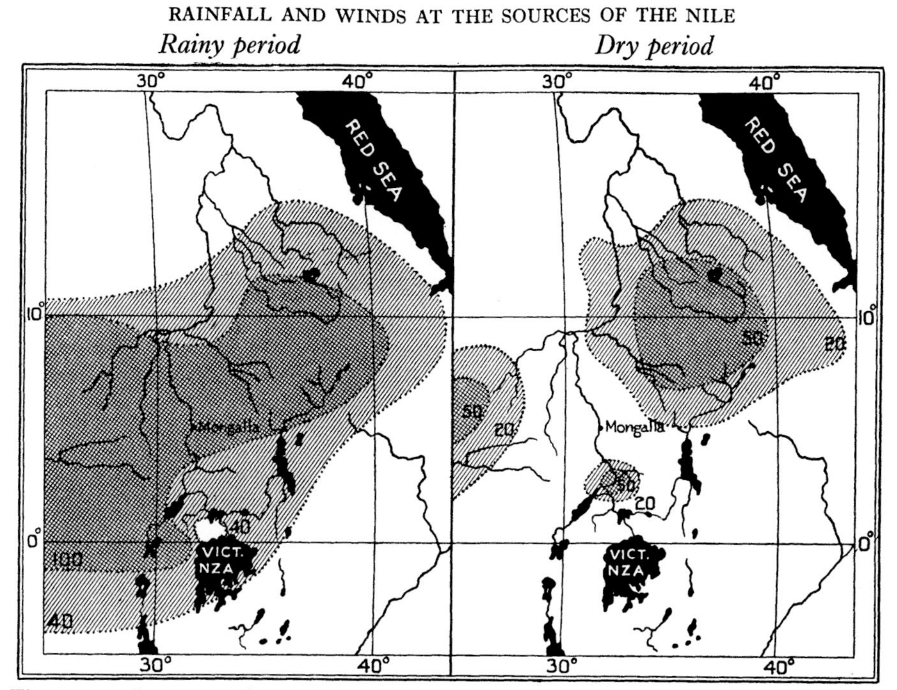

Fig. 179*a*. Isograms of rainfall in millimetres over Abyssinia and the Sudan in 1908 during two rainy weeks Aug. 7–13, Aug. 21–27 and during a dry week Aug. 14–20.

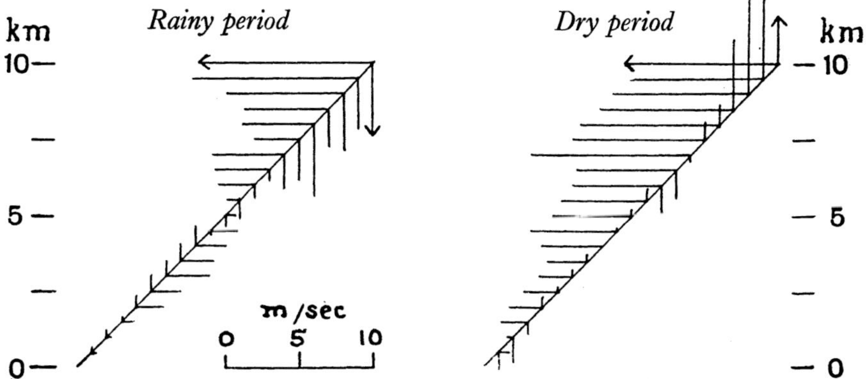

Fig. 179*b*. Winds in the upper air above Mongalla during the dry period Aug. 14–21 and the rainy period Aug. 7–13 and 22–27.

Perhaps the most instructive examples of those which we have selected for reproduction are contained in the brief investigation of the upper winds of Egypt and Abyssinia in the month of August 1908. The feature to which we wish to draw special attention is the association of rainy weather in the Sudan with winds from the West. Abyssinia itself seems to get it both ways. The West is certainly not a quarter from which we should expect rain in Egypt, and the re-assertion of the character of West as a rainy wind in that district is a suggestive indication.

The reversal of the trade-winds. The anti-trades.

With regard to the general results of the inquiry we may remind our readers that the reversal of the North East trade-wind in the upper levels has been recognised for a long time by observations on the Peak of Tenerife. The reversal of the South East trade at the level of about 1000 metres over the South Atlantic Ocean was clearly shown by pilot-balloons from H.M.S. *Repulse* during the journey of the Prince of Wales in 1925. The Peak of Tenerife rises to a height of 3707 metres. A careful investigation of the phenomena by R. Wenger[1] in 1910–11 by means of pilot-balloons showed that the winds in the region of the Canary Islands above the trade-wind of the surface were very variable. In winter the balloons mostly indicated an extension of the North East wind to the highest limits of the ascents and the continuity of the North East current extended over more than half the year. In the summer however the observatory at 2090 metres was in the anti-trade which is represented by a current from the South West. The earlier observations of Humboldt, Basil Hall, Buch and others, mostly summer observations, have been assumed to support the theory that the air which flows from North East in the trade-wind rises in the equatorial belt and flows back over the North Easterly current, taking its direction by the influence of the rotation of the earth exactly opposite to the influence at the surface. The return current is consequently called the anti-trade. But that facile explanation is not at all satisfactory. First, let us remark that the reversal is not by any means confined to the trade-wind, it was already noticed by the Greeks for the Mediterranean region, it is certainly common in certain conditions of weather over England as shown by a map compiled by C. J. P. Cave[2], and it seems not unlikely that it may happen in a similar manner in many parts of the world where the return of the original air overhead cannot be regarded as likely.

But the chief objection to any theory of direct reversal is set out in a paper in *The Air and its Ways*. With our present knowledge we must suppose that a current flowing from the North East has lower pressure on its left and higher pressure on its right, and any explanation of the reversal must also explain the change in the distribution of pressure which seems quite apart from the idea of a direct return of air overhead. It would occur if the general scheme of isobars for the levels above 2000 metres showed a deviation from the line of latitude with a return from the South West over the trade-wind region, but the best means of studying the subject is provided by Sverdrup's model (see p. 251). In the model the return current or anti-trade is shown as provided by a high-pressure over Africa round which air passes in the upper levels turning Northward over the Canaries and forming a common current with that which, further North, forms part of the general Westerly circulation. We conclude that the isobars of the upper air such as those of figs. 169–72 are the best guide to the circulation at the levels above 2000 metres.

[1] *Siebente Versammlung der Int. Komm. für wissenschaftliche Luftschiffahrt*, Wien, 1912; also H. von Ficker, *Festschrift d. Zentralanstalt*, Wien, 1926. For Mauritius, see p. 291.

[2] *The Structure of the Atmosphere in Clear Weather*, Cambridge, 1912.

THE NORMAL VELOCITIES OF THE CURRENTS IN THE GENERAL CIRCULATION

Individual observations of the velocity at different levels of the North East trade-wind are given by Sverdrup[1], Hergesell[2] and Peppler[3], from which mean values of the velocity of that primary current can be derived, but the observations are confined to the summer months.

Information about the intertropical current is given by occasional observations that have been taken in the West Indies, the Indian Ocean, Batavia and Victoria Nyanza and about polar currents by those of de Quervain in the North and Barkow in the South.

The surface-air.

The best values for the steady currents are those derived from the anemometer at St Helena, which is installed on the island heights at 632 metres above the sea in the heart of the South East trade. From the *Réseau Mondial*, 1917, following on the memoir published by the Meteorological Office in 1910, we obtain the following measures of the wind:

	Jan.	Feb.	Mar.	Apr.	May	June	July	Aug.	Sept.	Oct.	Nov.	Dec.
m/s	7·3	6·5	6·4	6·6	6·0	6·4	6·9	8·1	8·8	8·7	8·3	7·8

From the same memoir, measures of the velocity of the North East trade computed from ships' observations on the Beaufort scale in the region between the parallels 10–30° N, and between the 30th meridian W and the African coast, gave the following results:

	Jan.	Feb.	Mar.	Apr.	May	June	July	Aug.	Sept.	Oct.	Nov.	Dec.
m/s	5·3	5·8	5·6	6·0	5·5	5·0	4·6	3·7	4·3	3·3	4·4	5·2

The corresponding values of the South East trade obtained by the same method from the region 0–10° E and 10–20° S and 0–10° W and 0–20° S are:

	Jan.	Feb.	Mar.	Apr.	May	June	July	Aug.	Sept.	Oct.	Nov.	Dec.
m/s	5·9	6·3	6·0	6·7	6·4	6·7	5·8	6·7	6·4	6·1	6·7	6·4

It is noteworthy that the mean velocity of the two trades thus estimated shows very little variation within the year.

For the period 1855–1914 values of the surface-flow of the North East trade-wind in the region 15–25° N and 25–40° W have been given by P. H. Gallé[4]:

	Jan.	Feb.	Mar.	Apr.	May	June	July	Aug.	Sept.	Oct.	Nov.	Dec.
m/s	5·5	5·2	4·6	4·7	5·8	6·2	6·1	5·3	5·3	4·5	4·6	4·7

and for the South East trade in the region 5–10° S and 15–35° W the three-monthly mean values in metres per second are as follows: Dec.–Feb. 4·6, Mar.–May 4·3, June–Aug. 6·1, Sept.–Nov. 5·6.

[1] *Veröff. d. Geophys. Inst. der Univ. Leipzig*, Leipzig, 1917.

[2] 'Passatstudien in Westindien,' *Beitr. z. Physik d. freien Atmosphäre*, Band IV, Leipzig, 1912, pp. 153–87.

[3] 'Die Windverhältnisse im nordatlantischen Passatgebiet, dargestellt auf Grund aerologischer Beobachtungen,' *Beitr. z. Physik d. freien Atmosphäre*, Band IV, Leipzig, 1912, pp. 35–55.

[4] *K. Akad. v. Wetenschappen te Amsterdam*, vol. XVII, 1915.

Gallé[1] gives values also for the South East trade in the square 10–20° S and 80–90° E as follows:

	Jan.	Feb.	Mar.	Apr.	May	June	July	Aug.	Sept.	Oct.	Nov.	Dec.
Beaufort	3·6	3·0	2·9	3·9	4·1	4·0	4·8	4·7	4·8	4·6	4·3	4·1
Direction in degrees from N through E	123	111	128	126	137	147	143	147	140	133	129	134

We quote also a table of the monsoon region for the square 5–15° N and 50–60° E:

	Jan.	Feb.	Mar.	Apr.	May	June	July	Aug.	Sept.	Oct.	Nov.	Dec.
Beaufort	3·0	2·6	2·0	0·9	1·8	5·1	5·5	4·9	3·5	0·1	2·4	3·2
Direction in degrees from N through E	57	66	81	109	216	223	221	220	213	86	48	54

Wind-roses of one sort or other have been compiled for 10-degree squares or some smaller division for all oceans and for a number of stations on land; but the main features of the surface-circulation may best be inferred from the distribution of pressure except for the equatorial region.

The geostrophic winds of the upper air.

The mean velocity of geostrophic wind over England is represented by the curve of fig. 175 which shows a wind gradually increasing from the surface up to 13 kilometres and then falling off to an assumed zero at 20 kilometres. We can use a similar method to obtain an approximation to the motion of the air at various altitudes between 40° N and 60° N. See fig. 180.

For the intertropical belt we may conclude that there is a flow from East to West on the equatorial sides of the high-pressure systems which are shown in the maps of the different levels and in van Bemmelen's maps of the movement of cirrus. We may extend this conclusion to infer a general flow of air from East to West along the equator. The inference is supported by the careful observations that were made in 1883 of the motion of the clouds of dust ejected at the eruption of Krakatoa, a volcanic island to the west of Java and quite close to the equator. From those observations it was concluded that the motion of the air 80 kilometres above the equator formed a current carrying the cloud of dust from East to West at about 80 kilometres per hour. The dust spread gradually over the whole sky.

The velocity thus computed is greater than that assigned to the mean motion of any part of the atmospheric circulation within the region of computed isobars and the gradual spread over the whole earth has also to be accounted for.

Temperature in the lower stratosphere has a positive gradient from South to North; it follows that in the higher regions of middle latitudes which have been explored by sounding balloons the gradient of temperature is opposite to the gradient of pressure, and consequently the gradient of pressure Northward would be diminished with height. It is suggested by W. H. Dines that the pressure-gradient might be reduced to zero at a height of about 20 kilometres. We may go a step further and suggest that if the

[1] 'Klimatologie van den Indischen Oceaan,' *K. Ned. Met. Inst. Meded. Verhandl.* 29a, 1924.

scheme of distribution of temperature in isothermal columns were continued, the gradient of pressure over the West to East circulation would be reversed and a circulation from East to West would be maintained. In the equatorial region on the other hand, where there is a slight gradient for Easterly winds from the belts of high pressure towards the equator, the gradient of temperature would be in the same direction as that of pressure and consequently the gradient for Easterly winds would be increased.

SKETCH OF AN ATMOSPHERIC CLOCK TO ILLUSTRATE THE ROTATION OF LAYERS OF THE ATMOSPHERE; DERIVED FROM THE DISTRIBUTION OF PRESSURE REPRESENTED IN FIGS. 167, 172, 170.

JULY

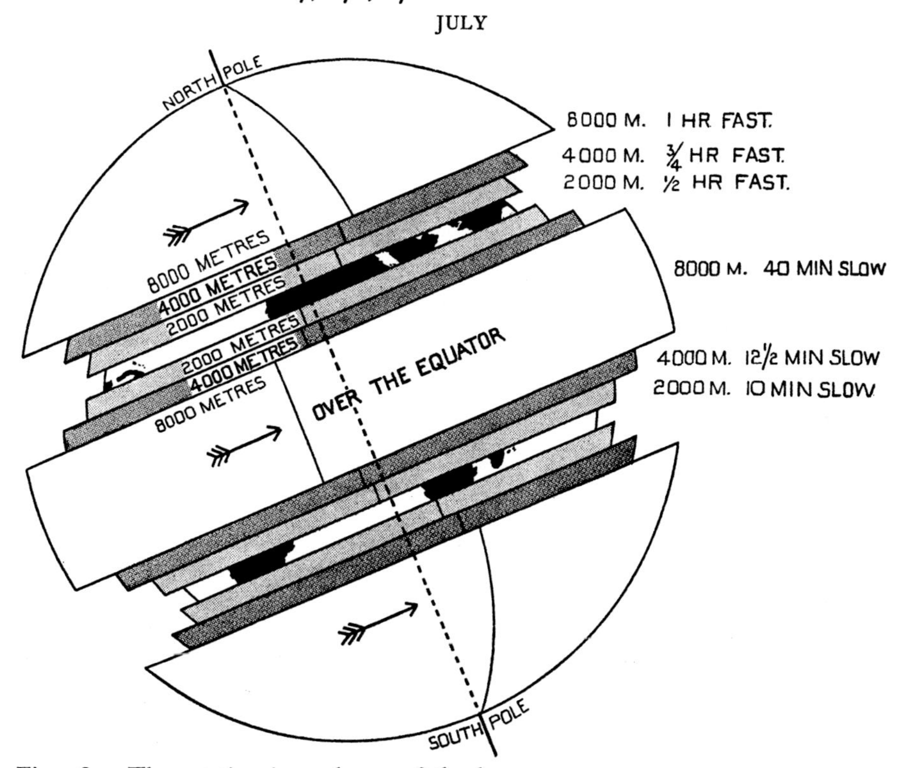

Fig. 180. The rotation in 24 hours of the layers at 2000 metres, 4000 metres and 8000 metres North and South of the "belt of calms" in each hemisphere is compared with the rotation of the solid earth which is represented by the black and white band. The lag in 24 hours of the layers in the intertropical belt and the advance in the same time of layers North and South of that belt are indicated by the positions of the cross-lines on the diagram with reference to the interrupted line which marks the meridian of Greenwich.

We have however not yet attained sufficient certainty in our knowledge of the upper stratosphere to draw final conclusions about the general circulation. Meanwhile we may exhibit a diagram, fig. 180, to give a general idea of the circulation of the atmosphere in relation to the permanent rotation of the solid earth at three levels, derived from the consideration of the pressure-maps of this chapter for those levels.

Clayton's or Egnell's law of variation of wind with height.

The profiles of the curves of fig. 175 will be recognised by those who are accustomed to work with pilot-balloons as essentially characteristic of the atmosphere in many parts of the world. In individual cases of pilot-balloons there is often complication at the bottom, but the upper part of the curve shows the general shape for winds from North or South in England. The gradual increase with height had been noticed quite independently of any observations of pressure-gradient, and both Egnell in France and Clayton in the United States have drawn attention to cases in which there is inverse proportion between the velocity and the air-density at different levels. Such a condition would exist if the gradient did not change with height (volume IV, p. 204) and it certainly does not hold beyond the tropopause.

THE DIURNAL VARIATION OF WIND AND PRESSURE

One of the most notable features of the mean values of wind recorded upon an anemometer is its diurnal variation. Turning at random for information on the subject to the *British Meteorological and Magnetic Year-Book* we find the following figures for yearly averages of wind-velocity in metres per second at the four observatories of the Meteorological Office. Mean values for the several months for the last two are included in the diagrams of isopleths, volume I, fig. 107; and the first two, with Ben Nevis and Falmouth in addition, in volume IV, p. 97.

Hour	2	4	6	8	10	12	14	16	18	20	22	24
Aberdeen	3·4	3·5	3·5	4·0	4·4	**4·6**	**4·6**	4·4	4·0	3·6	3·4	3·5
Eskdalemuir	4·3	4·3	4·5	5·1	5·9	6·2	**6·6**	6·3	5·3	4·7	4·4	4·4
Cahirciveen	5·5	5·6	5·5	5·5	6·0	6·4	**6·8**	6·7	6·2	5·7	5·6	5·4
Richmond	2·6	2·6	2·7	3·2	4·1	4·3	**4·5**	4·0	3·5	3·0	2·9	2·7

In these figures a maximum is clearly shown about two o'clock in the afternoon, a minimum at about two o'clock in the morning. The ratio of maximum to minimum is about 5 to 3 at Richmond in a flat plain, 3 to 2 at Eskdalemuir on high moorland, 5 to 4 at Aberdeen on the East coast, and at Cahirciveen on the West coast. The actual ratios are 1·73, 1·54, 1·35 and 1·26 respectively. This increase of wind in the day-time and falling off at night is a very well-known characteristic of land-stations. It is not the same at different levels and is reversed at a moderate height. It does not occur at sea where there is an indication of a 12-hourly variation[1]. In discussing the diurnal variation in the force of the South East trade-winds of the Indian Ocean, P. H. Gallé[2] writes: "The amplitude of the single-daily period is by far the largest and from May to October the strength of the trades reaches a maximum during night-hours. Sailors of a past generation used to say that by night the speed of their ships was better than by day."

[1] Report of H.M.S. *Challenger*, vol. II, pt. V, Plate II.

[2] 'Klimatologie van den Indischen Oceaan,' *K. Ned. Met. Inst.* No. 102, *Meded. Verhandl.* 29 a, 1924, p. 58. See also Kidson, 'Observations from the Willis Island Meteorological Station,' *A. A. Report*, Adelaide, 1926.

The semi-diurnal variation of pressure.

Pressure also is subject to well-recognised diurnal variation; it has been studied comprehensively by Hann[1], Angot[2], Simpson[3] and others. Its most characteristic feature is an oscillation with a period of twelve hours that has its maximum about ten o'clock in the forenoon and afternoon, at the same local time all over the intertropical and temperate zones changing in the polar region to an oscillation which has its maxima simultaneously along a circle of latitude. The 12-hourly oscillation is greatest in the equatorial and tropical regions and diminishes to a small and indefinite amount as the polar regions are approached. It can only be represented as a wave of pressure, like a tidal wave, passing round the earth two hours in advance of the sun. The oscillation which is manifest in the polar regions is, on the other hand, a standing wave between pole and equator.

By analysing the observed daily variation of the barometer we obtained the value of the amplitude and phase of the 12-hourly pressure-oscillation for 190 stations between equatorial regions and the North pole. These observations were grouped into eight zones, and those in each zone reduced to a common mean latitude, so that they can be considered as observations made on the same circle of latitude.

It was then assumed that the observed oscillations were due to two vibrations of the atmosphere, one of which [the equatorial vibration] occurs at every place on the circle of latitude at the same local time, while the other [the polar vibration] occurs at every place on the circle of latitude at the same Greenwich time.

From the observed values the amplitude and phase of each of these two vibrations were determined for each zone by the method of least squares.[3]

The values (transformed from mm to mb) are given in the following table:

	Equatorial vibration		Polar vibration	
Latitude N	Amplitude mb	Time of maximum Local time	Amplitude mb	Time of maximum G.M.T.
80	0·029	9 h 54 m	0·107	11 h 7 m
70			0·096	11 43
60	0·128	9 44	0·083	11 23
50	0·320	9 54	0·055	11 31
40	0·516	9 52	0·057	11 58
30	0·837	10 02	0·079	2 39
18	1·113	9 49	0·109	3 46
0	1·227	9 46	0·091	3 8

Besides the term of 12 hours, which can always be evaluated by harmonic analysis from a series of hourly values, there is a 24-hour term in the variation of pressure as well as others of shorter period, 8 hours, 6 hours and so on. The 24-hour term is different both in amplitude and phase at different places.

An interesting question arises as to the relation, if any, between the diurnal variation of wind and that of pressure in view of the fact that a close connexion between pressure-gradient and wind is already recognised. By way of inviting

[1] Hann-Süring, *Lehrbuch der Meteorologie*, 4 Aufl., Leipzig, pp. 194 et seq.

[2] A. Angot, *Ann. Bur. Cent. Météor. France*, 1887, pt. I, p. B. 237 and 1906, pt. I, p. 83.

[3] G. C. Simpson, 'The twelve-hourly barometer-oscillation,' *Q. Journ. Roy. Meteor. Soc.*, vol. XLIV, 1918, pp. 1–18.

the student of meteorology to take up this question we give a table of the seasonal and diurnal variation of wind over Lindenberg.

Diurnal variation of wind in the free air over Lindenberg[1] (cm/sec).

Summer (May–October).

West component.

m	0	2	4	6	8	10	12	14	16	18	20	22
4000	624	641	640	639	631	609	593	589	600	609	600	601
3000	560	477	524	587	581	511	456	441	555	528	501	548
2000	423	454	470	459	457	421	416	393	444	421	377	427
1000	301	360	394	380	380	390	372	364	326	287	277	290
500	261	314	333	285	285	316	320	332	293	295	246	219
130	29	20	17	48	111	162	196	191	150	87	46	28

South component.

m	0	2	4	6	8	10	12	14	16	18	20	22
4000	−7	45	52	2	−8	60	76	8	−19	−3	−8	−25
3000	−70	−57	−6	20	37	30	9	8	−5	−24	−33	−63
2000	−10	−25	−24	−11	−8	38	7	−5	−31	−44	−81	−28
1000	45	38	3	6	23	47	51	−3	−59	−56	−13	12
500	55	51	−8	43	108	80	53	14	−47	−71	−42	−4
130	31	44	59	74	73	68	57	34	−6	−21	−10	17

Diurnal variation of wind in the free air over Lindenberg (cm/sec).

Winter (November–April).

West component.

m	0	2	4	6	8	10	12	14	16	18	20	22
4000	507	429	453	385	352	470	544	467	516	616	552	557
3000	450	424	370	413	357	358	437	440	494	447	382	408
2000	347	367	367	370	271	341	435	365	446	387	321	357
1000	386	377	328	412	383	388	350	388	429	348	394	358
500	329	356	400	378	356	342	309	341	397	388	337	325
130	78	75	78	69	53	93	122	142	106	84	72	72

South component.

m	0	2	4	6	8	10	12	14	16	18	20	22
4000	−42	85	180	129	92	91	−85	−146	−26	−5	−22	−58
3000	6	48	−9	20	37	−28	−53	−63	−52	20	71	2
2000	19	45	−36	24	7	66	−4	−48	74	89	49	−67
1000	80	35	−15	17	52	108	67	−2	66	85	−1	4
500	118	123	9	52	114	166	118	44	100	83	55	45
130	84	80	73	74	88	100	100	94	90	77	86	93

These tables give the following resultant direction and velocity:

	Winter						Summer					
Height m	130	500	1000	2000	3000	4000	130	500	1000	2000	3000	4000
Direction, ° from N	225	256	264	267	270	268	249	266	269	272	271	269
Velocity cm/sec	123	365	381	365	415	488	97	292	343	431	522	615

The lowest level for which values are assigned, 130 m, is the ground-level of the observatory.

The number of observations at the several hours in winter as indicated in the tables ranges from 9 to 32 at 4 kilometres and from 107 to 230 at the surface. For the summer the numbers are not very dissimilar.

[1] 'Der tägliche Gang des Windes in der freien Atmosphäre über Lindenberg (1917–1919),' von Otto Tetens, *Arbeit. Preuss. Aeronaut. Obs. bei Lindenberg*, Band XIV, Braunschweig, 1922.

It should be noted first that the velocities of the two components, South to North and West to East, are given separately, and secondly that the figures in each case represent the difference of positive and negative values of each component; we cannot therefore draw any conclusions from the table as to the "mean force of the wind." If for example there were always an equally strong wind either from North or South, the result might be represented in the table as no wind at all from either of those directions.

We reinforce the invitation by a vector-diagram of the diurnal changes in the South East trade-wind obtained from the records of the anemometer at St Helena (fig. 181).

DIURNAL VARIATION OF WIND

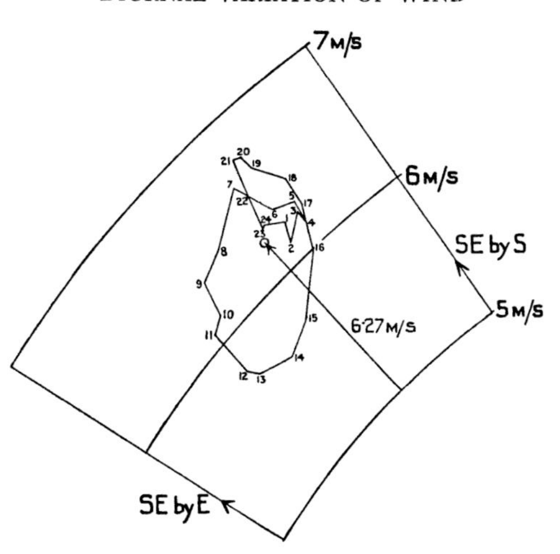

Fig. 181. Vector-diagram of the velocity and direction of the wind by the anemometer at St Helena (632 metres above sea-level) for each hour of the day, 1 to 24, in the month of April, 1892–1907.

An explanation of the marked diurnal change in the velocity of the wind at the surface has been sought first by Espy and subsequently by Köppen in the warming of the surface-layers in the day-time, and its consequent mixing with the upper layers by convection. And certainly if the normal diurnal variation of temperature be compared with the diurnal variation of wind as represented, for example, by curves for both adjusted to give the same range on a diagram or by the juxtaposition of the isopleths for temperature and wind as on p. 267 of volume I, the similarity will certainly be impressive, quite as much so as the curve-parallels of p. 286 or any others.

The justice of the explanation is further supported by demonstration of a diurnal change in the wind of opposite character at stations above the ordinary level of an anemometer. The diagrams of wind-velocity at the top of the Eiffel Tower and at the Bureau Central Météorologique (fig. 182), which is not far from the foot of the tower, illustrate this point.

It is fair therefore to conclude that the movement in the lower layers is speeded up and that in the upper layers slowed down by the mixing of the surface-air with the layers above; that the mixing is notably greater in the day-time than in the night. It may not be at all apparent at night when the upper wind is light.

But the process of mixing is no longer regarded as a simple effect of convection due to the warming of the ground. The variation of wind with height is now treated as the effect of turbulence or eddy-motion in the moving air consequent upon the retarding action of the surface. The effect of that action upon the speed and direction of motion of air or water has been studied by Ekman, Åkerblom, Hesselberg and G. I. Taylor, with the result that we are now rather more curious to inquire why there should be little mixing at night with an upper wind which is so effective in that sense during the day. The mixing depends indeed partly on the strength of the wind and also partly on the thermal conditions of the layers at the surface.

The diurnal variation of pressure has been treated from a different standpoint. The semi-diurnal wave preceding the sun leads to the idea of a solar tide as the impression of the mechanical conditions due to the sun's radiation and the rotation of the earth. We do not wish to examine the details of the question now. Under pp. 232–9, we have given a series of normal values of pressure at pairs of stations not far apart, with a considerable difference of level, in order that the curious student may have some evidence about the changes with height in the behaviour of the atmosphere with respect to diurnal variation of the barometer.

We leave the question as to the relation between the semi-diurnal oscillation of pressure and any corresponding change in the wind. Some light is thrown upon it by the examination of the question in relation to the trade-wind at St Helena in a memoir which has been referred to already more than once.

WIND-VELOCITY AND HEIGHT

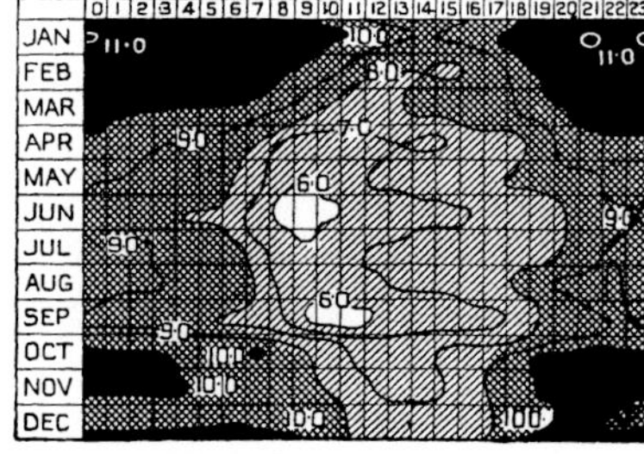

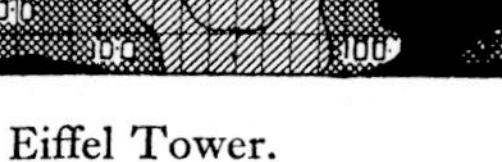

Eiffel Tower.

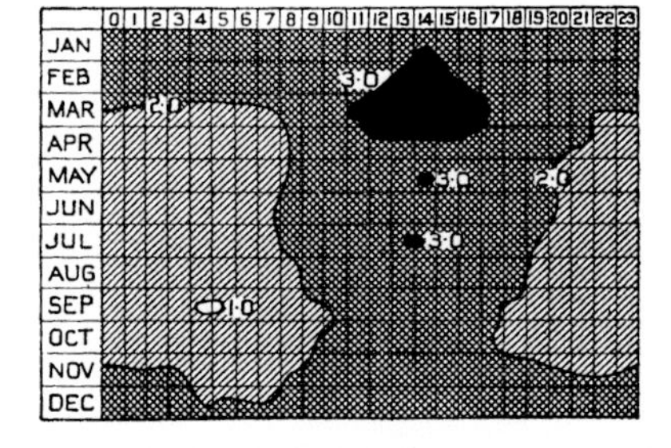

Bureau Central Météorologique.

Fig. 182. Isopleths of wind-velocity in metres per second at the Eiffel Tower and at the Bureau Central Météorologique (301 m) from observations 1890–5.

The figures indicate velocities in metres per second according to the following shades: *Eiffel Tower:* less than 6·0 m/s white, greater than 10·0 m/s black, intermediate lines 7·0 m/s, 8·0 m/s, 9·0 m/s. *Bureau Central Météorologique:* less than 1·0 m/s white, greater than 3·0 m/s black, intermediate line 2·0 m/s.

THE GENERAL CIRCULATION IN RELATION TO RAINFALL

We may close this chapter with a brief note on rainfall because the distribution of rain must be ordered and controlled by the general circulation of the air which carries it. The study of the tables of the seasonal variation of rainfall which we have given in the previous chapter ought therefore to afford us some guidance in checking the ideas of the general circulation which we derive from the study of the normal distribution of pressure.

In some cases the relationship is quite clear. The rainfall of the Indian peninsula, for example, is clearly dependent upon the air-currents which are associated with the monsoons. Over the greater part of the peninsula the rains come with the Westerly and South Westerly winds of the summer monsoon. In the Eastern region of Madras however the currents coming from the West still shed their moisture, but the East coast gets some rainfall from the North Easterly winds of the winter monsoon. The Island of Ceylon makes use of both to obtain its supply of water.

The situation is clear in some other cases. We cannot suppose that the rainfall of the Amazon valley is brought from the Pacific Ocean over the high lands of the chains of the Andes and Cordilleras in Peru and Bolivia; the suggestion that the water which flows down the Amazon finds its way previously up the valley as a continuation of the intertropical Easterly current can hardly be disputed, and a little consideration will show that the rain which supplies the Mississippi and Missouri with water comes from the Gulf of Mexico into which those rivers ultimately return a large part of it. We can easily reconcile these suggestions with currents which are in conformity with the distribution of pressure at the surface. With a little imagination we can provide, in like manner, for the rainfall of the Eastern United States, and we have no difficulty in associating the rainfall of British Columbia and the British Isles and Western Europe generally with the prevailing Westerlies of middle latitudes. The extension to the West coast of Norway is obvious enough. Equally well we may regard a Westerly current as carrying rain to the Mediterranean countries and to Northern India.

The supply of rain to Central and Eastern Europe and to Asia, North of the great mountain chains, is less easy. It seems likely that it is also carried primarily from the Atlantic by the extension of a Westerly current in the form of travelling depressions, and it may replenish its stock of water by evaporation on the way from lakes, rivers and wet land. It is a summer phenomenon.

Many questions of this kind can be suggested; two are of outstanding interest. One is the origin of the rain which supplies the Nile: the water comes by the White Nile from the equatorial lakes and by the Blue Nile from Abyssinia. We have noted a diversion of the South East trades over the Guinea coast and it seems quite probable that the air which forms that current passes Eastward along an isobar close to the equator on the North side and reaches the high lands of Abyssinia by that route. The observations of pilot-balloons quoted on p. 275 seem to confirm that view; but it is only fair to say that the meteorologists most nearly concerned prefer to rely upon the curvature across East Africa of the South East trade-wind of the Indian Ocean in the summer to give the Kenya rainfall on its way to help the South West monsoon.

The second of these outstanding questions is the origin of the snowfall of the Himalayas. We have no easy explanation at hand. The main difficulty is the very limited amount of water contained in saturated air at the low temperatures of great heights. An ordinary snowstorm seems quite unlikely. Before we take up such a question seriously we should like to know how much of the snow is really hoar-frost formed by the radiative cooling of the surface. The pictures of the Everest expedition suggest that the clouds of the highest regions are mainly, if not entirely, blown snow like that represented in the picture of Everest in volume I (fig. 20). Yet J. Barcroft has described a case of snow falling on January 14, 1922, at 5700 metres in the Andes without wind. The subject is of considerable interest, for the supply of snow in the higher regions offers a key to the uniformity of flow in rivers as Herodotus explained long ago; his shrewd observation is confirmed by the behaviour of the rivers of South Africa which, lacking the reserves that are provided by snow-fields, have a seasonal variation which may be very distressing for the country through which they flow.

Seasonal changes. Phase-correlations.

The occurrence of regular seasonal change has been a commonplace of meteorological study from the dawn of history, but having instrumental observations we can prosecute its examination in the hope of getting thereby a clue to the working of atmospheric processes.

There are a number of interesting questions that arise from an inspection of the series of normal values for the months which are set out at the foot of the maps of monthly distribution of rain. First among these may be mentioned the phase-correlations of the several seasonal changes in different parts of the world. We may take for example the month of maximum phase in the seasonal rainfall. These we have indicated for the successive zones of the globe in the panels of the maps of rainfall.

WIND-VELOCITY AND RAINFALL

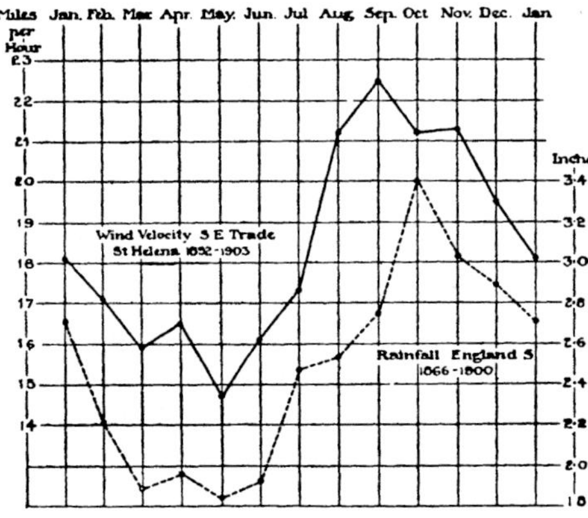

Fig. 183. Average seasonal variation of the velocity of the South East trade and of the rainfall in England S.

By way of reminiscence we reproduce a particular example, namely, the parallel between the seasonal variation of the South East trade at St Helena and that of the rainfall of the Southern district of England. The scale is, of course, chosen to make the parallelism conspicuous, but even when expressed as fractions of the mean value it is still notable. The diagram was reproduced in *Nature* in 1905, and some facts were added which supported the idea of a relation. Thereupon a London newspaper carried the idea to its limit and obtained monthly values of wind at St Helena as a means of predicting the rainfall of Southern England. That the result was disappointing need cause no surprise, for the diagram is based on the mean values of a series of years, and to use it for an individual year is to stretch the relationship beyond its limit. What is clear from the diagram is that normally the South East trade-wind reaches its maximum at about the same time of year as the Southern English rainfall, and as both these are elements of the general circulation they are almost certainly connected through that circulation; we may learn a good deal about the general circulation by studying the phases of the various elements of it. Of all the elements, wind and rainfall are of the greatest practical importance. Pressure-gradient might perhaps be better as an element than wind and more easily worked. Rainfall is difficult to deal with because its phases are subject to local influences. We may illustrate this point by the consideration of the rainfall of Abyssinia. We have already mentioned that the water which supplies the rainfall is thought to come from the equatorial Atlantic. The maximum phase of the Abyssinian rainfall is in July and August, that of the equatorial lakes in April, and that of the Nigerian coast in June, but at certain stations of the hinterland in August and September.

Disturbing causes affect the rainfall upon the coasts of any large land-area, and consequently the rainfall in the far interior is governed not immediately by the general circulation but rather by the influence of the orographic features upon the circulation which we have already illustrated in fig. 3.

* * *

As in the case of temperature, which is referred to on p. 47, or in that of the mass of air on the Northern Hemisphere as given in the heading of this chapter, so with the other meteorological elements rainfall, sunshine or cloudiness, pressure, all have marked seasonal variations which are characteristic of the general circulation. They are based upon monthly values because the great majority of meteorological data are arranged on that plan. On pp. 204–8 of *The Drama of Weather* we have shown a number of step-diagrams representing the seasonal variation of a number of elements arranged according to the periods at which the maxima are attained; and perhaps here we may remark that what is known as the seasonal variation is represented by the sequence of the monthly normals, but, as each year passes, the normal sequence is not followed except in the most general sense; the repetition is subject to considerable fluctuations from year to year. The next chapter will be devoted to the consideration of these disturbances of the normal circulation.

AUTHORITIES FOR THE DATA CONTAINED IN THE MAPS, DIAGRAMS AND TABLES OF CHAPTER VI

MONTHLY CHARTS OF MEAN ISOBARS OVER THE GLOBE

A Barometer Manual for the Use of Seamen. M.O. publication, No. 61. The maps therein published have been compared with:

MS normals prepared in the Meteorological Office for the Réseau Mondial.

The distribution of pressure and the air-circulation over Northern Africa. By H. G. Lyons. Q. J. R. Meteor. Soc., vol. XLIII, 1917, pp. 116–50.

Report on the barometry of the United States, Canada, and the West Indies. By F. H. Bigelow. Report of the Chief of the Weather Bureau, 1900–1901. Vol. II.

Monthly Meteorological Charts of Davis Strait and Baffin Bay. M.O. publication, No. 221. 1917.

PRESSURE IN THE ARCTIC

Sagastyr, Treurenberg Bay, Inglefield Bay, Angmagsalik. MS normals prepared in the Meteorological Office.

Franz-Josef Land. Einige Ergebnisse der meteorologischen Beobachtungen auf Franz-Josefs Land zwischen 1872 und 1900. Von J. Hann. Met. Zs. vol. XXI, 1904, p. 550.

Griffith Island. Contributions to our knowledge of the meteorology of the Arctic Regions. Vol. I. London, 1885.

PRESSURE IN THE ANTARCTIC

Kerguelen, Cape Horn, Port Charcot, Snow Hill, "Gauss," "Belgica," "Discovery." Deutsche Südpolar Expedition, 1901–1903. Band III. Meteorologie, I, 1. Erster Teil. Meteorologische Ergebnisse der Winterstation des "Gauss," 1902–1903. Von W. Meinardus. Berlin. S. 31.

The values are smoothed from the mean monthly values by means of the formula $\frac{1}{4}(a+2b+c)$. The values for Cape Horn are the mean of two years' observations at Cape Horn and Ushuaia, 1882–3, and two years at Staten I., 1888 and 1896. The values for Snow Hill are from the series March 1902 to February 1903 and November 1902 to October 1903.

S. Georgia. The climate and weather of the Falkland Islands and S. Georgia. By C. E. P. Brooks. Geophysical Memoir, No. 15; M.O. publication, No. 220 e. London, 1920.

Cape Adare, Framheim and McMurdo Sound. British Antarctic Expedition, 1910–13. Meteorology. Vol. I. Discussion. By G. C. Simpson. Calcutta, 1919. The figure for the pressure on the South Polar Plateau is from vol. III of the same publication.

"*Endurance.*" Réseau Mondial, 1915. M.O. publication, 222 g. London, 1924.

The position of the *Endurance* during the several months is as follows:

	Jan.	Feb.	Mar.	Apr.	May	June	July	Aug.	Sept.	Oct.	Nov.	Dec.
Lat. ° S	74	77	77	76	75	74	74	71	70	69	69	67
Long. ° W	27	35	37	40	44	46	48	49	51	51	52	52
Press. mb	988	988	988	992	997	991	995	1000	997	995	985	986

DIURNAL AND SEASONAL VARIATION OF PRESSURE

Arctic, Aberdeen, Calcutta, Batavia, Cape of Good Hope, Hut Point and Cape Evans, Ben Nevis and Fort William, Lahore and Leh. The authorities are the same as those given on p. 126 for the diurnal and seasonal variation of temperature.

Clermont Ferrand, Puy de Dôme, Eiffel Tower and Bureau Central Météorologique. Études sur le Climat de France. Par A. Angot. Ann. du Bur. Cent. Météor., 1906, pt. I. Paris, 1910.

Pike's Peak. A brief discussion of the hourly meteorological records at Pike's Peak and Colorado Springs from Nov. 1892 to Sept. 1894. By P. Morrill. U.S. Department of Agriculture. Report of the Chief of the Weather Bureau, 1895–1896. Washington, 1896.

Sonnblick. The table has been compiled from the monthly values published in the Jahrbücher der Zentral-Anstalt für Meteorologie und Geodynamik, Wien, for the years 1909 to 1918. In a paper 'Weitere Untersuchungen über die tägliche Oscillation des Barometers,' published in Band LIX of the Denk. der Mat. Natur. Classe der K. Akad. der Wiss., pp. 40 and 41, Wien, 1892, J. Hann gives diurnal values of pressure for the seasons at Sonnblick and Salzburg for the period 1887–1889.

Values for Vienna for a period of 19 years are given in the same publication, Band LV, Wien, 1889.

PRESSURE AT THE LEVEL OF 8 KILOMETRES

The authorities for the results of ballon-sonde data in the several countries are given on p. 127. In cases where the values of the pressure are not given in the original papers the values have been computed from the figures for the upper air-temperatures and the surface-pressure.

SEASONAL VARIATION OF DENSITY IN THE UPPER AIR

United States. Mean values of free-air barometric and vapour-pressures, temperatures and densities over the United States. By W. R. Gregg. Monthly Weather Review, Washington, vol. XLVI, 1918, pp. 11–20. The data are from sounding-balloon-observations at Fort Omaha, Indianapolis, Huron and Avalon.

Munich. Münchener Aerologische Studien, No. 11. Die Dichte der Atmosphäre über München. Von A. Schmauss. Deutsches Meteorologisches Jahrbuch für 1921, Bayern. München, 1922.

Pavia. Le caratteristiche dell' Atmosfera libera sulla Valle Padana. Per P. Gamba. Venezia, 1923.

Agra. Discussion of results of sounding balloon ascents at Agra during the period July 1925 to March 1928. By K. R. Ramanathan. Memoirs of the Indian Meteorological Department, vol. XXV, part 5, Calcutta, 1930, p. 163.

For *England* see chap. X, note 19.

TABLES OF CLOUD-MOTION

The data have been taken mainly from two papers by H. H. Hildebrandsson:

Rapport sur les observations internationales des nuages au Comité international météorologique. Upsala, 1903. Tr. in Q. J. Roy. Meteor. Soc., vol. XXX, 1904, pp. 317–43. See also Les Bases de la Météorologie dynamique, tome II.

Résultats des recherches empiriques sur les mouvements généraux de l'atmosphère. Nova Acta Regiae Soc. Scient. Upsaliensis, ser. IV, vol. V, No. 1. Upsala, 1918. A summary appears in the Monthly Weather Review, 1919.

The authorities for the additional data are as follows:

Karajac. A wind-rose based on data published by H. Stade is given in 'The rôle of the glacial anticyclone in the air-circulation of the globe,' by W. H. Hobbs, Proc. Amer. Phil. Soc., LIV, No. 218, 1915.

Potsdam. Photogrammetrische Wolkenforschung in Potsdam in den Jahren 1900 bis 1920. Von R. Süring. Veröff. des Preuss. Met. Inst., Nr. 317, Abhand. Bd. VII, Nr. 3. Berlin, 1922.

United States. (St Paul, Kansas, Abilene, Vicksburg, Louisville, Detroit, Cleveland, Buffalo, Blue Hill, Washington, Waynesville, Ocean City and Key West.) The average monthly vectors of the general circulation in the United States. By F. H. Bigelow. Monthly Weather Review, vol. XXXII, 1904, p. 260.

Barcelona. E. Fontseré. Movimiento de las Nubes altas y medias en el zenit de Barcelona. Memorias de la Real Academia de Ciencias y Artes de Barcelona, vol. XV, No. 8, pp. 251–9. Barcelona, 1919.

Atlantic Ocean, Hawaii, Cape Verde, Ascension, Johannesburg. Monthly Weather Review, LII, 1924, p. 445.

Pavia. Le osservazioni di Nubi. Per P. Gamba. Ann. dell' Uff. Centr. Meteor. Geodin., vol. XXXIV, parte I, 1912. Roma, 1914.

Zyosin. Results of the Meteorological Observations in Korea for the Lustrum 1916–20. Meteorological Observatory of the Government General of Chosen. Zinsen, 1922. Data for eleven other stations are also included in the publication.

India. (Lahore, Simla, Jaipur, Allahabad, Calcutta, Madras.) Cloud observations made in India between 1877 and 1914. By W. A. Harwood. Memoirs of the Indian Meteorological Department, vol. XXII, part 5. Calcutta, 1920, p. 565. Data for the years 1914–1919 are published in the same memoirs, vol. XXIV, parts 7 and 8.

India. (Kojak, Agra, Darjiling, Bombay, Bangalore, Kodaikanal.) The Free Atmosphere in India. Heights of clouds, and directions of free air movement. By W. A. Harwood. Memoirs of the Indian Meteorological Department, vol. XXIV, part 7. Calcutta, 1924. The mean direction has been computed from the tables of percentage frequency. Values based on less than 10 observations have been omitted.

Tsingtau. Results of the Meteorological Observations made at Tsingtau for the Lustrum 1916–1920. Published by the Tsingtau Meteorological Observatory, 1921.

Havana. Las Diferentes Corrientes de la Atmósfera en el Cielo de la Habana, 1892–1902. By L. Gangoiti.

West Indies. (Cienfuegos, Santiago, Santo Domingo, San Juan, Basseterre (St Kitts), Roseau (Dominica), Bridgetown (Barbados), Willemstad (Curaçao), Port of Spain (Trinidad).) Studies on the circulation of the atmospheres of the sun and of the earth. Results of the nephoscope observations in the West Indies during the years 1899–1903. By F. H. Bigelow. Monthly Weather Review, vol. XXXII, 1904. The vectors given in the original paper were obtained by taking account of both the direction and the velocity of the cloud-motion; they are published in graphical form. The values given in the table have been obtained by reading the direction to the nearest ten degrees.

Taihoku and Naha. Results of the meteorological observations in Japan, 1906–1910. Tokio, 1913. Data for 15 other stations are also available. Data for six stations in Formosa (not reproduced in the table) are published in Meteorological Observations in Formosa, 1896–1901.

Kingston. Cloud-drift at Kingston, Jamaica, 1907–1913. By J. F. Brennan. Jamaica, No. 463, 1916.

Samoa. Wolkenbeobachtungen in Samoa. Von G. Angenheister. Gesell. Wiss. Göttingen, Nachr. Math.-phys. Klasse, 1909, Heft 4, Sn. 363–70.

Pontianak and Batavia. The atmospherical circulation above Australasia. By W. van Bemmelen. Proc. K. Akad. Wetenschap. Amsterdam, vol. xx, 1918, pp. 1313–27.

Melbourne. Annual and seasonal variation in the direction of motion of cirrus clouds over Melbourne. By E. T. Quayle. Australian Monthly Weather Review, vol. I, 1910. Melbourne, 1912.

New Zealand. (Kaikoura.) Observations of upper clouds in New Zealand. By J. St C. Gunn. Q. J. Roy. Meteor. Soc., vol. XXXIII, 1907, p. 180.

Cape Evans and Cape Adare. British Antarctic Expedition, 1910–1913. Meteorology, vol. I, p. 133, 1919. By Dr G. C. Simpson.

Antarctic. (Belgica, S. Orkneys, S. Georgia, Cape Horn, Ushuaia, New Year's Island.) The Meteorology of the Weddell Quadrant and Adjacent Areas. By R. C. Mossman. Trans. Roy. Soc. Edin., vol. XLVII, part I, No. 5.

WINDS OF THE UPPER AIR

Batavia. W. van Bemmelen, as above *Pontianak and Batavia.*

Egypt. The upper currents of the atmosphere in Egypt and the Sudan. By L. J. Sutton. Physical Department Paper No. 17. Cairo, 1925.

India. The free atmosphere in India. By W. A. Harwood. Memoirs of the Indian Meteorological Department, vol. XXIV, parts VII and VIII. Calcutta, 1924.

Lindenberg. Die Windverhältnisse der freien Atmosphäre. Von W. Peppler. Die Arbeiten des Preussischen Observatoriums bei Lindenberg, Band XIII. Braunschweig, 1919.

Mauritius. Additional examples of the reversal of the trade-wind in the upper levels are given in an official report on Upper Air Investigations by A. Walter, Port Louis, 1926, from which we draw the following summary:

"The westerly drift is at times at a very much lower altitude than was formerly suspected."

The features of the S.E. trade stratum are:

(1) a steady increase in velocity up to 500 metres with a slight backing towards the North (probably a local effect);

(2) a rapid falling off in velocity after 750 metres until the wind falls practically calm at a height between 1000 and 2000 metres;

(3) the cessation of all easterly motion.

(4) A north westerly or south westerly current finally becoming almost due west at 6000 or 7000 metres. The height of the westerly current varies and is at times below 1000 metres. On rare occasions it is not encountered below 6000 metres, but the average height appears to be between 3000 and 4000 metres.

Samoa. Observations of upper air-currents at Apia, Western Samoa. (Second series.) By A. Thomson. Wellington, 1929.

United States. Washington, D.C., U.S. Department of Agriculture, Weather Bureau. An Aerological Survey of the United States. By W. R. Gregg. Part I. Results of observations by means of kites. Monthly Weather Review, Supplement No. 20, 1922. Part II. Results of observations by means of pilot balloons. Supplement No. 26, 1926.

CHAPTER VII

CHANGES IN THE GENERAL CIRCULATION. RESILIENCE OR PLASTICITY

Some long records of weather [see chap. x, note 20].

Austria
- Vienna, temperature from 1775
- Klagenfurt, rainfall from 1813

Denmark
- Copenhagen, temperature from 1798
  - rainfall from 1821

France
- Paris, temperature and pressure from 1757
  - number of rain-days 1698–1716, from 1752
  - amount of rainfall 1688–1717, from 1806
- Montdidier, rainfall from 1784
- Lille, rainfall from 1806

Germany
- Berlin, temperature 1719–1721, 1728–1751, from 1756

Great Britain
- London, temperature from 1763
  - rainfall from 1782
- Edinburgh, temperature from 1764
  - wind-direction 1732–1736, from 1764
  - pressure from 1769
  - rainfall 1770–1776, 1780–1781, from 1785
  - number of days T, ✱, ▲, ⁄, ≡° and ≡ from 1770
  - number of observations of aurora 1773–1781, 1800–1896
  - number of days of lightning 1807–1835, 1868–1896
- Kendal, rainfall from 1788
- Rothesay, rainfall from 1800
- Greenwich, rainfall from 1815

India
- Madras, rainfall from 1813

Italy
- Padua, rainfall from 1725
- Milan, rainfall from 1764
- Chioggia, rainfall 1771–1797, 1800–1814, from 186[illegible]
- Verona, rainfall 1788–1796, 1798–1816, 1822–1826, 1829–1841, from 1845
- Palermo, rainfall from 1806
- Pavia, rainfall from 1812
- Bologna, rainfall from 1813
- Florence, pressure, temperature, rainfall from 1813
- Naples, rainfall from 1821
- Rome, rainfall from 1825

Norway
- Oslo, temperature and pressure from 1816

Poland
- Warsaw, rainfall from 1813

Russia
- Archangel, temperature 1814
- Moscow, temperature 1780–1820, from 1821
- Leningrad, rainfall from 1836, incomplete from 1741, temperature from 1765, pressure from 1822

Sweden
- Upsala, temperature from 1739
- Lund, rainfall from 1748, temperature and wind-force from 1753
  - wind-direction from 1741
- Stockholm, temperature from 1756

United States
- New Bedford, Mass., rainfall from 1814
- Baltimore, rainfall from 1817
- Boston, rainfall from 1818 [1750]
- Philadelphia, rainfall from 1820

MEAN VALUES

With the sixth chapter we completed our picture of the normal general circulation of the atmosphere and its seasonal variation. It is based upon the mean values of observations which extend over a long series of years, and which in accordance with international agreement are grouped according to months. We shall now turn our attention to the changes disclosed by differences in the values from which the means of the several months have been obtained. Before doing so we digress for a moment to consider the intrinsic meaning of the picture which we have constructed. The late Sir Norman Lockyer used to cite, as though he were quoting from a copy-book, "La méthode des moyennes—c'est le vrai moyen de ne jamais connaître le vrai." He gave no reference for the aphorism; but it is well to remember it, because, as we have already learned, the "mode" is often different from the "mean"; the mean value of the measurements for any single point is not necessarily that which is of most frequent occurrence and for certain distributions might even be the least frequent. When we consider that that remark applies in some measure to every point on our maps we admit the possibility that the picture which we have elaborated may be a composite and have no real existence. That can hardly be true about certain main features like the East to West movement of the equatorial belt, the West to East rotation of the polar caps, or the winds and pressures of the Eastern sides of the "hyperbars" of the great oceans, but it may certainly be true of areas of low mean pressure which survive the treatment of statistical arithmetic; we know for example that the arithmetical process wipes out nearly all the centres of low pressure and of high pressure which are the most striking features of a synchronous chart, and the same may be true of many other details of the maps.

We shall not be able to write with confidence on such a question until we can examine the material from which our maps are constructed, with the aid of daily synchronous charts of the globe, suggested long ago by the assiduous scientific ambition of Teisserenc de Bort. That is now certainly within the range of possibility and almost within reach; but the prospect has terrors of its own. Since 1918 the British Meteorological Office has required ten quarto pages for the particulars of the weather of a single day over the part of the globe between 35° N and 80° N, and between 40° W and 30° E. For the whole globe twenty such areas may be necessary, and on the same scale of precision, for a single day, 200 quarto pages would be at our disposal and demand our consideration. To epitomise 50 years of such charts in a representation of the normal sequence of changes would be no light task.

It is clear that years will elapse before we can reach any conclusion concerning the effectiveness of the mean values as representing a real circulation, or find a proper substitute for the representation which we have given, if it is found to be defective; in the meantime, though the normal for the month may conceal a great variety of meanings, we may suppose our picture to represent reality if it is understood that we are using a very coarse brush, and do not attempt to press the accuracy of details. This is particularly to be remembered when we are considering such quantities as coefficients of correlation between the mean values of individual months. There is a certain danger in assuming that, being numerical, the figures of computation are necessarily exact, and that the conclusions drawn from them *prima facie* are not only accurate but exhaust the subject. We cannot fairly accept that position so long as the physical basis of the relationship is not understood.

We ought also to be on our guard about treating the most salient features of the monthly maps, high pressure or low pressure, as covering all that a reader would naturally understand by the expression "centres of action," which occurs frequently in the literature of this part of the subject. Teisserenc de Bort introduced the idea in 1881 with the following words[1]:

L'étude de la répartition des pressions et des vents sur le globe m'a conduit dans des travaux précédents, à définir et à classer les divers maxima et minima de pression

[1] 'Étude sur l'hiver de 1879–80 et recherches sur la position des centres d'action de l'atmosphère dans les hivers anormaux,' *Annales du Bureau Central Météorologique de France*, 1881, part 4. Paris, 1883.

que présente l'atmosphère à la surface du globe. J'ai été amené ainsi à désigner, sous le nom de *grands centres d'action de l'atmosphère*, ces aires de hautes et de basses pressions que l'on retrouve dans presque toutes les cartes moyennes, et dans celles qui représentent l'état de l'atmosphère à un moment donné.

It will be seen that the definition may be held to include either the extreme features of a map of mean values of pressure, or the centres of high and low pressure of a synoptic chart. It would be begging the question which we have in mind to regard the two conceptions as in effect identical. The expression "centres of action" is not infrequently used without any further definition. It is not altogether satisfactory: the attribution of "action" to an area of high pressure whether permanent or transitory does not consort well with what is usually connoted by "action." Such areas seem rather to be dumping-grounds for the air which has taken part in atmospheric action and has to be disposed of. In fact the enhanced pressure may perhaps be regarded as the result of atmospheric action which determines the accretion of air against the force of the existing gradient; otherwise the pressure could not rise; and, on the other hand, the locality of low mean pressure may be more a question of path than of local action. The violence of the chief action may be over by the time that the favourite rendezvous is reached. We shall therefore do well to keep an open mind about limiting the idea of centres of action to regions of high or low mean pressure.

Monthly means in a new light.

We must not be understood to imply that correlations between centres of high or low mean pressure and other features of the maps, or between each other, are meaningless or unimportant, but merely that they do not dispense with the necessity for further attention to the dynamical and physical conditions which are connoted by the word "action."

Before we consider the actual sequence from year to year, month to month, week to week, or day to day, we should like, indeed, to offer some examples of the use of graphic representation of monthly means to convey definite ideas of sequence in the events of the atmospheric circulation by incorporating the time-co-ordinate in the diagrams.

The examples chosen are:

First for fig. 184 step-diagrams of normal "sea-level" pressure at a number of stations in the British Isles which occupy very interesting geographical positions on the margin between the vast continent on the East and the active Atlantic on the West. This perhaps accounts for the marked increase of pressure in September shown at nearly all the stations.

Secondly, in fig. 185, pressure connotes the actual weight of air per unit area at the level of the barometer-cistern, and its changes from month to month suggest the export or import of air between the several localities, so we present a series of maps of the Northern Hemisphere with lines showing by "isallobars" the amount in millibars (tons per square dekametre) of import or export of air for the several localities between the so-called months.

Thirdly, we add a chart of step-diagrams (fig. 186) of monthly normals of rainfall at a number of stations in the British Isles which exhibit a notable characteristic of the difference of climate of the Eastern and Western sides.

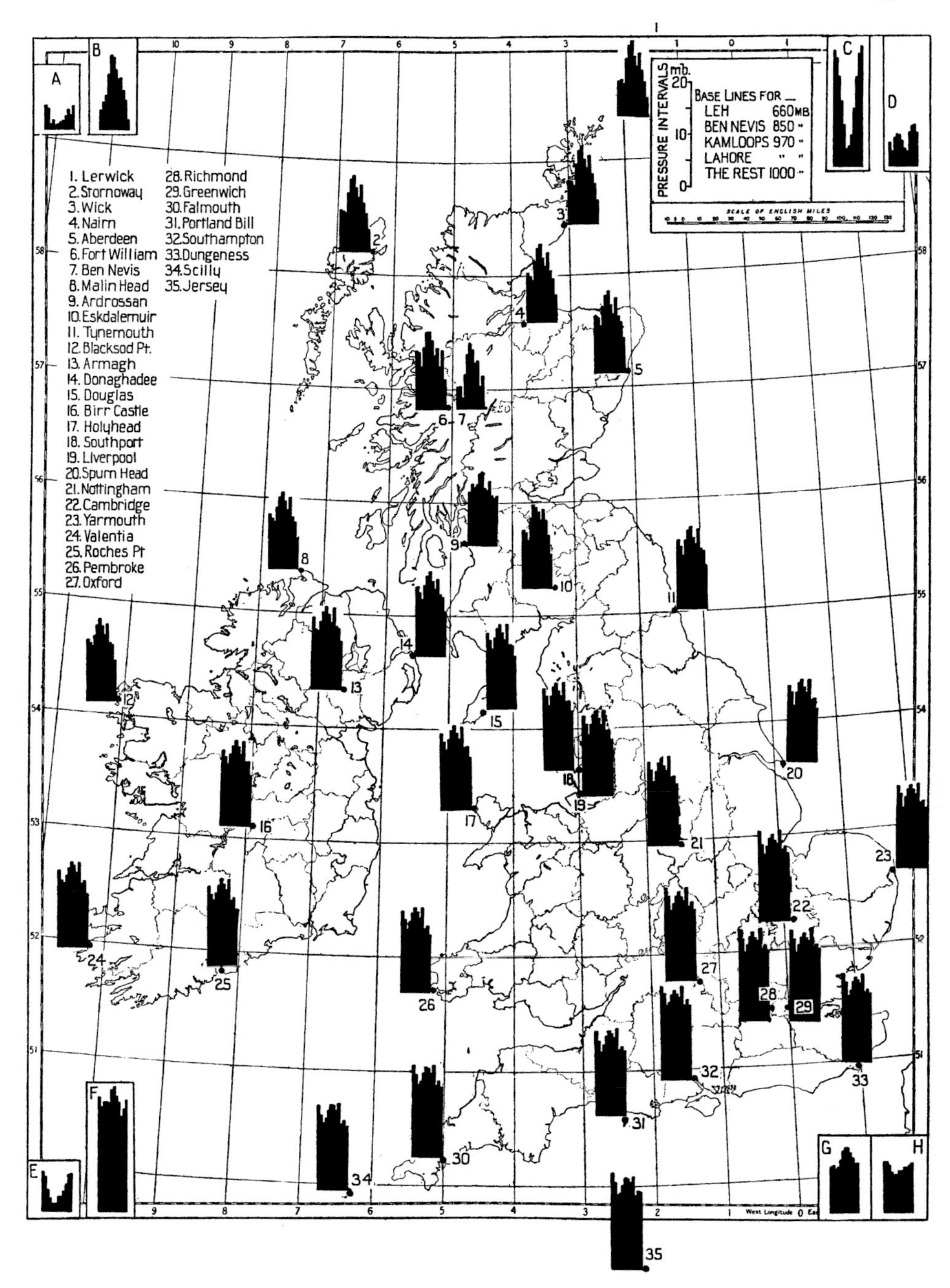

Fig. 184. Normals are given for thirty-five stations in the British Isles with eight examples of continental or oceanic stations elsewhere for comparison, viz. A. Kamloops, B. Thorshavn, C. Lahore, D. Leh, E. Little Rock, F. Azores, G. Lagos, H. Colombo.
Each diagram runs from January to December.

Fig. 185. MONTHLY IMPORT OF AIR (+) AND EXPORT (−)

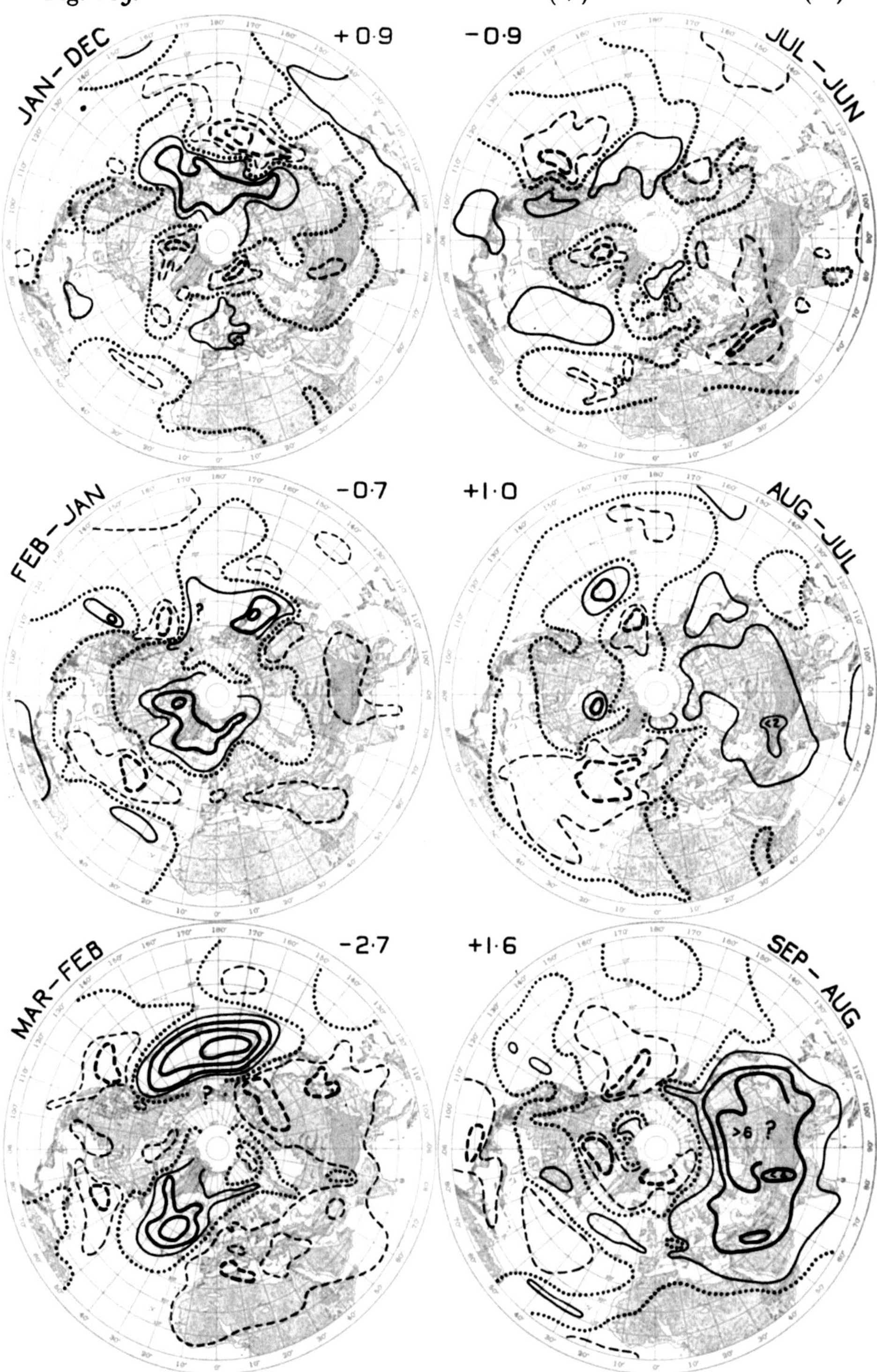

Isallobars: dotted lines .... for no change; full lines for import, — 2mb, — 4mb etc.; pecked lines for export of the same amounts. Data for land-stations are from *World Weather Records*.

FOR THE HEMISPHERE AND ITS PARTS

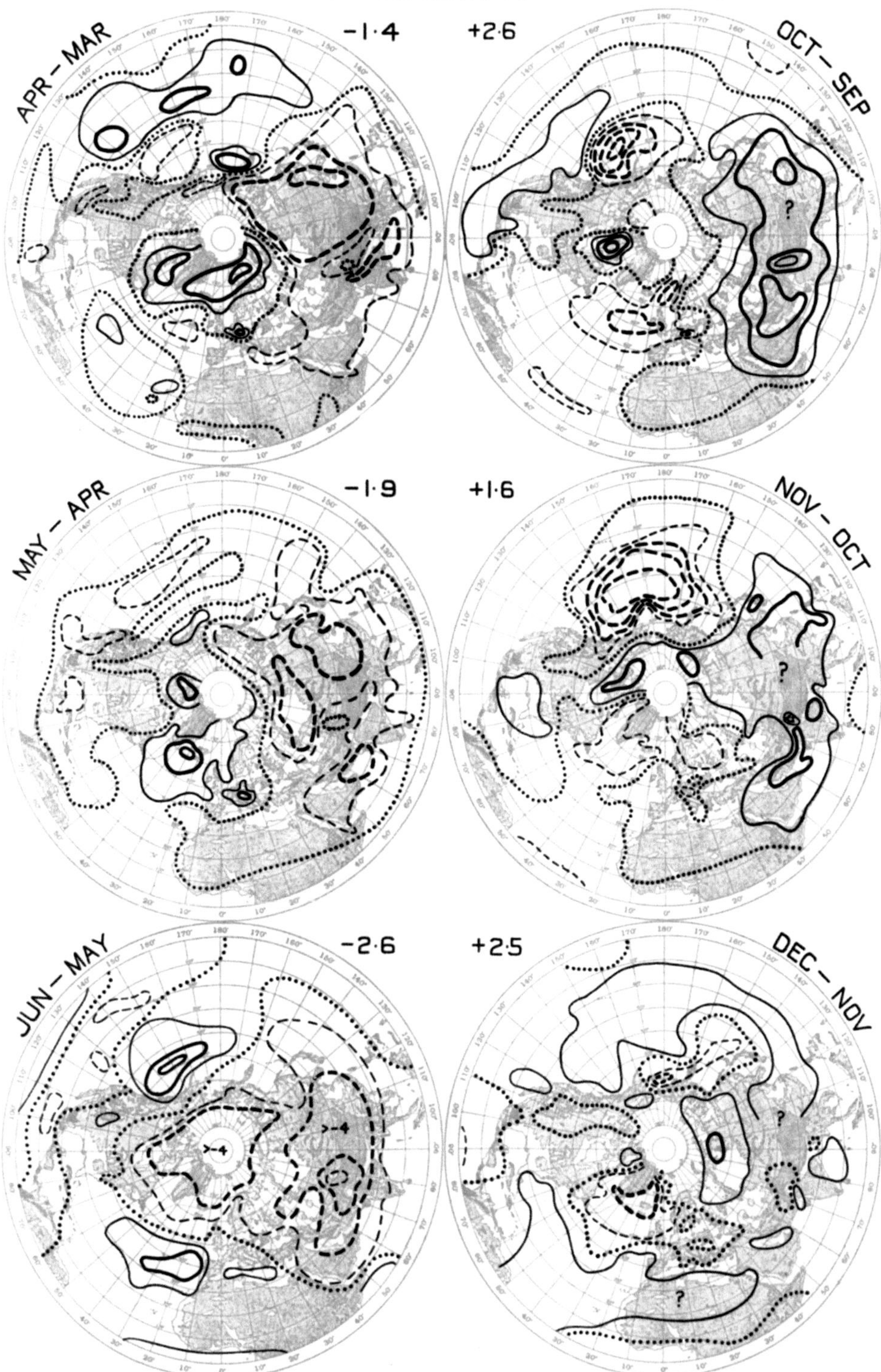

For the whole hemisphere the figures represent the monthly import or export in billions of tons (10^{15} grammes). [Over a 10° sector 1 mb would mean 75 thousand million tons.]

NORMAL MONTHLY FALL OF RAIN IN THE BRITISH ISLES

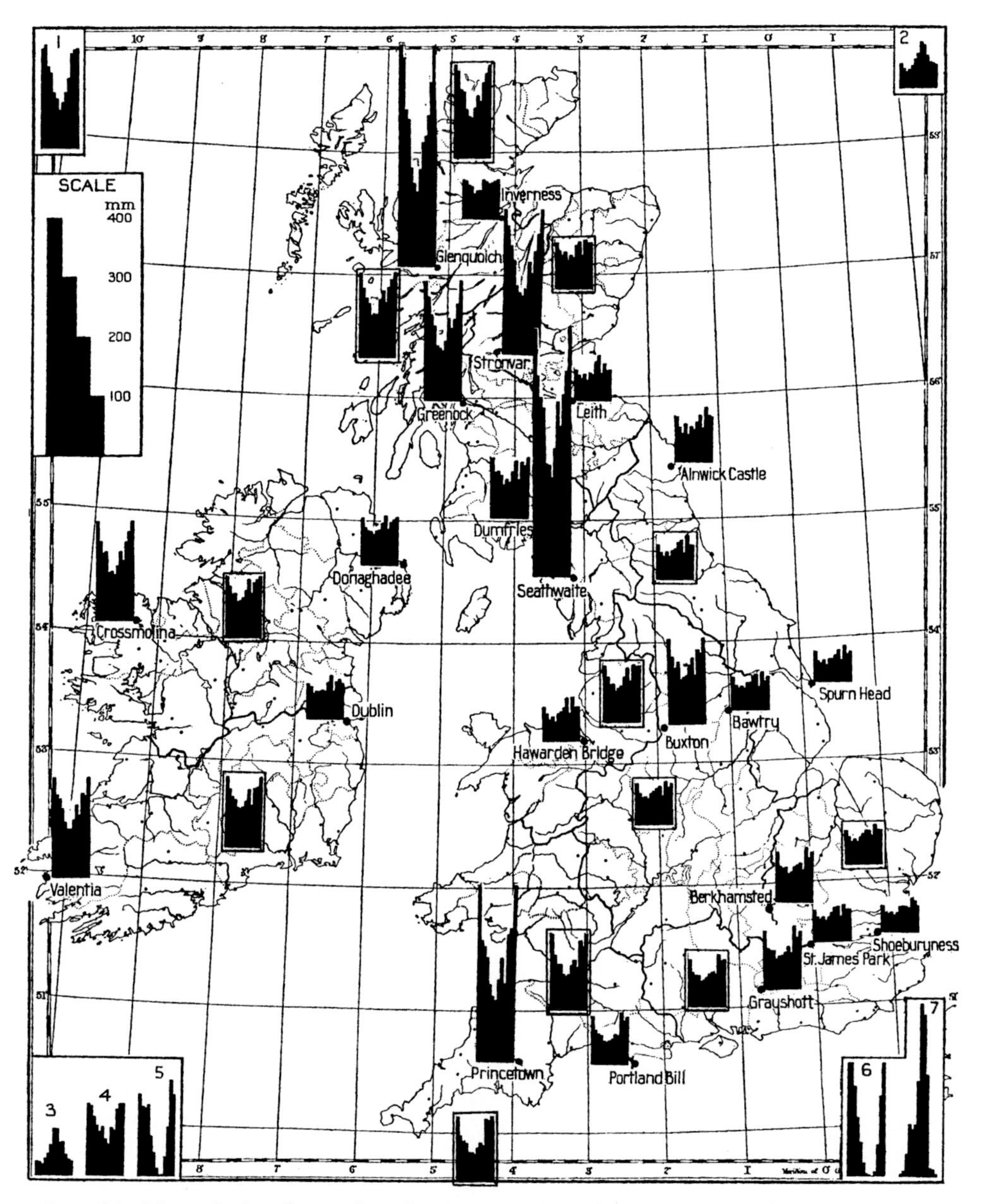

Fig. 186. Normals for the twelve districts are given by diagrams enclosed in frames. Normals for the stations of highest and lowest rainfall in each district are given by diagrams, the points indicating the positions of the stations. Data from the *Book of Normals*.

Each diagram begins and ends with the normal fall for the month of December.

In the four corners are examples of corresponding normals for: 1. Thorshavn, 2. Moscow, 3. Winnipeg, 4. Horta, 5. Gibraltar, 6. Beirut, 7. Addis Ababa.

The normal circulation.

In a paper in the *Philosophical Magazine* for February 1926, D. Brunt estimates the kinetic energy of the Westward motion of air of the equatorial belt of the atmospheric circulation at $1{\cdot}35 \times 10^{27}$ ergs (4×10^{13} kilowatt-hours). This refers approximately to one-half of the atmosphere, and allowing a somewhat greater mean velocity to the other half, which is represented by the two polar caps beyond 30° N and S, he obtains 3×10^{27} ergs, or 10^{14} kilowatt-hours, as a rough measure of the kinetic energy of the normal circulation, which in the form of electrical energy at one penny per unit might be valued at half a billion pounds sterling. As we have already hinted, the actual circulation is really never normal: there are always cyclonic depressions involved in it which are smoothed out by the process of "meaning." A computation, by the author, of the energy of a cyclone 10 mb deep and 1400 km in diameter as $1{\cdot}5 \times 10^{24}$ ergs, is compared by Brunt with the average of 10^{24} ergs for an equal area of the normal circulation, and accordingly he suggests 50 per cent. as an indication of the increase of energy of any part, in consequence of the disturbance of its area by a cyclone.

These are of course very rough figures. The whole question of the kinetic energy of motion on a rotating globe needs careful consideration, and certainly the calculations quoted make no claim to represent anything more precise than the order of magnitude of the quantities expressed. Accepting that position Brunt has gone further, and, assuming normal initial values for velocity and energy, has calculated the effects of frictional viscosity, or of turbulence due thereto, upon the energy of the circulation. Making certain further assumptions, which we need not now discuss in detail, he computes that, if left to itself, energy would be used up in turbulence at the rate of 50,000,000 ergs per square metre per second, and therefore, that the kinetic energy of the circulation would be reduced to one-tenth part of its original value within three days. It may seem of little moment that the energy should be reduced from 10^{14} kilowatt-hours to 10^{13} kilowatt-hours, which is what reduction to one-tenth means; the residue of 10^{13} kilowatt-hours may impress us still as a stupendous amount. But with the usual assumption of rate of loss being proportional to the total energy, reduction to one-tenth in three days means reduction to one-hundredth in six days; and that would correspond with reduction of velocity to one-tenth—from an assumed average value of ten metres per second to one metre per second. Such a reduction is therefore practically an annihilation of the circulation. Hence it appears that if six days were originally required to set up the machinery necessary to make the world fit for men to live in, in six days also, if left to itself, the machinery would run down again, and the world would be reduced to an uninhabitable condition.

The comparatively uneventful continuance of the conditions of life is clear evidence that the energy, as a matter of fact, does not run down in six days; it remains practically the same, year in and year out. On the whole, day by

The rainfall in inches for each month is shown by the length of a column as measured against the scale of inches at the side, and the fall in inches for each year is given in figures above the heads of the columns.

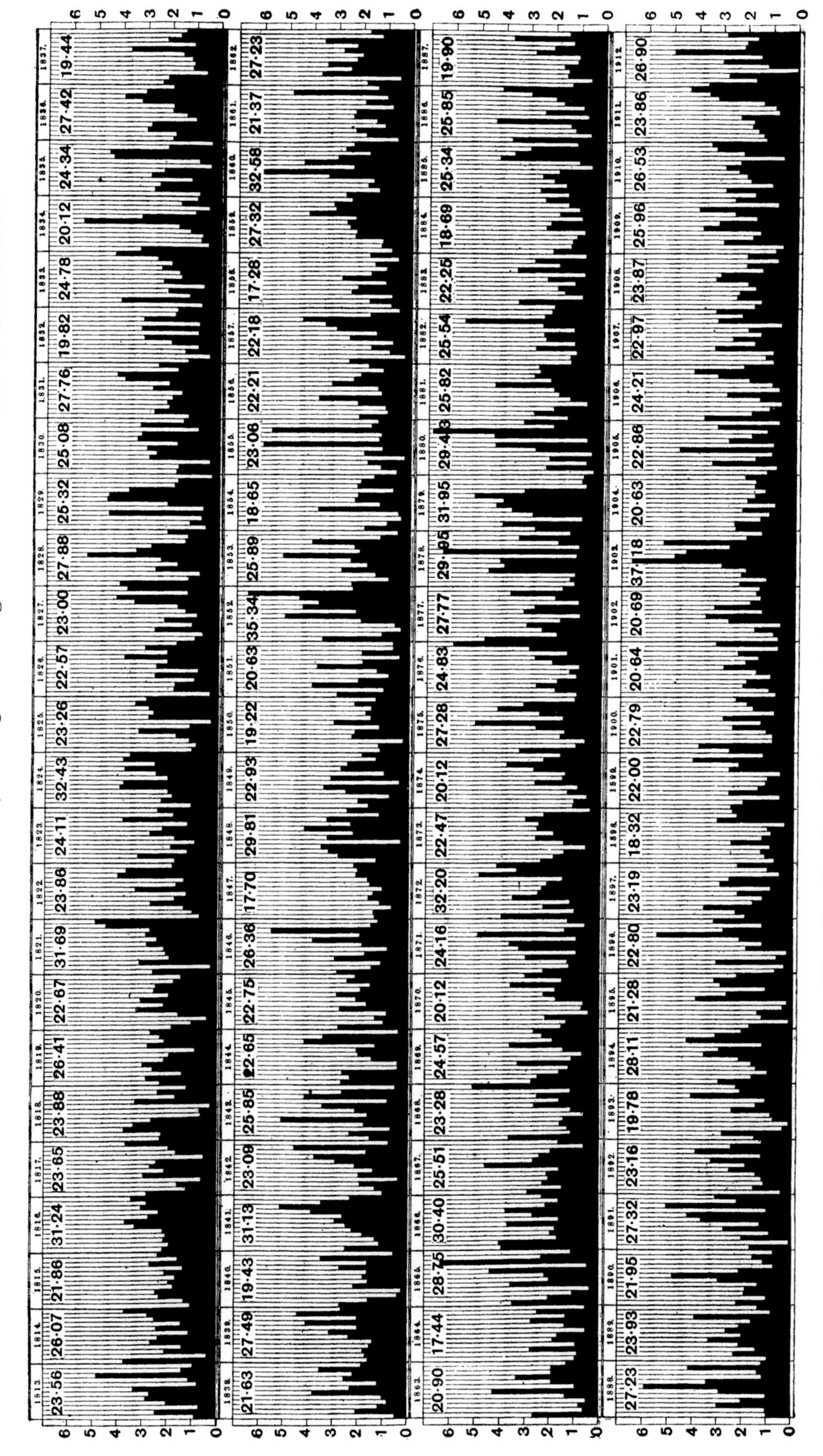

Fig. 187. One hundred years' rainfall over London.

day, our atmosphere gets as much as it loses. Although we have a clear idea of the decay of energy through turbulence we cannot yet describe with equal clearness the process by which the energy of the atmosphere is increased or built up. The unfailing supply comes from the sun in the form of radiant energy, acting in conjunction with the loss of energy by radiation from and through the atmosphere into space. Brunt estimates that 2 per cent. of the effective supply of solar energy, converted into kinetic energy, would keep up the working conditions. We will consider the process subsequently. Our immediate purpose is to point out that the circulation represents the balance between loss and gain of energy continuously operating one against the other, but each taking place independently of the other without any simple automatic relation between the loss of energy by turbulence and the gain of energy on a thermodynamic basis from solar and terrestrial radiation; for ever, one or other will be overrunning the adjustment. Hence we must conclude that constantly recurring changes or fluctuations are within the natural order of the general circulation. These changes find obvious expression in the monthly values, and it is to these that we now turn our attention. We shall be interested mainly by the irregular changes, represented by departures from the normal, which are shown by consecutive years or other periods, day, week, month, season, on the one side, or on the other side the lustrum of five years.

THE SEQUENCE OF CHANGES

We thus face the general problem of successive changes in the general circulation. We cannot introduce it better than by presenting the diagram of monthly rainfalls in London for 100 years which was completed by J. S. Dines at the Meteorological Office in 1913 by additions to the corresponding diagram prepared by his grandfather George Dines and published in the *Quarterly Weather Report* of the Meteorological Council. We have added overleaf fig. 188, in order to bring the information up to date so that the reader may examine for himself the question of normals and recurrence after a long period of years.

As examples, in quite different form, we have added tables of rainfall in Paris[1], Padua and Edinburgh, arranged in lustra, and a similar table for the United States which we owe to Professor A. McAdie of Blue Hill Observatory.

Rainfall is a peculiarly difficult element to deal with but it is not on that account an unsuitable one to choose for the presentation of the problem, which is indeed to find out laws which will account for the fluctuations portrayed in the diagram.

Many long series of values are available for pressure and temperature as well as rainfall, and there are a few for other meteorological elements. The table at the head of the chapter gives a list of records extending over more than a hundred years arranged according to countries. It is perhaps unfortunate that wind-directions at well-exposed stations figure so little in the compilations. Some particulars for Britain are given in a paper by C. E. P. Brooks and Theresa Hunt in the *Quarterly Journal Roy. Meteor. Soc.* 1933.

[1] *Met. Zeitschr.*, Bd. XXX, 1913, p. 45.

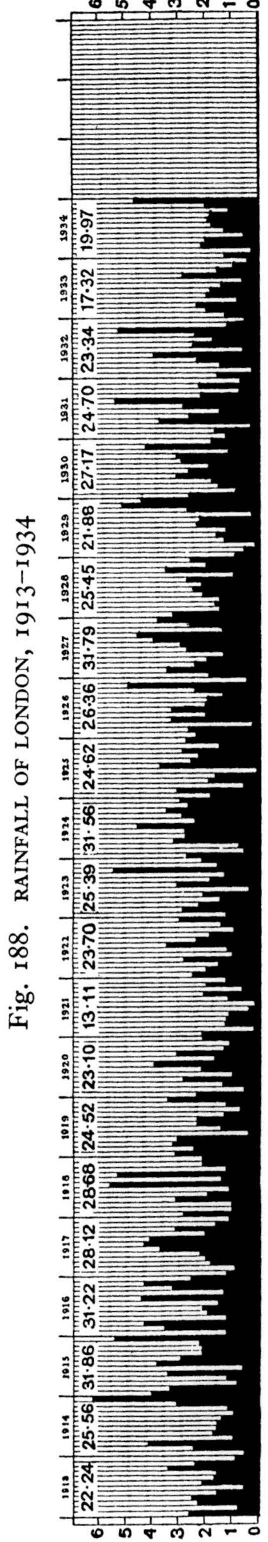

Fig. 188. RAINFALL OF LONDON, 1913–1934

ANNUAL RAINFALL

MEANS FOR LUSTRA AND FOR 35 YEARS

	United States	France	Italy	Scotland
	Cambridge Boston New Bedford New Haven	Paris: Observatoire to 1910 Montsouris 1911–30	Padua	Edinburgh Charlotte Sq. to 1895 Royal Obs. 1896–1930
Lustrum	• in mm	• in mm	• in mm	• in mm
1750	1072		821	
1751–55	1159		968	
1756–60	1110		942	
1761–65	850		939	
1766–70	965		988	
1771–75	1044		1086	
1776–80	1010		926	
1781–85	1269		859	
1786–90	966		812	668
1791–95	974		847	764
1796–1800	970		912	587
1801–05	987		1036	499
1771–1805	**1031**		**925**	
1806–10	1041	464	995	661
1811–15	1090	492	795	630
1816–20	979	509	667	640
1821–25	935	502	671	645
1826–30	1086	526	769	692
1831–35	1082	472	765	584
1836–40	987	548	831	710
1806–40	**1029**	**502**	**785**	**652**
1841–45	1062	515	872	581
1846–50	1070	547	835	648
1851–55	1098	514	934	615
1856–60	1203	543	833	697
1861–65	1127	462	772	711
1866–70	1154	525	857	666
1871–75	1119	540	852	732
1841–75	**1119**	**521**	**851**	**664**
1876–80	1096	540	898	761
1881–85	1036	478	772	625
1886–90	1268	470	809	609
1891–95	1059	467	752	623
1896–1900	1136	504	907	636
1901–05	1108	539	923	583
1906–10	956	558	823	710
1876–1910	**1094**	**508**	**841**	**650**
1911–15	993	608	849	600
1916–20	1093	627	943	697
1921–25	1022	614	793	664
1926–30		658	900	783

These all represent the succession of events by the sequence of monthly or annual values at individual stations. It need not be assumed that that selection is the best for the purpose of investigation. A year or even a month is a long interval for the integrated effect of weather; the same mean for a month may cover great differences between weeks or days. Looking at the subject from another point of view the values for an individual station are liable to vicissitudes due to local conditions which are not characteristic of the circulation, in the more general sense, in being at the time. When the Meteorological Council set up an organisation for the supply of statistics of weather for the problems of agriculture and hygiene in 1878 they chose the week as the fundamental period, with a daily map as an auxiliary on the one side; and, on the other side, they aggregated the weeks into quarters and years. Moreover, they grouped the stations of the British Isles in twelve districts instead of dealing with the long sequence at the several stations.

Homogeneous data on this plan are now available for 50 years. It has many advantages[1]. One of the easiest conclusions, for example, is that the district Scotland North belongs to a different meteorological province from that which includes the rest of the districts, a feature which is constantly in evidence[2].

Coincidence of the fiftieth anniversary of the weekly report with the corresponding anniversary of the International Congress at Rome in 1879, which confirmed the international practice of using the month as the time-unit for seasonal variation, is sufficient to suggest that it was not for lack of cognizance of the issues involved that the Meteorological Council chose the week as their phenological time-unit. It invites us to reconsider the question now, in view of the experience of the past half-century. The real advantage of a weekly report lies in the possibility of making an effective presentation of the seasonal variation of the elements by the successive weekly values.

In the original edition of this volume we reproduced diagrams made in 1911 which showed the differences from normal of successive quarterly and annual values of temperature and rainfall in each of the twelve districts of the British Isles from 1879 to 1910. We wished the reader to draw the conclusion that when the quarters are compiled into years a good deal of the vigour of the atmospheric action is lost. They are not reproduced here partly because the representation was by successive points for the quarterly or annual results connected by lines which indicate gradual increase or decrease and cannot have a real meaning.

In chapter IV, following the convention set out on p. 50, we have already given an indication of the manner in which the fifty-two values, for a year, can be represented week by week in a single line of print, and overleaf, on the same plan, we show on a single page the weekly weather of one district, the Midland Counties, for the twenty years 1907–1926. For any district the differences of weekly results from year to year are effective illustrations of the character of the British climate.

[1] *Journal of the Royal Statistical Society*, vol. LXXXVIII, part IV, pp. 489–512, 1925.
[2] J. Glasspoole, *British Rainfall*, 1922, London, pp. 260–66.

The Weather of the Midland Counties, Week by Week, for Twenty Years, beginning with the second week each year.

Temperature.

U very unusual, u unusual, m moderate, d deficient, D very deficient.

	Winter		Spring			Summer			Autumn			
			†			*			†			*
1907	umDd	dmmm	mmuUu	mmmd	umDd	dmdDD	Dumd	mmdm	duuuu	mmmm	uumm	muumd
1908	dmdm	muum	dmdmm	mmDm	umum	Umddu	mduu	udmm	DdmuU	UUmu	muuU	mmumdm
1909	umDm	dddD	ddmmU	uumu	muum	DmdDd	mmdm	UudD	dmmmu	uUDm	dDdu	mmdUm
1910	umDm	mumm	ududd	umdd	mUum	UuuDD	dddm	mumm	dmduu	ummd	DdDd	uUUmm
1911	mmud	muuu	mmmmd	muum	UuuU	Udmdu	uUUU	UUuu	Umdmd	mUmu	mumm	mmuuu
1912	mmdD	duuU	uumuu	muum	Uudm	dmumd	uumD	dDdd	dDmdd	uumm	mmum	mUuUu
1913	udmm	umdm	uummu	dmmm	dmmU	dmudm	Dmdm	dumu	ummUU	uuuu	uuUU	uuuddu
1914	dduU	Ummu	mmmum	uUum	mUdm	mUmUm	umdd	umuU	umdmm	ummu	uDmu	mmmmu
1915	umdu	mmdu	mmddm	mmuu	dmum	Umdmm	Dddm	ummD	mUuDm	umdD	mDDd	ummUU
1916	UUuu	mmdd	dmmdm	mmum	dUum	DDddd	dmuu	uuum	umduU	Ummm	ummd	dddmu
1917	mdDD	Ddmm	DmddD	Dmmu	muUu	uUuDd	muUd	uuum	umuum	dmmm	mmUU	dmDmd
1918	ddUm	uumm	muUmm	mdmm	mUUu	mmDDm	dumu	uuUm	mdmmd	mmmU	mddu	UUmmm
1919	mmdd	dDmm	mmddd	mmdd	uUuu	uUmDD	dmdm	uUud	uUmmm	ddmD	dDdm	umuumm
1920	Uumm	uumu	muuum	mmmm	muUm	dummD	mmDd	mddd	mmduU	ummd	uudu	mDmUU
1921	uUUm	muuu	uuuuu	mmum	umum	uuudu	UUuu	muum	umumU	UUmu	Ddmd	uuuu
1922	mddm	dmUU	mmddd	mddd	muUU	ummDD	Dddd	Dddd	mDmmu	mmdD	mmmm	uuumu
1923	uuuU	uudu	mmmuu	mddU	mDDD	ddddu	Uumm	uumd	mmDum	mmuu	DmdD	dmmmm
1924	duuu	uddd	dmdmd	dmum	duuu	dmmmd	umdm	mmdm	uuumm	mumu	mumU	uuuuum
1925	umuu	ummm	dudmu	mmdm	UUum	Uudum	UUdm	uumm	Dddum	duuu	DdDD	mddUu
1926	dduu	muUu	Ummuu	mmmd	ddUm	mmmmm	Uumm	uuuu	UUmmU	mDDd	mumd	mmmuu
	Jan.	Feb.	Mar.	Apr.	May	June	July	Aug.	Sept.	Oct.	Nov.	Dec.

Sunshine.

A very abundant, a abundant, m moderate, s scanty, S very scanty.

			†			*			†			*
1907	mmaa	mmaa	mmAAa	smmm	msSS	Smmss	sasm	mmsa	sAaam	mmms	sAsA	amssa
1908	mmmm	assm	smssm	mmsm	ssma	asmaa	SSaA	amsa	sassA	mmmm	sAmm	sasssm
1909	Ammm	sAas	smssA	maaA	aamS	smSss	Sama	Amsm	mSssm	mmms	Aams	msAas
1910	aAaa	mama	mamaS	ssaa	asam	samsS	mssm	smsm	sSmmm	ssSm	aams	smmam
1911	aama	smaa	mssss	mmmm	mSmA	Amssa	Aaaa	aAsa	Aamam	mSmm	Asms	Aamss
1912	sSma	mSss	mssmm	aaAs	smsm	SmmmS	smSs	sSSs	mSmmm	amsm	sssm	ssasm
1913	smSs	ssas	ammss	smms	Smma	ammsS	SSsm	sSsa	Smmms	smaa	aamm	ssmmam
1914	sssa	msmm	msssm	AAas	mass	mAaam	SsSs	aaam	maAas	Ssms	masm	smmmm
1915	mmss	mmam	smmma	ssam	mmAa	aAmSs	ssas	msam	AmmmS	sSmm	mAms	mmsmm
1916	mmms	aaSs	SSSma	aSas	Samm	ssSSS	Ssaa	assm	mssss	samm	mmmm	assmm
1917	sSSs	aSSs	Smmmm	mmsa	assm	aAmSa	mmmS	smas	assmA	saam	amma	Aammm
1918	assa	sSma	ssass	SSas	mmaA	amsma	msma	maaS	mmmas	msss	Aams	ssamm
1919	aasS	saSs	sssam	ssms	saAA	mAaSS	smss	aAms	sammA	Aama	Ssam	masmsm
1920	mmam	samm	amaSS	ssmm	ammm	mmsSS	smss	ssss	sssmm	smas	mmsm	mssms
1921	mmss	ssas	mmmaa	amam	mAAa	aamaa	amms	smsm	Ammam	AAam	Asss	smaam
1922	amss	asaa	msmsm	smmm	mmAA	AaSSS	msss	Smms	Ssmsm	aama	mmsm	ssaaa
1923	amms	msSm	SSmss	ssam	assS	Smsss	mmsa	amsa	ammmm	smmm	Aaam	amamm
1924	msma	sSsm	aAaSs	mass	smss	Ssmmm	aass	mmsS	Ssmam	mmms	msmm	msmmmA
1925	mmSa	mmsS	asmms	mmss	msmA	aasms	mmss	sssm	smmmm	msmm	AmaA	asmsm
1926	mssS	Smsa	mSsss	mmSm	smsm	msmaS	aSsa	msas	Samms	mmam	asmm	mmmmm
	Jan.	Feb.	Mar.	Apr.	May	June	July	Aug.	Sept.	Oct.	Nov.	Dec.

Rainfall.

H very heavy, h heavy, m moderate, l light, L very light, O no rainfall.

			†			*			†			*
1907	lLll	lhml	mmmOm	hmmh	mhhh	mhmmh	mOHh	lhlL	hLOLl	hHmh	mlmh	hHllL
1908	hlLl	Llhm	hmmHh	llhH	hhlL	hlmOL	hhLL	LlmH	HmmmL	lLhl	Lmmm	lmhlmm
1909	hlLl	mLlH	mmhhO	mhhO	llhh	mlHmh	mlHm	OHlm	mmmhm	hmmL	Lllh	hhHml
1910	hmhm	mhHl	mlOLm	hmhh	mhLm	HLhhH	Lmml	lhhh	lmlml	hmmm	hmhH	hHmlm
1911	mLLL	Lmhh	mhmml	Llhm	llmL	LmHmL	OLLm	LLmm	lmmmm	Lmhh	hmlm	hHHmh
1912	hHmL	hmmm	HhHll	lOOl	lmHm	Hhmhm	llHH	hmHH	llOlm	Olhh	lmlh	mmmHh
1913	hlhh	mmll	mhHhm	mhhH	hlLl	mmlLl	mLLL	llml	hmhmm	HLlh	lHml	mllmlm
1914	lLll	hhlm	hHmmh	OLLh	lmlL	HlmHm	mhmm	LLmL	mmllO	mmhh	mlmH	hhhHH
1915	hmlh	hhmm	llmmm	mlmm	HhLL	lOmhH	Hhmh	hlLl	OlmmL	Lmhm	HlLH	Hmhhm
1916	lmlh	hHhh	hHHhl	mhlh	hlml	hlmmh	mmLO	Lhmh	mlmlh	lmhh	mmmL	mlHmm
1917	HlLL	Llhl	hmmmm	hmOO	hhmh	mlmhL	mhlh	HhmH	lmhlh	hlmm	mLlm	llmml
1918	mHll	hmlm	lllhh	mhlm	mmhO	LlmlL	hHhm	mLLh	HHHmh	mlll	hLLm	mmhhH
1919	mhhm	llHl	hHHml	mhmh	mlLL	lmmlh	mhmL	LLhH	mmmml	Llhm	llmm	Hlhhhh
1920	mmHL	mhlm	mhhHH	Hmhm	mmhL	hhlHH	mmhm	llLl	lHlmh	mLLm	lmmh	llmHm
1921	mmlm	LLmm	lhmml	mmlm	mLLm	OOllL	LLmh	hmlm	lhOOm	lmlh	Lmlm	llmhm
1922	mhhh	OmhH	hLmmh	hmhm	LmmO	LmmmH	mmml	Hllm	lhmhl	LlLm	mLLl	LlHhh
1923	mlLm	HhHH	mmmhm	Hlmm	mmmm	llLLl	hlhh	Lmhh	lmhmm	mlml	mhlm	mlmhm
1924	hhhL	lllm	mOmmL	mmmH	hmHH	hhmLh	LmhH	lmhm	lhhhm	hlhH	Lmlh	hmmhHl
1925	mlhm	HlHl	lmllh	hmmH	lHhL	OLLll	lhhh	mmll	mHhlL	mHmh	LLll	mmhHm
1926	mhmh	mhlm	lllLm	hmhm	hLmm	hmlLm	Lhml	hmLm	mlmlm	hlmh	HHll	llLm
	Jan.	Feb.	Mar.	Apr.	May	June	July	Aug.	Sept.	Oct.	Nov.	Dec.

† Equinox. * Solstice. Black type means High Sun. Italic means Low Sun.

PLASTICITY AND RESILIENCE OF THE CIRCULATION

There are different ways in which we may regard the problem of explaining the sequence of changes. From one point of view the circulation may be looked upon as a plastic structure which, like clay, can be moulded by the external influences that are operative from time to time. It may be supposed to have no natural resilience but simply a capacity for taking the impress of disturbing causes. From this point of view the line of approach is to exercise our ingenuity in devising possible causes, estimating their magnitude, and tracing their effects in such a way that complete knowledge of the effects will follow a sufficient knowledge of causes. For this purpose we have to compile a history of the circulation in past time, as complete as possible, and to fit the recorded changes to ascertained variations in the causes. This is a fair description of the method by which the question of changes in climate has been, in fact, approached. We shall denote the underlying idea as that of the plasticity of the atmospheric circulation.

Another point of view to which we shall refer, under the name of resilience, depends ultimately on the idea that the circulation as a whole is a resilient structure. It may be disturbed by any temporary exceptional cause such as some peculiarity of the orbits of the sun, moon or planets, a change in the solar energy, or a loss of transparency of the atmosphere on account of dust or carbonic acid gas, an accumulation of ice in the polar regions or a recession of the glaciers of lower latitudes, and so on. When the temporary cause is removed and the conditions are restored to the normal (if that be possible) the circulation will, in virtue of its resilience, recover its normal condition; but it may oscillate about the normal in some period or periods of its own before resuming its normal state. If the source of the disturbance be continuously operative, like the effect of orographical features, or intermittently so, like solar radiation, the resilience will give rise to oscillations continually recurring but perhaps with irregularity of amplitude. If the cause of the disturbance is itself periodic, like sunspots or changes in the planetary orbits, the periodicity of the cause will certainly be forced upon the circulation; but the response to the exciting cause will be much enhanced if there is any natural tendency to periodicity in the circulation itself coincident with the period or periods of the cause. From this point of view it is of the first importance to find out the natural periodicity of the atmospheric circulation. Unfortunately we cannot calculate its natural periods of oscillation, the problem is too complicated. We can only approach the solution by examining the changes themselves with the view of detecting the periodicities that are latent therein, however distorted the changes may be by effects that must be regarded as local and irregular.

It is the idea of resilience either in the circulation itself or in the causes which affect it that is the justification of the efforts to discover the latent periodicities of the long records. Resilience is in fact the dynamical correlative or equivalent of oscillation or periodicity. Only a resilient structure is capable of oscillation of a natural kind. We have come across the same idea in chapter IV in considering the structure of the atmosphere which has a resilient stratification on account of its potential temperature. So also here periodicity implies resilience either in the circulation itself or in the causes which disturb it.

The underlying idea of this way of approaching the explanation of the sequence of changes is referred to as the resilience of the circulation, as distinguished from what we have called its plasticity. We shall deal separately with the changes, from these two different points of view.

THE GENERAL CIRCULATION AS A PLASTIC STRUCTURE

The study of the variations of the circulation from the point of view of its plasticity consists in the first place of compiling its life-history in order that the changes may be brought into comparison with certain known or supposed causes. We have already given some account of the process in chapter VI of volume I, and here we can only supplement the inquiry by a brief chronological table of salient meteorological events with an occasional reference to agents or presumptive causes. This is not by any means the first time that an attempt of this kind has been made. Otto Pettersson has studied the chronology from the point of view of oceanography in conjunction with meteorological records, and has sought to bring the salient features into relation with astronomical conditions. W. J. Humphreys has examined the chronology of weather in relation to volcanic eruptions, and indeed the voluminous inquiry into the causes of the glacial epochs of past ages, which is described by Humphreys, affords conspicuous examples of the method to which we are now referring. In the work on the fluctuations of climate, to be described later, Brückner has dealt with a chronology reaching back to the beginning of the eleventh century and taking into account a great variety of climatic and economic data.

There is a well-known chronology of meteorological and economic events compiled by E. J. Lowe in 1870, and more recently a chronology of weather in Britain from the eighth century onwards has been made the subject of a presidential address to the Geographical Association by Sir Richard Gregory. In this connexion it may be remarked that economic conditions are very effective indices of the weather of past times, when, owing to lack of transport, each locality had to bear the whole burden of the stress of its own weather. It is on that account that the price of corn can be regarded as a meteorological record.

The most striking chronological sequence of an indirect character is that given by the annual rings of growth of sequoias which, with the aid of modern records, A. E. Douglass[1] has interpreted in terms of annual rainfall from 1306 B.C. to the present time. We have already used the inferences from this

[1] *Climatic cycles and tree-growth*, Washington, Carnegie Institution, Publication 289, 1919.

investigation in chapter VI of volume I by reproducing the diagram in which Ellsworth Huntington compares them with the evidence which he has collected about apparent changes of climate from other sources[1].

In our table we have noted only the salient events. From the year 1860 onwards there is abundant material for the examination of all questions of presumptive causes, as of periodicities, from every meteorological point of view. We have already cited examples of long records of trustworthy observations which for certain selected areas will carry back the means of inquiry as far as 1725; beyond that time the records are scanty, and before 1650 there are only notes of weather without instruments, and the notes are chiefly concerned with the economic effects of weather. We shall get a more homogeneous table for our immediate purpose if we treat the last five half-centuries in the same way as those that preceded them, citing the salient features only, with the knowledge however that the material for a more elaborate form of investigation is at hand if required.

In a number of cases, we have mentioned at the end of the entry any "cause" that may have been suggested as explaining the particular feature referred to. It will be understood from what we have said elsewhere that we are not much disposed to regard any meteorological event, still less a sequence of events, as being dependent upon a single cause; but, until we are able to treat the circulation as an entity with known characteristics and properties, we must be content to explore what can be reached by hypotheses which in the end are quite likely to be found inadequate for the whole burden.

In order to keep the chronology in touch with the endeavour to represent the sequence of weather as the combination of regular oscillations of recognised periods which are set out in the table of p. 320, we have entered, with an asterisk, the dates for twenty-four intervals of thirty-one years each, counting backwards from 1925–6. A period of thirty-one years includes a number of well-marked periodicities, and twenty-four such periods make up the full 744 years of the grand lunar-solar cycle to which reference will be made later on. We have entered too, for the eleventh to the fourteenth centuries, the warm and cold periods, arrived at by Brückner by the collation of historical information for Europe or the world at large, which form part of the basis of his computation of a cycle generally regarded as having a period of thirty-five years and given on p. 316 as $34{\cdot}8 \pm {\cdot}7$ years.

The chronology provides some notable coincidences with these periodicities, but also some notable discrepancies. Local discrepancies may also be noted in the table in which some of the entries may appear to be self-contradictory. There is apparently no overpowering influence that obtrudes itself, at its own regular intervals, above all the combinations of cycles of shorter periods.

[1] Another remarkable type of evidence of the variation of climate in the past is that afforded by the "varves" in the deposits of glacial sediment worked out by Baron Gerard De Geer for Sweden and elsewhere, *Stockholms Högskolas Geokronol. Inst.*, Data 14–16, 1929–31. And solar radiation in relation to terrestrial climates is the subject of a paper by Sir George Simpson with discussion in *Q.J. Roy. Meteor. Soc.*, vol. LX, 1934, p. 425.

A Chronological Table of Salient Meteorological or Economic Events of the Historic Period with some indications of Agencies or Presumptive Causes.

Based mainly on the following authorities:

I. Albano, *O secular problema do nordeste*, Rio de Janeiro, 1918.
E. Brückner, *Klimaschwankungen seit* 1700, Wien, 1890.
F. Eredia, *La Siccità del* 1921, Rome, 1923.
Sir Richard Gregory, 'British Climate in Historic Times,' *Geographical Teacher*, 1924.
W. J. Humphreys, *Physics of the Air*, Philadelphia, 1920.
E. J. Lowe, *Natural Phenomena and Chronology of the Seasons*, London, 1870.

Dates at intervals of 31 years from 1925–6 are indicated by an asterisk *.

Dates before 1752 are liable to be misunderstood in consequence of the change in the commencement of the year with the new kalendar.

Twentieth Century.

*1925–6. Cold Nov.,Dec., Jan. in England, warm Feb. and early April, cold May and June. [Predicted as periodic.] Severe floods in Holland.

1921. Rainfall only 50 per cent. of normal in S and E England. Wet summer in NE of the United States.

1912, 1913. Feeble sunshine, Europe. [Eruptions: Katmai, 1912; Colima, 1913.]

1911. Record for warmth in summer, England. (*See* Q. J. Roy. Met. Soc. XLII, 1916, and The Air and its Ways, p. 19.)

1903. Very rainy year, London, 920 mm, chiefly in summer. Feeble sunshine. [Eruptions: La Soufrière, Mont Pelée, 1902.]

Nineteenth Century.

1897, 1899–1901. Famines in India.

*1894–5. Very cold in England, February. Rivers frozen, boat-races abandoned at Cambridge. [Periodic.]

1888–92. Feeble sunshine. [Eruptions: Bandaisan, Ritter, 1888; Bogosloff, 1890; Vesuvius, 1891–9; Awoe, 1892.]

1887. "Driest year on record," British Isles.

1887–9. Famine in China.

1884–5–6. Feeble sunshine and celebrated sunset-glows. [Eruptions: Mauna Loa, 1881–2; Krakatoa, St Augustin, Bogosloff, 1883; Falcon I., 1885; Tarawera, Niuafu, 1886.]

1881. Temperature much below normal in England, 2tt to 3tt Jan.–March.

1878–9. Agricultural disaster. Temperature 1tt to 2tt below normal, Oct. 1878 to Dec. 1879. Rainfall Apr. to Sept. 50 per cent. above normal all over Great Britain except Scotland N, less than 50 per cent. of normal Oct. to Dec. [Eruptions: Cotopaxi, 1877; Ghaie, 1878.]

1877–8. Severe famine in N. China.

1874, 1876 and 1877. Droughts and famines in India.

1872. "Wettest year on record," British Isles.

1868. Famine in India, N.W. Provinces.

1866. Famine in Bengal and Orissa.

*1863–4.

1861. Famine in N.W. India.

1858. Least rainfall of any year of the century, London.

1852. Greatest rainfall of the century, London.

1845–7. Potato-famine in Ireland.

1838. Famine, India, N.W. Provinces.

1834. Famine, India, N.W. Provinces. 1833. Drought in N. Bombay. 1832. In N. Madras.

*1832–3.

1824. Drought in Bombay, etc. 1823. In Madras.

1816. "Year without a summer," cold all over the world. [Eruption of Tambora which killed 50,000 people, 1815.]

1804. Famine in India, N.W. Provinces.

1802. Drought in S. Hyderabad and Deccan. 1803. In N.W. Provinces and Central India.

*1801–2.

Rain above 30 inches (normal 24 inches) in London, 1816, 1821, 1824, 1841, 1852, 1860, 1866, 1872, 1879, 1903, 1915, 1916, 1924.

A list of volcanic eruptions is given on pp. 21 and 25.

Eighteenth Century.

1790–92. Famine in Bombay, Hyderabad and N. Madras.

1783–4. Feeble sunshine, Europe and N. America. Winter more severe than for many years. (B. Franklin.) Asamayama the most frightful eruption on record, 1783.

1782. Drought in Bombay and Madras. 1783. In Upper India.

1778–9. Maximum number of sunspots ever recorded.

*1770–1.

1769–70. Great famine in Bengal.

1752. Many gales and floods, England.

1750. Unusually hot in February. Earthquake, Feb. 8, Lyndon, Rutland. The first three weeks in July the hottest ever known. The summer (except a few very hot days) exceedingly cold and scarcely a day without rain, London, Plymouth, Dublin.

1749. Snow and frost, May 16 to June 16 over England. Temperature 88° in shade, July 2.

1747–8. Cold winter, rivers dry in January.

*1740. Jan. 1 to Feb. 5. Great frost, lasted 9 weeks. Great fair and coaches on the Thames: known as the hard winter. December. Great snows, rains, storms and severe frosts. Similar in France, worse in Holland and Germany where whole territories were under water.

*1739. December, Rutland. Ice froze 3 inches thick in 24 hours, the most severe ever known.

1738. Jan. 14, Scotland. Violent gales, no such storm for many years. Dec. 29. Frost more severe than any since the memorable winter of 1715–16.

1737. Jan. 9. Great gale and flood at Chepstow, the water rose 70 feet.

1736. From the beginning of March continued rains and floods till July, 5 inches in 3 days.

1735. Jan. 8. Very severe gales, England, France and Holland. "The greatest gale ever heard of in the South of England."

1722. Cold, wet year. [Eruption: Kötlugia, 1721.]

1715–16. Great frost. Nov. 24–Feb. 9. Fair on the Thames.

*1708–9. "Hyems atrocissima," Europe except Scotland and Ireland. Frost in England, Dec. 1708 to Mar. 1709, in Edinburgh from early October to end of April. Wheat 78*s.* 6*d.* per quarter compared with 43*s.* in 1715, and 46*s.* 6*d.* in 1704. Mild winter in Constantinople but very severe in Eastern North America. [Eruption: Santorin, 1707.]

1703. Great gales in England, Nov. 26 to Dec. 1, "so disastrous as to fill a volume of the Philosophical Transactions with accounts of it." Probably surpassing all others on record. Eddystone Lighthouse destroyed.

Seventeenth Century.

1698–9. The coldest years between 1695 and 1742; the latest spring for the past 47 years. Apples in bloom, July 30. The first wheat cut in middle September and much barley uncut at Christmas, and in Scotland corn was reaped in January 1699. Wheat 64*s.* to 73*s.* per quarter compared with less than 30*s.* previously.

1697. Much rain and hail.

1696. Crops bad and dear.

1695. Cool summer.

1694–5. Severe frost, many forest trees split. [Eruptions: Celebes, Amboyna, Gunong Api, 1694.]

1683–4. Great frost from beginning Dec. to Feb. 4. Longest frost on record, nearly all birds perished. Very severe weather in Dorset about Christmas is referred to as a subject of discussion on Feb. 10 and March 10, 1684–5, in 'Early Science at Oxford' in *Nature*, 1925, but is not mentioned in the compilations. There may be confusion of dates because February 1684 might easily be classed with the winter of 1683–4 according to our reckoning, but should be classed with the winter of 1684 old style, or 1684–5 new style.

1681. Severe drought and cold spring. [Eruption: Celebes, 1680.]

*1677–8.

1671. An ice-storm, Dec. 9, 10, 11, followed by great heat. "An apple bloomed before Christmas."

1661. Great Indian famine, no rain for two years.

1660. Very cold, the price of wheat doubled.

1649. Great frost in January, the Thames frozen over.

1648. Prodigiously wet summer.

*1646–7.

1637. Long, severe winter. [Eruption: Hekla, 1636.]

1632–3. Severe autumn, winter and spring, in Europe. Hot, dry summer (1632) in Italy. Vesuvius more violent than at any time since 79.

1622. Frost, all the rivers of Europe frozen and also the Zuyder Zee.

1616. Great drought, Nottingham.

*1615–16.

1607–8. Great frost and snow from Dec. 5 to Feb. 14. Rivers frozen, including the Ouse, at York, Thames, at Lambeth.

Sixteenth Century.

1594. "Rain and flood ruined the harvest." Dearth owing to rain from the beginning of May to July 25.

"Therefore the winds...
...have sucked up from the sea
Contagious fogs; which falling in the land
Have every pelting river made so proud
That they have overborne their continents;
The ox hath therefore stretch'd his yoke in vain,
The ploughman lost his sweat, and the green corn
Hath rotted ere his youth attained a beard;
The fold stands empty in the drowned field,
And crows are fatted with the murrion flock.
...And thorough this distemperature we see
The seasons alter. Hoary-headed frosts
Fall in the fresh lap of the crimson rose,
And on old Hiems' thin and icy crown
An odorous chaplet of sweet summer buds
Is, as in mockery, set: the spring, the summer,
The chilling autumn, angry winter, change
Their wonted liveries, and the mazed world,
By their increase, now knows not which is which.".

1586. Famine in England.

*1584–5.

1581. River Trent dry, Dec. 21.

1571. Oct. 5. Gales and floods in Lincolnshire.

1565. Famine in Britain.

1564. Great frost in London at Christmas. Thames solid as a rock.

1556. Drought. Wheat rose from 8*s.* to 53*s.* per quarter.

*1553–4.

1541. Drought. Trent a straggling brook; Thames so low that the sea-water extended beyond London Bridge.

1540. Cherries ripe at the end of May.

1538–41. Hot and dry summers.

1538–40. Drought.

1537. "This year in December was the Thames of London all frozen over."

1531. "All description of corn gathered this year is still very moist owing to the quantity of rain that has fallen."

*1522–3.

1515. Intense frost, Thames frozen throughout January.

1506. Intense frost, Thames frozen.

Fifteenth Century.

1450–1500. Dry period in tree-growth.

*1491–2.

1473–4. Very hot summers.

*1460–1.

1447. Very hot summer.

1438. Famine. The people of England obliged to make bread of fern-roots.

1433–4. Thames frozen below London Bridge to Gravesend from Nov. 4 to Feb. 10. The price of wheat rose to 27*s.* per quarter but afterwards fell to 5*s.*

*1429–30.

1407. "There was a gret Wynter that dured both December, Januari, Februari and March that the most part of small birds were dead."

"The 1,600 pages of Mr. James Gairdner's edition of the Paston Letters, dated from 1422 to 1509, do not contain a single reference to the weather or to any kindred topic."

Fourteenth Century.

1300–1450. Rainy period in tree-growth (Douglass).

1300–1400. Period of violent changes (Pettersson).

*1398–9.

1396–1407. Great famine in India.

1393 and 1394. Excessively hot and dry summers.

1385–1405. Cold period (Brückner, see p. 315).

1370–85. Warm period (Brückner).

1369. The corn was greatly damaged by floods.

*1367–8.

1363. Severe frost from December 7 until March 19.

1354. Drought—no rain at Nottingham from end of March to end of July (Lowe).

1353. Great drought from the month of March until July (Gregory).

1350–70. Cold period (Brückner).

1350. The Black Death throughout Europe.

1348. "There were grate reynes which dured fro the Nativite of Seynt Jon Baptist (June 24) onto Christmas."

1348. Mild and serene year: very mild in winter.

1345. Drought—"the dry summer."

1344–5. Great famine in India.

1342. There was spring-like weather the whole time between September to the end of December.

1340. In this year there was not much cold after the autumnal equinox.

1339. Drought in Scotland, people were reduced to feed on grass while wheat in England was only 3*s*. 4*d*. per quarter.

1338–9. From the beginning of December came a very hard frost which lasted 12 weeks.

1337. In January there was warmth, with moderate dryness, and in the previous winter there had not been any considerable cold or humidity.

*1336–7.

1327. Famine owing to a succession of cold rainy harvests.

1325–50. Warm period (Brückner).

1315. Famine: people ate the bark of trees: 20,000 are said to have starved to death in London alone. It arose from continued rain destroying the corn and causing mortality among sheep and cattle. The famine lasted several years.

1310–25. Cold period (Brückner).

*1305–6.

Thirteenth Century.

1293–4. Excessively hot summers.

1292. February and winter very severe.

1290–1310. Warm period (Brückner).

1290. This year were frequently inundations of rain and especially in summer and autumn.

1283. All the summer and great part of the autumn vehemently and continually rainy.

1282. Floods in Holland. Zuyder Zee formed.

1281–2. From Christmas to nearly February 2 was so much cold and snow that five arches of London Bridge were thrown down by the force of the ice.

1281. Drought. Men passed over the Thames dry-shod at Lambeth and over the Medway between Strood and Rochester.

1276–8. Hot, long summers, drought.

*1274–5.

1270–90. Cold period (Brückner).

1269. Great frost from November 30 to February 2.

1260. Long, great and severe drought.

1258. All kinds of corn nearly destroyed by the autumn rain. When April, May and the principal part of June had past scarcely were there visible any shooting buds of flowers. Many thousand human beings died of hunger.

1257–8. Not a single frost or fine day occurred nor was the surface of the lake hardened by frost, but uninterrupted, heavy falls of rain and mist obscured the sky until Purification (February 2).

1257. Continuous rains in summer and autumn.

1256. Year very stormy so that the cornfields perished, blackened and putrified in autumn.

1255–70. Warm period (Brückner).

1254. Severe frost from January 1 to March 14. From the middle of autumn until spring continually stormy.

1248. The temperature of the winter was entirely changed to that of spring so that neither snow nor frost covered the face of the earth for two days together. Barns full of corn; very mild winter.

1245. Plenty of corn and but little frost.

1245. Exceptionally mild winter.

1245–55. Cold period (Brückner).

*1243–4.

1242. Stormy in winter: dry and very hot in summer.

1241. From March 25 to October 28 continuous dryness and incomparable heat dried up the deep marshes and broadest ponds and exhausted the rivers.

1240. In March, April and May the season was dry, but for the rest of the months rainy.

1239. In the four months preceding Easter the flooding rain-clouds did not cease.

1236. In January, February and part of March there were such floods of rain as no one remembered to have seen before. After a winter beyond measure rainy a constant drought attended by an almost unendurable heat succeeded which lasted for four months and more.

1234–5. Great frost from Christmas to February 2.

1230–45. Warm period (Brückner).

1222–3. Inundations of rain from September 14 to February 2.

*1212–3.

1205. Frost from January 15 until March 22.

1202. About this year great rains caused floods throughout the whole world.

1200–30. Cold period (Brückner).

Twelfth Century.

1193. Unusual storms, floods of water, fearful and wintry thunderstorms throughout the whole of this year.

1190–1200. Warm period (Brückner).

*1181–2.

1175–90. Cold period (Brückner).

1175–6. Snow and frost lasted from December 25 to February 2.

1173. There was an extraordinary season of fine weather throughout the winter and spring and the month of May until Ascension Thursday.

1165–75. Warm period (Brückner).

1157. Thames dry.

1149–50. Severe frost from December 10 to February 19.

1145–65. Cold period (Brückner).

1130–45. Warm period (Brückner).

1125. So bad a season as there had not been for many years before. Intense frost. The dearest year known for wheat.

1117. This was a very deficient year in corn through the rains that ceased not almost all the year.

1116. This year was a very gloomy winter, both severe and long. Excessive rains came immediately before August and greatly troubled and afflicted people until February 2.

1115. Nearly all the bridges throughout England were broken by the ice.

1114–15. In this year there was so severe a winter with snow and with frost that no man that then lived ever before remembered a severer.

1114. April 4, Thames dry; October 6, Medway dry.

1112. This was a very good year and very abundant in wood and in field.

1111. A very long and sad and severe winter.

1110. The earth-fruits were greatly injured through tempests and the tree-fruits over all this land almost all perished. Thames dried up.

1105–30. Cold period (Brückner).

1105. A very calamitous year.

1103. This was a very calamitous year in the land.

Eleventh Century.

1098. Great rains ceased not all through the year.

1097. This was in all things a very sad year, and over grievous from the tempests.

1093. There was such a great deluge of rain, so great a season of rain, as no one remembered before. Then the winter coming the rivers were so frozen that they were a road to riders and to wagons.

1092. Very severe frost.

1091. Gales and floods over England. London Bridge swept away.

1087. Through the great tempests there came a very great famine.

1086. This was a very heavy and toilsome year in England, and so great unpropitiousness in weather as no man can easily think.

1080–1105. Warm period (Brückner).

1076–7. Frost in England from November until April.

1075–6. A great frost.

1065–80. Cold period (Brückner).

1063. The Thames frozen 14 weeks.

1055–65. Warm period (Brückner).

1046. There was no man who could remember so severe a winter.

1041. All this year it was very sad in many things, both in bad weather and earth-fruits.

1040–55. Cold period (Brückner).

1035. Severe frost on June 24.

1020–40. Warm period (Brückner).

A list of severe winters according to the prevalence of ice in Danish waters since 1000 A.D. is given on p. 340.

Before 1000 A.D.

900–1050. Rainy period in tree-growth.

998, 987, 923, 908, 874, 827, 806, 759, 695, 545, 508, 359, 329, 291, 230, 220. Notable frosts are assigned somewhat dubiously to these years.

944. Gale.

836. Flood.

800. December 24 a violent wind and "an inundation of the sea flowed beyond its limits."

763. Drought.

761. "In this year was the great winter."

738. Flood.

721. "A very hot summer."

353. Flood.

100–200 A.D. Dry period in tree-growth.

ca. 80. Severe drought for several years in middle Asia.

79. Aug. 24, destruction of Pompeii by eruption of Vesuvius.

200–50 B.C. Dry period in tree-growth.

450–200 B.C. Rainy period in tree-growth.

The sequence of events which is very imperfectly represented in the foregoing table has been summarised by C. E. P. Brooks[1] according (*a*) to the number of years of storms and floods, and (*b*) the number of droughts, with the following results for successive half-centuries from 1000 to 1750 A.D.

Century	11th	12th	13th	14th	15th	16th	17th	18th
Years of storms and floods	2, 9	4, 1	8, 8	3, 3	3, 3	2, 10	8, 20	17
Years of droughts ...	3, 1	10, 3	7, 13	7, 7	1, 6	6, 12	15, 22	29

He has formed therefrom an opinion upon the rainfall of the century expressed in the form of a diagram which we may translate thus as a general representation of the rainfall of Britain century by century:

Eleventh century: transition from light to heavy rainfall which continued with slow diminution until 1750. Twelfth century: decline and recovery. Thirteenth century: continuance of recovery followed by slight decline and recovery. Fourteenth century: small decline and recovery. Fifteenth century: increase and decline. Sixteenth century: decline, recovery and decline. Seventeenth century: further decline. Eighteenth century: rapid decline to the middle of the century.

The past history of weather in China has been expressed by the analysis of Chinese records in a similar manner by giving the percentage ratio of years of flood to those of floods + those of droughts with the following results:

Century	2nd	3rd	4th	5th	6th	7th	8th	9th	10th	11th	12th	13th	14th	15th	16th
$\frac{\text{Flood years} \times 100}{\text{Floods + droughts}}$	34	38	11	33	20	23	43	36	36	37	49	36	49	31	34

The increase of the ratio from the eleventh century and its persistence for three centuries is thought to show concurrence with similar changes in Europe[2].

[1] *Meteorological Magazine*, vol. LX, June 1925, p. 108. The subject of changes of British climate is taken up again by Brooks in *Q. J. Roy. Meteor. Soc.*, vol. LIV, 1928, p. 309, with a revised diagram, and in vol. LVII, 1931, p. 13, for the Old World.

[2] *Meteorological Magazine*, vol. LXI, June 1926, p. 115.

THE GENERAL CIRCULATION AS A RESILIENT STRUCTURE. PERIODICITY.

To the unsophisticated reader the sequence of monthly values of London rainfall (fig. 187) may seem to be unpromising material out of which to extract evidence of regular periodicity, even when the process has been eased by smoothing the curve and smothering some of the awkward projections; nevertheless, it has been attempted by the methods described in chapter XIII of volume I, and if the reader will look at the way in which periodic changes combine in fig. 111 in that chapter, he will understand that the combination of periodicities may produce some astonishing variations. We shall quote the results of some of the efforts which have been made. We do not think the possibilities of the process have yet been exhausted; sufficient allowance has not been made for the fact that a natural resilience of the circulation implies the possibility of the decay of oscillations of an existing period and their recrudescence in a new phase. The phase is generally regarded as one of the constants of the oscillation but it is not necessarily so if new disturbances occur. We have not space to enter into the details but we may refer the reader to a series of papers by H. H. Turner[1], in which he treats of the variations of the meteorological elements as grouped into separate and more or less independent "chapters."

It must be remarked that the periodicities computed may be regarded as real, although, on any occasion, they may be obscured by disturbing influences which are outside the period and have not been accounted for. This circumstance may make an acknowledged periodicity of little use in practical meteorology such as forecasting. We may, for example, with proper regard for scientific truth, regard the seasonal variation of rainfall in South East England indicated in fig. 183 as real in spite of the fact that a glance at the series of values for London in fig. 187 shows that it does not appear in the record of every year, and that to make a general forecast for any particular October as a wet month could not be justified.

The best known example of this kind of real but concealed periodicity is that of the semi-diurnal variation of pressure which may be described as a wave of pressure going round the earth about 2 hours in advance of the sun followed by a similar wave 10 hours after the sun. We have already given an account of its amplitude in different latitudes and other features connected with it. On any particular day it may be completely masked at any station in the temperate zone, yet it is looked upon as real because it always becomes apparent in the means for a succession of days.

The universality of the semi-diurnal wave of pressure leads us to think that periodic changes are not likely to be local; they must be felt at neighbouring stations, not necessarily with the same amplitude but practically in the same phase. The detection of the same period at near stations is in consequence an assurance of the reality of the period even if its amplitude is

[1] 'Discontinuities in meteorological phenomena,' *Q. J. Roy. Meteor. Soc.*, vol. XLI, 1915, pp. 315–334; vol. XLII, 1916, pp. 163–171; vol. XLIII, 1917, pp. 43–57.

small. J. Baxendell has given close attention to the question of locality in relation to periodicity in his investigation of the periodicities of Easterly winds at Southport. The reader should therefore examine the evidence for identical periodicities in the values for the different localities referred to on p. 292 which are set out on pp. 320–4.

Periodicities elicited by the co-ordination and comparison of data.

The direct examination of a long succession of values of any meteorological element without any formal arithmetical manipulation generally suggests some periodic fluctuation, and the impression is strongly reinforced when the same period is suggested by the sequences of a number of different elements or of different localities. It may not be possible accurately to define the period, or the amplitude of the variation which may be expected from its influence, but a case may be made out for its existence not altogether dissimilar from the seasonal variation of rainfall or the semi-diurnal variation of pressure.

The eleven-year period.

The suggestion is particularly frequent with regard to a period which is often called an eleven-year period and which has been discerned in the annual frequency of occurrence of tropical revolving storms, especially in the South Indian Ocean, rainfall and barometric pressure in various localities, famines in India, depression in trade, the annual frequency of sunspots, the annual variations in the terrestrial magnetic elements and the frequency of aurora. There is a voluminous literature in relation to this subject to which contributions were made by Sir E. Sabine, Loomis, Wolf, Gautier, Balfour Stewart, Piazzi Smyth, E. J. Stone, W. de la Rue, C. Meldrum, Sir N. Lockyer, W. Köppen, H. F. Blanford, Sir J. Eliot, C. and F. Chambers, E. J. Pogson, Stanley Jevons, A. Buchan, A. Supan and many others. It has been one of the most attractive subjects of discussion on the geophysical side of meteorology. The actual periods have not been very rigorously defined for the different elements, nor is the sequence uniform in period or in amplitude, but the consensus of evidence is considerable. We shall include a number of the suggestions made with regard to a periodicity of about eleven years, often called the sunspot period, in the table which is given on pp. 320–4.

In like manner a period of nineteen years, associated with lunar changes, has been suggested for the weather of Australia by H. C. Russell, and pursued by W. J. S. Lockyer, H. E. Rawson and others. A period of 3·8 years is also suggested by the sequence of annual values of pressure in all parts of the globe[1] and has been regarded as associated with solar prominences which have a similar period. In fact the study of the relation of solar and terrestrial changes, induced by these suggestions, was a noteworthy feature of meteorological activity in the years of the current century before the war. Since the war there has been a general recognition of a period of 3 or 3·1 years in rainfall

[1] W. J. S. Lockyer, *Roy. Soc. Proc.*, vol. LXXVIII, p. 43, 1907.

and other elements at Oxford, in Australia and in other parts of the world. H. R. Mill drew attention to a periodicity of that character in *British Rainfall* in 1905.

The best known effort to establish a period in weather by the co-ordination of observations is that of E. Brückner, who found many indications of a period not rigorously uniform either in duration or in amplitude but quite notable in amount and with an average duration of about thirty-five years. A cycle of that duration had already forced itself upon public attention in the Low Countries in the sixteenth century according to the extract from Bacon's essay *On the vicissitudes of things* which is quoted in the biographical notice in volume I.

The Brückner cycle.

Brückner's contribution to the question of periodic changes in the weather of the world is contained in a volume entitled: *Klimaschwankungen seit 1700 nebst Bemerkungen über die Klimaschwankungen der Diluvialzeit*[1]. It originated in the idea of correlating the rhythm of the oscillations of level of the Caspian, the Black Sea and the Baltic with that of the variations in Alpine glaciers which had been set out by v. Lang in 1885, and which might be attributed to alternations of cool and rainy epochs with warmer and drier periods. It was natural to conclude that there would be similar fluctuations of weather outside the region of the Alps, associated with the changes in the levels of the three great inland seas, and the discussion of these relations showed that the changes affected not only Europe but the whole of the Northern Hemisphere and were not less apparent in the Southern Hemisphere. The investigation led to the suggestion of a period in world-weather of $34{\cdot}8 \pm {\cdot}7$ years.

Dass die gewonnenen Ergebnisse in keiner Weise abschliessend sind, brauche ich nicht hervorzuheben, handelt es sich doch um den Beginn der Discussion einer bisher nicht beachteten Frage. Manche Probleme konnten überhaupt nur gestreift werden, so unter anderem die wichtige Frage nach der Endursache der Klimaschwankungen. Nur aphoristisch ist die praktische Bedeutung der Klimaschwankungen behandelt, und nur wenige Worte sind den durch die Klimaschwankungen verursachten Schwankungen der Meere gewidmet, obwohl diese ursprünglich den Ausgangspunkt der Untersuchung bildeten; das in meinen Händen befindliche einschlägige Material ist noch nicht vollständig genug, um allgemeine Resultate zu liefern; es muss die Verwerthung desselben einer späteren Veröffentlichung vorbehalten bleiben.

Eine gewisse Schwierigkeit bot die Wahl eines passenden Titels. Ich schreibe Klimaschwankungen seit 1700, obwohl ich im Verlaufe meiner Untersuchungen auch Material für weiter zurückliegende Jahrhunderte fand und bis zum Jahre 1000 zurückzugehen versuchte. Allein thatsächlich ist das Material erst von 1700 an so reichhaltig und vielseitig, dass die Klimaschwankungen im Einzelnen verfolgt werden konnten. Die Ergebnisse für die früheren Jahrhunderte sind noch durchaus der Ergänzung bedürftig.

The ten chapters of Brückner's work furnish a large fund of information about the fluctuations of climate which are disclosed in records of the rainfall, temperature and pressure, and those which are indicated for Central Europe

[1] *Geographische Abhandlungen*, herausgegeben von Prof. Dr A. Penck in Wien, Bd. IV, Heft 2, ed. Hölzel, 1890.

in the accounts which have come down to us concerning the frequency of cold winters, the phenology of the vintages, the ice-conditions of rivers, the levels of inland seas or lakes and the variations of glaciers.

We have not space for more than the following summaries of the fluctuations of climate which are indicated by the data grouped according to lustra extending in one form or other from the beginning of the eleventh to near the end of the nineteenth century.

The fluctuations of climate which are disclosed by the information available for the eleventh, twelfth, thirteenth and fourteenth centuries concerning cold winters are included in the chronological tables of pp. 308–11.

Fluctuations of climate disclosed by frequency of cold winters (W), *the phenology of vintages* (V), *and ice-conditions in rivers* (I), *in lustra of the fifteenth and sixteenth centuries.*

Cold period	W	1385–1405	1425–55	1475–95	1505–20	1535–45	1555–70
	V	1391–1415	1436–55	1481–95	1511–15	1541–50	1561–80
	I	—	—	—	—	—	1556–65
Warm period	W	1405–25	1455–75	1495–1505	1520–35	1545–55	1570–90
	V	1416–35	1456–80	1496–1510	1516–40	1551–60	1581–90
	I	—	—	—	—	—	1566–85

Fluctuations of climate disclosed by the frequency of cold winters, the phenology of vintages, the ice-conditions of rivers, the fluctuations of lakes and of glaciers, in lustra of the seventeenth, eighteenth and nineteenth centuries.

	Winters	Vintages	Rivers	Lakes	Glaciers
Cold	1591–1600	—	1586–1600	1600	1595–1610
Warm	—	1601–1610	1601–20	—	—
Cold	1611–35	—	1621–25	1638	—
Warm	—	1636–45	—	1656	—
Cold	1645–65	—	1651–67	1674	1677–81
Warm	—	1665–90 or 85	—	1683	—
Cold	1685–1705	1691–1705	1702–20	1707–14	1710–16
Warm	1705–30	1706–35	1721–35	*ca.* 1720	—
Cold	1730–50	1736–55	1736–50	*ca.* 1740	—
Warm	1750–65	1756–65	1751–70	*ca.* 1760	1750–67
Cold	1765–75	1766–75	1771–90	*ca.* 1780	1760–86
Warm	—	1776–1805	1791–1805	*ca.* 1800	*ca.* 1800
Cold	—	1806–20	1806–20	*ca.* 1820	1800–15
Warm	—	1821–35	1821–30	*ca.* 1835	1815–30
Cold	—	1836–55	1831–60	*ca.* 1850	1830–45
Warm	—	1856–75	1861–80	*ca.* 1865	1845–75
Cold	—	1876–90	—	*ca.* 1880	1875–90

Fluctuations of climate disclosed in the records of rainfall, temperature and pressure.

Rainfall		Temperature		Pressure
Dry periods	Wet periods	Cold periods	Warm periods	
1716–35	1736–55	1731–45	1746–55	Each of the rainy periods is associated with a weakening of all pressure-differences, each of the dry periods with a strengthening of the same, and not only of the pressure-differences between one station and another but also of those between the seasons at the same station
1756–70	1771–80	1756–90	1791–1805	
1781–1805	1806–25	1806–20	1821–35	
1826–40	1841–55	1836–50	1851–70	
1856–70	1871–85	1871–85	—	

An examination of the summaries which we have noted will justify the remark that the cycle of thirty-five years does not represent a rigorous mathematical periodicity. The data for the different climatic factors support one another in a general way but the intervals indicated are not always equal nor are the fluctuations of different elements always similarly concurrent. The numerical value is arrived at by grouping the whole series of lustra into sequences of five fluctuations and computing a period for each group. We thus obtain:

For the interval	of	1020	to	1190	a periodicity	of	34	years
,,	,,	1190	,,	1370	,,	,,	36	,,
,,	,,	1370	,,	1545	,,	,,	35	,,
,,	,,	1545	,,	1715	,,	,,	34	,,
,,	,,	1715	,,	1890	,,	,,	35	,,

The final result is $34{\cdot}8 \pm {\cdot}7$ years.

The fact that for a succession of intervals the period should work out approximately the same is very striking. If there were a true single periodicity differing from thirty-five years by as much as one year the repetition of it twenty-five times would infallibly disclose the difference.

The fluctuation of thirty-one years already mentioned is not very different from the Brückner cycle, but over the range of centuries the two could not be mistaken the one for the other.

We shall also indicate a period of $12\frac{1}{3}$ months as the most noteworthy of all the periods in several long series of observations of temperature. An oscillation of this period would give a "beat" with the ordinary seasonal variation which would recur in thirty-seven years. That again is not far from the Brückner period but the difference could hardly escape detection in the long series of lustra.

The irregularities have to be accounted for. We may perhaps be concerned with the combination of periodicities of many denominations which express the Brückner cycle as a resultant effect. We proceed therefore to consider the periods which have been disclosed by rigorous mathematical analysis.

Periodicities elicited by mathematical analysis.

The idea of representing the observed changes in the general circulation of the atmosphere in whole or in part as a combination of changes of fixed and recognisable period is derived from Fourier's analysis of a curve of any shape into a series of curves each of which can be represented by a periodic function of the general type

$$y - y_0 = A \cos\left(\frac{t}{\tau} 360^\circ - \phi\right).$$

y_0 is the normal, $y - y_0$ the departure from the normal at the time t, τ is the period in the same unit as t, and ϕ is the angle which gives the maximum phase; ϕ is obviously given directly as a number of degrees of angle, but it can also be expressed in time by the ratio t/τ, the fraction of the whole period which falls between the epoch from which time is measured and the time of maximum.

I. *Selected natural cycles verified by harmonic analysis or by other examination of the available data.*

The idea of periodic oscillation is most easily applied to phenomena like diurnal and seasonal changes about the recurrence or periodicity of which we have no doubt; we are then only concerned with their regularity and the range of their operation. We can indeed represent the changes which have a periodicity of a day or a year by the constants of an algebraical equation which, with not more than four periods, gives a representation of the observed changes sufficiently exact for many meteorological purposes. Quite frequently two periods in the Fourier series are sufficient. Any set of twelve values equally spaced through the day or the year is sufficient for the determination of the amplitudes and phases of these components.

Seasonal variation of sea- and air-temperature and of barometric gradient.

$$=A_0+A_1\cos\left(\frac{t}{\tau}360^\circ-\phi_1\right)+A_2\cos 2\left(\frac{t}{\tau}360^\circ-\phi_2\right)+A_3\cos 3\left(\frac{t}{\tau}360^\circ-\phi_3\right)+A_4\cos 4\left(\frac{t}{\tau}360^\circ-\phi_4\right)$$

Atmospheric temperature.

Station	Mean value A_0 °F	1st order curve A_1 °F	1st order curve Max. date	2nd order curve A_2 °F	2nd order curve 1st max. date	3rd order curve A_3 °F	3rd order curve 1st max. date	4th order curve A_4 °F	4th order curve 1st max. date
Lerwick									
Sea	45	7·00	Aug. 12	0·89	Feb. 11	0·43	Jan. 11	0·04	Mar. 4
Air	[44·8]	7·55	Aug. 7	1·30	Jan. 29	0·25	Jan. 13	0·23	Feb. 1
Richmond									
Air	49·1	12·08	July 23	1·38	Feb. 3	0·04	Feb. 16	0·12	Feb. 19
Vienna									
Air	49·1	19·82	July 18	0·99	Mar. 24	0·32	Jan. 2	0·41	Feb. 12
Agra									
Air	78·8	15·51	June 29	5·22	Apr. 27	1·14	Feb. 4	0·38	Mar. 11

Barometric gradient.

Station	Mean value	A_1	Max. date	A_2	1st max. date	A_3	1st max. date	A_4	1st max. date
London—Valencia	mb [1·62]	mb 1·04	Jan. 1	mb 0·46	Jan. 15	mb 0·30	Jan. 15	mb 0·17	Feb. 4
London—Aberdeen	[5·48]	2·0	Dec. 5	1·15	Jan. 27	0·14	Apr. 19	0·27	Mar. 21

The periods of observation are as follows: Lerwick, Sea, 1880–1882, Air, 1871–1900; Richmond, 1871–1895; Vienna, 100 years; Agra, 1868–1874; Barometric gradient, 1871–1900.
The phases of maximum, ϕ, have been converted into days by multiplying by the factor 365/360.

We have already given an account of the use of these ideas to describe the diurnal variation of the barometer all over the world; temperature and other elements could be similarly treated. We do not enter further into that part of the subject at present, but we give some examples of the use of the method to represent the seasonal variation of temperature and of barometric gradient at separate stations[1], and thus provide material for expressing the seasonal variation of the general circulation. The full expression of these changes over the globe might be given in terms of Laplace's coefficients or spherical harmonics, but the analysis is too advanced for our present requirements.

[1] *Proc. Roy. Soc.*, vol. LXIX, 1902, pp. 65–6.

We quote these examples in order to call attention to the importance of the second term in temperature, 5·22° F at Agra, which represents a range of more than 10° F, as compared with Richmond (Kew Observatory), where the amplitude of the second term for a long period of years is little more than a degree, a range of only two degrees; and the sea at Lerwick which shows less than a degree. Further, we wish to point out that the regularity shown in the long series of years is itself liable to disturbance from year to year. At Richmond, for example, an amplitude of 3·57° F in the term of the second order was necessary to represent the seasonal changes of 1884. There are likewise changes in the time of maximum phase.

No meteorologist would be prepared to deny the reality of the existence of periodic changes in the general circulation which are regularly related to the day or the year, and in like manner the changes in any other natural interval might on analysis disclose a periodic change of a certain magnitude, but there is nothing which finds such general and easy recognition as the alternation of day and night and of summer and winter.

What in this way is done successfully for the easily recognised periods has been attempted for other natural astronomical periods. The somewhat vague period of about eleven years in sunspots is often used though it is not exactly what is understood by an astronomical period; endeavour has been made to correlate it with a combination of the rigorous periods of the planet Jupiter (11·86 years) and of Jupiter's opposition and conjunction with Saturn (9·93 years)[1].

One of the common methods of attacking the problem of the periodicity of weather is to select a period that corresponds with some cycle related to the sun or moon. A period which embraces almost exactly an integral number of days, years, sunspot periods and all the various lunar revolutions would include almost everything external which can be thought of as affecting the earth's atmosphere. Such a period is 372 years, 135870·1 days; this will be referred to later as one-half of a still longer period of 744 years.

The various lunar cycles have often been looked upon as likely to control the weather as they control the tides, but the relationships have not been generally accepted. The semi-lunar day, which produces such visible effects in the way of sea-tides, has very slight effect upon barometric pressure. The calculation has been made by S. Chapman with the following result:

Period	Subject and locality	Range of observations	Amplitude and time of maximum phase	
Semi-lunar day	Barometric pressure			
	Aberdeen	1869–1919	·0188 mb	11·53 h from lunar transit
	Greenwich	1854–1917	·0120 mb	11·2 h
	Batavia	1866–1895	·087 mb	·8 h

[1] E. W. Brown, 'A possible explanation of the sunspot period,' *Monthly Notices, R.A.S.*, vol. LX, pp. 599–606.

The semi-diurnal solar tide at Batavia, which is not regarded as having any influence on weather, is 1·00 mb, more than ten times that of its lunar counterpart.

New claims are made for lunar cycles in a recent discussion of hydrographic and meteorological data by Otto Pettersson.

Lunar cycles.

Lunar year, 355 days, 12 synodic, 13 tropic revolutions

8·004 *years*, 2923 days, 99 synodic, 107 tropic revolutions

18 *years* 11⅓ *days*. The Saros 223 synodic, 241 tropic, 239 anomalistic, 242 draconitic revolutions

3 *years less* 2 *hours*, 1095 days, node and apside of the moon's orbit coincident on the ecliptic

1800 *or* 1850 *years*, the conjunction of node and apside at perihelion

These periods are adopted by Otto Pettersson to account for changes in submarine tides and atmospheric tides which are represented in the Baltic herring fishery, the level of the Baltic, cyclical changes in temperature and other phenomena. The last conjunction of the long cycle in 1400–1425 was associated with exceptional submarine tides, an increase of the polar current, and of Greenland ice, and with inundations of the North Sea and submergence of part of Holland

The lunar-solar cycle of 744 years has been invoked by the Abbé Gabriel[1]. It combines 9202 synodic revolutions, 9946 tropical, 9986 draconitic, 9862 anomalistic, 40 revolutions of the ascending node of the lunar orbit and 67 periods of sunspots. It has harmonics of 372 years, 186 years. The last was relied upon for a prediction, made in the summer of 1925, of a cold winter to follow. The prediction was fulfilled in England by the occurrence of exceptionally cold weather in November, December and January. It must, however, be remarked that February, which is accounted as a winter month, brought the highest recorded temperature of that month for 154 years, and a spell of warm weather compared with which the first half of the month of May was winterly.

We record these events before proceeding to our second type of period, that which is disclosed by the arithmetical manipulation of a long series of observations arranged according to years, months or days. The method employed is generally that of the periodogram, an extension of the process of harmonic analysis, which has been described in chapter XIII of volume I; but authors often use other modifications.

The most notable contributions to this branch of the subject are those of Sir A. Schuster, Sir W. Beveridge, D. Brunt and J. Baxendell. We give a summary of the results of a number of investigations arranged according to the length of the period; with them we have included for the sake of completeness the periodicities which have been suggested by any form of examination of the data.

[A revision of our table should include information from a paper by W. Köppen in the *Meteorologische Zeitschrift*, 1930, expounding a paper by C. Easton on severe winters in Western Europe, and from H. P. Berlage (*K. Mag. en Met. Obs. Batavia, Verh.* 26, 1934) who lays stress on the 3-year period and a 7-year period special to Peru, and numerous contributions from A. Wagner.]

[1] *Comptes Rendus*, tome 181, Paris, 1925, p. 22.

II. *Empirical periods derived from the examination of long series of observations by arithmetical manipulation or by inspection.*

For the periods marked y the amplitude for rainfall is given in mm per year, and the computation is as a rule based on annual values of rainfall; for the periods marked m the computation is from monthly values and the amplitude is given in mm per month.

The periodicities marked * were used by Sir W. Beveridge to form a synthetic curve for the period 1850–1923 which is roughly indicative of the rainfall of Western Europe in that period. The amplitudes are expressed in a conventional unit.

• is used as an abbreviation for rainfall; mb for pressure; bb for barometric gradient; tt for temperature on the tercentesimal scale and ⊙ for sunspots.

Period in years	Range of observations	Amplitude and time of last maximum		Subject and locality	Author
260	0–1680 y	·2 mm	Maxima coincident	Tree-rings	Mohorovičić
	0–1680 y	38		Earthquakes (China)	Fotheringham
	640–1360 y	65 cm		Nile-flood minima	Turner
171	1737–1909 y	—	—	Nile-floods and Palestine droughts	Keele
	1726–1890	—	1875–83	• Britain	,,
155	1400–1910	·25 mm	[1860 app.]	Tree-rings	Huntington
130	400–1880	—	—	Winter-cold	Köppen
119	,,	—	—	,,	,,
108	1520–1907	—	1850–77	Wet periods, Chile	Brooks
93 (½ Gabriel's smaller cycle)					
89 (8 ⊙ cycles)	809–1878	—	1921	Winter-cold	Köppen
	760–1916	"Ingang eener nieuwe 89-jarige periode 1917"		—	Easton
80	—	—	—	Earthquakes (China)	Turner
*68·000	1545–1864	13·58	1863·43	Wheat-prices, Europe	Beveridge
57	1737–1909 y	—	—	Nile floods	Keele
	1726–1890 y	—	1875–1883	• Britain	,,
*54	1545–1864	26·09	1914·4	Wheat-prices, Europe	Beveridge
53	1726–1926 y	43 mm	? 1880	• All England	Baxendell
50 (multiple of 5)	—	—	—	Severe winters, England	Watson
44·5	760–1916	—	—	Winter cold, W. Europe	Easton
36	1884–1919 y	·44 mb	259° ‡	mb Bayern, 4 stations	Baur
	,,	·42 tt	206° ‡	tt Germany, 10 stations	
		‡ Phases on July 1, 1884.			
	1520–1907 y	—	1898–1907	• Chile	Brooks
	—	—	—	Brückner's data with additions	Clough
*35·5	1545–1704	15·64	1912·52	Wheat-prices, Europe	Beveridge
	1705–1864	46·98			
35	1725–1900 y	52 mm	1909·8	• Padua	Brunt
The approximate length of Brückner's	1782–1922 y	30 mm	1914·7	• London	,,
cycle with rainfall	1757–1878 y	·61 mb	1902·8	mb Paris	,,
maxima, 1815, 1846–	1756–1905 y	·31 tt	1900·6	tt Stockholm	,,
50, 1876–80, minima	1756–1907 y	·34 tt	1901·8	tt Berlin	,,
1831–5 and 1861–5	1757–1886 y	·25 tt	1916·7	tt Paris	,,
33·8	1730–1910	—	? 1900	Tree-rings	Huntington
33	1839–1909 y	73 mm	1917	• Ohio	Moore
31 (24th harmonic of 744 years)					
30	1757–1886 y	·26 tt	1920·7	tt Paris	Brunt
	1764–1900 y	49 mm	1908·0	• Milan	,,
25	1725–1900 y	64 mm	1926·2	• Padua	,,
	1770–1896 y	·51 mb	1911·1	mb Edinburgh	,,
	1756–1905 y	·27 tt	1910·0	tt Stockholm	,,
23	1763–1918 y	·39 tt	1918·2	tt London	,,
	1764–1863 y	·28 tt	1918·3	tt Edinburgh	,,
22	1785–1896 y	43 mm	1922·1	• ,,	,,
	1785–1896 y	·37 mb	1916·5	mb ,,	,,
	1757–1878 y	·53 mb	1917·0	mb Paris	,,
21	1400–1910	·1 mm, fluctuation 20 % of the mean		Tree-rings	Huntington
20	1725–1900 y	53 mm	1910·1	• Padua	Brunt
*19·9	1545–1704 y	50·07	1912·6	Wheat-prices, Europe	Beveridge
	1705–1864 y	23·97			
19	Provisional cycle for the Southern Hemisphere				H. C. Russell; W. J. S. Lockyer
	1726–1890 y	—	—	• Britain	Keele

Period in years	Range of observations	Amplitude	Time of last maximum	Subject and locality	Author
18·6 (lunar period)	1867–1906	—	—	mb, tt, •, wind Karlsruhe	F. Schuster
18	1884–1919 y	·30t	187° ‡	tt Germany, 10 stations	Baur
	—	39·1 mm	225° ‡	• ,, ,,	,,
	‡ Phases on July 1, 1884.				
17·5	1763–1918 y	·28tt	1916·5	tt London	Brunt
17·400	1545–1704 y 1705–1864 y	69·34 55·52	1921·30	Wheat-prices, Europe	Beveridge
17	1785–1896 y	43 mm	1912·3	• Edinburgh	Brunt
	1764–1863 y	·17tt	1914·1	tt ,,	,,
16	1764–1900 y	42 mm	1924·5	• Milan	,,
16	—	·74t	1923	tt Vienna, difference of summer and winter	Wagner
	—	2·5t to 3t	—	Difference of tt summer and winter Obir and Sonnblick	,,
15·225	1545–1704 y 1705–1864 y	86·11 111·03	1922·1	Wheat-prices, Europe	Beveridge
15·2	1899–1919	—	1926	N and NE wind, Southport	Baxendell
15	1756–1905 y	·27tt	1924·0	tt Stockholm	Brunt
	1763–1918 y	·22tt	1915·0	tt London	,,
	1756–1907 y	·25tt	1914·6	tt Berlin	,,
	1757–1886 y	·20tt	1926·5	tt Paris	,,
14·5	1770–1896 y	·34 mb	1926·9	mb Edinburgh	,,
	1757–1878 y	·4 mb	1923·5	mb Paris	,,
14	1756–1907 y	·25tt	1919·0	tt Berlin	,,
	1775–1874 y	·32tt	1920·3	tt Vienna	,
13	1725–1900 y	35 mm	1927·0	• Padua	,,
	1764–1900 y	40 mm	1918·0	• Milan	,,
	1785–1896 y	30 mm	1917·4	• Edinburgh	,,
	1764–1863 y	·17tt	1922·3	tt ,,	,,
*12·840	1545–1704 y 1705–1864 y	44·82 72·16	1917·23	Wheat-prices, Europe	Beveridge
12·050	1545–1704 y 1705–1854 y	35·10 31·73	1914·2	Wheat-prices, Europe	,,
12	1782–1922 y	21 mm	1926·5	• London	Brunt
11·4	1400–1900 y	Total variation 16 % of mean		Tree-growth	Huntington
	1863–1912	—	—	•, tt California	,,
11·21	—	—	—	Sunspots	,,
11·1	—	—	1917·6	Sunspots	N.S.
⊙ period	—	—	—	Level of Lake Victoria	M.O. Glossary
	—	—	—	• in tropical countries	Supan
	—	·72tt before ⊙ min.		tt tropics	Köppen
	—	·55tt min. at ⊙ max.		tt temperate latitudes	,,
	Period resulting from Jupiter's period 11·86 years, and the period of Jupiter's opposition and conjunction with Saturn, 9·93 years				E. W. Brown
	1878–1919	—	—	Weather, E. Indies	Braak
	30 years	—	—	,, Australia	Kidson
11 (app.)	–1866	—	—	tt, and wind-direction, England	J. Baxendell, *senr.*
11	1764–1863 y	·31tt	1922·0	tt Edinburgh	Brunt
	1884–1915	192 mm	—	• Bathurst	Brooks
11·000	1545–1704 y 1705–1864 y	96·01 3·47	1925·54	Wheat-prices, Europe	Beveridge
11 6 components	1885–1906 y	The whole range	1898	Wheat-yield, England, E	Shaw
9·75	1545–1704 y 1705–1864 y	38·44 29·72	1923·78	Wheat-prices, Europe	Beveridge
9·7	1764–1863 m	·55 mb	1922·06	mb Paris	Brunt
9·5	—	—	—	Mean position of high pressure belt of S. Hemisphere, E and W of S. Africa	Rawson
	1840–1906	—	—	mb Australia, Nile-floods	Lockyer
	1813–1912 m	3·3 mm	1927·17	• London	Brunt
8·33	1764–1863 m	4·3 mm	1920·15	• Padua	,,
8·2	1871–1922 y	33 mm	late 1925	• Southport	Baxendell
*8·050	1545–1704 y 1705–1864 y	9·52 42·98	1924·71	Wheat-prices, Europe	Beveridge
8	1865–1918	—	1920–21	mb Alps	Maurer
	—	—	—	mb U.S.A.	Bigelow
	—	—	—	mb Europe	Pettersson

Period in years	Range of observations	Amplitude and time of last maximum	Subject and locality	Author
8	1839–1910 y	105 mm 1922·6	• Ohio	Moore
7·9	1808–1863 y	— —	Winter-cold, Stockholm	Woeikof
7·66	1813–1912 m	2·8 mm 1919·97	• London	Brunt
7·5	1764–1863 m 1785–1884 m	4·7 mm 1922·97 3·0 mm 1920·73	• Padua • Edinburgh	,, ,,
7·417	1545–1704 y 1705–1864 y	29·06 29·54 } 1920·04	Wheat-prices, Europe	Beveridge
7·3	—	— —	Tree-growth, Arizona	Douglass
7·2	1884–1919 y	·25 t Phase on July 1, 1884, 104°	tt Germany, 10 stations	Baur
6·2	1774–1917 y	·4 mb 1925·3	mb London	Baxendell
6·17	1813–1912 m	3·3 mm 1921·9	• London	Brunt
6·0 ± (length of H. H. Turner's weather chapters)	— — — 1764–1863 m 1764–1863 m	— — — — — — 4·8 mm 1922·34 4·4 mm 1921·27	mb Europe mb Batavia tt Tropics • Milan • Padua mb Port Darwin	Schneider Angenheister Newcomb Brunt ,, Braak
*5·96 (1545–1844)	1545–1704 y 1705–1864 y	29·48 33·07 } 1924·06	Wheat-prices, Europe	Beveridge
*5·671	1545–1704 y 1705–1864 y	33·06 32·58 } 1924·00	,, ,,	,,
5·6 (half ⊙ cycle)	1851–1905 y	10 % on either side of mean • — 6 to 15 months after max. and min.	• Europe, NW	Hellmann
,,	,,	— 1 year before max. and min.	• Europe, E	,,
"Unaltered through several centuries"	—	— —	• Europe	Brooks
5·56 (half ⊙ cycle)	1774–1919 y —	·4 mb 1924·5 — —	mb London Heights of Thames, Elbe, Seine and Nile	Baxendell B. Stewart
5·423	1545–1704 y 1705–1864 y	3·14 65·69 } 1924·29	Wheat-prices, Europe	Beveridge
5·33	1764–1863 m	4·2 mm 1926·17	• Padua	Brunt
5·1	1871–1923 y	69 mm, 17 %, 1923, Feb.	• Southport	Baxendell
	1868–1924 y	— end of 1924	Violent gales, Lancashire and Cheshire	,,
	1899–1919 y	231 h, 40 %, 1926, end of Apr.	N and NE winds, Southport	,,
	1899–1924 y	215 h, 35 %, 1926, mid-May	E and NE winds, Southport	,,
	1841–1917 y	196–261 h, 19 %–29 %, 1926, mid-Apr.	N and NE, or E and NE winds, Greenwich	,,
5·1 (some change of phase)	1831–1923 y	50 mm, 9 %, 1923, mid-June	• Bolton	,,
	1774–1922 y	·2 to ·3 mb, 1925, mid-Sept.	mb London	,,
	—	— —	tt London	,,
*5·1	1545–1704 y 1705–1864 y	34·05 57·08 } 1922·26	Wheat-prices, Europe	Beveridge
5·1	—	"Well indicated"	Velocity of SE trades, St Helena	Brooks
5·1 ±	1841–1905	·02 tt (obscured by taking year instead of winter-season)	tt Greenwich	Brunt Brooks
5·0	1785–1884 m 1770–1869 m	4·1 mm 1927·08 ·51 mb 1924·58	• Edinburgh mb Edinburgh	Brunt ,,
5	1600–1900 y	? 80 % of frequency 1926	Winter-cold, England	Watson
4·8	—	— —	Ice, Arctic	Meinardus Brooks
4·67	1764–1863 m	5·3 mm 1925·72	• Padua	Brunt
*4·415	1545–1704 y 1705–1864 y	29·15 7·06 } 1925·12	Wheat-prices	Beveridge
4·37	1872–1919 y	41 mm, 9½ %, 1925, Apr.	• Southport	Baxendell
4·08	1764–1863 1764–1863	7·3 mm 1924·74 3·4 mm 1925·22	• Milan • Padua	Brunt ,,

Period in years	Range of observations	Amplitude and time of last maximum		Subject and locality	Author
3·98	—	—	—	tt London	Brooks Brunt
	—	—	—	Earthquakes, World	Turner
3·75 app.	—	—	—	Solar prominences	Lockyer
*3·415	1545–1704 y 1705–1864 y	23·55 11·56	1923·58	Wheat-prices, Europe	Beveridge
3·2	1882–1915	180 mm	—	• Bathurst	Brooks
3·155	—	—	—	• S. England	Jenkin
3·1	1871–1920 y	56 mm, 14 %,	1925, early Feb.	• Southport	Baxendell
	1832–1921 y	46 mm.	1925 mid-Feb.	• Bolton	,,
	150 years	1·4 mb	1924 mid-Feb.	mb London	,,
3·09 (not found in all data, probably has no equal)	—	Practically unalterable		Wind, mb and •, Oxford	Rambaut
	1878–1919 y	·7 mb	—	mb Dutch E. Indies	Braak
	30 years	1·3 mb	—	•, mb Australia	Kidson and Braak
3·08	1764–1863 m	·25 tt	1926·81	tt London	Brunt
	1764–1863 m	·42 tt	1926·78	tt Berlin	,,
	1764–1863 m	·25 tt	1926·84	tt Paris	,,
3·0	1884–1919 y	·68 mb	70° ‡	mb Bayern, 4 stations	Baur
	,,	39·8 mm	284° ‡	• Bayern, 10 stations	,,
	—	—	—	mb and • Batavia	Braak
		‡ Phase on July 1, 1884.			
2·92	1813–1912 m	2·5 mm	1924·02	• London	Brunt
2·85	1869–1920 y	76 mm, 18 %,	1926, July	• Southport	Baxendell
	1831–1877 y 1877–1919 y	23 mm 71 mm	1923 end July	• Bolton	,,
2·735	1545–1704 y 1705–1864 y	13·97 1·52	1924·69	Wheat-prices, Europe	Beveridge
2·66	1764–1863 m	4·1 mm	1924·95	• Padua	Brunt
	1807–1912	76 cm	—	Level of Vänersees	Wallén
	1777–1912	40 cm	—	Level of Mälarsees	,,
2·58	1764–1863 m	5·1 mm	1925·5	• Milan	Brunt
	1770–1869	·67 mb	1924·88	mb Edinburgh	,,
2·55	—	—	—	Weather and crops, U.S.A.	Jevons
2·42	1813–1912 m	2·8 mm	1925·72	• London	Brunt
	1785–1884 m	4·1 mm	1925·57	• Edinburgh	,,
	1770–1869 m	·51 mb	1926·75	mb Edinburgh	,,
	1764–1863 m	·69 mb	1926·61	mb Paris	,,
2·41	1831–1876 y 1877–1922 y	89 mm 56 mm	1924 mid-Nov.	• Bolton	Baxendell
2·4	1884–1919 y	·48 mb	48° §	mb Bayern, 4 stations	Baur
		·23 t	57° §	tt Bayern, 10 stations	,,
		§ Phase on July 1, 1884.			
2·33	1785–1884 m	4·1 mm	1925·67	• Edinburgh	Brunt
	1860–1912	200 mm	—	• Göteborg	Wallén
2·25	1874–1910	1·3 t	—	tt Tromsö	,,
	1860–1908	1·8 t	—	tt St Petersburg	,,
2·21	1764–1863 m	6·0 mm	1926·35	• Milan	Brunt
2·2 ($\frac{1}{5}$ ⊙ cycle)	—	—	—	tt Various parts of the world	Arctowski
	1831–1921 y	41 mm	1925 late May	• Bolton	Baxendell
	1789–1920 y	·6 mb	1924 mid-July	mb London	,,
	1899–1921 y	135 h	1923 end Oct.	bb cold winds Southport	,,
2·19	1871–1922	53 mm, 13 %,	1925, early July	• Southport	,,
2·17	1764–1863 m	4·5 mm	1924·93	• Padua	Brunt
	1874–1910	270 mm	—	• Tromsö	Wallén
	1860–1908	140 mm	—	• St Petersburg	,,
	1873–1908	·8 t	—	tt Thorshavn	,,
	1756–1912	1·4 t	—	tt Stockholm	,,
	1860–1908	2·1 t	—	tt Moscow	,,
2·1	1882–1915	102 mm	—	• Bathurst	Brooks
2·08	1868–1910	470 mm	—	• Bergen	Wallén
	1738–1912	120 mm	—	• Upsala	,,
	1860–1908	120 mm	—	• Moscow	,,
2·0	1868–1910	1·1 t	—	tt Bergen	,,
	1860–1912	1·4 t	—	tt Göteborg	,,
	—	—	—	Tree-rings	Douglass
	1757–1906 y	—	[1924–5]	Winter-cold, Stockholm	Woeikof
1·92	1764–1863 m	4·1 mm	1925·59	• Padua	Brunt
	1873–1908	240 mm	—	• Thorshavn	Wallén

Period in years	Range of observations	Amplitude and time of last maximum	Subject and locality	Author
1·72	1764–1863 m	4·4 mm 1925·93	• Milan	Brunt
	1764–1863 m	4·4 mm 1925·72	• Padua	,,
	1770–1869 m	·57 mb 1926·23	mb Edinburgh	,,
1·65	1898–1923 y	69 mm, 17 %, 1925, Apr.	• and mb Southport	Baxendell
1·63	1871–1897	58 mm, 14 %	• and mb Southport	,,
	—	— —	• Greenwich	Turner
	Lapsed after 1894	— —	tt London	Brunt
1·625	1764–1863 m	4·5 mm 1926·5	• Padua	,,
1·6	Old records	— —	• Europe	Turner
1·58	1813–1912 m	2·5 mm 1925·82	• London	Brunt
	1770–1869 m	·61 mb 1926·92	mb Edinburgh	,,
1·54	1764–1863 m	4·3 mm 1926·49	• Milan	,,
1·5	1764–1863 m	4·4 mm 1925·06	• Padua	,,
1·275	1764–1863 m	5 mm 1926·63	• Milan	,,
1·22	1764–1863 m	6·6 mm 1926·37	• Milan	,,
	1764–1863 m	5·5 mm 1926·33	• Padua	,,
	1813–1912 m	2·8 mm 1925·88	• London	,,
	1764–1863 m	·55 mb 1926·86	mb Paris	,,
1·11	1764–1863 m	5·1 mm 1926·49	• Padua	,,
	1813–1912 m	4·3 mm 1926·02	• London	,,
1·11	1910–1916	— —	Changes in the general circulation	Göschl
1·08	1770–1869 m	·51 mb 1926·08	mb Edinburgh	Brunt
1·05	1764–1863 m	6·0 mm 1926·39	• Milan	,,
1·03	1770–1869 m	·53 mb 1926·65	mb Edinburgh	,,
	1764–1863 m	1·24t 1926·66	tt Stockholm	,,
	1764–1863 m	·68t 1926·62	tt London	,,
	1764–1863 m	·80t 1926·63	tt Paris	,,
	1775–1874 m	·94t 1926·27	tt Vienna	,,
1·00	1764–1863 m	15·5 mm 1926·71	• Milan	,,
	1764–1863 m	9·6 mm ·64	• Padua	,,
	1813–1912 m	10·9 mm ·37	• London	,,
	1785–1884 m	13·7 mm ·69	• Edinburgh	,,
	1770–1869 m	2·3 mb ·45	mb Edinburgh	,,
	1764–1863 m	·91 mb ·59	mb Paris	,,
	1764–1863 m	6·2t ·56	tt Edinburgh	,,
	1764–1863 m	11·0t ·57	tt Stockholm	,,
	1764–1863 m	7·4t ·56	tt London	,,
	1764–1863 m	10·4t ·55	tt Berlin	,,
	1764–1863 m	8·3t ·55	tt Paris	,,
	1775–1874 m	8·8t ·54	tt Vienna	,,

In the preceding table we have given the amplitude and phase of what are regarded from various points of view as the chief cycles of oscillation of the meteorological elements between 93 years, which may be called the eighth harmonic of the comprehensive theoretical cycle of 744 years computed by the Abbé Gabriel, and one year, the one cycle which is universally recognised from innumerable observations as being operative in every meteorological element. Above the line we have cited some periods which have been suggested as being applicable in the centuries before numerical values were available.

Many of the cycles noted in the table may be regarded as harmonics of a primary cycle of 93 years, and others again may be regarded as multiples of 12⅓ months or of one year or of combinations of the two. It is an interesting matter of speculation whether the main features of the variation could be represented on one or other of these principles or whether the two are not in a sense identical.

The discussion in a paper by D. Brunt before the Royal Meteorological Society (*Journal*, January 1927) of prominent periodicities disclosed by the analysis of twelve of the records of long range has furnished a guide to the selection of periods quoted in the table. In a note added to the paper in the *Journal*, Brunt calls attention to a curious effect of "beats" between the oscillations with a period differing from a year by a fraction of a month, and the regular seasonal variation. He notes that in those summers when the two oscillations are in conjunction the daily temperature will be enhanced. Six months later they will still be approximately in conjunction in the opposite phase and the daily temperature will be depressed. The annual range will accordingly be expanded in conjunction and reduced in opposition. The beat will therefore appear as a real period in a long series of values of *annual range*, though it would not show in a series of values of the *annual mean*. In this way a period of 16 years in the difference of summer and winter temperature at Vienna, discovered by Wagner, is attributed to the beat of the seasonal variation with a period of $11\frac{1}{3}$ months, that of 19 years to a periodic term of 11·4 months, and that of 33 years to one of $11\frac{2}{3}$ months.

The results of the analysis of summer temperatures only, or winter temperatures only, have therefore to be received with due caution. On the other hand we may ask whether an indirect cause, such as freedom from cloud which enhances summer temperatures and lowers winter ones, is a real explanation of periodicity which differs from a year by a fraction of a month.

BIBLIOGRAPHY OF AUTHORITIES FOR THE TABLE OF PERIODICITIES.

ANGENHEISTER, G. *Gött. Nachr.*, 1914.

ARCTOWSKI, H. Washington, *Proc. 2nd Pan-Amer. Sci. Congress*, vol. II, p. 172. Washington, 1917.

BAUR, F. Bayern, *Deutsches Met. Jahrb.*, 1922. München, 1923, pp. D. 1–D. 8.

BAXENDELL, J. London, *Q.J.R. Meteor. Soc.*, vol. LI, 1925, pp. 371–90.

—— Fernley Observatory, Southport, *Report and Results of Observations for* 1921.

BEVERIDGE, W. *Journ. R. Stat. Soc.*, vol. LXXXV, part III, 1922, pp. 412–59.

BIGELOW, F. H. *Report of the Chief of the Weather Bureau*, 1900–1901, vol. II. Washington, 1902.

BRAAK, C. Batavia, *K. Magnet. en Meteor. Observ.*, Verh. V. 1919.

—— *Meteor. Zs.*, Braunschweig, Bd. XXXVII, 1920, p. 225.

BROOKS, C. E. P. Washington, *Monthly Weath. Rev.*, vol. XLVII, 1919, p. 637.

—— London, *Meteorological Magazine*, vol. LV, 1920, p. 205; vol. LVI, 1921, p. 113.

BROWN, E. W. *Monthly Notices, R.A.S.*, vol. LX, 1900, pp. 599–606.

BRUNT, D. *Phil. Trans. Roy. Soc.* A, vol. CCXXV, pp. 247–302, 1925.

—— London, *Q.J.R.Meteor. Soc.*, vol. LIII, 1927, pp. 1–25.

CLOUGH, H. W. *Astrophys. Journal*, vol. XXII, 1905, p. 42.

DOUGLASS, A. E. *Climatic Cycles and Tree Growth.* Carnegie Institution, Washington, 1919.

EASTON, C. *K. Akad. van Wetenschappen*, Naturk. Afd., 1904 u. 1905.

—— Amsterdam, *Proc. K. Akad. Wet.*, vol. XX, 1919, No. 8.

GÖSCHL, F. *Meteor. Zs.*, Bd. XXXIII, 1916, p. 230.

HELLMANN, G. *K. Preuss. Met. Inst.*, Abhand. Bd. III, No. 1. Berlin, 1909.

HUNTINGTON, E. *The Climatic Factor as illustrated in arid America.* Carnegie Institution, Washington, 1914.

JENKIN, A. P. London, *Q.J.R. Meteor. Soc.*, vol. XXXIX, 1913, p. 29.

JEVONS, H. S. *Contemporary Review*, August, 1909.

KEELE, T. W. *Journ. Roy. Soc., N. S. Wales*, vol. XLIV, 1910, p. 25.

KIDSON, E. 'Some Periods in Australian Weather.' Bureau of Meteorology, Melbourne, Bulletin No. 17, 1925.

KÖPPEN, W. *Zeit. Oesterr. Gesell. Meteor.*, Bd. VIII, 1873, pp. 241, 257; Bd. XVI, 1881, pp. 140, 183.

—— *Meteor. Zs.*, Bd. XXXV, 1918, p. 98.

LOCKYER, W. J. S. *A discussion of Australian Meteorology.* Publications of Solar Physics Committee, 1909.

MAURER, J. *Meteor. Zs.*, Bd. XXXV, 1918, p. 95.

MEINARDUS, W. Berlin, *Ann. Hydrog.*, Bd. XXXIV, 1906, pp. 148, 227, 278.

MOHOROVIČIĆ, S. *Meteor. Zs.*, Bd. XXXVIII, 1921, p. 373.

MOORE, H. L. *Economic cycles: their law and cause.* New York, The Macmillan Co., 1914.

NEWCOMB, S. *Trans. Amer. Phil. Soc.*, 1908.

PETTERSSON, O. 'Étude sur les mouvements internes dans la mer et dans l'air.' *Ur Svenska Hydrografisk-Biologiska Kommissionens Skrifter*, VI.

RAMBAUT, A. *Radcliffe Observations*, vol. LII, Oxford, 1921.

RAWSON, H. E. *Q.J.R. Meteor. Soc.*, vol. XXXIV, 1908, p. 165.

RUSSELL, H. C. *Notes on the Climate of New South Wales.*

—— *Journ. Roy. Soc., New South Wales*, vol. X, 1876, p. 151; vol. XXX, 1896, p. 70.

SCHNEIDER, J. *Ann. Hydrogr. Berlin*, Bd. XLVI, 1918, pp. 20–30.

SCHUSTER, F. *Meteor. Zs.*, Bd. XXX, 1913, p. 488.

SHAW, W. N. London, *Proc. R. Soc.* A, vol. LXXVIII, 1906, p. 69.

STEWART, BALFOUR. *Proc. Manchester Lit. Phil. Soc.*, vol. XXI, 1881–2, pp. 93–96.

SUPAN, A. *Grundzüge der physischen Erdkunde.* Leipzig.

TURNER, H. H. *Q.J.R. Meteor. Soc.*, vol. XXXVII, 1911.

—— *Monthly Notices, R.A.S.*, 1919, pp. 461–6, 531–9.

WAGNER, A. Wien, *Sitzber. Akad. Wiss.* II a, Bd. CXXXIII, 1924, pp. 169–224.

WALLÉN, A. *Meteor. Zs.*, Bd. XXXI, 1914, p. 209.

WATSON, A. E. *Q.J.R. Meteor. Soc.*, vol. XXVII, 1901, pp. 141–150.

WOEIKOF, A. *Meteor. Zs.*, Bd. XXIII, 1906, p. 433.

* * *

BERLAGE, H. P., Jr. East-monsoon forecasting in Java. K. Mag. en Meteor. Obs. te Batavia, *Verhandelingen*, No. 20, 1927, reconsidered in the subsequent paper noted on p. 319.

An example of the application of periodicity in Sweden is given in *Monthly Weather Review*, May 1927, 'Twelve years of long range forecasting of precipitation and water-levels' by Axel Wallén, Director of the Meteorological and Hydrographic Institute, Stockholm. Perhaps the most remarkable example is a sunspot periodicity ($11\frac{1}{3}$ years) detected in the thunderstorms of Siberia, see C. E. P. Brooks, *Q. J. Roy. Meteor. Soc.*, vol. LX, 1934.

The application of periodicity in practical meteorology.

The compilation of this table has been a work of no little labour and not much satisfaction because, notwithstanding the effort devoted to that object, the author is conscious that he will not satisfy the curious reader. It must be confessed that with few exceptions writers of papers on periodicity conceal from their readers the means of testing the results of the work in the only way that can carry conviction, namely, by compiling a synthetic curve to be compared with the actual sequence of events during and after the limited range of observations upon which the computations are based. In this matter the author would venture to make an appeal to the writers of papers. They must be aware from their knowledge of Fourier's theorem that a curve representing a sequence of magnitudes of any element can be resolved into periodic components. For every period, large or small, an amplitude and phase can be computed by a simple arithmetical process; and the questions that require answers are not whether the periodic changes exist in the set of observations, but, first, whether the combination of periods selected by the writer when taken together gives a fair approximation to the original curve, and, secondly, whether they are inherent in the observed phenomena or merely the result of arithmetic for the set of observations selected. If the reader is to find answers to these questions for himself he must know the amplitudes and the phases of the several components in a manner which enables him to trace the synthetic curve and examine its relation to current events. It has not always been easy to get even the computed amplitude of the vibration, and to set out the phase of the vibration at the close of the range has often proved to be too much for the patience of the compiler. For practical purposes the phase of the last stage of the range investigated ought to be given. It would be most agreeable to have the date of the last ascending node because a node is a much better defined point than a maximum, and concurrence of a number of components in a node a very important periodic event. Failing the date of the ascending node the date of the maximum phase is desired. In the table we have endeavoured to identify the phases in that manner but the computation in many cases is so complicated that errors are hardly avoidable. The original author, who has the whole series of vibrations so to speak at his fingers' ends, would be much quicker in detecting errors than any subsequent computer can be. Our appeal is that authors who detect mistakes in our compilation would be good enough to forward the necessary corrections.

With a passing sigh for what might have been, we pass over the solecism of measuring of time in so-called "months," instead of days as the astronomers do, which blunts the edge of numerical accuracy, the cardinal virtue of any arithmetical process. The blame for that rests with the International Organisation. We puzzle our brains with the effort, in such a table as we have compiled, to comprehend the practical meaning of monthly rainfall as a unit of amplitude in a periodic term of 1 year or 100 years, and wonder whether anything except archaic prejudice stands in the way of daily rainfall or daily sunshine as a unit. We can appreciate the utility of combining the observations in order to smooth out everything except the selected periodic terms when the hunt for periods is in progress; but we cannot condone the practice of leaving the reader, when the periods have been unearthed, to decide for himself whether they are of any importance in the sequence of actual observations.

For this and other reasons the reader whose interest is in practical meteorology may find little satisfaction in contemplating the accumulation of figures contained in this admittedly incomplete table. So far as we are aware there is no regular practice of using established periods as a guide to future weather, apart from occasional recognition of some noted periods such as the eleven years' period in West African rainfall or in terrestrial magnetism; nor are we able to suggest any practicable method of co-ordinating the whole set of periodicities which occupy the table. But the situation is not hopeless. There must be some reality underlying Brückner's cycle in the long run, though its recurrent phases are obscured by effects which are not yet accounted for; and ordinary folk are not satisfied with the realities of the long run when they are concerned about the weather of a particular month, season or year. The period of eleven years corresponding more or less accurately with a period of sunspots affords too many cases of agreement in more or less separate elements or regions for the conclusion as to reality to be set aside. It is noteworthy that in both these cases of periodicity which have achieved a qualified acceptation the conclusion is derived, notably so in the case of Brückner's cycle, from the general variation within a wide geographical region rather than from the direct application to a single station. In like manner Sir William Beveridge's cycles, marked in the table with an asterisk *, out of which he constructed a synthetic curve for rainfall for the years 1850 to 1921, are based upon information as to food-prices for Western Europe, and the rainfall curve is in like manner a general one. In our own experience the periodicity which was well marked for 22 years in the yield of wheat in Eastern England was not nearly so well marked in the yield of the counties which made up the area[1]. The probability seems to be that in general the changes in the circulation affect wide areas more clearly than individual stations; geographical smoothing may be a valuable step towards the identification of general periodic changes which are overlaid and obscured by local conditions. The same remarks apply almost equally to the other schemes of investigation of the changes in the circulation to which we refer in this chapter.

A notable example of prediction by cycle.

Within the year 1925 the Abbé Gabriel announced the probability of a cold winter as appropriate to the lunar-solar cycle of 186 years referred to on p. 319. The notice raised some discussion in the public press, mostly antipathetic. A private communication from J. Baxendell supported the suggestion of a period of cold winds, because the three cycles which he found

[1] *First Report Met. Com. to H.M. Treasury*, 1906, p. 21.

most pronounced for cold winds, viz. 15·2 years, 5·1 years and 3·1 years, were to come to their maxima together in April or May of 1926. The Northerly or North Easterly winds which he anticipated were certainly realised during the months of April, May and June of 1926 which, following the cold November, December and January, provided a spell of cold weather long enough to be considered in relation to a period of three years or more.

We remember the notable cold period of February 1895, when rivers were frozen and the water-supplies of many houses in London were cut off by frost. That occurred 31 years before, and we note that a period of 31 years would include three periods, 15·5 years, 5·17 years and 3·1 years, which agree with or are not far from J. Baxendell's selected periods. Moreover, the most conspicuous result of D. Brunt's analysis of the long series of observations in Europe is a period of $12\frac{1}{3}$ months which is so nearly one-third of 3·1 years as to suggest its association with the 31-year period.

It is remarkable that the Abbé Gabriel's period of 186 years, which is theoretical, covers six periods of 31 years or sixty periods of 3·1 years, or one hundred and eighty periods of approximately $12\frac{1}{3}$ months, and Baxendell's periods of 15·2 years and 5·1 years are not far away from one-twelfth and one-thirty-sixth respectively of 186 years.

In a recent paper (see p. 325) H. P. Berlage has a formula which "has yielded the right result the first time it was applied, viz. in 1926" for the rainfall in Java in the second half of the E. monsoon (July to Sept., or Aug. to Oct.) with reference to a 3-year cycle, a 7-year cycle and the sunspot cycle.

We are at present unable to say whether a multitude of cycles can be made up by a combination of a limited number of periods of natural resilience, approximately represented by 2·8 years, 3·09, 5·2, 15·5 years, which in the course of centuries find coincidences with the astronomical periods, or whether the short cycles are harmonic components of high order belonging to a dominant astronomical cycle such as that of 744 years; but we think the former suggestion offers an opportunity more likely to be practically productive. It is not difficult to understand that when a resultant curve is made up of a finite number of perfect sine curves the addition of selected groups of values will eliminate all variations that are not periodic in the selected interval, but it is more difficult to feel assured of the working of the process when each value is liable to errors of observation and special local influence which may spoil the symmetry.

Brückner's cycle of approximately 35 years may perhaps be explained as the result of interference of the fifth and sixth components of a period of 186 years. If we combine the two components of those orders of the master-period τ, supposing them to be of equal amplitude, we get as a resultant

$$\sin 2\pi \frac{5t}{\tau} + \sin 2\pi \frac{6t}{\tau} = 2 \sin 2\pi \frac{11}{2} . \frac{t}{\tau} . \cos 2\pi \frac{t}{\tau} .$$

The first factor has a period of 2/11 of 186 years, approximately 34 years, and the second factor represents the very slow variation with a period of 186 years.

In dealing arithmetically with a long series of observations the effect of an exceptional occurrence, such for example as the eruption of Krakatoa, upon an otherwise perfect sine curve would have to be represented by suitable harmonic components. We cannot at present distinguish between effects of that kind and genuine periods either of the resilience of the circulation or of the influencing cause.

We should like to notice further one of Sir W. Beveridge's tables in reference to the use of long periods for the calculation of the constants of a periodic change. He gives the following values for the intensity and phase of variations of 11 years and $5\frac{1}{2}$ years in successive groups of years, each group extending over 66 years.

Group ...	1500–65	1566–1631	1632–1697	1698–1763	1764–1829	1804–1869
11 years						
Intensity	57	71	214	14	29	5
Phase	338°	269°	228°	14°	245°	56°
5·5 years						
Intensity	104	26	1	59	135	77
Phase	40°	190°	296°	297°	177°	85°

The changes in intensity and phase suggest that a periodic change is not necessarily persistent throughout a long series of observations, but nevertheless may have at times a notable importance. With this may be associated (subject to the note at the end of the table on p. 324) a pertinent remark of J. Baxendell's that the amplitude of a variation of temperature may show larger in a series of values for a month or season of the same name than in that of the means for the whole year.

INTERCORRELATION OF THE ELEMENTS OF THE CIRCULATION

There is a third process of scientific inquiry into the changes, or departures from normal, of the elements of the general circulation which depends upon the statistical method of correlation. It is an arithmetical process and requires an ample series of accurate and homogeneous numerical values. It can therefore only be applied in those cases in which a sufficiently long series of values exists, but the great variety of elements and the enormous diversities of geographical conditions afford so vast an amount of material for the exercise of the method that the literature on the subject is already almost overwhelming.

Soon after the method had been introduced into meteorological practice W. H. Dines pointed out to me the necessity for an official compilation of the published coefficients as a concise representation of the results of much labour which would otherwise be lost to meteorologists with limited libraries. A compilation was accordingly printed in 1919 as a supplement to one of the sections of the *Computer's Handbook*. Even that compilation is bewildering for the inexpert, and since its publication the volume of available coefficients has been very much increased. In a publication of the International Research Council in the year 1926, C. E. P. Brooks[1] devoted thirty-

[1] *First Report of the Commission appointed to further the study of solar and terrestrial relationships*, Paris, 1926.

five pages to a summary and bibliography of the literature of the correlation of solar and meteorological phenomena dating from the contribution of Helland-Hansen and Nansen, so that the subject has passed beyond the range of the general reader and is not altogether easy for a modest expert. We cannot here reproduce a complete list, we must confine ourselves to a very rough indication of the general tenor of the results which have been achieved by these impressive masses of figures, themselves representing voluminous compilations of numerical results.

My own excursion into the arena of correlation was before the war in the early days of the application of the method to meteorology. I had been gratified by receiving a letter by an unknown correspondent dated from an impressive address. It conveyed the unusual wish to undertake as a hobby any computing that I might like to suggest. I was at the time seeking a computer to ascertain by correlation the parts of the circulation by which British air-pressure was governed. I accepted the offer gratefully and my correspondent supplied me with a large number of coefficients excellently worked out. I had indeed some difficulty in keeping pace with his appetite.

As the result of our joint labour I was struck with the resemblance between the related areas of correlation (positive and negative) and the disturbance caused by a stone thrown into a pond. The control seemed to be geographical rather than meteorological. Long explorations of the kind I was engaged upon lost their attraction for me; perhaps the month was too long an interval because its mean may represent by the same arithmetical figure a period of violent activity and one of undisturbed calm; or some suggestion of a physical connexion seemed wanting. But in that I am apparently exceptional, for the computation of correlations has been taken up *con amore* and even more than that, both in the old world and in the new.

Whatever else it has achieved, the new method, which enshrines Galton's memory much more visibly than any of his remarkable services to ordinary meteorology, has taken over the duty of correcting the vague impressions produced by curve-parallels; but our present inquiry includes the time before the wisdom of the correlation coefficient had dawned, we must therefore include the relationships indicated by Hildebrandsson and others with those of arithmetical computation.

Researches by curve-parallels and maps.

In his final memoir on the centres of action in the atmosphere in 1914 Hildebrandsson summarised the conclusions which he drew from the study of the graphs of the variations of the elements in all parts of the world. We translate his description because it deals with many questions to the investigation of which statistical methods have been subsequently applied.

In *winter* the deviation of the meteorological elements on the ocean between Iceland and Norway is almost always in accord with that observed over the North of Europe between the North Cape and Hamburg, but in opposition to the deviation of the same elements in the subtropical region Azores-Mediterranean. Between these two regions there is an intermediate zone which is associated sometimes with the Northern régime, sometimes with the Southern. This zone extends from Greenwich over the greater part of France and thence Eastward across Central Europe to Russia. In North America the same opposition is found between North and South. The run of the curves for the Northern region bounded on the West by California and British Columbia and on the East by Greenland and Newfoundland is in general opposite to that for the South which ranges from Mexico to Bermuda and from Toronto to Key West. We might expect an intermediate zone also. Adequate series of observations are wanting but it appears that Winnipeg is within that zone.

Comparing Europe and North America there is general agreement between the run of the graphs in Northern Europe and Southern North America and consequently opposition between the Northern parts of the two continents. If for example winter is cold in Northern Europe it is equally so in Mexico and the United States, but on the contrary mild in the South of Europe and in the North of America or *vice versa*.

Generally speaking, these results apply equally to pressure, temperature and precipitation. There are however exceptions. We know, for example, that in the North a spell of rain brings a rise of temperature in winter and a fall in summer; rain is associated with a lapse of temperature in Southern countries which have a high winter temperature.

In *summer* the thermal equator moves Northward with all the centres of action. The Azores maximum extends to the British Isles and the Iceland minimum passes to the glacial zone. In consequence, the conditions in Northern Europe are different, the maritime régime has no longer the same action. It is only on the North West coasts of Europe that the summer-temperature is determined by the sea-temperature of the same season. In the rest of Northern Europe, Scandinavia, the Baltic and North Germany summer-temperature depends on the greater or less depression of temperature during the preceding winter, in other words on the temperature over the sea between Norway and Iceland and the duration of the period of "snow-lying." O. Pettersson and Meinardus have proved that not only in winter is there an intimate relation between the surface-temperature of the sea between Norway and Iceland and that of the North West of Europe, but further the temperature of that sea in winter advances or retards the spring in Scandinavia and North Germany. This action continues even in summer. Vegetation and harvest round Berlin depend on the temperature of those seas during the preceding winter.

Equally in the region of the Baltic the temperature of the summer depends on the preceding winter in the Iceland region. There is also agreement between the precipitation in Thorshavn, January to March, and that of Berlin in the following April to September.

In the South of Europe temperature in summer as in winter is in opposition to that of the Northern region and consequently to that of the preceding winter in the Iceland region. The Southern régime extends further North in summer than in winter; it includes Paris and Switzerland. The intermediate zone is less broad, it passes by Greenwich, Vienna and Debreczin. Pola is still in the Southern régime.

In North America the same opposition is shown, the whole system is displaced Northward, the Southern régime includes Winnipeg. There is less regularity than in winter and a comparison of North America and Europe is hardly possible.

In Siberia the course of the graphs is in general opposite to those of Northern Europe. However, in mid-winter the influence of the sea at Thorshavn often extends as far as Barnaoul and even to Yenisseisk. In summer also there is opposition between the North of Europe on the one side and Southern Europe and Siberia on the other....

For the Southern Hemisphere few series of observations are available but comparing Punta Arenas with Cordoba a similar opposition is apparent.

In the Southern Hemisphere the types of season are propagated from West to East like waves. To show this clearly we have compared (for 1881–1903) precipitation in Java from October to March with barometric pressure (1) at the Cape in October to

March of the preceding year, (2) at Mauritius in the preceding April to September, (3) at Sydney and Melbourne for the same epoch as Java, (4) at Cordoba and Santiago de Chili in April to September of the following year. The opposition of the courses of the different curves is indeed very clear; from the Cape it would be almost possible to predict with sufficient probability the rain in Java a year in advance.

Finally, we must seek the principal cause of the different types of season in the state of the ice of the polar sea. In effect tropical climates are very regular and in the temperate regions no phenomenon presents variations from one year to another of sufficient importance to be the *cause* of the considerable differences between successive seasons. Only the extent and duration of "snow-lying" in winter and spring could be the cause to some extent of such variations. But we have seen that even that depends on the temperature of the Iceland-Norwegian sea....

The results are not perhaps sufficient to justify forecasts *à longue échéance* but they give in certain cases indications of coming seasons worthy of attention.

In order to deal with certain exceptional cases, two or three in the course of 25 to 30 years, when in spite of the ordinary opposition of the curve-parallels both curves showed values above normal or both below, Hildebrandsson compiled monthly maps of the world to show the grouping of the deviations and found that almost always the deviations of the same sign covered very large areas; thence he concluded that there must be variations of a higher order which tend to mask the normal conjunctions or oppositions of the curve-parallels. As a possible cosmical cause he turned his attention to sunspots and found that years of minimum sunspots (1888–9, 1902–3) show positive deviations over the tropical zone and negative deviations over a great extent of the temperate zones, while the deviations are in the opposite direction for years of maximum sunspots (1884, 1894–5, 1905). He concludes therefrom that there is less heat in the solar radiation during years of maximum sunspots than during those of minimum, a conclusion which is not borne out by the computed values of the solar constant.

The final paragraphs of the memoir that concluded the work to which Hildebrandsson had devoted so large a part of his life bring us face to face with the possibility of the effect of dust in the atmosphere due to volcanic eruptions such as those of Krakatoa, Mont Pelée and Katmai, which may exercise an influence powerful enough to override the effect of variations in the sun's surface. "S'il en est ainsi, une prévision quelle qu'elle soit, même la mieux fondée, sera toujours menacée d'un échec impossible à prévoir."

The investigation of changes of climate on the basis of curve-parallels with all the incidental questions of the influence of sunspots and solar radiation has been followed by B. Helland-Hansen and Fridtjof Nansen in a well-known volume on temperature-variations, of which the sub-title is "Introductory studies on the cause of climatic variations," published originally in 1917 and in translation for the Smithsonian Institution in 1920; to this reference must be made in the sequel.

Meanwhile the division of the earth into two regions between which there was maintained an oscillation of barometric pressure was described by Sir Norman Lockyer and Dr W. J. S. Lockyer as a barometric see-saw, Bombay and Cordoba being typical centres in which the reciprocal oscillation was most clearly marked. Fluctuations in the Trade-Winds with those of the Gulf Stream compared by C. F. Brooks (*Gerlands Beiträge*, 1931) lead to the conclusion

> When other factors controlling the seasonal weather of the Atlantic region are weak this...factor is sufficiently strong to permit successful seasonal weather-forecasting for a quarter of a year to nearly two years in advance.

Coefficients of correlation.

The method of correlation can certainly be used as a kind of searchlight for sweeping the meteorological horizon from some selected point. The principal features of the otherwise invisible landscape, which in this case extends over the whole globe, can thus be located. In this sense the method has been used by J. I. Craig for the relation of temperature between Egypt and North Western Europe; by Exner for the relations of pressure between Iceland and the rest of the world, between Lugano and the rest of the Northern Hemisphere; and by Sir G. Walker for a large number of selected points, and a correspondingly large number of elements. It was in that sense also that I tried to use it for British pressure. But the process was not so handy nor so speedy as a searchlight, and so much depends upon the point chosen for the centre that I have come to regard the method as more appropriate for verifying results, obtained otherwise, than for primary exploration. Sir Gilbert Walker was in a much more favourable position for an inquiry on these lines, because the time-scale of the meteorological sequence in India may fairly be regarded as a month, whereas in Britain a day is more appropriate. He is distinguished among workers on statistical methods as having applied it numerically in the calculation of the prospects of seasonal rains in India.

Experts in the method are punctilious about the reality of the relation which is disclosed by even a small coefficient provided that the probable error is not more than one-third of the coefficient. We are unable to persuade ourselves that a coefficient is of much value if it is insignificant, however real it may be. We recognise as correct all the things that may be said about the necessity for intense scrutiny; but at the same time we cannot forget that a correlation-coefficient is a very sensitive plant, it is much easier to kill one than to make one; whatever happens in the way of accidental errors, it must suffer. A few mistakes either in observation or transcription or bad luck in the selection of the time-unit, taking for example months instead of weeks, may batter a coefficient out of recognition. We shall probably alter our opinion when we can use the method for dealing with observations from the open sea; that presents a meteorological horizon so attractive as to overcome all the feeling of want of confidence inseparable from orographical vicissitudes. Meanwhile, of course, the big coefficients are a boon of the first importance. We have accordingly made a list of them on pp. 334–6. In order to make the study of the method of correlation more complete we have included in the table some coefficients of correlation for the upper air, though they deal with individual observations and not with monthly values and are not strictly pertinent to the subject of this chapter.

COEFFICIENTS OF METEOROLOGICAL CORRELATION

In the tables correlation may be represented by the figures of two places of decimals, full correlation would be represented by 100, a negative correlation will be underlined.

I. Relations between the synchronous variations of local atmospheric structure.

1. *Pressure, mb, and temperature, tt, at the same height; year.*

Height (km)	0	1	2	4	6	8	10	12	14
England, S	11	42	66	84	86	86	32	36	..
Canada	..	..	..	88	93	93	78	04	..
Batavia	..	..	2	38	64	72	88	86	76
Lindenberg	..	..	..	..	86	76	..	..	..
India (Gopal Rao)	68	..	14	56	83	88	89	79	22

2. *Pressure, temperature and wind-components at* 3·048 *km*, 10,000 *ft* (C. K. M. Douglas).

mb and tt +·68; mb and W to E component −·37; mb and S to N component −·26; tt and W to E component −·22; tt and S to N component +·10.

Wind at different levels as noted with the aid of pilot-balloons has formed the subject of correlation; but the method is not suitable for it unless the cases are previously classified in such a way as to limit the changes investigated to variations in the numerical measures of a comparable quantity. The changes with height in direction and velocity of wind, represented in a number of diagrams in vol. IV, will be sufficient to explain to the reader the difficulty with which in that case the statistical method is faced until a method of vectorial correlation is introduced such as J. I. Craig has suggested.

3. *Mutual relations between pressure, mb, at the surface, at nine kilometres; temperature, tt, the mean of* 1 *km to* 9 *km, at* 0 *km,* 4 *km, at the tropopause; the height of the tropopause, htp; vapour-content of the column, and the geostrophic-wind at the surface G* (W. H. Dines).

	mb surface	mb 9 km	tt mean 1–9 km	tt 0 km	tt 4 km	htp
mb surface	100	68	47	16	..	68
mb 9 km	68	100	95	28	82	84
tt mean, 1–9 km	47	95	100	..	..	79
tt 0 km	16	28	..	100	..	30
tt 4 km	..	82	..	..	100	64
tt tropopause	52	47	37	..	..	68
Height of tropopause	68	84	79	30	64	100
G, W to E	..	..	..	37	13	11
G, S to N	..	..	..	19	6	2
Vapour-content	8	..	..	..	..	39

II. Mutual relations between local variations of the surface circulation and the contemporaneous variation of the elements elsewhere.

1. *Centres of Action* (Sir G. Walker) *represented by names in column repeated as initials in the heading.*

Figures above the diagonal refer to northern winter, below to southern.

Northern Winter. December to February.

		mb Al.	mb Ic.	mb Si.	mb Ch.	mb Az.	mb In.	• I.Pe.	mb Ho.	• Ja.	mb P.D.	mb St H.	mb Ma.	mb Sa.	mb C.T.	mb Au.	mb Am.	mb S.O.
Alaska	mb	100	18	18	35	24	8	..	71	33	25	23	8	18	51	13	9	47
Iceland	mb	55	100	34	33	54	5	..	19	26	6	7	26	13	10	9	19	14
C. Siberia	mb	27	22	100	29	3	1	..	29	15	7	12	28	2	1	2	4	24
Charleston	mb	18	2	26	100	35	52	..	20	20	52	15	12	28	25	30	8	51
Azores	mb	33	49	27	25	100	2	..	15	15	7	27	4	8	9	10	5	57
N. W. India	mb	21	14	37	51	26	100	..	13	32	67	12	21	39	56	47	20	1
Indian Peninsula	•	11	3	33	1	10	11	100	..	..	..	..	..	..	..	..	..	..
Honolulu	mb	28	34	16	8	20	32	46	100	32	39	12	25	19	29	33	7	37
Java	•	..	..	..	..	..	..	..	..	100	46	6	5	45	33	42	28	13
Port Darwin	mb	25	23	7	12	8	3	41	67	..	100	11	11	57	59	75	15	30
St Helena	mb	10	13	22	49	22	14	27	12	..	16	100	6	53	20	28	9	11
Mauritius	mb	22	29	9	9	11	23	52	22	..	25	41	100	36	3	25	6	54
Samoa	mb	36	9	31	22	34	18	36	73	..	55	17	17	100	37	54	24	2
Cape Town	mb	3	9	27	16	3	17	10	7	..	17	29	15	0	100	52	12	44
S.E. Australia	mb	6	11	12	12	5	27	31	28	..	73	13	43	37	25	100	8	4
S. America	mb	7	5	31	4	17	25	44	52	..	46	6	38	49	10	33	100	14
S. Orkneys	mb	42	9	2	11	14	17	8	7	..	25	5	9	13	22	31	25	100

Southern Winter. June to August.

December to February: tt Seychelles, tt Mauritius +·54; tt Seychelles, tt Batavia +·50.

March to May: mb Batavia, tt Batavia +·52; mb Allahabad, tt Allahabad −·61; tt Seychelles, tt Batavia +·51.

September to November: tt Seychelles, tt Batavia +·54. June to August: tt S. Orkneys, • Evangelistas +·65.

Sir G. Walker[1] has also published additional tables of coefficients for spring and autumn.

"We have not hit upon any temperatures in the equatorial regions or Southern Hemisphere which are prime factors in controlling the general weather distribution."

[1] *Indian Meteorological Memoirs*, vol. XXIV, part IX, Calcutta, 1924.

General relations between the variations at a selected centre and the synchronous variations in the surrounding regions.

(i) EGYPTIAN CENTRE. (J. I. Craig, *Q. J. Roy. Meteor. Soc.*, vol. XLI, 1915, p. 89, and vol. XXXVI, 1910, p. 341.)

Mean temperatures: Egypt, North Africa and Europe.

Cairo and S.W. England. Quarters of the kalendar year.

First Quarter −·72; Second Quarter −·28; Third Quarter −·17; Fourth Quarter −·54; Year −·43.

The line of zero correlation runs as follows:

January from WSW bending towards E, Gibraltar, Gulf of Lyons, Venice, Odessa.

April from SSW bending towards NE, Tunis, Rome, Moscow.

July from SW, Algiers, Rome, Warsaw.

October from WSW bending towards N, Tunis, Palermo, Brindisi, Kiev.

(ii) ICELAND CENTRE. F. M. Exner, *Sitz. Ber. der Akad. der Wiss., Wien. Math. Nat. Klasse*, II a, Bd. 133, 325–408, 1924.

Pressure at Stykkisholm and the rest of the world.

Northern winter. Correlation is positive from the centre to a zero line which runs near the 55th parallel from Kamchatka to Alaska, and, after a dip to San Francisco, along the 50th parallel from Winnipeg to Valencia, thence to near the 70th beyond Archangel, then a dip to near the 40th at Tashkent and back to the 60th on the Siberian coast.

South of the zero line a zone of negative correlation with a second zero line about 10° N and gentle deviation Southward to St Helena, 16° S, and thence Northward to 30° N at Jeypur and back to 10° near Manila. There is a well-marked but narrow region of maximum −·5 extending from the Azores to Italy surrounded by oval curves of similar shape for −·4 and −·2, the latter extends from the Mississippi to the Black Sea and from the Atlantic steamer routes to the West Indies and Cape Verde, and a less well-marked centre of −·3 over Northern Japan.

Beyond the equator is a wide belt of positive correlation with regions of +·1, a zero line South of Cape Horn and Cape Pembroke and negative correlation −·1 at South Orkneys.

Northern summer. The distribution is on the same lines but the long region of high negative correlation is shrunk to a small oval of −·4 round the Azores, and at the same time negative correlations in the Southern Hemisphere are more pronounced with two centres −·2 off the East coast of South America, lat. 30° S, and off South Africa near the equator, respectively.

(iii) PRESSURE IN HIGH POLAR NORTHERN LATITUDES IN WINTER (Jacobshavn, Gjesvaer, Iakutsk) AND

(iv) LUGANO CENTRE.

In a previous communication Exner exhibited the correlations between pressure in winter in high Northern latitudes and (*a*) pressure, and (*b*) temperature in the rest of the Northern Hemisphere; the same for pressure at Lugano in the same season and pressure or temperature elsewhere. In the former the same negative correlation as in (ii) is shown for the Mediterranean and a reciprocal relation is shown for pressures at Lugano.

Other correlations (R. C. Mossman) are as follows:

mb Stykkisholm (May), 1902–11, mb Laurie Is. −·90.

• Malden Is. (May–Aug.), 1890–1911, tt Punta Arenas −·59.

• Baltimore (Jan.–Mar.), 1851–80, • San Fernando −·57.

• Baltimore (Jan.–Mar.), 1851–1904, • San Fernando −·21.

III. THE LARGER CORRELATIONS BETWEEN CONDITIONS AT SELECTED CENTRES AND SUBSEQUENT WEATHER ELSEWHERE.

1. *Conditions at selected stations during Northern summer, June–August, and weather elsewhere six months later, December–February.* (Sir G. WALKER.)

mb Alaska +·51 mb S. Orkneys, −·47 mb St Helena.

mb Siberia +·51 mb S. Orkneys.

• Indian Peninsula +·51 mb Alaska, −·46 mb Port Darwin, +·49 mb Samoa, −·62 mb S.E. Australia, −·51 mb Cape Town, −·52 mb S. Orkneys.

mb Honolulu −·64 mb Port Darwin, +·50 mb Samoa, −·54 mb S.E. Australia.

mb Port Darwin +·83 mb Port Darwin, +·46 mb Honolulu, −·50 • Java, +·53 mb Cape Town, +·58 mb S.E. Australia.

mb Mauritius +·60 tt Mauritius, +·49 mb S.E. Australia.

mb Samoa −·62 mb Port Darwin, −·46 mb Cape Town.

mb S.E. Australia −·52 • Java, +·48 mb Port Darwin, +·48 mb Cape Town.

mb S. America −·57 mb Port Darwin, −·57 mb S.E. Australia, −·46 mb N.W. India, −·49 mb Mauritius, +·48 mb Samoa, −·48 mb Cape Town.

mb S. Orkneys +·67 mb Iceland, −·75 mb Samoa.

• Zanzibar (May) −·50 • Indian Peninsula, June–Sept.

2. *Northern winter, December–February, and the weather there or elsewhere in the following June–August.*

mb N.W. India +·49 mb Port Darwin.

mb S. Orkneys −·65 mb Siberia.

mb Honolulu −·46 mb S. Orkneys.

mb Alaska −·51 mb S.E. Australia, −·47 mb Azores.

mb Cape Town −·47 mb Samoa.

3. *Conditions during March to May and the weather later.*

mb Mauritius +·65 tt Mauritius, Sept.–Nov.

mb India +·53 tt India, June–August.

tt Seychelles +·50 tt Batavia, June–August.

---

mb Batavia +·41 to +·63 tt Batavia in periods 3 to 9 months later.

mb Santiago, May +·65 tt S. Orkneys, July.

4. *Other correlations with future weather.*

• Santiago (May–Aug.) −·62 Nile flood, July–Oct. (R. C. Mossman.)

• Trinidad, Apr.–Sept. +·79 • Azo (Argentine) Oct.–Mar.

• Trinidad, Apr.–Sept. −·47 • Buenos Aires, Oct.–Mar.

• Java, Oct.–Mar. −·47 • Trinidad, Apr.–Sept.

• Habana, May–Oct. −·47 • Ireland S, Jan.–Mar. (A. H. Brown.)

• Habana, May–Oct. −·54 • England SW, Jan.–Mar.

bb Zikawei, Mar. +·68 to ·78 tt N.E. Japan, July–Aug.

bb Zikawei, Mar. +·61 to ·75 tt N.E. Japan, July.

bb Zikawei, Mar. +·37 to ·58 Duration of sunshine, N.E. Japan, July. (T. Okada.)

bb Zikawei, Mar. +·41 to ·47 Duration of sunshine, N.E. Japan, Aug.

bb Zikawei, Jan.–Mar. +·50 to ·61 tt N.E. Japan, Aug.

tt Irkutsk, Apr. −·53 tt San Francisco, July.

bb Nafa, Jan.–Apr. +·54 • Koti, S. Japan, July. (T. Akatu.)

bb Nafa, Jan.–Apr. −·45 tt Koti, S. Japan, July.

bb Nafa, Jan.–Apr. −·53 sunshine, S. Japan, July.

Sea-water and air-circulation of the N. Atlantic. (V. I. Pettersson.)

tt sea, annual, 36–37° N} ·86 tt air, Madeira. 12–15° W} ·67 tt sea 20–21° N, 20–23° W.

tt sea, 31–34° N, 73–77° W +·10 tt sea, iceberg region, 47–50° N, 30–33° W.

tt sea, 31–34° N, 73–77° W −·03 tt air, Bermuda.

tt sea, 47–50° N, 30–33° W +·50 tt sea, 54–56° N, 27–29° W one year later.

tt sea, 40–50° N, 39–48° W August, ·46 tt sea, 54–56° N, 27–29° W. Annual value one year later.

tt sea, 54–56° N, 27–29° W ·64 • Ireland one year later.

Wind or current and air-temperature, rainfall and sea-level. (P. H. Gallé.)

NE trade (May–Oct.) > +·70 tt Central Europe (Dec.–Feb.).

NE trade (May–Oct.) > −·60 tt North of Iceland (Dec.–Feb.).

NE trade (June–Nov.) > +·80 tt Central Europe (Dec.–Feb.).

NE trade (June–Nov.) > −·40 tt North of Iceland (Dec.–Feb.).

Monthly energy of wind in Bay of Bengal ·93 • Calcutta.

Monthly energy of wind in Bay of Bengal ·94 • Cherrapunji.

Monthly wind, 10–20° N, 80–90° E ·92 Height of Hooghly two months later.

Monthly wind, 10–20° N, 45–60° E ·96 Sea-level, Aden one month later.

Current, 15–25° N, 25–40° W. Maximum correlation with the level of the sea simultaneous at Washington and Baltimore, one month later Horta, three to four months later British and Norwegian coasts.

IV. Correlation between external influences and synchronous local variations in the general circulation.

1. *Correlation of sunspot numbers with meteorological phenomena.*

Water-level +·88. Victoria Nyanza. Meteorological Office.

Nile flood +·36. J. I. Craig.

Siberian thunderstorms +·88. E. Septer.

Ozone at Oxford −·62. G. M. B. Dobson.

•, *tt, mb at* 168 *stations in all parts of the world.* (Sir G. Walker.)

Rainfall •. Positive correlation at 69 stations; the largest coefficients are: ·51 Bathurst, Gambia; between ·30 and ·40 at New Caledonia; Wellington, N.Z.; Bahia, Brazil; Durban, Natal; St Louis, Senegal; Gouriev, Siberia; Leh, India; Victoria, B.C.

Negative correlation at 81 stations; the largest coefficients are: −·50 Edmonton, Canada; −·43 Punta Arenas, Chile; between −·30 and −·40 at Kasalinsk, Siberia; Ekaterinbourg, Russia; Tokio, Japan; Albany, U.S.A.; Suakin, Sudan; Newcastle, Jamaica; Alice Springs and Eyre, Australia; Pelotas, Brazil.

Temperature tt. Positive correlation at 18 stations, the largest coefficient is +·27 Auckland. Negative correlation at 76 stations: −·58 Pelotas and −·45 Recife, Brazil; −·55 Winnipeg; −·43 Agra; −·42 Brisbane; −·44 Calcutta; −·45 Hong Kong; −·49 Sydney (N.S. Wales); between −·40 and −·30 at Victoria, B.C.; Tashkent, Siberia; Newcastle, Jamaica; Santiago, Chile; Baghdad; Batavia; Bombay; Cordoba; Jakutsk; Sydney (Nova Scotia).

Pressure mb. Positive correlation at 39 stations: ·36 Pelotas, Brazil; ·35 Santiago, Chile; ·30 Galveston, Texas; between ·30 and ·24 Honolulu; Sydney (Nova Scotia); Albany, N.Y.; Ekaterinbourg; Hong Kong; Para, Brazil; Buenos Aires, Argentine. Negative at 48 stations: −·47 Cape Town, −·46 Zanzibar, −·43 Blumenau, Brazil; between −·39 and −·27 Bombay, Calcutta, Madras, Colombo, Leh, Port Darwin, Rio de Janeiro, Brisbane, Adelaide, Alice Springs, Batavia, Derby, W. A.

Storminess in the North Atlantic. (Huntington, *Earth and Sun*, p. 29.)

+·61 same year, −·03 three years before, ·003 three years after.

Minimum temperature, Colombo +·5. (G. T. Walker.)

2. *Correlation of solar constant with meteorological phenomena.*

Temperature. +·6 tt Santiago, June–Aug.; +·4 minimum tt Colombo; −·82 tt Sarmiento second day after. "No such relation in India." G. T. Walker.

Ozone −·54 Oxford.

Figures are now available for the correlation of sunspots with the monthly water-levels of Lake Michigan for the past century (S. M. Wood).

V. Sunspots and solar constant (G. T. Walker).

Synchronous (June to August) +·64.

Solar data are now included annually in the *Quarterly Journal* of the Royal Meteorological Society.

* * *

Correlation coefficients show some analogy to periodic terms in their variability in different groups of years. 'The correlation [of Nile flood] with St Helena pressure was ·6 for the first 16 years but only ·06 for 1892 to 1920.' *Q.J. Roy. Meteor. Soc.*, vol. LIII, p. 42.

The relations between Arctic Ice and the weather of Europe are put into correlational form in Geophysical Memoir, No. 41, by C. E. P. Brooks and Winifred A. Quennell. A coefficient of −·64 shown for the relation between ice off Iceland in winter and pressure for the following three months at Ponta Delgada is conspicuous in the table as a temporary effort. The subject is also referred to by E. H. Smith in a paper which we have noticed elsewhere.

For more complete information about the method of correlation and its achievements the reader should refer to Sir G. Walker's address to Section A of the British Association at Leicester in 1933 with the title of "Seasonal Weather and its Prediction," as well as to the papers by Sir G. Walker and E. W. Bliss in the publications of the Royal Society or Royal Meteorological Society. A number of papers on long period forecasting are also to be found under the name of F. B. Groissmayr in the *Meteorologische Zeitschrift*, the *Annalen der Hydrographie*, the *Monthly Weather Review*, Washington, and *Az Idöjárás*, Budapest.

The figures for the relation between the changes in the sun and the weather of different parts of the general circulation of the atmosphere are incomplete without reference to the work on the subject in the volume on *World-Weather* by H. H. Clayton which contains a large collection of data, some of them of novel character, and in the volume on *Earth and Sun* by Ellsworth Huntington in which Clayton also assisted. Included in those volumes are accounts of the direct relations of the variations in the sun and the weather of South and North America a few days after the event. We cannot do justice to that part of the subject by the citation of the pertinent coefficients of correlation, and we must therefore ask the reader to consult the works which we have mentioned.

The information compressed into the tables of pp. 334–6 is grouped into five classes, first, the synchronous relations between the elements of the local atmospheric structure, second, the synchronous relation between the surface-elements of the circulation in separate localities, third, the relations of consecutive values of the elements treated in a similar manner, fourth, the relation between the local elements of the surface-circulation and presumptive external causes and, fifth, a reference to sunspots and solar constant.

Reviewing the results in the groups into which the table of coefficients is divided we may note with regard to the first class that a good deal of ingenuity has been displayed in disavowing the inference with regard to the structure of cyclonic depressions which the large magnitude of the coefficients demands. It is unnecessary to follow the discussion: the real ground for criticism is different from what has been set out therein. The values upon which the coefficients are based are from consecutive observations in the same locality, and the direct inference that the consecutive changes in pressure and temperature, between four kilometres and eight kilometres, are very closely proportioned one to the other is undeniable. We make no such claim for levels below four kilometres or above eight kilometres. What is not fully justified is to assume that what is true about consecutive readings in the same locality applies equally to simultaneous readings in separate but not far distant localities. That is another matter; and the proper conclusion to be drawn is that observations should be so arranged, on a national or international basis, that the two cases which we have cited in chapter IV may be supplemented by a sufficient number of ascents, and in suitable meteorological conditions, for the question to be discussed in the only way that can really carry conviction.

With regard to the second class: in general terms it may be said that Exner's exploration of the world for correlation of pressure with the pressure in Iceland confirms the conclusions at which Hildebrandsson arrived in his researches on centres of action. Both are agreed as to the importance of the Mediterranean. Exner gives to that importance the numerical value of $-\cdot 5$ for the winter season but less than $-\cdot 2$ for the summer. This conclusion deserves special notice. The importance of pressure as the element selected is enhanced by the fact that so far as we know pressure can only be changed by removing air. The formation of a cyclonic depression over the Mediterranean means the removal of the corresponding quantity of air which has to go somewhere. Hence we may express Exner's result by saying that if cyclonic depressions are formed in the Mediterranean a considerable part of the air removed will be found over Stykkisholm. So far as the single map is concerned the converse is also true, we must look over the Mediterranean for the favourite place of deposit of air lifted from Stykkisholm in the winter. Which of these two operations should be regarded as the more natural one we cannot say; but in that connexion the difference between summer and winter is very striking because in the Mediterranean cyclones are winter phenomena and do not happen in summer. There is moreover a good deal of rainfall associated with the winter-depressions of the Mediterranean, and the correlations seem therefore to suggest that the Mediterranean sea is itself a "centre of action" from which vast quantities of air are removed from time to time, as depressions are formed or intensified, and one of the results associated therewith is a rise of pressure in the Icelandic region. Apparently the air which is removed goes North, not East or West.

This conclusion confirms the impression, mentioned elsewhere, that "centres of action" in the atmosphere are not adequately represented by the centres of high and low pressure on a map of monthly distribution. We should rather look for them in the regions where there is a considerable difference of temperature between sea and air and the sea is warmer than the air. (See fig. 16 of chapter IV.)

The third class, which deals with correlation between changes with a fixed interval of time, is very promising and it is a matter of some surprise that, so far as we know, its use is limited to the Meteorological Service of India. The relationships are well marked and could easily be applied. It requires, however, a well-defined climate, either continental or marine. A coastal region between the two, such as that of the British Isles, belongs to one or the other according to circumstances. As ocean climate has not yet been studied in this way the chance of further inquiry rests with continental countries.

The fourth and fifth classes represent only a small fragment of a vast and still growing literature on the relation of solar and terrestrial changes. The subject has gained in meteorological interest since G. E. Hale at Mt Wilson in California demonstrated the existence of magnetic fields in sunspots, and thereby classed the spots as forming rotating systems which might furnish an analogy to the cyclonic depressions of the earth's atmosphere.

Besides the subjects which are represented in the examples cited in the classes there is a voluminous and valuable accumulation of coefficients which express the relation between weather and crops. They have a close connexion with meteorology; crop-data are to a considerable extent weather-data. We have in fact already taken Sir W. Beveridge's periodicities derived from wheat-prices; but the space which is available is already exhausted and we must ask the reader to consult the volumes and memoirs which are specially devoted to this branch of the subject.

The general impression which we derive from the voluminous literature of correlation is as bewildering as our effort to represent the achievements of this method, with the necessary brevity, may prove to be to the reader. Relations have been established between various elements of the circulation, but the coefficients do not generally exceed ·5, which is a very moderate figure when we consider that it implies the direct control of a quarter of the one variate by the other. Moreover, the numerical value of the coefficient is often subject to change when the series of values is extended by additional observations. An appreciation of the present position of the subject has been contributed to the *Meteorologische Zeitschrift* for April, 1926, by A. Defant. He also finds himself oppressed by a larger collection of coefficients than can be understood at first reading.

The relation to solar changes through sunspot variations is similarly unimpressive; correlations are established for various elements in various parts of the world but the co-ordination is not clear.

One conclusion upon which all investigators seem to be agreed is that the intensity of solar radiation at the outer confines of the atmosphere increases with the sunspots, but at the same time the temperature of the earth is less, a state of things which W. J. Humphreys has expressed by the paradox of "a hot sun and a cool earth." Sir Gilbert Walker, in agreement with the suggestion of Helland-Hansen and Nansen, derives from the results the conclusion that there must be increased absorption by the upper air at the times of increased sunspots in order to explain the smaller amount of radiation which reaches the earth's surface at sunspot maximum than at other times. That is a conclusion which could be verified by measures of radiation at the earth's surface; observations to that end would make a greater appeal than any indirect method of approaching the question. Instruments are now available for the purpose. We await the results of observation with some curiosity because we cannot regard the readings of the dry bulb thermometer at land-stations, which are brought into the computations, as being an adequate expression of the effect of solar radiation upon the earth's surface. Land-area forms only a minor part of the earth's surface in any belt of latitude between the tropics and we are comparatively speaking ignorant of the effect of sunshine upon the sea. *A priori* we should expect the solarisation of a water-surface to result partly in the raising of temperature of the water and air but also partly and more markedly, perhaps almost exclusively, in increased evaporation of water. It would follow that with greater radiation there would be a greater amount of water-vapour in the atmosphere of the air over solarised waters; and we ought to inquire what effect that would have upon the atmosphere over the land-surfaces as well as over the sea itself. E. Kidson remarks about Australia "sunspot minimum years are relatively dry and maximum years wet."

Helland-Hansen and Nansen have pronounced against this suggestion because the depreciation of temperature in years of many sunspots is shown also in the observations of certain squares in the Indian Ocean for which data have been compiled by the Meteorological Service of the Netherlands but, if we have correctly understood the account, the values taken are annual. Without further investigation we should be unwilling to accept annual values as properly expressive of the effect of special conditions; and, of all parts, the North Indian Ocean, the region of the alternating monsoons, is perhaps least effectively represented by annual values. The general circulation is a very complicated scheme of currents in the upper regions and the lower regions, and we cannot arrive at final conclusions from the intensive study of comparatively few localities, nor can we afford to neglect the reflexion of solar radiation from clouds, which is a very considerable fraction of the incident energy. An increase in the area of the reflecting layer would involve a considerable loss of energy which otherwise would reach the earth.

[The whole subject of radiation in relation to temperature has been fully treated by Sir George Simpson in the Meteorological *Memoirs* or *Quarterly Journal* of the Royal Meteorological Society.]

ADDITIONAL INFORMATION CONCERNING THE SEQUENCE OF WEATHER IN HISTORIC TIMES

In a paper on the State of the Ice in Danish Waters by C. I. H. Speerschneider of the Danish Meteorological Institute, after a critical examination of all the available authorities and the rejection of many years on account of the chronicler having misunderstood the localities or misread the date, the following winters according to the more modern method of recording time are named as cold winters for the Danish seas: 1892–3, 1870–1, 1854–5, 1837–8, 1829–30, 1798–9, 1788–9, 1783–4, 1775–6, 1739–40, 1708–9, 1683–4, 1669–70, 1657–8, 1634–5, 1607–8, 1592–3, 1545–6, 1459–60, 1422–3, 1407–8, 1322–3, 1305–6, 1295–6, 1268–9?, 1047–8. February and March are the chief "winter months" for ice: "Every third winter ice forms in the principal waters."

Om Isforholdene i Danske Farvande i oeldre og nyere Tid Aarene 690-1860. Af C. I. H. Speerschneider. *Publikationer fra Det Danske Meteorologiske Institut, Meddelelser Nr.* 2, Kjøbenhavn, 1915. [For the years 1861–1906 see Nr. 6, 1927.]

The paper refers to compilations of salient features of weather in past times by Arago *Oeuvres complètes*, Pfaff *Ueber die strengern Wintern etc.*, Hennig *Katalog bemerkenswerter Witterungsereignisse etc.*, Berlin, 1904. The Danish Journal *Berlingske Tidende* is noted by Speerschneider as the most effective source of his information since 1750.

The reader may refer also to the *Chronique des Évènements Météorologiques en Belgique jusqu'en* 1924 by E. Vanderlinden, Bruxelles, 1924, and to a number of contributions by G. Hellmann, Berlin. For a satisfactory presentation of the whole subject, the earth should be divided into separate meteorological regions each with its own chronological table.

[The volumes of *Nature* 1930 contain week by week a calendar of 'Historic Natural Events'; we note also A. Réthly, 'Les calamités naturelles en Hongrie 930 à 1876,' *Matériaux pour l'étude des calamités*, vol. II, p. 77; P. Vujević, 'Documents historiques sur les variations de climat dans...Yougoslavie,' *Union Geog. Int. Beograd*, 1931; papers by A. Wallén, 'Om kalla vintrar i Europa,' 1929 and 'Klimatet förr och nu,' 1930; and a translation in the *Monthly Weather Review*, 1928, of a paper by H. Fritz, 1893 on 'The periods of solar and terrestrial phenomena.']

CHAPTER VIII

TRANSITORY VARIATIONS OF PRESSURE. CYCLONES AND ANTICYCLONES

The implications of some notable barometric ranges.

	Amount of air lost*	Diameter of depression	Rate of travel
Tropical hurricanes:	Million tons	km	m/s
Cocos Island, November 27, 1909	40,000	600	4·4
West Indies, Louisiana, Sept. 22–30, 1915, 8 a.m. Sept. 29		650	
8 p.m. (by chart)	6,600	1150	5
8 p.m. (by barogram) ...	3,300	800	
Stationary depressions:			
North Sea, July–Aug. 1917	86,000	1400	Nil
Atlantic, lat. 40°, Oct. 10–15, 1921	190,000	1900	Nil
Large Atlantic depressions:			
Lat. 50° N, February 6, 1899	2,000,000	3440	Nil
Lat. 50° N, January 31, 1926	2,100,000	3800	5
Tornado:			
S. Wales, Oct. 27, 1913, core only	0·4	2·5	15·7
Paris, Sept. 10, 1896, core	0·22	2	—
depression round the core	64	—	—
Waterspout, C. Comorin	—	0·05 at surface 0·17 at 1500 m	—

* Part of the loss of air may be expressed as the mass of the fallen rain

Rate of settlement of anticyclones.

Tropical North Atlantic, about 86 metres per day
British Isles, March 23, 1918, about 350 metres per day
British Isles, October 7, 1919, about 450 metres per day

In the discussion of the changes in the general circulation in the preceding chapter, we have suggested that a picture which is elaborated from mean values cannot be regarded as representing the actual condition of the atmosphere at any moment of its history. That office, however, is filled by the synchronous chart, based upon observations which are approximately simultaneous. It represents the distribution of the meteorological elements at the epoch for which the chart is drawn.

It is on this plan, each referring to its own limited region, that the Daily Weather Charts of the several meteorological services of the world are drawn. For some years now daily circumpolar charts have been issued in London, Paris and Toronto. A circumpolar chart for 1 h G.M.T. is published by the Deutsche Seewarte which has designed a synoptic chart for the whole hemisphere in connexion with the work for the polar year 1932–3. The remarkably effective outline of the chart with its contours is also used by the U.S.S.R. for a daily chart of the hemisphere to supplement a regional chart on a larger scale.

With Teisserenc de Bort we can easily imagine the system extended practically to the whole world, and the result which will be obtained when that step is reached can be foreshadowed by combining the weather-maps of different countries for the same day.

FIGURE 189

Millibars, inches, millimetres.

SYNCHRONOUS CHARTS OF THE WORLD

February 14, 1923

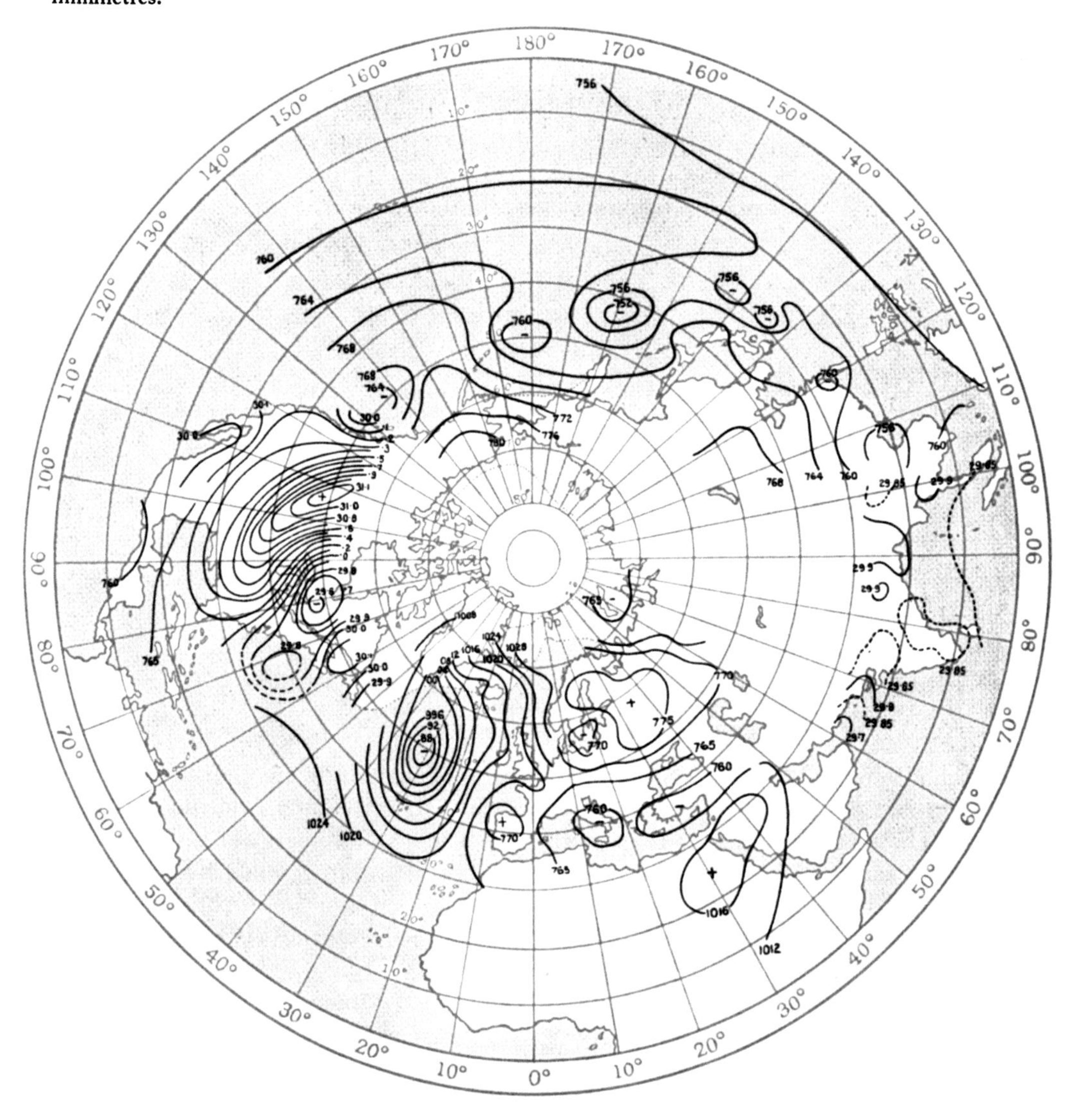

Pressure at sea-level along latitudes 23° N *and* 50° N, *February* 14, 1923.

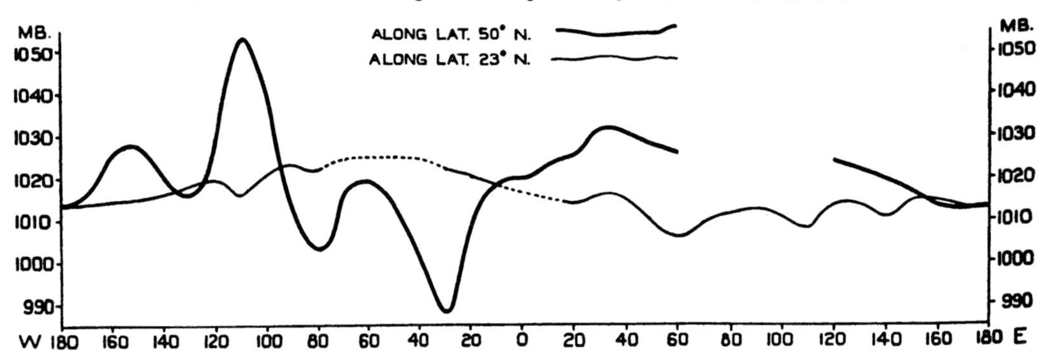

February 14, 1923

FIGURE 190

Millibars, inches, millimetres.

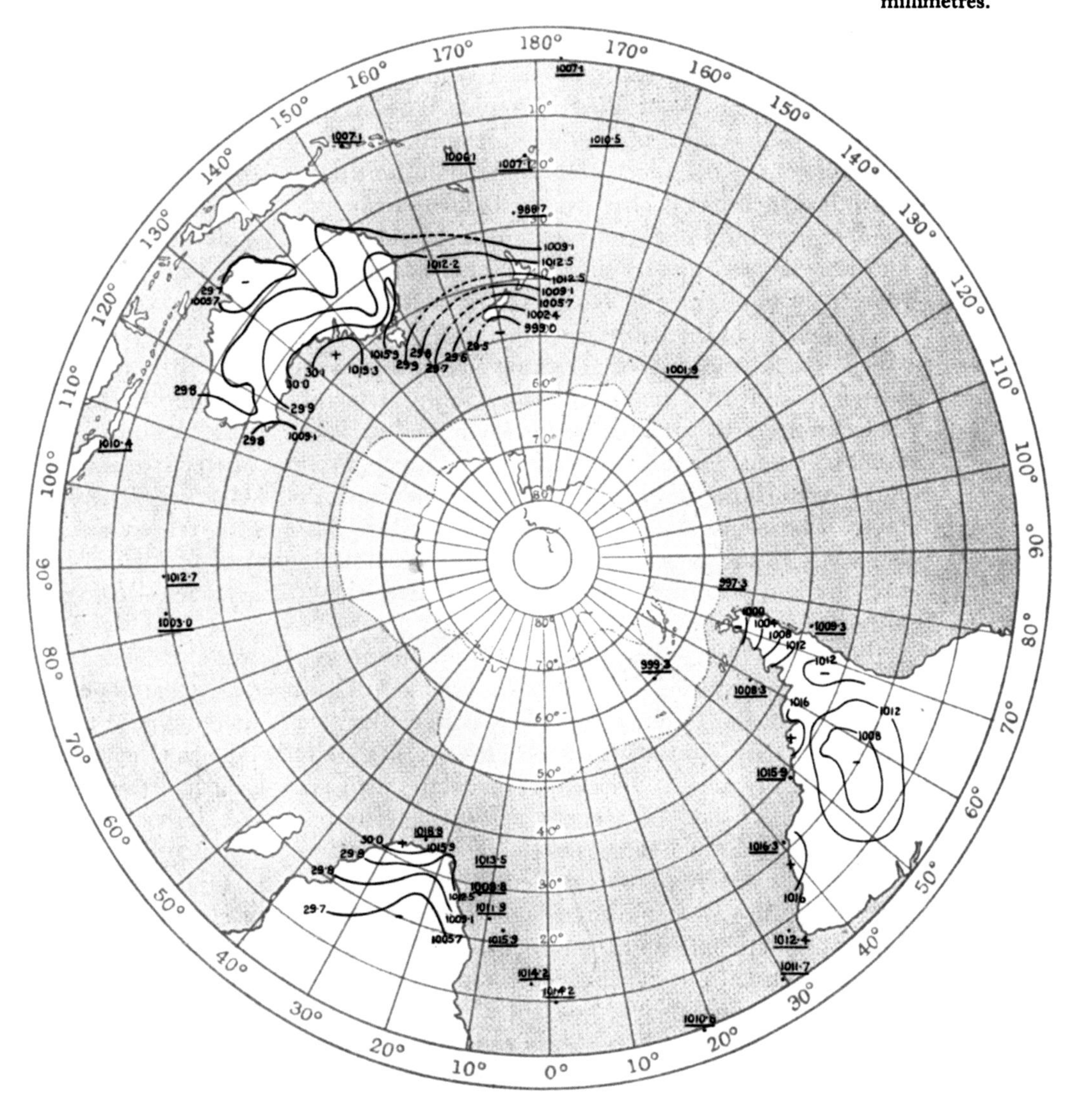

Coefficients of correlation between changes of pressure in the upper air of Batavia, England and Canada, and corresponding changes of temperature at the same levels.

Height km	Batavia	England						Canada	Height km
	Year	Year	Jan.–Mar.	Apr.–June	July–Sept.	Oct.–Dec.	Year	Year	
14	·76	—	—	—	—	—	—	—	14
12	·86	−·36	−·38	−·24	−·41	−·34	−·36	·04	12
10	·88	·32	·35	·20	·43	·29	·32	·78	10
8	·72	·86	·91	·81	·87	·86	·86	·93	8
6	·64	·86	·84	·92	·83	·85	·86	·93	6
4	·38	·84	·86	·89	·75	·83	·84	·88	4
2	·02	·66	·82	·49	·56	·76	·66	—	2
0	—	·11	−·02	·14	−·02	·33	·11	0	0

We use a combination of that kind in figs. 189 and 190, which represent the weather of February 14, 1923, so that we may call attention to the difference between the monthly normal distribution of pressure as represented for February in figs. 134 and 135 of chapter VI and the actual distribution in the various regions of the world on one day in February 1923.

The two are obviously different in type. The sweeping curves of the normals enclose a few centres of high or low normal pressure which constitute the "grands centres d'action" of Teisserenc de Bort.

In the synchronous chart we see that the temperate zone is occupied by a number of centres of high or low pressure, surrounded by numerous isobars enclosing larger or smaller areas. In some cases a single closed isobar envelops more than one centre of low pressure, a feature which is characteristic of weather-maps in the temperate zone.

It is the areas of high or low pressure on the synchronous chart to which we now wish to call attention as representing transitory variations of pressure. They are referred to also as anticyclones and cyclones. Those names are not at all felicitous because they connote too much. We want merely to indicate that a region is one of high pressure or low pressure, and, at this stage, nothing more. The simple adjectives "high" and "low" are employed in various languages in the maps of many countries to indicate centres of high or low pressure, and sometimes the centres are marked by the symbols + for high and – for low as in figs. 189 and 190.

The names high and low, simple as they are, are not quite suitable, because other things may be high and low as well as pressure. A. McAdie finds an attraction in the Greek-coined word "hyperbar" for high pressure, and a correlative term "infrabar" for low pressure, at the sight of which other people experience a repulsion. Discussing the matter with a classical friend we arrived at "hypsobar" and "hypobar" as possible alternatives in Greek coinage. Certainly some such names meaning high pressure or low pressure and nothing else are required for the satisfactory discussion of meteorological questions; then we could understand that with the exception of certain permanent anticyclones and the seasonally permanent cyclonic area of the East Indian monsoon, the hypsobars and still more markedly the hypobars, which are so easily distinguished in the charts of daily weather, are transitory variations of pressure, here to-day and gone to-morrow. For the past century they have been thought to hold the key to all the fluctuations of weather of middle latitudes.

THE CHARACTERS OF ANTICYCLONES AND CYCLONES

These transitory groups of isobars which form anticyclones and cyclones have one feature in common, in that Buys Ballot's law, which was originally an inference from the study of the barometric gradients associated with them, applies equally to either. In anticyclones as in cyclones in the Northern Hemisphere, the lower pressure is on the left-hand side of a person who has his back to the wind. The motion of the wind along the isobar is, however, compounded with a drift across the isobars from high to low. The drift is very irregular, more so in the case of the anticyclone than the cyclone. A whole volume of this work (vol. IV) is devoted to the examination of these irregularities and their implications.

The resultant air-movement is a local circulation, along the isobars, counter-clockwise in a cyclone with an irregular drift inwards, and clockwise in an anticyclone with a still more irregular drift outwards. Generally the drift is from high to low.

Another common property of these two characteristic groups of isobars, which has been elicited from the study of a multitude of examples during the last hundred years, is that they travel, but there is a difference between the two in this respect also. Both exhibit irregularities of travel but the irregularity is much more marked with anticyclones than with cyclones. The motion of anticyclones may often be described as spreading rather than travelling, and as a rule they travel more slowly than cyclones. They are more frequently stationary though stationary cyclones are by no means unknown. Another common characteristic is that both change their shape and internal arrangement while apparently preserving their identity; the changes in a cyclone are much more marked and more easily classified than those of anticyclones.

To sum up the behaviour of these two types of isobars, they have given the impression of definite entities, subject to travel and to change, but yet preserving a kind of individuality so that the use of the language of travel with regard to them seems almost as natural as to a passenger by an ocean-steamer. In both cases we are sure that at the end of the voyage the material is different though the identity has been preserved. So marked has been this impression during the past fifty years that Clement Wragge, in Australia, used to give names to those that appeared on his maps, and for the few days of their transient visit referred to them by name.

These are the cyclones and cyclonic depressions the development of the study of which we have traced in chapter XIV of volume I. They form the essential elements of the method of forecasting by means of weather-charts. We have already noted that Buchan is remembered for having been the first to trace the progress of systems of this kind across the Atlantic; and since that time the relation between the changes in the barometer and the travel of low and high pressure systems across the map, generally from West to East, has been accepted by meteorologists of all countries as a means of expressing the sequence of weather-changes. It has been the basis of explanation of the arrival of weather from the West which has been recognised as characteristic of European conditions throughout the ages. A low pressure system coming from the West indicated a fall of the barometer with a wind from South East or South; stripes of cirrus appeared in the sky, the cloud thickened to cirro-stratus or cirro-cumulus or alto-cumulus and then rainy weather, followed by a veering of the wind through South and South West, with strong winds or gales, to West. From that quarter or the North West came boisterous cool winds with clearing showers and ultimately fine weather as the barometer recovered its level.

Anticyclones, on the other hand, brought calm quiet weather, nothing violent, perhaps fog, but generally fine weather with the possibility of strong North Easterly winds as the barometer came back to its level; at most a light drizzle but sometimes persistent drizzle.

The explanation of the life-history of cyclones and anticyclones is still a part of the challenge of meteorological science. We have pointed out some common characteristics of the two groups of isobars, and also some differences. It has been customary since the weather-map came into use to regard the two almost as different aspects of a single vertical circulation of air, upward in a cyclone, downward in an anticyclone and horizontal between the two. We are disposed to regard the appearance of similarity or reciprocity as misleading, and to look upon the formation of an anticyclone as a process different altogether from that of a cyclone[1]. The explanation of the cyclone is the primary attraction of modern meteorology, but the problem is still asking for solution. In this chapter, regarding them rather from the point of view of geography than of dynamics, we shall confine our attention to the various types of cyclonic depression and their travel over the earth's surface, reserving for another chapter the consideration of the details of their horizontal structure and its relation to the upper air.

ENTITIES WHICH TRAVEL

Before going further we ought to explain to some extent the different ideas to which the name travel may be applied. We may take as two well-known examples a vortex-ring of smoke and a water-wave. At one time the vortex-ring was a very attractive subject of study on the part of physicists. Clerk Maxwell wrote:

> My soul's an amphicheiral knot
> Upon a liquid vortex wrought,

and Sir J. J. Thomson discussed a vortex-atom as the basis of the physical universe; meteorologists felt secure in regarding a cyclone as something of the same nature. The vortex-ring is still useful. It is a striking example of the transference of energy *by convection*; that is to say, a ring of smoke, having a certain store of energy of spin, makes its way through the air of a room, where it is created by sudden projection of smoky air through an orifice, and with surprisingly little loss on the journey carries both its smoke and its energy until it dissipates both upon some obstacle.

No one could refuse to regard a vortex-ring as a travelling entity, though one might find some difficulty in realising any change in its condition except a dissipation of its original store of energy.

A water-wave, on the other hand, obviously travels, children might give names to advancing waves as Clement Wragge did to cyclonic depressions. Nothing indeed is more suggestive of irresistible travel than an ocean wave.

[1] But see p. 399.

It may also transfer a vast amount of energy, but the transference is not by convection. The energy of a wave does actually consist in the displacement and motion of the material of which the wave is formed, and yet the material does not travel with the wave; after describing a limited orbit it may either repeat the process for a new wave or come to rest.

A third and still more obvious example of a travelling entity is the motion of a stream, an ocean-current or an air-current.

All sorts of modifications or combinations may occur in the atmosphere, and it is not generally possible to determine the true nature of the motion in any particular case by simple inspection. The steam from a locomotive labouring along a hillside appears to a distant spectator to be travelling with a velocity that can easily be identified. Its appearance is that of a travelling entity, but in reality the steam is not travelling at all, the apparent velocity is that of the locomotive which is ejecting it.

So with our cyclones, one of the essential elements of the problem is to decide what is the entity which travels. If it is a vortex we have to make out how it is protected against rapid dissipation, how the material of the vortex, or what part of it, is permanent, how it takes in or rejects material and energy during its travel. And, on the other hand, if it is a wave we have to explain why the section of it at the surface over which it travels is so much like that of a vortex with a vertical axis.

Hence the scrupulous analysis of examples of cyclonic disturbances of all kinds in order to determine their structure, not only at the surface but at different levels above the surface, is a matter of fundamental importance. We must not forget that a map gives us information about the section at sea-level, and the supplementary information necessary to form a true mental picture of the superstructure is even yet very fragmentary compared with the abundance of knowledge about the surface.

THE BAROGRAM AND ITS RELATION TO TRAVELLING CYCLONES

The essential groundwork of the weather-map to which we have been referring in considering cyclones and anticyclones as the key to the sequence of weather has been based upon readings of atmospheric pressure; we have regarded those entities as defined by isobars. The sequence of events at any point affected by the entity can in consequence be traced in the barogram, and indeed changes in the weather are so clearly indicated by the features of the curve of the barogram that the instrument has become indispensable as an index. We have seen in chapter VIII of volume I that we may be wrong in thinking that every change of wind or weather would be marked by a change in the barometric pressure in the locality. The converse proposition is much more certain, that if any variation is shown on the barograph there is some cause for it in the weather. Yet the relation between cause and effect is not by any means a fixed proportion. A heavy squall of wind or rain, even a thunder-storm, may be shown by an apparently insignificant notch in the barogram. We call these small changes that are sometimes to be found, the embroidery of the barogram, and among all the occupations of the meteorological student none is more fascinating than the relation between the perceptible changes in the barogram and the weather.

In tropical regions the embroidery of the barogram is as a rule all there is to watch in relation to weather; the normal trace is a uniform wavy line with two maxima and two minima in the twenty-four hours recurring with great regularity; when the regularity is disturbed by very small dislocations they generally mean showers of heavy rain with wind.

HURRICANES

And yet in those regions, where, as a rule, the disturbance of the barogram is seldom noticeable, it may on rare occasions, perhaps not more than once in a year at any single station, develop into a specimen of the destructive visitations which are known as tropical revolving storms. Most of the tropical oceans are liable to these appalling developments, which are all reasonably well included in Dampier's description of a typhoon[1]. They are known as hurricanes in the West Indies and the South Indian ocean in the neighbourhood of Mauritius, typhoons in the Western North Pacific, cyclones in the Arabian Sea and the Bay of Bengal, Willy-willies in Western Australia.

All these are included in the general designation of tropical revolving storms, and the general characters which they possess have led to a typical representation of the phenomena as a huge whirl or vortex, extending over a diameter of the order of 500 to 1000 km and having a vertical axis. The motion of the air can be represented as rotation in a circle with a drift of convergence towards the axis. The speed of the motion of the air is of the order of 100 miles per hour (44·7 m/s) more or less, and corresponds with a wind exceeding the lower limit of hurricane force. The higher limit cannot be given exactly.

Upon a map, a tropical revolving storm is represented by a system of circular isobars very close together, the gradient being balanced partly by the rotation of the earth but chiefly by the velocity of spin[2].

At the centre of the vortex there is a circular area of calm called the eye of the storm, a few miles in diameter. Round it is the ring of maximum wind-velocity; further away the velocity diminishes in something like the inverse ratio of the distance from the axis.

The whole system moves along comparatively slowly; the ordinary velocity of travel is about ten miles per hour, being governed as to direction and velocity by the field of pressure in which the circular isobars are set. Hence a stationary observer in the belt of 500–1000 kilometres along which the system sweeps will experience the phenomena which are appropriate to the line traced in the system during its passage. An observer in the path of the centre, supposed for the moment to be from East to West as many paths are in the initial

[1] Vol. I, p. 294.

[2] The representation of the thermal environment of a cyclone (tropical or otherwise) and of an anticyclone was represented in figs. 58, 59 of the original edition.

stages of the travel, will first experience a Northerly to North Westerly wind, increasing in violence until the eye of the storm is reached, then falling to calm during the passage of the central area, and suddenly resuming the full force of the hurricane as a Southerly to South Easterly wind in the rear of the storm.

The weather during the passage of the storm is generally characterised by torrents of rain with thunder and lightning, though these are not always present. The sea in the central area is very destructive because it is confused: the changing directions of the wind give confusing trains of waves. The waves caused by the wind which marks also the direction of travel, raise a swell that travels faster than the storm.

All these phenomena are consistent with the idea of spiral convergence of air from all sides towards a centre near which, for the convergence to be maintained, there must be some arrangement by which the air passes upwards and is disposed of after depositing its moisture as torrents of rain. It is natural to suppose that the air ascends in spirals which coil round the central area, so that the line of ascent is not immediately from the centre but in concentric tubes at some distance from it. The analogy to the vortex formed when water flows through a hole in the bottom of a circular basin and drags down a column of air round which the falling water circulates is too close to be disregarded.

It is upon Sir J. Eliot that we rely for a description of the beginnings of a tropical revolving storm as set out in his account of the cyclonic storms of the Bay of Bengal. The type of storm which is described has its seasonal maximum in July and must be regarded as being formed within the isobars of the South West monsoon which lie over the Bay. In anticipation of what will be discussed later in this chapter it should be remarked that the main current of the South West monsoon is regarded as being derived in large part from the Southern Hemisphere, where conditions are "winterly" in July.

It should be kept carefully in view by mariners of the Bay of Bengal that the formation of a cyclonic storm is a gradual process, and that it is only when the disturbance has passed beyond the initial stages that it becomes a storm in the proper sense of the word. The formation of a large storm is due to the prolonged continuance of actions, processes and changes of the same kind as those that are occurring in the atmosphere at all times when rain is falling and strongish humid winds are blowing. Whatever the causes and origin of cyclones may be, the history of all cyclones in the Bay shows that they are invariably preceded for longer or shorter periods by unsettled, squally weather, and that during this period the air over a considerable portion of the Bay is gradually given a rapid rotatory motion about a definite centre. During the preliminary period of change from slightly unsettled and threatening weather to the formation of a storm more or less dangerous to shipping, one of the most important and striking points is the increase in the number and strength of the squalls, which are an invariable feature in cyclonic storms from the very earliest stages. First of all, the squalls are comparatively light and are separated by longish intervals of fine weather and light variable or steady winds, according to the time of year. They become more frequent and come down more fiercely and strongly with the gradual development of the storm.

The area of unsettled and squally weather also extends in all directions, and usually most slowly to the north and west. If the unsettled weather advances beyond this stage (which it does not necessarily do), it is shown most clearly by the wind-directions

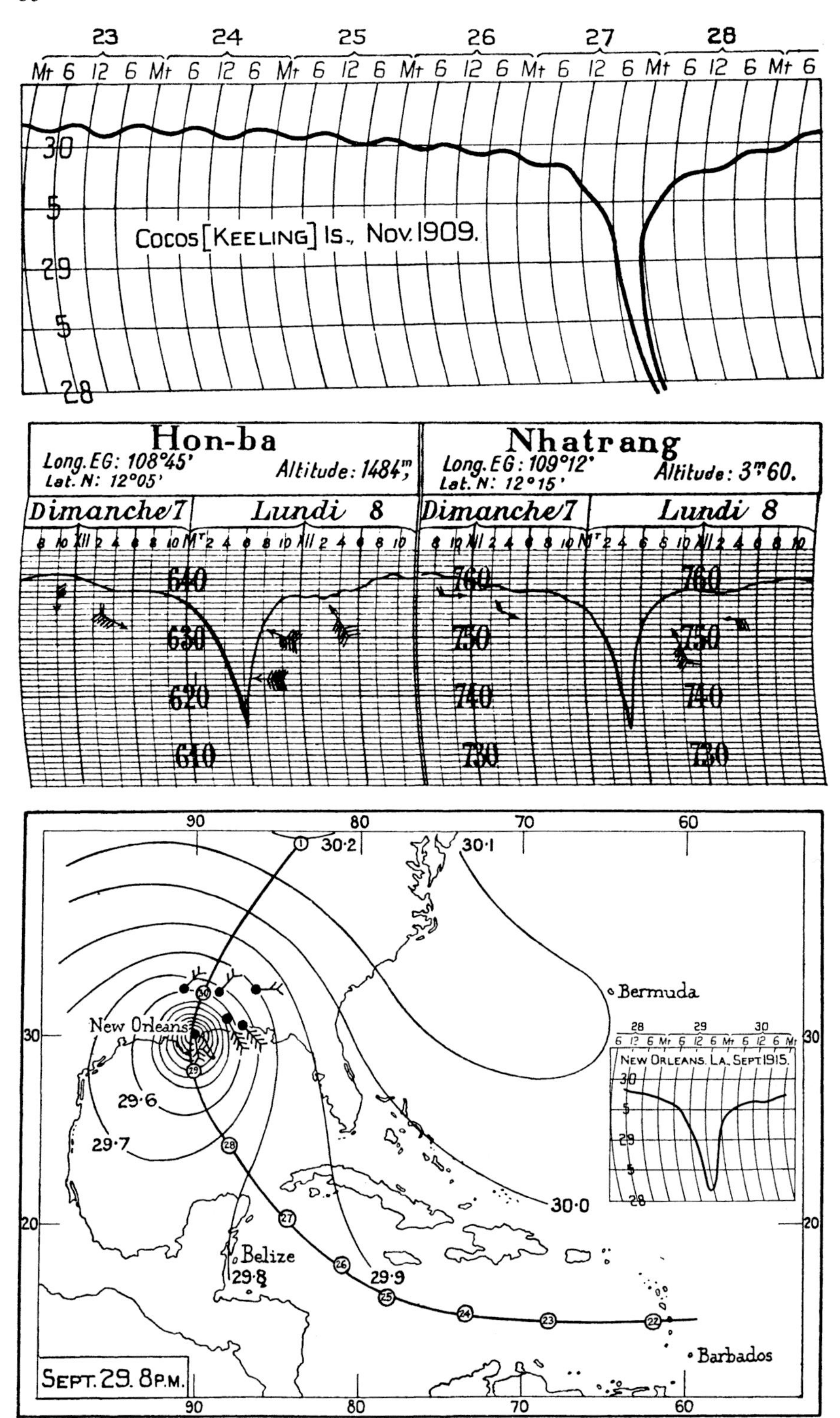

Figs. 191–3, Indian Ocean; Indo-China 8 Nov. 1926; West Indies 29 Sept. 1915.

over the area of squalls. The winds always settle down into those which invariably occur over an area of barometric depression or cyclonic circulation, or, in other words, are changed into the cyclonic winds of indraft to a central area of low barometer and heavy rain. As soon as the wind-directions indicate that a definite centre of wind-convergence has been formed in the Bay, it is also found that the centre never remains in the same position for any considerable interval of time, but that it moves or advances in some direction between north-east and west with velocities which not only differ very considerably in different storms, but also at different stages of the same storm.

This preliminary period of unsettled, squally weather may extend over several days, or may last only a few hours.... It should, however, be carefully noted that squalls more or less severe occur under several sets of conditions in the Bay, and it is hence desirable to discriminate between these. This is the more necessary in order that it may be fully realised that whilst squally weather is a necessary antecedent in time to the commencement of a cyclonic storm, squally weather is not necessarily followed by a cyclonic storm[1].

These phenomena are, of course, of special interest to seamen; and rules for navigating ships in circumstances which precede or accompany tropical revolving storms are given in all handbooks for sailors[2].

By way of illustration we give a reproduction (fig. 191) of the barogram of a hurricane of the Cocos Islands of November 1909. Barograms of such occurrences are not easily obtained because the wind force is of the order of 100 miles an hour (44·7 m/s) or more. We give too (fig. 192) the records[3] of two barographs for Nhatrang and Hon-ba, two stations, one high and the other low, apparently on the line of travel of the vortex.

We give also a barogram and corresponding map (fig. 193) of a West Indian hurricane of September 22 to 30, 1915, described by C. L. Mitchell[4].

It may be noted that Eliot writes of squally weather preceding the formation of a centrally organised storm, and the manner in which he treats the subject indicates that the resulting cyclones may be of different intensity on different occasions. We may arrive at a similar conclusion from C. L. Mitchell's memoir on "West Indian Hurricanes[4]." In his charts of the paths of hurricanes he differentiates between those which are known to have reached hurricane force, those which are known not to have reached that limit, with a third class of those about which it is doubtful whether they reached hurricane intensity.

We have apparently no criteria for classifying the intensity of severe tropical revolving storms of the several regions except by the destruction which they cause. And, on the other hand, we have found no information of the relation of the preliminary squalls to the barogram, whether these squalls are circulating systems or not. In 1927 we were in some doubt whether the phenomena which we know as cyclonic depressions occurred in the tropical regions subject to hurricanes; we learn from Dr Normand that depressions in all stages of development occur in the Bay of Bengal.

[1] *Handbook of Cyclonic Storms in the Bay of Bengal*, by J. Eliot, 2nd edition, Calcutta, 1900–1901, p. 137.

[2] *Barometer Manual for the Use of Seamen*, M.O. publication, No. 61, London, 1919, p. 77.

[3] E. Bruzon and P. Carton, *Le climat de l'Indochine*, Hanoi, 1929.

[4] *Monthly Weather Review*, Supplement, No. 24, Washington, 1924.

We gather from a paper by A. J. Bamford on "Cyclonic Movements in Ceylon[1]" that cyclonic depressions are part of the experience of that country, and indeed some of the cyclones of the Bay of Bengal which were supposed to be formed North of Ceylon had really traversed the island from the South before they were recognised as cyclones in the Bay.

We represent by the tables of p. 364 the frequency of these visitations in the different seas and, by charts, their place of origin and the tracks which they follow. We shall, if possible, represent only the tropical revolving storms of the year 1923 because we regard the information for a single solar period as being the most effective way of expressing the reality of the case.

CYCLONIC DEPRESSIONS OF THE TEMPERATE ZONE

Velocities of hurricane gusts in the British Isles.

Station	Date	m/s	Station	Date	m/s
Quilty	27. i. 1920	>50	Holyhead	2. i. 1899	42
Londonderry	28. i. 1927	49	Quilty	4. xii. 1914	41
Tiree	28. i. 1927	48	Pendennis Castle	27. x. 1916	41
Scilly	8. iii. 1922	48	,, ,,	29. i. 1927	41
Paisley	28. i. 1927	47	Lerwick	28. i. 1927	41
Pendennis Castle	14. iii. 1905	46	Southport	12. i. 1899	40
,, ,,	8. iii. 1922	46	Scilly	23. x. 1909	40
,, ,,	26. xii. 1912	44	Southport	14. ix. 1914	40
,, ,,	4. iii. 1912	44	Eskdalemuir	5. xi. 1911	40
Scilly	16. xii. 1917	43	Scilly	28. xii. 1900	40
Plymouth	8. iii. 1922	43	Eskdalemuir	25. x. 1917	40

The barograms of the British Isles and other parts of the temperate zone are marked by much greater versatility than those of the tropics. They show all kinds of changes small and great: the steadiness of the tropical barometer is a very transitory experience in the temperate zones and is soon replaced by fluctuations of a magnitude which is very rare in the tropics. The pen often travels over a wide range of pressure, sometimes as wide as that of a tropical storm, but the changes are less abrupt, and the winds which accompany them are seldom violent. On the other hand, the winds show much more fluctuation, strong winds and gales are much more frequent. A wind with a force of 8 on the Beaufort scale is called a gale, 9 a strong gale, 10 and 11 storm, and force 12 is reserved for a wind-force that no canvas can withstand, and is called "hurricane" force, to which a lower limit of 75 mi/hr, 33·5 m/sec, has been assigned for the "good exposures" of the British anemometers.

A few of the strongest gusts of wind in the British Isles as recorded by pressure-tube anemometers are enumerated in the table, with regard to which it may be remarked that the Quilty figure is not now regarded as authoritative. We may quote instead 49 m/sec at St Mary's, Scilly on 6 Dec. 1929 and perhaps 52 at Achariach on 27 Feb. 1902.

It may also be noted that these gusts were recorded during the occurrence of strong winds, and that the determination of the lowest limit of hurricane force refers to mean hourly velocity. Estimated in that way "hurricanes"

[1] *Ceylon Journal of Science*, Section E, including *Bulletins of the Colombo Observatory*, vol. 1, part 1, Colombo and London, 1926, p. 15.

were experienced on two occasions, namely, at Pendennis Castle on March 8, 1922, and at Holyhead on January 2, 1899.

Cyclonic depressions in the temperate zone are much more numerous than hurricanes in the tropics. The average number of tropical hurricanes per year all over the world (so far as it is known) is about 44, and if the life of the depression as a hurricane be counted at five days the average chances are three to two in favour of being able to find a hurricane somewhere on a map of the world on any day of the year, but the chance of finding a day when a map of the North Atlantic or the Southern Ocean shows no cyclonic depression is practically zero. [Visher counts 90 hurricanes a year, *M.W. Rev.*, 1930.]

The odds against a gale in the several parts of the British Isles are given in the following table which is taken from *The Weather of the British Coasts*:

Odds against the Occurrence of a Gale on any Section of the British and Irish Coasts on any day in the various months of the year.

Based upon records extending over the forty years 1876 to 1915. (The figures represent in each case the "odds against one.")

Coasts	Jan.	Feb.	Mar.	Apr.	May	June	July	Aug.	Sept.	Oct.	Nov.	Dec.	Year
Scotland:													
North East	5	6	7	15	27	74	102	43	17	8	5	5	10
East	8	10	10	26	43	74	77	61	20	11	8	10	15
North West	5	6	9	19	30	74	61	43	14	10	6	5	11
West	7	7	11	22	43	74	102	33	17	10	7	6	13
Ireland:													
North West	4	5	6	12	25	37	30	23	10	6	4	4	8
South West	4	5	7	14	25	59	61	21	15	8	5	4	9
Irish Sea	5	6	6	15	39	42	43	23	14	8	5	5	9
St George's Channel	6	7	8	20	38	74	77	30	19	8	6	5	11
Bristol Channel	5	5	8	14	33	42	43	17	13	6	5	4	9
England:													
South West	6	6	8	18	27	74	61	23	22	8	5	5	10
South	8	8	12	26	51	99	51	21	24	9	7	6	13
South East	9	11	14	32	61	149	77	30	29	10	7	7	16
East	12	12	16	29	61	149	154	51	42	12	9	9	19
North East	7	8	9	24	43	74	77	61	22	10	9	7	14

Many interesting features of the incidence of gales are submerged in the representation by months. The daily record during the period of 40 years in the stormiest part of the British Isles (N.W. Ireland), and in the least stormy (E. England) is represented by graphs in *The Weather of the British Coasts*, and shows remarkable prominences. The most conspicuous for North West Ireland are at the beginning of December and the middle and end of January; the maximum is 17 gale-days out of the 40 for January 28 and December 3. December 8 and 9 have 16, January 16 has 15 and November 11, 14½. December 28 has 13, whereas there is no day with more than 10 between December 19 and 25. In the summer half-year no day counts more than 5 in the 40 years except at the end of September and beginning of April perhaps a suggestion of "equinoctial gales."

On the East coast the stormiest day is November 10 with 9; October 15 and December 4 have prominences of 8. March is remarkably quiet with less than 4, and no day in the first three quarters of the year gets beyond 6.

Barograms.

Since the introduction of the portable barograph, which records on a weekly chart the pressure as indicated by an aneroid barometer, barograms are so familiar that it is not necessary to describe them. We give three examples (fig. 194) taken almost at random. The first, for September 28, 1925, represents an anticyclone; pressure was at the level of 1025 mb or above. Careful examination of the trace will disclose a reduced impression (appropriate to the latitude) of the double diurnal oscillation of pressure superposed upon the more gradual change. The second, February 9–16, 1925, shows traces of four well-marked cyclonic depressions and a very mild indication of a fifth. The third, February 23 to March 2, shows a succession of depressions terminated by a gradual rise of pressure from the lowest point of the week to 1015 mb, which continued without interruption in the following week until 1030 mb was reached on March 4th.

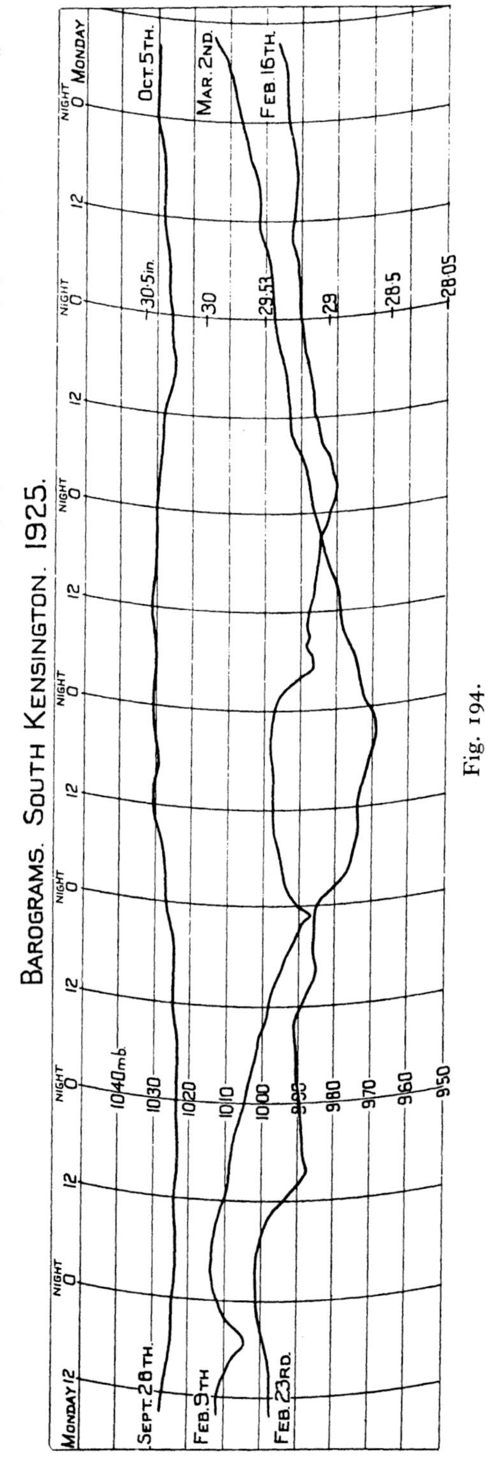

Fig. 194.

The travel of cyclonic depressions.

Some of the notable cyclonic depressions which cross the Atlantic and the British Isles have been regarded as the continuation of the existence of tropical revolving storms, and on that hypothesis the storms have been traced for very long distances.

Some doubt has, however, been expressed as to the identity of the organism in those cases. Modern inquiry, to which we shall refer later, draws special attention to a feature that was noted in *The Life History of Surface Air Currents*, namely, the growth of a new depression in the South East quadrant of an old one at the expense of its parent. It is clear that a process of that kind might be repeated time after time, and be represented as continuous travel when there are really successive new formations. We do not propose at this stage to enter into details concerning the identity or otherwise of the depression in this case. If the growth of the new centre is contingent upon the possession of the dying body of the old, we may regard the process as a change of capital rather than a change or dynasty. We reserve that interesting question as a subject for vol. IV, and in this chapter regard a depression as travelling, so long as it keeps a centre within its fringe of closed isobars.

The behaviour of cyclonic depressions can best be studied by examining the synoptic charts of the Atlantic issued by the Danish Meteorological Office and the Deutsche Seewarte jointly, or the corresponding charts for August 1882–August 1883, issued by the Meteorological Council in London.

These charts destroy the illusion that the travel of a cyclonic depression is a simple matter, as it was supposed to be in the early days of the weather-map. The changes which take place are infinite in their complexity and variety, and defy any simple classification.

There are, however, a large number of estimates of the tracks of depressions and the velocity with which they have travelled along them. We give a table of rates of travel of depressions across the British Isles, compiled by N Silvester.

Rates of travel of cyclonic depressions.

	Range m/s	Mean m/s		Range m/s	Mean m/s		Range m/s	Mean m/s
January	0–36	11	May	2–26	8	September	2–22	9
February	2–28	10	June	1–31	9	October	2–28	11
March	2–29	10	July	2–24	9	November	1–34	10
April	2–37	10	August	1–32	9	December	2–31	12

We add also the result of a special examination of depressions which were traced during the year 1913 in the North temperate zone. A column on the right of the table gives the number of days which would have been required for the depression to complete a circuit round the earth in its own latitude between 45° and 60°. [An additional table is given in chap. X, note 21.]

Date		Locality	Duration days	Speed m/s	Circuit days
1912	Jan. 15	N. Atlantic	8	7	44
1913	May 14	Europe	11	6	40
	June 10	Europe	7	6	38·5
	July 4	Europe	4	5·5	41
	July 27	N. America	11	4·5	60
	Sept. 10	Europe and Asia	16	5·5	45
	Oct. 18	Europe and Asia	12	5·5	43
	Nov. 1	N. America	13	5·5	49

Not all depressions travel. We have notes of two depressions that grew and filled up on the spot without arriving or departing as a traveller does;

the first over the North Sea on August 3, 1917, and the second over the Atlantic on October 10–15, 1921.

TORNADOES

Gales and sometimes severe gales accompany the cyclonic depressions of middle latitudes which have sufficient gradient. Occasionally gales are found over the whole area of a deep depression, but more often they are restricted to localities where the isobars are so deviated as to be congested.

Tropical revolving storms do not occur in the temperate zone but another variety of revolving storm, the tornado, is experienced, fortunately very rarely. The tornado belongs to the land, in the same general manner as the hurricane belongs to the tropical seas. It is smaller than its sea-brother, using perhaps for its formation not more than one-ten-thousandth part of the air that has to be disposed of in making a hurricane; yet it is more destructive than that agency. It is formed round a secondary centre in the South East quadrant of a large cyclonic depression[1], but only rarely. Being formed within an air-current, part of a system which already has a considerable velocity of South West wind, the vortex is carried in it and therefore we have to reckon with a considerable addition, one of 20, 30 or 40 miles an hour, to the velocity of the wind on the South East side of the centre and a corresponding deduction on the other. The added velocity makes a great difference to the force. This circumstance probably explains why tornadoes lay out a comparatively narrow belt of destruction in a business-like way that suggests the passage of a reaping or harrowing machine.

Briefly stated, a tornado is a very intense, progressive whirl, of small diameter, with inflowing winds which increase tremendously in velocity as they near the centre, developing there a counterclockwise vorticular ascensional movement whose violence exceeds that of any other known storm. From the violently-agitated main cloud-mass above there usually hangs a writhing funnel-shaped cloud, swinging to and fro, rising and descending—the dreaded sign of the tornado. With a frightful roar, as of "10,000 freight trains," comes the whirl, out of the dark, angry, often lurid west or south-west, advancing almost always towards the north-east with the speed of a fast train (20 to 40 miles an hour, or more); its wind velocities exceeding 100, 200, and probably sometimes 300 or more miles an hour; its path of destruction usually less than a quarter of a mile wide; its total life a matter of perhaps an hour or so. It is as ephemeral as it is intense. In semi-darkness, accompanied or closely followed by heavy rain, usually with lightning and thunder, and perhaps hail, the tornado does its terrible work. Almost in an instant all is over. The hopeless wreck of human buildings, the dead, and the injured, lie on the ground in a wild tangle of confusion. The tornado has passed by.

* * *

We may thus roughly classify the damage done by tornadoes as follows: (1) that resulting from the violence of the surface-winds blowing over buildings and other exposed objects, crushing them, dashing them against each other, etc.; (2) that caused by the explosive action; and (3) that resulting from the uprushing air movement close around the central vortex. Carts, barn-doors, cattle, iron chains, human beings,

[1] R. De Courcy Ward, 'The Tornadoes of the United States as Climatic Phenomena,' *Q. J. Roy. Meteor. Soc.*, vol. XLIII, 1917, pp. 317–329.

are carried through the air, whirled aloft, and dashed to the ground, or they are dropped gently at considerable distances from the places where they were picked up. A horse has been carried alive for over two miles. Iron bridges have been removed from their foundations. A cart weighing 600 lb. has been carried up in a tornado, torn to pieces, and the tyre of one wheel was found 1300 yards away. Beams are driven into the ground; nails are forced head-first into boards; cornstalks are driven partly through doors. Harness is stripped from horses; clothing is torn from human beings and stripped into rags. In one place the destruction may be complete, with every building and tree and fence levelled to the ground. A few feet away the lightest object may be wholly undisturbed. The damage is greater, and extends farther from the centre, on the right of the track than on the left, for the wind velocities are greater on the right, as in the "dangerous semi-circle" on the right of the track of tropical cyclones.

(R. De Courcy Ward, *loc. cit. supra.*)

Destructive tornadoes are most frequent and most violent in the plains of the United States, particularly of the Mississippi basin. They are also unpleasantly frequent in various parts of Australia.

The following table is taken from an article entitled, 'Tornadoes, windstorms and insurance,' in the *Bulletin of the American Meteorological Society* for April 1926. [For another census of tornadoes see chap. X, note 22.]

TORNADOES, 1916 TO 1924 INCLUSIVE. COMPARATIVE DATA FOR NINE STATES.

State	Number	Damage $\$ \times 1000$	State	Number	Damage $\$ \times 1000$
Florida	6	53	North Dakota	21	428
Indiana	22	5049	Ohio	28	15,742
Kansas	85	5764	Oklahoma	57	3,939
Kentucky	8	2080	Tennessee	28	1,066
Missouri	66	3811			

"On an average, about 50 a year are reported in the United States. Judging from the numbers recorded it appears possible that there are as many in Australia[1]."

The most violent local winds of the British Isles are of the same character; they are very rare and far less violent than those of the United States; but nevertheless so impressive as always to be the subject of notice in meteorological journals; they are found on one side of small secondary whirls which are formed in the already strong current of a large depression that covers the whole country with nearly straight isobars. Each is confined to a very limited area, so limited indeed that the normal weather-map shows little or no sign of them. A storm of this kind, which made a track of destruction from Devonshire to Shrewsbury on October 27, 1913, is described in a publication of the Meteorological Office.

The first signs of damage were noted at Dyffryn Dowlais to light structures such as hencots and outhouses. A haystack was partially removed. The width of the track was about 50 yards. The storm then bore slightly east of north to the house of Mr Rees, the coroner, at Llantwit Fardre, uprooting several trees on the way, but not carrying them any distance. Mr Rees' house was damaged, but the houses on either side did not suffer except for the removal of a few slates. The force of the wind had increased,

[1] 'Australian hurricanes and related storms,' *Bureau of Meteorology, Melbourne*, Bulletin No. 16, 1925.

as is shown by the fact that a heavy wooden stable was toppled over, and part of a shed on the tennis-lawn carried for a considerable distance over a mound 20 feet high situated to the north. The instruments at the Post Office were put out of order by the lightning. The path continued in the same direction past Lleiyon Cottages—where the top of a dog's kennel was blown 100 yards, to the vicarage, in a north-east direction—to Nant-yr-arian Cottages, close to the vicarage, which was undamaged and marks the eastern limit of the storm track.

The storm then passed over the hill to the north of the vicarage, and does not seem to have done much damage till it got to Treforest, about a mile distant in a direction a little east of north, where the first building to suffer was the generating station. This stands on rising ground to the east of the town. The iron stack at the south of the power house fell on to the roof. The western side of the building, which was made of corrugated iron sheets, was blown *outwards completely*. At the same time all the lights went out. The width of the storm track at this point was not more than 150 yards.

The subject has been fully treated by A. Wegener in a book *Wind und Wasserhosen in Europa*[1], and attention has been recently directed to it by E. van Everdingen[2] in a description of the effects of a remarkable visitation of that character which caused local devastation in the Eastern parts of Holland, in 1925. We make the following extract:

It is in the first place the capricious character of the destructions which forces us to ascribe them to whirlwinds. If the effects ought to be ascribed to nothing else but a gale, then the force of a hurricane of such strength would be required that over a wide area *everything* would have been destroyed. In whirlwinds indeed exceptionally high wind-forces occur locally and temporarily, even 100 m/s has been mentioned, though nobody can tell with certainty that such velocities have been reached. Indeed all calculations of the force of the wind from the pressure of the wind, estimated from its destructive effect, yield too high values, because it is certain that other forces of the same order of magnitude must have been present at the same time. These are the differences in atmospheric pressure, which in the rare cases where a whirlwind came across a barograph have proved to be able to reach 20 to 30 mm (27 to 40 mb). The mean diameter of a whirlwind is something of the order of 100 m. Hence mean pressure gradients of the order of 1 mm of mercury in 5 to 3 metres play a part, locally and near the axis, perhaps 10 times bigger. This causes forces of the order of 25 to 40 kg per m^2 on an object of one metre thickness, hence forces already equal to the wind-forces experienced in heavy gales, but differing from these in this respect, that they increase with the thickness of the object on which they are displayed. If such a whirlwind progresses with a big velocity—in this case there is reason to estimate that velocity at 20 to 30 m/s—then there is certainly no time to develop everywhere an equal distribution of pressure, and we may expect quasi-explosive effects of the not expanded air, which is present inside buildings.

That is why roof-constructions are tilted up, roof-coverings are blown off, window-panes and even walls are thrown outwards at the side opposite to that exposed to the strongest wind, even if the walls on that side remained intact, as has been observed in many cases. This also explains the possibility of very different directions of fall of trees at neighbouring places. This, lastly, explains why heavy objects may be carried through the air over rather large distances, borne by a diminution of pressure over and before them, and forced on by air streaming towards the depression and ascending at the same time[3].

[1] F. Vieweg & Sohn, Braunschweig, 1917.

[2] 'The cyclone-like whirlwinds of August 10, 1925,' *K. Akad. van Wetenschappen te Amsterdam, Proc.*, vol. XXVIII, No. 10, pp. 871–889, 1926.

[3] "I consider this explanation to be more acceptable than Wegener's supposition that transport over large distances would take place in a horizontal part of the vortex."

Another visitation of a somewhat similar character but of larger dimensions is described in *Forecasting Weather*[1]. It passed over the town of Cambridge from WSW to ENE early in the afternoon of Sunday, March 24, 1895, as a wind-storm with no rain. It did a vast amount of damage by uprooting trees and removing roofs within a belt of about 400 yards. Later in the day it reached the North coast of Norfolk and before leaving the shore it levelled a belt of trees, less than 100 yards wide, in a plantation on the neighbouring hills.

Barograms of tornadoes are not easily obtained, partly because the area affected by them is so limited and still more because the tornado at its worst destroys everything in its path on the one side. The few examples of records which we have seen are however very instructive; they show a sudden plunge downward of the pen and a recovery within the width of the descending trace.

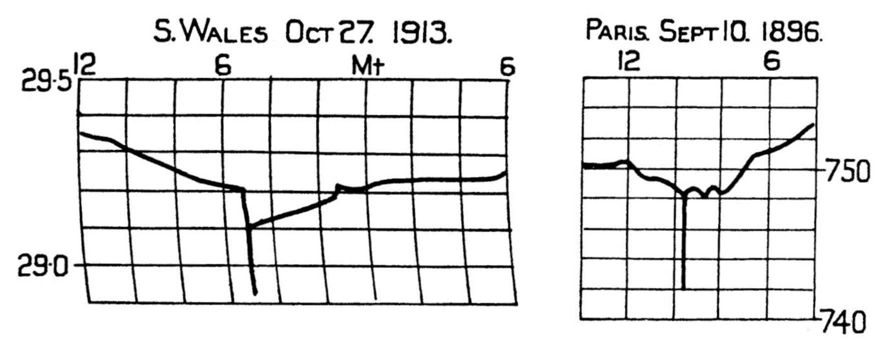

Fig. 195. Barograms of tornadoes.

We reproduce two examples (fig. 195), one for a tornado at Paris on September 10, 1896, and the other for that in South Wales on October 27, 1913. The depression and recovery must have been accomplished in a few minutes. There seems no limit to the possible descent of the pen in the more violent examples of the same character.

We have referred to tornadoes as occurring generally in extra-tropical land-areas in contradistinction to the revolving storms which are specially characteristic of the tropical seas to the eastward of the continents of America and Asia. It is therefore of some interest to note the occurrence of a tornado at Ireland-Island Dockyard, Bermuda, as early as 3h 25m a.m. on December 12, 1925, reported by the Commander of H.M. Surveying vessel *Ormonde*. The tornado is marked by a linear dip of 7 mb on the barogram of the Dockyard in the track of the tornado, and about 5 mb on that of the *Ormonde* at the dock-wall less than 100 metres away. It came apparently from the South West and went away northward after doing some damage. The breadth of the path was estimated at about 100 metres. (*Marine Observer*, December 1926.)

[1] Also *Proc. Roy. Soc.* A, vol. XCIV, 1918, p. 34

Many smaller whirls of the same kind are noticed but are not recorded because they do no damage. One reported to the author exhibited the conditions in a peculiarly effective manner. On a sultry summer afternoon a fête was arranged in the South of London, and the band after completing a "piece" was resting, when there came a breeze from the South West which lifted all the loose sheets of music from the stand and carried them up in a whirl to a height of some 1000 feet, and then left them. The life-history of such a whirl is presumably similar to that of the dust-devils of India and the African deserts.

DUST-WHIRLS

Fig. 196. Sketch by Raoul Pictet of a dust-whirl in the desert of Abassieh, three kilometres North East of Cairo, June 2, 1873 (*Les bases de la météorologie dynamique*, tome II).

DUST-WHIRLS, WHIRLWINDS AND WATERSPOUTS

Waterspouts which are sometimes seen at sea or on the coasts are somewhat akin to tornadoes, and the local whirlwinds which occur sometimes on hot summer days, and the dust-devils of dry sandy countries are still milder visitations of like character. They are very prevalent "in this part of India (Lahore) during the dry months, April, May and June[1]."

As we approach these smaller manifestations we are less and less helped by the barogram that may escape altogether from a visitation which produces violent results in its near neighbourhood. But, on the other hand, the sketch-book and the photographic camera can be used to give a permanent record of the visible features of these interesting phenomena. We reproduce from *Les bases* a sketch of a sand-column in Egypt by Pictet (fig. 196) and a sketch of dust-columns near Lahore by P. F. H. Baddeley (fig. 197), from whom we have already taken an illustration in the cloud-forms of vol. I.

Fig. 197. Sketch by P. F. H. Baddeley of dust-whirls in the neighbourhood of Lahore, with a dust-storm in the background (London, Bell & Daldy, 1860).

[1] *Whirlwinds and Dust-storms of India*, by P. F. H. Baddeley, Surgeon, Bengal Army. London, 1860. (Separate volume of plates.)

We have, besides, three photographs of waterspouts, two (fig. 198) at sea by Captain J. Allan Mordue, S.S. *Karonga*, San Francisco to Japan, August 1916, approximately in latitude 40° N and longitude 180° E, taken from the *Marine Observer*, August 1925, and one (fig. 199) on the Lake of Geneva, August 3, 1924, by P. L. Mercanton of Lausanne Observatory, taken from the *Meteorologische Zeitschrift*, June 1925.

In compensation for missing such interesting phenomena as tornadoes, the barogram sometimes discloses depressions of the duration of an hour and an intensity of ten millibars which are quite sufficient to produce dynamical results, but pass without notice except by the barograph. An example may be seen in the reproduction of the meteorographic traces at Stonyhurst for August 3, 1879, which appears as fig. 207 on p. 375.

WATERSPOUTS

Fig. 198. Pacific Ocean.

Fig. 199. Lake Geneva.

THE REMOVAL OF AIR

The reduction of pressure corresponding with the occurrence of a cyclonic depression can be expressed as the removal of a definite quantity of air. The main problems of the explanation of these striking phenomena are the questions of the method of removal of the air and its destination. Many theories of the origin and maintenance of cyclones and cyclonic depressions, some indeed of the most elaborate, disregard these questions. We are perhaps accustomed to take the removal of air from the locality in which the barometer falls as just a natural consequence of the travel of a tropical revolving storm or depression, but by that practice we merely evade the real question. In order that the barometer may fall air must be removed, and the problem of its removal by travel is hardly less a matter for inquiry than its removal from above the recording barometer by some process which is less obviously kinematical.

With the aid of a map, or of a barogram, and an estimate of the rate of travel of the barometric depression it is possible to compute the extent of removal of air which is represented by the formation of a depression. We have made an attempt at calculation of this kind for some examples of tropical revolving storms, cyclonic depressions of temperate latitudes and tornadoes. The results are given as a numerical table at the head of this chapter.

From the mere fact that the results of the calculations are expressed in millions of tons of air removed in the course of the development we infer that the processes of nature which produce these depressions are on a gigantic scale. The amounts range from 2,100,000,000,000 tons removed in the case of a vast Atlantic depression to 220,000 tons removed in the production of the barometric depression of a tornado at Paris in 1896 (fig. 195). It should be noted that the greater part of the excavation is needed in forming the outer circles of the depression which extend over a comparatively large area. The figures quoted for the Paris tornado represent only the narrow vertical column of the barogram. If the general depression is included from which the vertical column seems to descend the amount of the evacuation reaches 64,000,000 tons. In the case of the tornado of South Wales there is nothing that can be counted as representing a preliminary depression and consequently the eviction of 400,000 tons represents the vertical column of the barogram and nothing else.

It is interesting to note that the calculation of the eviction of the air to form the West Indian hurricane of September 1915 at 8 p.m. is carried out by two separate methods: the first by the map of the distribution of pressure shown in fig. 193 and the second by the depression of the barogram, taken together with the rate of travel of the depression estimated from the map. The result of the first is 6600 million tons, and of the second 3300 million. The difference arises from the different values assigned to the areas, and it is unremunerative at this stage to look more closely into the computation. It may, however, be remarked that the smaller value would have been reached if the areas of the isobaric curves had been taken from the map for 8 a.m. instead of that for 8 p.m., and that the time of deepest depression of the barogram was at 6 p.m., two hours before that of the map on which the areas were estimated.

With regard to the question of the destination of the air removed, we may remark that alternating with falls of the barometer we have periods of high pressure which are associated with what Galton called anticyclones. They also change their positions, but their travel is slower and even less regular than that of low pressures. It may, of course, be regarded as axiomatic that these areas of high pressure are the natural expression of the continuity of mass of the atmosphere—an inevitable consequence of reducing the quantity of air in one locality is its increase in some other locality—the problem which we have to think out is why, on any particular occasion, a particular locality should be chosen as the dumping ground for the air which has been removed in order to make the pressure low, and what form the process of dumping naturally takes.

THE ORIGIN AND TRAVEL OF HURRICANES AND CYCLONIC DEPRESSIONS

We have already pointed out that the most characteristic feature of tropical revolving storms and of cyclonic depressions is that they travel without loss of identity over considerable tracks, long or short. This feature is easily mapped; and in consequence among the most notable contributions to our knowledge of the meteorology of the globe are the many maps of the tracks of hurricanes and of cyclonic depressions. References to these maps will be found in the bibliography in vol. I, chap. XIV and in chap. X, note 23.

It may be said without fear of contradiction that the maps of the tracks of cyclonic depressions of middle latitudes lead only to very vague generalisations apart from the fact that the depressions are numerous in certain zones and occur seldom outside those zones. The tracks of hurricanes are more informing: nearly all of them originate over the ocean near the equator, advance first to the Westward until they reach the Western limit for the time being of the oceanic anticyclone, recurve round the high pressure, and travel thereafter in the zone of prevailing Westerly winds as cyclonic depressions.

These conditions for the year 1923 we have sought to represent upon the maps which are reproduced as figs. 200–203, with the view that individual examples are practically more informing than a scheme of mean results.

A natural inference from the travel of depressions is that if the track be traced back to the first source the conditions of formation must necessarily be disclosed. It is not easy to pursue that idea in the case of cyclonic depressions of middle latitudes. Most of them come already formed from beyond the confines of the map on which they are first shown; but there is a useful general conclusion that over the North Atlantic and North Western Europe a great number of cyclonic depressions originate as secondaries in the Southern half of a parent depression which is in process of decay; and, for the depressions of the American continent, many of them come as tropical depressions from the Gulf of Mexico or are "calved" as secondaries from the fluctuating area of low pressure over the North Pacific near Alaska, within the influence perhaps of Mount St Elias and its neighbours. The paths of cyclonic depressions from these two sources converge South of the St Lawrence and pass on to the Atlantic in that neighbourhood.

For tropical revolving storms no such generalisations are available: the paths can be traced back to some point not far from the equator and there the organism seems to have developed speedily and completely from next to nothing. We shall naturally return to the consideration of this vital question later on, but for the present we leave the actual origin of tropical revolving storms unexplained.

We introduce the maps of depressions of 1923 by a table of the frequency of hurricanes in all parts of the world arranged according to months, with an estimate of the average number per year for each district, and also a tabular summary of average annual frequencies by S. S. Visher, arranged according to the intensity of the phenomena, in order that the reader may place the year 1923 in proper relation to the normal. For that year observations of the upper air are the subject of an international publication. [The origin and travel of depressions are the salient features of practical meteorology and we may refer therefore to vol. I and vol. IV.]

FREQUENCY OF TROPICAL REVOLVING STORMS ARRANGED ACCORDING TO MONTHS.

Locality	Period	Jan.	Feb.	Mar.	Apr.	May	June	July	Aug.	Sept.	Oct.	Nov.	Dec.	Total	Year
West Indies	1493–1855	5	2	6	1	1	4	29	59	51	40	12	6	216	
,,	1887–1923	0	0	0	0	1	16	17	39	78	71	15	2	239	5
West Coast of Africa		Harmattan			Storm max.			Monsoon		Storm max.		Harmattan			
Arabian Sea	1648–1889	3	0	2	7	10	11	1	0	1	6	11	1	53	
,,	1890–1912	2	0	0	2	5	11	3	0	2	10	8	2	45	2
Bay of Bengal	1877–1912	0	0	0	7	21	42	65	55	70	51	37	17	365	10
Western N. Pacific	1880–1920	37	17	21	24	47	56	141	147	168	132	79	48	917	22
Eastern N. Pacific	1832–1923	1	1	2	0	1	5	10	14	31	22	7	3	97	2
S. Indian Ocean	1886–1917	42	54	39	18	6	0	0	0	0	2	8	25	194	6
40°–50° E	1851–1901	6	5	3	1	0	0	0	0	0	0	0	2	17 }	
50°–70° E	1848–1919	94	89	70	30	14	0	0	0	0	3	15	36	351 }	
70°–100° E	1848–1909	21	27	27	18	6	1	1	0	0	0	10	12	123 }	
S. Pacific:															
160° E–140° W	1789–1923	69	48	64	18	2	2	1	1	2	4	8	31	250	
Australia and adjacent waters:															
100°–160° E	1839–1922	54	49	58	29	7	7	6	0	5	4	10	22	251	5
W. Australia:															
Mainland	1812–1923	24	15	18	7	0	0	1	0	1	1	2	11	80 }	1–2
Ocean	1860–1924	3	7	7	2	2	0	0	0	0	0	1	1	23 }	
Queensland	1867–1925	33	27	32	14	6	8	7	0	5	4	2	10	148	1–2
N. Territory:															
Mainland	1839–1922	8	3	9	2	0	0	0	0	0	0	4	7	33 }	1
Near Timor	1778–1920	0	0	1	6	0	0	0	0	0	0	0	0	7 }	

AVERAGE ANNUAL FREQUENCIES OF RECORDED TROPICAL CYCLONES[1].

H = Severe hurricanes C = Lesser hurricanes and cyclones D = Cyclonic disturbances

Region	H	C	D
Western N. Pacific, 110 to 140 E	10	20	50
Central Pacific, 140 W to 140 E	2	4	—
Eastern N. Pacific, E of 150 W	2	3	—
Western S. Pacific, 130 W to 160 E	5	10	10
Australia, 110 E to 160 E	5	8	10
N. Atlantic	3	2	2+
Arabian Sea	2	2	—
Bay of Bengal	2	6	2+
S. Indian Ocean	8	5	—
Totals	39	60	74+

KEY TO THE MAPS OF TRACKS OF HURRICANES AND CYCLONIC DEPRESSIONS

The track of the centre of each tropical revolving storm or other well-marked depression in the year February 1923 to January 1924, inclusive, is shown in full line with an arrow at the extreme end. The dots on the lines show the positions of the centres day by day according to the morning maps. The dotted line of the February map indicates a specially interesting track for February 1924. The charts used for Siberia are those which were issued in Leningrad for 1925–6. Other departures from the primary method of representation are indicated in the margins of the maps.

[1] S. S. Visher, *Tropical Cyclones of the Pacific*, Honolulu, 1925.

The tracks which start from below latitude 25° and begin with motion westward are those of tropical revolving storms (cyclones, hurricanes or typhoons), the others are tracks of cyclonic depressions. When paths cannot be traced the positions of centres are indicated.

The land-areas within the contour of **2000 metres** are shown in black, for the Northern Hemisphere on one map of each quarter and for the Southern Hemisphere on two of the four quarterly maps.

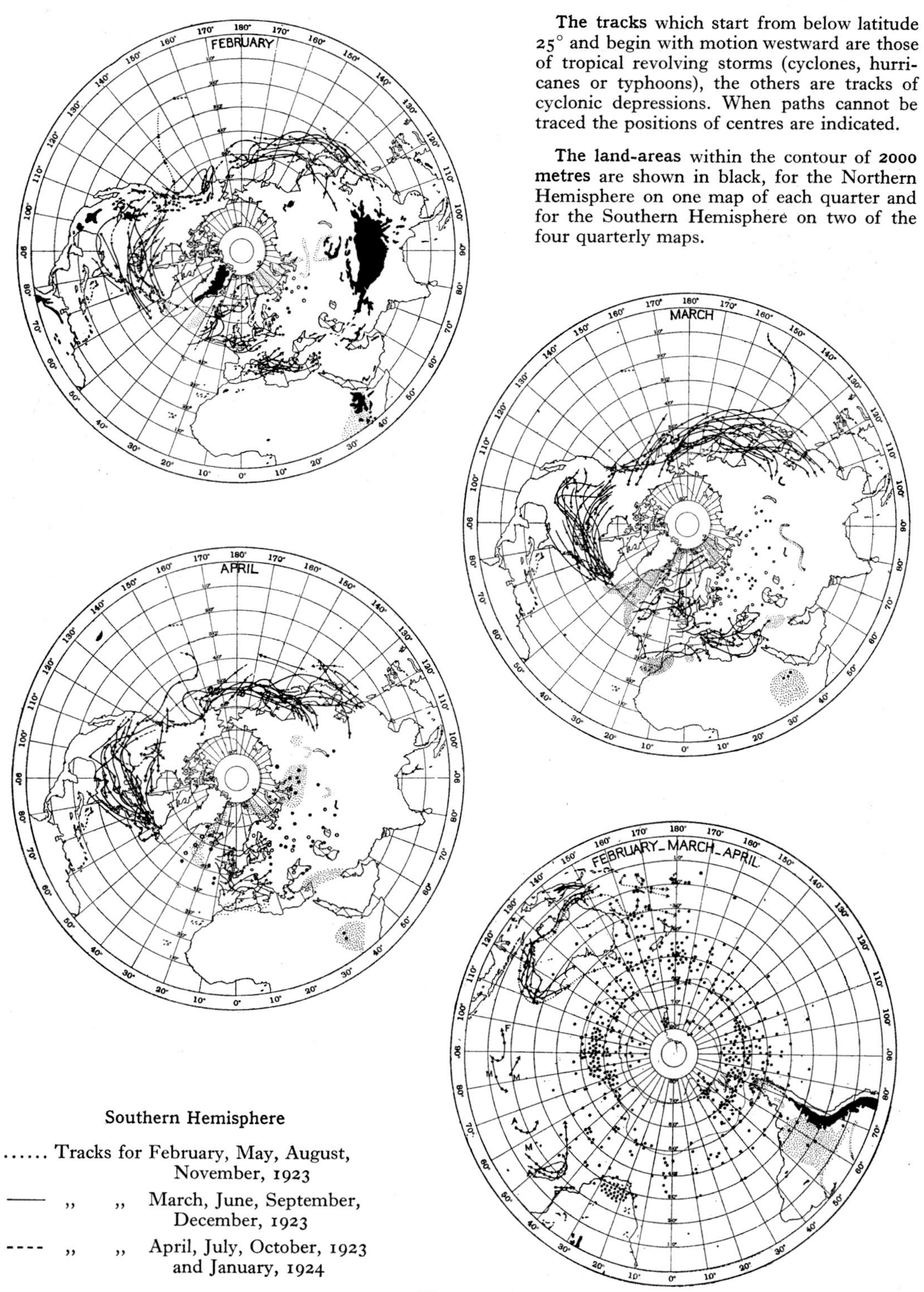

Southern Hemisphere

...... Tracks for February, May, August, November, 1923

—— ,, ,, March, June, September, December, 1923

---- ,, ,, April, July, October, 1923 and January, 1924

Fig. 200.

A separate dot indicates the position of the centre of a depression, well-marked on a map, which cannot be associated with previous or subsequent centres to form a track.

A small circle represents a closed isobar which surrounds a complex distribution of local centres.

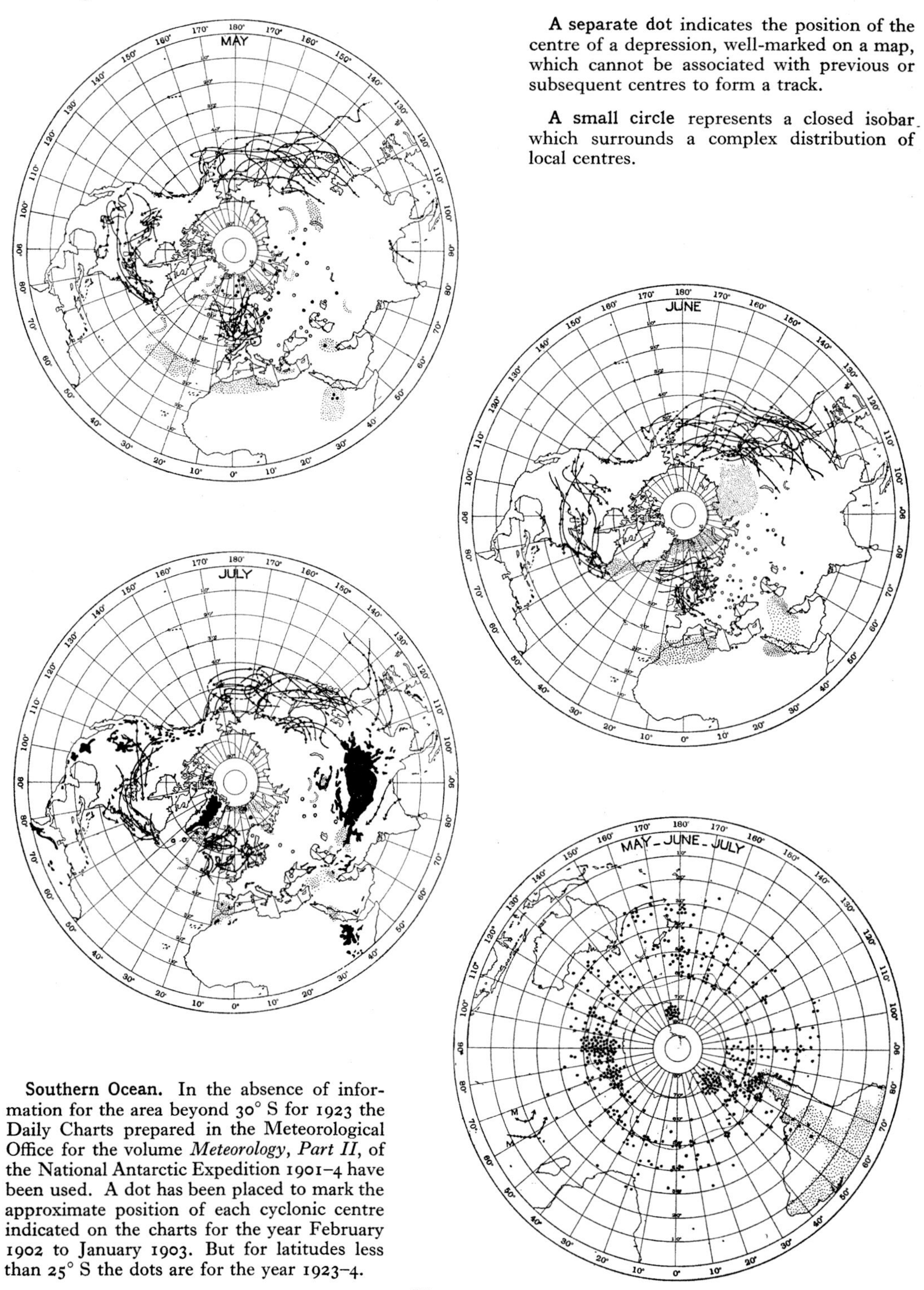

Southern Ocean. In the absence of information for the area beyond 30° S for 1923 the Daily Charts prepared in the Meteorological Office for the volume *Meteorology, Part II*, of the National Antarctic Expedition 1901–4 have been used. A dot has been placed to mark the approximate position of each cyclonic centre indicated on the charts for the year February 1902 to January 1903. But for latitudes less than 25° S the dots are for the year 1923–4.

Fig. 201.

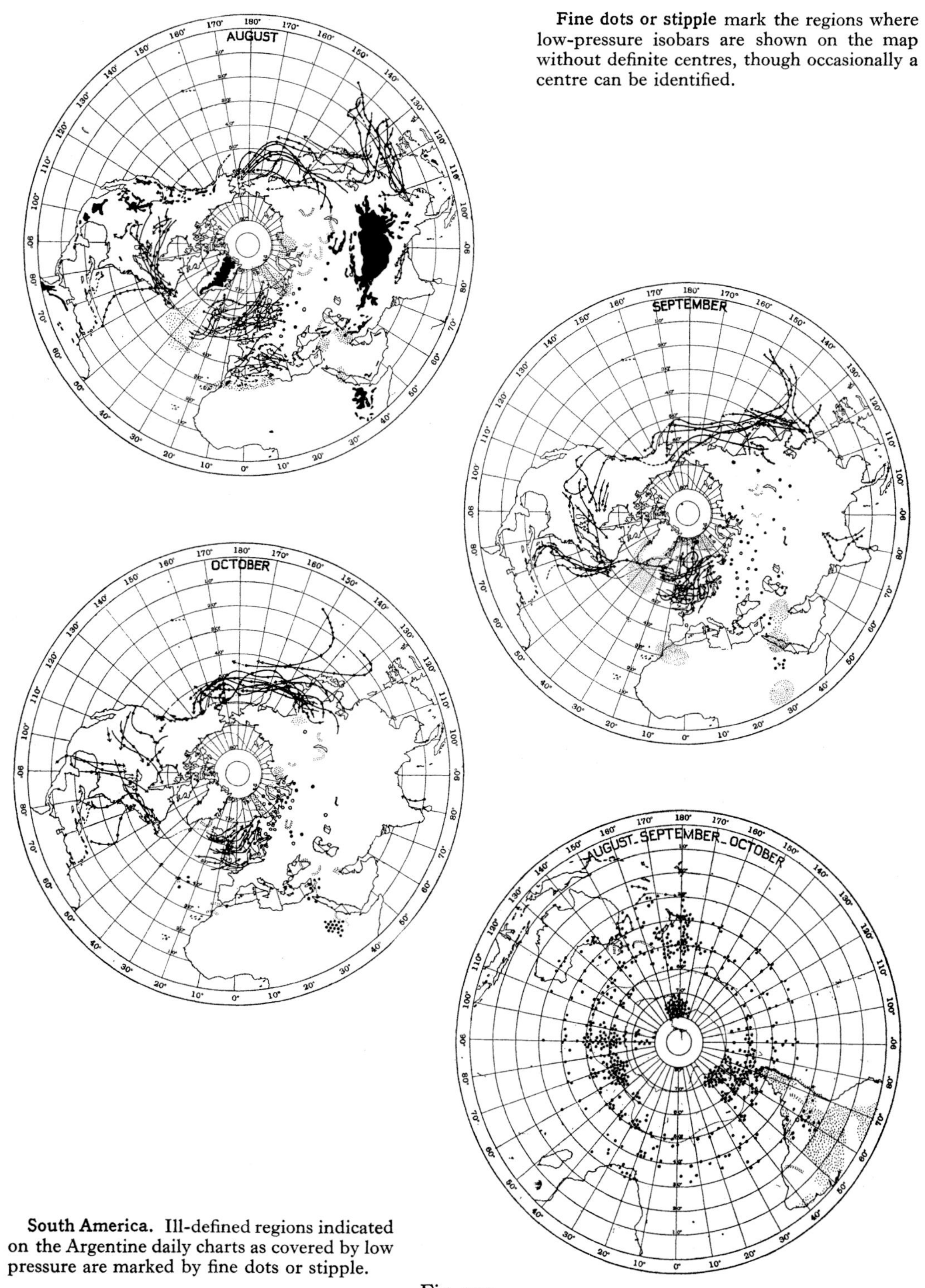

Fine dots or stipple mark the regions where low-pressure isobars are shown on the map without definite centres, though occasionally a centre can be identified.

South America. Ill-defined regions indicated on the Argentine daily charts as covered by low pressure are marked by fine dots or stipple.

Fig. 202.

Curved or serpentine stipple indicates a locality of cyclonic isobars on the margins of the maps consulted, with no information about a centre.

Short parallel lines are drawn to mark elongated **V**-shaped depressions. The lines are transverse to the axis of the depression.

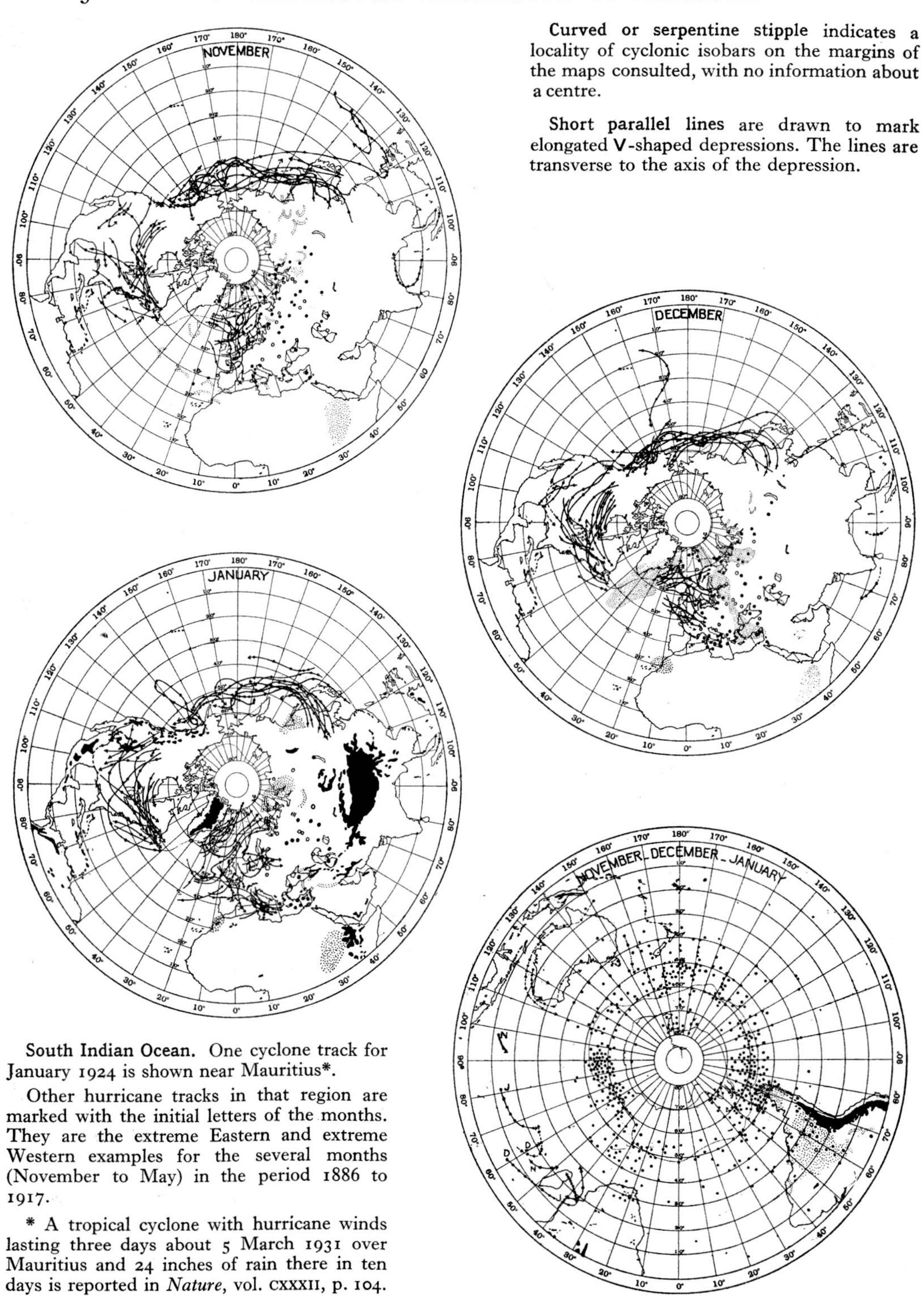

South Indian Ocean. One cyclone track for January 1924 is shown near Mauritius*.

Other hurricane tracks in that region are marked with the initial letters of the months. They are the extreme Eastern and extreme Western examples for the several months (November to May) in the period 1886 to 1917.

* A tropical cyclone with hurricane winds lasting three days about 5 March 1931 over Mauritius and 24 inches of rain there in ten days is reported in *Nature*, vol. CXXXII, p. 104.

Fig. 203.

ZONES OF BAROMETRIC DISTURBANCE

The most obvious conclusion to be drawn from the charts of the paths of hurricanes and the tracks of depressions is that the initial stages of the hurricanes, which are rare phenomena, are generally to be found not far from the equator, about ten degrees on either side of it, and therefore within the belt of the Easterly equatorial current; whereas the depressions, which are very frequent, belong to the region of prevailing Westerly winds in the temperate zone. This conclusion is rendered the more interesting by the fact that the first part of the paths of hurricanes is from the East, the paths of depressions from the West, and the complete track of a tropical barometric minimum is, first as a hurricane, along the hurricane path until the fluctuating Western margin of the permanent oceanic anticyclone is reached, then round the anticyclone to follow the path of depressions, as a depression.

The prevailing Westerlies may be described as occupying the Southern portion of the field of invasion of the equatorial winds by the colder polar winds; and the depressions may accordingly be regarded as expressing the result of the fluctuations of the recurrent and variable invasion.

If the juxtaposition of currents of different origin and of different temperatures is to be looked upon as the condition precedent for the formation of a depression, and at the same time tropical revolving storms and cyclonic depressions are to be regarded as manifestations of similar phenomena, we should have to look for a discontinuity in the Easterly equatorial current where a depression could be formed as the initial stage of a hurricane. Clearly discontinuity is not to be found in the Northern or Southern edge of the normal intertropical current; on those edges are the oceanic anticyclones, the most stable of all meteorological phenomena, and tropical storms are not formed there. It is near the equator, and therefore in the region where the air-supplies from the Northern and the Southern Hemispheres are brought into juxtaposition, that we find the origins of storms. The ordinary result of the juxtaposition is doldrums; very occasionally, however, perhaps on the edge of the doldrum region, conditions are favourable for the intense cyclonic circulation which causes the hurricane. The result is generally unexpected. It is the travel of hurricanes that forms the subject of warning, not their formation. The newspapers of September 20, 1926, report the visitation of the Bahamas and Florida by a West Indian hurricane which was sufficiently unexpected to cause the deaths of 1200 persons and the destruction of forty million pounds worth of property. Such an occurrence is so rare that in ordinary experience Florida is regarded as a place of steady barometer. It may be too bold a suggestion to regard the Bahamas, where this particular hurricane was first noticed, as the meeting place of air from the Southern with that of the Northern Hemisphere; but, for a reason already mentioned, namely, that the stream which forms the South West monsoon is a composite stream derived partly from the Southern Hemisphere, the suggestion is not improbable in the case of the depressions of Ceylon, or the summer storms of the Bay of Bengal.

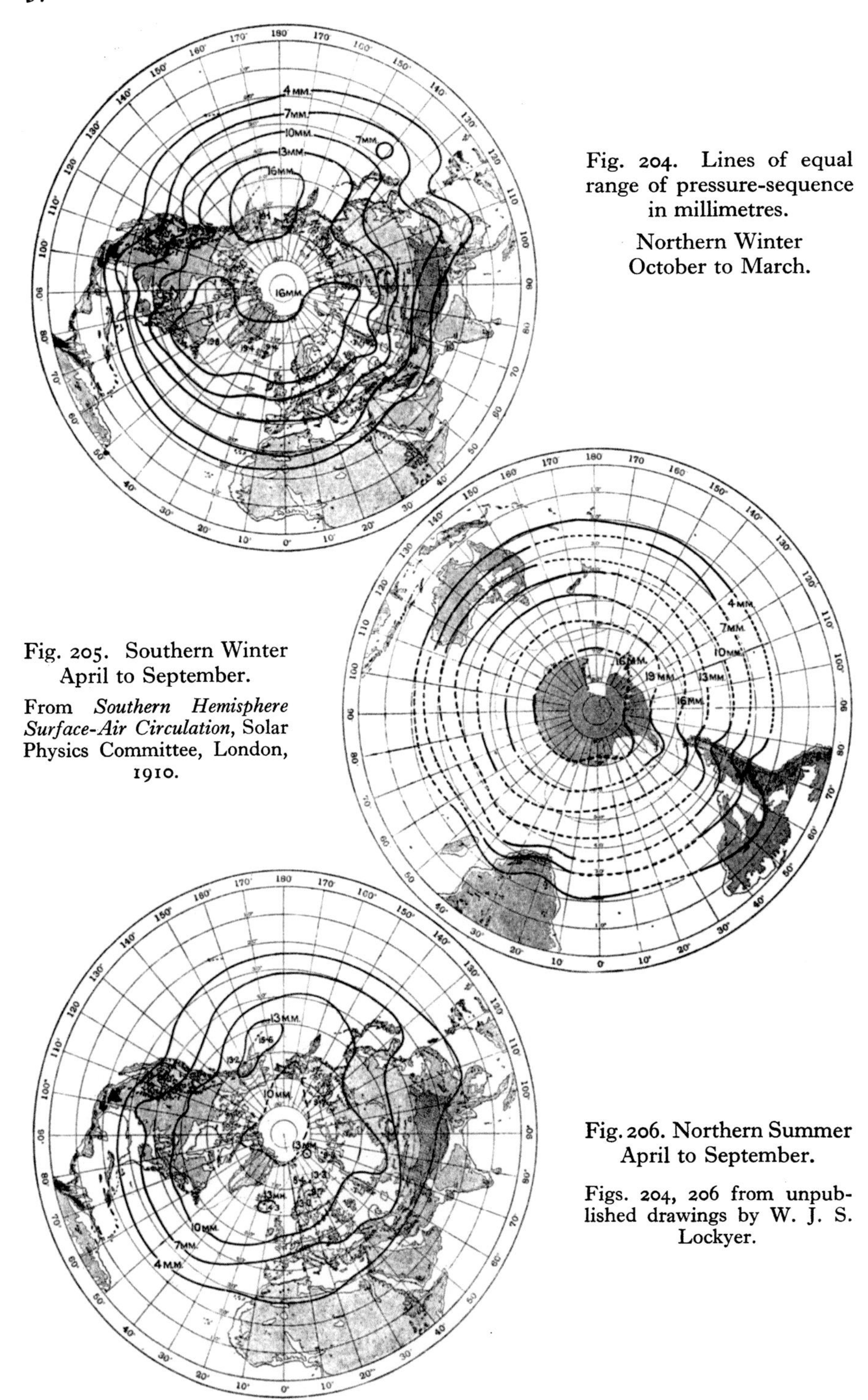

Fig. 204. Lines of equal range of pressure-sequence in millimetres.

Northern Winter October to March.

Fig. 205. Southern Winter April to September.

From *Southern Hemisphere Surface-Air Circulation*, Solar Physics Committee, London, 1910.

Fig. 206. Northern Summer April to September.

Figs. 204, 206 from unpublished drawings by W. J. S. Lockyer.

It should be noted that the only tropical ocean which is permanently free from hurricanes is the South Atlantic, and also that the doldrums of the Atlantic are always North of the equator, so that the whole of the South Atlantic may be regarded as always covered by Southern air. The same is not true of any other of the tropical oceans.

It can scarcely be possible that the development of a tropical revolving storm is dependent entirely on the surface conditions, the difference between storm and no storm is altogether out of proportion to any probable differences in surface conditions. We ought possibly to look for the explanation in differences in the conditions of the upper layers. In this connexion we should like to refer to the diagram of the upper air of India in fig. 63. If we could regard the summer and winter conditions of the air over India as roughly representative of the conditions of the upper air of the Northern and Southern Hemispheres respectively during the Northern summer, it would be pertinent to remark that the superposition of an upper layer of Southern air (above two kilometres) such as that of Agra in January above a moist warm layer of surface-air such as that of Agra in July would produce a condition of instability capable of supplying the energy of a hurricane.

Here, however, the subject becomes complicated by the question of "strays" or "atmospherics" in the reception of wireless messages. R. Bureau found that cold fronts are markedly associated with atmospherics, and Gherzi[1] has cited instances of steamers passing through the centres of typhoons and experiencing no atmospherics. Hence it may be inferred that there are no "cold fronts" in tropical storms. On the other hand "Reception of Arlington was impossible [owing to static] when the Nassau and Miami hurricanes were between Arlington and U.S.S. 'Kittery'" in the West Indies seas[2].

Information as to the behaviour of fronts as the initial stages of dynamic operations in the atmosphere is gradually accumulating in the East as well as in the West.

Isanakatabars.

Unlike the tropics the region of the prevailing Westerlies is one in which the barometer is persistently subject to fluctuation. The difference of character of these different regions can be studied by examining the curves of barometric variation in different localities. The range of variation between a minimum and the next succeeding maximum has been investigated by W. J. S. Lockyer, who finds that there is a belt of maximum variation both in the Northern and Southern Hemispheres in latitude 55–65° (figs. 204–206).

Classifying localities according to the average range of pressure from the minimum to the next succeeding maximum Lockyer has produced maps of the distribution of this form of barometric activity with lines of equal range, which are called isanakatabars, or lines of equal up-and-down of pressure.

[1] *Ondes électriques*, 1924; *Marine Observer*, vol. II, 1925, p. 132; vol. IV, 1927, p. 159.

[2] U.S. Pilot Chart of the Indian Ocean, March 1927, which gives a summary of results of "Static and its relation to navigation and communications."

They show for each Hemisphere the way in which the range works up to a maximum at or about latitude 60° and diminishes both towards the equator and towards the poles.

From the examination of the barograms for a large number of stations Lockyer has obtained the rate of travel of the pressure-waves in the Southern Hemisphere and has come to the conclusion that they would travel round the earth in thirty-eight days.

Curiously enough we find a similar period in the normal circulation of the upper air of the Northern Hemisphere as computed by A. W. Lee. His figures for the complete revolution of the Northern air-cap are 38 days at 4 km (392 gdkm), 33 days at 6 km (588 gdkm), and 23 days at 8 km (784 gdkm).

Moreover, in the table of the travel of depressions in the Northern Hemisphere quoted on p. 355 we note that the extreme maximum of the rates of travel over Europe, that of June 10, 1913, with seven days' observations, corresponds with the circumnavigation of the earth in 38·5 days; though the depressions of the Falkland Isles[1] might travel round the earth in 15 days.

Whether there is any real connexion between the travel of depressions and the normal velocity of the currents in the upper atmosphere it is not at the moment possible to say. It will, however, be remembered that Loomis and others have regarded the depressions as carried along in the main current which was indicated by the upper clouds. Bigelow[2] represents the air at 8 km above a cyclonic depression as being little disturbed from a West to East current.

Meanwhile tropical revolving storms are much more amenable to classification in respect of travel, the velocity of their winds is much greater than the velocity of travel which is, as a rule, about 10 miles an hour. They certainly seem to be carried along by the current in which they are formed, they occupy only a comparatively small part of it, whereas the cyclonic depressions of middle latitudes may sprawl over the whole area of the original current.

THE MEAN FOR A MONTH IN RELATION TO NORMAL DISTRIBUTION AND ISOCHRONOUS CHARTS

Reverting once more to the maps of the distribution of pressure representing the normal general circulation of the atmosphere in the light of Buys Ballot's law, we notice that the general circulation at the surface in the Northern Hemisphere can be related to certain centres of high and low pressure. In January there are areas of high pressure, 1020 mb or more, within closed isobars along the "desert" line of 30° to 35° North latitude over the sea and land with very slight interruptions constituting saddles or cols of lower pressure; and over Asia the high pressure area is extended far to the North and intensified to a central area of 1035 mb. On the other hand, there are two conspicuous areas of low pressure, one over the North Pacific Ocean

[1] *Meteorological Magazine*, vol. LXI, September 1926, p. 195.

[2] F. H. Bigelow, 'Report on the International Cloud Observations,' *Report of the Chief of the Weather Bureau*, 1898–99, vol. II, Washington, 1900.

from North Japan to British Columbia, the other over the North Atlantic extending from Labrador to Spitsbergen. In July the areas of high pressure over the oceans in the desert belt are intensified, the areas of low pressure over the oceans have become quite inconspicuous; but a vast region of low pressure has grown up over Southern Asia with an elongated line of minimum pressure along the Southern slope of the Himalaya (cf. fig. 185).

With these regions of high pressure and low pressure the prevailing trend of the surface-winds can be associated; and clearly the circulation is not very intense anywhere. It is most intense in the areas of low pressure over the oceans in January.

From the distribution of isobars in the upper levels we conclude that the closed isobars of the Northern map of surface-pressure belong only to the lower layers, possibly as a consequence of the distribution of land and water. In the upper air, isobars are closed by going round the pole and the complex circulation resolves itself into two simple circulations, one West to East in the half of the Hemisphere around the pole, and the other East to West along the equator.

In the Southern Hemisphere we have the suggestion of these conditions even at the surface. In the Southern winter there is a belt of high pressure over land and sea along the desert line of 35° S, and in the Southern summer the areas of high pressure over the sea become a little more isolated; but in both summer and winter between 40° S and 60° S the normal isobars are circumpolar and the circulation corresponds therewith; the winds indicated by the separation of the isobars about latitude 50° S are certainly not any less intense than those indicated in the limited areas of low pressure in the normal winter of the Northern Hemisphere.

When we transfer our attention to a synchronous chart of the weather for a definite epoch the impression which we get is quite different. Any weather-map will serve to explain our meaning. We are thus introduced to a type of motion related to local centres to which however Buys Ballot's law may still be applied in general terms but with somewhat distracting local variations.

If instead of taking normals for the month we took corresponding maps of mean values for the individual months, as Hildebrandsson did in his *Recherches sur les Centres d'Action de l'Atmosphère*, we should still find the features of the general circulation, modified somewhat differently perhaps in different parts, but still recognisable. In monthly maps it is the general circulation which preserves its identity, the local circulations are lost. The meteorological scheme would come to an end if, for example, there were no area of high pressure over the Atlantic and no area of low pressure over India in July, and equally if the high pressure failed over Asia in the winter. The changes represented on the maps for individual months would correspond with the resilience of the circulation, to the consideration of which chapter VII has been devoted. The recovery of pressure which attends the transition of a tropical revolving storm or a cyclonic depression may also be called resilience, but it is of a different character from that of the monthly distribution.

So far we have been dealing with the features of the barogram that are associated with the travel across the map of centres of low pressure or high pressure, the travelling entities of the synoptic chart. We may now recall that for the study of the charts the original designers were impressed with the necessity of using autographic records of the meteorological elements in order to account for the behaviour of the weather, and for the official meteorology of the British Isles a Committee appointed by the Royal Society arranged for self-recording instruments to be installed at seven observatories of the same type as those in use at the Kew Observatory of the British Association.

In the first report of the Committee in 1867 there appeared a copy of the records at Kew Observatory for an occasion when there was a sudden change simultaneously in pressure, temperature, wind-direction and force, as an indication of the important details that might pass unrecognised if they were not autographically recorded.

For twelve years from that date the Committee, or the succeeding Council, published in facsimile transcripts of the curves representing the records.

The reproduction was mainly the work of Warren de la Rue on the technical side and of Francis Galton on the mechanical and physical side. It extends over twelve complete years and each page covers five days with an ingenious combination of curves representing even in the smallest detail the changes in pressure, temperature of the dry bulb and wet bulb, vapour "tension," rainfall, wind-direction and velocity. Each five days' record is 12 cm long and the whole series would be 105 metres long. "It may be asserted without fear of contradiction that no record of a completeness and accuracy at all approaching that attained by the plates in question has yet been attempted in any other country, and that, moreover, the Meteorological Office is the only meteorological establishment which itself publishes the materials for testing the accuracy of its own numerical values[1]."

The possibilities of these records have never been explored but their use still remains an obvious responsibility of British meteorologists. It is rather pathetic to remember that after reproducing actual curves for twelve years with all the machinery for continuing the practice the experts at the time preferred the tabulation of five-day means of hourly values and the passing of the original curves through a mechanical harmonic analyser. Every page is suggestive to those who have the opportunity of studying it. We have given in fig. 207 on the opposite page a specimen of the whole in a portion from the sheet for July 30 to August 3 of 1879. The weather of that period culminated in a violent midnight thunder-storm which flooded the neighbourhood of Cambridge. We invite the reader in this case to compare the barograms of Falmouth, Stonyhurst and Kew and to note the difference of relation between barometric change and weather at the three observatories, also to note the similarity of the behaviour of the barogram with those of August 14 to 15, 1914 (fig. 209), which are figured in *Geophysical Memoirs*, No. 11, and with those of fig. 72 of vol. IV of this work.

[1] *Report, Proc. Roy. Soc.*, vol. XXIV, 1876, p. 206.

Five days of weather eventuating in a midnight thunder-storm in the Eastern counties of England 1879, July 30–August 3. (Quarterly Weather Report.)

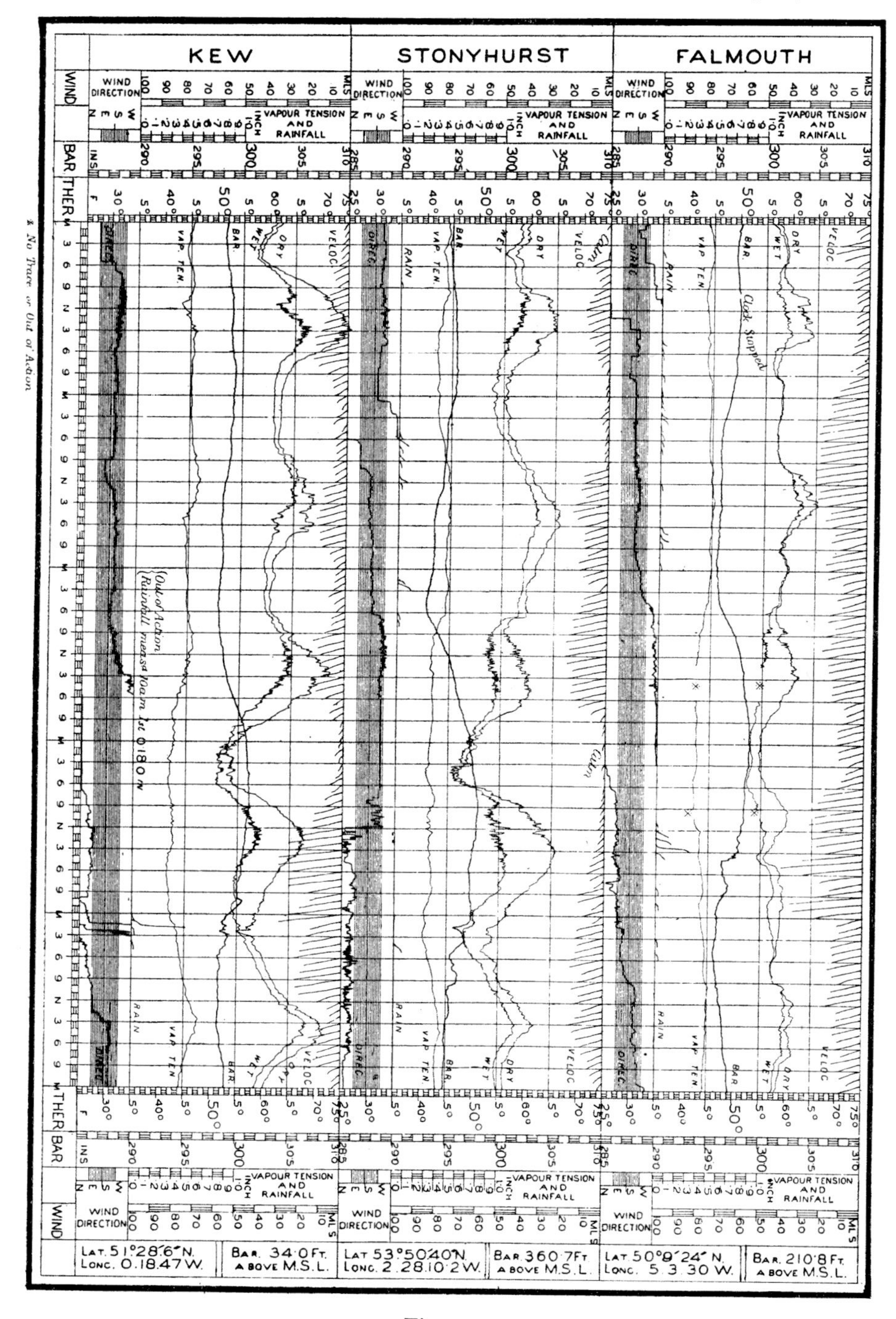

Fig. 207.

The embroidery of the barogram.

As an example of the relation of weather to the small fluctuations of pressure which we have called the embroidery of the barogram we have chosen a diagram compiled from the autographic records of the meteorological elements at Oxford.

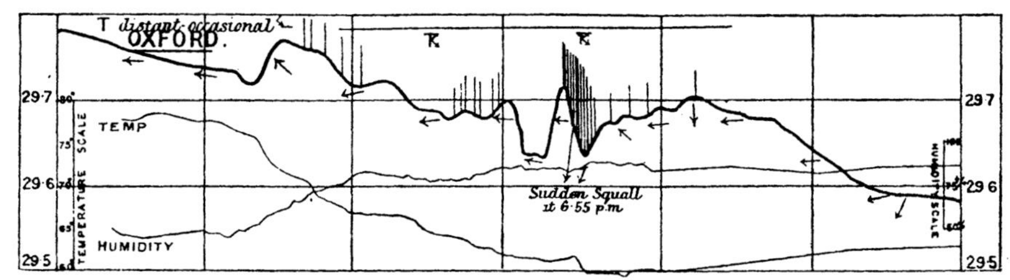

Fig. 208. Meteorogram of thunder-weather at Oxford, noon to midnight of July 27, 1900. Each horizontal space is two hours. Temperature and relative humidity are shown by fine lines; pressure by a thick line from which vertical lines are drawn one for each hundredth of an inch of rain. Wind-direction is indicated by arrows, the length indicates velocity on the scale of 5 m/s to 2 mm.

It is perhaps the fluctuations of wind and rain that are most intriguing, but the sudden and permanent fall of temperature corresponding with the sudden squall at 6.55 p.m. must be included. It will be seen that there is a gradual fall of the barometer in progress during the period represented by the diagram from noon to midnight on July 27, 1900. There was, in general, a light East wind with some occasional variation in direction and one squall from N of about 40 miles per hour (17·9 m/s) at 6.55 p.m. in the middle of heavy rain and a rapid fall of the barometer. It was during the half-hour of that particular small variation of pressure than the greater part of the rain at Oxford was concentrated. The rain which fell during the early part of the general fall of pressure was light and intermittent, whereas heavy rain came after a notable rise; the shower began when the temporary rise of pressure was complete.

Embroidery is generally very marked during the passage of a thunder-storm, but it is not a necessary accompaniment. On the other hand, there are many varieties of embroidery which are not accompanied by thunder. The subject is treated in a paper on 'The minor fluctuations of atmospheric pressure[1]' which deals with the relations of weather to the fluctuations shown by a microbarograph, an instrument which magnifies the rapid fluctuations twenty-fold, while allowing the slow changes to pass unregarded. The principal incidents referred to in the paper are sudden elevations of pressure during the clouding before the beginning of a shower of rain, and more especially of snow, with the subsequent remarkable fluctuations, waves of irregular type following the sudden rise of pressure before a thunder-shower, and waves of regular type having a period between ten and twenty minutes which have no apparent relation to weather. These phenomena are illustrated by figs. 209–212.

[1] Shaw, W. N. and Dines, W. H., *Q. J. Roy. Meteor. Soc.*, vol. XXXI, 1905.

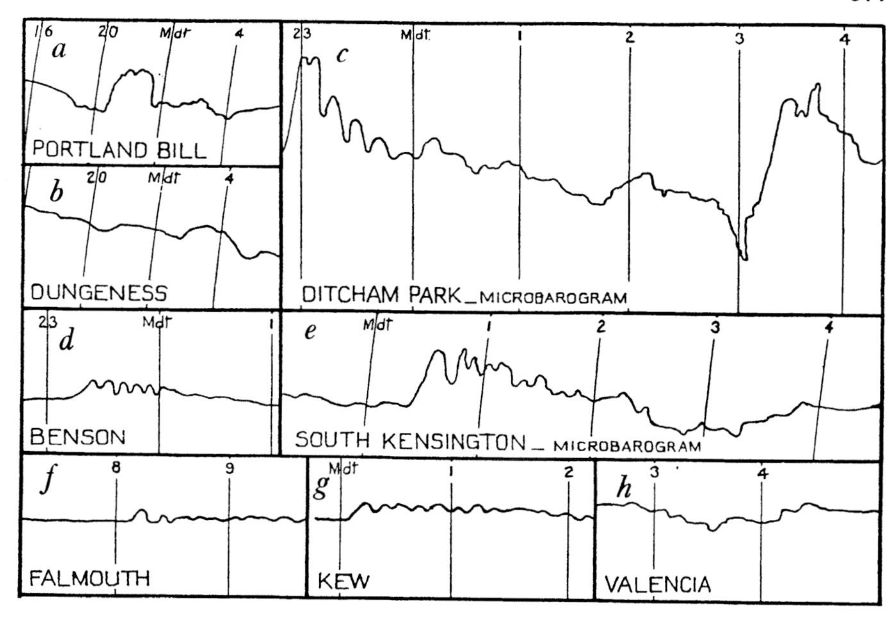

Fig. 209. Barometric disturbances associated with a sudden increase of pressure (3 mb) which occurred at 20h 30m and lasted till 23h 30m at Portland Bill on the occasion of thunder-storms in the vicinity during the night of August 14–15, 1914. (*a*) The "primary" disturbance; (*b*) the disturbance at Dungeness six hours subsequently; (*c*) the microbarogram of Ditcham Park, Hampshire, showing a disturbance of 1 mb which began at 22h 30m, with ten-minute oscillations and a renewal at 3h 5m; (*d*) eight-minute oscillations of pressure through ·5 mb at Benson, Oxford, beginning at 23h 20m; (*e*) microbarogram for London showing disturbance of ·5 mb at 0h 20m with ten-minute oscillations; (*f*) ten-minute oscillations [·5 mb] sixteen hours previously at Falmouth probably from a different primary disturbance; (*g*) ten-minute oscillations [·3 mb] at Richmond (Kew), beginning at 0h 10m; (*h*) irregular disturbances at Valencia contemporaneous with that at Dungeness.

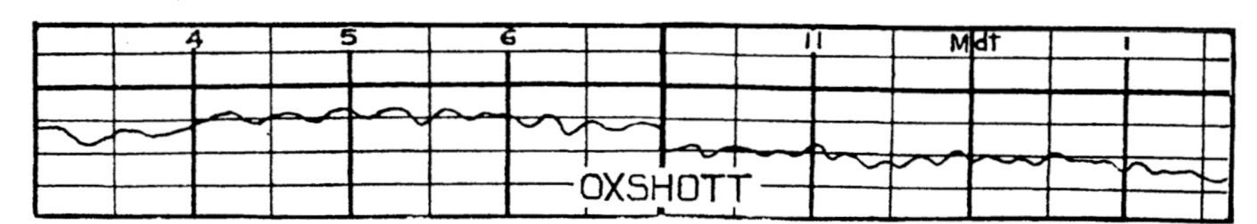

Fig. 210. Persistent oscillations shown on a microbarogram at Oxshott, Surrey, on February 22–23, 1904, twenty minutes period 3h to 7h, ten minutes period 22h to 2h.

One occasion is of peculiar interest. Observers in London, Reading, Cambridge and Ditcham Park on the look out for embroidery noticed a sudden oscillation of the barometer that died away after a short time. The records suggested a wave of pressure travelling across the country towards the South West and a member of the British Association at Dublin was asked to account for it. Many years afterwards there appeared in the newspapers an account of the destruction caused in Siberia by the fall of a great meteor. One of the observers remembered the date as that of the unexplained oscillation, and the record turned out to be that of pressure-waves transmitted through the air from the site of that meteor[1].

This explanation must be accompanied by an apology to the reader because the explanations of the observed changes belong to other volumes. In this we are concerned only with what an observer may expect to see, not with the explanation of it.

EMBROIDERY OF THE BAROGRAM

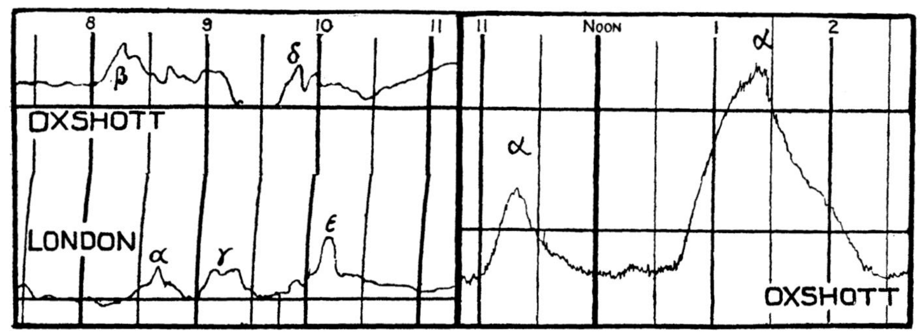

Fig. 211. On the left, contemporaneous records of similar microbarographs on May 27, 1904, at Oxshott, Surrey, and at London, South Kensington, about 14 miles to the North East. Taking the salient points in chronological order, β represents a thunder-storm with 7 mm of rain at Oxshott; α subsequent dark cloud with rain and hail, γ very dark cloud and rain-showers in London; δ thunder-storm with remarkable darkness and 10 mm of rain at Oxshott; ϵ total darkness with a few drops of rain in London. On the right α, α disturbances of the barogram for snow-showers at Oxshott, November 22, 1904.

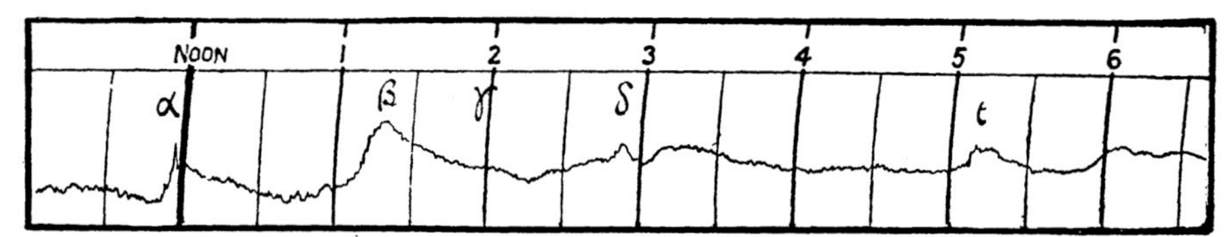

Fig. 212. Convective weather, South Kensington, July 2, 1904. α very sudden heavy rain with hail, β cloud and heavy shower, γ brilliant sunshine, δ dark cloud and rain-shower, ϵ squall, sudden cloud, slight shower.

As something intermediate between the two extremes we may reckon the setting out of the weekly results for twenty years on a single page, 304, or a diagram derived from daily observations which may prove of considerable use in agricultural meteorology and which exhibits on one page a curve of variation of daylight within the year, each day's duration of sunshine, a vertical line for each day's range of temperature from minimum to maximum, a corresponding line for the variation of dew-point and each day's fall of rain. The corresponding figures within each week, which might include the temperature of the soil, are also a possible help to trace the connexion between weather and growth and exhibit such features as a blackthorn winter or a St Luke's or St Martin's summer.

[1] F. J. W. Whipple, *Q. J. Roy. Meteor. Soc.*, vol. LVI, 1930, p. 287, and subsequent papers.

Perhaps the most notable feature of the embroidery of the barogram is the sudden but permanent rise of pressure that sweeps as a well-marked line across the country and is the prelude to a sudden veer of wind and fall of temperature. These phenomena belong exclusively to the right-hand side of the path of a depression and are indeed the mark of a transition from the equatorial air of the "warm sector" of a depression to the colder polar air of the rear which will be referred to later as characteristic of the trough.

CONTEMPORARY RECORDS OF THE *EURYDICE* LINE-SQUALL, IN THE AFTERNOON OF MARCH 24TH, 1878

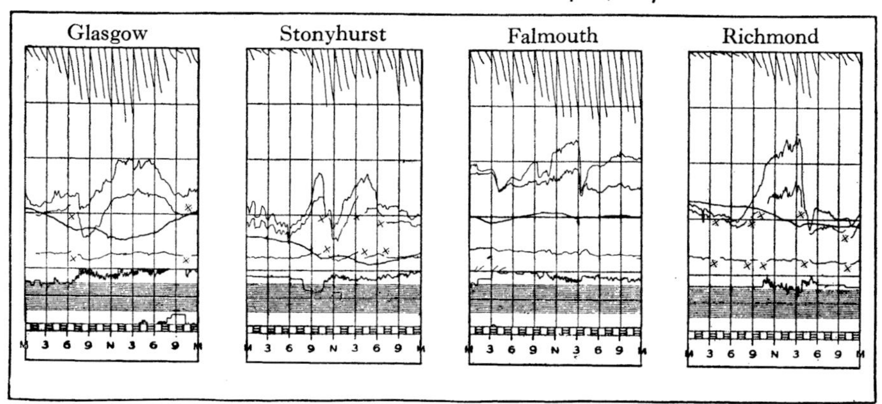

Fig. 213. The records from midnight of March 23 to midnight of March 24 are reproduced from the Quarterly Weather Report of the Meteorological Office and show the variation during the 24 hours of wind-velocity at top, temperature of the dry bulb and wet bulb, barometric pressure, vapour-pressure and rainfall in the middle panel, wind-direction at the bottom, the shaded band covering the directions with Southerly component. The scales are shown in fig. 207, and the whole incident recalls the remark about the report of 1867 on p. 374.

The most noteworthy feature of the squall, common to all the stations, is the sudden drop of temperature, and the onset of the squall may be timed thereby.

These phenomena may however be repeated several times in the South Western and Western quadrants of a depression. The result of the change may be an increase of wind represented by the new distribution of pressure, but in many cases the transition is from a field of close isobars and strong wind from the South West to a field of more widely spaced isobars with light winds from West or North West. At the transition there is generally a violent squall from the new direction and of short duration, which is called a line-squall because its onset is marked by an arched cloud advancing often with a line front which may extend across the whole sky. Occasionally line-squalls may be very destructive. A well-known instance is that of the *Eurydice* squall in 1878, so named from the fact that the heavy squall which marked the transition from the fine warm weather of a spring day to that of a wintry afternoon of March 24 capsized the training-ship of that name when she was with open ports. Another case is that of February 8, 1906, in which the line of advance ranged across the British Isles and was traced from the Hebrides to the Puy-de-Dôme. A third, that of October 14, 1909, took a more Easterly direction and exhibited marked violence on the East coast of Scotland. The phenomena were discussed at the request of the author by R. G. K. Lempfert and R. Corless[1], and have taken their place in the common knowledge of meteorological practice. Their relation to the interaction of polar and equatorial air in the polar front will be sufficiently apparent.

In studying these examples we can hardly fail to be struck by the analogy between the advance of the line which marks the transition from equatorial to polar air with its squall, and the advance of a breaking wave on a shelving shore as figured in a well-known diagram in Maxwell's *Heat*. We wonder whether there is any real analogy between the rise of pressure with its temporary squall and the advance of the wave with slightly deeper water behind it carrying the breaker as a sort of cap. An interesting feature is that the breaking of the wave is due to the fact that the upper part is travelling faster than the lower because the lower is retarded by the friction of the shelving shore, and yet the mere fact of breaking ensures that the rate of advance is controlled by the general mass of the wave and not by the more rapidly advancing crest. This peculiar characteristic is well seen at sea when a wave carries a breaking cap for a considerable distance and leaves it behind as a long patch of foam.

In the aerographic analogy the water would be replaced by a lower layer of polar air with a strong current of equatorial air above it. A suggestive example of the phenomena referred to is shown in the records of wind, pressure, temperature and rain for March 25, 1927, at South Kensington.

Something similar may be held to account for the embroidery of the barogram in the region of a line-squall. An attempt to study the phenomena in the line of junction is given in *Forecasting Weather*, 1923, p. 338, where also a fuller account of the embroidery of the barogram will be found. From the point of view of forecasting weather it is rather disappointing that the analogy of the breaking wave seems to point to the advance of the coming wave commencing nearer the base than the crest; with the line-squall also it is apparently not in the upper air that the first intimation is given but near the ground. We understand that, in like manner, waves of cold air which invade the region of the foothills of the Canadian Rockies make themselves felt in the plain beneath, before they affect the more elevated regions of Banff or the still higher level of Sulphur Mountain.

There are still a considerable number of features of the embroidery of the barogram which are unexplained and which, in view of the new interest in the structure of cyclonic depressions developed by the doctrine of the polar front, are very attractive for those who are curious about the physical processes of weather.

[1] *Q. J. Roy. Meteor. Soc.*, vol. XXXII, 1906, and vol. XXXVI, 1910.

CHAPTER IX

ANALYSIS OF CYCLONIC DEPRESSIONS

Such as of late o'er pale Britannia passed. J. ADDISON, 1704.

WE revert to the study of depressions by a diagrammatic analysis of a cyclone by Abercromby and Marriott, showing the distribution of cloud and weather in different parts of the area, divided into right and left halves by a line through the centre along the path, and into front and rear by another line through the centre at right angles to the path. The part of the cross-line which lies on the right of the path marks what is called the trough, it corresponds with the minimum of pressure on the barogram for any station over which the depression passes, and its passage is commonly the occasion of a sudden

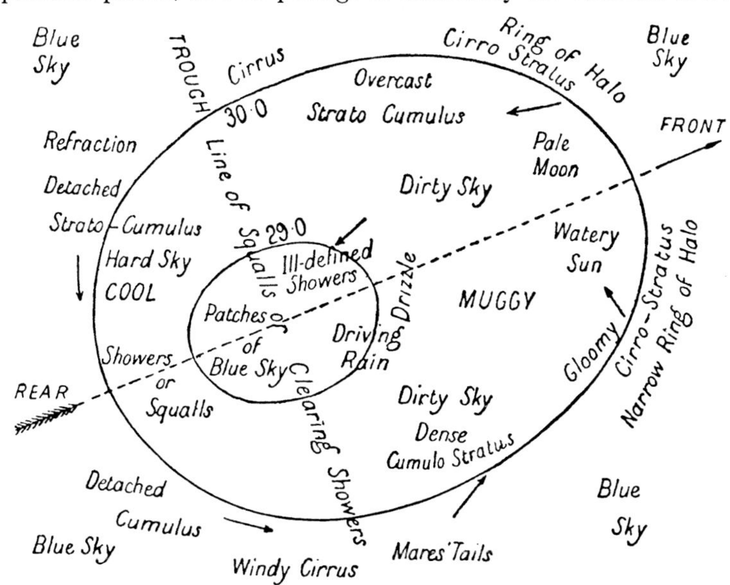

Fig. 214. Cyclone prognostics (Abercromby and Marriott 1882–7).

veer of wind and often of the change of the weather from a rainy to a showery type with a considerable fall of temperature. The fall of temperature is a very marked feature of the cyclonic depressions of the American winter and spring, in which the thermometer may suddenly drop through 40° F or more[1]. [50° on one occasion in Texas.]

In front of the advancing depression are shown the halo-forming cirro-stratus, followed by pale moon, watery sun with muggy weather on the right of the path, thereafter a belt of drizzle, heavy rain, dirty sky and dense "cumulo-stratus" to be followed by the trough of squalls and clearing showers, a "hard sky" within (clouds with sharply-defined edges), and detached

[1] *Descriptive Meteorology*, by Willis L. Moore, New York and London, 1911; *Climatology of the United States*, by A. J. Henry, Washington, 1906; *Bull. Amer. Met. Soc.*, 1932, p. 78.

CLOUDS IN RELATION TO DISTRIBUTION OF PRESSURE

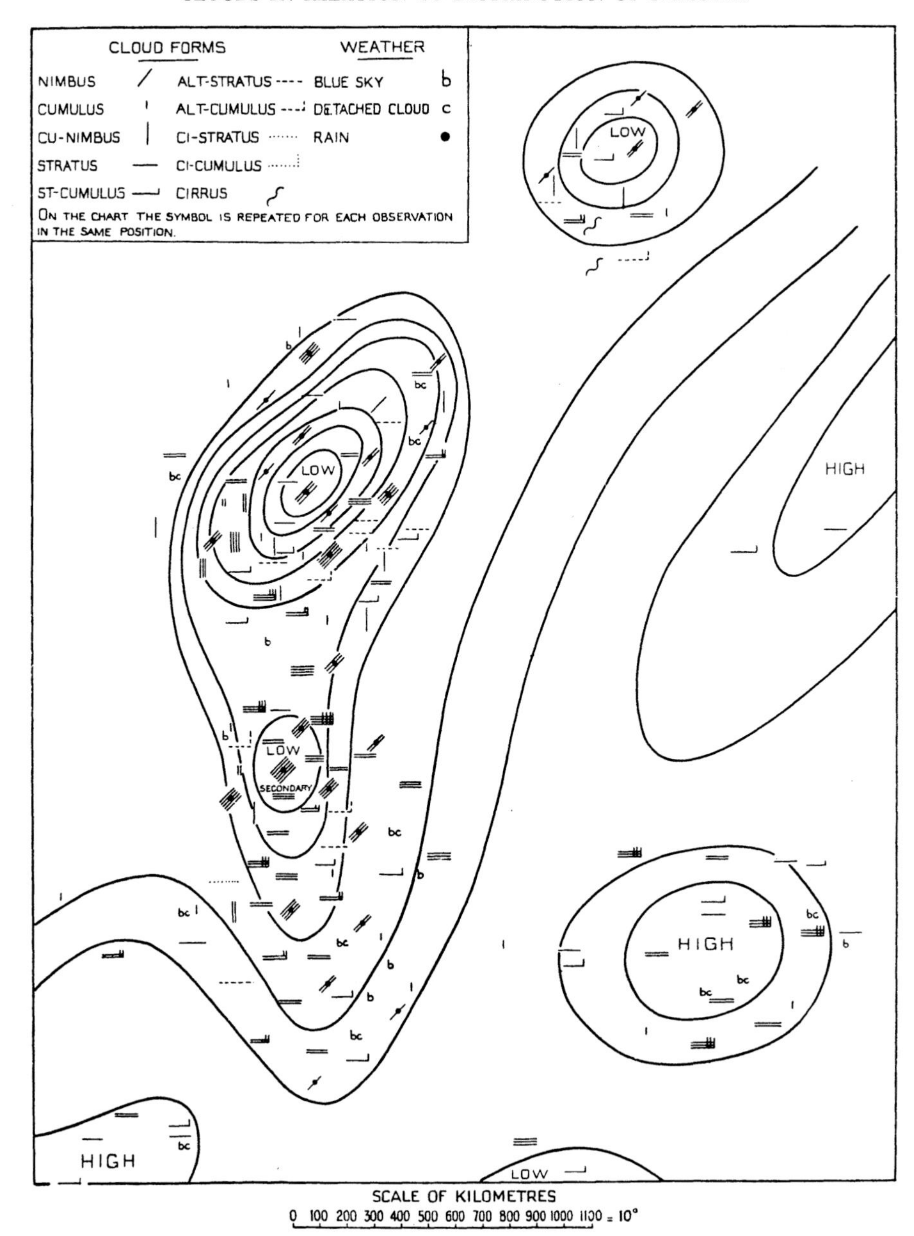

Fig. 215. Observations of cloud-forms in the Daily Weather Reports of December 1922, allocated to their positions in a general scheme of distribution of pressure.

cumulus on the margin. Cloud was bounded by an outer isobar, rain by an inner one. The plan was constructed to illustrate the relation of the weather-map to the prognostics that were at the time well known to seamen. The distribution was based rather upon Abercromby's experience at sea than any formal statistical grouping of observations.

The concentration of rainfall within an inner isobar and along the trough is not an effective representation of all the facts, but its implication conveys a very telling result of experience. The most impressive incidents of British weather within the author's recollection, apart perhaps from such violent experiences as thunder-storms and deluges of rain, are the prolonged snow-storms on the Northern side of the path of depressions the centres of which made their way up the English Channel: one on January 18, 1881, and the other very similar to it on March 28, 1916. These may be called winter experiences; another equally impressive one was an occasion when a depression made its way up channel in like manner but in the summer, and as it moved slowly and London lay on the left of the path the rain was practically continuous for three consecutive days.

An obvious reflexion for those who are accustomed to use weather-maps is that the well-formed cyclonic depression is not always present and the environment of a depression on the map is not an unlimited extent of blue sky and fine weather, but a scheme of isobars including all the varieties of grouping to which meteorologists have given names, such as straight isobars, secondary depression, V-shaped depression, col or saddle between two depressions, wedge or projection of high pressure from either side to meet at the col, as well as anticyclones from which the wedges spread out.

As a class-exercise we have endeavoured to put the assignment of weather upon a more definite observational basis by preparing a general scheme of isobaric distribution and setting thereon the observations of weather on such occasions as afforded the opportunity. We reproduce the result in a diagram based upon maps of nineteen days of December 1922 for which we are indebted to Captain L. G. Garbett.

Something of the same kind is achieved by Hildebrandsson and Teisserenc de Bort in a chapter on "Distribution des éléments météorologiques autour des minima et des maxima barométriques[1]."

We do not however lose sight of the fact that little stress can be laid upon statistical results of this kind; we are doubtful if one can reach a real expression of the structure of a depression in that way. Our reason is that the data have been collated by mapping observations of details of a number of cyclones in relation to the centre at the surface. The features of a cyclone portrayed by the results are like those of a composite photograph, they are only effective in so far as the different examples included are similar in their structure and dimensions. Each cyclone is, in actual fact, quite as much a separate individual as any one of the persons included in a composite photograph and probably

[1] *Les bases de la météorologie dynamique*, by H. H. Hildebrandsson and L. Teisserenc de Bort, tome II, chapitre I.

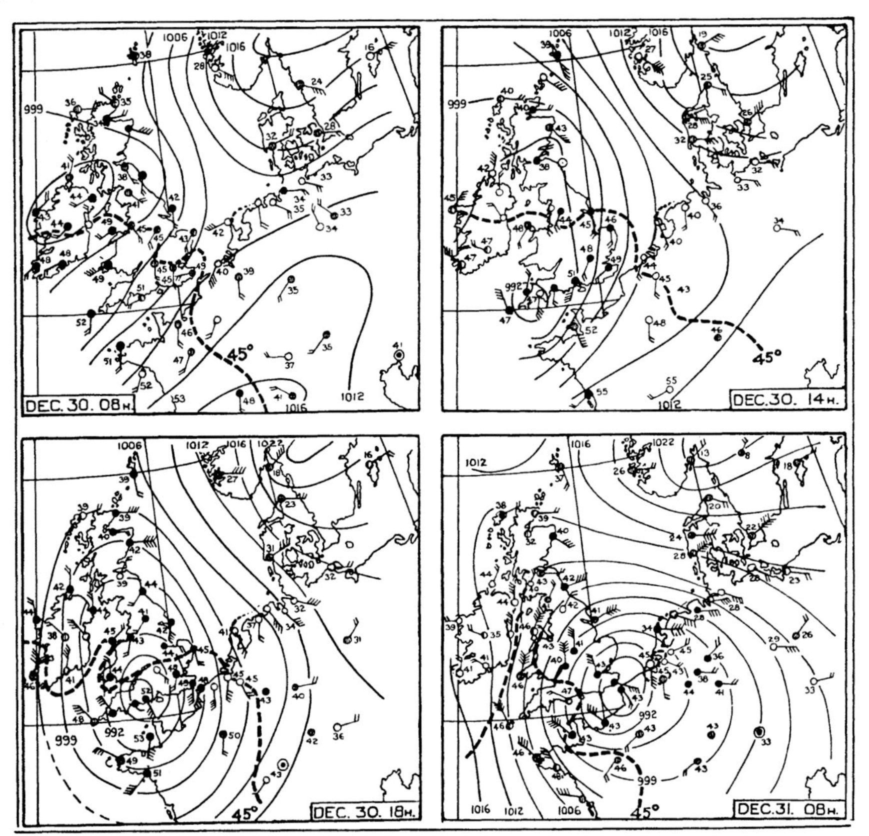

Fig. 216*a*. Maps showing pressures, winds, and temperatures between 8h Dec. 30 and 8h Dec. 31, 1900, from *The Life History of Surface Air-Currents*.

much more so. What we map, for example, gives us the pressure, temperature, circulation, etc. in the surface-layer. The whole atmosphere above it contributes to the pressure of the surface-layer and the upper structure must be regarded as consisting of a succession of layers each of which may have its own centre, pressure-distribution, circulation and temperature. Some cyclones may extend right through the troposphere though perhaps with different positions of centre, etc. for different layers; others in process of formation may not yet have reached the surface or the stratosphere as the case may be, depending upon whether the initiative in the formation is taken by the surface or the upper layers. It is therefore not quite effective for our purpose to make out the structure of the average of many cyclones. An intensive investigation of a single one, if only it were within our reach, would give a better foothold for our imagination.

That takes us back to the challenging suggestion of the original Meteorological Committee that the information derived from self-recording instruments should be used for tracing the sequence of conditions shown on the synchronous charts. In 1900 an endeavour was made to carry out the sug-

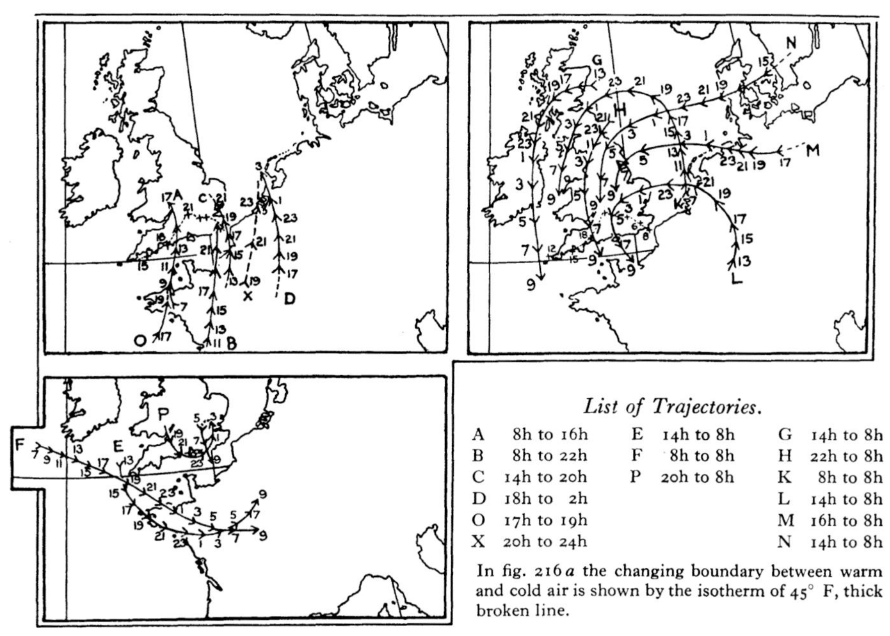

Fig. 216*b*. Trajectories between 8h Dec. 30 and 8h Dec. 31, 1900.

gestion which eventuated in the publication of a memoir on the *Life History of Surface Air-Currents*. The charts of the weather-map gave only the situations at 8 a.m., 2 p.m. and 6 p.m. and it became apparent that the seven observatories could not be expected to furnish all the details that were necessary. By that time portable barographs had become a common addition to the equipment of meteorological stations, anemometers had been erected in some important positions and a few thermographic records were also available. It was therefore decided to collect all the recorded information and prepare maps for successive hours during the passage of selected examples of cyclonic depressions. Those pictured in the *Life History* include two fast-travelling depressions, two slow travellers, two "circular storms," a V-shaped depression and a sudden decrease accompanying a veer of wind.

From the numerous maps thus obtained trajectories of air were drawn to show as far as possible the actual paths of the air during the experience of the depression and to these were added estimates of trajectories of air on selected occasions over the North Atlantic, some of which are shown on pp. 244 and 246 in order to obtain some idea of the flow of air over the ocean.

As an example of this mode of analysis of a cyclonic depression we give in fig. 216*a* the synchronous charts for 8h, 14h and 18h of December 30 and for 8h of December 31, 1900, on which the isotherm of 45° F shows by pecked lines the separation between warm and comparatively cold air. At 8h on the 30th there is a closed isobar 996 mb embracing Northern Ireland and South West Scotland, defining the central region of a depression. Over the mouth of the English Channel there is a slight opening-out of the isobars

THE TRAVELLING CYCLONE—WEATHER IN RELATION TO THE CENTRE

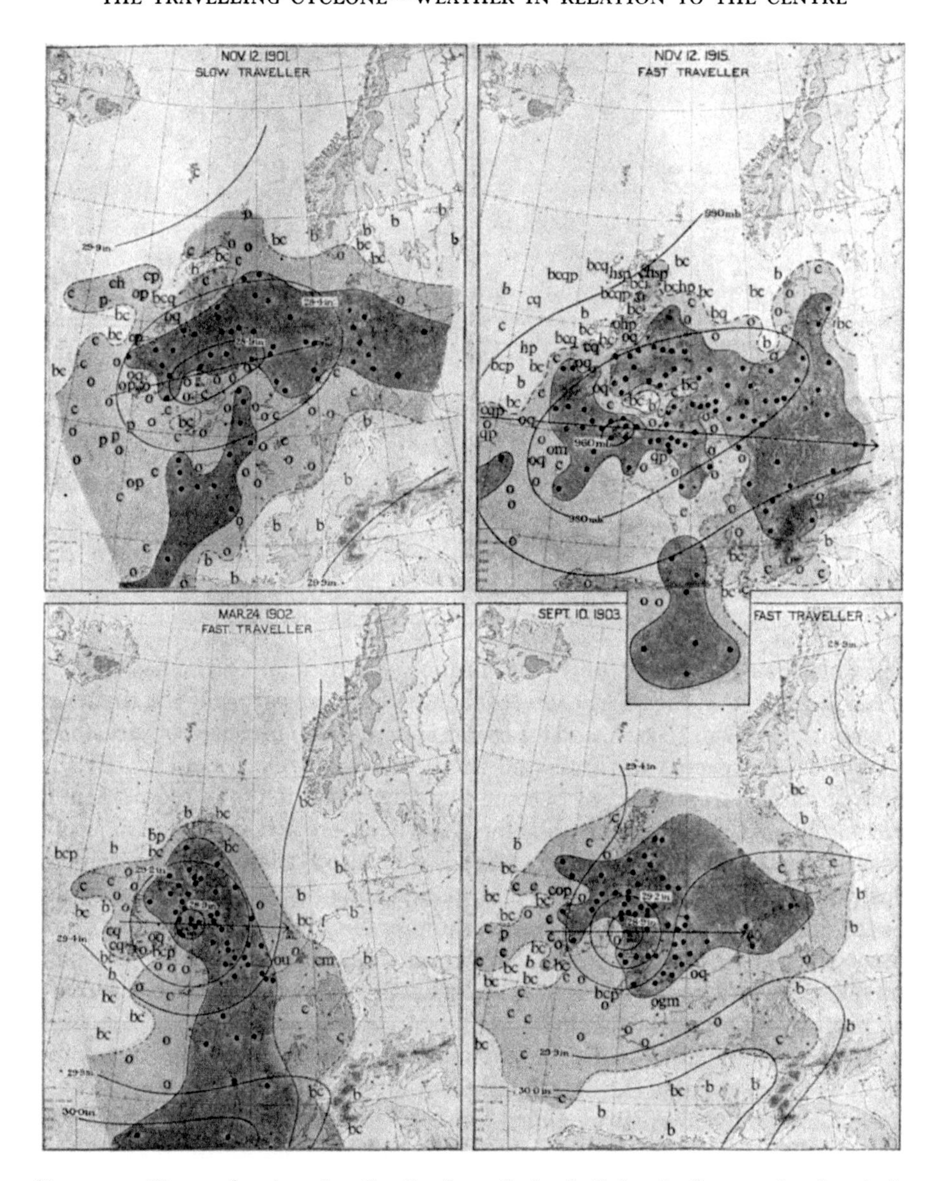

Fig. 217. Charts showing the distribution of cloud, light shading, and rain, dark shading, with reference to the centres in selected travelling cyclones. Scale 1 cm to 400 km.

Nov. 12, 1901. Slow traveller 700 km in 24 hrs, 8 m/s, Strongest winds 22 m/s
Nov. 12, 1915. Fast traveller 650 km in 13 hrs, 14 m/s, Strongest winds 26 m/s
Mar. 24, 1902. Fast traveller 550 km in 14 hrs, 11 m/s, Strongest winds 13–20 m/s
Sept. 10, 1903. Fast traveller 650 km in 14 hrs, 13 m/s, Strongest winds 20–32 m/s

The reader is asked to draw the conclusion that if the four charts were combined into a single chart for a typical cyclone the most noteworthy characteristics of each cyclone would be lost.

and by 14h that has developed into an isobar of 992 mb embracing Cornwall which in its turn becomes the centre of a notable depression advancing up the Channel as far as Dover by 8h on the 31st. Intermediate maps which confirm the development are shown in the original publication.

For the appropriation of the dignity of an important centre by the slight disturbance of the isobars of 8h on the 30th the courses of the several air-currents are shown in fig. 216*b*. The first tablet shows the behaviour of winds from the south, trajectories A to D, O and X; the second the origin of the winds which were strong from the North on the 31st, trajectories G to N, and the third E, F, P the winds which, coming from the West, had taken the place of the Southerly winds and had flanked the centre.

The process of formation of new centres by the interaction of winds from different quarters was represented by J. Bjerknes as an ordinary mode of progression of centres which had up to that time been supposed to move in continuous lines.

Another method of compiling information about individual cyclones which are moving, is to refer the weather conditions especially the rainfall to the centre instead of the fixed frame of the map. We give in fig. 217 a reproduction of four diagrams of this kind which are shown in the *Weather-map*; three of them are taken from *The Life History of Surface Air-Currents*, the other was contributed by Lempfert and Geddes. The noteworthy point about the diagrams is the distribution of rain which is notably different in different cases. These are examples of the intensive investigation of the relation of weather to isobars which was the guiding principle of the *Life History*; and, as a result, a new view of the general character of a cyclonic depression as shown on British weather-maps, with its attendant rainfall, is disclosed in a diagram taken from the author's work on *Forecasting Weather*.

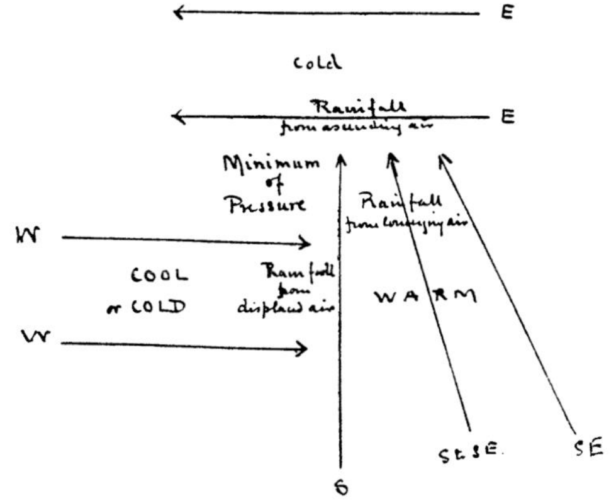

Fig. 218. Original sketch for *Forecasting Weather* (1911); the distribution of wind, temperature and rain with reference to the centre of a cyclonic depression.

The diagram shows that there were discontinuities of wind along the line of the path and along the trough-line, and that each was responsible for a rainy area. A further investigation by J. Bjerknes showed that there is a discontinuity of temperature also and that the region between the two lines of discontinuity is composed of warmer or equatorial air nearly surrounded by colder or polar air; the boundary, between the two, forms what is called the polar front.

The subject has been developed by J. Bjerknes in a series of papers which have so far revolutionised the practice of forecasting that their author was invited to spend the autumn and winter of 1925–6 in London in order to give the Meteorological Office the benefit of his personal experience[1].

Another intensive investigation of the relation of weather to isobars has been carried out by the Office National Météorologique de France, in which the transition of cloud-forms is the special subject of investigation. It is treated in two brochures published by the Office National: "Les systèmes nuageux," by Ph. Schereschewsky and Ph. Wehrlé, Paris, 1923.

A summary of the results of these important investigations was given by E. Gold for the Catalogue of the Exhibit of the Royal Society of London at the British Empire Exhibition in 1924. We are indebted to the Royal Society for permission to reproduce it here.

WEATHER FORECASTING

By Lt.-Col. E. GOLD, D.S.O., F.R.S.

This note treats only of weather forecasts for relatively short periods, and leaves entirely aside forecasts of the character of seasons or years. Broadly speaking, there are three processes which a weather forecaster must take into account in his prediction.

(*a*) The travel of weather from one place or region to another.

(*b*) The development of weather in moving air.

(*c*) The development of weather at a place or in an area consequent upon orographical features, such as hills or coast-lines.

Hills may produce cloud and rain; they may get snow when lowlands get rain; they may be in fog when the lowlands are clear; they may get gales when the lowlands get only fresh winds. At the coast, temperature may be low in summer when it is very hot inland, or mild in winter when there is frost inland; there may be fog in summer when it is clear inland, or thunder-storms in winter when there are none inland. The coast-line is also a line of discontinuity so far as the resistance to atmospheric motion is concerned, and on that account it modifies the normal development of a cyclone.

The meteorologist at a central institution is better situated for dealing with the problems presented by (*a*) and (*b*) than a meteorologist at a local sub-office; but the latter has a very definite advantage in regard to (*c*) so far as the locality in which he is situated is concerned.

The idea underlying the method by which the problem of forecasting has been attacked is that the problem is one of mathematical physics, and that if we knew exactly what the physical condition of the atmosphere is at a given instant we could, theoretically at least, deduce what the conditions would be at subsequent times. Accordingly as much information about the weather existing at a given instant at different places over as wide an area as possible is collected rapidly at a central institute.

The actual problem cannot be solved in exact mathematical fashion; but its explicit formulation and the application in present-day forecasting of the general mathematical and physical principles involved in the solution of the idealised problem do mark a definite advance beyond the pure empiricism of the earlier forecasters.

The organisation in Europe by which the collection of information is made has grown rapidly since the War. In each country the observations from a selection of stations are issued by wireless telegraphy according to a pre-arranged time-table; these issues are received at the Air Ministry, and the observations are plotted on maps to enable the forecaster to get a rapid view of the existing situation. This is done three times daily (7 a.m., 1 p.m. and 6 p.m. Greenwich time), and a smaller collection of

[1] *Geophysical Memoirs*, No. 50, London, 1930.

information is also made in the night (1 a.m. Greenwich time). The information from each observing station includes pressure (barometer); temperature; humidity; direction and force of the wind; existing weather and general character of the weather since the preceding report; form, amount and approximate height of cloud; visibility or distance at which objects can be seen; and the manner in which the barometer has been changing in the preceding three hours.

Early Forecasting.

The dominant idea in British forecasting used to be the travel of weather, and efforts were mainly directed to discovering how the path of weather could be foretold. It was found from the earliest studies of synoptic weather charts, some 50 years ago, that the two outstanding features were cyclones, or regions of low barometer, and anticyclones, or regions of high barometer. Broadly, the first were associated with windy and wet weather, and the second with quiet dry weather. If the coming of windy and wet weather can be forecast correctly, then, *ipso facto*, quiet and dry weather can also be forecast. It seemed therefore fundamental to understand the distribution of wind and weather in the area of a cyclone, and this was formulated by Abercromby some 40 years ago in a generalised way. Upon this and the analysis by van Bebber of the paths of depressions at different seasons, forecasts were mainly based. A cyclone was indicated by readings of the barometer, its path and speed of translation were estimated, and the sequence of weather to be expected at any place could then be written down more or less accurately.

It was indeed recognised that deformation of the cyclone might occur, and that a bulge of the isobars, usually at the southern extremity, might have the features of wind and weather appropriate to a cyclonic entity without necessarily having the closed isobars, although in some cases it might have them, too. Such bulges came to be called secondaries. It was also recognised that cyclones differed considerably among themselves in regard to the actual distribution of weather, but there was no satisfactory physical explanation of the variations: they could only be forecast for a region to which the cyclone was moving, when their existence was indicated in the actual reports from the region where the cyclone was then situated.

The manner in which the barometer changed was always recognised to be of importance, but it was not until the introduction of relatively cheap self-recording aneroid barometers (barographs) that the significance of these changes was more fully appreciated. The first great step forward in modern forecasting since 1885 came in 1908 with the introduction into the reports from observing stations of a precise indication of the character and rate of change of the barometer. From that time the forecaster knew, not merely if the barometer was rising or falling, but also the rate at which it was rising or falling; he also knew the character of the curve of variation for the preceding three hours; he knew, for example, if the barometer had been steady and had commenced to fall, or if it had been falling continuously, or if it had been falling and had commenced to rise.

This information facilitated considerably the task of estimating the travel of weather, but it contributed also to the solution of problem (*b*) and made the prevision of the development of secondaries more certain; or, what is, perhaps, a more important point, it enabled longer warning of these developments to be given. Still it was mainly from the point of view of the "travel" of weather that this and other improvements were effected and only a little light was thrown on development.

Significance of Cloud Movements.

Another advance was made about the same time when it was proved that the direction and speed of the wind at heights between about 2000 ft. and 5000 ft. could be found very simply from the charts of isobars with practically the same degree of accuracy as that reached by wind-measuring instruments. It had been proved that if the motion of the air were "steady" this would be the case; but it was not until the present writer verified in 1907 the agreement of the hypothetical values with actual records at heights of 1500 ft. and 3000 ft., that it was imagined that the motion could be steady enough for the calculation to be of practical value.

This result has applications in the whole field of practical meteorology. Its bearing on forecasting may be seen by considering that clouds travel, and are most important features of weather, and that a means of knowing the direction and speed of the air at heights comparable with those of clouds, permits of greater precision in regard to the time at which weather changes will arrive at a place.

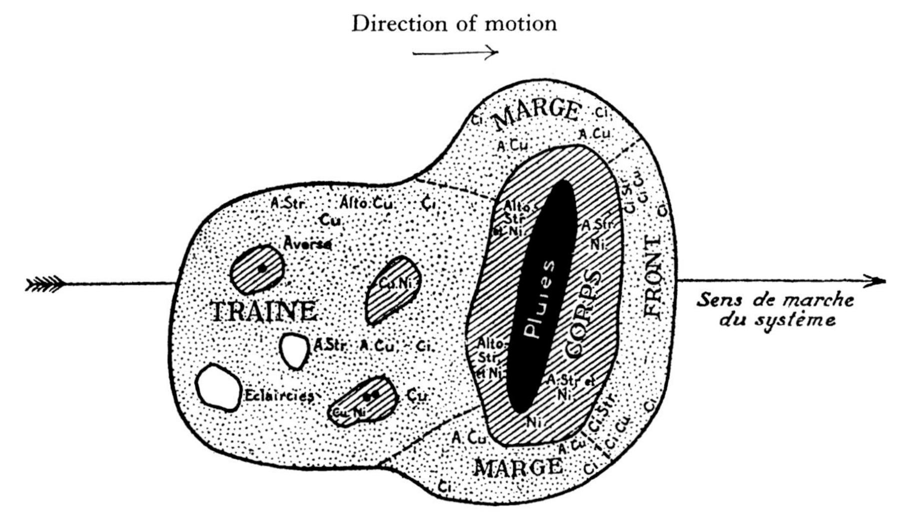

Cloud regions. After Fig. 1, Royal Society Handbook.

A further step forward was made when observations of temperature at different altitudes came to be available in time for use by the forecaster. Such observations are essential to a proper understanding of the general meteorological situation; they are especially valuable in connexion with the problem of "development" (*b*). Their usefulness in two special directions may be illustrated as follows:

(1) The temperature of the upper air practically sets an effective limit to the maximum temperature which can be reached at the ground in the daytime. If, for example, the temperature at 5000 ft. is 25° F, then the temperature at the surface cannot rise above 55° F, as a result of a day's sunshine; and if the air is damp as well as cold, the temperature at the surface will not rise above 45° F.

(2) A high temperature in the upper air limits the development of vertical motion and the consequent production of cloud. If, for example, the temperature at 10,000 ft. is 50° F, and decreases only slowly at higher levels, then there can be no development of cloud above the 10,000 ft. level unless the surface temperature rises above 85° F; if the temperature at 5000 ft. is also 50° F, then the surface temperature cannot rise above 80° F, and it is impossible to get the thick cumulo-nimbus clouds of thundery weather without the influx of entirely different air.

One of the most recent developments, a contribution from France, is essentially an extension of the principles enunciated by Abercromby. The atmosphere, at least in temperate latitudes, may be divided into regions of cloudless weather and regions of cloud. A region of cloud consists of a central core of thick cloud and wet weather; in passing from the core to the fine weather area, the cloud experienced depends upon the direction in which the transition is made.

In front of the "core," on the outer edge of the cloud region, are cirrus clouds (mares' tails); nearer the "core" these are replaced by cirro-stratus or a veil of whitish cloud covering the whole of the sky; just in front of the rain area is alto-stratus or a thick veil of grey cloud. On either side of the "core" are cirrus and cirro-cumulus (mackerel sky) or alto-cumulus. Behind the "core" are again occasionally cirrus clouds, but generally alto-cumulus and cumulo-nimbus (thunder cloud), giving showers or thunder-showers and bright intervals. The diagram (fig. 1) is a rough illustration of a cloud region as conceived in the French system. Actually there are, in individual cases, many divergences from this generalised representation.

The "core" is the dominating feature and, once it has been identified and its limits observed, forecasting becomes a simple matter if, as the French believe, the motion of the "core" is not attended with the uncertainty of the motion of the centre of a depression, but is practically identical with the speed and direction of the wind at about 10,000 ft. The "core" may be associated with a cyclonic depression, but it may also arise wherever there is a tendency for the formation of a secondary—i.e. wherever there is an area over which the change of the barometer relative to the change in surrounding regions is generally downwards.

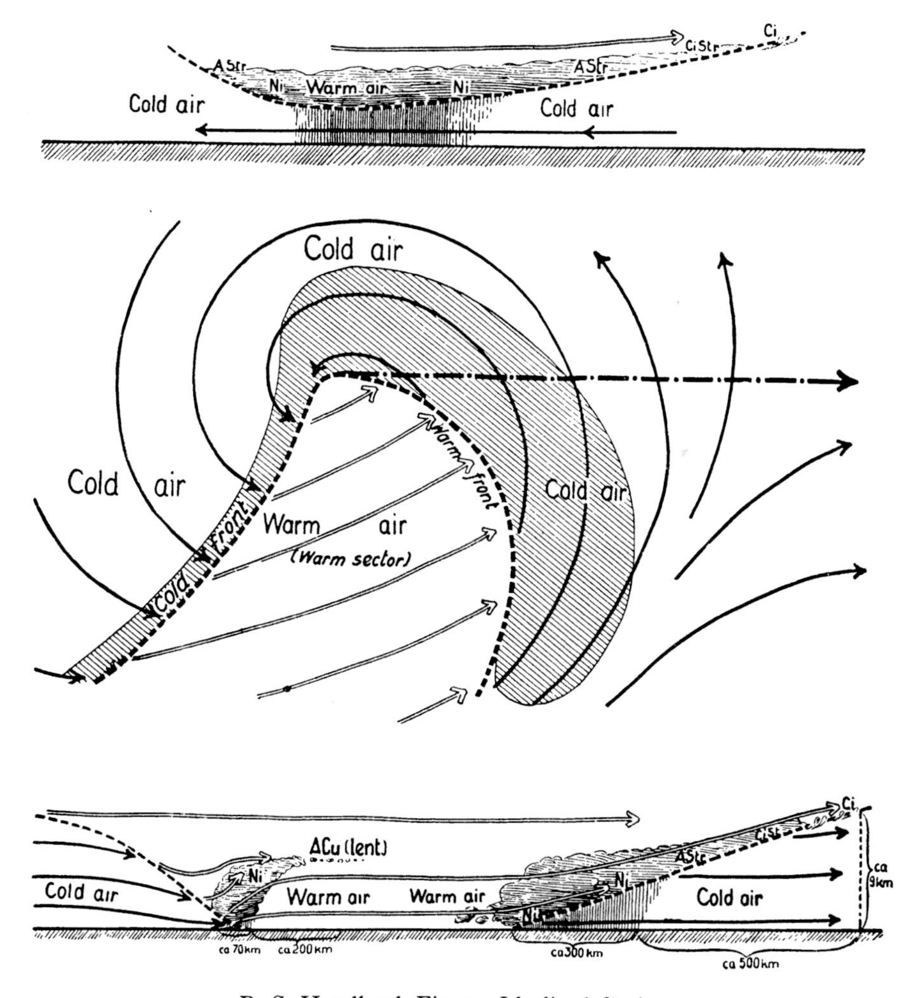

R. S. Handbook Fig. 2. Idealised Cyclone.

The Theory of the Polar Front.

Undoubtedly the greatest recent advance is that associated with the name of V. Bjerknes. Like a poet, he has turned to shape the ideas, latent in the minds of forecasters and physicists, or suggested tentatively, as in Sir Napier Shaw's diagram of the constituent parts of a cyclone, in which he shows a cold east current, a cool or cold west current, and a warm south current, as making up a cyclone.

Bjerknes' theory is briefly as follows:

The polar regions are covered by a mass of cold air, and the tropical regions by a mass of warm air. There is not a continuous change from the cold air to the warm air, but the two masses are separated by a surface of discontinuity—the *polar front*. Cyclones develop at this surface of discontinuity and constitute the mechanism by which interchange takes place between the cold air and the warm air. Each cyclone consists of two sectors, a warm sector of tropical air and a cold sector of polar air. The warm air pushes the cold air in front of it, and at the same time rises over the cold air; the cold air behind the warm sector pushes underneath the warm air, so that normally the warm sector is being reduced in area and is lifted upwards.

In the inner area of a cyclone, the cold air may, and often does, get right round, with the result that the warm sector is cut in two. The cyclone itself then begins to diminish in intensity. The result is a transposition of the polar front and a new cyclone or secondary usually forms with the remainder of the warm sector and the transposed polar air as its constituents.

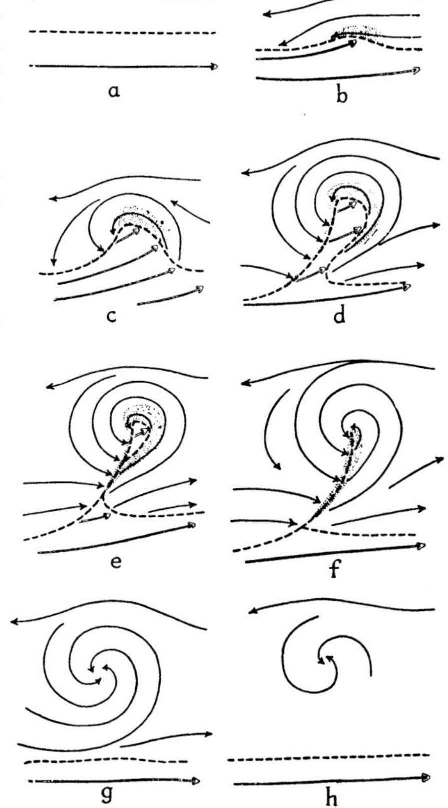

R. S. Handbook Fig. 3 (reduced facsimile). Life Cycle of a Cyclone.

The diagram (fig. 2)[1] illustrates the structure of a cyclone. The shaded part is the area of rain (or snow), and the sections at the top and bottom of the diagram are vertical sections through the cyclone to the left and right respectively of the path at the centre.

The normal birth, life and death of a cyclone are illustrated in fig. 3.

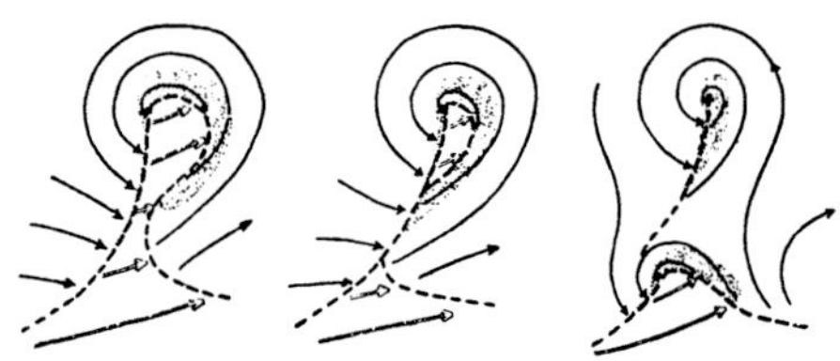
R. S. Handbook Fig. 4 (reduced facsimile). Formation of a secondary cyclone simultaneously with the occlusion of the "mother cyclone."

Fig. 3 (*a*) shows a cold easterly current and a warm westerly current in juxtaposition. The cold air begins to bulge southwards, and the warm air northwards in (*b*). In (*c*) a cyclone has formed, and (*d*), (*e*), (*f*) show successive stages, until the whole of the warm air has

[1] Figs. 2, 3 and 4 are reproduced from a paper by J. Bjerknes and H. Solberg, entitled, "Life Cycle of Cyclones and the Polar Front Theory of Atmospheric Circulation."

been lifted above the cold air. The cyclone gradually dissipates, as indicated in (*g*) and (*h*).

In fig. 4 the formation of a secondary is shown. The cold air in front of the line of discontinuity cannot move away quickly enough; it gets dammed up, and some of it begins to flow backwards, and so produces the conditions requisite for the development of cyclones.

In its broad features the theory of Bjerknes is in accordance with the facts of observation and the laws of physics. Further investigations of the upper air will elucidate some points of difficulty, and will possibly necessitate modification in details. The outstanding advantage of the theory is the light which it throws on "development." By its aid the forecaster can now see how an existing cyclone will develop; whether it will go on increasing in intensity, or if it will become stationary and fill up; he can also see where a new cyclone is likely to develop *before any actual development has commenced.* Our knowledge of process (*b*), development of weather, is now nearly, if not quite, as good as our knowledge of process (*a*), the travel of weather. Bjerknes has acknowledged his debt to previous investigators, notably Margules and Sir Napier Shaw; but the glory of the architect's work is increased rather than diminished by the excellence of the bricks which others had made ready for the building.

Addendum 1934.

In the brief description of the theory of the polar front on p. 392, the atmosphere is treated as being broadly divided into two types of air mass, one cold and the other warm: the cyclone developing at the surface of separation between two air masses of warm and cold type. In most cases of cyclones which reach the British Isles the cold air is by no means a homogeneous mass: that in front of the cyclone is usually different both in temperature and humidity from that in the rear of the cyclone. Thus when the cold front in the diagram on p. 391 catches up the warm front, producing thereby fig. 3 (*f*) on p. 392, the result is not uniformity of conditions at or near the surface; there still remains a discontinuity along the line where the two fronts have coalesced.

This line of coalescence of the two fronts is called the line of occlusion, or simply an occlusion, and in the neighbourhood of this line the weather partakes of the character of the weather of the fronts. The sky is usually covered with low cloud, and there is precipitation.

It is not infrequent for the cold front near the centre of the cyclone to catch up the warm front at a comparatively early stage in the life of the cyclone. The occlusion formed in this way starts originally from the centre of the cyclone, but as the cyclone progresses, the centre moves along the occlusion towards the end of the remaining warm sector. The occlusion then projects behind the centre of the cyclone and forms a secondary front: usually a secondary *cold* front. The special characteristic of this front, however, is that it is of limited length and runs for a comparatively short distance, only, from the centre of the cyclone, in the manner illustrated in the diagram. It is, however, of substantial practical importance, because it brings, behind the cold front and after the precipitation of the cold front itself has passed, a new region of low cloud and precipitation, instead of the fair, bright, weather which would otherwise be expected.

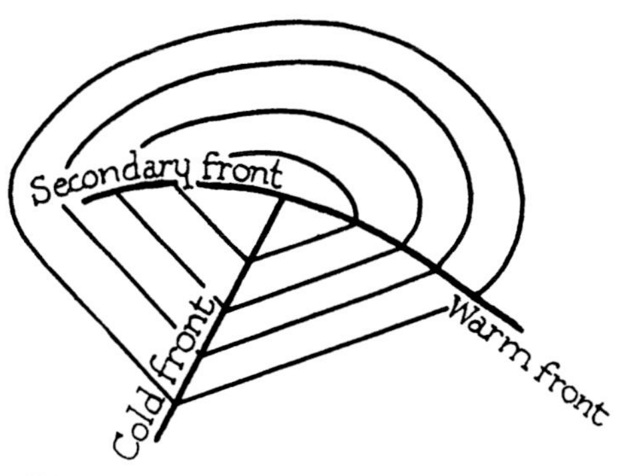

Diagram of cyclone with secondary front (a back-bent occlusion)

The process just described is one way in which a secondary cold front may be produced. There is another way in which such cold fronts are produced, viz. by descent or subsidence of the cold air in the neighbourhood of the surface of separation between cold air and warm air. This downward-sliding cold air, although colder near the surface than the air of the warm sector, is warmer than the horizontally moving cold air in the cold sector. There thus appears a first cold front dividing this descending air from the warm air, and a second cold front dividing the surface cold air from this descending cold air.

A development of this character will result in a diffusion of the phenomena of the cold front over a greater horizontal distance, and naturally, also, in some modification of the phenomena. This development appears most likely to arise with fast moving cold fronts. Sometimes it results not in two distinct and separate cold fronts but in a "diffuse" front, a zone of gradual change instead of a surface of discontinuity.

For a fuller description of the phenomena associated with fronts and the theory of their development, the reader may consult Geophysical Memoir, No. 50, 'Practical Examples of Polar Front Analysis' by Dr J. Bjerknes.

LOOKING BACK ON THE ANALYSIS

We may gather from Colonel Gold's account of recent progress in forecasting weather, following the lead of the Norwegian meteorologists, that the traditional idea of the cyclonic depression of middle latitudes as a travelling vortex with air flowing from all sides towards the centre and consequently disposed of by rising and causing rain, with the correlative idea of an anticyclone described by Francis Galton, who invented the name, as a region of descending cold air flowing outwards along the surface, is, for the time being at least, practically obsolete. Travelling vortices still exist perhaps as dust-devils, tornadoes, the historic whirlwinds, and perhaps also as tropical hurricanes for which indeed the name cyclone was originally coined; but the travelling depression with which we are familiar is now regarded as the region of interaction of fronts of various categories, named according to their probable region of origin and their past history. The finer analysis of their structure and inferences as to their behaviour confirms the effectiveness of the new practice.

Material for the inferences is obtained by the association of observations of the upper air by aeroplanes and pilot-balloons with those which are exhibited in the surface chart. We cannot here estimate their validity because they depend upon a recognition of the effect of physical processes of which the reader has not yet been reminded.

We have, besides, the note that most of the depressions which were regarded as coming to Europe from the Atlantic have suffered occlusion by the time they reach the land. By that we understand that the tropical air which at one time filled a sector of the depression, with one cold "front" actually in front of it and another on its flank, has been completely displaced by the latter so that at the surface, by the flanking movement, the cold front from the North West has now before it the other cold front over which the air of the warm sector has passed in the course of its persistent flow. That air is left to meet the fate in store for it in the region above ground. We have seen that the centre is observed to travel along the occlusion line, which at one time was the "steering line" and becomes a new element in the complication.

There is also the conception of subsidence which means the gradual descent of upper air along the sloping surface of an air-mass beneath it so that it reaches the recording instruments of the balloons or airplanes warmed and relatively dried by the gradual increase of its pressure. What exactly it is that subsides, whether there is a settlement of the superjacent ton-per-square-foot of the atmosphere above it or the slipping downward of a few thousand tons of intermediate air is not very clear.

Perusal of the recent literature of the subject brings to mind the necessity for care in the terminology to which we have called attention in the introduction to this volume. In modern meteorological literature it is of great importance to know whether "gradient" means slope along the horizontal or steps in the vertical; and the kinematic processes envisaged are so complicated that one would be glad to be quite sure about what the writer regards as "inverted" when he introduces an inversion. If it is the "temperature gradient" that is inverted the word "counterlapse" might help those of us who have to stop to think about the meanings of words.

And apart from terminology it is not easy to get definite ideas about the extent of the kinematics that is being invoked. We may read a great deal about fronts but find little or nothing about their backs or the extent of what are called air-masses. We may ask what is the real kinetic importance of any air-masses the fronts of which play their part on the weather-stage. We have estimated the energy of a square of surface-air one kilometre thick stretching across from one isobar to another on a map with isobars two millibars apart as meaning energy of 280 million kilowatt-hours. On January 25, 1935, the strip of kilometre air on the way from near Londonderry to London would mean about five thousand million, with momentum equivalent to more than 100 million 500-ton trains. So we should like to know how far horizontally the back of the air-mass is from its operating front, how high up in the atmosphere it extends and how it has obtained its immense store of energy. So far as we know it must rely upon gravity for its supply, or more precisely upon the disturbance of gravitational equilibrium caused by differences of its temperature and of that of the water-vapour which it carries.

These quantities of energy or momentum are seldom defined; authors mostly concentrate their attention on the operating front and we are left to form our own impression from descriptions which seem to regard air-masses as extending up through the atmosphere and moving over the face of the earth without any obvious guidance until they meet other air-masses in conflict with which travelling depressions are formed.

In the original edition of this volume the remarks of this ninth chapter were followed by a report of a lecture in Manchester on November 14, 1921, with the title of "New views about cyclones and anticyclones" and a series of 45 "articles" in syllogistic form as an indication of the physics and dynamics of the two volumes which were then to follow. These volumes are now

TWO MODELS OF CYCLONIC STRUCTURE

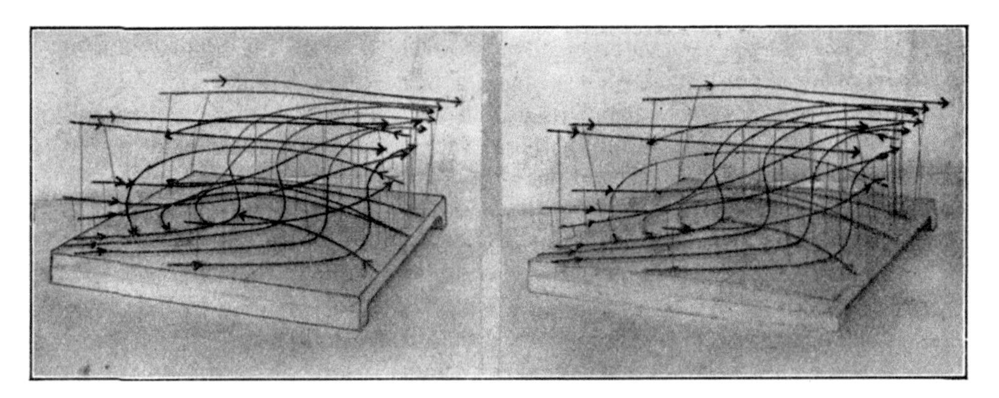

Fig. 219. Stereo-photograph of a wire model constructed by J. Bjerknes to represent the flow of air in successive layers of a cyclonic depression. In the highest layer the West current is represented as undisturbed: in the lower layers, warm currents from the West or South West are shown ascending over cold currents from the East, and cold currents are shown descending from the East, North East, North and North West to undercut the air of the warm sector of the cyclone. The model may be regarded as accounting for the discontinuity of fig. 218.

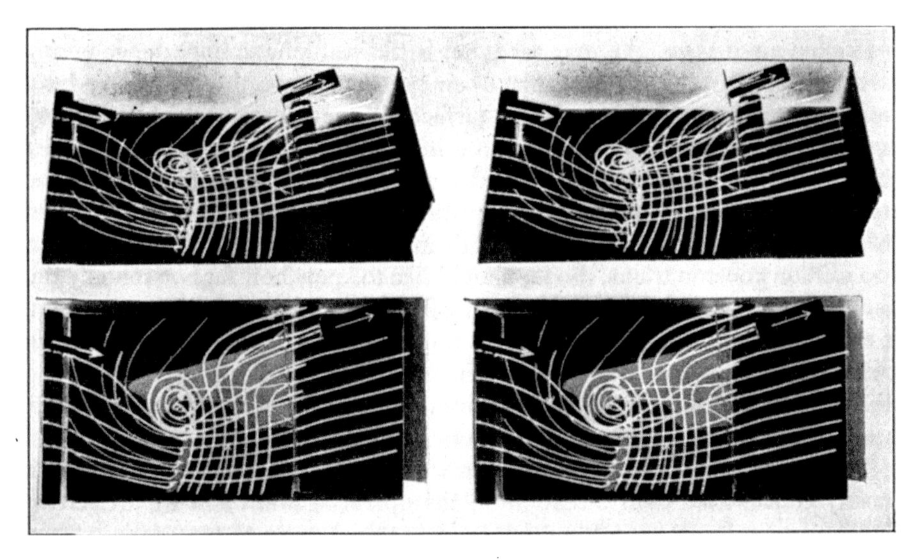

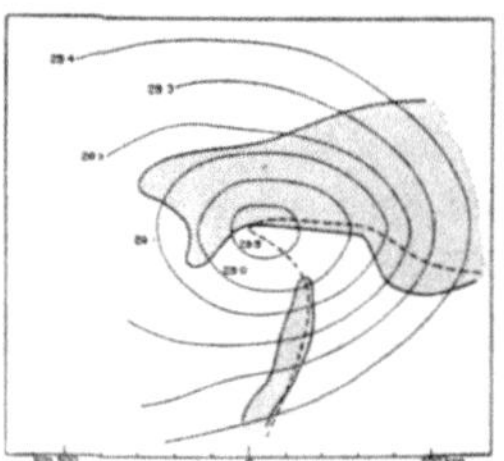

Fig. 221. Surface-plan of cyclone of 11–13 November, 1901.

Fig. 220. Photographs of a model with stream-lines of convection in a vortex superposed upon a uniform current of air at the level of the fourth kilometre, with retardation of the motion at the surface, "featuring" the cyclonic depression of fig. 221 (11–13 Nov. 1901).

The lower pair is taken from vertically above the model, the upper pair a front view from about 30°. The lighter area in the ground shows the distribution of rainfall with reference to the centre of low pressure at the surface, see fig. 217.

themselves available and the forty-five articles of Definitions, Axioms, Postulates and Propositions can be withdrawn for our private information. We may, however, consider the relation of the surface-conditions of a depression to the free air above it.

Many years ago the writer had the whole atmospheric process explained to him by an eminently successful school-master, and physical geographer, as the result of the obviously natural flow of air from high pressure to low pressure diverted to the right (in the Northern Hemisphere) from its primary objective like the trade-winds, by the rotation of the earth and so having a component of its motion along the isobars with the low pressure on its left. It was scarcely possible to persuade the exponent that the flow from high to low was really a minor consideration due to the interference of the surface-friction with the "natural" flow along isobars. We require, therefore, to take into consideration not only the motion at the surface but also the information about the pressure, temperature, humidity and motion of the upper air which is now available.

The relation of the surface to the upper air.

The height of cyclones or the behaviour of the air above them has been a subject of curiosity from the time of their invention; anybody may be interested in deciphering the way in which the layers of upper air act as scavengers for the layers beneath when they are in a convective humour. F. H. Bigelow devoted a good deal of attention to it in his Report on the International Cloud Observations in 1900 and in 1903 Teisserenc de Bort pointed out that when the temperature-gradient is opposite to the pressure-gradient, as it may be at the surface on the Northern side of depressions, the gradient for Easterly winds will diminish with height and the circulation may be lost within a small range above the surface. If, however, the gradient of temperature is everywhere in the same sense as that of pressure, increase of height does not annihilate the circulation.

Sir John Eliot gave an estimate of the height of a cyclone over the Bay of Bengal as about a mile, because the cyclone collapsed under the influence of a range of mountains of that height. That reasoning, however, does not seem to be conclusive because the dynamics of the crossing of mountains by a cyclone, whatever it may be, is difficult. The travelling meteor has not only to suffer the loss of its heaviest layer but has also to remake its lower limb when it has passed over the hills. T. Kobayashi[1] has examined very closely the crossing of a range of hills in Korea and the crossing of the Rocky Mountains has been studied to some extent.

The height of the typhoons of the China Seas is the subject of a paper by Rev. M. Selga[2] which suggests that those meteors may extend upwards to six or eight kilometres.

[1] 'On a cyclone which crossed the Corean peninsula and the variation of its polar front,' *Q. J. Roy. Meteor. Soc.*, vol. XLVIII, 1922, p. 169.

[2] *Proc. Third Pan. Pacific Science Congress*, Tokyo, 1926. The process of capping a vortex by the stratosphere is suggested in 'Illusions of the Upper Air,' *R. I. Proc.*, vol. XXI, Part III, London, 1928.

In the early days of the study of fronts J. Bjerknes constructed a wire model of a cyclonic depression which is represented stereographically in fig. 219 from which we might infer an upper wind from the West over all, to be reinforced in what used to be called the front of the depression by the air of the warm sector flowing along the surface as a South or South West wind and then rising over an air-current which flows from the east and curls round the centre to join the polar air coming from North or North West and help to produce the phenomena of the trough in an attack on the western side of the warm sector.

The model suggests a notable feature of the new analysis, namely that the centre of the depression has lost a good deal of the importance which it carried for fifty years; yet dynamically, or at least kinematically, it needs consideration because it is apparently a point of junction of all three fronts of a complete depression, the warm front of the warm sector, the cold front in front of that and the cold front on its flank all join apparently at the centre.

At such a point the process of replacement of warm air by cold air must be kinematically complicated. We may perhaps represent the conditions by that of the model which is shown in stereo-photograph in fig. 220 as including a spiral ascent of the warm air round the vertical drawn from the centre.

There is evidence in favour of this view in certain trajectories shown in the *Life History* as "looped," that is to say the air appears to be travelling *round the centre*, approaching from the South in front of the centre and receding from the North in the rear and crossing its own path by turning Eastward South of the centre. An example is shown in trajectory P of fig. 216*b*, difficult to see on the reduced scale but notable for the dotted connexion of the two sides of the loop beyond the centre.

A notable feature of these looped trajectories is that the receding air is cold compared with the advancing air. It suggests that the advancing warm air has disappeared and cold air has taken its place to maintain the apparent regularity of the flow.

So it may be possible that an area of low pressure at the surface with all its irregularities, aided by vigorous convection and the accompanying eviction, may evoke the development of a travelling vortex in the upper layers of the atmosphere, and there are some indications of its probability. There is no special prevalence of winds from any particular quarter in the upper air of middle latitudes at eight kilometres, and from the high coefficients of correlation shown on p. 334 for the layers from 4 km to 8 km the suggestion has been made that at those levels isobars are also generally isotherms and there may be symmetry about an axis.

Why, so far, such inferences have been disregarded is not generally clear. A relation with a coefficient exceeding ·8 can hardly be casual; it looks more naturally causal.

Anticyclones.

In modern weather-study attention is very much concentrated upon our depressions; very little light is thrown upon the formation of anticyclones. They seem to be regarded as the dumping-ground for air not required for the more interesting features of our weather such as are exhibited by the areas of loss of pressure where rain may occur with its antecedents and its consequences. And yet for the meteorological beginner the problem of the anticyclone must appear more difficult than that of the cyclone; he may easily imagine with our school-master air flowing from high pressure to low. We have given some estimates in the heading of chapter VIII of the rate of settlement which is necessary to feed the out-flowing stream at the bottom of some anticyclones. The basis of the estimates is explained in *The Air and its Ways*, p. 127. It is, of course, a very speculative calculation, but it is sufficient to account for the distribution of temperature which is usually noted over anticyclonic areas, and to suggest that if the pressure of an anticyclone is maintained and increased there must be a regular contribution to its mass at some level in the upper air. But to get it back again from the low to form the high "Hoc opus, hic labor est." How can it come about? So far as we can tell unless anticyclonic layers are always surmounted by cyclonic layers so that the gradient may be reversed in the upper layers the final stages of replacement and supply must be by motion against the gradient. The looped trajectories of the *Life History* certainly pass through a minimum of pressure and show higher pressures in the subsequent steps of their path; but whether the increase of the gravitational energy of the pressure is the equivalent of the arrest of their motion is not explored.

It seems possible that we owe the solution of that difficulty to the simple effect of the rotation of the earth if we are permitted to subject our school-master's theory to an "inversion" and claim that, instead of the winds being the result of the distribution of pressure, that distribution itself is created and controlled by variations in the distribution and strength of the flow of air which at the surface we call winds and regard as horizontal.

The flow is different at different levels and for each level the rotation of the earth implies a displacement to the left or right unless the proper relation between pressure-gradient and velocity is maintained. Wherever the velocity is too great for the gradient the drift will be to the right, where it is too small, to the left[1].

It is rather disappointing that no one has yet undertaken a systematic investigation of the upper air of a permanent anticyclone such as that of the North Atlantic Ocean. Investigations over the sea have been devoted in preference to the trade-wind.

For the investigation of anticyclones attention has been directed mainly to the more transitory specimens that are to be found alternating with, replacing

[1] "St Martin's summer in England in 1931," *Beitr. Phys. fr. Atmosph.* (Bjerknes-Festschrift) XIX, 1932, p. 113.

or surrounding, the cyclonic depressions of temperate latitudes. From the investigation of these cases, conclusions have been drawn as to the relatively higher temperature in the troposphere of the anticyclone compared with a depression and lower temperature in the stratosphere.

A more elaborate investigation by Hanzlik published in the Denkschriften of the Vienna Academy, 1908, reveals a difference between a stationary anticyclone, which has higher temperature than its environment, and a travelling anticyclone which is cold so long as it is moving. This difference may perhaps justify us in regarding a travelling anticyclone as one which is always in process of formation. Years ago when investigating the life-history of surface air-currents we were unable to make any satisfactory representation of the trajectories of air in the case of the travelling anticyclone.

Here we are dealing simply with what has been observed in recognised examples of the cyclone or anticyclone. The dynamics of their behaviour belongs to vol. IV and is the subject of a work by Prof. Brunt.

The point which impresses us most strongly is the need for the investigation of the life-history of the upper levels of areas of high pressure and of the debatable regions between high pressures and low pressures which we call cols.

Concerning cols.

The air-mass which is taking part in the maintenance or alteration of any barometric distribution must perhaps be regarded as extending backward until we find its back generally leaning against a col, the peculiar region where two "highs" and two "lows" face each other with an intermediate region that is suggestive of the geographer's col, a depression in a mountain ridge between two valleys. If we regard the energy which is to do the work we are contemplating in our own neighbourhood as that which is possessed by the air between the "front" and the col behind it, for all the levels up to the stratosphere, we have a vast store of energy to deal with, and that being provided, we may leave the rotation of the earth to adjust the pressure-distribution to the velocity at every level and so give us the picture of the motion of a cyclonic or anticyclonic centre carried along by the wind when it is really the motion of the air which is developing those entities on its flanks as it travels.

There is still room for further study of the facts. For some reason or other the col has escaped the detailed study that is necessary for the appreciation of its influence; but now that we recognise the part that a col takes as backing the front, an analysis of its features and particularly the distribution of winds in the air above might be helpful in the study of the anticyclone and lead us to the ultimate analysis of the cyclone.

CHAPTER X

NOTES AND SUPPLEMENTARY TABLES

The end must justify the means. M. PRIOR.

Note 1, Chap. 1, p. 6. THE SOLAR CONSTANT

RECENT observations lead to a value of 1·94 gramme calories per square centimetre per minute as the "normal" value of the solar constant, the equivalent in kilowatts per square dekametre is 135·2 so that the value of 135 kw/(10 m)^2 which has been used throughout the volume holds good within a quarter kilowatt.

Observations of the solar constant are now made regularly at three stations under the direction of the Astrophysical Observatory of the Smithsonian Institution: Mt Montesum a (22° 40′ S, 68° 56′ W, 2711 m) in N. Chile; Table Mt., California (34° 22′ N, 117° 41′ W, 2286 m), to which the station previously at Mt Harqua Hala was transferred in 1925, and Mt Brukkaros in S Africa (25° 52′ S, 17° 48′ E, 1586 m) where a station was established in 1926. [A station is now in operation at Mt St Katherine, Sinai Peninsula.]

The series of observations at Montesuma extends back to 1920; definitive values of the "constant" from 1920 to 1930 have recently been published[1] and for 1921–30 are given in the following table in place of the provisional values of p. 6:

Year	Year	Jan.	Feb.	Mar.	Apr.	May	June	July	Aug.	Sept.	Oct.	Nov.	Dec.
		Thousandths of gramme calories per square centimetre per minute											
1921	1900+52	55	56	49	44	43	39	56	44	69	62	51	53
1922	27	48	43	38	31	25	14	12	18	24	27	29	15
1923	37	46	30	32	32	36	28	36	41	44	40	41	33
1924	46	42	39	45	46	48	55	46	40	46	49	48	42
1925	46	43	43	39	47	50	45	51	45	50	46	46	45
1926	38	41	38	39	34	39	44	44	44	42	34	29	32
1927	43	38	43	42	44	45	46	45	41	44	44	44	42
1928	38	40	43	46	42	47	48	42	37	27	30	29	26
1929	34	38	29	31	37	38	34	33	31	28	29	36	40
1930	42	36	38	39	40	44	43	47	45	37	40	44	47
Means	1900+40	43	40	40	40	41½	40	41	39	41	40	40	37½ g cal/1000
	130+ 5·2	5·3	5·2	5·2	5·1	5·3	5·1	5·2	5·1	5·2	5·2	5·1	5·0 kw/(10 m)^2

Monthly mean values of "preferred solar constants" obtained after weighting the values observed at the several stations are given in the same volume. The values agree well with those for Montesuma alone. The maximum difference is in January 1923 and amounts to ·021 g cal/cm^2 min. In most months the differences are less than ·005.

The maximum and minimum values of the daily values of "preferred solar constants" from 1926–30 are shown in the following table in thousandths of gramme-calories per square centimetre per minute:

	1926	1927	1928	1929	1930
Maximum	1954	1954	1960	1948	1955
Minimum	1918	1930	1905	1914	1926
Range	36	24	55	34	29

[1] *Annals of the Astrophysical Observatory of the Smithsonian Institution*, vol. V, 1932, p. 258.

Note 2, Chap. II, p. 11. THE DURATION OF SNOW-COVER

The distribution of snow as a covering of the surface for considerable periods of the year, sometimes including the whole year, depends ultimately upon the height above sea-level but also partly on latitude and partly on the distance from the coast-line, with the understanding that the snow extends much farther Southward along the Eastern shores of the Continents than along the Western shores. A typical contrast is that presented in lat. 50° N by British Columbia on the Western side and Labrador on the Eastern side of America, and between the British Isles on the West and Eastern Siberia on the East of the Eurasian continent.

The contrast is shown also in the run of the isothermal lines. On our chart for January (fig. 19) the isothermal line for a "normal" temperature of 30° F cuts the Western coast of America North of British Columbia in about lat. 55°; it bends Southward over the Continent reaching 40° N in the central regions, then Northward again and skirting the Southern shores of Lake Erie cuts the Eastern coast of the Continent near Boston in lat. 45°. From off the SE coast of Newfoundland it runs North-Eastward to Iceland in lat. 65°, then after passing Eastward across that island it follows the West and South coasts of Scandinavia and turns South-Eastward to the Northern shores of the Black Sea whence it runs more or less due Eastward along lat. 35° reaching the Pacific just North of Shanghai. With a discontinuity of about 10° of latitude at the coast, it crosses the Pacific North of the Japanese islands, cuts across the Alaska peninsula and follows the Alaskan coast-line to enter the American continent again in about lat. 55°.

The isotherm of 40° F follows an almost parallel course over N. America and the Atlantic keeping about five degrees of latitude South of the 30° isotherm. It then runs from North to South over the British Islands and France, turns through a right angle to run roughly from West to East along the Northern shores of the Mediterranean, bends Southwards over the Balkans and crossing the Black Sea and the Caspian it emerges into the Pacific in lat. 30°, passes between the North and South islands of Japan and enters the American continent just South of lat. 50°.

About the snowfall of the United States, De Courcy Ward[1] writes:

"From a practical point of view it may be said that snow does not occur in sufficient amount to lie unmelted on the ground south of San Francisco (38° N) on the lowlands of the Pacific coast, or south of Cape Hatteras (35°) on the Atlantic coast. This statement does not hold for inland districts or for elevated areas. The southern boundary of a regular winter snow-cover in ordinary winters may be put at about lats. 41–42° [the latitude of Madrid] in the eastern United States but occasional winters carry the snow-cover a good deal farther south....

"Most of the snow falls from December to March; but at the higher elevations and in the northern states it begins as early as October or even September and falls as late as April or even May."

[1] *M.W. Review*, vol. XLVII, Washington, 1919, p. 695 (contains maps of annual snowfall and days of snow-cover).

The Canadian Atlas contains a map of the annual snowfall in the inhabited regions of Canada. The snowfall is "very heavy" on the Western slopes of the Rocky Mountains, it reaches 60 inches in parts of Saskatchewan and Manitoba, rises to 90 inches over the Eastern half of Ontario and exceeds 120 inches over the North-Eastern parts of Quebec and the Northern shores of New Brunswick. Snow falls from about October to May.

In the North snow may occur in any month except July, but is infrequent in June and August[1]. The snowfall varies from about 45 to 70 inches, equivalent to 5 to 7 inches of rain.

Of the countries in **Europe** we have maps for Germany[2] and Scandinavia[3], and memoirs on snow-cover are given by E. Bénévent for the French Alps in *Mémorial* No. 14 de l'Office National Météorologique de France and for the Austrian Alps by V. Conrad and M. Winkler in *Gerlands Beiträge zur Geophysik*, Bd. XXXIV, 1931. Snow is included in some of the charts of Note 24. The "Atlas climatologique de l'Empire de Russie" gives a chart extending to about 45° E of the number of days of snow on the ground, based on five years' observations. The number of days is 20 over the Northern shores of the Black Sea in latitude 45°; it increases Northwards reaching more than 190 days North of the Arctic circle. At Leningrad the number of days is about 140 and slightly more at Moscow.

In the Balkan hills snow may fall at any time between November and April and occurs frequently from December to February; in some winters the hills may be snow-covered from December to March, and the Roumelian Plain in January and February. At sea-level and on the islands of the Aegean snow falls occasionally but rarely lasts long.

The snow-cover in **Eastern Siberia** has been investigated by W. B. Shostakovitch[4] who gives a map of the average thickness of snow-cover in February and figures for the average thickness at a large number of stations for the months from September to June.

"The period during which the snow-sheet covers the ground lasts about 25–70 days in the Sea-side district (Primorskaya Oblast) and attains 250–259 days in the extreme N. regions. Almost all E. Siberia lies under an uninterrupted snow-cover during more than 150 days in the year, the duration of the snow-cover gradually increasing with the latitude."

Elsewhere in Asia the complication due to elevation is distracting.

For **Japan** T. Okada[5] gives maps of the date of first and last snowfall and

[1] A. J. Connor, 'The Temperature and Precipitation of N. Canada,' *Canada Year Book*, 1930.

[2] E. Hebner, 'Die Dauer der Schneedecke in Deutschland,' Stuttgart, 1928 (containing a map of the number of days of snow-cover 1900–14).

[3] 'La durée de la couche de neige sur la plaine en Suède,' *Geog. Ann.*, Bd. X, 1928, pp. 178–80 (containing a map of the mean duration of snow-cover 1910/11–1919/20), the number runs from a minimum of 34 days in the South to 254 in the North; 'Nedbøriagttagelser i Norge... Middelvaerdier Maksima og Minima,' *Tilleggsh. til Årb.*, vol. XXV, 1919 (containing seasonal maps of snow-depth and number of days of snow-cover and of snowfall). Particulars of the snow-cover in Scandinavia, Finland and the East Baltic region are also given by A. Peppler and F. Hummel in *Petermanns Mitteilungen*, 1934, Hefte 3 and 4.

[4] *Recueil Géophy.*, vol. IV, fasc. 3, Leningrad, 1926, pp. 111–23.

[5] 'The climate of Japan,' Tokyo, *Bull. Cent. Meteor. Obs.*, vol. 4, No. 2, 1931.

tables of the average depth of snow in each month and of the number of days of snowfall.

"Snow falls over the whole of Japan except in Formosa and the Ryûkyû islands and it remains unmelted in the provinces facing the Japan Sea, Tôhoku, Hokkaido and Sakhalin during the whole of winter and early spring."

Snow begins to fall in the extreme north in September and remains on the ground in those regions in October. In December the provinces facing the Japan Sea, N. Japan and Sakhalin become covered with snow, and N. Korea is also snow-covered. The depth of snow increases in January and in some places also in February, it begins to decrease in March, and melts away by May except in elevated regions.

For the Southern Hemisphere our information is even more fragmentary.

South Africa. "Falls of snow are comparatively infrequent and are usually confined to the mountainous districts. Snow is rarely met with on the coastal flats...on Table Mountain on only 7 occasions in 57 years; on the High Veld (Transvaal) falls were recorded in only 11 out of 57 years."...Snow "is much more frequent on the mountains in the east of the Cape Province where it may be seen at least 3 or 4 times each winter." (*Official Year-Book of the Union of S. Africa.*)

Australia. "Light snow has been known to fall occasionally so far north as 31° S and from the western to the eastern shores of the continent....During the winter for several months snow covers the ground to a great extent on the Australian Alps where also the temperature falls below zero Fahrenheit during the night. In the ravines around Kosciusko and similar localities the snow never entirely disappears." (*Official Year-Book of the Commonwealth of Australia.*)

New Zealand. "Snow is rare at sea-level, especially in the North Island. In the interior and at high altitudes it occurs more frequently. On the summits of the ranges in the whole length of the South Island and on the highest peaks in the North Island snow falls on the average on over 30 days per annum....

"Data regarding snow lying are scanty. In the North Island any snow falling on the low levels almost invariably melts as it falls, but on the high plateaux it may lie, especially in the hollows, for from one to three weeks during the year. In the South Island it practically never lies at low levels on the north or west coasts, but on the east coast does so on a few days in some years. At altitudes between 500 ft. and 1000 ft. in the interior of the South Island the average number of days appears to be between 7 and 14." (*The New Zealand Official Year-Book.*)

South America. Snowfall extends on the eastern side of the Cordilleras to very low latitudes. At Ouro Preto, in only 20½° S, snow certainly falls at a height of 1000 m. On the highlands of the Brazilian provinces Parana, Santa Catharina and Rio Grande do Sul snow frequently falls at heights above 800–400 m, also at Curityba (25·4° S, 900 m) where it even remains lying for part of the day. At Palmeira (27·9° S, 580 m) snow fell in August 1879 to a depth of 5–6 cm and remained for several days; at the same time at Passo Fundo (100 km to the SE) the snow lay 10 cm deep and at Vaccaria 80 cm. The great snowfall at Lages on 26–31 July, 1858, cost the lives of more than 30,000 head of cattle and lay on the surface in many places for 14 days. Farther to the south in the German colonies of Sao Leopoldo and Santa Cruz snow falls sometimes even at 100 m above sea-level. In the interior of the Argentine Republic it snows at Cordoba though very seldom.

On the west coast of S. America the snowfall begins about the latitude of Valdivia, 40° S. The snow does not lie there (in 25 years June had 2 snow days, July 4, August 1) but only on the plateaux of the interior. In Ancud (42° S) snow is said to be unknown probably because it is more exposed to the sea-air than Valdivia. Farther to the north in Santiago de Chile (33·4° S, 570 m) on the other hand snow falls occasionally inland. (Translated from J. Hann, *Handbuch der Klimatologie*, Bd. III, p. 436.)

Number of snow-days at stations in South America (Köppen-Geiger, *Handbuch der Klimatologie*).

	Lat.	Long.	Ht.	No. of days		Lat.	Long.	Ht.	No. of days
Los Andes	32° 50′ S	70° 37′ W	816 m	1	Evangelistas	52° 24′ S	75° 6′ W	55 m	28
Santiago	33° 37′	70° 42′	520	1	Punta Arenas	53° 10′	70° 54′	28	30
El Teniente	34° 6′	70° 38′	2134	39	San Isidro	53° 47′	70° 58′	20	33
Lonquimay	38° 26′	71° 24′	970	38	Bahia Douglas	55° 9′	68° 8′	5	58
Punta Dungeness	52° 24′	68° 26′	5	6	C. Pembroke (Falkland Is.)	51° 41′	57° 42′	21	54

Note 3, Chap. II, p. 12. THE ANTARCTIC CONTINENT

The outline of the Antarctic is still only imperfectly known. A "map of the Antarctic compiled by the American Geographical Society of New York" was published in 1928 in four sheets. It incorporates the results of expeditions in recent years and includes all known physical features—inland ice, glaciers, shelf-ice, pack-ice, elevations and depths as well as lines of equal magnetic variation and, on the margin of the charts, wind-roses compiled from observations made by expeditions that wintered in the several localities.

Heights are given along the routes to the pole followed by Scott and Amundsen, and a contour of 2000 m is indicated running between the magnetic pole (lat. $72\frac{1}{2}°$ S, long. 158° E) and the coast at a distance of about 100 miles from the magnetic pole. The height of the geographical pole computed from Scott's values is 2765 m, the highest point reached by Scott is given as 3005 m and by Amundsen 3168 m.

A contour line of 2000 m on the Antarctic continent would be an instructive addition to the information shown on the map of p. 12. It would include a large part of the area of the Continent but the details available are not yet sufficient to justify a line on the map.

Note 4, Chap. II, p. 15. ARCTIC ICE

A useful introduction to the study of ice in Northern waters is given by E. H. Smith[1] in the report of the scientific results of the *Marion* expedition to Davis Strait and Baffin Bay (between the Western coast of Greenland and the Arctic Islands which, with interruptions, continue the line of the Labrador coast Northwards). From this memoir the greater part of the information here given has been compiled.

The ice is classified into **sea-ice** which is produced by the freezing of sea-water, and **land-ice** which appears in the form of icebergs delivered to the sea by the glaciers that run down to the coast-line. So far as composition is concerned there is not much difference between the two, ordinary sea-water contains about 3·4 per cent. of salt and has a freezing-point of 271·15. When it freezes the water crystallises out of the solution and leaves the salts behind; but the large masses of pure ice-crystals may carry with them some separate salt-crystals and some relics of the brine that has lost some of its water in the freezing. So sea-ice is mostly pure ice, its salinity varies with age.

[1] U.S. Treasury Dept. Coast Guard, Bulletin No. 19, *Scientific Results*, Part 3, Arctic Ice Washington, 1931.

The land-ice on the other hand, formed by the consolidation of snow in the glaciers, carries mechanically a good deal of imprisoned air and its density is in consequence not exactly defined.

Sea-ice includes the North polar ice-cap which occupies the central portion of the North Polar Basin. About two million square miles of the Arctic is permanently covered with sea-ice, and in addition to the polar cap we have to take account of what is known as **fast ice**, formed by the freezing of the coastal border. This freezing may begin about September, reach its maximum area in December and then the ice gradually thickens until May. The largest areas of fast ice are the Siberian shelf and the American Arctic archipelago. The 12-fathom contour line is approximately the outer limit to which the fast ice spreads along the open coast-line. In spring this fast ice and parts of the polar cap-ice begin to melt and drift down with the currents into the N. Atlantic Ocean. Such ice which has drifted from its original position is known as **pack-ice**, or by seamen frequently as **field-ice**. It is estimated that this annual discharge and the summer melting amounts to about 1·1 million sq miles of northern sea-ice, over $2\frac{3}{4}$ million sq km.

The outer margin of the polar-cap-ice, on the Siberian side at least, shows a definite anticyclonic movement;...its main escape towards the Atlantic is through the opening between Spitsbergen and Greenland. Ice streams...distinctly secondary in size emerge around the southern side of Spitsbergen, through the Ellesmere-Greenland Strait to Baffin Bay and by still more minor passages through the sounds of the American archipelago. Through the Bering Sea the Polar Sea has no outflow.

Land-ice supplies the icebergs which find their way on to the steamer routes of the Atlantic. The *Marion* report gives a chart of the position and drift-tracks of icebergs around the Grand Bank (1900–1930) and also a chart of exceptional drifts extending in some cases nearly to latitude 30°. The general drift in open water is with the water that carries them. It is only in the later stages of disintegration when the height is mainly in the air that the wind takes control.

Icebergs are calved from the glaciers on the coast; the calving goes on all the year round but frequently the icebergs are held in position by the fast ice and may be so held for a single season or for several years so that a large number of icebergs in a single year may be due to exceptional melting of fast ice which sets free an accumulation of bergs.

The travel of icebergs into Baffin Bay from the Eastern side of Greenland is represented in fig. 5. Some go North of Greenland to Ellesmere Land.

Greenland contains about 90 per cent. of the land-ice of the N. polar regions and is the main source of the icebergs of the N. Atlantic. No bergs are found in the Bering Sea and only small bergs off Spitsbergen. It is estimated that Greenland produces 10,000 to 15,000 sizable bergs a year. The icebergs of the Arctic are much more irregular and picturesque than the box-shaped bergs of the Antarctic—this is due to the fact that in the Antarctic the ice-sheet extends beyond the coast, whereas in Greenland it stops short of the coast.

The wastage for the coast of W. Greenland is about 7 to 10 cubic miles, whereas the ice melted annually in Baffin Bay and Davis Strait is estimated at 467 cubic miles, the ratio of icebergs to sea-ice is therefore insignificant.

Icebergs have been measured up to heights of about 450 ft, those from the E. coast of Greenland are apparently not so high as those from the W. coast. The average berg in the Davis Strait area is estimated at about 50 million cubic feet, they gradually diminish in size as they drift Southwards and an average berg on the Grand Bank is estimated at about 6 to 8 million cubic feet. The density of the ice varies considerably, but probably about one-eighth of the mass is above water and seven-eighths below water. The relation of height exposed to height submerged is less than this and varies with the shape of the berg. For a block-shaped berg with precipitous sides it may be 1 : 5 whereas for a berg in the last stages of disintegration it may be 1 : 1. This must be borne in mind when considering the effect of wind and current on the drift of bergs.

"A berg of average size, 70 to 90 ft in height, in mixed waters South of the Grand Bank will survive as a menace to navigation for 12 to 14 days during April, May and June, but will not survive longer than 10 to 12 days thereafter. An equal sized berg... within the confines of the Gulf Stream, 65° to 70° F, will survive approximately 7 days, with considerable variation."

A good deal of definite information has been accumulated about Arctic ice from the Ice-patrol and other sources. More recent charts of the limits of ice in the Western North Atlantic than those of figs. 4*a* and *b* are published in the *Marine Observer*.

While this note has been in hand sea-ice has been the subject of an Atlas with 40 illustrations by Alf Maurstad, *Geofysiske Publikasjoner*, vol. X, No. 11, Oslo, 1935; and two other publications have been noted, viz., J. Richter, "Die Eisverhältnisse des Weissen Meeres," *Annalen der Hydrographie*, vol. LXII, 1934, p. 89, and Petersen and Oellrich, "Die Eisverhältnisse an den deutschen Küsten, einschliessig Memel und Danzig nach 25 jährigen Beobachtungen vom Winter 1903–4 bis 1927–8," vol. LVIII, 1930, p. 25.

Seasonal variation.

C. E. P. Brooks and W. A. Quennell[1] have made out the seasonal variation of the drift of pack-ice past Iceland on a special scale, and Mecking gives the drift of pack-ice off Newfoundland in percentage of the annual total. The seasonal variation in the number of icebergs South of Newfoundland and South of the Grand Bank has also been summarised from records of the U.S. Hydrographic Office[1] and of the International Ice-patrol 1900 to 1926. We give the figures in the following table.

Pack-ice	Jan.	Feb.	Mar.	Apr.	May	June	July	Aug.	Sept.	Oct.	Nov.	Dec.
Off Iceland	4·0	4·3	8·2	7·5	7·3	7·3	3·8	1·0	0·0	0·2	0·1	0·2
Off Newfoundland %	9	37	18	13	14	5	2	1	0	0	0	0
Icebergs												
S of Newfoundland	3	10	36	83	130	68	25	13	9	4	3	2
S of Grand Bank	0	1	4	9	18	13	3	2	1	0	0	0

[1] Meteorological Office, Geophysical Memoirs. No. 41, London, 1928.

Variation from year to year.

The comparative amounts of ice from year to year from 1880 and the numbers of icebergs observed from 1900 are shown in the two tables which follow.

Pack-ice S of Newfoundland 1880–1927 (on an arbitrary scale, 0 to 20).

Year	0	1	2	3	4	5	6	7	8	9
188–	9	5	12	10	13	15	8	10	9	7
189–	17	6	8	9	12	6	7	12	10	11
190–	6	6	5	14	8	15	10	12	8	18
191–	6	9	18	11	14	10	6	5	7	7
192–	10	14	11	8	3	6	9	10		

No. of icebergs S of Newfoundland (48th parallel). E. H. Smith.

	0	1	2	3	4	5	6	7	8	9
190–	89	88	41	802	265	845	405	638	222	1024
191–	50	396	1019	550	731	468	54	38	199	317
192–	445	746	523	236	11	109	345	389	515	1351
193–	475									

A table of the state of ice in the Davis Strait on the scale 0–10 for the years 1820–1930 is given by C. I. H. Speerschneider, *Det Danske Met. Inst., Med. Nr.* 8, Copenhagen, 1931.

Ice in the Baltic is also a subject of continuous and careful observation. The ice-season in that region usually runs from January or February to April. There is a lag of about two months upon the annual rhythm of the duration of daylight, a result which is generally characteristic of the surface temperature of the sea. Each of the Baltic states publishes data; the Deutsche Seewarte issues a daily chart. In 1928–9 the issue began on 19 December 1928 and in 1929–30 on 3 Feb. 1930.

Bibliography. Particulars of the extent of ice year by year are published as follows:

ARCTIC: *Isforholdene i de Arktiske Have*, Det Danske Met. Inst. Aarbøger, Copenhagen (reprinted from *Nautisk-Meteorologisk Aarbog*).

BALTIC AND NORTH SEA: Hamburg, Deutsche Seewarte, Eisübersichtkarte; *Ann. Hydrog. Marit. Met.* 'Die Eisverhältnisse a. d. deutschen Küsten einschl. d. Danziger und Memeler Gebiets in Winter'; 'Die Eisverhältnisse des Winters...in den ausserdeutschen europäischen Gewässern.'

Tartu Ulikooli Eesti veekogude uurimise komisjoni väljaanne, "Sea-ice observations made in Esthonia."

Helsingfors, Merentutkimuslaitoksen Julkaisu,' Übersicht der Eisverhältnisse im Winter...an den Küsten Finnlands.'

'Is- og Besejlingsforholdene i de Danske Farvande' (The state of ice and the navigational conditions in the Danish waters). Ministeriet for Søfart og Fiskeri, København. (Contains a comparison of winters from 1906–7 onwards.)

NORTH ATLANTIC: 'International ice-observation and ice-patrol service in the N. Atlantic Ocean,' U.S. Treasury Dept. Coast Guard Bulletins.

'Ice in the Western North Atlantic,' *Marine Observer*.

Note 5, Chap. II, p. 25. VOLCANIC ERUPTIONS SINCE 1914

E. H. Vestine in an article on 'Noctilucent clouds[1]' gives the following list of "the years in which explosive volcanic eruptions threw an appreciable amount of volcanic material well into the atmosphere." The list is based on information contained in 'The Volcano Letter' for the years 1913–33:

1914	Sakurajima, Komagadake
1919	Komagadake
1924	Komagadake
1926	Tokachi, Santorin [? 1925]
1929	Asama (2 major explosions, one throwing out an estimated amount of 3·4 million m³ of dust)

1931	Gorely (Kamchatka)
Mar.	Kluchevskaya (120 million m³ of ash)
May	Aniakchak (Aleutian)
1932	
Jan.	Mt Cleveland (Aleutian)
Feb.	Shishaldin (Aleutian)
Jan.–Feb.	Kluchevskaya (Kamchatka)
April	Quizopú (Andes)

[1] *Journ. R. Ast. Soc.*, Canada, vol. XXVIII, 1934, pp. 310–11.

Additional particulars of volcanic eruptions can be obtained from 'Matériaux pour l'étude des calamités,' Société de Géographie, Genève. 'Bulletin volcanologique,' Section de Volcanologie de l'U.G.G.I., Naples, Nos. 3–4, 1925, contains a paper by H. Tanakadate on 'The volcanic activity in Japan during 1914–24'; Nos. 7–8 give an 'Enumeration of active volcanoes in the Netherlands East Indian Archipelago' with the dates of the last eruptions.

Note 6, Chap. II, p. 29. ATMOSPHERIC ELECTRICITY AND THUNDER-STORMS

To the information of pp. 29–31 about the frequency of thunder we add here a bibliography of contributions to the subject since the date of Dr Brooks' memoir in 1925; and to supplement the information about the diurnal variation of thunder which represented only Edinburgh and Batavia we add a line of step-diagrams numbered 1 to 9 (fig. 222) showing the diurnal variation in the frequency of thunder at six stations on land and three for the West Indian seas, because in chap. II the sea is represented only by the information from H.M.S. *Challenger*.

Fig. 222. Diurnal variation of thunder on land and sea.

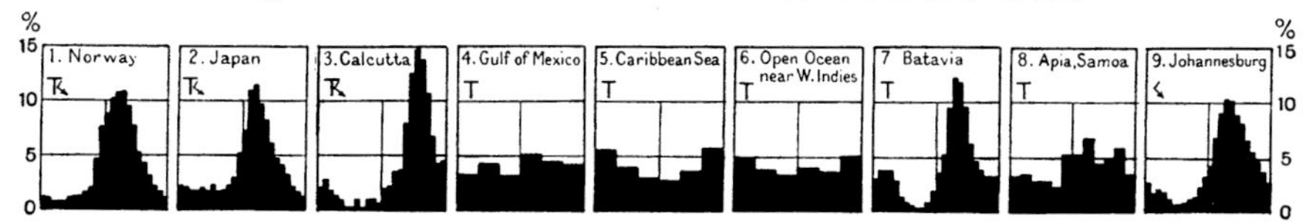

Each panel represents 24 hours; noon is marked by a vertical line.

1. Thunder-storms in the summer half year in Norway in the district Ostland which is the district in which thunder is most frequent. The mean number of thunder-storm-days a year in this district varies from 5·8 to 10·4 according to locality; almost all the thunder-storms (98 per cent.) occur between May and October. (G. Schou.)

2. Diurnal distribution of thunder-storms in Japan proper 1926–30. (H. Noto.)

3. Thunder-storms at Calcutta based on records of 511 storms between 1900 and 1926. Most of the storms occur between March and June with a maximum frequency of 6·1 in May. (V. V. Sohoni, *India Met. Dept., Scientific Notes*, vol. I, No. 3, Calcutta, 1928.)

4–6. Thunder in the open sea, (4) Gulf of Mexico, (5) the Caribbean Sea, (6) the open ocean 10°–30° N in the neighbourhood, from ships' observations. (W. Kloster, *Arch. d. Deutsch. Seewarte*, vol. XL, Heft 1, Hamburg, 1922.)

7. Thunder at Batavia representing 370 observations. Thunder is recorded on 133 days a year; on over 10 days a month from October to May and under 10 days a month from June to September. (C. Braak, *Natuur. Tijdschr. Ned. Ind.*, vol. LXXX, p. 275.)

8. Thunder at Apia, Samoa, Jan. 1891 to Dec. 1899 (with breaks amounting to 1½ years). The time of thunder was recorded to within ¼ hour on 235 occasions and less accurately on 41 occasions. (A. Thomson, *M.W. Rev.*, vol. LVIII, Washington, 1930, p. 327.)

9. Frequency of lightning as recorded by the lightning-recorder at the Union Observatory, Johannesburg during the five years July 1906 to June 1910. (*Official Year-Book of the Union of S. Africa*, No. 11, 1928–9, Pretoria.)

The diurnal variation of thunder is a thorny subject about which different inferences may be drawn according as we limit ourselves to personal observation or rely upon a theoretical relation with the sign of the potential gradient of atmospheric electricity or the theories of the origin of lightning[1]. In this volume we are concerned only with observed natural phenomena and for the time being take the observations at their face value. Vol. III should provide an opportunity for an endeavour to explore their theoretical implications.

A summary of data of Terrestrial Magnetism and Atmospheric Electricity is given by W. F. G. Swann in *International Critical Tables*, vol. VI, pp. 442–50, 1929: potential gradient, ionic content, conductivity of atmosphere, miscellaneous data, etc.

Bibliography. POLAR REGIONS: N. Kallio, 'Die Erstreckung des Gewitters nach dem Nord- und Südpol,' *Com. Physico-Math., Soc. Sci. Fenn.*, Helsingfors 2, 1924, No. 10, pp. 1–11.

EUROPE: G. Schou, 'Gewitter in Norwegen,' *Geofys. Publ.*, vol. IX, No. 7, Oslo, 1932; W. Köppen, 'Jahrliche Häufigkeit der Gewitter zwischen Felsengebirge und Ural,' *Ann. Hydr.*, 1927, p. 355; A. Eredia, 'Sui temporali in Italia,' *Elettrotecnica*, Milano, 19, 1932, No. 6; K. Keil, 'Die mittlere Zahl der Tage mit Gewittern auf verschiedenen deutschen Flugstrecken,' *Berlin, Erfahrb. d. Flugw.*, 7, 1932, No. 14, pp. 153–9; W. A. L. Marshall, 'The mean frequency of thunder over the British Isles and surrounding areas,' *Q.J. Roy. Meteor. Soc.*, vol. LX, 1934, p. 413.

ASIA: Poona, India Met. Dept. 'Frequency of thunderstorms in India,' *Sci. Notes*, 1, No. 5, Calcutta, 1929; H. Noto, 'Statistical investigations on thunderstorms in Japan,' *Tokyo, J. Astr. Geoph.*, vol. IX, 1932, pp. 207–43, vol. X, pp. 51–79.

N. AMERICA: W. H. Alexander, 'The distribution of thunderstorms in the United States, 1904–33,' *M. W. Rev. Washington*, 63, 1935, p. 157.

S AMERICA: W. Knoche, 'Über die Zahl der Gewitter in Chile,' *Santiago, Mitt. Deutsch. Chil. Bundes*, No. 3.

NEW ZEALAND: E. Kidson and A. Thomson, 'The occurrence of thunderstorms in New Zealand,' *Wellington N.Z., J. Sci. Tech.*, vol. XII, 1931, pp. 193–206.

Note 7, Chap. III, p. 35. THE UPPER ATMOSPHERE

Ozone. A summary of the present position of our knowledge with regard to ozone is given by G. M. B. Dobson and A. R. Meetham in the *Quarterly Journal of the Royal Meteorological Society* for 1934, p. 265.

We may perhaps note that ozone is a molecular modification of oxygen according to the chemical formula $3O_2$ (oxygen) $= 2O_3$ (ozone).

The main facts are stated to be as follows:

Distribution in the vertical. The ozone is distributed through all the atmosphere between the ground and 40 km or above; its average height in Switzerland is about 22 km above sea-level, the maximum amount per kilometre being found between 25 and 30 km. The proportion of ozone to air or oxygen increases steadily from the ground upwards and reaches a maximum at about 35 km, after which apparently it falls off again....Roughly half the total amount is above 20 km. When the total amount of ozone increases or decreases the change takes place chiefly between 5 km and 25 km.

Results at Tromsø (lat. 69½ N) in May and June 1934 show that there "the ozone is more concentrated in a region centred at a height of 21 km."

Seasonal variation. The amount of ozone shows a regular seasonal variation different in different latitudes. There is a maximum in local spring (March in the Northern Hemisphere and September in the Southern) and a minimum in local autumn. Near the equator there is little change throughout the year.

Geographical distribution. In spring there is a rapid increase from equator to pole, the amount over polar regions probably being more than double that at the equator. In autumn the polar regions have little more than the equator.

The Heaviside-Kennelly layer (p. 38). The present state of knowledge with regard to the highly ionised layers of the upper atmosphere was the subject

[1] F. J. W. Whipple, *Q.J. Roy. Meteor. Soc.*, vol. LV, 1929, p. 1; G. R. Wait, *Terr. Mag. and Atmos. Elec.*, vol. XXXIV, 1929, p. 237.

of discussion at the Royal Society in 1933. Two regions of high ionisation are now recognised, known as E and F. We quote the following:

"There can be little doubt that the maximum of Region E ionisation is reached at a height of about 100 km above the ground. The height of the corresponding Region F maximum is much more difficult to estimate, but a figure of 180 km is probably not greatly in error." (E. V. Appleton.)

"Throughout the discussion we have been dealing with equivalent heights... calculated on the assumption that the group travels with the velocity of light....It might well be, for instance, that the two main layers instead of being separated by a hundred kilometres or so, as is generally assumed *sub silentio*, are really quite close together and merely represent more or less typical changes in the ionic density-gradient." (F. A. Lindemann.)

Note 8, Chap. III, p. 37. Composition of the atmosphere

On p. 36 we left the discussion of the composition of the atmosphere above the level of 100 kilometres with the suggestion that the best guide to knowledge was the diagram of Chapman and Milne exhibited on p. 37 which represented the atmosphere above 50 kilometres as mainly nitrogen, changing to mainly helium at about 100 kilometres and above 100 kilometres exclusively so; the alternative for the upper atmosphere was mainly hydrogen or helium with the hypothetical geocoronium.

In the light of further investigation and the increase of knowledge of ozone the picture has lost its charm. Prof. Chapman has introduced the possibility of the dissociation by radiation of molecular oxygen and nitrogen into the atomic form. While allowing that recent flights by manned balloons into the stratosphere in U.S.S.R. show "that the composition of the air is the same at the lowest pressure reached (50 mm) as at the ground," he adds:

"The principles of my theory of the ozone-oxygen equilibrium still seem appropriate. They imply that ozone decreases and atomic oxygen increases in concentration above a certain height (perhaps 30 km in the light of the new observations), and that atomic oxygen and molecular nitrogen are the most important constituents above some level (whose precise value depends on various physical and chemical constants not yet independently determined). Nitrogen also will be dissociated by radiation of wave-length (<1370 Å) much shorter than that which corresponds to oxygen dissociation....The level at which this nitrogen dissociation occurs depends on the absorption-coefficient of the radiation concerned." (S. Chapman, *Q.J. Roy. Meteor. Soc.*, vol. LX, 1934, p. 127.)

Meanwhile from studies of the aurora Vegard has reached the following conclusions about the composition of the higher reaches of the atmosphere:

(*a*) The auroral spectra give no indication of the existence of an upper limiting atmospheric layer mainly consisting of the light gases hydrogen and helium.

(*b*) Nitrogen is a dominating component in the auroral region and exists largely in the form of positive ions.

(*c*) The auroral line is not emitted from a gas lighter than nitrogen.

(*d*) Spectroscopically the green auroral line can be followed to the very top of the auroral ray streamers to several hundred kilometres in height. This fact...shows that nitrogen is a predominant component of the atmosphere to its extreme limit towards empty space. (*Geofysiske Publikationer*, vol. IX, No. 11, Oslo 1932.)

He concludes that "the nitrogen which auroral investigations have shown to exist several hundred kilometres above the ground cannot be carried to these altitudes merely by the thermal motion, but is lifted towards larger altitudes by electric forces." That part of the subject however is not for this volume.

Spectrum analysis has also some new suggestions about the temperature at these high levels which include one of 1000° in the region of the aurora. Vegard is content with 225tt.

Note 9, Chap. III, p. 40. DUST-STORMS

Meteorology is quite familiar with the transport of vast quantities of dust by winds blowing across arid regions. A picture of a sand-storm in the Sudan or Iraq is an ordinary accompaniment of collections of cloud photographs and desert-sand has been regarded as a sort of snow that never melts.

The raising of small clouds of dust by wind in temperate regions where there is irregular rainfall is certainly not unknown but is not generally important. Interest in the subject has been stimulated by the raising of enormous clouds of dust over the North American continent during a notable drought in the spring of 1934 by a North-Westerly wind that carried vast quantities of valuable soil Eastward and South-Eastward from the Canadian prairie and neighbouring states. About the same time it is noted in the *Monthly Weather Review* that a well authenticated velocity of 231 miles per hour was recorded on an anemometer at the Observatory on Mount Washington in New Hampshire at a height of 6284 feet above sea-level—the highest wind-velocity ever recorded.

The London *Times* of 12 May, 1934, reports:

> "New York experienced to-day what for it was a rare meteorological phenomenon when a great cloud of dust enveloped the city in a grimy and eerie haze.
>
> "The dust cloud which originated in the parched plains of Western Canada and the neighbouring States of the Union reached the eastern seaboard early to-day, 11 May, and by noon had spread a grey pall over a wide belt of country extending from New England as far as Washington." A prominent geologist stated that so far as he knew it was the first time that prairie dust had been noticed in New York.

At 3 p.m. on May 11 the number of particles per cubic centimetre shown by the Owens dust-counter at Key Bridge, Georgetown[1], was more than 12,000. It exceeded the previous maximum by 72 per cent.; the average diameter in microns (·001 mm) of the particles, reduced to cubes, was 1·0+ and the visibility was $\frac{1}{8}$ to $\frac{1}{4}$ mile. At American University, Washington, at 3.22 p.m. the count was 11,445 per cc and the visibility $\frac{1}{4}$ mile; at 10 p.m. the count was 3150 and visibility (of lights) 10 miles.

The height of the cloud was estimated at 2 kilometres and the weight of dust carried as 101 "short tons" per square mile. The diminution of solar radiation amounted to about 75 per cent.

[1] *Monthly Weather Review*, vol. LXII, Washington, 1934, p. 156.

Severe dust-storms over Idaho, Washington, and Oregon on March 19, 1930, are described in the *Monthly Weather Review* of that month.

"The dust weighed 2·03 g/ft^2 which equals 62 tons/sq. mile. Assuming that the inland empire had 50,000 sq. miles affected by the storm and that the fall of dust at Cheney was an average of the whole region some 3 million tons of dust were deposited. It would take a freight train 500 miles long to move the dust that the storm moved in about 8 hours."

Similar storms occurred on 21–24 April, 1931, and on 13 November, 1933[1]. We note some examples in other parts of the world.

Africa. The transport of dust by the harmattan has already been referred to (p. 40). A note 'Sur l'extension de la zone dans laquelle souffle l'harmattan' was presented by H. Hubert to the meeting of the Meteorological Section of the U.G.G.I. at Prague in 1927 and appears in the Procès-Verbaux of that meeting. A recent instance of a protracted dust-storm 2 miles east of Tenerife is cited in the *Quarterly Journal of the Royal Meteorological Society*, vol. LX, 1934, p. 331. Dust of extreme fineness fell for more than a week, 18–24 February, 1934.

Europe and Asia. A storm of 26–27 April, 1928, which was most intense over the Ukraine, deposited dust in rain over Poland and Roumania and caused a darkening of the atmosphere over an area of about a million square kilometres. In South Poland the visibility was reduced to 300 m. Information was collected from nearly 1000 correspondents.

"It appears that from an area of 400,000 km^2 was raised about 15 million tons of the best cultivated soil. The author holds that within the Ukraine the quantity of dust deposited would total 9,600,000 tons." The finer dust was carried to the WNW. "Within Roumania fell about 100,000 tons. The quantity deposited in Poland the author estimates at 1,400,000 tons." (A. V. Voznesensky, *Trans. Bur. Agrometeor.*, vol. XXI, Leningrad, 1930.)

"The depth of accumulation of dust on the ground was from 3 mm in the SE to less than 1 mm in Central Poland. The total mass falling on Poland was computed to be 1,139,725 tons." (*Bull. Amer. Met. Soc.* 1928, p. 195, quoting B. Bonasewicz, *Bull. Meteor. Hydrog.*, Poland, April, 1928.)

Sand having its origin in the deserts of North Africa is often carried by the scirocco to the countries of Southern Europe. Notable occasions in the last ten years are: 26–7 March, 1929, 4–5 April, 1932, 24 April, 1926, 2 May, 1933, 23 July, 1931, 30–31 October, 1926, and 27 November, 1930. On these last two dates the dust was subjected to a special mineralogical analysis under the direction of E. Fontseré[2].

New Zealand is not exempt from these destructive storms. The supply of dust to those islands comes from Australia. A storm of 6–9 October, 1928, is described by P. Marshall and E. Kidson.

The largest particles were found to have a diameter of 0·08 mm, there were few particles larger than 0·04 mm....The indications are that the dust fell over an area

[1] *Monthly Weather Review*, Washington, 1931, p. 195; 1934, pp. 12–15; *Bull. Amer. Met. Soc.*, 1931, pp. 112, 127; *Geog. Rev.*, New York, 1935, p. 152.

[2] 'Sur les pluies de boue et de poussière,' *U.G.G.I. Stockholm, Association de Météorologie II, Annexes*, p. 103; F. Pardillo, 'Les pluges de pols,' *Notes d'Estudi*, No. 50, Barcelona, 1932. On 27 November the quantity of dust in Catalonia amounted to 32 tons/km^2.

of at least 10,000 square miles; consequently the total deposit must have amounted to something approximating 100,000 tons. (*N.Z. Journ. Sci. Tech.*, vol. x, No. 5, pp. 291–9, *see also* vol. xi, No. 6, 1930, pp. 417–18.)

Note 10, Chap. III, p. 44. Supersaturated air in clouds

At the meeting of the British Association at Leicester in 1933 L. H. G. Dines submitted a paper giving a number of examples of the record on the hair-hygrometer used with his sounding-balloons which showed, by extrapolation of the reading of the scale, relative humidities exceeding 100 per cent. of saturation by as much on occasions as 20 per cent.; and from the records he reached the conclusion that the air which carries the water-drops is itself "super-saturated," that is to say it contains more water in the form of vapour than that which corresponds with the saturation-pressure of water-vapour at the temperature of the air. He supported the conclusion by a number of references to the literature of the subject. Jaumotte devoted three pages of *Ciel et Terre*, 1925, to a case observed in Belgium.

The subject is an extraordinarily complicated one[1]. There is the difference of the vapour-pressure at saturation over water and over ice and the curious behaviour of water-drops in an atmosphere below the freezing-point of water referred to as surfusion. And further the humidity is recorded by the elongation of a hair in consequence of the remarkable capacity of hair apparently to appropriate from the surrounding atmosphere the amount of water which corresponds with the fraction of saturation of the air at the temperature. What the exact mechanism of the process is by which the structure of the hair claims possession of its modicum of water we do not know and can only express our admiration for the way in which it generally keeps the rules we have made for it.

Add to that the general turmoil of air in a moving cloud and the fact that the measuring instrument is travelling through it so that the whole set of conditions is dynamic or at least kinematic.

The difficulty of the situation is very clearly illustrated by the fact that some time ago the question of determining the actual amount of water in a given volume of fog was attempted by endeavouring to ascertain the actual amount of water-vapour in the foggy air by the determination of its dew-point, and after a number of trials the effort was abandoned apparently for the reason that the successive results were so variable.

In the circumstances we can only record the fact that the hair-hygrometer does show the peculiarity mentioned and leave it to the reader to decide by judicious experiment whether the peculiarity depends on the nature of the air which carries the cloud, or the thermal or dynamical conditions of the observations, or the inability of the hygrometer to face them successfully.

The behavour of the wet and dry bulb hygrometer at low temperatures is the subject of an investigation by J. H. Awbery and Ezer Griffiths which is published in the *Proceedings of the Physical Society*, 1935, p. 684.

[1] It may be associated with the deposit of ice on aeroplanes. "Surfusion" is regarded by Fontseré and Patxot i Jubert as responsible for "givre" on a sunshine recorder.

Note 11, Chap. IV, p. 46. DIURNAL VARIATION OF TEMPERATURE OF SEA AND AIR AT THE SURFACE

The determination of the actual diurnal variation of temperature of the air and water at sea is not an easy problem because the instruments are on a moving vessel and change of temperature may be caused by change of position. For the temperature of the water from the observations of H.M.S. *Challenger* we get a daily range of ·6° C in the N. Pacific, ·4 in the S. Pacific, ·6 in the N. Atlantic (vol. III, p. 191) and in Krümmel's *Handbuch*, G. Schott gives a mean range in the tropical seas 0·39 with overcast sky and 0·71 with clear sky in moderate to fresh breeze; and in calms or light airs 0·93 for overcast sky and 1·59 for clear sky. The late P. H. Gallé in a work on the 'Climatology of the Indian Ocean[1]' summarises his conclusions about diurnal variation of temperature by a harmonic formula from which we gather the following:

			Diurnal range of temperature in ° C			
			January		July	
	Lat.	Long.	Water	Air	Water	Air
Monsoon	5–14° N	50–59° E	·48	1·02	·42	1·18
Trade-wind	10–19° S	80–89° E	·50	1·56	·32	·78
West wind	35–44° S	70–79° E	·48	1·58	·14	·42

The range shows increase with decrease of cloudiness and wind-force and *vice versa*.

Corresponding information is given in a paper on the Scientific Results of the Norwegian Antarctic Expeditions 1927–8 by H. Mosby. Generalised results for the diurnal range of air-temperature are:

·75° C, 20°–50° N, 2 to 17 Oct. 1927, 5 to 15 Ap. 1928.
·8° C, 20° N–20° S, 19 Oct. to 5 Nov. 1927, 25 Mar. to 4 Ap. 1928.
about 1° C, 20°–54° S, 6 to 14, 19 to 30 Nov. 1927, 14 to 24 Mar. 1928.

The range of temperature of the sea in the period 14 March to 15 April, 1928, was about ·53° C.

Information for the water-surface of the Malay archipelago is given in a paper by Dr H. P. Berlage[2] on 'Sea-surface temperatures on some steamer routes.' For Singapore to Pontianak we have the ranges for each month of the year, for example of ·74° C in January, 1·08 in April, ·76 in July, ·89 in October, giving ·70 for the year with the differences from the mean in each month at the even-numbered hours.

"The ranges are highest during the months of monsoon change April (1·08° C) and November (·92° C) when low wind-velocities and stagnant currents are favourable to the development of maximum temperature-differences caused by the sun's radiation." The temperature of the surface is generally lowest about 3h and highest about 13h local time.

[1] *K. Ned. Met. Inst., Med. en Verh.* 29 b, 's Gravenhage, 1928.
[2] *K. Mag. en Met. Obs. te Batavia, Verh.* No. 21, Weltevreden, 1928; see also No. 25, Batavia, 1933.

For the other side of the world C. F. Brooks[1] gives us some information about the diurnal range of the temperature of the Gulf Stream not far from its emergence from the Caribbean Sea:

> From night to day in sunny quiet weather the sea-temperature at the surface rises 3° or 4° F, and at a depth of 6 ft about 2°. In windy weather the diurnal range is reduced by stirring to 1° or less.

The relations of temperature in the Gulf Stream and its environment are considered in a paper by P. E. Church, *Geographical Review*, 1932 (April).

For the range of temperature of the air over the sea we have some valuable information from ships equipped with thermographs. In many cases where the thermograph is exposed in a screen not far from the deck of a ship the records are not very different from those at land-stations; but when the instrument is at the mast-head and the ship is at sea the regular diurnal variation of temperature is within 3° F. The conspicuous features of the record in these circumstances are associated with showers and squalls and other incidents of weather rather than with the regular changes in the sun's radiation. A detailed analysis is complicated by the fact that changes in temperature due to changes in the geographical position of the ship may mask altogether the small diurnal variation.

A few examples from a thermograph exposed in a screen on the foremast of a ship about 117 ft above sea-level will illustrate the experience in respect of the temperature of air and sea. In the table are given some results for intertropical regions on "undisturbed" days, and these are followed by some notes of the changes associated with showers and squalls. The data are communicated by Captain L. G. Garbett, R.N., from information contained in logs of H.M. ships.

Ranges of air- and sea-temperatures on undisturbed days.

	Position at noon		Air-temperature					Sea-temperature				
			Minimum		Maximum		Range	Minimum		Maximum		Range
Date	Lat.	Long.	° F	h	° F	h	° F	° F	h	° F	h	° F
9. xii. 32	23½ N	58½ E	71	7	73	16	2		—		—	—
14. i. 33	14	68½	77½	4	79½	15½	2		—		—	—
10. ii. 33	Trincomali		78	3–7, 23–24	80	10	2		—		—	—
19. vii. 33	4½ S	49 E	74	3h 40m*	76	15–24	2	77	1–7, 9–14, 17–24	78	8, 15, 16	1
25. vii. 33	2½	60 E	78	0–5	80	9, 21–24	2	78	0–3	82	16–17, 19–22	4
23. x. 33	20 N	64 E	78	3–5, 22–24	80	16	2	77	5–8	82	15	5
30. i. 34	4	99½	78	20–24	81	2–5	3	79	1–4	82	5–15	3

* A fall of 2° F with a light shower.

Some notable changes of temperature in Fahrenheit degrees associated with showers and squalls.

Date	Position	Changes observed
28. xi. 32	Bombay	84° F at 17h 50 to 74° at 18h 10 during heavy rain. "The rain advanced in a solid wall of water on a line N'W to S'E."
16. xii. 32	Persian Gulf 25½ N, 53 E	73° at 18h 20 to 68° at 19h during the passage of a cold front. Wind: 17h 30 S'E 2; 18h SW 4; 19h NW 9.
3. v. 33	Equator 73 E	Sudden falls of about 7° F on five separate occasions associated with heavy showers. The temperature rose slowly after each fall and at midnight was only 2° F lower than at noon.
18. vii. 33	4½ S, 46¾ E	75½° F at 23h to 71½° at 23h 10 during a heavy squall, regaining its origina value at 24h.
10. i. 34	9 N, 95½ E	79° at 2h 45 to 72° at 2h 55. Heavy rain, T, ⚡, •, 79° at 6h.

[1] *Monthly Weather Review*, vol. LVIII, Washington, 1930, p. 148.

A record of transition from the NE trade to the SE trade off the East coast of S. America.

Date		Position		Changes observed				
11. ii. 31	4h	1° 22′ S	31° 44′ W	NE 2, 1008·1mb,	80° F (dry),	78° (wet),	Cu. 7, bc,	Swell from ENE
	8h	1° 53′	31° 50′	ESE 5, 1009·5	75°	75°	St-cu 10, c	,, SE

During the interval between the two observations the ship logged weather l,cqpl,c. The thermograph shows a fall from 79° F at 6h 30 to 72½° at 7h during the rain-squall which accompanied the transition, but it regained its original value at 11h. The temperature before and after the transition differed only by about 1° F.

In the Mediterranean between Sardinia and Malta observations[1] in May 1926 with an Assmann psychrometer held over the extreme bow of the ship 7·5 m above sea-level showed a diurnal variation of 1·05° C with maximum at 14h 30 G.M.T.

Observations on the *Carnegie* during its seventh cruise are noted in the *Monthly Weather Review*, 1931, p. 183.

A range of between 1° F and 1° C over the sea seems to be a normal variation but it comes to shipwreck at the smallest island.

Note 12, Chap. IV, p. 54. Desert plants

"The experiments described in this paper indicate that some plants of the Egyptian desert can, in an atmosphere of high humidity, increase in weight (presumably by the absorption of water-vapour by their aerial organs). In the Egyptian desert, owing to the great difference between day and night temperatures, it is often found that during the night the air-humidity approaches saturation even in summer, and hence may be the source of an appreciable part of the water-supply in plants like *Reaumuria histella*, which have salt-crystals on their leaves. These crystals apparently form part of the mechanism of absorption of water-vapour, as without them the plants do not increase in weight in atmospheres of high humidity.

"In this way one-sixth of the plant's loss by transpiration may be replaced by absorption of water-vapour at night." (R. E. Chapman, 'The absorption of water-vapour by the aerial parts of Egyptian desert plants,' B.A. meeting, Aberdeen, 1934.)

Note 13, Chap. IV, p. 57. Weather conditions in Greenland

It has been noted quite appropriately in a review of this volume in 1928 that in the monthly charts of mean temperature, figs. 17–41, the lines shown over the plateau of Greenland, where observations are very few, are more detailed than those over the Indian peninsula where observations are abundant. That is certainly remarkable; but from the meteorological point of view the shape and position of Greenland are also remarkable and at the time there was considerable interest in polar glacial areas. A volume on Greenland published about that time contains a chapter on the Climate of Greenland by H. Petersen with data for the stations on the coastal fringe, seven on the Western and four on the Eastern side, and some information about the plateau based partly on what is "well known" about the inland ice and the observations of Nansen 1889, de Quervain 1912 and of Koch-Wegener 1913.

It makes perfectly clear that the inland plateau is the only part of Greenland for which such conceptions as mean values of meteorological elements for months have any effective application and the idea of a mean is severely

[1] N. K. Johnson, *Q.J. Roy. Meteor. Soc.*, vol. LIII, 1927, p. 59.

strained even there. The surrounding sea is firm ice in winter and floating ice in summer and perverts ones ideas of marine influences. With the deep contorted indentations of the fjords the climate of any part of the coastal fringe is a chapter of accidents but the plateau is a comparatively smooth area thoroughly glaciated with perpetual snow, and with a ridge running North and South 2500 to 3000 metres high which acts as a "divide" for the cold air which flows downward to the East on the one side and the West on the other. By that arrangement the prevalent föhn winds with their variable temperatures and notable dryness are explained. The coastal fringe, with a coast that extends from lat. 60° to 80° N and has romantic experiences of duration of daylight, can set at nought almost all the customary inferences of our plains of the temperate zone.

From 1915 at least Prof. W. H. Hobbs of Ann Arbor has striven to develop the idea of a "glacial anticyclone" over cold high places like Greenland and the Antarctic continent and has defended his ideas in a volume on *Glacial Anticyclones*, 1926, and has also maintained expeditions to Greenland based on Mount Evans, off the Western side, for the study of the subject[1].

Recently the meteorological results of the long sojourn of the British Arctic Air-Route Expedition on the inland ice 1930–1, discussed by S. T. A. Mirrlees[2], have added greatly to our knowledge of the weather of the plateau which has vicissitudes like those of elsewhere. A summary by its author in the *Meteorological Magazine*, October, 1934, seems to confirm at least the general idea of the inferences which may be drawn from the distribution of temperature shown in the charts if we admit that the map-maker has omitted to reduce the plateau values to sea-level which the air of the fjords has often done of its own accord. The extremes of temperature observed on the ice-cap in lat. 67° N, long. 42° W, elevation 8200 ft. are shown in the accompanying table.

Temperatures on the ice-cap of Greenland in °F.

(140 miles NW of Angmagssalik)

	Maximum		Minimum	
1930–1	Highest	Lowest	Highest	Lowest
September	+29	− 1	+15	−27
October	+22	−38	+ 3	−48
November	+ 7	−36	− 9	−51
December	+ 7	−45	+ 2	−56
January	+20	−43	− 4	−52
February	− 6	−41	−15	−60
March	+ 8	−33	+ 1	−59

Observer snowed up after 17th

The "Beiträge zur Meteorologie des Luftweges über Grönland[3]" add evidence in support, with the caution of a footnote:

"Es muss ausdrücklich bemerkt werden, dass mit der Bezeichnung 'antizyklonaler Charakter' nicht die Behauptung aufgestellt werden soll, dass über Grönland stets eine

[1] Reports of the Greenland Expeditions of the University of Michigan (1926–31), Part I, Aerology, Ann Arbor, 1931.

[2] Meteorological Office, Geophysical Memoirs, No. 61, M.O. 356 d, 1934.

[3] *Arch. Deutsch. Seewarte*, Bd. LII, Nr. 4, Hamburg, 1933.

ausgeprägte stationäre 'Inlandeisantizyklone' vorhanden ist. Nur die auf jeden Fall wirksame Tendenz zu einer solchen ist gemeint."

So we may sum up the conclusion that whether the distribution of pressure over the Greenland plateau can properly be called an anticyclone or not the effect of the glaciation at that high level makes the lowest layers of the air behave as one would expect if there were an anticyclone.

In his short summary Mr Mirrlees remarks that "outflowing winds from Greenland may...attain hurricane force [wind velocities up to 129 mi/hr were recorded] but such outflow is limited in extent, appears to have no direct influence for disturbed weather in the North-East Atlantic and is probably not on a large enough scale to provide motive power for the atmospheric circulation." One would like to see the probability for Greenland as a whole expressed in numerical dimensions. With cold air descending on either side of a ridge some 1200 miles long and capable of attaining anywhere very high velocities one cannot lightly set aside its influence for the behaviour of the surface-air of the Northern Hemisphere.

In the days when the sequence of weather was expressed in terms of centres of low pressure, two areas where such centres were developed were noteworthy, one was to the South of Greenland and another South of the high mountains of Alaska. Any sequence of maps would afford some support for the suggestion; and if one were making an experimental model to illustrate the thermal behaviour of the atmosphere, Greenland for the Atlantic and Mt St Elias and its fellows for the Pacific could scarcely be left out.

Note 14, Chap. IV, p. 107. Additional particulars of the conditions of the upper air

Within the text of this edition we have given some particulars of temperature and pressure of the upper air of both the Eastern and Western Hemispheres. They do not by any means exhaust the information about the upper air which is now available. Observations of pressure, temperature and humidity by means of airplanes, and of the horizontal direction and velocity of the motion of the air by the use of pilot-balloons are now matters of daily duty at many stations associated with the various air-services, as are also visibility and the type, height and motion of clouds. We cannot give the data within the limits of this volume. They are not yet summarised for the whole world as are those which we have represented in some preceding chapters; we must be content with an indication of where an industrious reader may seek additional information when he desires it. This is the purpose of an abbreviated bibliography which as Note 25 concludes this volume.

One other subject may be illustrated here which is not covered by the diagrams of figs. 58 and 59, namely the variation of the temperatures and heights of the tropopause within the range of observations already represented in those figures. Fig. 223 gives this information for Sealand (Chester), SE England, Canada and Pavia as well as for Abisko (N. Swedish Lapland),

Lindenberg (SE of Berlin), Vienna, Agra and Batavia (Java) in order to illustrate the extent and limitation of the geographical as well as the seasonal variation.

TEMPERATURES IN THE TROPOPAUSE: AND THE RANGE OF LEVEL ASSOCIATED THEREWITH ILLUSTRATING VARIATION WITH LONGITUDE, COLUMNS 1–3, 4–5 AND WITH LATITUDE, COLUMNS 6–10.

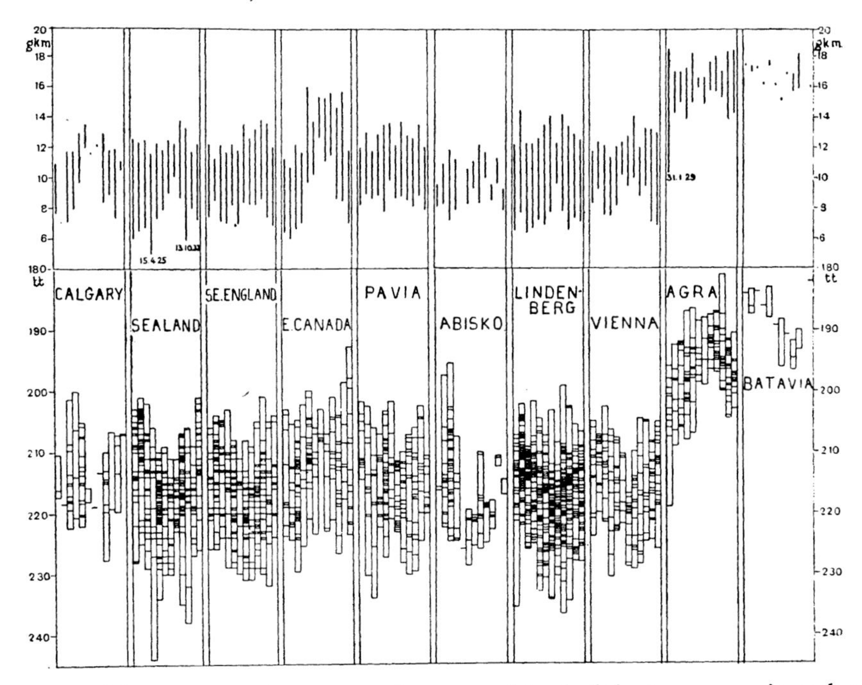

Fig. 223. The upper panel represents the range of level of the tropopause in each of the twelve Calendar months according to a scale of geokilometres. The lower panel, with a tercentesimal scale, shows temperatures in the tropopause as recorded in the several soundings.

We can detect very little difference in what we may estimate as the "normal" between Abisko and Sealand nor between Agra and Batavia but the large range of the tropopause at Agra in the "winter months" compared with the summer is very noticeable. We must however not forget that the definition of the tropopause becomes uncommonly vague when the transition from the troposphere to the stratosphere is made by a succession of counterlapses of temperature instead of the single change from high lapse-rate to none at all or negative.

And we are unwilling to allow the information derived from pilot-balloons to pass without some token of respect. We have accordingly set out in fig. 224 a representation by means of perspectives of the direction and velocity of the horizontal motion of the air at the surface and at successive heights above sea-level as determined by pilot-balloons at Plymouth, Holyhead and Leuchars near St Andrews.

FOOT-STEPS OF BRITISH WINDS BRINGING IN THE NEW YEAR 1936.

Plymouth Holyhead Leuchars

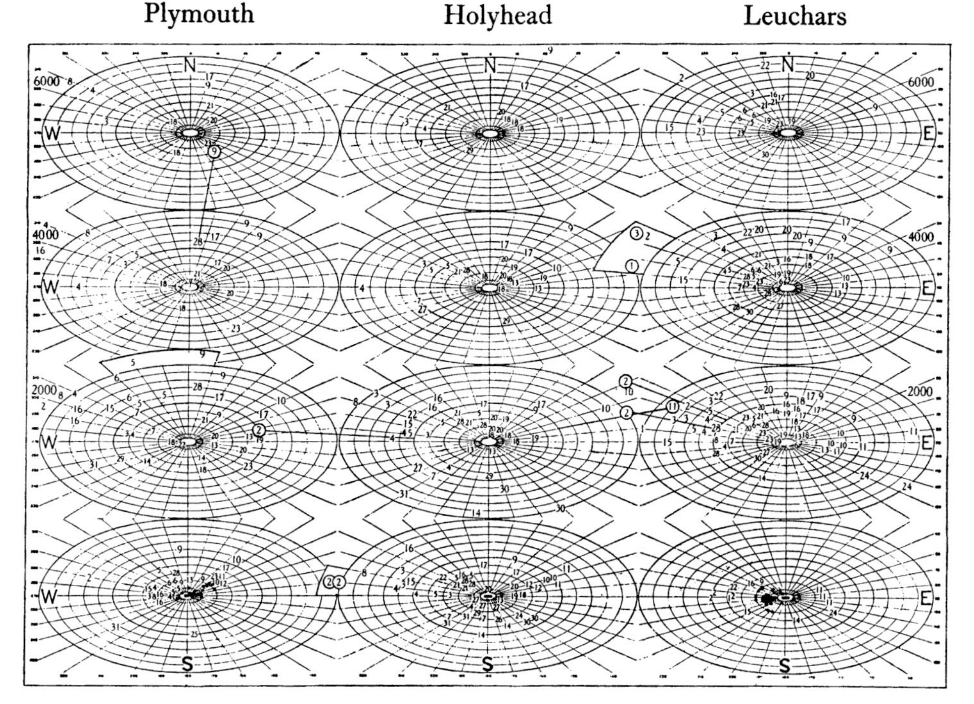

Fig. 224. Horizontal velocity of the winds at the surface and at the levels of 2000 ft., 4000 ft. and 6000 ft. for three British stations as recorded in the *Daily Weather Reports* for December, 1935. The numbers are the dates of the ascents, 47 at Plymouth, 53 at Holyhead and 60 at Leuchars. The position of the number in its diagram indicates the direction and speed of the motion. The outer curve of each diagram corresponds with a velocity of 40 miles per hour, the lower limit of "gale-force."

At Plymouth ascents on Dec. 7, 10, 11 (2), 12 (3), 13, 19, 23, 25, (eleven in all) failed to reach 2000 ft.; no ascents were made on Dec. 1, 24, 26, 27, 30. For Holyhead ascents on Dec. 2, 7, 11, 12 (2), 14, 24, 26, 27 (2), 31, also eleven in all, failed to reach 2000 ft.; no ascents on Dec. 1, 6, 23, 25. For Leuchars one ascent on Dec. 14 failed to reach 2000 ft.; no ascents on Dec. 12, 25, 26, 29, 31. Some of the light winds at the surface are indicated by dots.

Note 15, Chap. v, p. 143. VISIBILITY AND FOG

On account of the importance for aviation of a knowledge of the conditions of visibility the subject has received considerable attention in recent years. The International Commission for Aerial Navigation has recommended a special form of publication for monthly summaries. It gives the frequency of the various degrees of visibility at the three hours of observation.

We give below a list of countries which publish summaries in this standard form together with the number of stations for which information is given and the dates of the first and last issues received in the library of the Meteorological Office in London up to the end of 1934.

Yugo-slavia	18 stations 1927 (Jan.)	1931 (Dec.)	41 stations
*British Isles	25 ,, 1927 (Jan.)	1934 (Oct.)	25 ,,
Czecho-slovakia	5 ,, 1927 (July)	1933 (Dec.)	13 ,,
Italy	13 ,, 1928 (Jan.)	1933 (Dec.)	14 ,,
Poland	25 ,, 1928 (Sept.)	1933 (Aug.)	25 ,,
France	20 ,, 1929 (Jan.)	1932 (Dec.)	20 ,,
Siam	1 ,, 1929	1931	1 ,,
†Chile	1 ,, 1930 (Jan.)	1933 (July)	12 ,,
Roumania	5 ,, 1930 (July)	1933 (Mar.)	5 ,,
Azores	4 ,, 1931 (Jan.)	1933 (Dec.)	4 ,,
Sweden	11 ,, 1932 (Jan.)	1934 (June)	11 ,,
South Africa	18 ,, 1933 (Mar.)	1934 (Sept.)	17 ,,
Portugal	11 ,, 1934 (July)	1934 (Oct.)	11 ,,

* From Jan. 1933 observations at Malta and at 11 stations in Egypt, Palestine, Trans Jordania or Iraq are included in the report.

† Series incomplete.

We add also a list of some recent publications dealing with the frequency of occurrence of fog in different regions:

SWEDEN: E. Lindskog, 'On the geographical distribution of fog in Sweden,' *Geog. Ann., Stockholm*, vol. XIII, pp. 1–94, 1931. (Charts of monthly fog-frequency as a percentage of the total number of days and also curves of monthly frequency at a large number of stations.)

DENMARK: K. H. Soltau, 'Die geographische Verbreitung und Bedeutung des Nebels in Schleswig-Holstein und Dänemark,' *Breslau, Veröff. Schleswig-Holstein Univ. Ges.*, Nr. VII, 1927. (Tables of monthly values and an annual chart.)

GERMANY: 'Wetterstatistik des Deutschen Flugwetterdienstes, 1926–30.' (Observations of frequency of the various degrees of visibility at 10 stations in 1926 and 24 stations in 1930.)

H. Grimm, 'Nebelhäufigkeit an den deutschen strahlungsklimatischen Stationen,' *Ann. Hydrog.*, Bd. LXI, pp. 36–8, 1933.

W. Peppler, 'Ein Beitrag zur Kenntnis des Nebels besonders in Südwestdeutschland,' *Wetter*, Bd. XLI, p. 143, 1924. (Normals at 35 stations.)

G. Hankow, 'Sicht, Windrichtung und Windstärke in NW und W Deutschland,' *Berlin, Erfahrb. D. Flugw.*, Bd. VII, No. 10, pp. 65–103, 1932.

A. Mey, 'Die Sichtverhältnisse des Flugplatzes von Bremerhaven,' *Ibid.*, Bd. IX, No. 12, pp. 61–8, 1934.

GREAT BRITAIN: 'Wind, fog and mist, SW approaches to Great Britain and Ireland,' *Marine Observer*, vol. VII, London, 1930.

EUROPE IN GENERAL: V. A. Berezkin, 'Synoptical conditions of formation of fogs. Geographical distribution of fogs,' *Recueil Géophys. Leningrad*, vol. VII, fasc. 2, 1930. (Monthly maps of isograms of mean frequency of synoptic fogs.)

MALTA: J. Wadsworth, 'A study of visibility and fog at Malta,' *Geophysical Memoirs*, No. 51, London, 1930.

EAST ASIA: H. Seilkopf, 'Grundzüge der Flugmeteorologie des Luftwegs nach Ostasien,' *Arch. Deutsch. Seew.*, Bd. XLIV, Nr. 3, Hamburg, 1927. (Tables of monthly fog-frequency.)

JAPAN: T. Okada, 'The climate of Japan,' *Bull. Cent. Met. Obs.*, vol. IV, No. 2, Tokyo, 1931. (Tables of monthly data of frequency of fog at a large number of stations.)

S. AFRICA: *Official Year Book of the Union of South Africa*, No. 13, pp. 25–33, 1930–1. (Tables of mean monthly visibility and frequency of fogs at 30 stations.)

TABLE BAY: 'Wind, fog and mist, approaches to Table Bay,' *Marine Observer*, vol. VII, London, 1930.

SAHARA: J. Tilho, 'Sur la fréquence des brouillards dans le Sahara oriental,' *Comptes Rendus*, tome 170, pp. 1435–8, 1920.

AUSTRALIA: 'Monthly percentage of fogs in Australia,' Bureau of Meteorology, Melbourne, 1928. (Diagrams for about 50 stations based on observations from 1915–19.) Visibility 1923–4, *Bulletin*, No. 17.

N. AMERICA: W. E. Hurd, 'Fog at sea.' Pilot charts of the North Pacific, November, 1931, Washington, Hydrographic Office. (Tables of percentage frequency of fog at lighthouses on the Newfoundland coast, number of days with fog at stations in Newfoundland and Canada, on the Pacific coast of Canada and the United States, and at stations on the Pacific coast of Asia and in neighbouring islands.) R. G. Stone, 'Fog in the United States and adjacent regions,' *Geog. Rev. New York*, 1936, p. 111.

CHILE: W. Knoche, 'Nebel und Garúa in Chile,' *Zs. Ges. Erdk. Berlin*, 1931, p. 81. (Tables of monthly data for 54 stations, with an annual chart.)

OCEANS: Information about fog over the oceans is contained on the Pilot Charts published by the several countries and summaries are given also in some of the publications referred to in the bibliography of note 24.

We note also: K. Tsukuda, 'On the sea-fog in the N. Pacific Ocean,' *Geoph. Mag.* vol. VI, p. 147, Tokyo, 1932. (Tables of monthly data of percentage frequency for 5° squares, and maps for the months May to September.) An example is included in fig. 72.

'Fog at sea,' *Marine Observer*, vol. VII, London, 1930.

Note 16, Chap. v, p. 145. CLOUD IN THE ARCTIC

Much information about the details of cloud in the Arctic is contained in the results of the Norwegian North Polar expedition with the *Maud* which have been discussed by H. U. Sverdrup.

"A dull and monotonous appearance is characteristic of the arctic sky and is caused by the prevalence of the stratus clouds. In summer contourless fog or low stratus generally covers the sky without leaving a patch of blue, and days are rare with both lower and upper clouds visible. The same applies to the autumn and partly to the winter when, however, thin alto-stratus clouds which form a milky veil also are frequent. In spring the sky has a more varied appearance partly because the cloudiness is small and partly because the sky itself often shows beautiful colours immediately before sunset or after sunrise, or when the sun is low. However even in this season the stratus forms are by far the most dominating when the sky is cloudy."

Over the pack-ice stratus (which includes fog and drifting snow) accounts for about half the observations from February to April and for about three quarters in the summer months.

Seasonal variation. Cloud-amount over the pack-ice is greatest in July and August and least in the late winter and early spring, the cloudiness increases rapidly from April to May and decreases rapidly from October to November. The difference between summer and winter is about 5·3, from a minimum of about 3·5 in winter to a maximum of nearly 9 in summer.

At coastal stations the greatest cloud-amount is in autumn and the least in winter and early spring with a secondary maximum in May. The difference is about 3·2, considerably less than over the pack-ice.

Diurnal variation. In summer, conditions vary from month to month and on the average the cloudiness shows no diurnal variation. In the dark winter months, October to March, cloud-amount appears to show a maximum in the day at about 10h and a minimum at about midnight; but the amount of daylight has an important effect on estimates of cloud-amount and the apparent diurnal variation may be caused by errors of observation.

Height of clouds. On the occasions when pilot-balloons were lost in cloud the type of cloud was noted together with the height at which the balloon disappeared. The results are summarised in the following table:

Type of cloud	Ci	Ci-st	Ci-cu	A-st	A-cu	St-cu	Fr-st	St
Height of lower limit, km	Up to 9	$4\frac{1}{2}$–8	$4\frac{1}{2}$–8	$2\frac{1}{2}$–$4\frac{1}{2}$	$2\frac{1}{2}$–$4\frac{1}{2}$	$\frac{1}{2}$–$2\frac{1}{2}$	$\frac{1}{5}$–2	0–$1\frac{1}{2}$
Height of greatest frequency, km	—	—	—	$3\frac{1}{2}$–4	3	$1\frac{1}{2}$	—	$\frac{1}{2}$

Note 17, Chap. VI, p. 243. SURFACE-WINDS

In considering the details of distribution of the flow of air over the surface we may remind our readers that we have declined to draw continuous isothermal lines across the coast-line and we should do the same with winds. We would regard as separate, for the time being, the winds over the oceans which are duly represented by wind-roses in many volumes of oceanic charts and are effectively summarised for the months of highest and lowest sea-temperature by W. Köppen (pp. 244–7). The winds of the comparatively level land-surfaces below 200 metres we have already noted as being generally indicated by the isobars; but the coast-lines are special regions subject to the differences of the effect of solar radiation upon sea and land and should be treated separately.

It is a matter of common knowledge that in regions where there are steep slopes and fjords, air flows down the fjords and the cold air which they deliver has to be disposed of when it reaches the coast-line. We are not surprised therefore that some authors regard an Easterly wind along the Atlantic coast-line of the Antarctic as a notable feature of that region while others are impressed by the fact that at some distance from the coasts Westerly and North-westerly winds are prominent. We may support this conclusion by a reference to the chart of the winds of the Atlantic Antarctic by H. Mosby[1] in which Easterly winds are shown along the coast-line itself but a variety of other winds in more Northerly latitudes.

Corresponding with Mosby's discussion of the winds of the Antarctic we may take the results of the *Maud* expedition as set out by Sverdrup for the Arctic.

"In the pack-ice, 1922–24, the easterly winds were dominating in all months except in September 1923 and February and March 1924 when the directions NW, SW and S respectively were the most frequent. In the greater number of cases the most frequent wind came from due east."

Except for minor details the winds of coastal plains may share the general behaviour of the neighbouring ocean winds but in hilly and mountainous regions the behaviour of winds can only be regarded as better represented by micro-climatology. The considerable number of notes in the *Meteorologische Zeitschrift* and elsewhere on mountain and valley winds in the Alpine regions makes it clear that different localities must be separately treated.

For detailed charts of the wind in the several countries we may refer to the bibliography of note 24.

Note 18, Chap. VI, p. 258. NORMAL ISOBARS AT HIGH LEVELS

Since the publication of the charts of normal pressure in the upper air over the Northern Hemisphere more detailed charts for special countries have been issued which may be used for the correction or modification of our hemispherical charts.

UNITED STATES: W. R. Gregg, 'An aerological survey of the United States, Part I. Results of observations by means of kites,' *M.W. Rev. Washington Supplement*, No. 20, 1922. The memoir contains maps of mean summer and winter pressure at sea-level, 500 m, 1000 m, 2000 m, 3000 m and 4000 m, over the eastern and central United States; the isobars are drawn for steps of 2mb.

JAPAN: U. Nakaya, 'Monthly normals of isobars in Japan at the height of 3000 m,' *Tokyo, J. Faculty Sci.*, Sec. I, 1, 1927, pp. 301–12; K. Sinoda, 'The monthly normal isobars at 4000 and 6000 metre levels over Japan and its vicinity,' *Ibid.*, pp. 313–37.

INDIA: 'Monthly normal isobars and wind-roses at 0·5, 1, 2 and 3 km above sea-level over India and neighbourhood,' *India Met. Dept. Scientific Notes*, vol. I, No. 8, Calcutta, 1931.

H. C. Banerjee and K. R. Ramanathan, 'Upper air circulation over India and its neighbourhood up to the cirrus level during the winter and the monsoon,' *Ibid.*, vol. III, No. 21, Calcutta, 1930. The paper contains maps of approximate isobars at 4, 6, 8, 10 km for Dec.–Jan and July–Aug.

GREENLAND: S. T. A. Mirrlees, 'Meteorological results of the British Arctic air-route expedition 1930–1,' *Geophysical Memoirs*, No. 61, London, 1934. The memoir contains maps of computed pressure at mean sea-level, 2½, 3, 4 km for the quarter Oct.–Dec. 1930.

Note 19, Chap. VI, p. 258. DENSITY IN THE UPPER AIR OVER ENGLAND

We quote from Articles 21 and 32 of chap. X of the first edition.

At the level of the eighth kilometre the density is normally constant all the year round and apparently uniform all over the world, with a notable exception at Agra, June to November.

[1] 'The sea-surface and the air,' *Scientific results of the Norwegian Antarctic Expeditions 1927–1928 et sqq.*, No. 10, Det Norske Videnskaps-Akademi i Oslo, 1933.

At any level above the eighth kilometre the density of air increases from the pole to the equator and at any lower level it increases from the equator to the pole. Tables of the density at different heights and of its seasonal variation are given on p. 263 for U.S.A., Bavaria, Italy and British India. We add a table for England:

Seasonal variation of density at different levels over England.

Height	Jan.	Feb.	Mar.	Apr.	May	June	July	Aug.	Sept.	Oct.	Nov.	Dec.
km	g/m³	g/m³	g/m³	g/m³	g/m³	g/m³	g/m³	g/m³	g/m³	g/m³	g/m³	g/m³
15	184	188	184	188	191	195	198	198	195	195	191	188
14	219	219	216	219	226	226	230	230	230	226	223	219
13	254	254	254	258	261	261	268	268	268	264	261	258
12	299	299	296	303	306	310	313	310	313	310	306	299
11	348	348	348	351	358	362	362	362	362	358	351	351
10	404	407	407	411	414	414	418	414	414	411	407	404
9	459	466	466	466	470	470	470	470	466	466	463	463
8	529	529	525	525	529	529	529	529	529	525	525	525
7	595	595	588	588	588	588	588	588	592	592	592	592
6	672	668	665	665	665	661	654	654	654	654	658	665
5	748	748	741	738	734	731	727	727	731	731	738	741
4	832	828	828	821	814	811	807	807	811	814	818	821
3	926	919	922	908	912	905	898	898	901	905	912	915
2	1034	1027	1027	1016	1013	999	999	999	1002	1002	1013	1023
1	1152	1148	1145	1131	1124	1110	1107	1103	1114	1114	1128	1138
Surface	1274	1281	1270	1260	1246	1239	1228	1225	1235	1242	1260	1270

The range over the globe of the normal values of density at any level between 4 km and 9 km is within 4 per cent. of the mean value. (*Meteorological Glossary*, p. 53, and S. N. Sen, *Q. J. Roy. Meteor. Soc.*, vol. L, 1924, p. 29.)

The variability of density in local circulations.

In *normal* distributions densities are approximately uniform along the horizontal at 8 km; in local circulations there is considerable variation. For changes of pressure in the sequence of weather over England they are as follows (W. H. Dines, *Meteorological Glossary*, 'Density'):

	Density				
Height km	High pressure g/m³	Low pressure g/m³	Range g/m³	Percentage of mean %	Range of monthly means %
9	472	446	26	5·7	2·2
8	530	516	14	2·7	0·7
7	589	582	7	1·2	1·2
6	658	648	10	1·5	2·6
5	734	724	10	1·4	2·8
4	818	808	10	1·2	3·0
3	906	898	8	0·9	3·0

In the last column of the table the range of mean values for all observations for the month expressed as a percentage of the mean for the whole year is given for comparison.

Note 20, Chap. VII, p. 292. LONG PERIOD RECORDS

Records of pressure, temperature and rainfall month by month at a large number of stations were published in a volume of 'World Weather Records' (*Smithsonian Misc. Coll.*, vol. LXXIX, Washington, 1927) shortly after the text of chap. VII had been compiled, and a supplementary volume containing data for the ten years 1921–30 was issued in 1934.

In addition to the data of many of the long records already referred to in the caption of chap. VII the volume contains also records of pressure, mb, temperature, tt, or rainfall, •, going back to 1830 or earlier (but not in all cases complete), at the following stations:

Finland, Helsingfors	tt	Italy, Rome	tt, •	United States:	
Germany, Königsberg	•	Netherlands, Zwanenburg	tt	Albany	tt •
Trier	•	Norway, Bergen	mb, tt	Charleston	tt •
India, Bombay	•	Poland, Wilno	tt	New Haven	tt •
Calcutta	•			New York	tt •

Among other long period records we may note:

GREAT BRITAIN: Oxford, Radcliffe Observatory, vol. LV, tt, •. Some early wind records in the British Isles are summarised in *Q.J. Roy. Meteor. Soc.*, vol. LIX, 1933, p. 375; vol. LX, 1934, p. 62.

GERMANY: W. Naegler, 'Hundertjährige Temperaturreihen von Dresden und Leipzig,' *Zs. a. Met.*, Berlin, Bd. XLVII, 1930, pp. 60–2. O. Egert, 'Aus dem Archiv der Sachsischen Landeswetterwarte,' 16 *Tagung der deutschen Meteor. Ges.* Dresden, 1929, pp. 3–6 (mb, tt, • Dresden).

HUNGARY: A. Réthly, '100-jährige Temperaturmittel von Budapest,' *Met. Zs.*, Bd. LI, 1934, p. 43.

NORWAY: Bergen, Ullensvang, Vardö, compiled by B. J. Birkeland in *Geofysiske Publikasjoner*, vol. V, No. 8, 1928; vol. IX, No. 6, 1932; vol. X, No. 9, 1934 (mb, tt).

UNITED STATES: E. R. Miller, 'A century of temperatures in Wisconsin,' *Trans. Wisconsin Acad. Sci. Arts and Letters*, vol. XXIII, 1928, pp. 165–77.

Note 21, Chap. VIII, p. 355. RATE OF TRAVEL OF CYCLONES AND ANTICYCLONES

In a paper[1] on 'The extratropical cyclones of Eastern China and their characteristics' Shio Wang Sung quotes the following table, which he attributes to Dr C. Okada, of the velocities of movement of the cyclones of the world. For comparison with the tables of p. 355 we have converted the velocities from kilometres per hour to metres per second.

	Japan m/s	United States m/s	Western Europe m/s	European Russia m/s	North Atlantic m/s	Bering Strait m/s	Eastern China m/s
Winter	13·0	15·6	8·0	10·8	8·2	8·5	12·5
Spring	11·4	12·3	7·2	9·2	8·3	8·5	10
Summer	9·6	10·9	6·6	8·0	7·4	10·3	9·5
Autumn	12·5	12·3	8·2	9·6	8·3	9·3	10·5
Year	11·6	12·8	7·5	9·3	8·0	9·1	11

E: Kidson[2] in 'Notes on the general circulation in the New Zealand region' gives the Eastward velocity of anticyclones over the distance from Perth to New Zealand as about 9° per day (equivalent to 9 m/sec in lat. 40°). The velocity is said to be remarkably uniform throughout the year, the minimum, $1\frac{1}{2}°$ per day below the mean, occurs in May and the maximum, 1° per day above the mean, from October to December.

[1] *Memoir Nat. Res. Inst. of Met.*, No. III, Pei-chi-ko, Nanking, 1931.
[2] *Gerlands Beiträge z. Geophysik*, Band XXXIV, Köppen-Band III, 1931.

Note 22, Chap. VIII, p. 357. ADDITIONAL INFORMATION ABOUT TORNADOES

Tornadoes in the United States. In a paper read at the fourteenth Annual Meeting of the American Geophysical Union, Section of Meteorology, in 1933, C. W. Brown discussed the distribution of tornadoes in the United States in respect of time and locality, with a distinction of type between cyclonic and convectional tornadoes according to whether the tornado was associated with a well-formed cyclonic depression or not. He accounts for a gross total of nearly 5000 tornadoes from 1880 to 1931 with a distribution ranging from a minimum of none at all in 1902 to a maximum of 215 in 1930, and subsidiary maxima of 175 in 1885, 160 in 1887 and 130 in 1917. There was a period of very low frequency from 1900 to 1906.

The distribution and character of the tornadoes from 1917 to 1931, about 2000 in all, are discussed in greater detail and particulars are given about seasonal variation. For the United States as a whole the numbers in the successive six-day periods between the middle of March and the middle of June range from 30 to 120. He notes four belts of country running more or less North and South which are frequented by tornadoes. Kansas, Iowa and Texas are the most frequented states with 225, 220 and 190 respectively in the 15 years; 21 states have less than 50 in that period.

The *Report of the Chief of the Weather Bureau* gives year by year a description of all tornadoes that have occurred during the year with a chart of their paths. From the data there set out we obtain the figures given in the table below for the total number of tornadoes in each of the twelve months during the nine years 1923–31.

It should be noted that when conditions are favourable a large number of tornadoes may occur on the same day. Thirty-six were reported in the three days 7–9 May, 1927, and no fewer than twenty-eight on 1 May, 1930. "In some cases tornadoes have a tendency to form in groups and to move in parallel paths in a north-easterly direction."

To the figures for the total number of tornadoes in the United States as a whole we have added figures for the three states Kansas, Alabama and Virginia of which special studies have been published, and also data of "outstanding tornadoes" in the period 1804–1925.

State	Period	Jan.	Feb.	Mar.	Apr.	May	June	July	Aug.	Sept.	Oct.	Nov.	Dec.	Total
All states	1923–31	23	27	86	216	288	256	112	79	99	36	62	27	1311
Kansas	1914–28	0	1	15	22	30	65	10	8	13	7	5	0	176
Alabama	1794–Mar. 1925	22	34	80	69	29	8	3	2	0	1	15	9	272
Virginia	1814–Jan. 1925	0	1	2	6	11	5	13	11	11	3	0	0	63
"most important tornadoes"														
All states	1804–1925	1	1	13	14	7	6	4	4	1	1	2	0	54

For the United States as a whole the time of maximum occurrence is in the afternoon. Out of 452 tornadoes 80 per cent. occurred between noon and 6 p.m., and 15 per cent. between 6 p.m. and midnight. In Kansas "the late afternoon from 3 p.m. to 6 p.m. [15h to 18h] has been the usual time of occurrence though some tornadoes have been sighted shortly after the noon hour and some have formed almost as late as midnight. Three occurred between 3 a.m. and 5 a.m. but none during the forenoon." In Alabama the greatest frequency is from 16h to 19h, the frequency increases from about sunrise to sunset and then gradually decreases. In Virginia the "danger-hours" are said to be from 15h to 18h, 27 tornadoes out of 42 records examined occurred between those hours.

The track of a tornado is frequently remarkably straight. In Kansas tornadoes seldom travel more than 25 miles before breaking up, but in the Mississippi valley some travel 50 to 150 miles and tracks exceeding 300 miles have been recorded.

Bibliography: C. J. Root, 'Some outstanding tornadoes,' *M.W. Rev.* Washington, vol. LIV, 1926, p. 58; S. D. Flora, 'Kansas Tornadoes 1914–1928,' *Ibid.*, vol. LVI, 1928, p. 412; vol. LVII, 1929, p. 97; W. R. Stevens, 'Tornadoes in Alabama,' *Ibid.*, vol. LIII, 1925, p. 437; A. W. Giles, 'Tornadoes in Virginia 1814–1925,' *Ibid.*, vol. LV, 1927, p. 169; H. C. Hunter, 'Tornadoes in the United States 1916–23,' *Ibid.*, vol. LIII, 1925, p. 198.

Tornadoes in other countries.

With reference to tornadoes in other countries than America we may offer a specimen from North-West India published by Flt.-Lt. Veryard[1] from a photograph by Ldg. Aircraftsman Russel Pleasants.

Fig. 225. Tornado at Peshawar, 5th April 1933. Photographs at approximately five-minute intervals from 11.55 to 12.25 local time.

For India Dr C. W. Normand in a personal letter in 1929 writes as follows:

"On p. 359 tornadoes are indicated to be extratropical land-phenomena. While this is true as a general rule, tornadoes probably do occur sometimes within the tropics and near if not on the sea. In India severe thunderstorms (which in Bengal are called nor'westers) are of fairly common occurrence. A few of these attain to the severity and probably to the form, of tornadoes; on the latter point more reliable evidence is

[1] India Met. Dept., *Scientific Notes*, vol. v, No. 56, Delhi, 1934.

needed. In Bengal the severe local storms generally occur in April or May. In South Burma there are accounts of severe local storms in the middle of the monsoon season, after a lull in the monsoon itself and when a fresh advance is taking place. Such a storm, possibly tornado, at Moulmein in South Burma (lat. 16½° N) was described in the Annual Summary for 1923."

The destructive winds referred to by Dr Normand as the Nor'westers are characteristic winds of South Bengal. They are associated with thunder-storms carrying occasionally very heavy hailstones. Some account of the subject is given in a paper by V. V. Sohoni[1] on 'Thunderstorms of Calcutta,' and they have since been the subject of special inquiry. They seem to originate especially in March, April and May in a surface of discontinuity between the warm moist air of the Southerly to South-westerly current up to about 2 km from the Bay and the relatively cold and dry drift from the North West of the Gangetic plain supplied from the ranges to the North and East.

"Occasionally a Nor-wester rivals the tornado in its violence and destructive effects. But in it there is no whirling motion of the air such as is found in a tornado." For these reasons they may perhaps be classed with the line-squalls of North Western Europe and perhaps with the Southerly busters of Southern Australia.

As not unlike the Nor'westers comes a note about tornadoes in West Africa during the polar year described by D. E. Smith in the *Meteorological Magazine* of February 1934 as occurring in Lagos due to the interaction of the dry North Easterly and warm moist South Westerly winds in the transition seasons March to May and September to early November.

Each tornado has its squall line with a discontinuity of wind-direction and speed and a definite drop of temperature.

We have noted on pp. 357–60 some occurrences in the British Isles which are at least analogous to tornadoes in their destructiveness and its peculiar limits. We may add some other examples:

1729 May 20, Bexhill, *Q.J. Roy. Meteor. Soc.*, vol. LX, 1934, p. 186.
1850 April 18, Dublin, *Meteor. Mag.*, vol. LXIV, 1929, p. 266.
1916 October 26, Writtle, Essex, *Q.J. Roy. Meteor. Soc.*, vol. XLV, 1919, p. 147.
1918 July 26, Gosfield, Essex, *Ibid.*
1928 October 22, London, *Meteor. Mag.*, vol. LXIII, 1928, p. 238.
1931 June 14, Birmingham, *Ibid.*, vol. LXVI, 1931, p. 125.
1934 January 13, Carbis Bay, *Ibid.*, vol. LXIX, 1934, p. 71.
1934 June 6, Laindon, Essex, *Q.J. Roy. Meteor. Soc.*, vol. LX, 1934, p. 536.

"Tornadoes though rare in these islands, are by no means unknown. A full description of one which occurred on October 27th 1913 in south Wales and the west of England, was published by the Meteorological Office as *Geophysical Memoirs*, No. 11. It is there stated that the *Meteorological Magazine* gives references to forty tornadoes, more or less violent, which occurred in the British Isles in the years 1866 to 1895." (R. Corless.)

And we conclude our reference to these destructive occurrences with a note of an investigation at the National Physical Laboratory of the effect of strong wind on a building, illustrated by the effect on a model in the air-channel.

[1] India Met. Dept., *Scientific Notes*, vol. I, No. 3, Calcutta, 1928.

It is recorded in the report of a lecture at the Royal Institution on 2 March, 1934, by H. J. Gough and shows that the flow of wind on the front of the building reduces the external pressure over the whole of the roof, and the lifting of the tiles or the whole roof is a natural consequence.

Note 23, Chap. VIII, p. 363. TROPICAL HURRICANES. ADDITIONAL BIBLIOGRAPHY

WEST INDIES: C. L. Mitchell, 'West Indian hurricanes and other tropical cyclones of the N. Atlantic ocean,' *M.W. Rev.*, vol. LX, Washington, 1932, p. 253. (Charts for 1924–32 bringing up-to-date *M.W. Rev.*, Supp. 24.)

S. Sarasola, 'Los huracanes de las Antillas,' *Bogota, Obs. Nac. S. Bartolomé, Notas Geofis. Met.*, No. 2, 1925, Segunda edición, Madrid, 1928.

I. M. Cline, 'Tropical cyclones,' New York, London, Toronto, 1926.

A. J. Henry, 'The frequency of tropical cyclones that closely approach or enter continental United States,' *M.W. Rev.*, vol. LVII, 1929, p. 328.

INDIAN OCEAN: C. W. B. Normand, *Storm tracks in the Arabian Sea*, India Met. Dept., 1926; *Storm tracks in the Bay of Bengal*, India Met. Dept., Calcutta, 1925.

C. Poisson, 'Les cyclones de Madagascar,' *Matériaux pour l'étude des calamités*, vol. V, Genève, 1928.

PACIFIC OCEAN: E. Gherzi, 'Piccolo atlante dei tifoni dei Mari della Cina,' Genova, *Ann. Idrog., Ist. Idrog. Regia Marina*, vol. XII, 1928.

E. Bruzon and P. Carton, *Le climat de l'Indochine et les typhons de la Mer de Chine*, Hanoi, 1929.

T. F. Claxton, *Isotyphs showing the prevalence of typhoons in different regions of the Far East for each month of the year*, Hong Kong, 1932.

Manila, Bureau of Agriculture, *The tracks of remarkable typhoons which affected general weather conditions in the Philippines between* 1903 *and* 1925, Manila, 1926.

Coching Chu, 'A new classification of typhoons of the Far East,' *M.W. Rev.*, vol. LII, 1924, p. 570; 'The place of origin and recurvature of typhoons,' *Ibid.*, vol. LIII, 1925, pp. 1–5.

M. Selga, 'Tropical revolving storms within eight degrees of the equator,' 'The birthplace of typhoons,' *Melbourne, Proc. Pan-Pacific Sci. Congr.* 1923, pp. 665–80; *see also* Meteorological Notes, Nos. 5, 6, 7 of the Philippine Weather Bureau. *Bulletin*, 1931.

W. E. Hurd, 'Tropical cyclones of the eastern N. Pacific Ocean,' *M.W Rev.*, vol. LVII, 1929, p. 43.

Waterspouts. W. E. Hurd, 'Waterspouts,' *M.W. Rev.*, vol. LVI, 1928, pp. 207–11.

Dust-Devils. R. G. Veryard, *Meteor. Mag.*, vol. LXIX, 1934, p. 268.

L. Weickmann, *Zum Klima der Turkei*, Erstes Heft, p. 96.

Note 24. BIBLIOGRAPHY OF METEOROLOGICAL CHARTS

The charts of the meteorological elements reproduced in the first edition of this volume were compiled shortly after the end of the war and were based on the most recent data available at that time in the Meteorological Office in London. During the last fifteen years a large number of data and charts have been published giving details of one or more of the elements in different regions of the globe. For the scale, however, on which they are drawn the charts in the published volume probably give a fairly accurate picture of the general distribution and no attempt has been made to modify or correct them.

For readers who require more recent information about any particular region a bibliography has been compiled of the most recent charts available; in the case of a few regions for which no charts have been published references are given to compilations of data.

In order that the charts for a given element may be easily referred to, columns are given for the following elements:

tt temperature	p pressure	⌓ cloudiness
Y wind	• rainfall	

An entry is made in the appropriate column to show whether the charts included in the publication are for the months, the seasons or the year:

M charts for each of the 12 months	y charts for the year
m charts for a selection of months	* charts of special character
s charts for the seasons	x data only

Italic letters in the temperature column indicate the *temperatures of the sea*.

tt	☁	●	p	Γ	
					The World as a Whole
M y					W. Gorczynski. *Nouvelles isothermes de la Pologne, de l'Europe et du globe terrestre.* Warsaw, Pamietnik Fizyograficzny, 1918.
			M y		W. Gorczynski. *Pression atmosphérique en Pologne et en Europe* (avec 54 cartes contenant les isobares mensuelles et annuelles de la Pologne, de l'Europe et du globe terrestre). Warsaw, 1917.
*					C. E. P. Brooks and G. L. Thorman. 'The distribution of mean annual maxima and minima of temperature over the globe.' *Geophysical Memoirs,* No. 44, M.O, 307 d. London, 1928.
		y			M. Jefferson. 'A new map of world rainfall.' *Geog. Rev.*, 16, pp. 285–90. New York, 1926.
		y			E. Ekhart. 'Eine neue Regenkarte der Erde (1911–20).' *Petermanns Mitt.*, 76, p. 57. Gotha, 1930.
		y			W. Meinardus. 'Eine neue Niederschlagskarte der Erde.' *Ibid.*, 80, pp. 1–4. Gotha, 1934.
					Europe in General
	M y				K. Knoch. 'Die Verteilung der Bewölkung über Europa.' *Veröff. Preuss. Met. Inst.*, Abhand. Bd. VII, Nr. 5. Berlin, 1923.
			s y *		E. Alt. Klimatologie von Süddeutschland. IV. Teil, 'Die Luftdruckverteilung über Europa, dargestellt nach Pentadenmitteln (1880–1909).' *D. Met. Jahrb. Bayern,* 41, 1919, pp. E 1–12. (Charts of 5-day means.)
			*		G. v. Elsner. 'Die Verteilung des Luftdrucks über Europa und dem Nordatlantischen Ozean dargestellt auf Grund zwanzigjähriger Pentadenmittel (1890–1909).' *Veröff. Preuss. Met. Inst.*, Abhand. Bd. VII, Nr. 7. Berlin, 1925.
					Scandinavia (Sweden, Norway, Denmark, Iceland and the Baltic)
		M y			A. Wallén. 'Nederbördskartor över Sverige (1881–1920).' *Stat. Met. Hydrog. Anst., Medded.* 2, No. 3. Stockholm, 1924.
		y			H. W. Ahlmann. 'Karta över den årliga nederbördens fördelning på Skandinaviska halvön.' *Ibid.*, 3, No. 4. Stockholm, 1925.
m y	m	m y	m y	m	A. Wallén. 'Climate of Sweden.' *Stat. Met. Hydrog. Anst.*, Nr. 279. Stockholm, 1930. (Contains also charts of absolute humidity, sunshine and snow.)
M y	s		M	M	H. Mohn. 'Atlas de climat de Norvège.' Nouvelle ed. par A. Graarud et K. Irgens. *Geofys. Pub.*, II, No. 7. Kristiania, 1922. (Contains also seasonal charts of wind-force, gales, fog and thunderstorms; charts for four months of absolute and relative humidity and much additional information.)
		M y			Kristiania, Norske Met. Inst. 'Nedbøriagttagelser i Norge. Middelvaerdier, maksima og minima 1876–1915.' *Tillaegsh. til aarg.*, 24, 1918. Kristiania, 1920. (Normals for 1896–1925, but no charts, are given in *Tilleggshefte til årg.*, 33, 1927. Oslo, 1928.)
m y		m y	m y		G. Kunze. *Beiträge zur Klimatologie des Europäischen Nordmeeres (Skandik).* Inaug.-Diss. Univ. Breslau, Schramberg (Württ.), 1933.
		y			W. Petersen. 'Bidrag til et Nedbørskort over Danmark.' Kjøbenhavn *Geogr. Tids.*, 33, pp. 209–15. 1930.
M y		M y	m y		Copenhagen, Danske Met. Inst. *Danmarks Klima, Tabeller og Kort.* København, 1933.
m					N. H. Jacobsen. 'Isotermekort over Island for Januar og Juli.' København *Geogr. Tids.*, 35, pp. 54–66. 1932.
		*			T. Thorkellson. 'Um úrkomu á Islandi.' *Reykjavik, Bunadarritinu*, 38.
		y			H. Renier. 'Niederschlag und Bewölkung auf Island.' *Ann. Hydr.*, 61, pp. 252–4. 1933.
					European Russia, Finland and the Baltic States, Poland
M y					Leningrad, Geophys. Zent. Obs. *Klima der U.S.S.R.*, Teil I, 'Die Lufttemperatur.' Lief. I, 'Monatsmittel der Lufttemperatur im Europäischen Teil der U.S.S.R.,' by E. Rubinstein. Leningrad, 1927.
			M y	M y	Teil II, 'Luftdruck und Wind in der U.S.S.R.,' Lief. 1 und 2, 'Luftdruck nach Monatsmitteln und Windrichtung in der U.S.S.R.,' by A. Kaminsky. Leningrad, 1932. (Isobars at 500 m, 1000 m and 1500 m are given for the mountainous regions.)
		M y			S. Nebolsin. 'Quantités moyennes des précipitations atmosphériques en Russie d'Europe....' *Recueil de Géophysique*, Tome III, Fasc. 1. Petrograd, 1916.
M y		M y			Kiev, Ukraine Met. Service. *Climatological atlas of the Ukraine.* Kyiv, 1927.
		M y s			W. W. Korhonen. 'Niederschlagskarten aus Finnland.' *Met. Zentr. Anst. finn. Staat.*, *Mitt.* No. 1, Helsinki, 1925; also No. 9, Helsinki, 1921.
		s y			W. W. Korhonen. 'Niederschlagskarten von Fennoskandia.' *Ibid.*, No. 16. Helsinki, 1925.
M y					J. Keränen. 'Temperaturkarten von Finnland.' *Ibid.*, No. 17. Helsinki, 1925.
M y					R. Meyer and G. Bauman. 'Beiträge zur Klimakunde des Ostbaltischen Gebietes.' I, Mittelwerte der Temperatur, 1886–1910. *Riga, Arb. Meteor. Inst. Univ.*, No. 8. 1927. (For other temperature data see also *Ibid.*, No. 14. Riga, 1930.)
	y				H. Liedemann. 'Über die Sonnenscheindauer und Bewölkung in Eesti.' *Dorpat, Acta Comm. Univ. Tartuensis*, A, XIX 3. 1930.
					R. Gumiński. 'L'humidité de l'air en Pologne.' *Etudes météor. hydrog.*, fasc. 3. Warsaw, 1927. (My of absolute humidity and sy of relative humidity.)
		M y			St Kosińska-Bartnicka. 'Les précipitations en Pologne.' *Ibid.*, fasc. 5, p. 41. 1927.
				m y	L. Bartnicki. 'Les courants atmosphériques en Pologne.' *Ibid.*, fasc. 9. 1930. (For temperature and pressure in Poland *see* The world as a whole.)
					Germany and the Rhine
M y	M y	M y	M y	M y	Berlin, Preuss. Met. Inst. *Klima-atlas von Deutschland.* Berlin, 1921. (My of vapour pressure and relative humidity.)
		y			K. Schneider. 'Normalwerte des Niederschlags für Thüringen und benachbarte Gebiete.' *Jena, Mitt. Thür. Landesw.*, Heft 4. 1932. (Monthly data.)
M y					E. Alt. 'Die mittlere Temperaturverteilung in Süddeutschland....' *München, D. Met. Jb. Bayern*, 42. 1920.
		M y			J. Haeuser. 'Die Niederschlagsverhältnisse in Bayern...1901–25.' *Veröff. Bayer. Landesst. Gewässerk.* Atlas. München, 1930.
y	x	x		x	P. Polis. *Klima und Niederschlagsverhältnisse im Rheingebiet.* Berlin-Grunewald, 1928. (Chiefly tables.)
m y		y		y	M. G. Barbé. 'Climatologie de la région du Rhin.' *Mem. Off. Nat. Météor. France*, No. 8. Paris, 1924.
		y			A. Dieckmann. 'Die Niederschlagsverhältnisse der Deutschen Nordseeinseln.' *Ber. Strahlungs-Klimat. Stationsnetzes im Deutschen Nordseegebiet*, II, 1928 (1930), p. 55.
		s y			E. Kleinschmidt. 'Neue Niederschlagskarten von Württemberg.' *Württemberg. Jahrb. Stat. Landeskunde*, 1923–4. Stuttgart, 1926.
		M y			W. Peppler. 'Die Niederschlagsverhältnisse in Baden.' *Veröff. Bad. Landesw.*, Abhand. Nr. 1. Karlsruhe, 1922.
y	s y	y	s y	s y	E. Thost. 'Das Klima des nördlichen Württemberg.' *Stuttgarter Geog. Studien*, Reihe A, Heft 24/25. 1930.
					Netherlands
m y		M y	M	M	C. Braak. 'The climate of the Netherlands.' De Bilt, K. Ned. Met. Inst., *Meded. en Verh.*, 32–34. 1929–33.
		y			E. Vanderlinden. 'Sur la distribution de la pluie en Belgique.' *Brussels, Inst. Roy. Met., Mémoires*, 2. 1927. (Tables of monthly data.)

tt	⌂	●	p	ϒ	
					British Isles
* M y		M y			London, Meteorological Office. *The book of normals of meteorological elements for the British Isles for periods ending* 1915. Section III. Maps of the normal distribution of temperature (max. and min.), rainfall and sunshine. M.O. 236. London, 1920. (For more recent temperature data *see* M.O. 364, 1933.)
		M y			London, Royal Meteorological Society. *Rainfall atlas of the British Isles.* 1926.
					France
s y		m y		s y	E. Bénévent. 'Le climat des Alpes françaises.' *Paris, Off. Nat. Météor., Mémorial* No. 14. 1926.
					Iberian peninsula
m y					W. Semmelhack. 'Temperaturkarten der Iberischen Halbinsel.' *Ann. Hydr.*, 60, p. 327. 1932.
		y			W. Semmelhack. 'Niederschlagskarte der Iberischen Halbinsel.' *Ibid.*, p. 28.
		M y			J. Febrer. 'Atlas pluviomètric de Catalunya.' *Barcelona, Memòries Patxot*, vol. I. 1930.
x		x	x		J. M. de Almeida Lima. *O clima de Portugal continental.* Lisboa, 1922. (Contains also data of relative humidity, wind-velocity and evaporation.)
m		s y		m	H. Lautensach. 'Portugal: auf Grund eigener Reisen und der Literatur.' *Peterm. Mitt. Ergänzh.*, No. 213. 1932.
					Italy and Sardinia
x					F. Eredia. *La température moyenne mensuelle en Italie.* Rome, 1931.
m y					F. Eredia. 'La distribuzione della temperatura media dell' aria nella Sardegna.' *Ann. Lav. Pubb.* 1932, fasc. 12. Rome.
m y	y	y	y s		G. Crestani. *Climatologia.* Torino, 1931. (Tables of monthly data, charts chiefly from Eredia and De Marchi.)
					Switzerland
		y			H. Brockmann-Jerosch. 'Die Niederschlagsverhältnisse der Schweiz.' *Vegetation der Schweiz*, Heft 12. Zürich, 1925.
*	y				'Klimakarten der Schweiz.' By W. Brückmann. Zürich, Schweiz. Meteor. Zentr. Anst. *Annalen*, 68. 1931.
					Czecho-Slovakia, Austria, Hungary
m		y			N. Krebs. *Die Ostalpen und das heutige Oesterreich.* Stuttgart, 1928.
m y	m	m y			E. Biel. 'Klimatographie des ehemaligen oesterreichischen Küstenlandes.' *Wien, Denksch. Akad. Wiss.*, 101, pp. 137–93. 1927.
m y					Vienna, Hydrog. Zentralbureau im Bundesministerium für Land- und Forstwirtschaft. 'Temperaturmittel 1896–1915 und Isothermenkarten von Oesterreich.' *Wien, Mitt. Geog. Ges.*, 72. 1929.
				m	A. Defant. 'Die Windverhältnisse im Gebiete der ehemaligen Österr.-Ungar. Monarchie.' *Jahrb. Zentr.-Anst. Meteor. Geodyn.*, 1920 Anhang. Wien, 1924.
	*				J. Loidl. 'Die Bewölkung von Österreich.' *Ibid.*, 1924 Anhang. Wien, 1927. (Frequency of cloudiness ⩽2.)
		M y			K. Knoch and E. Reichel. 'Verteilung und jährlicher Gang der Niederschläge in den Alpen.' *Berlin, Preuss. Met. Inst.*, Abh. 9, Nr. 6. 1930. (The monthly charts are given as percentages of annual fall.)
					Balkan peninsula
		y			H. Renier. 'Die Niederschlagsverteilung in Südosteuropa.' *Mem. Soc. Geog. Beograd.*, vol. I. 1933.
		y			S. Škreb. 'Niederschläge in...Kroatien und Slavonien 1901–10.' Zagreb Kgl. Landesanstalt für Meteor. und Geodyn. (Monthly data but no monthly charts.)
m y	m y				E. Otetelişanu. 'Die Temperaturverhältnisse von Rumäniei mit einem Atlas.' *Bucharest, Inst. Met. Cent. al României, Memorii şi Studii*, vol. I, No. 1. Bucureşti, 1920. (For other climatic data see *Bucharest Inst. Met. Cent.*, Date climatologice, vol. I, No. 1. 1931.)
		y			E. Otetelişanu and G. D. Elefteriu. *Consideratiuni generale asupra Regimului precipitaţiunilor atmosferice in Romania.* Bucureşti, 1921.
M y		s y			E. Otetelişanu and A. Dissescou. 'Climat de la Dobroudja et du Littoral de la Mer noire.' *Inst. Met. Cent. al României, Memorii şi Studii*, vol. I, No. 3. Bucureşti, 1928.
	m y				A. N. Livathinos. 'Sur la nébulosité en Grèce.' *Ann. Obs. Nat.*, 8. Athens, 1926.
		s y y			E. G. Mariolopoulos. *Etude sur le climat de la Grèce. Précipitation.* Paris, 1925. (Monthly data.) 'Etude des régimes pluviométriques de la Grèce.' *Ann. Obs. Nat.*, 12. Athens, 1934.
		y s			E. Reichel. 'Die Niederschlagsverhältnisse der Türkei.' *Ann. Hydr.*, 60, p. 353. 1932.
					Mediterranean
m y			M	m	*Zum Klima der Türkei.* Erstes Heft: L. Weickmann, 'Luftdruck und Winde im östlichen Mittelmeergebiet; Zweites Heft: P. Zistler, 'Die Temperaturverhältnisse der Türkei.' München.
					Asiatic Russia (*see also* European Russia)
M y					Leningrad, Geophys. Zent. Obs. *Klima der U.S.S.R.* Teil 1, 'Die Lufttemperatur,' Lief. 3, 'Monatsmittel der Lufttemperatur im Asiatischen Teil der U.S.S.R.,' by E. Rubinstein (with atlas). Leningrad, 1931–2.
M y	s y	s y			Leningrad, Acad. des Sci. de l'U.R.S.S. *Travaux de la commission pour l'étude de la république autonome sov. soc. Jakoute.* Tome VI, 'Contributions à l'étude du climat de la république Jakoute et des pays contiguës de l'Asie septentrionale,' by V. B. Schostakowitsch. Leningrad, 1927. (Contains chart of wind-velocity but no direction.)
		x			Vladivostok Met. Obs. *Precipitations of Maritime Province.* Vladivostok, 1925.
m y					H. von Ficker. 'Untersuchungen über Temperaturverteilung, Bewölkung und Niederschlag in einigen Gebieten des mittleren Asiens.' *Geog. Ann.*, 5, p. 351. Stockholm, 1923.
			M y		P. I. Koloskoff. 'A climatological sketch of the Kamtchatka peninsula.' *Vladivostok. Rec. Far East Geophysical Inst.*, No. 2 (9), p. 119. 1932. *See also* 'A climatological sketch of the central part of the far eastern district.' *Ibid.*, p. 165.
				M y	A. Kaminsky. 'Über die Windrichtung auf dem Kaspischen Meere....' *Leningrad, Nachr. Zentr.-bur. Hydrometeor*, 5, p. 112. 1925.
					China
M y					H. Gauthier. *La température en Chine et à quelques stations voisines d'après des observations quotidiennes.* Shanghai, 1918.
		M y			E. Gherzi. *Etude sur la pluie en Chine 1873–1925.* Shanghai, 1928.
M *					E. Gherzi. *Atlas thermométrique de la Chine.* Zikawei, 1934.
			m	m	Coching Chu. 'Circulation of atmosphere over China.' *Mem. Nat. Res. Inst. Met.*, No. 4. Nanking, 1934.
				M	E. Gherzi. *The winds and upper air-currents along the China coast and in the Yangtze valley.* Zikawei Obs., Shanghai, 1931.
					Korea
		M y			'Rainfall in Chôsen (Korea).' *Met. Obs. of Govt. Gen. of Chôsen.* Zinsen (Chemulpo), 1925.

tt	⌓	●	p	Γ	
M y *					'Air-temperature in Tyôsen (Korea).' *Met. Obs. of Govt. Gen. of Chôsen*, Zinsen, 1928. (Also monthly and annual charts of mean daily max. and min.; absolute max. and min., etc.)
					Japan
M y	M y	M y	M y	M y	Tokyo, Central Meteorological Observatory. *Climatic atlas of Japan and her neighbouring countries*. Tokyo, 1929. (Contains also charts of humidity and other elements.) *See also* T. Okada, 'The climate of Japan.' *Bull. Cent. Meteor. Obs.*, 4, No. 2. Tokyo, 1931.
		M y			Taihoku, Met. Obs. *The rainfall in the island of Formosa*. 1920.
	M y				K. Tsukuda. 'The mean cloudiness in the Far East.' *Mem. Imp. Marine Obs., Kobe*, 4, No. 2. 1931.
					Indo-China and Siam
M		M y	M	s	Phu-Liên, Service Météorologique. *Atlas*. Hanoi, 1930.
		M y		s	E. Bruzon and P. Carton. *Le climat de l'Indo-Chine....* Hanoi, 1929.
		s			V. D. Iyer. 'Rainfall of Siam....' *India Met. Dept. Scientific Notes*, 4, No. 38. Calcutta, 1931. (Refers to monthly maps for the period 1901–21 published in the Administrative Report of the Royal Irrigation Dept. of Siam.)
					India
		x			G. T. Walker. 'Monthly and annual normals of rainfall and of rainy days.' *Ind. Met. Memoirs*, 23. Calcutta, 1924. (Tables only, no charts.)
					Ceylon
		M			A. J. Bamford. 'Ceylon rainfall, monthly averages.' *Mem. Colombo Obs.*, No. 3. Ceylon Survey Dept., Colombo, 1921.
					Malay Archipelago, Dutch East Indies, Philippines
		m			C. D. Stewart. 'The rainfall of Malaya.' *Malayan Agric. J.* 1930. (Data for each of 12 months.)
		M y			J. Boerema. 'Regenval in Nederlandsch-Indie.' *K. Mag. en Met. Obs.*, Batavia, Verh. No. 14, Deel II Kaarten. Weltevreden, 1925.
		M y			J. Boerema. 'Regenval in Nederlandsch-Indie.' *Ibid.*, Verh. No. 24. 1931–3, Deel II Sumatra, Deel III Borneo, Deel IV Celebes.
M *M*			M	M	C. Braak. 'The climate of the Netherlands Indies.' *Ibid.*, Verh. No. 8. 1921–9.
x		x			C. E. P. Brooks. 'Meteorology of British North Borneo.' *Q.J. Roy. Meteor. Soc.*, 47, p. 294. 1921. (Monthly data.) *See also* J. M. Hall. 'Ten years' rainfall in British N. Borneo.' *Brit. N. Borneo Herald*, 50, No. 8, pp. 80–4. 1932.
y		y		s	J. Coronas. *The climate and weather of the Philippines*. Manila, 1920. (Contains also monthly data for many elements.)
	x			M y	M. Selga. 'Wind-roses of ideal marine stations in and near the Philippines.' *Manila Obs. Publ.* 3, No. 3. 1931.
					Persia and Asia Minor (*see also* Balkan peninsula and Mediterranean)
x	x	x	x	x	B. N. Banerji. *Meteorology of the Persian Gulf and Mekran*. India Met. Dept. Calcutta, 1931. (Data only.)
x	x	x		x	K. Schneider. 'Klimatologie von Armenien.' *Peterm. Mitt.*, 77, p. 69. Gotha, 1931. (Data only.)
		M y			D. Ashbel. *Die Niederschlagsverhältnisse im südlichen Libanon, in Palästina und im nördlichen Sinai*. Berlin, 1930.
		y			Bagdad Weather Bureau. *Climate and Weather of Iraq*. 1919.

tt	⌓	●	p	Γ	
					Africa
		*			H. Schmidt. 'Der jährliche Gang der Niederschläge in Afrika.' *Arch. Deutsch. Seew.*, 46, No. 1. Hamburg, 1928.
		m	m y	m	C. E. P. Brooks and S. T. A. Mirrlees. 'A study of the atmospheric circulation over tropical Africa.' *Geophysical Memoirs*, No. 55. London, 1932.
m	y	y	m	m	C. E. P. Brooks. 'Le climat du Sahara et de l'Arabie.' *Le Sahara*, T. 1, pte 1. Paris, 1932.
m y					C. E. P. Brooks. 'Charts of mean temperature in Africa.' *Q.J. Roy. Meteor. Soc.*, 45, p. 251. 1919.
		y			H. L. Shantz and C. F. Marbut. *The vegetation and soils of Africa*, pp. 245–6. New York, 1923.
					Morocco, Algeria, Tunis and Tripoli
		y			L. Embérger. *Carte des pluies de Maroc*. Rabat, 1933. (See *Météorologie*, 3, p. 442, Paris, 1927, and 10, p. 491, 1934.)
x		x			Rabat, Institut scientifique Chérifien. 'Pluies et températures moyennes mensuelles.' *Bull. Offic.*, No. 749, p. 437. Rabat, 1927.
				m	Gouvernement Général de l'Algérie. *Les territoires du sud de l'Algérie*. Aperçu météorologique. By A. Lasserre. 1922.
		s y			*Le régime des pluies en Tunisie*...1901–30. Tunis, Service Météorologique, 1931.
m y *		s y		s	G. Ginestous. *Résumé de climatologie tunisienne*. Tunis, 1922; *Etude climatologique du Golfe de Tunis*. Tunis, 1925.
m y		s y		s	*Atlante meteorologico della Libia*. Tripoli, Ministero delle Colonie. Tripoli, 1930.
m		y			L. Wittschell. *Klima und Landschaft in Tripolitanien*. Hamburg, 1928.
					Egypt, Eritrea
m	m	m y	m	m	Cairo, Survey of Egypt. *Meteorological Atlas of Egypt*. Giza, 1931.
m y	m	m y	m	m	Cairo, Ministry of Public Works, Physical Dept. 'The Nile Basin,' vol. I, by H. E. Hurst and P. Phillips, ch. III, *Phys. Dept. Paper*, No. 26. 1931.
		y			F. Eredia. 'Le precipitazioni acquee nell' Eritrea dal 1923 al 1931.' *Rome, A. Lav. Pubbl.*, fasc. 9. 1932. (Contains also monthly data.)
m					F. Eredia. *Sui caratteri climatologici della Somalia Italiana meridionale*. Genova, 1927.
					West Africa
*		M y			H. Hubert. 'Cartes météorologiques de l'ouest africain.' *Paris, A. Phys. Globe France d'outre-mer*, 1, No. 1, p. 25. 1934. (Monthly charts of tt max. and min.)
		M y			R. Pignol. 'Die Niederschlagsverhältnisse des alten deutschen Schutzgebietes Togo.' Hamburg, *Arch. Deutsch. Seewarte*, 49. 1930.
		M y			R. Rousseau. 'Les pluies au Sénégal.' *Bull. Com. Etudes Hist. Sci.*, 14, p. 157. 1931. (See also *Météorologie*, 11, p. 153, Paris, 1935.)
y		y			N. P. Chamney. 'The climatology of the Gold Coast.' *Accra, Dept. Agric. Bull.*, No. 15. 1928.
		y			*Mean distribution of rainfall (to end of* 1927). Accra, Dept. of Agric.
M y	x			x	C. E. P. Brooks. 'The distribution of temperature over Nigeria.' *Q.J. Roy. Meteor. Soc.*, 46, p. 204. 1920. (For data of winds and cloudiness *see* 47, p. 203, 1921.)
		x			*Graphs showing amount of rainfall recorded at 34 stations for various years from* 1907–31. Lagos, Survey Dept. 1933.
		x			P. Gasthuys. 'Observations pluviométriques faites dans la province orientale (Congo Belge) Bruxelles.' *Bull. Agric. Congo Belge*, 13, p. 260. 1922.
					East Africa
		y			C. E. P Brooks. 'The distribution of rainfall over Uganda with a note on Kenya colony.' *Q.J. Roy. Meteor. Soc.*, 50, p. 325. 1924. (Monthly data, seasonal charts for Uganda as percentage of annual.)

tt	⌓	●	p	Γ	
					East Africa (*cont.*)
		x			'Available rainfall averages to end of 1925.' *Tanganyika Blue Book*, 1926.
		y			*Map of Uganda showing average annual rainfall in inches.* Uganda Survey Dept. Entebbe, 1925.
		s			*Rainfall map. Summer and autumn climate.* Zomba, Agric. Survey of Nyasaland. London (1934).
		y			*Normal rainfall map of S. Rhodesia* 1898–1926. Salisbury, Dept. Agric.
		M y			W. Paap. 'Die Niederschlagsverhältnisse des Schutzgebietes Deutsch-Ostafrika.' *Arch. Deutsch. Seewarte*, 53, Nr. 3. Hamburg, 1934.
					South Africa
		M y			J. R. Sutton. 'A contribution to the study of the rainfall map of S. Africa.' *Trans. R. Soc. S. Africa*, 9, p. 367. Cape Town, 1921.
		y			A. D. Lewis. *Rainfall normals up to the end of* 1925. Pretoria, Meteorological Office. Cape Town, 1927.
			m	m	*Official Year Books of the Union of S. Africa* contain normal values. No. 14, 1931–2, mean pressure and air-flow at 4350 ft. in Jan. and July with relative wind frequencies at surface and high levels in summer and winter.
		m y			P. Heidke. 'Die Niederschlagsverhältnisse von Deutsch-Südwestafrika.' *Mitt. deutsch. Schutzg.*, Bd. 32. Berlin, 1920.
					Madagascar
		x			F. Loewe. 'Die Niederschlagsverhältnisse von Madagaskar.' *Met. Zs.*, 43, p. 107. 1926.
		y s			*Cartes des pluies, total annuel, et saison sèche.* Tananarive, Dépt. de Géographie. 1930.
					North America as a Whole
			x		P. C. Day. 'Monthly normals of sea-level pressure for the U.S., Canada, Alaska and the W. Indies.' *M.W. Rev.*, 52, p. 30. 1924.
					Alaska
m y		M y	m	m	E. M. Fitton. 'The climates of Alaska.' *M.W. Rev.*, 58, p. 85. 1930.
					Canada
* x		s y			A. J. Connor. *The temperature and precipitation of Alberta, Saskatchewan and Manitoba.* Ottawa, Met. Service of Canada. 1920. (Mean max. and min. tt.)
m		x			A. J. Connor. 'The temperature and precipitation of N. Canada.' *Canada Year Book.* Ottawa, 1930.
m	y	y	m	m y	C. E. Koeppe. *The Canadian Climate.* Bloomington, 1931. (Monthly data of tt ⌓ and ●.)
					United States. [Detailed maps for individual states are not included in the bibliography.]
M *				m	Washington, D.C., U.S. Dept. Agric. *Atlas of American agriculture*, Part II, Climate: A, Precipitation and humidity, by J. B. Kincer; B, Temperature, sunshine and wind, by J. B. Kincer. Washington, 1928. (Contains also monthly charts of average daily max. and min. and highest and lowest monthly mean temperatures and monthly charts of hours of sunshine.)
		M s			J. B. Kincer. 'The seasonal distribution of precipitation and its frequency and intensity in the U.S.' *M.W. Rev.*, 47, p. 624. Washington, 1919.
		x			P. C. Day. 'Daily, monthly and annual normals of precipitation in the U.S.' *M.W. Rev. Supp.*, No. 34. Washington, 1930.
				M	E. R. Miller. 'Monthly charts of frequency-resultant winds in the U.S.' *M.W. Rev.*, 55, p. 308. 1927.
m	y	s y		m	R. de C. Ward. *The climates of the United States.* Boston, 1925: 'Note on atmospheric humidity in the U.S.' *M.W. Rev.*, 50, p. 575. 1922. (Charts of relative humidity and vapour-pressure in Jan. and July.)
					Mexico
M y		M y	m		*Atlas termopluviométrico de la República Mexicana.* Tacubaya, 1924. (Period 1906–10; also chart of humidity for the year.)
M y	*	M y	m y	M y	*Atlas climatológico de la República Mexicana.* Tacubaya. (Period 1921–25; also charts of relative humidity for 4 months and the year, and annual chart of frequency of clear and cloudy days.)
M y					J. Hernandez. 'The temperature of Mexico.' *M.W. Rev. Supp.*, No. 23. Washington, 1923. (Charts for surface and sea-level, and much additional information.)
m		y	m	m	J. L. Page. 'Climate of Mexico.' *M.W. Rev Supp.*, No. 33. Washington, 1930.
					Central and S. America
		s			E. van Cleef. 'Rainfall maps of Latin America.' *M.W. Rev.*, 49, p. 537. 1921.
		y y *			B. Franze. 'Die Niederschlagsverhältnisse in Südamerika.' *Peterm. Mitt. Erganzh.* No. 193. 1927. (Step-diagrams of monthly rainfall.)
	x				R. C. Mossman. 'Cloud amount in Brazil and Chile.' *Q.J. Roy. Meteor. Soc.*, 46, p. 294. 1920.
x	x	x		x	W. W. Reed. 'Climatological data for N. and W. tropical S. America.' *M.W. Rev. Supp.*, No. 31. Washington, 1928; '...for Southern S. America,' *Ibid.*, No. 32. 1929 (M data); '...for Central America,' *M.W. Rev.*, 51, p. 133. 1923.
		y			H. J. Spinden. 'The population of ancient America.' *Geog. Rev.*, 18, p. 641. 1928.
					Central America (Guatemala, Salvador)
		x			K. Sapper. 'Regenfall in den Republiken Guatemala und El Salvador.' *Met. Zs.*, 38, p. 279. 1921. (M data.)
		y			A. K. Botts. 'The rainfall of Salvador.' *M.W. Rev.*, 58, p. 459. 1930.
					West Indies and Bermuda
	s y	y *		m	W. Kloster. 'Bewölkungs-, Niederschlags- und Gewitterverhältnisse der westindischen Gewässer und der angrenzenden Landmassen.' *Arch. Deutsch. Seew.*, 40. Hamburg, 1922.
x	x	x		x	W. W. Reed. 'Climatological data for the W. Indian Is.' *M.W. Rev.*, 54, p. 133. 1926. (M data.)
x		x	x	x	'Estudio climatológico de Cuba.' Habana, *Bol. Obs. Nac.*, 24, p. 269. 1928. (M data.)
x		x			O. L. Fassig. 'Rainfall and temperature of Cuba.' *Tropical Plant Research Foundation, Bull.* 1. Washington, 1925.
		M y			E. J. Foscue. 'Rainfall maps of Cuba.' *M.W. Rev.*, 56, p. 170. 1928.
		y x			*The rainfall of Jamaica from about* 1870 *to end of* 1929. Jamaica, 1934. (M data.)
		x			C. E. P. Brooks. 'The rainfall of San Domingo.' *Met. Mag.*, 56, p. 73. 1921.
		y			O. L. Fassig. 'A tentative chart of annual rainfall over the island of Haiti-Santo Domingo.' *M.W. Rev.*, 57, p. 296. 1929.
					Venezuela, Trinidad
		x			W. Köppen. 'Neue Regenmessungen in Venezuela und Surinam.' *Met. Zs.*, 51, p. 108. 1934.
		y			P. E. James. 'The climate of Trinidad.' *M.W. Rev.*, 53, p. 71. 1925. (See also *Scot. Geog. Mag.*, 42, p. 84. Edinburgh, 1926.)
		y			F. M. Bain. *The rainfall of Trinidad.* Department of Agriculture. Trinidad, 1934.
					Brazil
y		y			H. Morize. *Contribuição ao Estudo do Clima do Brasil.* Rio de Janeiro, 1922. (*See also* M. Jefferson, *Geog. Rev.* 1924, p. 127.)
x	x	x	x	x	*Boletim de Normaes.* Rio de Janeiro, Directoria de Meteorologia, 1922.

tt	⌓	●	p	Γ	
		y			C. M. Delgado de Carvalho. 'Dados pluviometricos relativos ao nordéste do Brazil.' *Ministerio da Viação e Obras publicas, Publ.* No. 47. Rio de Janeiro, 1922; 'Atlas.' *Ibid., Publ.* No. 53. 1923.
y		y			*Boletim de Normaes de temperatura, chuva e insolação*... 1914–21. Commissão geographica e geologica do Estado de Minas Geraes. Bello Horizonte, 1923.

Argentine Republic

tt	⌓	●	p	Γ	
y		y			G. Hoxmark. *Las condiciones climatológicas y el rendimiento del trigo.* Ministerio de Agricultura. Buenos Aires, 1925.
*		m y			G. Hoxmark. *El maiz en la Argentina.* Buenos Aires, 1927.
		s y			'El clima de la República Argentina.' *Almanaque del Min. Agric.*, p. 441 Buenos Aires, 1927.

Uruguay

tt	⌓	●	p	Γ	
		y			*Mapa pluviométrico* 1914–27. Observatorio Nacional, Montevideo.

Chile

tt	⌓	●	p	Γ	
		m y			W. Knoche. 'Jahres-, Januar- und Juli-Niederschlagskarte der Republik Chile.' *Zs. Ges. Erdk.*, p. 208. Berlin, 1929.
	m				W. Knoche. 'Karten der Januar- und Juli-Bewölkung in Chile.' *Ibid.*, p. 220. 1927.
		y			M. Jefferson. 'The rainfall of Chile.' *Amer. Geog. Soc.*, Research Series, No. 7. New York, 1921.

Australasia

tt	⌓	●	p	Γ	
		x			R. C. Mossman. 'Southern hemisphere decadal and mean monthly and annual rainfall.' *Q.J. Roy. Meteor. Soc.*, p. 355. 1919.

Australia

tt	⌓	●	p	Γ	
			M	M	*Mean monthly pressure and gradient wind over Australia at 9 a.m. and 3 p.m.* Melbourne, Bureau of Meteorology. 1928.
		y			*Average annual rainfall map of Australia revised to* 1924. Melbourne, 1925.
			M y		*Charts showing the mean monthly isobars.* Melbourne, Bureau of Meteorology.
		x			Monthly rainfall figures for the several stations are given in:
					I. Jones. *Tables of rainfalls in Queensland.* Brisbane, 1933.
		M y			*Results of rainfall observations made in S. Australia and the N. Territory.* Melbourne, 1918.
		M y			*Results of rainfall observations made in W. Australia.* Melbourne, 1929.
x		x			*Meteorological data for certain Australian localities.* Melbourne, Council for Scientific and Industrial Research. Melbourne, 1933. (Max. and min. tt, humidity and ●.)
m y		m y	m y		Köppen-Geiger. *Handbuch der Klimatologie*, Bd. IV, Teil S, Griffith Taylor, 'Climatology of Australia.' 1932.

New Zealand

tt	⌓	●	p	Γ	
		y			E. Kidson. 'Average annual rainfall in New Zealand for the period 1891 to 1925.' *Wellington, Dept. Sci. Ind. Research.* 1930.
M y					E. Kidson. 'Mean temperatures in New Zealand.' *M.O. Note*, No. 7. Wellington, 1932.
			x		E. Kidson. 'The annual variation of pressure in New Zealand.' *M.O. Note*, No. 6. Wellington, 1931.
m y		m y			Köppen-Geiger. *Handbuch der Klimatologie*, Bd. IV, Teil S, E. Kidson, 'Climatology of New Zealand.' 1932.

Polar regions

tt	⌓	●	p	Γ	
m					R. N. Rudmose Brown. *The polar regions.* London, 1927.

Arctic and Greenland

tt	⌓	●	p	Γ	
m					*The geography of the polar regions.* New York, American Geog. Soc., 1928.
M	m	y	M	m	F. Baur. 'Das Klima der bisher Erforschten Teile der Arktis.' *Arktis*, 2. Jahrg. 1929, Heft 3, pp. 77–89; Heft 4, pp. 110–20.
m					H. U. Sverdrup. 'The Norwegian N. Polar expedition with the *Maud* 1918–25.' *Scientific results*, vol. II, Meteorology, Part I, Discussion, 1933.
x		x			H. Petersen. 'The climate of Greenland.' *Greenland*, I. 1927.

Oceans as a Whole

tt	⌓	●	p	Γ	
m *m*			m	m	'Meteorological charts for the world.' London, Admiralty. 1928. (Information of tt (sea) and Γ is lacking for the Pacific.)
x *x*	x	x	x	x	'Monthly meteorological data for 10° squares in the Oceans.' *K. Ned. Met. Inst.*, No. 107. Data for 1913–16, Atlantic and Indian Oceans only; 1917–30, all oceans. (Monthly data published year by year.)
				x	'Percentage frequency of gales in 5° squares in the four quarters.' *Marine Observer*, 2. 1925.

Atlantic Ocean
(*see also* W. Indies)

tt	⌓	●	p	Γ	
M *M*			M	M	'Oceanographische en meteorologische waarnemingen in den Atlantischen Oceaan.' Kaarten, *K. Ned. Met. Inst.*, No. 110, Dec.–Feb. (1870–1914), Mar.–May (1856–1920), June–Aug. (1870–1922), Sept.–Nov. (1870–1925). Amsterdam and Utrecht, *c.* 1920–31.
M					'Charts showing mean sea-surface temperature.' N. Atlantic, *Marine Observer*, 3, 1926; S. Atlantic, 4, 1927; Mediterranean and Black Seas, 7, 1930; Coastal regions of B. Isles, 10, 1933.
M *y*					'Atlas de température et salinité de l'eau de surface de la Mer du Nord et de la Manche.' *Conseil permanent international pour l'exploration de la mer.* Copenhagen, 1933.
M					'Monthly mean temperature of the surface water in the Atlantic, N of 50° N lat.' *Det Danske Met. Inst.* Kjøbenhavn, 1917.
m					'Atlas für Temperatur, Salzgehalt und Dichte der Nordsee und Ostsee.' *Deutsche Seewarte.* Hamburg, 1927.
M					P. M. van Riel. 'Surface temperature in the NW part of the Atlantic ocean.' *De Bilt, K. Ned. Met. Inst. Meded. en Verh.* No. 35. 1933.
m y	y	y	m y	m	G. Schott. *Geographie des Atlantischen Ozeans.* Hamburg, 1926. (Also tt sea.)
			*		G. v. Elsner. 'Die Verteilung des Luftdrucks über Europa und dem Nordatlantischen Ozean dargestellt auf Grund zwanzigjähriger Pentadenmittel (1890–1909).' *Preuss. Met. Inst.*, Abhand. 7, Nr. 7. Berlin, 1925.
m y	y		m y	M	'Monatskarte für den Südatlantischen Ozean, Jan.–Dez.' *Hamburg, Deutsche Seewarte.*
				M	'Monatskarte für den Nordatlantischen Ozean.' *Ibid.* (Issued monthly.)
M			M	M	'Pilot charts of the Central American waters.' *Washington, Hydrographic Office.* (Issued monthly.)
M			M	M	'Pilot charts of the N. Atlantic Ocean.' *Ibid.* (Issued monthly.)
s			s	s	'Pilot charts of the S. Atlantic Ocean.' *Ibid.* (Issued quarterly.)
M					J. Janke. 'Strömungen und Oberflächentemperaturen im Golfe von Guinea.' *Arch. Deutsch. Seew.*, 38, Nr. 6. Hamburg, 1920.

tt	⌓	●	p	Γ	Indian Ocean
M *M*	x		M	M	'Oceanographische en meteorologische waarnemingen in den Indischen Oceaan.' Utrecht, *K. Ned. Met. Inst.*, No. 104, Dec.–Feb. (1856–1910); Mar.–May (1856–1912); June–Aug. (1856–1908); Sept.–Nov. (1856–1914). With supplementary information for N of 5° to 1923.
M					'Mean values of sea-surface temperature for two-degree squares.' *Marine Observer*, 5. 1928.
m *m*		*	m	m	P. H. Gallé. 'Climatology of the Indian Ocean.' *K. Ned. Met. Inst.*, No. 102; *Meded. en Verh.*, 29. 1924–30. (● probability Dec.–Feb., and June–Aug.)
M			M	M	'Pilot charts of the Indian Ocean.' *Washington, Hydrographic Office.* (Issued monthly.)
					Indian and Pacific Oceans
	m	y			G. Schott. 'Die jährlichen Niederschlagsmengen auf dem Indischen und Stillen Ozean.' *Ann. Hydr.*, 61, p. 1. 1933. 'Die Bewölkung über dem Indischen und Stillen Ozean.' *Ibid.*, p. 280.

tt	⌓	●	p	Γ	Pacific Ocean
x *x*	x		x		'The mean atmospheric pressure, cloudiness, air and sea-surface temperature of the N. Pacific Ocean and neighbouring seas for the lustra 1916–20, 1921–25, 1926–30.' *Imperial Marine Observatory.* Kobe, 1925, 1929, 1932. (tt air, 1926–30 only.) '1911–30' Kobe, 1935.
M *y*					T. Okada. 'On the surface temperature of the Japan Sea.' *Mem. Imp. Marine Obs.* 1, p. 66. Kobe, 1923.
M y	M y		M y		K. Tsukuda. 'On the mean atmospheric pressure, cloudiness and sea-surface temperature of the N. Pacific Ocean.' *Mem. Imp. Marine Obs.*, 2, p. 163. Kobe, 1929. 'The mean cloudiness in the Far East.' *Ibid.*, 4, p. 75, Kobe, 1930.
			M	M	'Maps showing the mean atmospheric pressure and wind-direction and force over the China Sea for each month of the year.' *Hong Kong, Royal Observatory*, 1925.
x	x	x		x	W. W. Reed. 'Climatological data for the tropical islands of the Pacific Ocean (Oceania).' *M.W. Rev. Supp.*, No. 28. Washington. 1927.

Some Recent Additions.

tt	⌓	●	p	Γ	
					Italy
		s y			F. Eredia. *Le precipitazioni atmosferiche in Italia nel decennio* 1921–30. Ministero dei Lavori Pubblici, Servizio Idrografico, Roma, 1934. (Includes Sardinia and Sicily.)
					Hungary
		M y			F. Hajósy. 'Die Verteilung des Niederschlages in Ungarn (1901–30).' *K. Ung. Reichanst. Met. und Erdmag, Publikation* Bd. XI. Budapest, 1935.
					Greece
M y	M y	M y	M y	M y	E. G. Mariolopoulos and A. N. Livathinos. *Atlas climatique de la Grèce.* Observatoire National d'Athènes, 1935. (Contains also charts of humidity, snow and other elements.)
					Syria and Lebanon
y		y			Ch. Combier. 'La climatologie de la Syrie et du Liban.' *Rev. Géog. Phys. Géol. dyn.*, vol. VI, pp. 319–46. Paris, 1933.
					Rhodesia
m y		m y			Köppen-Geiger. *Handbuch der Klimatologie*, Bd. V, Teil X, C. L. Robertson and N. P. Sellick, 'The climate of Rhodesia, Nyasaland and Moçambique.' Berlin, 1933.
					Canada
m *			m	m	W. E. Knowles Middleton. 'The climate of the Gulf of St Lawrence and surrounding regions in Canada and Newfoundland, as it affects aviation.' *Toronto, Canad. Met. Mem.*, vol. I, No. 1. Ottawa, 1935. (Includes fog and snow.)

tt	⌓	●	p	Γ	
					Central and South America
m	m	m y	m	m	Köppen-Geiger. *Handbuch der Klimatologie*, Bd. II, Teil G, K. Knoch, 'Klimakunde von Südamerika.' Berlin, 1930; Teil H, K. Sapper, 'Klimakunde von Mittelamerika.' 1932.
					West Indies
		y			St Lucia. *Rainfall Map*, 1907–34. Department of Agriculture, St Lucia, 1935.
		M y			C. Braak. 'The climate of the Netherlands West Indies.' De Bilt, *K. Ned. Met. Inst.*, *Meded. en Verh.*, 36, 1935. (Suriname.)
					Argentine Republic
		M y			'El Régimen pluviométrico de la República Argentina.' *Ministerio de Agricultura, Dirección de Meteor. Geofis. e Hidrol.*, Serie F, Publ. No. 1. Buenos Aires, 1934.
					North Sea
M *M*				M	*Monatskarte für der Nord- und Ostsee.* Hamburg, Deutsche Seewarte, 1935.
					Mauritius
		y			M. Herchenroder. *La pluie à l'île Maurice.* Port Louis, 1935.
					Pacific Ocean
M *M*		*	M	M	Oceanographic and meteorological observations in the China Seas and in the Western part of the North Pacific Ocean. I. Monthly Charts for January to June (1910–30). *K. Ned. Met. Inst.* No. 115. 's-Gravenhage, 1935. (Includes also fog, gales and paths of depressions and typhoons.)

Note 25. Bibliography of Data for the Atmosphere below the 20 km level.

It may be remembered that in presenting the results of the ordinary observations of temperature and pressure for the whole world in chapters IV and VI, wisely or unwisely (with the exception of the charts of import and export of air) the values were "reduced to sea-level," in the case of temperature by an empirical table and, in the case of pressure, with the assistance of the readings of temperature, by a simple dynamical formula.

With the observations of the upper air no such reduction to a uniform level is available. Indeed, the variation of temperature, pressure, wind, or other element, with change of level is the subject of inquiry, for the representation of which results of observations in all parts of the world, over sea as well as land, are required.

In the chapters referred to we have given some results of actual observations of temperature, pressure and wind in various parts of the world. We have provisionally regarded them as typical and in combining them into a diagram to illustrate the variation of temperature with latitude we have disregarded longitude and used the winter values of stations in the Northern Hemisphere as substitutes for stations in the Southern Hemisphere during the Northern summer.

That, of course, can only be regarded as a temporary expedient, and as in ten years' time the jubilee of regular international co-operation in the study of the upper air may be celebrated, we may begin to think of the problem which confronts us. Beginning may be regarded as noting the existence of available material. In the course of an examination of the lists of a meteorological library, with the aid of Miss L. D. Sawyer's experience, we have noted the titles of three hundred publications which contain data for one or more meteorological elements in the upper air. Many of them are periodical, and as a periodical may have been discharging the duty for nearly forty years, a complete list of references connotes thousands of volumes.

And meanwhile the best way of using the material to represent effectively the meteorological conditions of the different parts of the atmospheric circulation is undecided. We have, therefore, to be content with indicating by means of an abbreviated bibliography the extent of the material available for different parts of the world, which a reader may co-ordinate in any way he pleases.

In preparing the list of data the order in which the countries are arranged needs some consideration. The atmospheric circulation is not exactly alphabetical. On a planet so nearly spherical as the earth, we might expect the zones of special meteorological significance to be controlled by parallels of latitude which regulate the duration of daylight and we should, therefore, consider separately the polar areas, Arctic and Antarctic, where daylight ranges during the year from twenty-four hours in a day to none at all, and the intertropical zone where daylight differs little from twelve hours per day throughout the year, with an intermediate zone where the variation of daylight during the year is the most conspicuous feature of climate but where the daylight never reaches either twenty-four hours or zero.

But for the difference of response of the land and the water to the influence of solar radiation, some simple arrangement of that kind would probably be effective, and with that possibility we may examine the conditions represented by the charts of pressure and temperature for the globe with the assurance that, as the elementary text-books of physical geography have told us, the motion of air going towards the equator will be complicated by an Easterly component and that going away from the equator by a Westerly component as exhibited in fig. 180.

Actually we may sum up as follows the zonal arrangements exhibited by the charts:

The Arctic circle with its associated ice and snow naturally retains its claim for separate treatment and the middle latitudes between 45° and 66° N are already identified as the favourite latitudes of cyclonic depressions travelling Eastward, apparently guided by the great West winds. Let us call them the great Westerlies. This zone, the natural home of fronts, has its counterpart in the Roaring Forties of the Southern Hemisphere.

Then in the Northern Hemisphere in the lower latitudes between 25° N and 45° N comes a curious zone which provides regions of permanent high pressure over the great oceans intersected by trade-winds on the Eastern sides conveying air to the equator. On the Continents are desert regions, and within the same belt we find the Mediterranean Sea, from the shores of which comes so great a part of our sciences. For lack of rainfall even the Mediterranean may be regarded as a desert in the summer months when the development of the oceanic anticyclones is most marked. And if we pursue the lines of the anticyclonic zone along the Mediterranean and across some of the Asiatic deserts we leave a region which we regard as semi-tropical and which provides for us a genial winter climate and come to the countries bordering on the Pacific which in spite of the favourable latitude have severe winters and so exhibit the marked difference in the weather of identical latitudes on opposite sides of the great Continents as noted in chapter IV.

Below this remarkable zone, which in the Western Hemisphere includes the desert regions of Northern Mexico and the Southern United States, we come to the equatorial zone which is characterised by a general drift of air Westward, a region of what we may call the great Easterlies, although, East of the African and Asiatic continents, it exhibits the remarkable reversals of the monsoons, and in or about the West Indies the reversal of the travel of the winds on the Northern side of the equator. We cannot deal with the Southern Hemisphere below 25° S as we would, for lack of observations over the sea.

So, after an enumeration of the publications in which data are to be found arranged according to countries in alphabetical order, our exposition of the available data begins with what information we can give about the air over the ordinary navigable sea, because, though not by any means the most abundant, it must be regarded from the point of view of meteorological science as ultimately of the greatest importance. With the zonal arrangement in mind, we begin with the ice-covered sea and snow-covered land North of the Arctic circle and then, dealing exclusively with land-areas, we step by way of Alaska to the few islands of the Pacific which contribute observations from the extremes of longitude East and West. Then we deal with the Americas and with the West Indian Islands, the scene of the defeat of the great Easterlies and the locus of origin of the great Westerlies.

We then consider the islands of the Atlantic, North of the equator, and proceed by the European countries on the Eastern shore of the Atlantic from Portugal to Norway. Then we take in order the Baltic States and the Central European countries, followed by the Russian lands South of the Arctic circle and leading across Central Asia to the far Eastern countries of Siberia below the Arctic circle, Mongolia and Manchukuo.

Next we revert to Europe and North Africa for the sub-tropical anticyclonic zone of 25° to 45°, taking the Mediterranean Islands as a sort of central region with the continental lands first on the North side leading to Greece and Asia Minor and then on the South side from Morocco through French and Italian North Africa to Egypt, Syria and Palestine, and so to Iraq and Iran with the Persian Gulf and onward behind the Himalaya to China, Korea and the Japanese Islands.

From there we pass to the equatorial zone of great Easterlies, roughly 25° N to 25° S, beginning with British India and Ceylon, through Burma, Siam and French Indo-China to the Philippine Islands, and so by Malaya to the Netherlands East Indies, leaving Australia for the time being but including Borneo.

On our way back to the West, we take the islands of the Indian Ocean, the Andaman Islands, Mauritius, Réunion and Madagascar, and reach Africa by Somaliland, after considering Southern Arabia and the Red Sea. Abyssinia and the Sudan come next, then equatorial Africa, British or French or Belgian, to the South Atlantic Islands, St Helena and Fernando Noronha which might perhaps be better associated with Brazil.

For the Southern Hemisphere we pass by Southern Rhodesia and South Africa to Australia and New Zealand and finally treat the Antarctic as nearly as possible like the commencement of our survey in the Arctic. The strip from South Africa to New Zealand, lat. 25° S to 45° S, should be correlated with the Anticyclonic or Mediterranean zone of the Northern Hemisphere, with special attention to the Roaring Forties on the Southern border.

The index of data to be covered by the programme thus sketched requires references to the publications in which the several data are to be found. These are provided by the consecutive enumeration of the items in the list of publications to the consideration of which we now proceed.

THE LIST OF PUBLICATIONS

International Publications.

The collection and dissemination of information about the upper air are parts of the duty of two international commissions.

One is the *International Aerological Commission* appointed in 1896 and referred to as the *International Commission for the Exploration of the Upper Air.* It publishes data for days selected year by year as "international days." Results for pressure, temperature, humidity and wind obtained by unmanned balloons (ballons-sonde, radio-sonde, captive balloons, kites or pilot-balloons) are included as well as observations of the form and motion of clouds. The latest issue, giving data for 12 and 13 November 1928, contains contributions from the following countries, twenty-nine in the Northern and five in the Southern Hemisphere, viz.: Argentine Republic, Australia, Austria, Belgium, Brazil, British India and the Andaman Is., Bulgaria, Ceylon, Colombia, Cyrenaica, Czecho-Slovakia, Danzig, Egypt, Esthonia, Finland, France, Georgia, Germany, Great Britain, Holland, Hungary, Italy, Japan, Malta, Mauritius, Norway and Jan Mayen, Persian Gulf, Poland, Portugal (Azores), Spain, Sweden, Tenerife, Russian Turkestan, Yugo-Slavia. A circular of the Commission dated 20. xii. 1935 says: The volume of 1928 of the results of aerological ascents has been finished now at a cost of 11,424 RM. It contains 16 parts with 1486 pages in total. "As the commands for the volume are in total 204 and this number has been expedited, the price of each will be RM 56" (about £4. 10*s*.).

The other is the *International Commission for Aerial Navigation* appointed by Governments after the War as a supplement to the Peace Treaty and now referred to as I.C.A.N. This institution collects from the co-operating governments monthly frequencies of the direction and velocity of the horizontal motion of the air, as well as information about visibility and clouds, and arranges for the distribution of the information to co-operating countries. At the moment data come from: Azores, Bermuda, Brazil, Chile, Czecho-Slovakia, Finland, France, Great Britain, Italy, Poland, Portugal, Roumania, Siam, South Africa, Sweden, Uruguay, Yugo-Slavia.

The regulations for these publications were subjects of discussion and revision at the meeting in September 1935 of the International Conference of Directors of Meteorological Services at Warsaw.

Quite recently Dr K. Keil, Secretary of the International Aerological Commission, has initiated the issue, in manifold, of a continuous bibliography of observations of the upper air arranged in sheets which can be cut to separate the items and form a card-index of the published data for recent years.

National Publications.

Next, we may note that data for the upper air in various forms are now included regularly in the official publications of the meteorological institutes of many countries. The commencement of publication may be different for different countries and before the practice of official publication became general the results of observations were published separately as official memoirs or as contributions to scientific journals.

We may regard as definable types of data, first those obtained by sounding balloons, which may include particulars of pressure, temperature, humidity and wind at specified elevations geometric or geodynamic. These we may notify in the bibliography by the Greek letter Θ. Secondly, observations obtained upon airplane flights or by means of kites or captive-balloons, which may give data for pressure, temperature, etc., also at specified elevations, generally within the range of eight kilometres. These contributions we may indicate with the symbol ⤻, and thirdly, the observations of pilot-balloons, which give the direction and velocity of the horizontal motion of air, also at specified elevations. These are identified by a foreshortened view of a "perspective" of the results ⊗.

So far as this volume is concerned we have regard more particularly to the individual data, which a reader may co-ordinate in any way he thinks desirable; but, in passing, we may note the publication of summaries which should appear in any survey of the literature of the subject. Contributions of this kind are identified by the symbol Ⓢ.

As already suggested the general plan of our bibliography, having noted the countries which contribute to the International Commissions, is to give a list of the publications numbered consecutively for the several countries taken in alphabetical order, and to follow that with an indication of the available data for each of the countries arranged according to the zones of latitude which we have already described.

Here we may remark that the discussion of the observations of the upper air has been the subject of the following memoirs:

'Die Temperaturverhältnisse in der freien Atmosphäre,' by A. Wagner, *Beiträge zur Physik der freien Atmosphäre*, Bd. 3, p. 57, 1909.

'The international kite and balloons ascents,' by E. Gold, *Geophysical Memoirs*, No. 5, 1913.

Résultats...sur les mouvements généraux de l'atmosphère,' by H. H. Hildebrandsson, *Nova Acta Reg. Soc. Scient. Upsal.* Ser. IV, vol. 5, 1918.

'Etude préliminaire sur les vitesses du vent et les températures dans l'air libre à des hauteurs différentes,' by H. H. Hildebrandsson, *Geografiska Annaler*, vol. 2, 1920, p. 97.

'Circulation générale de l'atmosphère,' by V. Khanevsky, *Bulletin de l'Institut de Physique cosmique de Moscou*, fasc. 1, 1923, pp. 30–67.

Köppen-Geiger Handbuch der Klimatologie, Band I, Teil F, 'Klimatologie der freien Atmosphäre,' by A. Wagner, Berlin, 1931.

Authorities for the Issue of Publications or Reports in which the Results of Observations of the Upper Air are included, with Reference Numbers for the Several Publications.

International Authorities.

Organisation Météorologique Internationale. Publications du Secrétariat, **1**.

Commission Aérologique Internationale (I.A.C.). Résultats des ascensions aérologiques internationales: Dec. 1900–June 1912, Strassburg; 1923–24, London; 1925 + Berlin, **2**.

International Commission for Aerial Navigation (I.C.A.N.).

Union Géodésique et Géophysique Internationale. Association de Météorologie. Procès-Verbaux, **3**.

+ This symbol placed after the year of commencement of the reports of data means that reports have been continued in the publication referred to from that year to the present time.

National Authorities or local Institutes, for the countries of the world taken in alphabetical order, with the titles of the publications which they issue.

Africa, East

British East African Meteorological Service, Nairobi. Annual Report, **4**; Results of meteorological observations made at...Chukwani Palace (Zanzibar)... Kabete (Kenya)...Kololo Hill (Uganda), **5**; Meteorological Report for Northern Rhodesia, **6**.

Statistical Research Department, East African Governor's Conference, Nairobi. Quarterly Bulletin of Statistical Research, **7**.

Africa, French equatorial and French West

See Algeria.

Africa, South

Meteorological Office, Department of Irrigation, Pretoria. Monthly frequency tables (I.C.A.N.), **8**; Report (annual), **9**; Memoirs, **10**; Report on...the polar year, **10*a***.

*South African Journal of Science**, Pretoria, **11**; Union of South Africa. Official Year Book, **12**.

Algeria

Institut de météorologie et de physique du globe. Bulletin quotidien, **13**; Bulletin mensuel, **14**; Observatoire Jules Carde (Tamanrasset). Observations, **15**.

Argentine Republic

Dirección de Meteorología, Buenos Aires. Daily Weather Report, **16**; Anales, **17**.

Australia

Commonwealth Bureau of Meteorology, Melbourne. Bulletin, **18**.

Australian Association for the Advancement of Science, Adelaide, Report, **19**.

'The frequency and velocity of the wind above Sydney,' by H. E. Camm and H. E. Banfield (typescript), Sydney, 1934, **20**.

Austria

Zentralanstalt für Meteorologie und Geodynamik, Vienna. Monatliche Mitteilungen, **21**; Jahrbuch, **22**.

Bundesministerium für Handel und Verkehr, Vienna. Aerologische Beobachtungen des österreichischen Flugwetterdienstes, **23**.

Sonnblick-Verein, Vienna. Jahresbericht, **24**.

Akademie der Wissenschaften, Vienna. Sitzungsberichte, Mathematische-Naturwissenschaftliche Klasse, **25**.

Österreichische Gesellschaft für Meteorologie; Deutsche Meteorologische Gesellschaft. Meteorologische Zeitschrift, **26**.

Azores

Service Météorologique, Angra do Heroismo. Tableaux mensuels des fréquences (I.C.A.N.), **27**; Résumé d'observations, **28**

Belgium

Institut Royal Météorologique de Belgique, Uccle. Bulletin quotidien, **29**; Annuaire météorologique, **30**.

Bermuda

Meteorological Office. Pilot-balloon ascents and monthly frequency tables (I.C.A.N.), **31**; 'Daily meteorological observations taken at St George's, Bermuda, during the polar year, 1932–3,' 1935, **32**.

Brazil

Instituto de Meteorologia do Departamento de Aeronautica Civil, Rio de Janeiro. Boletim Diario, **33**; Boletim Mensal, **34**.

Bulgaria

Service météorologique de l'aviation bulgare, Sofia. Annuaire aérologique, **35**; Résultats des observations avec ballons-pilotes en Bulgarie, **36**.

Canada

Meteorological Service of Canada, Toronto. 'Upper air investigation in Canada,' by J. Patterson, Ottawa, 1915 (with supplementary sheets), **37**; Canadian Meteorogical Memoirs, **37*a***.

Ceylon

Survey Department, Colombo Observatory. Annual Report, **38**; Bulletins and Memoirs (subsequently included in Ceylon Journal of Science, Section E), **39**.

Chile

Oficina Meteorológica de Chile, Santiago. Anuario Meteorológico, **40**.

Estudio Meteorológico de Chile. Ruta: Arica-Santiago-Magallanes by J. B. Navarrete, Santiago, 1930, **41**.

China

Observatoire de Zi-ka-Wei, Shanghai. Bulletin aérologique, **42**; 'The winds and the upper air currents along the China coast,' by E. Gherzi, 1931, **43**.

National Research Institute of Meteorology, Nanking. Bulletin of upper air current observations, **44**; Memoirs, **45**.

Royal Observatory, Hong Kong. Meteorological results, **46**; 'The upper winds of Hong Kong,' by G. S. P. Heywood, 1933, **47**; 'Meteorological information for aviation purposes,' 1935, **48**.

Cuba

Observatorio Nacional, Habana. Boletín, **49**.

* When the publication referred to is an independent journal its title is printed in italics.

Czecho-Slovakia

Institut Météorologique de la République Tchécoslovaque, Prague, Tableaux mensuels (I.C.A.N.), **50**.
Institut Aéronautique Militaire, Section Météorologique, Prague, Zpravy, **51**; Résumé des observations des stations météorologiques d'aviation militaire, **52**.

Danzig

Staatliches Observatorium. Ergebnisse der meteorologischen Beobachtungen, **54**.

Denmark

'Travaux de la Station Franco-scandinave de sondages aériens à Hald, 1902–3,' by L. Teisserenc de Bort, Viborg, 1904, **55**.
Danmark-Ekspeditionen til Grønlands Nordøstkyst 1906–8, Meddelelser om Grønland, vol. 42, by A. Wagner, **56**.

Egypt

Meteorological Service of Egypt, Physical Department, Cairo. Daily Weather Report, **57**; Annual Meteorological Report, **58**; Physical Department Papers, **59**.
Cairo Scientific Journal, **60**.

Esthonia

Tartu Ulikooli Meteoroloogia Observatoorium, Tartu. Meteorologische Beobachtungen angestellt in Dorpat, **61**; continued as Meteorologisches Jahrbuch für Eesti, **62**.

Finland

Ilmatieteellinen Keskuslaitos, Helsinki (Meteorologische Zentralanstalt des finnischen Staates). Månadsoversikt av våderlaken i Finnland, **63**; Meteorologisches Jahrbuch, Teil 3, **64**; Mitteilungen, **65**. Monthly Frequency Tables (I.C.A.N.), **65***a*.
Meteorologisches Institut der Universität, Helsinki. Mitteilungen, **66**.
Societas scientiarum Fennica. Commentationes Physico-Mathematicae, Helsinki, **67**.

France

Office National Météorologique, Paris. Bulletin quotidien d'études, **68**; Bulletin quotidien de l'Afrique du Nord, I. Maroc, Algérie, Tunisie, **69**; Bulletin quotidien de l'Afrique du Nord, II. Sahara, **70**; Climatologie aéronautique (I.C.A.N.), **71**.
Institut de Physique du Globe, Strasbourg. Annuaire, **72**.
Société météorologique de France, Paris. La Météorologie, **73**.
Académie des Sciences, Paris. Comptes Rendus, **74**.
Annales de Physique du Globe de la France d'Outre Mer, Paris, **75**.
Association française pour l'Avancement de Science. Comptes Rendus, **76**.
L'Aérophile, Paris, **77**.
Observatoire de Trappes. Travaux scientifiques, **78**.
'Préparation météorologique des voyages aériens,' by J. Rouch, 1920, **79**.

Georgia

Observatoire Physique de la Géorgie, Tiflis. Monthly Bulletin, **80**.

Germany

Meteorologisches Observatorium, Aachen. Deutsches Meteorologisches Jahrbuch, **81**.
Deutsche Meteorologische Gesellschaft, Berlin. Meteorologische Zeitschrift, see Austria, **26**.
Flugwetterdienst, Berlin (subsequently Reichsamt für Flugsicherung). Erfahrungsberichte, **82**; Aerologische Berichte, **83**.
Preussische Akademie der Wissenschaften, Berlin. Sitzungsberichte, **84**; Abhandlungen, **85**.
Preussisches Meteorologisches Institut, Berlin (now Reichsamt für Flugwetterdienst). Abhandlungen, **86**; Deutscher Witterungsbericht (now published in *Wirtschaft und Statistik*), **87**.
Reichs-Marine-Amt, Berlin. Forschungsreise S.M.S. "Planet" 1906–7, Bd. 2, Aerologie, Berlin, 1909, **88**.
Bremische Landeswetterwarte, Bremen. Deutsches Meteorologisches Jahrbuch, Bremen, **89**.
Forschungs-Institut der Rhön-Rossitten Gesellschaft, Darmstadt. Veröffentlichungen, **90**.
Sächsische Landeswetterwarte, Dresden. Dekaden Monatsberichte, **91**; Deutsches Meteorologisches Jahrbuch, Sachsen, **92**.
Physikalischer Verein, Meteorologisch-Geophysikalisches Institut, Frankfurt am Main. Wetterkarte, **93**; Berichte, **94**.
Wetterdienststelle, Frankfurt am Main. Synoptische Bearbeitung, **95**.
Deutsche Seewarte, Hamburg. Täglicher Wetterbericht, **96**; Monatskarte für den südatlantischen Ozean, **97**; Aus dem Archiv, **98**; Annalen der Hydrographie und maritimen Meteorologie, **99**.
Bädische Landeswetterwarte, Karlsruhe. Deutsches Meteorologisches Jahrbuch, Baden, **100**.
Geophysikalisches Institut der Universität, Leipzig. Veröffentlichungen, **101**; Zum Klima der Turkei, **102**.
Sächsische Akademie der Wissenschaften, Leipzig. Berichte der Mathematisch-Physikalischen Klasse, **103**.
Preussisches Aeronautisches Observatorium, Lindenberg. Mitteilungen 1920–1, Kurzer Monatsbericht 1922–32, Aerologische Monatsübersicht 1933+, **104**; Arbeiten (previously Ergebnisse der Arbeiten), **105**.
Ebeltofthafen-Spitzbergen, Deutsches Observatorium, Veröffentlichungen, **106**.
Bayerisches Landeswetterwarte, München. Deutsches Meteorologisches Jahrbuch, Bayern, **107**.
Wetterwarte, Saarbrücken. Meteorologische Beobachtungen, **108**.
Statistisches Landesamt, Stuttgart. Deutsches Meteorologisches Jahrbuch, Württemberg, **109**; Drachenstation am Bodensee, Ergebnisse der Arbeiten, **110**; Jahrbuch für Statistik und Landeskunde, **111**.
Beiträge zur Physik der freien Atmosphäre, Leipzig, **112**.
Zeitschrift für angewandte Meteorologie (Das Wetter), Leipzig, **113**.
Gerlands Beiträge zur Geophysik, Leipzig, **114**.
Zeitschrift für Geophysik, Braunschweig (Deutsche Geophysikalische Gesellschaft), **115**.
Petermann's Mitteilungen, Erganzungsband, Gotha, **116**.
Zeitschrift der Gesellschaft für Erdkunde, zu Berlin, **117**.
Köppen-Geiger, Handbuch der Klimatologie, Band I, Teil F. 'Klimatologie der freien Atmosphäre,' by A. Wagner, Berlin, 1931, **118**.
Deutsche-Grönland Expedition, Wissenschaftliche Ergebnisse, Band IV, Meteorologie by A. Wegener, Leipzig, 1935, **119**.
Deutsche Atlantische Expedition "Meteor," Wissenschaftliche Ergebnisse, Berlin and Leipzig, 1933, **120**.
K. Gesellschaft der Wissenschaften, Göttingen. Nachrichten Mathematisch-Physikalische Klasse, **121**.
'The temperature of the air above Berlin,' by R. Assmann, 1904, **122**.
'Die Winde in Deutschland,' by R. Assmann, Braunschweig, 1910, **123**.

[Note: The official meteorological observatories are now incorporated in the Reichsamt für Wetterdienst, Berlin.]

Great Britain

Meteorological Office, London. Daily Weather Report, Upper Air Section (with monthly supplement), **124**; Weekly Weather Report, 1907–11, **125**; Geophysical Journal, 1912–21, **126**; Observatories' Year Book, 1922+, **127**; Monthly frequency tables (I.C.A.N.), **128**; Meteorological Magazine, **129**; Geophysical Memoirs, **130**; Professional Notes, **131**; Computer's Handbook, **132**; Airship Meteorological Reports (manifolded), **133**; 'Barometric Gradient and Wind-force,' by E. Gold, **134**; 'The free atmosphere in the region of the British Isles,' by W. H. Dines, M.O. 202, **135**; Aviation Meteorological Reports (manifolded), **136**.
Royal Meteorological Society, London. Quarterly Journal, **137**; Memoirs, **138**; Collected Scientific Papers of W. H. Dines, **139**.
Scottish Meteorological Society, Journal, **140**.
Royal Society, London. Proceedings A, **141**; Philosophical Transactions A, **142**.
Royal Aeronautical Society. Journal, **143**.
Advisory Committee for Aeronautics. Reports and Memoranda, **144**.
Board of Trade. Report of the work carried out by the ss. Scotia, 1913, **145**.

Great Britain (*contd.*)

British Antarctic Expedition 1910–13, Committee for the publication of the scientific results. Meteorology, vol. I, Calcutta, 1919, **146**.

'The structure of the atmosphere in clear weather,' by C. J. P. Cave, C. U. Press, 1912, **147**.

London, Admiralty. Chart showing the distribution of observations of upper winds and upper air temperature over the oceans. London, 1931, **148**.

Nature, **149**.

Holland

Koninklijk Nederlandsch Meteorologisch Instituut, De Bilt. Daily Weather Report, **150**; Ergebnisse aerologischer Beobachtungen, **151**; Résumés des observations aérologiques à Soesterberg et De Bilt (monthly), **152**; Mededeelingen en Verhandelingen, **152a**.

Koninklijke Akademie van Wetenschappen. Proceedings, **153**.

Hemel en Dampkring, Groningen, **154**.

'De gemiddelde hoogtewind boven De Bilt...(1922–31)' by W. Bleeker, Zutphen, 1936, **154a**.

Hungary

Meteorológiai és Földmágnességi Intézet, Budapest. Daily Weather Report, **155**; Jahrbuch, **156**; Az Idöjárás, **157**.

Meteorologisches Institut der Universität, Debrecen. Jahrbuch, **158**; Mitteilungen, **159**.

Geographisches Institut der Universität, Debrecen. Höhenwindmessungen...in Debrecen, **160**.

Iceland

Reykjavik Vedurstofan. Höhenwindmessungen (manifolded), **161**.

India, British

India Meteorological Department, Poona. Upper Air Data, **162**; Monthly Weather Report, **163**; India Weather Review, **164**; Memoirs, **165**; Scientific Notes, **166**; 'Meteorology of the Persian Gulf and Mekran,' by B. N. Banerji, **167**; 'Meteorological conditions affecting aviation over the NW frontier,' by R. G. Veryard and A. K. Roy, Delhi, 1934, **168**.

Asiatic Society of Bengal. Proceedings, **169**.

Indo-China

Service Météorologique de l'Indo-Chine. Annales, **170**.

Italy

R. Ufficio Centrale di Meteorologia e Geofisica, Rome. Annali, **171**; Memorie, **172**; Rivista Meteorico Agraria, **173**.

Ufficio Presagi, Ministero dell' Aeronautica, Rome. Bollettino meteorologico e aerologico (D.W.R.), **174**; Annali, **175**; Riassunto Mensile (I.C.A.N.), **176**; Sondaggi aerologici eseguite nei...giorni stabiliti dalla Commissione internazionale, **177**.

Società Meteorologica Italiana, Perugia. Bollettino bimensuale, continued as Meteorologia pratica, **178**.

Pontificia Accademia della Scienze Nuovi Lincei. Atti, **179**.

R. Accademia Nazionale dei Lincei. Rendiconti, **180**.

Comitato Nazionale Italiano per la Geodesia e la Geofisica, Milano. Bollettino, **181**.

R. Comitato Talassografico Italiano. Memorie, **182**.

Aeroclub. Bollettino tecnico, **183**.

Aerotecnica, Rome, **184**.

Rivista aeronautica (Ministero dell' Aeronautica), Rome, **185**.

Le caratteristiche dell' Atmosfera libera sulla Valle Padana, by P. Gamba, Venezia, 1923, **186**.

Lanci di Palloni-Sonda effetuati nel R. Osservatorio Geofisico di Pavia nell' anno 1925. Mendovi, 1926, **187**.

La meteorologia e l' aerologia degli Oceani. L' Oceano Atlantico Nord, by F. Eredia, Roma, 1935, **188**.

Spedizione Italiana De Filippi nell' Himalaia, 1913–14, ser. I, vol. III, Bologna, 1931, **189**.

Japan

Central Meteorological Observatory, Tokyo. Report of Upper Air Observations, **190**; Geophysical Magazine, **191**.

Aerologia Observatorio, Tateno. Raporto, **192**; Bulteno, **193**; Observadoj meteorologiaj kaj aerologiaj ...en la internaciaj tagoj dum la polusa jaro 1932–3, **194**.

Meteorological Observatory, Osaka. Annual Report, **195**; Bulletin of upper air current, **196**.

Imperial Marine Observatory, Kobe. Weekly Weather Report, **197**; Monthly Bulletin, **198**.

Meteorological Society of Japan, Tokyo. Journal, **199**.

International Latitude Observatory, Mizusawa. Soundings with pilot-balloons...concurrently with latitude observations, **200**.

Korea

Meteorological Observatory of the Government-General of Tyosen, Zinsen. Upper air-current observations, **201**.

Latvia

Bureau Central Météorologique de Lettonie, Riga. Daily Weather Report, **202**.

Madagascar

Service Météorologique, Tananarive. Observations météorologiques (monthly), **203**.

Malaya

Malayan Meteorological Service. Upper wind-roses from pilot-balloon observations, **204**.

Mauritius

Royal Alfred Observatory. Results of Magnetic and Meteorological Observations, **205**; Miscellaneous publications, **206**; 'Report on upper air investigations,' by A. Walter, 1926, **207**.

Morocco

Service de Physique du Globe et de Météorologie de l'Institut Scientifique Chérifien, Casablanca. Daily Weather Report, **208**.

Netherlands Indies

K. Magnetisch en Meteorologisch Observatorium te Batavia. Verhandlingen, **209**.

Natuurkundig Tijdschrift, Batavia, **210**.

New Zealand

Meteorological Office, Wellington. Notes, **211**.

Lands and Survey Department, Records of the Survey of New Zealand, **212**.

Nigeria

Survey Department, Lagos. 'Meteorological observations in Nigeria during the polar year 1932–3,' **213**.

Norway

Det Norske Meteorologiske Institutt, Oslo. Daily Weather Report, **214**; Jahrbuch, **215**; Geofysiske Publikationer, **216**.

Geofysisk Institutt, Bergen. Norwegian North Polar Expedition, s.s. "Maud," 1918–25, Scientific Results, II, III, Bergen 1933, 1930, **217**.

Det Norske Videnskaps-Akademi i Oslo. Norwegian Novaya-Zemlya Expedition, 1921. Report No. 39. Meteorologische und aerologische Beobachtungen, by O. Edlund, 1928, **218**.

Persian Gulf

See India, British.

Peru

Servicio Meteorológico Nacional, Lima. Meteorología de la Estratosfera en el Perú, by Sergio Korff and G. A. Wagner, **219**.

Philippines

Manila Observatory. Publications, **220**.

Poland

Państwowy Instytut Meteorologiczny Warszawa (Institut National Météorologique, Varsovie). Daily Weather Report, **221**; Résumé mensuel (I.C.A.N.), **222**; Guide météorologique à l'usage de la navigation aérienne, Warsaw, 1934, **223**.

Observatoire astronomique de Wilno, Bulletin, II, Météorologie, **224**.

Société géophysique de Pologne. Etudes géophysiques, **225**.

Institut Géophysique et Météorologique de l'Université, Lwów. Communications, **226**.

Portugal

Serviço Meteorológico, Ministerio da Marinha, Lisbon. Boletim meteorológico (D.W.R.), **227**.

Serviço Meteorológico do Exercito. Résumé des observations (I.C.A.N.), **228**.

Escola Militar, Lisbon. Sondagens aerológicas com balão piloto, **229**.

Rhodesia, Northern

See Africa, East.

Rhodesia, Southern

Department of Agriculture, Salisbury. Meteorological Report, **230**.

Roumania

Institutul Meteorologic Central al Romaniei, Bucharest. Daily Weather Report, **231**.

Ministère de l'Industrie et du Commerce, Aviation Civile et Navigation aérienne, Section Météorologique. Tableaux mensuels (I.C.A.N.), **232**.

Samoa

Apia Observatory. Annual Report, **233**; Upper air observations, by A. Thomson, 1st and 2nd series, Wellington, 1925, 1929, **234**.

Siam

Ministry of Agriculture and Commerce, Royal Irrigation Department. Monthly frequency tables (I.C.A.N.), **235**.

Spain

Servicio Meteorológico Español, Madrid. Daily Weather Report, **236**; Resumen de las observaciones aerológicas, **237**; Publicación, Serie A, **238**; Anuario, **239**.

Servei Meteorològic de Catalunya, Barcelona. Daily Weather Report, **240**; Notes d'Estudi, **241**.

Institut de Ciencias, Barcelona. Treballs de l'Estació Aerològica de Barcelona, **242**.

Observatorio del Ebro, Tortosa. Boletín Mensual, **243**.

Observatorio de Igueldo, San Sebastian. Trabajos, Publicación, **244**; Resumen Mensual de las observaciones meteorológicas, **245**.

Sweden

Statens Meteorologisk-Hydrografiska Anstalt, Stockholm. Daily Weather Report, **246**; Monthly frequency tables (I.C.A.N.), **247**; Observations météorologiques suédoises, **248**; Aerologiska iakttagelser i Sverige, Årsbok, **249**; Meddelanden, **250**.

Svenska Sällskapet för Antropologi och Geografi. Geografiska Annaler, **251**.

R. Societatis scientiarum Upsaliensis, Upsala. Nova Acta, **252**.

Artilleri-tidskrift, Stockholm, **253**.

Switzerland

Schweizerischen Naturforschenden Gesellschaft u.s.w. Ergebnisse der Schweizerischen Grönlandexpedition 1912–13, by A. de Quervain and others, Zürich, 1920, **254**.

Trinidad

Department of Agriculture. 'The rainfall of Trinidad,' by F. M. Bain, 1934, **255**.

Tunis

Service Météorologique de la Tunisie, Tunis. Statistique (annual), **256**; 'Etude climatologique du Golfe de Tunis,' by G. Ginestous, Tunis, 1925 (see also Résumé de climatologie tunisienne, Tunis, 1922), **257**.

Turkey

Konstantinopel, Feldwetterzentrale. Seewind im Golf von Smyrna (Imbad). Sonderveröffentlichungen No. 1, 1919, **258**.

United States

Weather Bureau, Washington. Monthly Weather Review, **259**; Monthly Weather Review, Supplements, **260**; Bulletin of the Mount Weather Observatory, **261**.

Hydrographic Office, Washington. Pilot charts of the upper air. North Atlantic, **262**; Pilot charts of the upper air. North Pacific, **263**.

Scripps Institute of Oceanography, Berkeley Institute of California, **264**.

American Meteorological Society. Bulletin, **265**.

Massachusetts Institute of Technology. Professional Notes, **266**.

National Research Council, American Geophysical Union. Bulletins, **267**; Transactions, **268**.

National Advisory Committee for Aeronautics. Reports, **269**.

Astronomical Observatory of Harvard College, Cambridge, Mass. Annals, **270**.

Pacific Science Congress, **271**.

Pan-American Scientific Congress. Proceedings, **272**.

University of Michigan, Greenland expeditions 1926–31. Part I, Aerology, by W. H. Hobbs, Ann Arbor, 1931, **273**.

Geographical Society, Baltimore. 'The Bahama Is.,' by O. L. Fassig, 1905, **274**.

U.S. Department of Commerce, Aeronautics Branch, Washington. Airways Bulletins, **275**.

Uruguay

Observatorio Nacional (Servicio Meteorológico del Uruguay) Montevideo. Sondeos de la Atmósfera en Montevideo, **276**.

Escuela Militar de Aviación, Servicio Meteorológico Aeronáutica, Montevideo. Sinopsis meteorológica, **277**.

U.S.S.R.

Central Geophysical Observatory, Leningrad. Works of the Aerological Observatory, Sloutzk, Pavlovsk, **278**; Journal of geophysics and meteorology, **279**; Recueil de géophysique, **280**.

Observatoire Constantin, St Pétersbourg Etude de l'atmosphère, **281**.

Administration de l'Hydrographie, Section Hydro-Météorologique, Leningrad. Matériaux de l'Expedition hydrographique de l'Océan glacial du Nord, 1910–15, by N. Evgenov, **282**; Résultats des observations effectuées au moyen des ballons pilotes en 1926 aux stations aérologiques de l'Administration de l'Hydrographie, 1930, **283**.

Zentralbureau für Hydrometeorologie, Leningrad. Nachrichten, **284**.

Academy of Sciences, Leningrad. Travaux de la Commission pour l'étude de la république autonome soviétique socialiste Iakoute, **285**.

Institut de Physique Cosmique, Moscow. Bulletin, **286**.

Observatoire Aérologique, Moscow. Etudes scientifiques de l'atmosphère, **287**.

Central Weather Bureau Committee, Moscow. Journal of Geophysics, **288**; Daily Weather Report, **289**.

Aerodynamical Institute, Koutchino. Bulletin, **290**.

Arctic Institute, Leningrad. Bulletin, **291**.

Ukrainian Meteorological Service, Urkmet, Kiev. Geophysical characteristic of the Ukraine. Upper air observations, **292**.

Académie des sciences d'Ukraine. Mémoires de la classe des sciences naturelles et techniques, **293**.

Meteorological Institute of Middle Asia, Tashkent. Meteorological data, 1924, **294**; Bulletin of the Geophysical Observatory of Tashkent, **295**.

U.S.S.R. *(cont.)*

Magnetic and Meteorological Observatory, Irkutsk. Bulletin, **296**.

Turkoman Meteorological Bureau, Ashkabad. Meteorological data for 1928, **297**; 'A few characteristics of the distribution of the wind over Ashkabad from data obtained from pilot-balloon observations, 1928,' by N. P. Bizov, 1929, **298**.

Far East Geophysical Institute, Vladivostok, Records, **299**.

Yugo-Slavia

Service Météorologique, Ministère de la Guerre et de la Marine, Direction de l'Aéronautique. Novi Sad. Tables mensuels de fréquence (I.C.A.N.), **300**.

Index of Data for the World arranged in Zones or Sectors

The information available for different parts of the globe Θ by ballon-sonde or radio-sonde, ⟋ by aeroplane or kite, ⊗ by pilot-balloon.

The numbers in black type refer to the list of publications. When the issue of data is periodical the number precedes the entry; when the issue is occasional the number follows the entry.

Immediately following the data-symbol are figures for the number of stations and the periods covered by the observations there.

The number of the year, if any, at the end of the entry gives the date of publication.

The symbol + following the number of a year means that the publication is continued from that year onwards.

The symbol ‡ following the name of the country indicates that no regular publication of upper air data has been traced.

The Sea

The nearest approach to a regular publication of upper air data over the sea is in the *Archiv der Deutschen Seewarte*, which records the results of pilot-balloon observations on successive voyages of certain merchant ships across the Atlantic in the period from September 1928 (**98** vol. 48, No. 3; vol. 50, No. 1; vol. 51, No. 3). The same publication contains also a series of observations on occasional voyages from March 1922 onwards (**98** vol. 40, No. 4; vol. 41, No. 4; vol. 42, No. 2; vol. 43, No. 3; vol. 45, No. 3; vol. 46, No. 2; vol. 49, Nos. 1, 3, 4; vol. 51, Nos. 1, 2; vol. 52, No. 3; vol. 54, Nos. 2, 3). Others are included in the *Annalen der Hydrographie und maritimen Meteorologie* (**99** 1910, p. 201; 1912, p. 454; 1929, p. 353; 1930, p. 369; 1932, p. 49).

A summary of observations over the Atlantic Ocean subsequent to those recorded in chap. IV is included in the report of the expedition of s.s. "Meteor" (**120**), which gives also the results of a large number of observations of pilot-balloons, kites and ballons-sonde during the voyage of the expedition Jan. 1925–March 1927.

The most recent volume of data for the upper air of Holland (**151** 1934) includes data from pilot-balloons over the sea. Similar data for the Mediterranean in four months of 1930 are given by F. Eredia (**175** vol. IV). The data available from British ships are indicated in a report by Commander L. G. Garbett (**1** No. 21; **148**).

The references for observations from earlier expeditions are as follows:

Θ ⟋ ⊗ 1905–7, **78** vol. IV; 1906, **88**.

Θ 1908–10, **2**; Dec. 1909, **112** vol. 4, p. 153; 1927 (July 5), **129** vol. 62, p. 235; Ⓢ **101** vol. 2; **112** vol. 4, pp. 17, 224; **251** vol. 2, p. 97.

⟋ 1905, **112** vol. 1, p. 205; 1903–4, **78** vol. 3; 1906, **105** vol. 2, p. 132; 1911, **98** vol. 37, No. 1; 1913, **145**; 1915, **260** No. 3.

⊗ Dec. 1909, 1910, **112** vol. 4, p. 153; Ⓢ **101** vol. 2; Ⓢ **112** vol. 4, p. 35.

The Arctic zone, land and ice North of lat. 66°.

Arctic Sea‡

Θ ⟋ ⊗ 1906–7, 1910, **112** vol. 6, p. 224.

Θ July 1931 (Graf Zeppelin), **103** vol. 83, p. 333; **116** No. 216, p. 58; **26** vol. 48, p. 409.

⟋ 1913–15, **282**.

⟋ ⊗ 1919–25, **217** (see also Ⓢ **26** vol. 51, p. 401; **118**).

2 stations 1912–13, **254**; 1930–1 (7 ascents), **130** No. 61; 2 stations 1930–1, **119**.

⊗ 1909, **112** vol. 5, p. 132; 4 stations 1912–13, **254**; 1 station (mean of 24 ascents), **98** vol. 52, No. 4; 1 station May-Aug. 1933, **151** 1934; 3 stations Aug. 1930–July 1931, **119**.

Alaska‡

⊗ 3 stations, polar year 1932–3, Ⓢ **259** vol. 62, p. 244; **268** 1934, p. 14.

Greenland

214 ⊗

Θ (2 ascents), ⊗ one station 1926–9, **273**; Θ ⊗ Thule, polar year, data not yet published.

⟋ 1906–8, **56** No. 1, Ⓢ **112** vol. 3, p. 33 and **118**;

Spitzbergen‡

Θ ⟋ ⊗ 1911–12, **106** Heft 1–4; Ⓢ **118**.

Arctic Russia‡

Θ Alexandrovsk 13–16 Jan. 1931, **114** vol. 34, p. 36; polar year 1932–3, **291** 1933, p. 457 (note only, no data).

⊗ 3 stations Feb.-Sept. 1919, Ⓢ **131** No. 32.

Novaya Zemlya‡

⊗ July-Sept. 1921, **218** p. 13.

Far Eastern and Western Sector. Pacific Islands.

Sandwich Islands‡

⊗ 1 station April-Aug. 1923, Ⓢ **259** vol. 51, p. 525; 1922–4, Ⓢ **259** vol. 55, p. 222; 1924–7, Ⓢ **259** vol. 56, p. 496; Ⓢ **118** and **263**.

Guam‡

⊗ 1921–4, Ⓢ **259** vol. 55, p. 224; Ⓢ **263**.

Palau‡

⊗ Feb.-Dec. 1925, **190** vol. 1, No. 2.

Samoa, Apia

233 ⊗ 1932+.

⟋ May-Aug. 1906, **121** 1906; 1908–9, **121** 1910.

⊗ Apia 1923–8, **234** (see also **259** vol. 57, p. 255); 1923–4, Ⓢ **118**.

The American Sector with the West Indies: the sources of the Great Westerlies.

CANADA‡

Θ 1–2 stations 1911–15, **37**; 1–2 stations 1923+, **2** (height of tropopause 1932–3, **129** vol. 69, p. 166); (§) 1911–15, **118**.
⊗ *c.* 14 stations 1920+ (data available but not published); 1 station 1920–31, 10 stations 2–3 years, (§) **133** Nos. 57–69; 6 stations, 1929–31, (§) **37** *a* No. 1.
A note on data available for the polar year 1932, **268** 1934, p. 13.

UNITED STATES

259 Θ Misc. stations, varying periods, 1926+ (vol. 55, p. 293; vol. 57, p. 231; vol. 58, p. 235; vol. 59, pp. 297, 417; vol. 60, p. 12; vol. 62, pp. 45, 121; vol. 63, p. 49).
⟋ 6 stations 1922–31, *c.* 20 stations 1931+ (monthly means).
⊗ 20–24 stations March 1929+ (monthly mean resultant winds).

Θ St Louis 1904–7, **270** vol. 68, part 1; 2 stations 1909–11, **261** vols. 3, 4; Misc. stations, short periods, **259** vol. 44, p. 247, vol. 42, p. 410; **260** No. 3; (§) 4 stations combined, vol. 46, p. 17; St Louis 1906–8, **112** vol. 2, p. 35.
Θ ⟋ ⊗ (§) **267** Physics of the Earth, vol. III; periods to 1929, (§) **118**.
⟋ 1–5 stations 1915–18, **259** vol. 47, p. 367; **260** Nos. 3, 5, 7, 8, 10–15; Mt. Weather 1907–13, **261** vols. 1–6; Blue Hill 1894–1905, **270** vols. 42, 43, 58; Mt. Washington 1932–3, **268** 1934, pp. 114, 118. (§) Mt. Weather 1907–10, **261** vol. 4; 1907–12, **261** vol. 6, and **259** vol. 46, p. 11; Drexel 1915–18, **259** vol. 48, p. 1; 8 stations, varying periods 1896–1920, **260** No. 20.
Notes of special investigations: **259** vol. 56, p. 221, vol. 61, pp. 61, 170, 321, vol. 62, p. 195; **260** No. 21; **270** vol. 58, part 1, vol. 68, part 2; **272** 2nd meeting, p. 632.
⊗ (§) *c.* 50 stations in 9 groups, varying periods, **260** No. 26; 14 stations, varying periods, **260** No. 35; 3 stations 1918–22, **259** vol. 51, p. 448; Lansing 1919–22, vol. 50, p. 642; San Diego 1924–7, 1933, vol. 56, p. 221, vol. 62, p. 195; Havre 1927–30, vol. 59, p. 189; Phoenix W. Bureau 1930, vol. 59, p. 270; Reno 1930–2, vol. 61, p. 171. 14 stations, **262**; 5 stations, **263**; Cyclones and anticyclones, (§) **270** vol. 68, part 1; **259** vol. 47, p. 647, vol. 54, p. 195, vol. 56, p. 47; **272** 2nd meeting, p. 632.
Wind on air routes: **259** vol. 51, p. 111; vol. 52, p. 153; vol. 54, p. 10; **269** No. 245.

WEST INDIES, AND JAMAICA‡

(The scene of the defeat of the Great Easterlies)

49 ⊗ Cuba 1926+.

Θ ⊗ Misc. stations Dec. 1909, 1910, **112** vol. 4, p. 153.
⟋ Bahamas, 5 ascents, June-July 1903, **274**; **259** vol. 31, p. 582.
⊗ Barbados, a few ascents, 1907, 1909–12, **137** vol. 34, p. 265, vol. 37, p. 17, vol. 39, p. 295; Porto Rico 1920, 1923, **259** vol. 52, p. 22; 1920–31, (§) vol. 59, p. 414; 1920–9, (§) **268** 1933, p. 69; Jamaica, observations 1925+ not yet published; Caribbean Sea, summer 1925, **112** vol. 12, p. 183; Trinidad (Port of Spain) 1933, (§) **255**; 5 stations, (§) **262**. Curaçao, July 1923–Jan. 1924, (§) **152** *a* No. 36.

PANAMA‡

⊗ (§) Coco Solo, **262**.

PERU‡

⊗ Lima Dec. 1934–March 1935, **219**.

BRAZIL

33 ⊗ 13–16 stations Sept. 1928+.
34 ⟋ 1–3 stations 1924+.
⊗ 10–20 stations. 1924+ (Sept. 1929–Jan. 1934, summaries in I.CA.N. form).

⟋ Flight over S. Brazil, Jan.-Feb. 1928, **98** vol. 49, No. 4; **82** 2 Sonderband p. 88; Rio de Janeiro 1930–5, (§) **259** vol. 63, p. 190.
⊗ Jan.-Feb. 1928, **98** vol. 49, No. 4; (§) 1921–34, **259** vol. 62, p. 164 (short note, no data).

ARGENTINE REPUBLIC

16 ⊗ *c.* 8 stations 1933+.
17 (Tomo XIX, Volumen I) ⊗ Buenos Aires 1928.

⊗ Feb. 1928, **98** vol. 49, No. 4.

URUGUAY

276 ⊗ Montevideo Aug. 1927+.

⟋ ⊗ Montevideo 1930, **277**. (The data are referred to but not published.)
⊗ Aug. 1927–Dec. 1928, (§) **118**.

CHILE

40 Santiago 1931+.
⊗ Santiago (§) **41**.

The Northern Atlantic Sector. Atlantic Islands: Iceland, Bermuda, Azores, Madeira.

ICELAND‡

Θ Reykjavik July-Aug. 1933, ⟋ ⊗ Sept. 1932–Aug. 1933, **151** 1933 (see also **113** vol. 51, p. 317; **154** vol. 32, p. 12).
⊗ 3 stations 1909–28, **98** vol. 51, No. 5; 1 station 1927–8, **161**; 1 station 1912–13, **254**; 1909, **112** vol. 5, p. 132; July-Sept. 1928, **98** vol. 49, No. 3; (§) 1909, 1912–13, 1926–7, **115** vol. 4, p. 352; **118**.

BERMUDA

31 ⊗ 1 station 1934+.
⊗ 1 station March-Aug. 1933, **32**.

AZORES

27 ⊗ 2 stations 1931+.
28 ⊗ 1–2 stations (international days) 1922+.
Θ 3 ascents, July 1907 (§) **112** vol. 4, p. 17.
⊗ Angra do Heroismo (§) **262**.

The zone of the Great Westerlies. The Atlantic coast of Europe: Portugal, (Spain), France, Great Britain, Belgium, Holland, Germany, Denmark, Norway.

PORTUGAL

227 ⊗ a few stations *c.* 1926+.
228 ⊗ 10 stations July 1934+.

⊗ Lisbon 1930–2, **229**.

SPAIN (see Mediterranean zone)

FRANCE

68 ⟋ *c.* 6 stations; ⊗ *c.* 40 stations 1923+.
71 ⊗ 11 stations 1929+.
72 Θ ⊗ 1 station 1922+.

Θ Trappes 1905–12, 1923+, **2**; Strasbourg 1905–12, **2**; (§) Trappes 1899–1903, **74** vol. 138, p. 42; Strasbourg 1922–8, **72** 1928, p. 98.
(§) Paris, Strasbourg, **118**, **130** No. 5.
⟋ Paris, (§) **133** Nos. 72–83; Berck 1918, **131** No. 8.
⊗ (§) Bayonne 1918, **77** 1922; Paris 1916–18, **79**; Aubagne, 3 years, **76** 1922, p. 262; Lyon-Bron, **73** vol. 3, p. 205; N. France (diurnal variation), **131** No. 15; Angoulême, 1921–30, **73** vol. 11, p. 468.

The zone of the Great Westerlies—contd.

GREAT BRITAIN

124 ⟋ 1–2 stations 1919+; ⊗ 20–30 stations 1918+.
128 ⊗ 7 stations 1927+.
127 Θ Kew 1923+; Sealand 1924+. (A preliminary issue is made in manifold.)

Θ Pyrton Hill, Benson, Manchester and occasional stations 1907–11, **125**; 1912–21, **126**; 1922, **127**; Pyrton Hill, 1907–8, **135**.
(§) SE. England 1907–11, **142** vol. 211; 1908–10, **118**; British Isles 1908–15, **132** Section II (see also **130** Nos. 5, 13); England 1921–9, **127** 1929; England (humidity) 1927–31, **137** vol. 59, p. 157.
Discussions of ascents on special occasions, **137** vol. 36, pp. 7, 127; vol. 37, pp. 1, 11, 23. Temperature of stratosphere: **138** No. 18; Cyclones and anticyclones, **140** ser. 3, vol. 16. (See also **139**.)
⟋ Crinan 1902–4, **142** vol. 202, p. 123; **137** vol. 30, p. 155; **137** vol. 31, p. 217; **141** vol. 77, p. 440; **139**; Oxshott 1904–5, **141** vol. 77, p. 440; 1906, **125**; c. 4 stations 1906–13, **125**, **126**; 1–2 stations 1917–20, **126**; (§) **126** 1920; **131** No. 4.
(§) Oxshott 1906, **134**; 4 stations 1906–7, **137** vol. 34, p. 15; 1917–18, S. England, **133** Nos. 72–83.
⊗ 3–4 stations 1907–20, **125**, **126**; Scilly Nov.-Dec. 1911, **130** No. 14, **137** vol. 45, p. 21; S. Farnborough (diurnal variation) Oct.-Nov. 1912, **144** No. 92; Shoeburyness (exceptionally high ascents), **131** No. 13; no lift balloons: Blackpool 1910, **137** vol. 37, p. 33; Upavon 1914, **144** New series, No. 325.
(§) 11 stations, varying periods, 1920–32, summaries available but not published; 5 stations **262**; Mt. Batten (wind-roses 1500 ft.), **129** vol. 66, p. 112; Worthy Down 1928–32, **137** vol. 59, p. 408; Upavon 1913, **137** vol. 40, p. 123; Aberdeen 1912–13, vol. 41, p. 123; Ditcham Park 1907–10, **147**; Pyrton Hill 1910–11, **144** No. 47; Calshot, **133** No. 69. (A summary of many of these data is given in vol. IV of this Manual.)

BELGIUM

29 ⟋ 1 station; ⊗ c. 2 stations.

Θ Uccle 1906–12, **2** (see also **30** 1907–14, lists of ascents and data for some years); 1906–13, (§) **30** 1914, p. 53; 1906–11, (§) **130** No. 5; **118**; Dec. 26–28, 1928 (ascents at short intervals), **216** vol. 9, No. 9; April-Oct. 1933 (temp. of stratosphere), **112** vol. 21, p. 208.
⟋ ⊗ 2 stations 1915–18, **98** vol. 47, Nos. 3, 4 (monthly means) (see also **118**).

HOLLAND

150 ⟋ Soesterberg, ⊗ De Bilt (before 1931).
152 ⟋ Soesterberg, ⊗ De Bilt 1931–2.
151 Θ De Bilt 1914+; ⟋ 1–3 stations 1909+; ⊗ 2–3 stations, 1911+.
154 ⟋ 2 stations (isopleths) 1926+.

⟋ 2 stations, periods to 1927, (§) **112** vol. 14, p. 52; (§) to 1930, **1** No. 8; (§) 1–2 stations 1911–28, **118**.
⊗ De Bilt, (§) **262**; 1922–31, **154a**.

GERMANY

96 ⟋ ⊗ c. 1902+.
83 ⟋ c. 9 stations, ⊗ c. 35 stations, 1926+.
104 ⟋ 1 station 1920+.
105 Θ Lindenberg 1905–17, 1921, 1923+; ⟋ (kites) 1905–16, 1923–4, (aeroplanes) 1922+; ⊗ 1908–16, 1923+.
107 Θ München 1906–14, 1922+.
100 Θ Karlsruhe 1926+, ⊗ 1932+.
92 Θ Dresden and Leipzig 1926+.
108 Θ ⊗ Saarbrucken 1928+.
113 ⟋ (isopleths or monthly summary) Lindenberg 1905+; Friedrichshafen 1913+; 9 stations 1934+.
87 ⟋ 2 stations 1921+.

Θ Hamburg 1905–12, **2**; 1926–30, **98** vol. 50, No. 5.
(§) Lindenberg 1906–16, **105** vol. 13; 1906–27, vol. 16; München 1906–14, **107** 1915; 1906–14, 1922–8, **107** 1928; Lindenberg, München, Hamburg to 1911, **130** No. 5; Lindenberg 1906–27, München 1906–28, **118**.
Studies of conditions at high levels: see **112** vol. 14, p. 65, vol. 17, p. 176, vol. 20, p. 274; **105** vol. 7, p. 253.

⟋ Berlin 1902–3, **122**; 1903–4, **113**; 1903–5, **98** vol. 31; Brocken Jan.-Feb. 1906, **105** vol. 2; Darmstadt 1926–8, Wasserkuppe i Rhön, summer 1928, 1929, **90** Nos. 1, 3, 4; Frankfurt am Main 1914–19, **94** Nos. 2–4; Friedrichshafen June-Dec. 1903, **112** vol. 1, p. 1; 1908–15, **110**; 1916–17, **111**, 1921–2; (§) 1911–20, **118**; **112** vol. 7, p. 218, vol. 14, p. 278; **26** vol. 48, p. 382; Fürth 1925–6 **107**; Hamburg 1903–5, **98** vol. 31; 1927–8, **98** vol. 49, No. 10; (§) polar year 1932–3, **99** vol. 62, p. 406; **98** vol. 34, No. 5; (§) 1904–9, **118**; Königsberg 1932–3, (§) **112** vol. 21, p. 313; Lindenberg (§) **105** vols. 2, 5, 8; 25 year mean, **104** Mar. 1930; **118**; Taunus Obs. (§) **94** No. 4; 5 stations 1930–1, 1932–3, **26** vol. 50, p. 104, vol. 51, p. 193; 6 stations (polar year) **93**.
Papers dealing with special subjects, see **105** vols. 5–9, 14 and **107**.
⊗ Berlin (§) 1926–32, **112** vol. 20, p. 291 (see also **113** vol. 50, p. 212; **82** vol. 8, p. 187); Bremen 1926–9 (annual (§) only), **89** part IV; Dresden 1914–17, **91**; 1923–4, **92**; Frankfurt 1913–15, **94** Nos. 1, 2; (§) 1909–12, **94** No. 1, p. 51; Friedrichshafen 1908–15, **110**; (§) **109**; Fürth 1925–6, **107**; Hamburg 1932–3, (§) **82** vol. 9, No. 9; Karlsruhe 1921–6, **100**; (§) 1922–6, **82** vol. 1, No. 6; Königsberg 1932–3, (§) **112** vol. 21, p. 313; **82** vol. 9, No. 9; (§) 1926–32, **82** vol. 8, No. 5; Lindenberg (§) **105** vol. 13; Mannheim May 1927–April 1928, (§) **82** vol. 3, No. 2; München 1910–14, **107**; Zugspitze (§) Sept. 1913–July 1915, **107** 1931.

DENMARK‡

Θ ⟋ Hald 1902–3, **55**.

NORWAY

214 ⟋ Kjeller c. 1931+; ⊗ c. 3 stations c. 1923+.
215 Θ Ås 1932+; Bergen 1933+; ⟋ Kjeller 1928+.

The Baltic countries: Sweden, Finland, Esthonia, Latvia, Lithuania, Danzig, Poland.

SWEDEN

246 ⟋ 1–2 stations c. 1925+; ⊗ 1–8 stations 1921+.
247 ⊗ c. 9 stations 1932+.
249 Θ Riksgränsen 1930+; ⟋ 1–5 stations 1928+.

Θ Kiruna 1907–9, **252** ser. IV, vol. 3, No. 7, (§) **118**; Abisko, 1921–9 and Kiruna 1907–9, **250** vol. 5, No. 5.
(§) Abisko 1921–9, **1** No. 8.
⟋ Malmstätt 1924–5, **250** vol. 3, No. 10; 1924 only, **253** vol. 54, p. 51; (§) **118**.
⊗ Abisko 1913–15, **248** vol. 57, App. 2, (§) **112** vol. 8, p. 200; 6–7 stations 1928–30, **249**; 6 stations 1912–29, (§) **250** vol. 6, No. 3.

FINLAND

63 ⊗ 4–8 stations 1921+ (mean monthly velocity).
64 ⟋ Ilmala 1919+; ⊗ 1–5 stations 1918+.
65 *a* ⊗ 4 stations 1934+.

⟋ Ilmala 1922–4, **63** (monthly summary only); 1911–14, (§) **65** No. 5 (see also **118**).
⊗ Petsamo July-Aug. 1926, **66** No. 7; Ilmala 1920–2. (§) **66** No. 3.

RUSSIA (see Arctic zone, also U.S.S.R. in the Eastern extension of the zone of the Great Westerlies).

ESTHONIA

62 ⟋ Dorpat 1928+; ⊗ Dorpat 1925+.
61 ⊗ Dorpat 1916–17.

LATVIA

202 ⊗ Riga 1930+.

DANZIG

54 ⊗ 1 station 1930.

The Baltic countries—contd.

POLAND

221 ⊗ 1–6 stations *c.* 1922+.
222 ⊗ 7 stations April 1928+.
224 (Nos. 6–10) ⊗ Wilno 1925+.

⊗ Warsaw 1919–26, Ⓢ **225** vol. 7, fasc. 8, p. 15; Warsaw 1919–26, 4 stations 1926–30, Ⓢ **223**; Staraja Jablonna (nr. Warsaw) Feb.-July 1915, Ⓢ **280** vol. 3, fasc. 2, p. 41.

Central Europe: Switzerland, Austria, Czecho-Slovakia, Hungary, Roumania, Bulgaria.

SWITZERLAND‡

Θ Zürich 1905–8, 1910–11, **2**; Ⓢ to 1911, **130** No. 5.
⊗ Arosa, winter 1929–30, **112** vol. 18, p. 81.

AUSTRIA

22 Θ Vienna 1924+; ⊗ Vienna 1910+.
23 *c.* 8 stations 1930+.

Θ Vienna 1904–17, **21**; Ⓢ to 1911, **130** No. 5; Ⓢ 1906–16, 1924–8, **22** Beihefte 1928, pp. 23–57; Ⓢ 1907–27, **118**.
↗ Vienna 1904–*c.* 1914, **21**.
⊗ Vienna 1911–19, **21**; Ⓢ **112** vol. 7, pp. 77, 174, 197; **26** vol. 34, p. 70; **113** vol. 37, pp. 19, 185; Hochobir 1913–14, **22** vol. 57; Ⓢ **25**, Abt. IIa, Bd. 132, p. 233; 16 stations, Ⓢ **26** vol. 45, p. 131.

CZECHO-SLOVAKIA

50 ⊗ 5 stations 1927+.

↗ 3 stations 1926 (international days), **51** vol. 2, No. 7.
⊗ 11 stations 1926 (international days), **51** vol. 2, No. 7; 12 stations 1927, **52** (monthly mean force and direction); 10 stations 1921–5, Ⓢ **51** vol. 1, No. 1.

HUNGARY

155 ⊗ 1–2 stations *c.* 1928+.
156 Θ Budapest 1913–14, 1927+; ↗ 1–2 stations 1925+; ⊗ 1–4 stations 1913–14, 1927+.

Θ Budapest 1913–14, 1927, Ⓢ **118**.
⊗ Debrecen Aug. 1932–Dec. 1933, **160**; Debrecen 1934, **158**.

ROUMANIA

231 ⊗ 1 station.
232 ⊗ 5 stations 1930+.

↗ Temesvar 1915–17, Ⓢ **26** vol. 36, p. 197, **118**.

BULGARIA

35 ⊗ 2–3 stations 1932+.

⊗ 1–2 stations 1925–31, **36**.

The extension of the zone of the Great Westerlies to the Far East: Georgia, U.S.S.R.

GEORGIA‡

Θ Tiflis 1910–15, **80**.
⊗ Tiflis 1910–15, 1926–7, **80**.

U.S.S.R.

289 ↗ *c.* 2 stations; ⊗ *c.* 20 stations.
278 Θ ↗ ⊗ 1–2 stations 1920–7.

Ⓣ Leningrad and Moscow Jan.-April 1901, **78** vol. 3; Moscow (Koutchino) 1905–14, **290** fasc. 5; 1–4 stations, 1905–12, 1923+ [Pavlovsk 1905–12, 1923+, Ekaterinbourg 1908–12, Nijni Oltchédaeff 1907–12, Omsk 1907–8, 1910 (1 ascent)], **2**; Sloutzk Jan. 1933, **26** vol. 50, p. 429; Kharkiv 1912–17 (data incomplete), **292** vol. IV, part 1; Ichan Bazar Oct. 1908 (4 ascents), **280** vol. 1, p. 23.
Ⓢ 1–3 stations 1902–9, **26** vol. 28, p. 1, **118**; 2 stations to 1911, **130** No. 5; 3 stations, **251** vol. 2, p. 97; Koutchino 1905–14 (thermoisopleths), **279** vol. 1, p. 68.

Θ ↗ Pavlovsk and Leningrad 1901–3, **281**.
↗ Leningrad 1903–6, **98** vol. 31, No. 1; Aug.-Oct. 1911 (wind), **280** vol. 1, fasc. 2, p. 23; Tashkent 1907, **280** vol. 1, p. 32.
Ⓢ Pavlovsk 1904–11, **280** vol. 1, fasc. 1, p. 75; 1920–5, **278** vol. 6, pp. 152, 153, vol. 7, p. 147.

⊗ Sloutzk and Leningrad, Aug-Oct. 1911, **280** vol. 1, fasc. 2, p. 23; Moscow, 1925, **287** fasc. 4; Kharkiv, 1915–26, **292** vol. IV, part 1; Ukraine, 6–9 stations, 1924–6, **292** part II; Kiev, 1918–30, **293** No. 3; Markhot, Aug.-Oct. 1926–7, **284** part 8, p. 211; 7 stations, 1926, **283**; Tashkent, 1907–8, **280** vol. 1, p. 20; 1924, **294**; 1928, 1930, **295** Nos. 1, 2, 5, 6; Ⓢ **118**; Jakoutsk and Petropavlovsk, 1925–6, Verkhoiansk, 1926, **285** vols. 8, 16; Ashkabad, 1928, **297**; Vladivostok, winter 1929–30, **299** No. 1 (VIII); Irkutsk, Aug.-Nov. 1915, **296**.
Ⓢ Sloutzk and Leningrad, 1920–4, **112** vol. 14, p. 127; **278** vol. 6, p. 150, vol. 7, p. 148; 1921–5, **279** vol. 5, No. 1; Ashkabad, 1928, **298**.

The Mediterranean zone (a zone of permanent anticyclones, deserts or winter rainfall) and its Eastern extension, 25° N to 45° N: Spain, Italy, Yugo-Slavia, Greece, Turkey, Mediterranean Is., Canary Is., French Morocco, Algeria, Tunisia and the Sahara, Tripolitania, Egypt, N. Arabia, Syria, Palestine, Iraq, Iran, Persian Gulf, Tibet, Mongolia, China, Korea, the Japanese Is.

SPAIN

236 ⊗ 20 stations *c.* 1921+ (charts).
240 ⊗ 1–2 stations *c.* 1923+.
237 Θ Madrid, ⊗ 14–20 stations 1925, 1932, 1934+.
241 ⊗ Barcelona 1919+.
243 ⊗ Tortosa 1930+. (Annual summary only 1920–9.)
245 ⊗ San Sebastian 1933+.

Θ Madrid 1924+; Guadalajara 1905–8, **2**; Ⓢ **238** No. 1 (discussion only); **118**.
⊗ Madrid 1913–20, **239** vols. 1–4; Ⓢ **262**; Barcelona 1914–15, **242**; 1915–18, **239** vols. 3, 4; Ⓢ 1914–30, **241** No. 54; La Coruña 1917–20, **239** vol. 4; San Sebastian 1927–30, **244** Nos. 1, 3, 5.

ITALY

174 ⊗ 20–40 stations 1915–18, 1921+.
176 ⊗ 14 stations 1928+.
177 Θ ↗ Vigna di Valle 1932; ↗ Montecelio 1928+; ⊗ *c.* 25 stations 1926+.
175 (vols. 2–5) Θ Vigna di Valle 1926+; ↗ Vigna di Valle 1915+; Montecelio 1928+; ⊗ Vigna di Valle 1926+.

Θ ⊗ Pavia 1905–6, **2**; 1907–12, **171** vols. 30, 32–6; 1913–16, **172** ser. III, vols. 1, 2, 4; 1923–4, **173** 1925; 1925–30, **178**; 1931–2, **172** vol. 4; Ⓢ 1906–20, **186** and **118**.

The Mediterranean zone—contd.

ITALY—*contd.*

Θ Milan Sept.-Oct. 1906, **105** vol. 2; Vigna di Valle (temp. at high levels), **181** ser. 2, II, No. 1; Milan, Pavia to 1911, (§) **130** No. 5.

↗ Vigna di Valle 1915–26, (§) **179** vol. 81, p. 85 **175** vol. 2, **118**; Montecelio, 1930, **181** ser. 2, II, No. 4.

⊗ 16–20 stations May 1912–April 1913, **182** Nos. 14, 22, 25, 33; Genoa, **172** vol. 3; Etna and Catania, a few months 1921–4, **175** vol. 3, p. 189; 4 stations July 1912, **171** vol. 34.

(§) 10 stations, **112** vol. 14, pp. 21–5; Capo Sperone, **175** vol. 5, p. 114; Ciampino 1917–27, **179** vol. 82, p. 94; Florence 1926–31, **175** vol. 5, p. 96; Florence, Livorno and Milan, **185** vol. 9; **82** vol. 8, No. 3; Ischia, **178** vol. 11, p. 233; 1916–26, **179** vol. 81, p. 391; Montecelio June 1928, **185** vol. 5; Palermo 1927–9, **184** vols. 10, 11; Pavia 1908–28, **172** ser. 3, vol. 3; Perugia, **178** vol. 11, **179** vol. 81, p. 413; Trieste, **26** vol. 33, p. 64; Trapani 1914–18, **175** vol. 5, p. 114; **185** vol. 8; Varignano, **178** vol. 9, p. 129; **180** vol. 7, p. 244; Venice 1913–15, **184** vol. 11, p. 21; Vigna di Valle 1910–11, **182** vol. 8; 1911–25, **175** vol. 1; **180** vol. 5, p. 675, vol. 6, p. 49; 1911–13, **183** Nos. 3–4; 1912–28 (tramontana), **185** vol. 5, No. 12; Zara 1927–32, **175** vol. 5, p. 87; *c.* 15 stations, varying periods, **133** Nos. 9, 72–83.

YUGO-SLAVIA

300 ⊗ 4–12 stations 1927 +.

ALBANIA‡

⊗ Valona, **133** Nos. 9, 72–83.

GREECE‡

⊗ 3 stations 1916–18, (§) **98** vol. 38, No. 5; 23 stations (Balkan peninsula), **98** vol. 41, No. 3.

TURKEY‡

↗ Waniköy 1917–18, (§) **102** Heft 2.

MEDITERRANEAN ISLANDS

124 ↗ ⊗ Malta 1925+.
128 ⊗ Malta 1933 +.
237 ⊗ Mahon Aug. 1925, Jan. 1934 +.
177 ⊗ Leros 1928 +.

⊗ Malta 1918–24, (§) **130** No. 37, **133** Nos. 72–83.
↗ Malta 1926–30, (§) **133** Nos. 18, 72–83.

CANARY ISLANDS

237 ⊗ 1–2 stations 1925, 1932, 1934 +.

⊗ Tenerife 1912–16, **239** vols. 1–3; (§) 1912–13, **84** 1928, No. 23; 1912–23, **85** 1930, No. 1.

FRENCH MOROCCO, ALGERIA, TUNIS AND THE SAHARA

69 ⊗ 10–25 stations 1929 +.
70 ⊗ 2–10 stations 1929 +.
208 ⊗ *c.* 3 stations 1922 +.
13 ⊗ 1–2 stations 1930 +.
14 ⊗ Algiers 1921 + (monthly frequencies).
15 ⊗ Tamanrasset 1932 + (monthly frequencies).
256 ⊗ 2 stations 1926 + (monthly frequencies).

⊗ Algiers 1918–28, (§) **73** vol. 5, p. 214; Tunis 1920–3, (§) **257**; Tamanrasset 1932–3, (§) **73** vol. 11, p. 80.

TRIPOLITANIA AND CYRENAICA

177 ⊗ Tripoli, Benghazi 1928 +.

⊗ Tripoli Jan.-April 1913, **182** No. 33; Tripoli and Tobruk, **133** Nos. 9, 72–83, (§) Benghazi, 1931–4, **178** vol. 17, p. 1.

EGYPT

57 ⊗ 1–4 stations.
58 ⊗ Helwan 1911–14, 1920 +.
128 ⊗ 2 stations 1933 +.

↗ Ismailia (monthly means) May 1922, **129** vol. 57, p. 215; June-Aug. 1922, **131** No. 41; Aug.-Oct. 1925, March-July 1926, **130** No. 56; Helwan Oct.-Nov. 1907, March 1908, **60** vol. 2; Oct. 1923–April 1924, **131** No. 41; (§) 1922–4, **118**; Heliopolis July 27–Aug. 3, 1920, **131** No. 20.

(§) Lower Egypt **133** Nos. 72–83.

⊗ (§) Helwan 1920–3, **59** No. 17; 1920–8, **59** No. 27 (see also **59** No. 26); 4 stations, **133** No. 17; Helwan 1907–9, (§) **118**.

SYRIA AND PALESTINE AND TRANSJORDANIA

128 ⊗ 2 stations 1933 +.

↗ Jordan valley, 1 ascent June 1931, **137** vol. 58, p. 58.
⊗ (§) 3 stations 1917–19, **137** vol. 46, p. 15; 3 stations 1918, **102** Heft 3; 2 stations April-Sept. 1916, **112** vol. 8, p. 170; 2 stations, varying periods 1921–6, **133** Nos. 17, 72–83.

IRAQ

128 4 stations 1933 +.

↗ Hinaidi July, Sept. 1925, **131** No. 59; 26 ascents 1926–7, **143** vol. 32, p. 901.
⊗ (§) 2 stations June-Sept. 1921–30, **131** No. 64; 8 stations (mean direction), **112** vol. 15, p. 63; 5 stations, varying periods 1921–6, **133** Nos. 17, 72–83; 2 stations, **136** No. 3.

IRAN (PERSIA) AND THE PERSIAN GULF

162 2 stations 1928 +.

↗ 1928 (a few ascents), **112** vol. 17, p. 126.
⊗ (§) 3 stations, **167**; Jask 1924–5, **166** No. 8; 3 stations, varying periods 1924–9, **166** No. 17; 3 stations (mean direction), **114** vol. 30, p. 196; 4 stations, varying periods, **136** No. 3.

MONGOLIA AND CHINA

42 ↗ data for air routes 1930 + ; ⊗ 2–3 stations 1931 + (see also Supp. 1932).
44 ↗ 2 stations 1932 + ; ⊗ Nanking 1930 +.
46 ⊗ Hong Kong 1933 +.

⊗ 5 stations (China), varying periods 1928–30, **43**; Mongolia 1925–7, **296** Nos. 2–3; (§) June-July 1927, **112** vol. 17, p. 23 (short note); Hong Kong 1921–32, **47**; 1921–34, **48**; 3 stations, (§) **263**.

KOREA

201 ⊗ 6–8 stations 1930 +.

THE JAPANESE ISLANDS

192 Tateno Θ 10 July 1926 (No. 3, p. 153); ↗ 1923 + (Nos. 1, 3, 6–9); ⊗ 1923 + (Nos. 1, 4, 6–9).
193 ↗ ⊗ Tateno 1927 +.
196 ⊗ Osaka 1928 +.
190 8–11 stations 1924–5.

↗ ⊗ Tateno 1932–3, **194**; ↗ (§) 1923–7, **192** No. 3; ⊗ Tateno (§) 1923–5, **192** No. 1 (see also **118**); 1923–7, **192** No. 4 (see also **26** vol. 49, p. 42). Kobe 1920–2, **197**; April-Dec. 1924, **198** vol. 3; (§) **199** 1924; Mizusawa 1922–5, **200**; Osaka 1922–7, **195**; Misima 1930–1, (§) **191** vol. 6, p. 69; 3 stations, (§) **263**. Summary of upper winds in typhoons, **191** vol. 6, p. 98.

The zone of the Great Easterlies. British India and Ceylon, Burma, Siam, French Indo-China, Philippines, Malaya, Netherlands East Indies, the Indian Ocean to Madagascar (the region of the defeat of the Great Easterlies), Somaliland and Southern Arabia with the islands of the Red Sea, Abyssinia, the Sudan, East Africa, French equatorial Africa, Nigeria, French West Africa, South Atlantic Islands.

BRITISH INDIA

162 Θ 2–4 stations 1928+; ⟋ 3 stations 1929+; ⊗ *c.* 30 stations 1928+.

Θ Agra 1925, **165** vol. 25, part 5; 1926–7, **164** Annual supplement.

Ⓢ 1915–18, **165** vol. 24, part 6, 1925–8, vol. 25, part 5 (see also **136** No. 1 and **118**); Poona and Hyderabad 1928–31, Ⓢ **165** vol. 26, part 4.

Tropopause and stratosphere: **259** vol. 57, p. 64; **114** vol. 25, p. 266; **166** Nos. 10, 48.

⟋ 1–2 stations, summer months 1905–7, **165** vol. 20, parts 1, 2, 7.

Ⓢ **165** vol. 24, part 6; 1–3 stations 1927–30, **166** Nos. 42, 50, **168**.

⊗ 13 stations, short periods 1910–19 (resultant winds), **165** vol. 23, part 3; 4–18 stations 1918–27 (resultant winds), **164**; *c.* 11 stations 1926–7, **163** supplement; Kashmir, June 1914, **189**.

Ⓢ Up to 26 stations, periods to 1929, **165** vol. 22, part 4, vol. 24, parts 7, 8; **166** Nos. 7, 8, 17, 21, **168**; 25 stations (mean wind), **114** vol. 30, p. 196; 5 stations, varying periods, **136** No. 1; 8 stations, varying periods **118**.

Special investigations, **166** Nos. 1, 13, 15, 28, 30, 41.

CEYLON

38 ⊗ Colombo 1927+ (monthly wind-roses).

⊗ Colombo 1922–3, **39** No. 5; Ⓢ 7 years, **39** vol. 1, p. 173 (see also **166** No. 17; **114** vol. 30, p. 234).

SIAM

235 ⊗ Bangkok July 1931+.

⊗ **136** No. 1.

FRENCH INDO-CHINA‡

⊗ 2–7 stations 1932–3, **75** Feb. 1935; Ⓢ (notes only, no data, **74** vol. 194, p. 1595; vol. 197, p. 1677; vol. 198, p. 1055).

PHILIPPINES‡

⟋ Manila Oct. 1930–Feb. 1932, **220** vol. 2, No. 5.
⊗ Manila Ⓢ **263**.

MALAYA

204 ⊗ 3 stations 1932+ (monthly wind-roses).

NETHERLANDS EAST INDIES‡

Θ Batavia 1910–15, **209** No. 4; Ⓢ 1910–15, **118**.
⟋ Batavia and Java Sea 1909–10, 1912–14, **209** Nos. 2, 3 and 8, pp. 316, 320–3.
⊗ Batavia 1909–11, **209** No. 1; Batavia and occasional stations 1911–18, **209** No. 6; Ⓢ 1909–17, **153** vol. 20, p. 1313; **209** No. 8 (vol. 1, pp. 74–83, 104; vol. 2, pp. 146–50, 392, 426–7, 483); **210** vol. 80, p. 265 (note only); Ⓢ **118**; **263**; **136** No. 2.
⊗ *c.* 7 stations 1929+ (data available but not published).

THE ISLANDS OF THE INDIAN OCEAN

162 Andaman Is. ⊗ 1928+.
206 Mauritius ⊗ July 1927–Dec. 1929 (Nos. 9, 11).
203 ⊗ Madagascar, 4 stations 1935+.

⊗ Andaman Is. 1926–7, **164** (resultant wind); Ⓢ 1926–9, **166** No. 17c (see also Nos. 8, 16); Mauritius (Vacoas), **207**; July–Sept. 1925, **205**; July 1925–April 1926, **130** No. 39; Ⓢ 1925–8, **118**; Réunion (data available but not published); Madagascar, 5 stations 1933, **74** vol. 198, p. 1250 (note only, no data).

SOMALILAND AND SOUTHERN ARABIA, WITH THE ISLANDS OF THE RED SEA

175 ⊗ Cape Guardafui 1930+ (vols. IV, V) (see also **177**).

⊗ Cape Guardafui 1930–2, Ⓢ **112** vol. 21, p. 193; Mogadiscio polar year 1932–3, Ⓢ **115** vol. 10, p. 360; Berbera April 1931–July 1932, Ⓢ **131** No. 65; Aden Oct. 1929–31, **162**; Kamaran Is. Ⓢ Nov. 1927–Jan. 1928, **73** vol. 5, p. 519; Aug. 1927–April 1928, **166** No. 8; 1928–30, **133** No. 70.

ABYSSINIA‡

⊗ Bahrdar Georgis March 1923–April 1924, **59** No. 17.

THE SUDAN

57 ⊗ *c.* 2 stations.

⊗ 4 stations, varying periods 1908–9, 1924, **59** No. 17 (see also **60** vols. 2, 3; **26** vol. 26, p. 565, vol. 27, p. 227); Khartoum 1926–8, **59** No. 27 (see also No. 26).

EAST AFRICA AND N. RHODESIA

5 ⊗ 3 stations (Kenya, Uganda, Zanzibar) 1931+.
6 ⊗ Broken Hill Jan. 1932+ (No. 9) monthly mean values.

Θ ⟋ ⊗ Victoria Nyanza, Mombasa, Zanzibar Aug.–Dec. 1908, **105** Bericht über die aerologische Expedition...nach Ostafrika; Ⓢ **118**.
Θ ⊗ Zanzibar July–Aug. 1908, **171** vol. 30.
⊗ 1–2 stations Dec. 1926–March 1927, **7** Jan.–March 1927.

FRENCH EQUATORIAL AFRICA, SAHARA

See French Morocco, Algeria and Tunis.

NIGERIA‡

⊗ Kaduna Aug. 1932–June 1933, **213**.

FRENCH WEST AFRICA

70 ⊗ 1929+.

⊗ 1–3 stations 1932–3, **75** Feb. 1935 (see also **74** vol. 183, p. 229; vol. 194, p. 902).

SOUTH ATLANTIC ISLANDS

34 ⊗ Fernando Noronha Dec. 1930+.

⊗ St Helena: observations available from 1928 but not published.

Towards the Roaring Forties.

S. RHODESIA

230 ⊗ Salisbury Oct. 1928+.

S. AFRICA

8 ⊗ 6–8 stations 1932+.

Θ Pretoria (1 ascent, May 1922), **12** No. 7; 5 ascents, 1932, **10***a*.

⊗ Ⓢ 13 stations, varying periods, **9** 1929; **10** No. 1 (see also, shorter periods, **11** vol. 23, p. 103; **12** No. 14); 9 stations, polar year, **10** *a*.

AUSTRALIA‡

Θ 1913–14, **18** No. 13; Ⓢ **118**.
⊗ Ⓢ 4 stations, varying periods 1922–8, **112** vol. 15, p. 163 (see also **118**); Sydney 1926–32, **20**; see also **136** No. 2; Willis Is. 1922–4, **19** vol. 17, p. 155; Melbourne (vertical velocities), **137** vol. 52, p. 415, vol. 55, p. 153.

NEW ZEALAND‡

⊗ Christchurch 1926–7, **212** vol. 4, App.; Ⓢ **118**; Ⓢ Wellington 1929–31, **211** No. 15; for exceptional values 7 Feb. 1933 see **137** vol. 59, p. 238.

[Data of a few Θ, and of ⊗ at 3 stations are available but have not yet been published.]

ANTARCTIC‡

Θ Cape Evans, Aug., Nov.–Dec. 1911, **146**; Ⓢ **118**.
⟋ ⊗ Weddel Sea 1912, Ⓢ **86** vol. 7, No. 6; **118**; **117**, 1927, p. 50.
⊗ Little America Jan. 1929–Feb. 1930, Ⓢ **268** 1932, p. 124.

General summary for Antarctic, **73** vol. 3, p. 337.

ITEMS OMITTED FROM THE ORIGINAL TEXT ON REVISION

Pp. 104–7, figs. 58–59. A model of the normal distribution of temperature from equator to pole at the level of the tropopause, showing cyclone, anticyclone and tropical hurricane.

P. 220. Geostrophic scales for hemispherical maps.

Pp. 299–303, figs. 185–8. Quarterly values of temperature and rainfall for the twelve districts of the British Isles in the period 1878–1910.

Pp. 396–404. Some new views about cyclones and anticyclones. Lecture at the University of Manchester, 14 November, 1921.

Pp. 405–24. Chapter X. The earth's atmosphere. The foundation of meteorological theory.

Forty-five articles expressing the meteorological conditions of the middle atmosphere between the levels of 400 and 800 geodekametres and their connexions with the layers above and below.

(i) *Some preliminary notions.* Geodynamic or level surfaces. The significance of pressure-difference at any level. Geostrophic wind. Atmospheric shells.

(ii) *Some properties of the general circulation.* Graphic projection by adiabatic lines. Stratification and resilience. Height of the homogeneous atmosphere. Deviation from the geostrophic wind. Penetrative convection. Vortex-motion and its analogy with the general circulation. Uniformity of lapse-rate and its consequences. Variation of geostrophic wind with height.

(iii) *The origin of local circulations.* Conditions for penetrative convection. Eviction of air and resulting vortex-motion.

(iv) *Vortices in a uniformly flowing air-current.* The combination of uniform motion with vortex-motion and the resulting fields of stream-function and pressure.

(v) *The properties of cyclonic systems derived from observation.* The correspondence of isobars and isotherms in the middle atmosphere. The adiabatic change of pressure and temperature at 9 km. The creation of regions of convective equilibrium. The effect of convection on the variation of wind with height.

(vi) *The layers of the atmosphere outside the limits of* 400 *and* 800 *geodekametres.* The analysis of the normal pressure-distribution. The effect of turbulence on geostrophic wind. The ceiling of a vortical column.

(vii) *The wave-theory of cyclonic depressions.*

INDEX

*** A subject-index of the authorities for the charts, tables and diagrams is given under the heading Bibliography. Entries have not been made under the author's name in the case of papers referred to only in the bibliographies.

Printed in the United States
By Bookmasters